BIOLOGICAL WASTEWATER TREATMENT

POLLUTION ENGINEERING AND TECHNOLOGY

A Series of Reference Books and Textbooks

EDITOR

PAUL N. CHEREMISINOFF

*Associate Professor
of Environmental Engineering
New Jersey Institute of Technology
Newark, New Jersey*

Additional Volumes in Preparation

BIOLOGICAL WASTEWATER TREATMENT

Theory and Applications

C. P. Leslie Grady, Jr. and Henry C. Lim

School of Civil Engineering
Purdue University
West Lafayette, Indiana

School of Chemical Engineering
Purdue University
West Lafayette, Indiana

MARCEL DEKKER, INC.   New York and Basel

Library of Congress Cataloging in Publication Data

Grady, C   P   Leslie
    Biological wastewater treatment.

    (Pollution engineering and technology ; 12)
    Includes bibliographies and indexes.
    1. Sewage--Purification--Biological treatment.
I. Lim, Henry C.   joint author.
II. Title. III. Series
TD755.G72   628.3'51   80-20171
ISBN 0-8247-1000-2

MARCEL DEKKER, INC.

270 Madison Avenue, New York, New York 10016

Current printing (last digit):
10 9 8 7 6 5

PRINTED IN THE UNITED STATES OF AMERICA

To Art and Tony
with much appreciation
for a good beginning.

CPLG

The components in wastewater treatment processes may be conveniently categorized as physical, chemical, and biochemical unit operations and a thorough understanding of the principles governing their behavior is a prerequisite for successful process design. This "unit operations approach" to the study of process engineering has been widely accepted in the fields of chemical and environmental engineering and a number of books are available which delineate the fundamental theories explaining physical and chemical operations and illustrate their applications in practice. The purpose of this book is to provide similar coverage for the biochemical unit operations. Although a certain amount of introductory material is necessary in any book, the authors have worked from the premise that the reader is familiar with the process trains commonly used in wastewater treatment and that he has come to this book to gain additional insight into the theory and applications of the biological operations in them. There are inherent interactions among the operations in a process train, however, and these interactions will be noted and discussed when appropriate.

An understanding of biological wastewater treatment operations requires knowledge in two fundamental areas: microbiology and reactor engineering. Students in environmental engineering programs generally have adequate training in the former but have little exposure to the latter. For students in chemical engineering the situation is generally reversed. Thus when we set out to prepare a text which could be used to teach advanced undergraduate and beginning graduate students in either type of program it was apparent that the two groups had different fundamental needs and that the needs of each would have to be met before we could proceed with the main topic of interest. Our answer was to include the material needed by each. Consequently, the text material is suitable for use by a broad range of individuals, including those who need training in both of the fundamental areas.

The book is organized into five parts:

        Part   I - Introduction
        Part   II - Fundamentals of Reactor Engineering
        Part III - Biochemical and Microbiological Background
        Part   IV - Theory: Modeling of Ideal Biochemical Reactors
        Part   V - Applications: Named Biochemical Operations

The first problem faced by a student of biological wastewater treatment is in under-standing the meanings of the sometimes curious names which have been given to the various biochemical operations. Part I overcomes this by describing the "named bio-chemical operations" in terms of the treatment objective, biochemical environment, and reactor configuration. Having learned that both the biochemical environment and the reactor configuration play an important role in the performance of a biological reactor the reader progresses to either Part II or Part III to prepare for the theory of Part IV. In Part II, after kinetics and stoichiometry are introduced the concepts of reactor engineering are developed in enough detail to allow the reader to use them. This is particularly important because it provides the groundwork from which to build an understanding of how reactor configuration and kinetics interact to determine the extent of reaction that will occur in any system. In Part III, the major elements of biochemistry and microbiology are presented in order to give the reader an under-standing of how biological wastewater treatment operations function, thereby providing an appreciation of their limitations. This is important because in Part IV mathe-matical modeling is used to illustrate the theoretical performance of the operations and lack of appreciation of their limitations could cause the reader to make extra-polations which were not valid. Having obtained the needed background, however, the reader is prepared to learn in Part IV how various perfect reactors would respond to changes in operational conditions thereby forming the basis for understanding why biological wastewater treatment systems behave as they do. In this part performance equations are developed and investigated for simple continuous stirred tank reactors, reactors in series, reactors with recycle, and reactors containing organisms growing in fixed films. Finally, in Part V the theory is applied to the common named bio-chemical operations introduced in Part I. In that application, however, considerable care is taken to point out when the theory is not valid and when practical constraints must be applied to ensure that the system will function properly in the real world. In this way the reader will attain a rational basis for the design of biological wastewater treatment operations which incorporates the knowledge that has been ob-tained through practice.

The organization of the book gives instructors a great deal of latitude in its use. If students were unprepared in both reactor engineering and microbiology the entire book could be used in a course spanning two semesters. Conversely, if the students were well prepared in both areas the instructor could emphasize Parts IV and V in a 3 semester-hour course, referring to Parts II and III as needed for review. For chemical engineering students, Parts III, IV, and V could be covered in either a 3 or 4 semester-hour course while for environmental engineering students Part II would be used instead of Part III. Finally, if an instructor so desired, he could teach a 3 semester-hour course around Part V, drawing on the previous parts as needed to improve understanding. Of course it is not necessary to use the book in a formal

course. The key points, study questions, and references will all help the individual
reader who wishes to use the book either for self-paced study or as a reference
guide.

A large number of people have contributed to the preparation of this book and
the authors are deeply indebted to them all. First, this book would not have been
possible had it not been for the hundreds of researchers who contributed the knowledge upon which it is based. In addition, we know that we have overlooked or
neglected many articles of merit which could have been included; for that we ask your
forbearance. We are also indebted to the many students who read earlier versions of
the manuscript for their constructive criticism and helpful comments; we would
especially like to acknowledge Anne Schwartz and Glen Daigger in this regard. We
also appreciate the comments of our colleagues and would particularly like to thank
Ed Kirsch, Joe Sherrard, and Bob Sykes for reviewing portions of the manuscript. A
work of this size involves the efforts of many people. We appreciate the patience
and skill of Ché Tiernan, Marilyn Weston-Kit, Sylvia Ryland, and Janet Taylor who
deciphered our handwriting to type the original draft of the manuscript. We would
especially like to recognize the hard work and dedication of three wonderful people
without whom the final production would not have been possible: Barbara Steel for
drawing the artwork; Shirley Velten for typing both the final draft of the manuscript
and the camera-ready copy; and Joni Grady for editorial assistance, proofreading, and
preparation of the indexes. Finally, we would like to thank Bob Greenkorn who introduced us to each other and encouraged us to pursue joint teaching and research
activities.

C. P. Leslie Grady, Jr.
Henry C. Lim

CONTENTS

_____

BIOLOGICAL WASTEWATER TREATMENT

PART I

INTRODUCTION

CHAPTER 1

CLASSIFICATION OF BIOCHEMICAL OPERATIONS

Biochemical operations, in which chemical transformations are carried out by living microorganisms, have many diverse uses. Brewing and wine making are two of the oldest and perhaps most widely appreciated examples. The medical revolution has been due largely to the production of wonder drugs using biochemical operations in the pharmaceutical industry. Finally, of course, biochemical operations play a large role in the treatment of wastewaters although much of the nomenclature associated with them evolved without much consideration for the basic mechanisms involved. In fact, the most common operation, activated sludge, was named before its biochemical nature was even recognized. This situation has often caused confusion in those attempting to learn about these operations. Consequently, before starting a study of the operations it would be wise to establish clearly what they are and what they do.

1.1 CRITERIA FOR CLASSIFICATION

The classification of biochemical operations may be approached from three points of
view: (1) the biochemical environment, (2) the nature of the biochemical transformation, and (3) the reactor configuration. If all three are considered together, the
result is a detailed classification system which will aid the engineer in choosing
the operation most appropriate to his needs.

1.1.1 The Biochemical Environment

There are two major environments in which biochemical operations are carried out:
aerobic and anaerobic. An aerobic environment is one in which dissolved oxygen is
present in sufficient quantity so as to not be rate-limiting. In such an environment oxygen serves as the terminal electron acceptor in microbial metabolism and
growth occurs in an efficient manner. An anaerobic environment is one in which
dissolved oxygen is either not present (and is not entering the system) or one in
which its concentration is so low as to limit aerobic metabolism. In an anaerobic
environment some substance other than oxygen serves as the terminal electron acceptor. If that substance is itself an organic molecule, the system is termed a fermentation. If the terminal electron acceptor is an inorganic substance, the culture
is said to be undergoing anaerobic respiration.

   The biochemical environment has a profound effect upon the ecology of the
microbial population. Aerobic operations tend to support entire food chains from
bacteria at the bottom to rotifers at the top. Anaerobic operations on the other
hand tend to contain predominantly bacterial populations although they too can be
ecologically complex. The biochemical environment influences the biochemistry of
the cells because aerobic organisms use many metabolic pathways which are different
from anaerobic ones. This means that an aerobic operation can carry out some transformations that cannot be accomplished anaerobically and vice-versa.

1.1.2 Nature of the Biochemical Transformation

   REMOVAL OF SOLUBLE ORGANIC MATTER. One major use of biochemical operations is
in the removal of soluble organic matter which can be used as a food source by the
microorganisms present. When this occurs, a portion of the carbon is converted to
carbon dioxide and the rest is incorporated into new cell material. The carbon
dioxide is evolved as a gas and the cell material is removed in a physical operation
leaving the wastewater free of the original organic matter.

   Aerobic cultures are particularly suitable for the removal of organic matter
in the concentration range between 50 and 4000 mg/l as biodegradable chemical oxygen demand (COD). At lower concentrations carbon adsorption often proves to be
more economical. Anaerobic biochemical pretreatment is frequently used for concentrations above 4000 mg/l in order to reduce the quantity of oxygen which must be
provided in the subsequent aerobic operation. If the concentration of organic
matter to be removed is above 50,000 mg/l then evaporation and incineration or wet

combustion may be more economical. It should be emphasized that the concentrations
listed are for soluble organic matter. If the organic waste is either suspended or
colloidal it is often cheaper and easier to remove it by physical or physical-
chemical means, although mixtures of soluble and colloidal wastes are routinely
treated by biochemical means.

  STABILIZATION OF INSOLUBLE ORGANIC MATTER. In aqueous systems the organic
solids are generally in the form of a slurry with high solids concentrations. Traditionally, stabilization has been carried out anaerobically although within the
last ten years numerous uses have been made of aerobic reactors. The end products
of such stabilizations are inorganic solids and insoluble organic residues which are
relatively resistant to further biological activity and have characteristics similar
to humus. An additional product of anaerobic operations is methane gas. In addition to slurry systems, stabilization reactions are also carried out in environments
which are essentially moist solids, i.e., the solids concentration exceeds 30%.
These are generally aerobic and the end product is a humus-like material which finds
use as a soil conditioner and fertilizer.

  CONVERSION OF SOLUBLE INORGANIC MATTER. Since the discovery during the 1960's
of the effects of eutrophication upon lakes more attention has been turned toward
methods of removing inorganic compounds from wastewaters. One of the prime causes
of inorganic pollution is soluble phosphate and aerobic biochemical operations have
been proposed for their removal. Currently there is considerable controversy over
whether this is indeed a biochemical operation. A more clearly biochemical operation involving inorganic compounds is the one for the conversion of ammonia nitrogen
to nitrate nitrogen through the metabolism of two classes of aerobic autotrophic
bacteria. Complete removal of the nitrogen can then be accomplished in a balanced
anaerobic respiration operation utilizing nitrate as the terminal electron acceptor,
releasing nitrogen gas as the end product. Other inorganic transformations occur in
nature and an understanding of their mechanisms by engineers will aid in their control. Among these are the anaerobic conversion of sulfate ions to sulfide ions resulting in both noxious and toxic conditions. Equally important is the aerobic conversion of sulfide ions to sulfate ions with the resultant corrosive conditions.

1.1.3 Reactor Configuration

Herbert [1] has suggested a classification of biochemical operations according to
reactor type. This has been modified as shown in Fig. 1.1. The importance of
classifying with respect to reactor type is that a given reactor type will perform
in much the same way regardless of the biochemical transformation that is being
carried out in it. Therefore before turning to a study of operations capable of
carrying out specific transformations it is important to get a clear picture of
the many reactor types available.

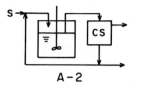
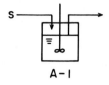

A-1

A-2

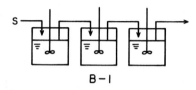

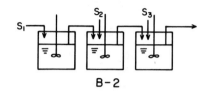

B-1

B-2

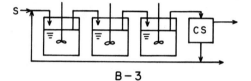

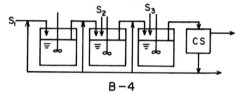

B-3

B-4

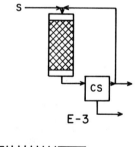

C

D

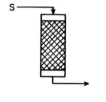

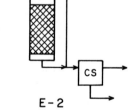

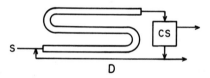

E-1

E-2

E-3

F-1

F-2

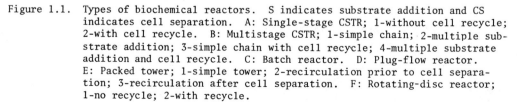

Figure 1.1. Types of biochemical reactors. S indicates substrate addition and CS
indicates cell separation. A: Single-stage CSTR; 1-without cell recycle;
2-with cell recycle. B: Multistage CSTR; 1-simple chain; 2-multiple substrate addition; 3-simple chain with cell recycle; 4-multiple substrate
addition and cell recycle. C: Batch reactor. D: Plug-flow reactor.
E: Packed tower; 1-simple tower; 2-recirculation prior to cell separation; 3-recirculation after cell separation. F: Rotating-disc reactor;
1-no recycle; 2-with recycle.

The continuous stirred tank reactor (CSTR) (A) is the simplest possible continuous flow slurry reactor system, with a liquid stream containing a substrate
entering and a stream of microbial slurry leaving. The liquid volume is maintained
constant by some sort of liquid level device and the performance of the reactor is
controlled by the hydraulic residence time. Mixing is sufficient to make the concentration uniform throughout and consequently these reactors are also called completely mixed reactors. The mixing maintains the culture in a constant average
physiological state. Considerable flexibility may be gained by the addition of a
cell separator which allows recycle of a portion of the microorganisms. This is because performance of the reactor will then depend upon the cell residence time
rather than the hydraulic residence time. As normally operated the liquid stream
leaving the top of the cell separator is free of microorganisms while that leaving
the bottom contains a concentrated slurry. Part of that concentrated stream is recycled and the remainder constitutes cell wastage.

Connecting two to four stirred reactors in series affords another variation
(B). Feed may be added to only the first reactor or to each one. In addition, cell
recycle may be employed about the entire chain or any portion of it. The behavior
of systems such as these is complex because the physiological states of the cells
change as they pass from reactor to reactor. One of the advantages of multistage
systems is that a different reaction may be carried out in each stage. This may
even occur when all substrate is added to the first stage because of normal metabolic control mechanisms. When multiple substrate addition is employed it is often
possible to introduce a different substrate into a later stage in order to take
maximum advantage of the physiological state of the microorganisms.

In a batch reactor (C) there is no flow; instead, a "batch" of material is
placed into a vessel, inoculated, and growth is allowed to occur. As growth progresses, reaction conditions change and consequently so does the growth environment.
This can lead to the familiar "bacterial growth curve" wherein the cells present at
different times will be in different physiological states and no steady state is
possible. The perfect plug-flow reactor (D) may be considered a moving batch culture because each increment of flow moves through the reactor unaffected by those on
either side. Recycle is necessary to provide the inoculum needed to start growth
and there will be an increasing concentration of cells from inlet to outlet. The
single most important difference from a CSTR is that the cells at different points
in the reactor will be in different physiological states, thus a steady state with
respect to the entire reactor cannot be achieved. A sort of steady state can be
obtained, however, wherein the conditions at a particular point in the reactor do
not change with time. In spite of the fact that the "batch growth curve" has been
recognized much longer than the character of growth under constant conditions

(i.e., CSTR) our ability to model the batch or plug-flow system is much less advanced because of the extreme complexity introduced by changing physiological
conditions.

Whereas reactors A-D in Fig. 1.1 may all be classified as slurry reactors, E
and F are referred to as "fixed films." In a packed tower (E) the organisms grow
as a film on an immobile support, such as rock or plastic media. The medium is not
submerged, but instead the fluid runs in a thin sheet over it. If no recirculation
is employed there is considerable change in reaction environment from top to bottom
of the tower as the bacteria remove the substrate. Recirculation of fluid around
the tower tends to reduce the severity of the change in reaction environment. The
higher the recirculation flow, the more homogeneous the reaction environment becomes. The kinetics of this reactor type can be strongly influenced by the manner
in which the flow is recirculated. In order to attain a stable condition organisms
are continually sloughed from the solid surface. If those organisms are removed
prior to recirculation, then substrate removal is caused primarily by the activity
of the organisms attached to the support media. On the other hand, if flow is
recirculated prior to the removal of the sloughed-off organisms the fluid stream
will resemble that of a slurry reactor and substrate removal will be due to both
the fixed and suspended organisms.

In the rotating-disc reactor (F) the microorganisms grow attached to plastic
discs which are rotated in the liquid. If the reaction vessel is relatively small
with respect to the flow the environmental conditions will be relatively uniform
throughout. If, however, a long rectangular basin is employed, then reaction conditions will change along the tank length so that the physiological state of the
organisms on one disc may be different from that of those on another. The recycle
of sloughed-off organisms makes the reactor behave in a manner intermediate between
slurry and fixed-film systems.

There are a number of other reactor systems which could be employed. Some are
just modifications of those listed here, while others are totally different. Familiarity with the reactor types shown in Fig. 1.1 will serve the needs of most situations, however.

1.2 COMMON "NAMED" BIOCHEMICAL OPERATIONS

In almost all fields certain operations have gained common names through years of
use and development. Although such names are not always logical, they are recognized and accepted because of their historical significance. Many of the common
biochemical operations were developed prior to an understanding of the principles
underlying them. Consequently, some of the names are applied to more than one
reactor configuration and in some cases to more than one operational objective.
For purposes of discussion nine common names have been chosen and these are listed
in Table 1.1. In order to relate these names to the classification scheme discussed

TABLE 1.1
Common Named Biochemical Operations

---

1. Aerated Lagoon
2. Activated Sludge
3. Aerobic Digestion
4. Trickling Filter
5. Rotating Biological Contactor
6. Nitrification
7. Anaerobic Contact
8. Anaerobic Digestion
9. Denitrification

---

previously, Table 1.2 was prepared. Table 1.2 is in the form of a matrix labeled
across with operational objectives and down with reactor configurations. Each block
of the matrix is placed into one of three categories. If the combination of objective and reactor type fits into one of the named operations then that name is given.
Combinations which are possible but which are seldom if ever seen are denoted by a
single asterisk. Combinations which have been tried but which are not yet in common
use are marked with two asterisks.

1.2.1 Aerated Lagoons

Aerated lagoons can generally be classified as completely mixed reactors which do
not employ cell recycle. Usually they are large earthen basins which are mixed and
aerated by surface aerators. When the hydraulic retention time is long, very good
removal of soluble organic matter is achieved. If the treatment provided by the
lagoon is the only treatment that the wastewater is to receive then some provision
must be made for the removal of the microorganisms. This is often accomplished in
a large quiescent pond. If the lagoon is used only as a pretreatment device then
the cells are discharged with the liquid. The primary purpose of the operation is
the removal of soluble organic matter through its conversion to microbial cells.
The main difference between it and activated sludge, however, is that the organisms
in the lagoon usually grow in the dispersed state rather than as a flocculent mass.

1.2.2 Activated Sludge

"Activated Sludge" appears seven times in Table 1.2, indicating that it is not a
very descriptive term. It is generally used to denote an aerobic flocculent slurry
of microorganisms which remove organic matter from wastewater and are then removed
themselves, usually by sedimentation. Activated sludge is best suited for the removal of soluble organic matter because insoluble organic matter can usually be
removed more economically by physical-chemical means. Often, however, wastewaters
will contain both soluble and insoluble organic matter. If the concentration of
soluble organic matter is greater than about 50 mg/l of biodegradable COD then the
activated sludge process may be used. Under those conditions it is usually more
economical to use the activated sludge for the removal of the insoluble organics as
well, in which case they are removed by adsorption and physical entrapment within

Table 1.2
Classification of Named Biochemical Operations

Objective	Removal of Soluble Organic Matter		Stabilization of Insoluble Organic Matter		Conversion of Soluble Inorganic Matter	
Reactor Configuration	Aerobic Environment	Anaerobic Environment	Aerobic Environment	Anaerobic Environment	Aerobic Environment	Anaerobic Environment
A1	Aerated Lagoon	Anaerobic Contact **	Aerobic Digestion **	Anaerobic Digestion **	Oxidation Pond	**
A2	Activated Sludge	Anaerobic Contact *	Aerobic Dig. or Act. Sludge *	Anaerobic Digestion *	Nitrification	Denitrification
B1	Aerated Lagoon	**	**	**	*	*
B2	Aerated Lagoon	*	*	*	*	*
B3	Activated Sludge	**	**	**	Nitrification	*
B4	Activated Sludge	*	*	*	Nitrification	*
C	Activated Sludge	*	Aerobic Digestion	Anaerobic Digestion	*	*
D	Activated Sludge	*	Activated Sludge	*	Nitrification	*
E	Trickling Filter	**	--	Anaerobic Filter	Nitrification	Denitrification
F	Rotating Biol. Contactor	*	--	--	Nitrification	*

* = Possible, but very rare; ** = Possible, but not common.

the floc. The kinetics of the operation are therefore still governed by the concentration of soluble organic matter and are reasonably well defined.

The first uses of activated sludge were on a batch basis. At the end of each
aeration period a sludge was present which was left in the reactor when the clear
wastewater was withdrawn after settling. As this batch procedure was repeated the
sludge built up giving more complete removal of organic matter within the allowed
reaction time. Although this increase in removal activity was due to the growth of
a viable microbial culture its reason was unknown to the early researchers who
characterized the sludge as being "activated" thereby giving it its name [2]. This
batch or "fill and draw" technique is still used in some very small installations,
although it is rare.

As the need for continuous operation increased, the batch operation was converted to continuous flow through the use of long aeration chambers similar to plug-
flow reactors. Various modifications of the plug-flow reactor were tried, among
them introduction of the substrate at various points along the tank in a manner
similar to the introduction of feed into CSTR's in series. In the middle 1950's
various engineers began advocating the CSTR with cell recycle as an alternative to
the conventional activated sludge reactor because of its inherent stability. That
stability, plus the advantages regarding the physiological state of the cells, has
caused an increase in the use of the CSTR in the last decade. More recently, attempts to use pure oxygen for oxygen transfer have been successful. The efficacy
of the oxygen transfer system is dependent upon the use of completely mixed reactors
in series, and therefore this innovation in oxygen transfer has caused increased
interest in the use of multistage CSTR systems.

1.2.3 Aerobic Digestion

Aerobic digestion is the name given to the aerobic destruction of insoluble organic
matter in a slurry reactor. Generally aerobic digesters employ a CSTR with a long
solids retention time allowing ample time for the conversion of much of the organic
carbon to CO_2. When the solids being destroyed are waste bacteria their reduction
occurs primarily as a result of endogenous decay. Aerobic digestion is often used
to get rid of the excess activated sludge formed in the treatment of soluble wastes.
It is also often used at small "package plant" installations. Sometimes the aerobic
digestion is allowed to go on in the same reactor in which the soluble organics are
being removed, in which case it is given the name "extended aeration activated
sludge." In that particular case, nitrification usually occurs as well. In some
very small installations, such as dairy farms, aerobic digestion of solids is
carried out as a fill and draw semibatch operation.

1.2.4 Trickling Filter

Trickling filter is the common name used to describe a fixed-film reactor in the shape of a packed tower. Until the mid 1960's trickling filters were made of stone, which limited their height to around six feet. More recently plastic media have been used which resemble egg crates and allow excellent passage of air through the tower. Because of their greater void space and lighter weight, plastic media trickling filters are self-supporting to heights of 20 ft. The wastewater runs in a thin sheet over the packing material upon which a microbial film is growing, and the organisms extract the soluble organic matter as a source of carbon and energy. Traditionally, trickling filters have been used to treat domestic sewage in small to medium size installations desiring minimum operational expense. Since the introduction of the plastic media many trickling filters have been built as pretreatment devices preceding other biochemical operations. This is because they have the ability to reduce the waste concentration at relatively low operating cost, a bonus when aerobic treatment is being employed. Trickling filters cause relatively little degradation of suspended organic matter and thus should not be used for its degradation.

1.2.5 Rotating Biological Contactor

The rotating biological contactor is a relatively new application of an old idea for the removal of soluble organic matter. Microorganisms growing attached to the rotating discs convert the organic matter into energy and new cells. Its applications are similar to those of a trickling filter, ranging from pretreatment to complete removal of soluble organic matter.

1.2.6 Nitrification

Nitrification, a type of reaction rather than a distinct operation, occurs when ammonia nitrogen in a wastewater is converted to the nitrate form through the action of autotrophic nitrifying bacteria. This is becoming more popular as the need for nitrogen control increases. It can be accomplished in several types of reactors as long as the environment is aerobic. In slurry reactors it is possible for nitrification to occur simultaneously with the removal of soluble organics, consequently many activated sludge reactors are designed to carry out both conversions. Recently, however, interest has shifted to the use of packed towers.

1.2.7 Anaerobic Contact

Operations used to remove soluble organic matter under anaerobic conditions are called anaerobic contact. It is also used to treat wastes containing a mixture of soluble and suspended organics just as the activated sludge process is. Although Table 1.2 indicates the use of a CSTR with or without cell recycle the vast majority employ cell recycle. The anaerobic contact operation is well suited as a pretreatment method for wastes containing more than 4000 mg/l COD but less than 50,000 mg/l [3], because it is less expensive than either activated sludge or evaporation.

Its main advantages over activated sludge are lower power requirements and the production of fewer excess solids. Further treatment is required for the effluent
from anaerobic contact, however, because many hydrolysis products will be left in
solution. ·

1.2.8 Anaerobic Digestion

By far the biggest use of anaerobic cultures is in the stabilization of suspended
organic matter. Two major groups of bacteria reside in the culture. One is responsible for the hydrolysis of the solids, with the major end products being
soluble, short chain fatty acids and a stable insoluble residue similar to humus.
The other group is responsible for the conversion of the fatty acids to methane gas.
Anaerobic digestion is one of the oldest means of wastewater treatment and yet because of the complex ecosystem involved the control of the process is still a subject of much research. Designers currently favor the CSTR, because of its uniform
environmental condition. Most have been built without solids recycle but recycle
is gaining in popularity because it allows the use of smaller reactors. There is
also considerable interest in two-stage reactors because each reactor could then be
operated in the optimum manner for one of the groups of organisms. Finally, consideration is also being given to the use of the plug-flow reactor although its
efficacy has not been proven under full scale conditions.

1.2.9 Denitrification

The conversion of nitrate ion to nitrogen gas is called denitrification, which is a
type of reaction rather than a named operation. Because nitrification is relatively
easy to accomplish, if denitrification could be carried out, then the resultant
nitrogen gas could be stripped from the system leaving the water free of excess
nitrogen. This conversion is carried out by organisms undergoing anaerobic respiration. When facultative bacteria are placed under anaerobic conditions some will
use an inorganic ion as a terminal electron acceptor if it is present. In a denitrification reactor anaerobic conditions are maintained and organic matter is added in
the stoichiometric amount required to cause the nitrate to be converted to nitrogen
yet not allow sulfates to be converted to sulfides. Most investigations have been
centered around the CSTR with cell recycle and the fixed-film reactor.

1.2.10 Miscellaneous Operations

The large number of asterisks in Table 1.2 indicates the potential for development
of biochemical operations. No doubt many of those possibilities will be proved to
be feasible and thus join the group of "named operations." One point should always
be kept in mind however, and that is: all of these operations are dependent upon
living systems. Consequently all follow the fundamental laws of biochemistry and
microbiology. Because reactor configuration and reaction environment can strongly
influence the kinetic response of the organisms involved, the environmental engineer
should always strive to choose the optimum conditions for the particular objective

to be accomplished. The truly innovative engineer will exploit the flexibility of microbial systems in order to achieve the best possible design.

1.3 KEY POINTS

1. Biochemical operations may be carried out in either an aerobic or an anaerobic environment, and the choice of environment has a profound effect upon the ecology of the microbial population.

2. Three major biochemical transformations may be performed with biochemical operations: removal of soluble organic matter; stabilization of insoluble organic matter; and conversion of soluble inorganic matter.

3. The major reactor types are: single-stage continuous stirred tank reactor; multistage continuous stirred tank reactor; batch reactor; plug-flow reactor; packed tower; and rotating-disc reactor.

4. The major named biochemical operations are: aerated lagoon, activated sludge, aerobic digestion, trickling filter, rotating biological contactor, nitrification, anaerobic contact, anaerobic digestion, and denitrification.

1.4 STUDY QUESTIONS

1. List and define the three major biochemical transformations which may be performed with biochemical operations.

2. Describe each of the six major reactor types which find use in biochemical operations.

3. List the nine named biochemical operations.

4. Describe each of the named biochemical operations in terms of the biochemical transformation involved, the reaction environment used, and the reactor configuration employed.

REFERENCES AND FURTHER READING

1. D. Herbert, "A theoretical analysis of continuous culture systems," Society of Chemical Industry, Monograph No. 12, 21-53, London, 1960.

2. C. N. Sawyer, "Milestones in the development of the activated sludge process," *Journal of Water Pollution Control Federation,* 37, 151-170, 1965.

3. G. G. Cillie, *et al.* "Anaerobic digestion-IV. The application of the process in waste purification," *Water Research,* 3, 623-643, 1969.

Further Reading

Annual reviews of the water pollution control literature appear in the June issue of the *Journal of the Water Pollution Control Federation.* Reviews which are particularly pertinent to this text include "Biological Filters," "Activated Sludge," "Lagoons and Oxidation Ponds," and "Sludge Treatment, Utilization, and Disposal."

PART II

FUNDAMENTALS OF REACTOR ENGINEERING

Two major factors determine the quality of the effluent leaving a biochemical operation: the kinetics of the reactions occurring within the operation and the characteristics of the reactor within which they occur. "Reactor engineering" is the name
given to the study and exploitation of the interactions of those two factors and
through its use, a biochemical engineer may choose a reactor system that will accomplish the desired goals in the most expeditious fashion. Consequently, an understanding of the basic concepts of reactor engineering is a prerequisite to the
proper design of biochemical operations. The purpose of this part of the book is to
provide those concepts. First we will review stoichiometry and mass balance equations. Next we will investigate the various forms that reaction rate expressions
can take and then we will combine them with mass balance equations for ideal continuous stirred tank and plug-flow reactors in order to see how reactor type and
reaction rate influence system performance. After the importance of reaction rate
expressions has been established, we will investigate the techniques by which rate
data can be acquired and analyzed to give us the needed rate expressions. Finally,
because real-world reactors are not ideal, we will briefly consider the effects of
nonideal flow patterns and the techniques by which the characteristics of nonideal
reactors may be evaluated.

CHAPTER 2

STOICHIOMETRIC RELATIONSHIPS, REACTION RATES, AND MASS BALANCE EQUATIONS

The starting point for the design of any reactor is the writing of the mass and energy balance equations. The mass balance equations keep track of various chemical and microbial species which enter and leave the reactor, whereas the energy balance equations keep track of energies associated with the influent and effluent streams as well as the reacting system. In general, the mass and energy balance equations are coupled and must be solved simultaneously. In addition to the number of moles of reactants consumed for each mole of products formed it is also necessary to know the rates at which the reactants are consumed and the products generated. In this chapter we begin by developing the stoichiometric relationships among the reactants and products. Then the rates of generation of product and consumption of reactant are defined, as well as the generalized reaction rate. For introductory purposes we will consider only isothermal reaction conditions and hence the energy balance equation is not discussed here although it is in later chapters. In this chapter we will concentrate on how the rates enter into the mass balance equations around three common types of industrially important reactors: batch reactors, continuous-flow stirred tank reactors, and plug-flow reactors.

2.1 STOICHIOMETRIC EQUATIONS

Consider a simple reaction in which a reactant A reacts to form a product B
according to the stoichiometric equation

$$aA \rightarrow bB \tag{2.1}$$

where a and b are stoichiometric coefficients and therefore, are positive numbers.
Equation 2.1 states that for every a moles of A consumed b moles of B are generated.
Quantitatively this is equivalent to

$$(N_{Ao} - N_A)/a = (N_B - N_{Bo})/b \tag{2.2}$$

where N_A and N_B stand for the number of moles of A and B at any time and N_{Ao} and N_{Bo}
are the initial number of moles of A and B. The term $(N_{Ao} - N_A)$ represents the
number of moles of A consumed while $(N_B - N_{Bo})$ represents the number of moles of B
generated. In terms of Δ, which stands for the difference between the present (or
final) and the past (or initial) states (i.e., $\Delta N_A = N_A - N_{Ao}$ and $\Delta N_B = N_B - N_{Bo}$),
Eq. 2.2 can be put into the following form

$$\Delta N_A/-a = \Delta N_B/b \tag{2.3}$$

Note that in this form the denominator associated with ΔN_A is the negative of the
stoichiometric coefficient of A, -a, while the denominator associated with ΔN_B is
the stoichiometric coefficient of B, b. Equation 2.2 or 2.3 represents an equation
in four unknowns, N_A, N_B, N_{Ao} and N_{Bo}, and therefore, may be used to calculate any
one unknown, provided that the other three are known.

In general, many reactions involve more than one reactant and more than one
product so that a more general stoichiometric equation involving two reactants and
two products is

$$aA + bB = cC + dD \tag{2.4}$$

where a, b, c and d are stoichiometric coefficients and therefore, are positive
numbers. Since Eq. 2.4 implies that for every a moles of A and b moles of B consumed c moles of C and d moles of D are generated, it gives the following quantitative relationships

$$\Delta N_A/-a = \Delta N_B/-b = \Delta N_C/c = \Delta N_D/d \tag{2.5}$$

Equation 2.5 represents three independent equations ($\Delta N_A/-a = \Delta N_B/-b$, $\Delta N_B/-b = \Delta N_C/c$
and $\Delta N_C/c = \Delta N_D/d$) in four unknowns ($\Delta N_A$, ΔN_B, ΔN_C, and ΔN_D), so that once the
change in the number of moles of any one species (A,B,C, or D) is known, the corre-
sponding changes for all other species are fixed. Since the denominators in Eq.
2.5 are the negative values of the stoichiometric coefficients of the reactants and
the positive values of the stoichiometric coefficients of the products, it is convenient to write the stoichiometric equation (Eq. 2.4) in modified form:

$$cC + dD - aA - bB = 0 \tag{2.6}$$

Note that the stoichiometric coefficients of the products retain their original
signs while the coefficients of the reactants are given negative signs. In terms
of these modified stoichiometric coefficients the stoichiometric relationships of
Eq. 2.5 state simply that the change in the number of moles of one species divided
by its modified stoichiometric coefficient is equal to the change in the number of
moles of any other species involved in the reaction divided by its modified
stoichiometric coefficient.

EXAMPLE 2.1-1

Given the following stoichiometric equation:

$$2A + B = 3C + 4D$$

write the modified stoichiometric equation and establish the quantitative relationships for the changes in the number of moles of each of the species.

The modified stoichiometric equation is:

$$3C + 4D - 2A - B = 0$$

The stoichiometric relationships are

$$\Delta N_C/3 = \Delta N_D/4 = \Delta N_A/-2 = \Delta N_B/-1$$

or

$$(N_C - N_{Co})/3 = (N_D - N_{Do})/4 = (N_A - N_{Ao})/-2 = (N_B - N_{Bo})/-1$$

2.2 REACTION RATE

To design a chemical reactor to produce a desired amount of product in a given time,
we need to know not only the stoichiometric relationships, but also the rates at
which the reactants are consumed and the products formed. Since the rate of generation of product and the rate of consumption of reactant are used in the mass
balance equations for the product and the reactant respectively, it is convenient
to define a rate associated with a particular species. Consider the simple reaction
in which for every mole of A consumed two moles of B are formed:

$$A \rightarrow 2B \tag{2.7}$$

The notation r_B denotes the rate of generation of species B in a control volume of
reactor and has the units of moles of species B formed per unit time per unit
volume. In general, r_j stands for the rate of generation of species j in a control
volume of reactor. Thus, r_A is the rate of generation of A. However, since A is
consumed in the reaction and not generated, a more meaningful term for this situation is the rate of consumption of A. This, by convention, is $-r_A$. With this sign

convention, and when the rates are expressed in molar units, the stoichiometric
equation, Eq. 2.7, implies the rate of consumption of A, $-r_A$, is equal to one half
of the rate of production of B, r_B, i.e.,

$$-r_A = r_B/2 \qquad (2.8)$$

It should be emphasized that the relationship between the rates given by Eq. 2.8 is
valid only when the rates are expressed in molar units, moles per unit volume per
unit time. When it is difficult to establish the molecular weight because the exact
molecular structure is unknown, as in the case of microbial cells, the rates are
usually expressed as mass per unit time per unit volume, for example grams/(hr·liter).
In this situation the stoichiometric equations are not known exactly and the relationship between rates cannot be established a *priori,* but must be determined
experimentally.

EXAMPLE 2.2-1

Reactant A, having a molecular weight of 100, reacts according to Eq. 2.7 to yield
product B with a molecular weight of 50. If the rate of generation of B in mass
units is known to be 10 grams/(hr·liter) what would be the corresponding rate of
consumption of A?

The rate of generation of B in molar units is:

r_B = [10 grams/(hr·liter)]/50 grams/mol

r_B = 0.2 mol/(hr·liter)

According to Eq. 2.8

$-r_A = r_B/2 = 0.2/2$

$-r_A$ = 0.1 mol/(hr·liter)

In mass units,

$-r_A$ = (0.1)(100) = 10 grams/(hr·liter)

It is also convenient to define the generalized rate of a reaction, in addition
to the rates associated with the individual species. This is done through the modified stoichiometric equation. For example, rewriting Eq. 2.7 in modified form gives

$$2B - A = 0 \qquad (2.9)$$

Inspection of Eqs. 2.8 and 2.9 suggests that the rate of generation of one species
divided by its modified stoichiometric coefficient is equal to the rate of generation

of any other species in the reaction divided by its modified stoichiometric coefficient, i.e.,

$$r_B/2 = r_A/-1 \qquad\qquad (2.10)$$

Note the analogy between the relationship among individual species rates, Eq. 2.10, and the stoichiometric relationships, Eq. 2.3. In view of Eq. 2.10, which implies that the reaction rate for only one species needs to be known, we can define the generalized reaction rate, r, by

$$r = r_B/2 = r_A/-1 \qquad\qquad (2.11)$$

For the general reaction of Eq. 2.4 with the modified stoichiometric equation of Eq. 2.6, the generalized rate of reaction is

$$r = r_C/c = r_D/d = r_A/-a = r_B/-b \qquad\qquad (2.12)$$

EXAMPLE 2.2-2

Determine the generalized reaction rate for the following reaction:

$A + 2B \rightarrow 3C$

First, write the stoichiometric equation in modified form

$3C - A - 2B = 0$

Then, write the generalized rate in accordance with Eq. 2.12.

$$r = r_C/3 = r_A/-1 = r_B/-2$$

Thus far, we have defined the individual species rates, r_j, and the generalized reaction rate, r, but nothing has been said about the nature of these rates. In general, r_j or r is an intensive quantity which depends on the temperature and composition of the homogeneous phase in which the reaction takes place. Unfortunately, the algebraic expression for the rate cannot generally be predicted *a priori* so that each reaction must be studied experimentally to establish it. In fact, the development and verification of rate expressions are two of the major concerns of applied kineticists. In later chapters we shall consider systematic ways for hypothesizing rate expressions and verifying them experimentally. In this chapter we are primarily concerned with the use of rate expressions in mass balance equations. Therefore, for the present we shall assume that the rate expressions are known and proceed to show how they are used in the design of various reactors.

2.3 MASS BALANCE EQUATIONS

The starting point for a mass balance on any system is the specification of the
control volume, or the boundary of the system. For reactors the control volume may
be either the entire reactor volume or a differential portion of it. When the
reaction conditions, including the composition, are uniform over the whole reactor
volume, then the entire volume may be taken as the control volume. Otherwise, a
differential element of the reactor volume should be taken as the control volume.
Having picked an appropriate control volume the mass balance on each species
(reactants and products) is made around the control volume by keeping track of the
amounts entering or leaving, and generated or consumed. As we saw in Section 2.1,
for a reaction with known stoichiometry it is only necessary to write a mass balance
on any one species (usually the reactant) because once this mass balance has been
completed, giving the number of moles of the reactant consumed, the numbers of moles
of other species participating in the reaction may be accounted for through the
stoichiometric relationships, such as Eq. 2.5. For multiple reactions, such as
parallel reactions, two or more mass balance equations must be written to define the
system completely. In addition, if the reaction is carried out nonisothermally, an
energy balance equation is also necessary.

The general unsteady-state mass balance equation for species j around the
control volume is

$$
\begin{bmatrix} \text{Rate of flow of species } j \\ \text{into the control volume} \\ \text{(mol/time) or (mass/time)} \end{bmatrix} - \begin{bmatrix} \text{Rate of flow of species } j \\ \text{out of the control volume} \\ \text{(mol/time) or (mass/time)} \end{bmatrix} +
$$

$$
\begin{bmatrix} \text{Net rate of generation of} \\ \text{species } j \text{ in the control volume} \\ \text{(mol/time) or (mass/time)} \end{bmatrix} = \begin{bmatrix} \text{Rate of accumulation of species } j \\ \text{within the control volume} \\ \text{(mol/time) or (mass/time)} \end{bmatrix} \quad (2.13)
$$

or simply

$$
\text{Input - Output + Generation = Accumulation} \qquad (2.14)
$$

When the stoichiometry is known, as in most chemical reactions, it is more convenient to use molar units in the mass balance equation. However, when the molecular weights and stoichiometry are not known exactly, it may be necessary to use
mass units. In this situation the rate expression must also be expressed in mass
units.

In molar units Eq. 2.14 can be written as

$$
n_{jo} - n_j + r_j V_c = \frac{\partial}{\partial t} (N_j) = \frac{\partial}{\partial t} (C_j V_c) \qquad (2.15)
$$

where n_{jo} and n_j are the inlet and outlet molar flow rates of species j, r_j the rate
of generation of species j in molar units, V_c the control volume, N_j the total moles

of the species j in the control volume, C_j the molar concentration of the j species, and t the time. Note that n_{jo}, n_j, C_j and V_c may all be functions of time.

Equation 2.14 can also be expressed in mass units as

$$m_{jo} - m_j + r_j' V_c = \frac{\partial}{\partial t} (M_j) = \frac{\partial}{\partial t} (C_j' V_c) \tag{2.16}$$

where m_{jo} and m_j are the inlet and outlet mass flow rates of species j, r_j' the rate of generation of species j in mass units, M_j the total mass of the species j in the control volume, and C_j' the mass concentration of the species j. Note that the balance equation in molar units, Eq. 2.15, may be obtained from the balance equation in mass units, Eq. 2.16, by dividing both sides of it by the molecular weight of species j. Conversely, Eq. 2.16 may be obtained by multiplying Eq. 2.15 by the molecular weight.

In most chemical reactions the stoichiometry is known so that it is convenient to work with the mole balance equation, Eq. 2.15. For microbial reactions for which the exact stoichiometry is not known, we need to use Eq. 2.16. The use of Eq. 2.15 will now be illustrated for three types of reactors: batch, continuous stirred tank and plug-flow reactors. For this purpose the simple reaction of Eq. 2.7 will be used. Furthermore, the reaction is assumed to take place isothermally in an aqueous phase without a catalyst. Under these assumptions the rate expression is expected to be a function of only the concentration of A, C_A. If one needs to calculate the heat exchange requirement for this reaction it would be necessary to write an energy balance equation. We assume here that no heat exchange requirement needs to be determined and hence, will dispense with the energy balance equation.

2.3.1 Continuous Stirred Tank Reactors (CSTR)

A CSTR, also known as a continuous-flow stirred tank reactor (CFSTR), backmix reactor, or completely mixed reactor, is used very frequently in industry. As shown in Fig. 2.1 it has a feed stream called the influent and an exit stream called the

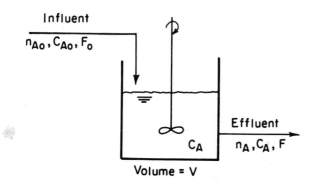

Figure 2.1. Continuous stirred tank reactor (CSTR).

effluent. It is usually equipped with baffles and a powerful mixer which is operated at a sufficiently high speed so that the mixing is assumed to be perfect, i.e., it is homogeneous and instantaneous so that any reactant carried into the reactor by the feed is dispersed evenly throughout the reactor without any time delay. Thus, samples taken from all parts of the reactor have the same composition. In addition, the reaction is assumed to take place only in the reactor so that the effluent composition is the same as the reactor composition.

Under the above assumptions the concentrations of A and B are uniform throughout the reactor and hence it is logical to pick the entire volume occupied by the fluid within the reactor as the control volume. Application of Eq. 2.15 on the reactant A around the entire reactor volume yields

$$n_{Ao} - n_A + r_A V = \frac{d}{dt}(N_A) = \frac{d}{dt}(C_A V) \tag{2.17}$$

where V stands for the reactor volume. This type of reactor is normally operated at steady state, which means that the time rate of change of any component is zero. Thus, the steady-state balance on A around a CSTR is obtained by setting $d(C_A V)/dt$ equal to zero in Eq. 2.17:

$$n_{Ao} - n_A + r_A V = 0 \tag{2.18}$$

or

$$F_o C_{Ao} - FC_A = -r_A V \tag{2.19}$$

where F_o and F are the volumetric flow rates of the influent and effluent respectively, and C_{Ao} and C_A are the concentrations of A in the influent and effluent (or reactor), respectively. For aqueous reactions it is safe to assume that the flow rates of the influent and effluent are the same, i.e., $F_o = F$, so that Eq. 2.19 may be written as

$$-r_A = F(C_{Ao} - C_A)/V \tag{2.20}$$

Since A is consumed, $r_A < 0$ and $-r_A > 0$. Since the right-hand side is the difference in the mass flow rates of A in the influent and effluent per unit volume of reactor, and the left-hand side is the rate of consumption of A in the reactor, Eq. 2.20 simply states that the difference in the mass flow rates of A into and out of the reactor is due to the consumption of A in the reactor. Equation 2.20 can be used to calculate the rate of consumption of A as a function of C_A, and thus it may be used to obtain a rate expression experimentally. By varying the flow rate F, the reactor volume V, or the influent concentration C_{Ao}, the effluent concentration C_A can be varied experimentally and the corresponding rates calculated using Eq. 2.20. This will provide data on the effects of C_A upon r_A, which can be used either to test a proposed rate expression or to obtain an empirical rate expression by using standard correlation techniques to obtain a functional relationship between C_A and r_A.

The major utility of Eq. 2.19 does not, however, lie in obtaining the rate expression as a function of C_A. The major use is for reactor design and evaluation. Rearrangement of Eq. 2.19 yields

$$V = F(C_{Ao} - C_A)/-r_A \qquad\qquad (2.21)$$

Equation 2.21 can be used to calculate the required reactor volume once the flow rate, F, the influent concentration, C_{Ao}, the desired outlet concentration, C_A, and the rate, r_A, are known. Conversely, it can also be used to calculate the volumetric flow rate that can be handled by a reactor of volume V,

$$F = -r_A V/(C_{Ao} - C_A) \qquad\qquad (2.22)$$

In summary the mass balance equation can be used to recover the rate expression from a series of steady-state runs. In addition, the reactor volume necessary to process a given flow rate of the feed or the flow rate that a reactor with a given volume can handle can also be calculated, provided that the rate expression and the feed and effluent concentrations are known.

EXAMPLE 2.3.1-1

For the aqueous reaction A → B the rate expression is given by $r_A = -10C_A$, mol/(liter·hr). This reaction is to be carried out in a CSTR, and the feed concentration C_{Ao} is 2 mol/liter.

(a). Determine the volume of the reactor necessary to convert 50% of A in the influent, which is flowing at a rate of 100 liter/hr.

Solution:

Given information;

F = 100 liters/hr

C_{Ao} = 2 mol/liter

$C_A = C_{Ao}(0.5) = 2(0.5) = 1$ mol/liter

$r_A = -10C_A$ mol/(liter·hr)

Equation 2.21 gives

$V = F(C_{Ao} - C_A)/-r_A = 100(2 - 1)/-(-10)(1) = 10$ liters

(b). Assuming the same conversion, if the reactor volume were 50 liters what would be the influent flow rate it could handle?

Solution:

Given information;

V = 50 liters

C_{Ao} = 2 mol/liter

C_A = C_{Ao}(0.5) = 2(0.5) = 1 mol/liter

r_A = -10C_A mol/(liter·hr)

Equation 2.22 gives

F = $-r_A V/(C_{Ao} - C_A)$ = 10(1)(50)/(2 - 1) = 500 liters/hr

(c). What is the rate of reaction, $-r_A$, in b?

Solution:

Equation 2.20 gives

$-r_A$ = $F(C_{Ao} - C_A)/V$ = 500(2 - 1)/50 = 10 mol/(liter·hr)

which, of course, must be equal to the rate calculated from the rate expression

$-r_A$ = 10C_A = 10(1) = 10 mol/(liter·hr)

2.3.2 Batch Reactors

Another industrially important type of reactor is a batch reactor. Typically, a
batch reactor is initially charged with a feed containing the reactants, sealed, and
then the required reaction conditions (e.g., temperature) are imposed upon it.
After allowing the reaction to take place for the period of time necessary to achieve
a desired conversion, the reaction is stopped (e.g., by cooling) and the reactor is
emptied for another run. Thus, during the reaction period a batch reactor has no
in nor out terms in the mass balance equation. This is depicted in Fig. 2.2. As in
a CSTR the batch reactor is assumed to be perfectly mixed so that the entire reactor
content is homogeneous at any given time. Hence, the rate is independent of reactor
position and the concentration varies only with time. Therefore, it is appropriate
to take the whole reactor volume as the control volume. The mass balance on A
around the batch reactor is

$$0 - 0 + r_A V = \frac{d}{dt}(C_A V) \tag{2.23}$$

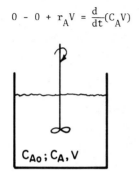

Figure 2.2. Batch reactor.

or

$$r_A = \frac{d}{dt}(C_A V)/V = (dN_A/dt)/V \qquad (2.24)$$

Equation 2.24 states that the rate of generation of A, r_A, in a batch reactor is
equal to the time rate of change of the total number of moles of A in the reactor
divided by the reactor volume. When the reactor volume is constant, as in aqueous
systems, this simplifies to

$$r_A = dC_A/dt \qquad (2.25)$$

which states that the rate is equal to the time rate of change of the reactant
concentration.

There has been considerable confusion in the chemical and environmental
engineering literature [1,2,3] concerning Eqs. 2.24 and 2.25 because many authors
have used them as general definitions of reaction rate. This is not correct,
however [4]. They are simply mass balance equations on a batch reactor, and as such
are not valid for any other type of reactor. In fact, some rather ridiculous conclusions can be reached if Eq. 2.24 or 2.25 is used to define the reaction rate.
For example, we know that at steady state in a CSTR, dC_A/dt is equal to zero, i.e.,
there is no change in the concentration with time. If we were to use Eq. 2.25 to
define r_A we would conclude that the reaction rate in the CSTR was zero, so that the
effluent concentration was equal to the influent concentration. This is, of course,
absurd. Consequently Eqs. 2.24 and 2.25 should never be used as a general definition
of reaction rate. Instead, they should always be thought of as the mass balance
equations for a batch reactor.

The above arguments should not be interpreted to mean that Eqs. 2.24 and 2.25
cannot be used to generate rate data, for they can. In fact, one of the main uses
of batch reactors is for the determination of rate equations. Because chemists
usually handle only batch reactors, however, Eqs. 2.24 and 2.25 are presented in most
chemistry textbooks as the definition of reaction rate [5,6,7]. Thus, care must be
exercised to keep the true definition of reaction rate in mind when applying the
resulting rate expression to other systems.

Equation 2.25 implies that by running a batch reactor experiment, measuring the
concentration of A in the reactor as a function time, and taking the time derivatives
of the concentration history, one can obtain the numerical values of the rate, r_A,
at various concentrations of A, C_A. Having obtained the data of rate versus reactant
concentration it is then possible to obtain a correlation between r_A and C_A. Once
this algebraic relationship is established and verified experimentally, it can then
be used in mass balance equations for any other reactors. Note that if the reactor
volume is not constant, it is necessary to take the time derivatives of the total
number of moles of A (Eq. 2.24) rather than of the concentration.

Perhaps the primary use of the mass balance equation is in the determination of the required reaction time. Rearranging Eq. 2.24 one obtains

$$\int_o^t V dt = \int_{N_{Ao}}^{N_{Af}} \frac{dN_A}{r_A} = \int_{C_{Ao}}^{C_{Af}} \frac{d(C_A V)}{r_A} \qquad (2.26)$$

This requires knowledge of the reactor volume as a function of time. For the aqueous reactions with which we are dealing here, the reactor volume may be safely assumed to be constant and Eq. 2.26 reduces to

$$t = \int_{C_{Ao}}^{C_{Af}} \frac{dC_A}{r_A} \qquad (2.27)$$

It is necessary to know the rate expression as a function of C_A to integrate the right-hand side of Eq. 2.27, thereby giving the reaction time necessary to convert the initial charge (C_{Ao}) to a desired concentration C_{Af}. With a known reaction time, down time (time necessary to stop the reaction, to empty the product and to recharge with a new batch), and production rate, one can then calculate the size of batch reactor necessary.

EXAMPLE 2.3.2-1

For the aqueous reaction A → B with a given rate expression, $r_A = -C_A$, (C_A in mol/liter, r_A in mol/(liter·hr));

(a). Determine the reaction time necessary to convert 50% of the initial charge with C_{Ao} = 1 mol/liter

Solution:

Given information;

C_{Ao} = 1 mol/liter

$r_A = -C_A$ mol/(liter·hr)

$C_A = C_{Ao}(0.5)$ = 0.5 mol/liter

Equation 2.27 yields

$$t = \int_{C_{Ao}}^{C_{Af}} \frac{dC_A}{r_A} = \int_{C_{Ao}}^{C_{Af}} \frac{dC_A}{-C_A} = \ln(C_{Ao}/C_A)$$

t = ln (1/0.5) = 0.693 hour

(b). If the down time is one hour and the desired production rate is 2400 moles of B/day, determine the size of batch reactor necessary under the conditions of (a).

Solution:

Given information;

t_r = reaction time = 0.693 hr

t_d = downtime = 1 hr

n_B = 100 mols/hr

Since 0.5 mol/liter of B is produced ($C_B - C_{Bo}$ = 0.5 mol/liter) in 1.693 hours
(0.693 hour of reaction time plus one hour of down time) the required reactor volume
is obtained by the following relationship

$$\text{Reactor volume} = \frac{\text{Desired production rate for B}}{\text{Avg. prod. rate for B per unit volume of reactor}}$$

$$= \frac{2{,}400 \text{ moles/24 hours}}{(0.5 \text{ mole/liter})/1.693 \text{ hours}} = 3386 \text{ liters}$$

2.3.3 Plug-Flow Reactors (PFR)

The last important flow reactor is the plug-flow reactor (PFR), which can be either
a simple tube or one packed with a catalyst or some other sort of packing. A feed
containing the reactants is fed continuously to the reactor inlet while the effluent
containing the products and unreacted reactants is removed from the outlet. For an
ideal plug-flow reactor it is assumed that the flow pattern inside has uniform
velocity and concentration in the radial direction at any point along the length of
the reactor and no longitudinal (axial) diffusional mixing of either reactants or
products along the reactor. This type of reactor is also known as a tubular, ideal
tubular, or piston-flow reactor.

Because of the assumption of plug flow and because the reaction takes place all
along the reactor length, the concentrations of reactants and products vary with the
axial distance x only. Therefore, it is appropriate to consider as the control
volume an infinitesimal volume, ΔV, in which the concentration may be considered
uniform. This control volume is shown in Fig. 2.3.

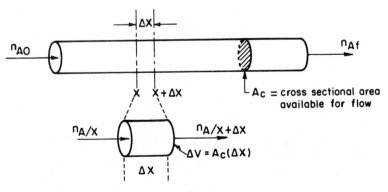

Figure 2.3. Plug-flow reactor (PFR).

The mass balance on the reactant A around the control volume is

$$n_A\big|_x - n_A\big|_{x+\Delta x} + r_A A_c \Delta x = \frac{\partial}{\partial t}(A_c \Delta x C_A) \tag{2.28}$$

or

$$-\frac{(n_A\big|_{x+\Delta x} - n_A\big|_x)}{\Delta x} + r_A A_c = A_c \frac{\partial C_A}{\partial t} \tag{2.29}$$

where A_c is the cross-sectional area of the reactor, x the distance from the reactor entrance, Δx the length of infinitesimal reactor volume, and $n_A\big|_x$ and $n_A\big|_{x+\Delta x}$ the molar flow rates of A evaluated at the distances x and x + Δx from the reactor entrance. In the limit as $\Delta x \to 0$, the first term in Eq. 2.29 becomes the partial derivative of n_A and Eq. 2.29 reduces to

$$-\frac{\partial n_A}{\partial x} + r_A A_c = A_c \frac{\partial C_A}{\partial t} \tag{2.30}$$

At steady state there is no change with respect to time in concentration at any point within the reactor so this equation reduces to

$$-\frac{dn_A}{dx} + r_A A_c = 0 \tag{2.31}$$

The molar flow rate may be taken as the product of the flow rate and molar concentration, i.e., $n_A = FC_A$. For aqueous solutions the flow rate may be assumed constant, i.e., it does not vary with the axial distance, so that Eq. 2.31 reduces to

$$-F\frac{dC_A}{dx} + r_A A_c = 0 \tag{2.32}$$

or

$$r_A = \frac{F}{A_c}\frac{dC_A}{dx} \tag{2.33}$$

Hence, it is theoretically possible to calculate the rates. If samples of the reacting fluid can be taken at various points along the reactor length without perturbing the concentration profile, one can obtain the concentration profile along the length, $C_A(x)$, from which the derivatives, dC_A/dx, can be evaluated all along the distance, thereby allowing computation of r_A. We then have a series of values of r_A at various values of C_A, thereby allowing correlation of the two. In practice, however, it is difficult, if not impossible, to sample at various points along the reactor length without disturbing the concentration profile. Consequently, plug-flow reactors are seldom used to generate rate data in this fashion.

The main purpose for which the mass balance equation is used is to determine the size of reactor required to achieve a desired conversion from a feed stream of given composition and flow rate. If the rate expression is known, Eq. 2.32 can be rearranged and integrated over the length of reactor L, to give

$$\int_0^L A_c dx/F = A_c L/F = V/F = \int_{C_{Ao}}^{C_{Af}} dC_A/r_A = \int_{C_{Af}}^{C_{Ao}} dC_A/(-r_A) \qquad (2.34)$$

Hence, the ratio of reactor volume to the volumetric flow rate of the feed, V/F, can be obtained by performing the indicated integration.

EXAMPLE 2.3.3-1

For the reaction considered in Example 2.3.2-1 determine the size of plug-flow reactor necessary to achieve a 50% conversion of A. The aqueous feed stream contains 1 mol/liter of A and is available at a flow rate of 1000 liters/hr.

Solution:

Given information;

F = 1000 liter/hr

C_{Ao} = 1 mol/liter

$r_A = -C_A$ mol/(hr·liter)

$C_A = C_{Ao}(0.5) = 1(0.5) = 0.5$ mol/liter

Equation 2.34 gives

$$V/F = \int_{C_{Af}}^{C_{Ao}} dC_A/(-r_A) = \int_{C_{Af}}^{C_{Ao}} dC_A/C_A = \ln(C_{Ao}/C_{Af})$$

$$= \ln(1/0.5) = 0.693 \text{ hour}$$

$$V = (0.693)(F) = (0.693)(1000) = 693 \text{ liters}$$

In the previous sections we have taken a simple isothermal reaction A → B in ideal reactors and developed mass balance equations on the reactant A. Naturally, one may ask the question, is it necessary to write another mass balance on the product B in each reactor in order to keep track of it? The answer to this question is no, because the stoichiometric relationship provides a molar relationship, $N_{Ao} - N_A = N_B - N_{Bo}$ which can be used to determine what happens to B, once the fate of A is known.

Let us now consider the stoichiometric relationships for multiple reactions involving one or more reactants and products.

2.4   STOICHIOMETRIC RELATIONSHIPS FOR MULTIPLE REACTIONS

For multiple reactions involving one or more reactants and products each individual reaction must be considered both separately and simultaneously to develop the overall stoichiometric relationships. For example, for the reactions

(1) $F + 2G \rightarrow H$ (2.35)

(2) $H + 2F \rightarrow 3I$ (2.36)

in which the species H and F are involved in both reactions (1) and (2), we can write the individual stoichiometric relationships for reactions (1) and (2). Denoting reactions (1) and (2) by subscripts 1 and 2 we have

$$(\Delta N_F)_1/-1 = \Delta N_G/-2 = (\Delta N_H)_1/1$$ (2.37)

$$(\Delta N_H)_2/-1 = (\Delta N_F)_2/-2 = \Delta N_I/3$$ (2.38)

Since one cannot measure $(\Delta N_F)_i$ nor $(\Delta N_H)_i$ individually, but only the overall ΔN_F and ΔN_H, it is necessary to combine Eq. 2.37 with Eq. 2.38

$$\Delta N_F = (\Delta N_F)_1 + (\Delta N_F)_2 = (1/2)\Delta N_G - (2/3)\Delta N_I$$ (2.39)

and

$$\Delta N_H = (\Delta N_H)_1 + (\Delta N_H)_2 = -(1/2)\Delta N_G - (1/3)\Delta N_I$$ (2.40)

Similarly the following relationships hold among the individual rates

$$(r_F)_1/-1 = (r_G)/-2 = (r_H)_1/1 = r_1$$ (2.41)

and

$$(r_H)_2/-1 = (r_F)_2/-2 = r_I/3 = r_2$$ (2.42)

The net rates of generation of F, G, H, and I can then be expressed in terms of the generalized rates, r_1 and r_2.

$$r_F = (r_F)_1 + (r_F)_2 = -r_1 - 2r_2$$ (2.43)

$$r_G = (r_G)_1 + (r_G)_2 = -2r_1$$ (2.44)

$$r_H = (r_H)_1 + (r_H)_2 = r_1 - r_2$$ (2.45)

and

$$r_I = (r_I)_1 + (r_I)_2 = 3r_2$$ (2.46)

Note that there are four species involved (F,G,H, and I) in the reaction and Eqs. 2.39 and 2.40 provide two equations. Therefore, two mass balance equations on two species are needed to describe the system completely.

For reversible reactions the stoichiometric equations for the forward and reverse reactions should be written separately and the resulting multiple reactions should be handled as was done above. For example

$$2A \overset{\rightarrow}{\leftarrow} B \tag{2.47}$$

can be written as

(1) $2A \rightarrow B$ (2.48)

(2) $B \rightarrow 2A$ (2.49)

so that

$$(r_A)_1/-2 = (r_B)_1/1 = r_1 \tag{2.50}$$

$$(r_B)_2/-1 = (r_A)_2/2 = r_2 \tag{2.51}$$

Therefore,

$$r_A = (r_A)_1 + (r_A)_2 = -2r_1 + 2r_2 \tag{2.52}$$

$$r_B = (r_B)_1 + (r_B)_2 = r_1 - r_2 \tag{2.53}$$

Likewise

$$\Delta N_A/-2 = \Delta N_B/1 \tag{2.54}$$

Unfortunately, in the wastewater treatment field the stoichiometric relationships are very seldom known and consequently it is necessary to write mass balance
equations for all components of interest. This situation presents no particular
difficulty except that the mass balance equations must be solved simultaneously.
In other words simultaneous algebraic equations need to be solved for reactor design.

2.5 KEY POINTS

1. The change in the number of moles of one species divided by its modified stoichi-
    ometric coefficient is equal to the change in the number of moles of any other
    species involved in a reaction divided by its modified stoichiometric coefficient.

2. The rate of generation of any species divided by its modified stoichiometric
    coefficient is equal to the rate of generation of any other species divided by
    its modified stoichiometric coefficient.

3. The mass balance equation for a CSTR can be used to determine the reactor volume
    necessary to process a given influent flow rate or the flow rate which can be
    handled by a reactor of known volume. The equation can also be used to recover
    the rate expression from a series of steady-state runs.

4. For a constant volume batch reactor the reaction rate is equal to the time rate
    of change of the reactant concentration. This relationship may be used to deter-
    mine the reaction rate as a function of reactant concentration, which may then
    be used in the mass balance equation for any other reactor type.

5. The main purpose of the mass balance equation for a plug-flow reactor is to
    determine the size of reactor required to achieve a desired conversion from a
    feed stream of given composition and flow rate.

6. For multiple reactions and reversible reactions each individual reaction must be considered both separately and simultaneously when developing the stoichiometric relationships.

2.6 STUDY QUESTIONS

The study questions are arranged in the same order as the text:

Section	Questions
2.1	None
2.2	1,2
2.3	3-14
2.4	15-20

1. Define in words the terms r_A and $-r_A$. What are typical units?

2. For the reaction A → 2B

    (a) is $r_A < 0$? Why?

    (b) if $r_A = -1$ mol/(hr·liter), what is r_B?

    (c) if r_B is given as 10 mol/(hr·liter), what is the value of r, the generalized rate of reaction?

    (d) if $r_B = kC_A$, what would be the expression for r_A?

3. What are the assumptions which are made in deriving the mass balance equation for a CSTR? Can you do away with any of the assumptions? Explain.

4. For an aqueous-phase reaction, A → B, which is to be carried out in a CSTR, it is known that $r_A = -2C_A$ mol/(hr·liter).

    (a) What size reactor do you need to convert 50% of a feed stream containing one mol/liter of A at a flow rate of 100 liters/hr? What is the numerical value of r_A in the reactor?

    (b) Suppose there is a 10 liter reactor available for your use. Given a feed containing 2 mol/liter of A, what volumetric flow rate is required to convert 50% of A so that the effluent concentration would be 1 mole/liter of A?

    (c) Suppose that the feed contains 1 mol/liter of A and is available at 100 liters/hr to a reactor with a volume of 10 liters. What would be the concentration of the effluent? What would be the rate, r_A?

5. Describe in detail how you may use a CSTR to determine the rate r_A as a function of C_A for reaction A → B.

6. State the assumption made in deriving the mass balance for a batch reactor.

7. Under what conditions is $r_A = dC_A/dt$?

8. For the aqueous reaction A → B with $r_A = -C_A$ mol/(liter·hr), it is desired to produce a product containing 25% A and 75% B from a feed containing 1 mol/liter of A only. How long should the reaction be carried out in a batch reactor? What would be the value of r_A at the terminal time?

9. For the aqueous reaction A → B the rate of consumption of A is given as
    $-r_A = kC_A$, where k is the rate constant which is unknown. This reaction was
    carried out for one hour in a batch reactor, at which time the concentration
    of A was 36.79% of the initial concentration. What would be the concentration
    of A if the reaction were carried out for 2 hours?

10. What are the usual assumptions involved in writing the mass balance equation
     for a plug-flow reactor?

11. Describe how a plug-flow reactor may be used to determine the rate as a function
     of composition.

12. There is an analogy between a constant volume batch reactor operating under
     unsteady-state conditions and a plug-flow reactor operating under steady-state
     conditions. Show the analogy by writing mass balance equations.

13. For the reaction A → B with $r_A = -2C_A$ mol/(hr·liter) what size plug-flow
     reactor is needed to convert 50% of feed at a flow rate of 1000 liters per
     hour containing 5 mol/liter of A?

14. Write an unsteady-state mass balance for a plug-flow reactor in which an
     aqueous reaction A → B is taking place.

15. For the aqueous phase reaction 2A → 3B, the generalized rate is given by
     $r = 3C_A$ mol/(hr·liter). The volume of a CSTR is 100 liters and feed con-
     taining 2 mol/liter of A is available at 1000 liters/hr. Determine the
     concentration of B in the effluent by:

     (a). Writing the mass balance on A and using the stoichiometric relationship,
           and by

     (b). Writing the mass balance on B.

16. Repeat Study Question 15 using a plug-flow reactor.

17. For aqueous reaction 2A $\rightleftarrows$ 3B (reversible) the forward and reverse reaction
     rates are given by $r_1 = k_1 C_A$ and $r_2 = k_2 C_B$, respectively. What is the
     expression for the *net* rate of formation of A?

18. The reaction of Study Question 17 is carried out in a batch reactor. Write a
     mass balance on A. You will note that the mass balance equation contains both
     C_A and C_B. Eliminate C_B in terms of C_A, C_{Ao} and C_{Bo}. Do not integrate.

19. For the reaction

     A + 2B → C $r_1 = k_1 C_A C_B^2$

     C + A → 2D $r_2 = k_2 C_C C_A$

     Write the expressions for the rates of formation of A,B,C, and D.

20. The reaction of Study Question 19, was carried out in a batch reactor starting
     with one mole of A and 2 moles of B. At the end of reaction it was found that
     there was one mole of B and 1/2 of D in the reactor. How many moles of A and
     C would you expect in the reactor?

REFERENCES AND FURTHER READING

1. S. Aiba, *et al. Biochemical Engineering,* 2nd Ed. Academic Press, New York, N.Y., 131, 1973.

2. Metcalf and Eddy, Inc., *Wastewater Engineering,* 2nd Edition, McGraw-Hill Book Company, New York, N.Y., 142-179, 1979.

3. O. Levenspiel. *Chemical Reaction Engineering,* John Wiley & Sons, Inc., New York, N.Y., 5, 1972.

4. D. C. Dixon. "The definition of reaction rate," *Chemical Engineering Science,* 25, 337-338, 1970.

5. F. Daniels, and R. A. Alberty. *Physical Chemistry,* 3rd ed., John Wiley & Sons, Inc., New York, N.Y., Chapter 10, 1966.

6. W. Moore, *Physical Chemistry,* 4th Ed. Prentice-Hall, Inc., Englewood Cliffs, N.J., 325, 1972.

7. C. F. Prutton, and S. H. Maron. *Fundamental Principles of Physical Chemistry,* 2nd ed., Macmillan Company, New York, N.Y., 614, 1956.

Further Reading

Boudart, M., *Kinetics of Chemical Processes,* Prentice-Hall Inc., Englewood Cliffs, N.J., 1968. See Chapter 1.

Fogler, H. S., *The Elements of Chemical Kinetics and Reactor Calculations (A Self-Paced Approach),* Prentice-Hall Inc., Englewood Cliffs, N.J., 1974. See Chapters 1 and 2, which cover the topics from a self-paced approach.

Himmelblau, D. M., *Basic Principles and Calculations in Chemical Engineering,* 2nd Ed., Prentice-Hall Inc., Englewood Cliffs, N.J., 1967. See Chapters 2 and 3 for further discussion of stoichiometry and the development of general balance equations.

Smith, J. M., *Chemical Engineering Kinetics,* Mc-Graw-Hill, Inc., New York, N.Y., 1970. See Chapters 2 and 3.

CHAPTER 3

REACTION RATE EXPRESSIONS

In the previous chapter we have shown how the rate expression r fits into the mass balance equations for various types of reactors. In this chapter the equations for the rate expression will be classified according to the reaction types which they represent and will be put into forms suitable for use in the mass balance equations. For those complex reactions whose rate expressions require involved derivations, for example catalytic and chain reactions, only a cursory treatment will be given here. A more detailed coverage will be given in Chapter 5 where experimental methods for verification of proposed rate expressions and for determination of kinetic parameters will be covered. In this and subsequent chapters it will be assumed that the reactions take place with no volume change.

3.1 TERMINOLOGY

3.1.1 Reaction Order, Molecularity, and Reaction Rate Constant

In general, the rate of a chemical reaction depends on the concentrations of the reactants. It may also depend on the concentrations of one or more products or other species not involved in the stoichiometric equation. When the rate expression involves powers of concentrations, the order of the reaction with respect to a particular species is equal to the exponent on that species in the rate expression. The generalized rate of reaction r, involving species A, B, and C is often found to be well approximated by the expression

$$r = kC_A^a C_B^b C_C^c \tag{3.1}$$

in which k is the reaction rate constant, C_A, C_B, and C_C are the concentrations of species, A, B, and C, and a, b, and c are exponents which may or may not be numerically equivalent to the stoichiometric coefficients of the corresponding species in the stoichiometric equation. This reaction is ath order with respect to A, bth order with respect to B, and cth order to C. The overall order of the reaction is the sum of the exponents. Thus, it is $(a + b + c)^{th}$ order. Furthermore, the overall order of a reaction need not be an integer. It may be zero or a positive quantity which seldom, if ever, exceeds three.

EXAMPLE 3.1.1-1

The rate of reaction for A + B → C is given by

$$r = kC_A C_B^{\frac{1}{2}}$$

The order of reaction is first with respect to A and one-half with respect to B. The overall order is 3/2.

The term molecularity refers to the number of molecules involved in the step leading to reaction and hence it must take on only positive integer values. For example, if a single reactant molecule leads to a product, the reaction is unimolecular. Bimolecular and termolecular reactions involve two and three molecules of reactants to form products. For example A + A → Product, and A + B → Products are bimolecular reactions, while A + A + A → Products, A + B + C → Products, and A + B + B → Products, are termolecular reactions.

The rate constant, which is the proportionality constant in the rate expression (Eq. 3.1), is also known as the specific rate constant. In homogeneous reactions the rate constant, k, is usually a function only of temperature although it can also be a function of other parameters, such as pH, ionic strength, or solvent. However, in practice these parameters are usually kept constant and therefore, we may assume that k depends only on temperature. Because the units of r are (concentration)/(time) the units of k are easily found from the units associated with each term in Eq. 3.1 and are equal to $(\text{concentration})^{1-a-b-c} (\text{time})^{-1}$.

3.1.2 Elementary and Nonelementary Reactions

Elementary reactions are those for which the molecularity and order are identical so that the exponents in the rate expression correspond to the coefficients in the stoichiometric equation. For example, if the stoichiometric equation is given by

$$A + 2B \rightarrow \text{products} \tag{3.2}$$

and the rate is given by

$$-r_A = kC_A C_B^2 \tag{3.3}$$

there is a correspondence between the stoichiometric coefficients and the exponents to which the concentrations are raised so that the reaction is elementary. Conversely, if the reaction characterized by the stoichiometric equation

$$aA + bB + cC \rightarrow \text{products} \tag{3.4}$$

is known to be elementary, the rate expression must be given by

$$-r_A = kC_A^a C_B^b C_C^c \tag{3.5}$$

Those reactions which do not show any correspondence between the coefficients in the stoichiometric equation and the exponents to which the concentrations are raised in the rate expression are called nonelementary. Generally, nonelementary reactions can be explained by a sequence of elementary reactions. The reason for observing what appears to be a single reaction rather than the sequence of elementary reactions is that the intermediates formed in the sequence of reactions are either so low in concentration or so short-lived that they are difficult to detect. Nonelementary reactions are often characteristic of catalytic reactions, chain reactions, and biochemical reactions catalyzed by enzymes.

3.1.3 Fractional Conversion and Yield

Fractional conversion represents the fraction of a reactant that has gone through chemical transformation at a particular stage of the reaction process, i.e., the ratio of the amount of the reactant that has been consumed to the amount that was subjected to reaction. Thus, the value of fractional conversion ranges from zero to one. In molar units the fractional conversion for the reactant A, x_A, is defined as

$$x_A = (N_{Ao} - N_A)/N_{Ao}$$
$$x_A = (C_{Ao}V_o - C_A V)/C_{Ao}V_o \tag{3.6}$$

Almost all wastewater treatment systems are of the constant density, constant volume type so that $V_o = V$. In this situation Eq. 3.6 reduces to

$$\chi_A = (C_{Ao} - C_A)/C_{Ao} \tag{3.7}$$

or

$$C_A = C_{Ao}(1 - \chi_A) \tag{3.8}$$

Sometimes per cent conversion is used instead of the fractional conversion, which, of course, is equal to $100\chi_A$.

When only part of a reactant is converted into the desired product and the remaining is converted into undesirable byproducts, the term yield is used to describe the situation. Let us suppose that in the reaction system given by the stoichiometric equations:

$$A \diagup\!\!\!\overset{\textstyle D + E}{\diagdown U + W} \tag{3.9}$$

a_o moles of A are fed to the reactor, of which a moles are reacted to give d moles of desired product D and u moles of undesirable side product U. The yield for D is either d/a_o or d/a depending upon whether one chooses as a basis the total amount of A fed to the reactor or that part of A which reacted. Similarly, the yield for U is either u/a_o or u/a. A molar yield of 30% for D may imply either that 0.3 mole of D is formed for every mole of A reacted or that 0.3 mole of D is formed for every mole of A charged. It is obvious that these two interpretations are different unless all A charged has been reacted. Therefore, it is important to state the basis used to compute the yield. Generally, within environmental engineering, the yield is computed on the basis of the amount reacted. Sometimes, it is more convenient to use mass units. For example, a 50% mass yield based on the amount reacted implies a yield of 0.5 gram of D per gram of A reacted.

3.2 ELEMENTARY REACTIONS

3.2.1 Single Irreversible Reactions

FIRST-ORDER (UNIMOLECULAR) REACTIONS. Consider an elementary reaction involving a single molecule of reactant A,

$$A \rightarrow Product \tag{3.10}$$

Since this is unimolecular and elementary, the rate expression may be written by examination of the stoichiometric equation:

$$r = -r_A = kC_A \tag{3.11}$$

The reaction rate constant, k, has the units of reciprocal time because the units of r_A and C_A are (concentration)/(time) and (concentration), respectively. This expression may now be substituted into the appropriate mass balance equations to calculate the time required to achieve a desired concentration in a batch reactor (Eq. 2.27), or the volume needed to convert a given influent into a desired effluent concentration in either a plug-flow reactor (Eq. 2.34) or a CSTR (Eq. 2.21).

SECOND-ORDER (BIMOLECULAR) REACTIONS. A bimolecular reaction can be represented by

$$A + B \rightarrow \text{Products} \qquad (3.12)$$

Since this is an elementary reaction, it follows from the stoichiometric equation that

$$-r_A = kC_A C_B = -r_B = r \qquad (3.13)$$

The rate constant k has the units of $(\text{concentration})^{-1}(\text{time})^{-1}$. Now, if Eq. 3.13 is used in the mass balance equation for species A for any type of reactor, it will be seen that there are two concentration terms, C_A and C_B, which are unknown. Therefore, to facilitate the use of Eq. 3.13 in the mass balance equation for species A we need to eliminate C_B in terms of C_A. Alternatively, we could eliminate C_A in terms of C_B, and use Eq. 3.13 in the mass balance for species B. In either case the stoichiometric equation allows us to do this. From Eq. 3.12

$$\Delta N_A = \Delta N_B \qquad (3.14)$$

so that

$$C_A V - C_{Ao} V_o = C_B V - C_{Bo} V_o \qquad (3.15)$$

Since we are dealing with reactions with no change in volume, this reduces to

$$C_B = C_{Bo} + C_A - C_{Ao} \qquad (3.16)$$

and the rate (Eq. 3.13) can be written as

$$-r_A = kC_A(C_{Bo} + C_A - C_{Ao}) \qquad (3.17)$$

or

$$-r_B = kC_B(C_{Ao} + C_B - C_{Bo}) \qquad (3.18)$$

The rate expression given by Eq. 3.17 can be used readily in the mass balance equation for species A, while Eq. 3.18 can be used for species B.

A considerable simplification results when A and B are present in equal molar concentration initially. In this situation A and B are numerically indistinguishable so that it is equivalent to a situation in which the species A and B are the same, as in the reaction

$$2A \rightarrow \text{Products} \qquad (3.19)$$

For this reaction the rate expression is

$$-r_A = kC_A^2 = 2r \tag{3.20}$$

which is the same as the result when C_{Ao} is set equal to C_{Bo} in Eq. 3.17.

   THIRD-ORDER (TERMOLECULAR) REACTIONS. The most general type of third-order reaction is one in which three different molecules react,

$$A + B + C \rightarrow \text{Products} \tag{3.21}$$

Inspection of Eq. 3.21 suggests that

$$-r_A = kC_A C_B C_C = -r_B = -r_C = r \tag{3.22}$$

In this case the units of k, the reaction rate constant, are (concentration)$^{-2}$ (time)$^{-1}$. If the reaction rate expression is to be used in the mass balance for species A it must be expressed in terms of C_A alone. Therefore, C_B and C_C must be expressed in terms of C_A. This is done through the stoichiometric equation $(\Delta N_A = \Delta N_B = \Delta N_C)$ to obtain

$$-r_A = kC_A(C_A - C_{Ao} + C_{Bo})(C_A - C_{Ao} + C_{Co}) \tag{3.23}$$

When two of the initial concentrations are the same, say $C_{Bo} = C_{Co}$, Eq. 3.23 reduces to

$$-r_A = kC_A(C_A - C_{Ao} + C_{Bo})^2 \tag{3.24}$$

When two of the species are identical, say C = A, so that

$$2A + B \rightarrow \text{Products} \tag{3.25}$$

the rate is given by

$$-r_A = kC_A^2 C_B = -2r_B \tag{3.26}$$

which simplifies to

$$-r_A = kC_A^2(C_{Bo} + C_A/2 - C_{Ao}/2) \tag{3.27}$$

Finally, when all three species are identical the stoichiometric equation becomes

$$3A \rightarrow \text{Products} \tag{3.28}$$

and the rate expression which follows from Eq. 3.28, is

$$-r_A = kC_A^3 \tag{3.29}$$

   AUTOCATALYTIC REACTIONS. An autocatalytic reaction is one in which a product of the reaction catalyzes the reaction. The simplest type is

$$A + B \rightarrow 2B + C \tag{3.30}$$

which has the rate:

$$-r_A = kC_A C_B \qquad\qquad\qquad (3.31)$$

The stoichiometric equation, Eq. 3.30, gives

$$C_B = C_{Bo} + C_{Ao} - C_A \qquad\qquad\qquad (3.32)$$

so that Eq. 3.31 simplifies to

$$-r_A = kC_A (C_{Ao} + C_{Bo} - C_A) \qquad\qquad\qquad (3.33)$$

Compare Eq. 3.33 with Eq. 3.17. In an autocatalytic reaction, a quantity (however small) of product which acts as the catalyst must be present initially for the reaction to proceed as implied by the rate expression of Eq. 3.31. It is very important to note that unlike all previous rate expressions of elementary reactions, which are monotonically increasing functions of the reactant concentration, the rate expression for an autocatalytic reaction, Eq. 3.33, exhibits unusual characteristics. For example, it is not difficult to show that $-r_A$ exhibits a maximum. Equation 3.33 can be rewritten as

$$-r_A = -k\{ [C_A - (C_{Ao} + C_{Bo})/2]^2 - (1/4)(C_{Ao} + C_{Bo})^2 \} \qquad (3.34)$$

which is the equation for a parabola with the axis at $C_A = (C_{Ao} + C_{Bo})/2$. Thus, the maximum rate occurs when $C_A = (C_{Ao} + C_{Bo})/2$ and is equal to $k(C_{Ao} + C_{Bo})^2/4$. This is shown in Fig. 3.1. This type of behavior is unusual and special attention is given in Chapter 4 to the design of reactors for autocatalytic reactions.

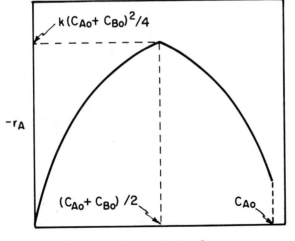

Figure 3.1.   Effect of reactant concentration on
             the rate of an autocatalytic reaction.

3.2.2 Single Reversible Reactions

Reversible reactions are those in which both forward and reverse reactions proceed simultaneously so that the products of the forward reaction are also the reactants for the reverse reaction. In principle, all reactions are reversible but in practice the rate of the reverse reaction is sometimes so small that the overall reaction may be approximated as an irreversible one.

REVERSIBLE FIRST-ORDER REACTION. This is the simplest reversible reaction and can be described by

$$A \underset{k_{-1}}{\overset{k_1}{\rightleftarrows}} B \tag{3.35}$$

where k_1 and k_{-1} are the rate constants for the forward and reverse reactions, respectively. This reaction can be described by two irreversible reactions:

$$A \overset{k_1}{\rightarrow} B \tag{3.36}$$

$$B \overset{k_{-1}}{\rightarrow} A \tag{3.37}$$

Therefore, application of the concept of net rate, as done in Eq. 2.52, yields

$$r_A = -k_1 C_A + k_{-1} C_B \tag{3.38}$$

This can be expressed in terms of C_A only by application of stoichiometric relationships to obtain

$$r_A = -(k_1 + k_{-1})C_A + k_{-1}(C_{Ao} + C_{Bo}) \tag{3.39}$$

At equilibrium the rate is zero, i.e., $r_A = 0$, and the concentrations of A and B may be denoted by C_{Ae} and C_{Be} so that Eq. 3.38 reduces to

$$r_A = 0 = -k_1(C_{Ae}) + k_{-1}(C_{Be}) \tag{3.40}$$

Making use of the stoichiometric relationships this may be rearranged to yield

$$C_{Be}/C_{Ae} = (C_{Ao} + C_{Bo} - C_{Ae})/C_{Ae} = k_1/k_{-1} = K \tag{3.41}$$

where K is the equilibrium constant. Thus, the forward rate constant k_1 and the reverse rate constant k_{-1} are not independent of each other but are related by the equilibrium constant, K. In terms of K, Eq. 3.39 can be written as

$$r_A = -k_{-1}[(K + 1)C_A - (C_{Ao} + C_{Bo})] \tag{3.42}$$

REVERSIBLE SECOND-ORDER REACTION. A general stoichiometric equation for a reversible second-order reaction is

$$A + B \underset{k_{-1}}{\overset{k_1}{\rightleftarrows}} C + D \qquad (3.43)$$

The rate expression for species A is

$$-r_A = k_1 C_A C_B - k_{-1} C_C C_D \qquad (3.44)$$

which suggests that the equilibrium constant is

$$K = k_1/k_{-1} = C_{Ce} C_{De}/C_{Ae} C_{Be} = \frac{(C_{Co} + C_{Ao} - C_{Ae})(C_{Do} + C_{Ao} - C_{Ae})}{C_{Ae}(C_{Bo} - C_{Ao} + C_{Ae})} \qquad (3.45)$$

The stoichiometric equation can be used to eliminate C_B, C_C and C_D in terms of C_A in Eq. 3.44, yielding

$$r_A = -k_1 C_A (C_A - C_{Ao} + C_{Bo}) + k_{-1}(C_{Co} + C_{Ao} - C_A)(C_{Do} + C_{Ao} - C_A) \qquad (3.46)$$

Equation 3.45 may then be used to eliminate k_1 in Eq. 3.46 as was done in Eq. 3.42.

When only one reactant and one product are involved, i.e., A = B and C = D, the stoichiometric equation becomes

$$A + A \underset{k_{-1}}{\overset{k_1}{\rightleftarrows}} C + C \qquad (3.47)$$

and the expression reduces to

$$-r_A = k_1 C_A^2 - k_{-1} C_C^2 = k_1 C_A^2 - k_{-1}(C_{Ao} + C_{Co} - C_A)^2 \qquad (3.48)$$

MIXED-ORDER REVERSIBLE REACTIONS. We restrict the treatment to the reversible reactions of first- and second-order

$$A + B \underset{k_{-1}}{\overset{k_1}{\rightleftarrows}} C \qquad (3.49)$$

The rate of consumption of A is

$$-r_A = k_1 C_A C_B - k_{-1} C_C \qquad (3.50)$$

Eliminating C_B and C_C in terms of C_A gives

$$r_A = -k_1 C_A (C_{Bo} - C_{Ao} + C_A) + k_{-1}(C_{Ao} + C_{Co} - C_A) \qquad (3.51)$$

The equilibrium constant is

$$K = k_1/k_{-1} = C_{Ce}/C_{Ae}C_{Be} = (C_{Co} + C_{Ao} - C_{Ae})/C_{Ae}(C_{Bo} - C_{Ao} + C_{Ae}) \qquad (3.52)$$

Thus the rate given by Eq. 3.51 reduces to

$$r_A = -k_{-1}\{KC_A^2 + [K(C_{Bo} - C_{Ao}) + 1]C_A - (C_{Ao} + C_{Co})\} \qquad (3.53)$$

3.2.3 Multiple Reactions

PARALLEL REACTIONS. Suppose that the reactant A can be consumed in a number of concurrent reactions to yield several products. For example, consider the following elementary reactions:

$$A \xrightarrow{k_1} P \qquad (3.54)$$

$$A + A \xrightarrow{k_2} Q \qquad (3.55)$$

The net rate for A is the sum of the two individual rates and is given by

$$-r_A = k_1 C_A + k_2 C_A^2 \qquad (3.56)$$

which can be used in the mass balance equation for A. The stoichiometric relationship (Section 2.4) gives

$$C_A - C_{Ao} = C_{Po} - C_P + 2(C_{Qo} - C_Q) \qquad (3.57)$$

The mass balance on species A using the rate given by Eq. 3.56 keeps track of species A. The stoichiometric relationship given by Eq. 3.57 provides another equation relating species A to species P and Q simultaneously but not individually. Thus, to describe the system it is necessary to follow experimentally C_P or C_Q as well as C_A.

Consider now parallel reactions in which two reactants are involved:

$$A \xrightarrow{k_1} P \qquad (3.58)$$

$$A + B \xrightarrow{k_2} Q \qquad (3.59)$$

Since each reaction is assumed to be elementary, the net rate of consumption of A is given by

$$-r_A = k_1 C_A + k_2 C_A C_B \qquad (3.60)$$

The usual approach would be to use the stoichiometric relationships to eliminate C_B in terms of C_A so as to express $-r_A$ in terms of C_A only. Using the method developed in Chapter 2, the stoichiometric relationships under constant volume resulting from Eqs. 3.58 and 3.59 are

$$C_{Ao} - C_A = C_P - C_{Po} + C_Q - C_{Qo} \tag{3.61}$$

and

$$C_{Bo} - C_B = C_Q - C_{Qo} \tag{3.62}$$

Inspection of Eqs. 3.61 and 3.62 reveals that C_B cannot be expressed in terms of C_A alone. Therefore, when the mass balance equation for species A is written for a reactor, it will contain both C_A and C_B so that it cannot be solved for C_A. Another equation involving C_A and C_B is needed which can be solved simultaneously with the mass balance equation for species A. The needed equation is the mass balance equation for the other reactant, B, which may be written using the reaction rate for B:

$$-r_B = k_2 C_A C_B \tag{3.63}$$

After the two mass balance equations are solved simultaneously for C_A and C_B, C_Q may be calculated from Eq. 3.62 and C_P from Eq. 3.61.

  SERIES REACTIONS. Let us consider a reaction in which a reactant A is irreversibly converted into a final product C through an intermediate B,

$$\overset{k_1 \quad k_2}{A \rightarrow B \rightarrow C} \tag{3.64}$$

Assuming each step is elementary the rate expressions are

$$r_A = -k_1 C_A \tag{3.65}$$

$$r_B = k_1 C_A - k_2 C_B \tag{3.66}$$

and

$$r_C = k_2 C_B \tag{3.67}$$

The rate expression for A, Eq. 3.65, can be used in the mass balance equation for species A in a particular reactor to determine C_A in that reactor. In the case of a CSTR, C_A is a fixed value, while in batch and plug-flow reactors C_A is a function of time and reactor volume, respectively. Once C_A is known it can be substituted into Eq. 3.66 and the resulting rate expression for r_B may be used in the corresponding mass balance equation for species B to determine C_B. Finally, C_C can be obtained from the stoichiometric relationship,

$$C_{Ao} - C_A = C_B - C_{Bo} + C_C - C_{Co} \tag{3.68}$$

Even when the order of each reaction is other than one, the same procedure may be followed. For series reactions the desired product frequently is the intermediate, B, which needs to be obtained in good yield.

3.3 NONELEMENTARY HOMOGENEOUS REACTIONS

Nonelementary reactions are those whose rate expressions cannot be written from inspection of the stoichiometric equation, because there is no correspondence between the exponents in the rate expression and the stoichiometric coefficients, or the rate expression is not made up of simple powers of the concentrations of reactants. Under this category come many reactions that are important from an environmental engineering viewpoint, i.e., biochemical reactions, chain reactions, catalytic reactions, etc.

3.3.1 Fractional- and Zero-order Reactions

A reaction is said to be zero-order if the rate expression is independent of the concentrations of the reactants involved in the reaction. For example,

$$r_A = -kC_A^o = -k \qquad\qquad\qquad (3.69)$$

In reality there are very few reactions which are truly zero-order overall. Instead, it is frequently found that reactions behave in a zero-order manner with respect to one or more reactants, such as in the case of excess reactants. For example, consider an irreversible enzymatic reaction involving a reactant and a product. The reaction takes place at the active centers of the enzyme, which can become saturated when the concentration of reactant reaches a certain level. At reactant concentrations above that level the reaction seems to proceed at a rate independent of the reactant concentration (see Eq. 3.81). Hence, it behaves as a zero-order reaction when the concentration is above the critical level.

Sometimes the order of a reaction turns out to be fractional, say 3/2 or 5/2. For these fractional-order reactions the rate may be given as

$$r_A = -kC_A^q C_B \qquad\qquad\qquad (3.70)$$

where q is a fraction.

3.3.2 Catalytic Reactions

Catalytic reactions occur very frequently and constitute an important type of non-elementary reaction. However, they are very complex and to treat them in depth would require more space than can be justified in this text. Consequently, a rather brief treatment will be given here and those whose interests are so inclined should consult the texts listed under Further Reading.

To illustrate the basic concepts involved in the development of the rate expression for a catalytic reaction let us consider the case of a biochemical reaction catalyzed by a biological catalyst (enzyme, E), in which a single reactant (usually called the substrate, S) is irreversibly converted to a product

$$S \xrightarrow{E} P \tag{3.71}$$

The first step is the postulation of a series of elementary reactions leading to the product:

$$(1) \quad S + E \underset{k_{-1}}{\overset{k_1}{\rightleftarrows}} ES \tag{3.72}$$

$$(2) \quad ES \xrightarrow{k_2} P + E \tag{3.73}$$

Step (1) states that an enzyme-substrate complex ES is formed reversibly from the substrate S and enzyme E, while step (2) states that the enzyme-substrate complex is irreversibly transformed into the product P and free enzyme E. The enzyme can again react with S to form ES to repeat the cycle. Since steps (1) and (2) are elementary by assumption, the rates are:

$$r_{ES} = k_1 C_E C_S - (k_{-1} + k_2) C_{ES} \tag{3.74}$$

$$r_E = -r_{ES} \tag{3.75}$$

$$r_p = k_2 C_{ES} = -r_S \tag{3.76}$$

The objective is to express the rate of consumption of substrate, $-r_S$ (or the rate of formation of product, r_p), in terms of concentrations of readily observable species, such as the total (initial) enzyme concentration, C_{Eo}, and the substrate concentration, C_S. The other alternative would be to use all three rate expressions, r_{ES}, r_E, and r_S, in their respective component mass balance equations and then solve them simultaneously, as was proposed in Section 3.2.3, but this approach quickly becomes unwieldly. The way to represent $-r_S$ in terms of C_S and C_{Eo} (constant) is to use the pseudosteady-state hypothesis (PSSH) which allows us to set the rate of formation of enzyme-substrate complex to zero, i.e., $r_{ES} = 0$. This assumption yields one algebraic equation in two unobservable variables, C_E and C_{ES},

$$r_{ES} \cong 0 = k_1 C_E C_S - (k_{-1} + k_2) C_{ES} \tag{3.77}$$

Another equation in C_E and C_{ES} is needed to provide two equations in two unknowns. This equation is the mass balance on the enzyme, which states that the enzyme added initially, E_o, must exist either in the form E or ES so that

$$C_E + C_{ES} = C_{Eo} \tag{3.78}$$

Rearranging Eqs. 3.77 and 3.78 into a matrix form and applying Cramers rule to solve for C_{ES} results in

$$C_{ES} = \frac{k_1 C_{Eo} C_S}{k_{-1} + k_2 + k_1 C_S} \qquad (3.79)$$

Substitution of Eq. 3.79 into Eq. 3.76 yields the desired rate expression

$$-r_S = k_2 C_{Eo} C_S / [(k_{-1} + k_2)/k_1 + C_S] \qquad (3.80)$$

which is a function of C_S so that it can be readily used in the mass balance equations. Equation 3.80 is usually written as

$$-r_S = V_m C_S / (K_m + C_S) \qquad (3.81)$$

where $V_m = k_2 C_{E_o}$ and $K_m = (k_{-1} + k_2)/k_1$. The quantity V_m is called the maximum rate for the enzymatic reaction, and K_m is called the Michaelis constant or saturation constant for the substrate. Equation 3.81 is widely known as the Michaelis-Menten equation [1], and sometimes as the Briggs-Haldane equation [2].

Whenever the substrate concentration is significantly higher than the Michaelis constant, i.e., $C_S \gg K_m$, the denominator in Eq. 3.81 approaches C_S so that the rate expression approaches an asymptotic value, V_m, effectively becoming independent of C_S. Thus, in this high substrate concentration range the rate is zero-order. On the other hand, in the range of substantially low concentration in which $C_S \ll K_m$, the rate is effectively first-order, i.e., $-r_S \rightarrow (V_m/K_m)C_S$. It is, then, in the intermediate range of concentration that the rate is nonlinear with respect to C_S.

Let us now go back to the assumption of PSSH. A rigorous justification is somewhat mathematically involved, therefore suffice it to state that in a batch reactor when the initial enzyme concentration is much smaller than that of the substrate ($C_{Eo} \ll C_{So}$) the assumption is valid [3,4,5]. This hypothesis, which allows one to set the net rate of formation of all enzyme-substrate complexes (or active intermediates) to zero, is one of the most important developments in chemical kinetics and was proposed as early as 1919 [6,7,8] to explain the kinetics of the reaction between hydrogen and bromine, which had been studied by Bodenstein and Lind [9] in 1906.

Recapitulating the process leading to the rate expression for the enzyme reaction, the steps are: (1) postulation of a sequence of elementary reactions, (2) writing of all rate expressions for the reactant and complexes, (3) expression of the concentration of each enzyme complex, which is usually unobservable, in terms of the concentrations of observable species by solving simultaneously the algebraic equations resulting from the PSSH and the mass balance equation on the enzyme species, and (4) expression of the rate in terms of the concentration of observable species using the results of (3). For other reactions which involve active complexes the same procedure may be used as shown below for chain reactions.

3.3.3 Chain Reactions

We now consider chain reactions. These reactions are similar to catalytic reactions since active complexes (or sites) are involved, although they are not provided by a catalyst, but rather by ultraviolet light, gamma rays, etc. The chain reaction is characterized by the following sequence of elementary reactions

(1)   Reactant → (Intermediate)*

(2)   Reactant(s) + (Intermediate)* $\overset{\rightarrow}{\leftarrow}$ Product(s) + (Intermediate)*

(3)   (Intermediate)* → Reactant (or stable product)

where the asterisks are used to denote the active complexes. Step (1) initiates the chain reaction by producing the intermediate which then participates in subsequent reactions. Therefore, it is appropriately called the initiation reaction. Step (2) propagates the chain reaction by having the initiated intermediate react to produce other intermediates and products, and it is called the propagation reaction. The last step, step (3) is called the termination reaction because the intermediates are terminated (annihilated) in it. Frequently, the initiation and termination steps are called also head and tail respectively.

A classical example of chain reactions is the formation of hydrogen bromide from hydrogen and bromine [9],

$$H_2 + Br_2 \overset{\rightarrow}{\leftarrow} 2HBr \tag{3.82}$$

This reaction may be explained by the following free-radical chain mechanism [6,7,8]

$$Br_2 \overset{k_1}{\rightarrow} 2Br* \qquad\qquad INITIATION \tag{3.83}$$

$$\left.\begin{array}{l} Br* + H_2 \overset{k_2}{\underset{k_{-2}}{\overset{\rightarrow}{\leftarrow}}} HBr + H* \\[2em] H* + Br_2 \overset{k_3}{\rightarrow} HBr + Br* \end{array}\right\} PROPAGATION \tag{3.84}$$

$$2Br* \overset{k_4}{\rightarrow} Br_2 \qquad\qquad TERMINATION \tag{3.85}$$

Each of these is assumed to be elementary. The procedure to be used to develop the rate expression is same as that used in Section 3.3.2. The net rate of formation of HBr is

$$r_{HBr} = k_2 C_{Br*} C_{H_2} - k_{-2} C_{HBr} C_{H*} + k_3 C_{H*} C_{Br_2} \tag{3.86}$$

Equation 3.86 needs to be freed of the concentrations of free radicals, C_{H*} and C_{Br*},

both of which are difficult, if not impossible, to measure. The net rates of
formation of the free radicals are

$$r_{Br*} = 2k_1 C_{Br_2} - 2k_4 C_{Br*}^2 - k_2 C_{Br*} C_{H_2} + k_{-2} C_{HBr} C_{H*} + k_3 C_{H*} C_{Br_2}$$

or

$$r_{Br*} = 2k_1 C_{Br_2} - 2k_4 C_{Br*}^2 - r_{H*} \tag{3.87}$$

and

$$r_{H*} = k_2 C_{Br*} C_{H_2} - k_{-2} C_{HBr} C_{H*} - k_3 C_{H*} C_{Br_2} \tag{3.88}$$

Note that k_1 and k_4, as used in Eq. 3.87, are defined on the basis of Br_2, i.e.,

$$[r_{Br_2}]_t = [-r_{Br*}]_t/2 = k_4 C_{Br*}^2$$

and

$$[r_{Br_2}]_t = [-r_{Br*}]_t/2 = k_4 C^2 Br*$$

The PSSH allows us to set Eqs. 3.87 and 3.88 to zero to obtain

$$k_2 C_{H_2} C_{Br*} - (k_{-2} C_{HBr} + k_3 C_{Br_2}) C_{H*} = 0 \tag{3.89}$$

and

$$k_1 C_{Br_2} - k_4 C_{Br*}^2 = 0 \tag{3.90}$$

Solving the last equation for C_{Br*} yields

$$C_{Br*} = (k_1 C_{Br_2}/k_4)^{\frac{1}{2}} \tag{3.91}$$

Substitution of Eq. 3.91 into Eq. 3.89 yields

$$C_{H*} = \frac{k_2 C_{H_2} (k_1 C_{Br_2}/k_4)^{\frac{1}{2}}}{k_{-2} C_{HBr} + k_3 C_{Br_2}} \tag{3.92}$$

The desired rate of formation of C_{HBr} is obtained by substituting Eqs. 3.91 and 3.92
into Eq. 3.86 followed by simplification.

$$r_{HBr} = \frac{2k_2 (k_1/k_4)^{\frac{1}{2}} C_{H_2} C_{Br_2}^{\frac{1}{2}}}{1 + (k_{-2}/k_3) C_{HBr}/C_{Br_2}} \tag{3.93}$$

If HBr is absent initially, the initial rate is proportional to $C_{H_2} C_{Br_2}^{\frac{1}{2}}$, i.e., it is
first order with respect to hydrogen concentration and one-half order with respect
to bromine concentration. On the other hand, if a sufficient amount of HBr is added

initially so that $k_{-2}C_{HBr}/k_3C_{Br_2} \gg 1$, the initial rate is proportional to
$C_{H_2}C_{Br_2}^{3/2}/C_{HBr}$. Thus, in either situation the rate of formation of HBr is seen
to be of fractional order with respect to the concentration of bromine.

3.3.4 Autocatalytic Reactions

In this section we consider nonelementary autocatalytic reactions. As was done with
an elementary autocatalytic reaction in Section 3.2.1, it is best to consider an
example. A typical microbial process involves a batch reactor which is charged with
an aqueous solution of substrate, salts, and nutrients, and then inoculated with a
small amount of microorganisms called the inoculum. The microbial cells contain the
necessary enzymes to carry out complex sequences of reactions to generate more cells,
energy, and metabolic products. Under a proper environment the cells begin to grow
in number and as more cells are formed more enzymes are made available for the
reactions. Thus, the rate of cell production increases rapidly, reaching a maximum
and then gradually dropping to zero with the depletion of available substrate. Although the reaction sequences are very complex, we shall consider a highly simplified
version which may be represented by

$$S \xrightarrow{X} aX + b_1P_1 + b_2P_2 + \ldots + b_mP_m \tag{3.94}$$

in which X stands for the cells, S the substrate, P_i metabolic products, and a and
b_i the stoichiometric coefficients. The reaction scheme as represented by Eq. 3.94,
is very similar to the enzyme reaction scheme given by Eq. 3.71 in that the microbial
cells play a role equivalent to the enzyme in Eq. 3.71. Therefore, it is not too
surprising to find that the rate expression empirically proposed by Monod [10] for
this situation resembles the rate expression for an enzyme reaction proposed by
Michaelis and Menten. Monod proposed a rate equation of the form

$$r_X = \mu_m C_X C_S/(K + C_S) \tag{3.95}$$

where μ_m and K are constants. This has been shown to represent remarkably well a
wide variety of biological kinetic data. Let us assume that the stoichiometric
coefficient a is a known and constant (the difficulty associated with this assumption
will be discussed fully in Section 9.4). Then, the rate of consumption of substrate
is

$$-r_S = r_X/a = (\mu_m/a)C_S C_X/(K + C_S) \tag{3.96}$$

The stoichiometric relationship implied by Eq. 3.94 is used to eliminate C_X in terms
of C_S,

$$-r_S = \mu_m(\sigma - C_S)C_S/(K + C_S) \tag{3.97}$$

where

$$\sigma = C_{Xo}/a + C_{So} \tag{3.98}$$

The rate expression as given by Eq. 3.97 shows a maximum, as can be demonstrated by differentiation of $-r_S$ with respect to C_S and setting it to zero. The maximum occurs when

$$C_S = (K^2 + K\sigma)^{\frac{1}{2}} - K \tag{3.99}$$

and the maximum rate is

$$-r_S = \mu_m [2K + \sigma - 2(K^2 + K\sigma)^{\frac{1}{2}}] \tag{3.100}$$

Thus, the rate of consumption of substrate, $-r_S$, is not a monotonically increasing function of substrate concentration. It increases with the substrate concentration first, reaches a maximum, and decreases eventually. A similar phenomenon was also noted for autocatalytic reactions that are elementary. A typical plot of $-r_S$ versus C_S is very similar to Fig. 3.1. That the rate expression is not a monotonically increasing function, has very important implications in reactor design as we shall see in Chapter 4.

3.4 DEPENDENCE OF REACTION RATE CONSTANTS ON TEMPERATURE

As we have seen in this chapter, rate expressions are written as a product of a temperature dependent rate constant, k_i and concentration dependent terms. The temperature dependence of the reaction rate constant has been found to be well re-presented by the Arrhenius equation [11].

$$k = k_o \exp(-E/RT) \tag{3.101}$$

where k_o is the preexponential factor (or sometimes called a frequency factor), R the gas constant, T the absolute temperature, and E the activation energy. The pre-exponential factor k_o is assumed to be a constant independent of temperature and has the same units as the rate constant. Taking the natural logarithm of Eq. 3.101 yields

$$\ln k = \ln k_o - E/RT \tag{3.102}$$

According to this equation a straight line with slope of $-E/R$ should be obtained when the natural logarithms of the rate constants are plotted against the recip-rocals of the absolute temperatures. The value of k_o may be calculated from a particular value of temperature, T_1, and a corresponding value of k, k_1, which lies on the line,

$$k_o = k_1 \exp(E/RT_1) \tag{3.103}$$

Since two parameters k_o and E describe the temperature dependence of the rate constant in Eq. 3.102, it is necessary, in principle, to have only two experimental determinations of the rate constant at two different temperatures. However, in practice it is advisable to run additional determinations in order to ensure that the plot indeed yields a straight line.

Once a plot of Eq. 3.102 is available, the rate constant at any temperature may be estimated simply by locating the reciprocal of the absolute temperature on the abscissa and reading off the corresponding ordinate value from the straight line. Alternatively, one can use Eq. 3.101 to calculate the value of k at any other temperature once -E/R and k_o are known. A graphical determination of the rate constant and the parameters -E/R and k_o is depicted in Fig. 3.2. Reactions with high activation energies are much more sensitive to temperature than reactions with low activation energies. Whenever there is a fair amount of scatter in the plot, it is difficult to draw the best straight line by eye. In this situation the method of least squares should be used to obtain a reliable straight-line equation.

There are theories which predict the temperature dependency of the rate constant, for instance the collision theory of gas reactions and the transition theory. A detailed explanation of these is beyond the scope of this book, however, the results predicted by them are summarized below. A detailed coverage may be found elsewhere [12,13].

Collision Theory: $k \propto T^{\frac{1}{2}} \exp(-E/RT)$ (3.104)

Transition Theory: $k \propto T \exp(-E/RT)$ (3.105)

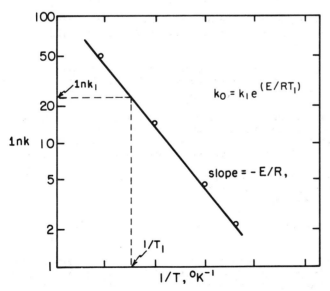

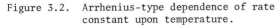

Figure 3.2. Arrhenius-type dependence of rate
             constant upon temperature.

Thus, the expression

$$k = k_o T^n \exp(-E/RT) \qquad n = 0, 1/2, 1 \qquad\qquad (3.106)$$

represents the temperature dependency of the rate constant from both a theoretical
and an empirical viewpoint. In practice, the exponential term is usually much more
sensitive to temperature than the T^n term, so that the change in k caused by the
latter is masked by the change in the exponential term. Therefore, k can be represented in effect as

$$k = k_o \exp(-E/RT) \qquad\qquad (3.101)$$

Thus, if the temperature range is not too great, it may be concluded that the
Arrhenius equation is a good approximation of the temperature dependency of the rate
constant predicted by both the collision and transition theories.

3.5 KEY POINTS

1. The order of a reaction with respect to a particular species is equal to the
    exponent on that species in the rate expression. The overall order of a
    reaction is the sum of the exponents in the rate expression.

2. The term molecularity refers to the number of molecules involved in the step
    leading to reaction.

3. Elementary reactions are those for which the molecularity and order are identi-
    cal so that the exponents in the rate expression correspond to the coefficients
    in the stoichiometric equation. Therefore, the rate expression may be obtained
    from the stoichiometric equation.

4. The yield of a reaction can be expressed either as the amount of product formed
    per unit amount of reactant charged or the amount of product formed per unit
    amount of reactant reacted.

5. The reaction rate expression for a single species in a single irreversible
    reaction may be written in terms of the concentration of that species alone
    through use of stoichiometric relationships.

6. In reversible reactions, the forward rate constant, k_1, and the reverse rate
    constant, k_{-1}, are not independent of each other but are related by the
    equilibrium constant K. This allows the reaction rate to be written in terms
    of only one rate constant and the equilibrium constant.

7. When parallel reactions involve more than one reactant, a mass balance equation
    must be written for each reactant and then the set must be solved simultaneously
    to determine the concentrations.

8. The mass balance equations for the reactants in series reactions may be solved
    sequentially.

9. Enzyme reactions are nonelementary reactions. The process leading to the rate
    expression for an enzyme reaction involves the following steps: (1) postu-
    lation of a sequence of elementary reactions, (2) writing of all rate expres-
    sions for reactants and complexes, (3) expression of the concentrations of each
    enzyme complex in terms of the concentrations of observable species by solving
    simultaneously the algebraic equations resulting from the pseudosteady-state

hypothesis and the mass balance equation on the enzyme species, and (4) expression of the rate in terms of the concentration of observable species using the result of (3).

10. Chain reactions are similar to catalytic reactions since active sites are involved. Thus the steps listed above may also be used to determine the rate expression for chain reactions.

11. The rate of substrate consumption during microbial growth is not a monotonically increasing function of substrate concentration, but instead reaches a maximum value before decreasing.

12. The temperature dependence of the reaction rate constant is well represented by the Arrhenius equation:

$$k = k_o \exp(-E/RT) \qquad\qquad (3.101)$$

3.6 STUDY QUESTIONS

The study questions are arranged in the same order as the text:

Section	Questions
3.1	1-5
3.2	6-11
3.3	12-13
3.4	14-17

1. If the rate expression is given as $r = kC_E C_F^2$, what is the order of the reaction with respect to E? What is the overall order?

2. If the rate expression is given as $r = kC_A/(K + C_A)$, what is the order of the reaction?

3. Define unimolecular and bimolecular reactions.

4. If a certain reaction has a rate given by $-r_A = 2.5C_A^2$, mole/(liter·hr) what would be an equivalent rate expression in units of milligram/(liter·min)? The molecular weight of A is 50. What would be the units of the rate constant?

5. For an aqueous reaction the rate expression is reported as $-r_A = 0.05C_A^2$ mg/(liter·hr)

    (a) What are the units of the rate constant and C_A?

    (b) The molecular weight of A is 100. Express the rate in units of mol/(cm^3·sec). What are the units of the rate constant?

6. If the stoichiometric equation of an irreversible elementary reaction is given by B + 2C = D, what is the rate expression for this reaction? What is the rate expression for species C?

7. The rate expression of an elementary reaction is given as $r_B = kC_C C_D^2$. Write the stoichiometric equation for this reaction.

8. Derive Eq. 3.34 from Eqs. 3.30 and 3.31.

9. Derive Eq. 3.53 from Eqs. 3.49 and 3.50.

10. The forward and reverse reactions involved in the reversible reaction

$$A + 2B \overset{k_1}{\underset{k_{-1}}{\rightleftarrows}} C + 2D$$ are known to be elementary. The rate constants k_1 and k_{-1} are defined on the basis of the species A and C, respectively. Derive the net rate of generation for species C.

11. Consider the following parallel and series reactions in which each reaction is elementary. Develop the rate expression $-r_A$.

$$A \overset{k_1}{\underset{k_{-1}}{\rightleftarrows}} B \overset{k_2}{\rightarrow} 2C ; \qquad A \overset{k_3}{\rightarrow} D$$

12. For the reaction $A + B \rightarrow C$ the rate expression was found to be given by $-r_A = kC_A C_B/(K_1 + C_A)$. Is this an elementary reaction? Why or why not? If the units of C_A and C_B are mole/liter and $-r_A$ is given in units of mole/(liter·hr), what are the units of k and K_1?

13. A more general reaction scheme for Eq. 3.71 can be represented by

$$E + S \overset{k_1}{\underset{k_{-1}}{\rightleftarrows}} ES \overset{k_2}{\underset{k_{-2}}{\rightleftarrows}} EP \overset{k_3}{\rightarrow} E + P$$

Develop the rate expression r_p by following the procedure developed in Section 3.3.2.

14. What are the typical units of the reaction rate constant k for the termolecular reaction $3A \rightarrow$ product?

15. It was found that the rate constant of a reaction at 47°C was twice that at 37°C. What is the activation energy associated with this reaction?

16. The rate constant of an elementary reaction was determined at 27°C and 57°C to be 10/min and 100/min, respectively. Determine the rate constant at 42°C.

17. The rate constant of a first-order reaction was measured at various temperatures as given in Table SQ3.1. Determine $-E/R$ and k_o.

TABLE SQ3.1
Effect of Temperature on Rate Constant

Temperature °C	Rate Constant min^{-1}
40	0.32
40	0.31
53	0.77
53	0.72
64	1.30
77	2.80
77	2.60
80	3.00

REFERENCES AND FURTHER READING

1. L. Michaelis and M. L. Menten, "Die Kinetik der Invertinwirkung," *Biochemische Zeitschrift,* 49, 333-369, 1913.

2. G. E. Briggs and J. B. S. Haldane, "Note on the kinetics of enzyme action," *Biochemical Journal,* 19, 338-339, 1925.

3. W. G. Miller and R. A. Alberty, "Kinetics of the reversible Michaelis-Menten mechanism and the applicability of the steady-state approximation," *Journal of the American Chemical Society,* 80, 5146-5151, 1958.

4. J. R. Bowen, *et al.* "Singular perturbation refinement to the quasi-steady-state hypothesis in chemical kinetics," *Chemical Engineering Science,* 18, 177, 1963.

5. F. G. Heineken, *et al.* "On the mathematical status of the pseudo-steady-state hypothesis of biochemical kinetics," *Mathematical Biosciences,* 1, 95-113, 1967.

6. J. A. Christiansen, *Kgl. Dansk Videnskab, Selskab Mat.-Fys. Medd.,* 1, 14, 1919.

7. K. F. Herzfeld, "Zur Theorie der Reaktionsgeschwindigkeiten in Gasen," *Zetschrift für Elektrochemie,* 25, 301-304, 1919.

8. Von M. Polayni, "Reaktionsisochore and Reaktionsgeschwindigkeit vom Standpunkte der Statistik," *Zeitschrift für Elektrochemie,* 26, 49-54, 1920.

9. M. Bodenstein and S. C. Lind, "Geschwindigkeit der Bildung des Bromwasserstoffs aus seinen Elementen," *Zeitschrift für Physikalische Chemie* (Leipzig), 57, 168-192, 1906.

10. J. Monod, "The growth of bacterial cultures," *Annual Review of Microbiology,* 3, 371-394, 1949.

11. S. Arrhenius, "Uber die Reaktiongeschwindigkeit bei der Inversion von Rohrzucker durch Säuren," *Zeitschrift für Physikalische Chemie* (Leipzig), 4, 226-248, 1889.

12. E. A. Moelwyn-Hughes, *Kinetics of Reactions in Solution,* Oxford University Press, New York, N.Y., 1947. See Chapters 1 and 2.

13. S. Glasstone, *et al. The Theory of Rate Processes,* McGraw-Hill Book Co., New York, N.Y., 1941. See Chapter 1.

Further Reading

Bensen, S. W., *Foundations of Chemical Kinetics,* McGraw-Hill Book Co., New York, N.Y., 1960. See Chapter 13.

Dixon, M. and Webb, E. C., *Enzymes,* Academic Press, New York, N.Y., 1958. See Chapter 4.

Frost, A. A. and Pearson, R. G., *Kinetics and Mechanisms,* Second Edition, John Wiley and Sons, New York, N.Y., 1961.

Laidler, K. J., *Chemical Kinetics,* Permagon Press, New York, N.Y., 1963.

Levenspiel, O., *Chemical Reaction Engineering,* Second Edition, John Wiley and Sons, New York, N.Y., 1972.

Moore, W. J., *Physical Chemistry,* Fourth Edition, Prentice-Hall, Inc., Englewood Cliffs, N.J., 1972. See Chapter 9.

CHAPTER 4

REACTOR DESIGN

In Chapter 2 we introduced the subject of reactor design for a simple liquid phase reaction, the role of stoichiometric equations, and the need for rate expressions. In Chapter 3 we considered various forms of the rate expressions for several named reactions. In this chapter we shall consider the design of three ideal reactors of industrial importance; batch, continuous-flow stirred tank, and plug-flow reactors. We shall first look at the design of individual single reactors for single reactions. A size comparison of the three ideal reactors will then be made to emphasize the volume efficiency of each. Many biological reactions behave like autocatalytic reactions and these exhibit some unusual characteristics. Therefore, the design of reactors for autocatalytic reactions will be treated next. After that multiple reactors for single reactions will be considered. Finally, the design of reactors for multiple reactions will be discussed briefly to illustrate how the type and size of reactor affects the distribution of products.

4.1 SINGLE REACTORS FOR SINGLE REACTIONS

Single irreversible and reversible reactions were considered in Sections 3.2.1 and 3.2.2, respectively. We now want to take the reaction types discussed in those sections and see how they behave in each of the ideal reactors, using the principles introduced in Chapter 2. It will be recalled from Chapter 2 that the mass balance equations can be simplified when the reactions are carried out in a liquid phase of approximately constant density. Since the vast majority of biological reactions utilized in wastewater treatment occur under such conditions, the development presented herein will be restricted to constant density (and volume) liquid phase reactions.

4.1.1 Batch Reactors

For a constant volume, liquid phase reaction the mass balance on reactant A around an ideal batch reactor was performed in Section 2.3.2, yielding

$$r_A = \frac{dC_A}{dt} \tag{2.25}$$

This may be integrated analytically or numerically between the initial concentration C_{Ao} and the final concentration C_{Af} to yield

$$t = \int_{C_{Ao}}^{C_{Af}} \frac{dC_A}{r_A} = \int_{C_{Af}}^{C_{Ao}} \frac{dC_A}{-r_A} \tag{2.27}$$

As we saw in Example 2.3.2, Eq. 2.27 can be used to calculate the reaction time necessary to achieve a desired concentration, from which the batch reactor volume necessary to meet a desired treatment rate or the concentration that can be reached in a given time may be calculated. The use of this equation for various types of reactions is illustrated below.

ANALYTICAL INTEGRATION. When rate expressions are available in analytical (algebraic) form, it is often possible to perform analytically the indicated integration in Eq. 2.27. We shall consider irreversible reactions first.

For the first-order reaction, A → Product, the rate of consumption of A was given by Eq. 3.11

$$-r_A = kC_A \tag{3.11}$$

which may be substituted into Eq. 2.27 and integrated to give

$$t = \int_{C_{Af}}^{C_{Ao}} \frac{dC_A}{kC_A} = \frac{1}{k} \ln(C_{Ao}/C_{Af}) \tag{4.1}$$

or

$$C_{Af} = C_{Ao} \exp(-kt) \tag{4.2}$$

Equation 4.2 predicts an exponential decay of the concentration of A.

For the second-order reaction

$$A + B \rightarrow \text{Products} \tag{3.12}$$

the rate was given by Eq. 3.17:

$$-r_A = kC_A(C_{Bo} - C_{Ao} + C_A) \tag{3.17}$$

This may be substituted into Eq. 2.27 and integrated using the method of partial fractions to obtain

$$(C_{Bo} - C_{Ao})kt = \int_{C_{Af}}^{C_{Ao}} (\frac{1}{C_A} - \frac{1}{C_{Bo} - C_{Ao} + C_A}) dC_A$$

$$= \ln[C_{Ao}(C_{Af} - C_{Ao} + C_{Bo})/C_{Af}C_{Bo}] \tag{4.3}$$

or

$$C_{Af}/C_{Ao} = (C_{Bf}/C_{Bo}) \exp[-(C_{Bo} - C_{Ao})kt] \tag{4.4}$$

where

$$C_{Bf} = C_{Af} - C_{Ao} + C_{Bo} \tag{4.5}$$

When $C_{Ao} = C_{Bo}$ or A = B the rate expression simplifies to Eq. 3.20

$$-r_A = kC_A^2 \tag{3.20}$$

and the integrated mass balance equation is

$$kt = 1/C_{Af} - 1/C_{Ao} \tag{4.6}$$

or

$$C_{Af} = C_{Ao}/(1 + ktC_{Ao}) \tag{4.7}$$

For the third-order reaction

$$A + B + C \rightarrow \text{Products} \tag{3.21}$$

the rate was given by Eq. 3.23,

$$-r_A = kC_A(C_A - C_{Ao} + C_{Bo})(C_A - C_{Ao} + C_{Co}) \tag{3.23}$$

The mass balance equation, Eq. 2.27, may be integrated using the method of partial fractions to yield

$$(C_{Ao} - C_{Bo})(C_{Bo} - C_{Co})(C_{Co} - C_{Ao})kt$$

$$= \ln[(C_{Ao}/C_{Af})^{(C_{Co} - C_{Bo})}(C_{Bo}/C_{Bf})^{(C_{Ao} - C_{Co})}(C_{Co}/C_{Cf})^{(C_{Bo} - C_{Ao})}] \tag{4.8}$$

where

$$C_{Bf} = C_{Af} - C_{Ao} + C_{Bo} \tag{4.9}$$

and

$$C_{Cf} = C_{Af} - C_{Ao} + C_{Co} \tag{4.10}$$

When only two reactants are involved so that

$$2A + B \rightarrow \text{Products} \tag{3.25}$$

and

$$-r_A = kC_A^2 C_B \tag{3.26}$$

Eq. 4.8 can be shown by l' Hopital's rule to reduce to

$$(2C_{Bo} - C_{Ao})^2 kt = \ln(C_{Bo}C_{Af}/C_{Ao}C_{Bf}) + (2C_{Bo} - C_{Ao})(C_{Ao} - C_{Af})/C_{Ao}C_{Af} \tag{4.11}$$

When the third-order reaction involves only one reactant species, so that

$$3A \rightarrow \text{Products} \tag{3.28}$$

and

$$-r_A = kC_A^3 \tag{3.29}$$

the mass balance equation can be integrated by the power rule to obtain

$$C_{Af} = C_{Ao}/(1 + 2C_{Ao}^2 kt)^{\frac{1}{2}} \tag{4.12}$$

This may be also shown to result from Eq. 4.11 when l' Hopital's rule is used to obtain the limit.

Zero-order reactions show an unusual characteristic which needs to be considered. The rate expression was given by Eq. 3.69

$$-r_A = k \tag{3.69}$$

so that the mass balance equation, Eq. 2.27, may be integrated to yield

$$t = \int_{C_{Af}}^{C_{Ao}} \frac{dC_A}{k} = \frac{1}{k}(C_{Ao} - C_{Af}) \tag{4.13}$$

According to Eq. 4.13 the reactant A will be consumed completely ($C_{Af} = 0$) when $t = C_{Ao}/k$. Any reaction time greater than this has no physical meaning so that Eq. 4.13 should be restricted to $t \leq C_{Ao}/k$.

For fractional- and n^{th}-order reactions with the rate of consumption of A given by $-r_A = kC_A^n$, the integrated mass balance equation is

$$C_{Af} = [(1 - n)kt + C_{Ao}^{(1 - n)}]^{1/(1 - n)} \tag{4.14}$$

Thus, we see that regardless of the reaction order, the design equations for constant-density batch reactors are obtained by substituting an appropriate rate equation for the consumption of A into the constant-density batch reactor equation, Eq. 2.27, and integrating between the starting concentration C_{Ao} and the desired concentration C_{Af}.

EXAMPLE 4.1.1-1

A first-order aqueous reaction was carried out in a batch reactor. It took five hours to reduce the reactant concentration from 1 mol/liter to 0.5 mol/liter. How long would it take to reduce the concentration to 0.25 mol/liter?

Solution:

  Given information;

  $-r_A = kC_A$, $C_{Ao} = 1$, $C_{Af} = 0.5$ when $t = 5$ hrs

Substitution of the given information into Eq. 4.1 and solving for k yield

  $k = \ln(1/0.5)/5 = (\ln 2)/5 = 0.139 \text{ hr}^{-1}$

The time to reach a concentration of 0.25 mol/liter is again determined by Eq. 4.1

  $t = \ln(C_{Ao}/C_{Af})/k = \ln(1/0.25)/0.139$

  $= \ln(4)/0.139 = 10$ hours.

Now consider reversible reactions. For reversible first-order reactions

$$A \underset{k_{-1}}{\overset{k_1}{\rightleftharpoons}} B \tag{3.35}$$

the rate expression was given by Eq. 3.42

$$-r_A = -k_{-1}[(K + 1)C_A - (C_{Ao} + C_{Bo})] \qquad (3.42)$$

This may be substituted into Eq. 2.27 and integrated to yield

$$(k_1 + k_{-1})t = \ln\left(\frac{KC_{Ao} - C_{Bo}}{KC_{Af} - C_{Bf}}\right) = \ln\left(\frac{C_{Ao} - C_{Ae}}{C_{Af} - C_{Ae}}\right) \qquad (4.15)$$

where the equilibrium constant and concentration, K and C_{Ae}, were given by Eq. 3.41. Equation 4.15 may be rewritten as

$$C_{Af} - C_{Ae} = (C_{Ao} - C_{Ae})\exp[-(k_1 + k_{-1})t] \qquad (4.16)$$

which is analogous to Eq. 4.2, the design equation for the first-order irreversible reaction. In the case of an irreversible reaction the equilibrium concentration, C_{Ae}, and the reverse reaction rate constant, k_{-1}, are zero, and therefore, Eq. 4.16 reduces to Eq. 4.2. This suggests that for reversible reactions the deviations of reactant concentrations from their equilibrium values, rather than the concentrations, should be used (e.g., $C_{Af} - C_{Ae}$ and $C_{Ao} - C_{Ae}$).

EXAMPLE 4.1.1-2

A first-order reversible reaction, A $\underset{\longleftarrow}{\longrightarrow}$ B was carried out in aqueous phase using a batch reactor. The forward reaction rate was determined to be 2.0 hr^{-1} and the reverse reaction rate 0.5 hr^{-1}. It is desired to reduce the concentration of species A from 1 mol/liter to 0.5 mol/liter. How long would it take if no B is present initially?

Solution:

   Given information

   $k_1 = 2.0\ hr^{-1}$

   $k_{-1} = 0.5\ hr^{-1}$

   $C_{Ao} = 1$ mol/liter

   $C_{Af} = 0.5$ mol/liter

   $C_{Bo} = 0$

Equation 3.41 gives the equilibrium constant

   $K = k_1/k_{-1} = 2.0/0.5 = 4$

The equilibrium concentration is determined from Eq. 3.41

   $(C_{Ao} + C_{Bo} - C_{Ae})/C_{Ae} = (1 - C_{Ae})/C_{Ae} = 4$

or

$$C_{Ae} = 0.20 \text{ mol/liter}$$

The required time is obtained from Eq. 4.15

$$t = \ln\left(\frac{1.0 - 0.20}{0.5 - 0.20}\right)/(2.0 + 0.5) = 0.392 \text{ hour}$$

How long would it take if the initial concentration of B is 1.0 mol/liter? The equilibrium concentration is again obtained from Eq. 3.41

$$(1 + 1 - C_{Ae})/C_{Ae} = 4 \text{ or } C_{Ae} = 0.4 \text{ mol/liter}$$

and the required time from Eq. 4.15

$$t = \ln\left(\frac{1.0 - 0.4}{0.5 - 0.4}\right)/(2.5) = 0.717 \text{ hour}$$

Note that the presence of B initially increases the equilibrium concentration of A, C_{Ae}, and also the time required to reduce the concentration of A.

For the reversible second-order reaction

$$A + B \underset{k_{-1}}{\overset{k_1}{\rightleftarrows}} C + D \tag{3.43}$$

the rate expression was given by Eq. 3.46. When combined with Eq. 2.27 this gives

$$\frac{dC_A}{dt} = - k_1 C_A (C_A - C_{Ao} + C_{Bo}) + k_{-1}(C_{Co} + C_{Ao} - C_A)(C_{Do} + C_{Ao} - C_A) \tag{4.17}$$

Equilibrium conditions dictate that

$$K = k_1/k_{-1} = C_{Ce}C_{De}/C_{Ae}C_{Be} = \frac{(C_{Ao} + C_{Co} - C_{Ae})(C_{Do} + C_{Ao} - C_{Ae})}{C_{Ae}(C_{Bo} - C_{Ao} + C_{Ae})} \tag{3.45}$$

and that

$$C_{Ao} - C_{Ae} = C_{Bo} - C_{Be} = C_{Ce} - C_{Co} = C_{De} - C_{Do} \tag{4.18}$$

Letting $x = C_A - C_{Ae}$ and using Eqs. 3.45 and 4.18 we may rewrite Eq. 4.17 as

$$\frac{dx}{dt} = - k_1(x + C_{Ae})(x + C_{Be}) + k_{-1}(C_{Ce} - x)(C_{De} - x) = ax^2 - bx \tag{4.19}$$

where

$$a = k_{-1} - k_1 \tag{4.20}$$

and

$$b = k_1(C_{Ae} + C_{Be}) + k_{-1}(C_{Ce} + C_{De}) \tag{4.21}$$

Equation 4.19 may be integrated by the method of partial fractions to give

$$t = \frac{1}{b} \ln \left[\frac{(x_f - b/a)x_o}{x_f(x_o - b/a)} \right] \tag{4.22}$$

When substitution is made for x this becomes

$$(C_{Af} - C_{Ae} - b/a)/(C_{Af} - C_{Ae}) = [(C_{Ao} - C_{Ae} - b/a)/(C_{Ao} - C_{Ae})]\exp(bt) \tag{4.23}$$

The computational scheme for the second-order reversible reaction is as follows. Suppose we are given the rate constants, k_1 and k_{-1} and the initial concentrations C_{Ao}, C_{Bo}, C_{Co} and C_{Do}. Using Eq. 3.45 the equilibrium concentration of A, C_{Ae}, is determined, and the corresponding equilibrium concentrations of all other species are calculated from Eq. 4.18. Equations 4.20 and 4.21 allow us to calculate the constants a and b which are then used in the integrated mass balance equations, Eq. 4.22 or 4.23, to calculate either the time required to reach a final concentration of A, or the final concentration of A at a given time.

EXAMPLE 4.1.1-3

The forward and reverse reaction rate constants for a reversible reaction
A + B $\underset{\longleftarrow}{\longrightarrow}$ C + D are given as

$$k_1 = 3 \text{ (liter)}/(\text{hr·mol})$$

$$k_{-1} = 1 \text{ (liter)}/(\text{hr·mol})$$

The reaction is to be carried out in a batch reactor starting initially with 1 mol/liter each of A and B and no C or D. How long would it take to reduce the concentration of A to 0.5 mol/liter? What would be the concentrations of B, C, and D? What would be the concentration of A at t = 1 hour?
Solution:

Given information;

$$C_{Ao} = C_{Bo} = 1 \text{ mol/liter}$$

$$C_{Co} = C_{Do} = 0$$

Equation 3.45 gives

$$K = \frac{3}{1} = \frac{(1 + 0 - C_{Ae})(0 + 1 - C_{Ae})}{C_{Ae}^2} = \left[\frac{1 - C_{Ae}}{C_{Ae}}\right]^2$$

which may be solved for C_{Ae}:

$$C_{Ae} = 0.366 \text{ mol/liter}$$

Equation 4.18 allows us to calculate the equilibrium concentrations:

$$C_{Be} = C_{Ae} = 0.366$$

$$C_{Ce} = C_{Co} + C_{Ao} - C_{Ae} = 0 + 1 - 0.366 = 0.634$$

$$C_{De} = C_{Ce} = 0.634$$

The parameters a and b are calculated from Eqs. 4.20 and 4.21, respectively

$$a = 1 - 3 = -2$$

$$b = k_1 (C_{Ae} + C_{Be}) + k_{-1}(C_{Ce} + C_{De})$$

$$= 3(0.366 + 0.366) + 1(0.634 + 0.634)$$

$$= 3.464$$

The time required to achieve $C_{Af} = 0.5$ mol/liter is calculated from Eq. 4.22,

$$t = \frac{1}{3.464} \ln[\frac{(0.5 - 0.366 + 1.732)(1 - 0.366)}{(0.5 - 0.366)(1 - 0.366 + 1.732)}] = 0.38 \text{ hour}$$

The stoichiometric relationships are

$$C_{Ao} - C_{Af} = C_{Bo} - C_{Bf} = -(C_{Co} - C_{Cf}) = -(C_{Do} - C_{Df})$$

from which we can obtain the final concentrations of B, C, and D:

$$C_{Bf} = C_{Af} = 0.5 \text{ mol/liter and } C_{Cf} = C_{Df} = 1 - 0.5 = 0.5 \text{ mol/liter}$$

At t = 1 Eq. 4.21 gives

$$(C_{Af} - 0.366 + 1.732)/(C_{Af} - 0.366) = [(1 - 0.366 + 1.732)/(1 - 0.366)]\exp(3.464)$$

$$(C_{Af} + 1.366)/(C_{Af} - 0.366) = 119.21$$

or

$$C_{Af} = 0.3807 \text{ mol/liter}$$

When only two species are involved, for example, as in Eq. 3.47

$$2A \underset{\longleftarrow}{\longrightarrow} 2C \qquad\qquad (3.47)$$

the above mass balance equation is still valid provided that we set $C_{Ae} = C_{Be}$, $C_{Ce} = C_{De}$, $C_{Ao} = C_{Bo}$ and $C_{Co} = C_{Do}$ in Eqs. 4.18 through 4.23 and Eq. 3.45.

When the rate constants for the forward and reverse reactions are equal so that the equilibrium constant is unity, i.e., K = 1, Eq. 4.19 reduces to

$$\frac{dx}{dt} = -bx \qquad\qquad (4.24)$$

and Eq. 4.23 becomes

$$C_{Af} - C_{Ae} = (C_{Ao} - C_{Ae}) \exp[-2k_1(C_{Ae} + C_{Ce})t] \tag{4.25}$$

For the mixed-order reversible reaction

$$A + B \underset{k_{-1}}{\overset{k_1}{\rightleftharpoons}} C \tag{3.49}$$

which is second order in the forward direction and first order in the reverse direction, substitution of the rate equation as given by Eq. 3.50 into the mass balance equation gives

$$\frac{dC_A}{dt} = -k_1 C_A C_B + k_{-1} C_C$$

$$= -k_1(x + C_{Ae})(x + C_{Be}) + k_{-1}(C_{Ce} - x) \tag{4.26}$$

or

$$\frac{dx}{dt} = -k_1 x^2 - [k_1(C_{Ae} + C_{Be}) + k_{-1}]x \tag{4.27}$$

Since Eq. 4.27 has the same form as Eq. 4.19, the result of integration is readily obtained from Eq. 4.23 as

$$(C_{Af} + C_{Be} + 1/K)/(C_{Af} - C_{Ae})$$

$$= [(C_{Ao} + C_{Be} + 1/K)/(C_{Ao} - C_{Ae})] \exp\{[k_1(C_{Ae} + C_{Be}) + k_{-1}]t\} \tag{4.28}$$

where

$$K = (C_{Co} + C_{Ao} - C_{Ae})/C_{Ae}(C_{Bo} - C_{Ao} + C_{Ae}) = k_1/k_{-1} \tag{3.52}$$

and

$$C_{Ao} - C_{Ae} = C_{Bo} - C_{Be} = C_{Ce} - C_{Co} \tag{4.29}$$

Likewise, the results for $A \rightleftharpoons C + D$ are

$$\frac{C_{Af} - C_{Ae} - C_{Ce} - C_{De} - K}{C_{Af} - C_{Ae}}$$

$$= \left[\frac{C_{Ao} - C_{Ae} - C_{Ce} - C_{De} - K}{C_{Ao} - C_{Ae}}\right] \exp\{[k_1 + k_{-1}(C_{Ce} + C_{De})]t\} \tag{4.30}$$

where

$$K = (C_{Co} + C_{Ao} - C_{Ae})(C_{Do} + C_{Ao} - C_{Ae})/C_{Ae} = k_1/k_{-1} \tag{4.31}$$

and

$$C_{Ao} - C_{Ae} = C_{Ce} - C_{Co} = C_{De} - C_{Do} \qquad (4.32)$$

From these results we see that for second- and mixed-order (first- and second-order) reversible reactions the integrated mass balance equations for batch reactors are identical in form so that the computational scheme used for second-order reactions (Example 4.1.1-3) may also be used for mixed-order reactions.

NUMERICAL INTEGRATION. When it is not possible to perform the integration in the mass balance equation analytically, it must be done numerically. Two different situations are possible. In one the rate expression is unusual in form and does not lead to analytical integration. In the other the rate expression is not available, but reaction rate data at various values of reactant concentration are. In these situations one has to rely upon a numerical integration scheme to evaluate the integral. Equation 2.27 suggests that the reciprocal of $-r_A$ should be plotted against C_A so that the area under the curve between the limits of C_{Af} and C_{Ao} becomes the value of the integral and therefore, the required reaction time. This is illustrated in Fig. 4.1. The area under the curve may be evaluated either by the trapezoidal method [1a] or by Simpson's rule [2].

In the trapezoidal method the concentration interval, (C_{Af} to C_{Ao}), is divided into a large number of subintervals (n) and the area under each subinterval is approximated by the area of the trapezoid, with the sides equal to the heights of the curve at the left- and right-hand sides of the interval and the base equal to

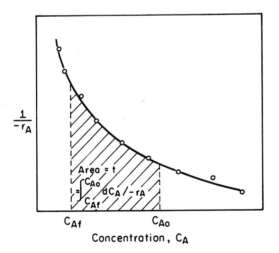

Figure 4.1.   Graphical representation of the equation
              for a constant density batch reactor (Eq. 2.27).

the subinterval length. The value of the integral is the sum of the areas of the trapezoids,

$$t = \int_{C_{Af}}^{C_{Ao}} \frac{dC_A}{-r_A} = [\frac{1}{-r_A(C_{Af})} + \frac{1}{-r_A(C_{A1})}](C_{A1} - C_{Af})/2$$

$$+ \sum_{j=2}^{n} [\frac{1}{-r_A(C_{Aj})} + \frac{1}{-r_A(C_{Aj-1})}](C_{Aj} - C_{Aj-1})/2 \qquad (4.33)$$

where $-r_A(C_{Aj})$ represents $-r_A$ evaluated at the jth interval, C_{Aj}, and C_{An} is equal to C_{Ao}. When the rate expression is available in analytical form, it is most convenient to break up the interval into n equal subintervals and the area is then determined from a reduced form of Eq. 4.33,

$$t = [\frac{1}{-2r_A(C_{Af})} + \frac{1}{-2r_A(C_{Ao})} + \sum_{j=1}^{n-1} \frac{1}{-r_A(C_{Aj})}](C_{Ao} - C_{Af})/n \qquad (4.34)$$

The use of Eq. 4.34 is illustrated in Fig. 4.2.

To apply Simpson's rule, the interval between C_{Af} and C_{Ao} should be divided into any number, n, of subintervals of arbitrary length, Δ_i. Simpson's rule then fits a quadratic equation to the three points on the curve of $1/-r_A$ versus C_A formed by the ends (C_{AL} and C_{AR}) and the midpoint (C_{AM}) of each subinterval, and evaluates the area under each quadratic analytically. The value of the integral being evaluated is the sum of the areas within each subinterval:

$$t = \int_{C_{Af}}^{C_{Ao}} \frac{dC_A}{-r_A} = \sum_{i=1}^{n} [\frac{\Delta_i}{6}(\frac{1}{-r_A(C_{AL})} + \frac{4}{-r_A(C_{AM})} + \frac{1}{-r_A(C_{AR})})]_i \qquad (4.35)$$

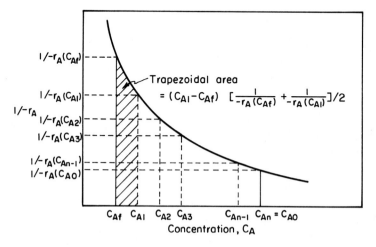

Figure 4.2. Trapezoidal method of evaluating an integral.

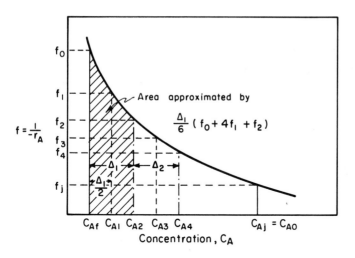

Figure 4.3. Simpson's rule for evaluating an integral.

where Δ_i is the length of a subinterval $(C_{AR} - C_{AL})$, $1/-r_A(C_{AL})$ is the negative reciprocal of the reaction rate corresponding to the concentration C_A on the left boundary of the subinterval, $1/-r_A(C_{AR})$ is the value on the right boundary, and $1/-r_A(C_{AM})$ is the value in the middle.

Equation 4.35 may be simplified to make its application easier. Referring to Fig. 4.3, let us choose subintervals of equal length so that all values of Δ_i are the same and equal to Δ. Furthermore, let f_0 equal the value of $1/-r_A$ at $C_A = C_{Af}$ and f_j equal the value at $C_A = C_{Ao}$. Then f_1 is the value of $1/-r_A$ at the center of the first subinterval and f_2 is the value at the right side of the first subinterval. However, f_2 is also the value at the left side of the second subinterval. Continuing in this manner it can be seen that the even numbered values of f correspond to the boundaries of the subinterval whereas the odd ones correspond to the values at the midpoints. Applying this concept to Eq. 4.35 reveals that it reduces to:

$$t = \frac{\Delta}{6}(f_0 + 4f_1 + 2f_2 + 4f_3 + 2f_4 + \ldots + 4f_{j-1} + f_j) \qquad (4.36)$$

Let us now illustrate the two methods of numerical integration with an example.

EXAMPLE 4.1.1-4

A reaction involving one reactant has been studied and the rates at various reactant concentrations are given in Table E4.1. Estimate the time required to reduce the reactant concentration from 1 mol/liter to 0.2 mol/liter in a batch reactor using (a) the trapezoidal method and (b) Simpson's rule

TABLE E4.1
Reaction Rates at Various Reactant Concentrations

C_A mol/liter	$1/-r_A$ (hr·liter)/mol	C_A mol/liter	$1/-r_A$ (hr·liter)/mol
1.2	0.48	0.5	0.57
1.1	0.49	0.4	0.59
1.0	0.5	0.3	0.62
0.9	0.51	0.2	0.66
0.8	0.52	0.1	0.72
0.7	0.54	0.05	0.77
0.6	0.55	0.03	0.80

Solution:

Given information; Rate data are given at equal concentration intervals from 0.2 to 1.0 mol/liter. Therefore, it is convenient to divide the interval into 8 equal subintervals of 0.1 units each and label them as shown in Table E4.2.

TABLE E4.2
Subintervals for Numerical Integration

C_A Label	Value	$1/-r_A(C_{Ai})$ Label	Value
C_{Af}	0.2	f_0	0.66
C_{A1}	0.3	f_1	0.62
C_{A2}	0.4	f_2	0.59
C_{A3}	0.5	f_3	0.57
C_{A4}	0.6	f_4	0.55
C_{A5}	0.7	f_5	0.54
C_{A6}	0.8	f_6	0.52
C_{A7}	0.9	f_7	0.51
$C_{A8}=C_{Ao}$	1.0	f_8	0.50

(a) Trapezoidal Method

Equation 4.34 is used to calculate the necessary reaction time, in which n = 8, C_{Ao} = 1, and C_{Af} = 0.2.

$$t = [(0.66/2) + (0.50/2) + (0.62 + 0.59 + 0.57 + 0.55 + 0.54 + 0.52 + 0.51)]$$
$$(1 - 0.2)/8$$
$$= 0.448 \text{ hour}$$

(b) Simpson's Rule

An interval for Simpson's rule, Δ, is twice the length of the subinterval used above, thus $\Delta = 0.2$. Substituting the appropriate values into Eq. 4.36 yields

$$t = \frac{0.2}{6}[0.66 + 4(0.62) + 2(0.59) + 4(0.57) + 2(0.55) + 4(0.54) + 2(0.52)$$

$$+ 4(0.51) + 0.5]$$

$$= 0.448 \text{ hour}$$

In this situation both the trapezoidal method and Simpson's rule give the same
numerical value, 0.448 hour. However, in general, these two techniques will give
different values because one fits the data with a series of straight lines whereas
the other uses curves.

4.1.2 Continuous Stirred Tank Reactors

The general mass balance equation on species A for a CSTR is given by Eq. 2.18

$$n_{Ao} - n_A + r_A V = 0 \tag{2.18}$$

where n_{Ao} and n_A are molar flow rates for the input and output, respectively and the
rate has to be evaluated at the reactor conditions which are the same as the exit
conditions. For constant volume reactions it is more convenient to work with concentrations and flow rates, i.e., Eq. 2.19

$$F_o C_{Ao} - F C_A = -r_A V \tag{2.19}$$

Since the influent and effluent flow rates are the same at steady state, Eq. 2.19
may be rewritten as

$$V/F = (C_{Ao} - C_{Af})/(-r_A)_f \tag{4.37}$$

where C_{Af} is the effluent concentration of A (same as the concentration in the CSTR
due to the perfect mixing assumption) and $(-r_A)_f$ is the rate of consumption of A
evaluated at the conditions within the reactor (same as the effluent conditions).
The quantity V/F has the units of time; it represents the time needed to process
one reactor volume of feed. Just as the reaction time for a batch reactor represents
a measure of processing rate (see Example 2.3.2-1), V/F is also a measure of processing rate for flow reactors and deserves special attention (to be given at the
end of this section). It is called space time [3a]:

$$\tau = V/F \tag{4.38}$$

For example, if τ = 10 hours, it means that every 10 hours one reactor volume of
feed is treated by the reactor.

   In terms of space time the mass balance equation for a CSTR can be written as

$$\tau = (C_{Ao} - C_{Af})/(-r_A)_f \tag{4.39}$$

Note that for a CSTR, unlike a batch reactor, it is not necessary to know the rates of reaction at various reactant concentrations. Rather, it is necessary to know the rate at only one concentration - the effluent concentration. Thus, if the initial concentration C_{Ao}, the effluent concentration C_{Af}, and the rate of consumption of A at the effluent concentration $(-r_A)_f$ are known, the space time τ can be calculated from Eq. 4.39. From τ, either the volumetric flow rate that can be handled with a given size reactor, or the size of reactor to handle a given flow rate, can be calculated. It is also possible to calculate the effluent concentration C_{Af} if τ is given and the rate expression is known. This will be illustrated through an example.

EXAMPLE 4.1.2-1

The second-order irreversible reaction $A + B \rightarrow C$ is to be carried out in aqueous phase. The rate constant at 30°C is given as 1.5 liter/(mol·hr). The feed rate is 10 liters/hr and the feed contains 2 mol/liter of A and B.

(a) Determine the size of CSTR necessary to reduce the concentration of A in the effluent to 0.2 mol/liter.

Solution:

Given information:

$C_{Ao} = C_{Bo} = 2$ mol/liter

$F = 10$ liters/hr

$k = 1.5$ (liters)/(mol·hr)

$C_{Af} = 0.2$ mol/liter

Since $C_{Ao} = C_{Bo}$, this reaction is equivalent to $2A \rightarrow C$ and $-r_A = kC_A^2$.

The mass balance equation, Eq. 4.39, gives

$$\tau = (C_{Ao} - C_{Af})/(-r_A)_f = (C_{Ao} - C_{Af})/kC_{Af}^2 = (2 - 0.2)/[1.5(0.2)^2]$$

$$= 30 \text{ hours}$$

and

$$V = \tau F = (30)(10) = 300 \text{ liters}$$

(b) If the CSTR volume is 100 liters, what are the concentrations of A, B, and C in the effluent?

$\tau = V/F = 100/10 = 10$ hours. Equation 4.39 gives

$$10 = (2 - C_{Af})/1.5C_{Af}^2$$

or

$$C_{Af} = 0.333 \text{ mol/liter}$$

The CSTR equation is an algebraic equation and can be solved when the rate of consumption of reactant is known at only one concentration, i.e., the effluent concentration. Therefore, the rate expression need not be known in analytical form. For example, if the rate data are available in tabular form, say $-r_A$ vs C_A, one may obtain by interpolation the rate at a desired effluent concentration. As in the case of a batch reactor it is possible to construct a graphical representation of the CSTR equation. Let us rearrange Eq. 4.39

$$\tau = (C_{Ao} - C_{Af})(-1/r_A)_f \qquad\qquad\qquad (4.39)$$

The first term $(C_{Ao} - C_{Af})$ represents the abscissa from C_{Af} to C_{Ao} and the second term $(-1/r_A)_f$ represents the ordinate at C_{Af}, as indicated in Fig. 4.4. Thus, the product of these two terms represents the area of the rectangle of base $(C_{Ao} - C_{Af})$ and height $(-1/r_A)_f$. It is important to remember that the ordinate is evaluated at the conditions within the reactor, which are the same as those in the effluent.

EXAMPLE 4.1.2-2
The decomposition of A was carried out in aqueous phase in a 10 liter experimental CSTR. The results from several experiments are summarized in Table E4.3. Determine the space time necessary to reduce the concentration from $C_{Ao} = 1$ mol/liter to $C_{Af} = 0.4$ mol/liter in a CSTR.

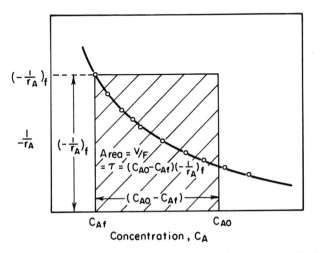

Figure 4.4. Graphical representation of the equation
            for a constant density CSTR (Eq. 4.39).

TABLE E4.3
Data from a CSTR at Steady State

C_{Ao} mol/liter	C_{Af} mol/liter	Flow rate, liters/hr	τ hr
2.0	1.20	15	0.67
2.0	1.00	10	1.00
2.0	0.66	5	2.00
1.0	0.66	20	0.50
1.0	0.50	10	1.00
1.0	0.33	5	2.00

Solution:

We need to construct a plot of $-1/r_A$ vs C_A. Equation 4.39 may be used to calculate the rate for each run:

$$\frac{1}{(-r_A)_f} = \frac{\tau}{C_{Ao} - C_{Af}}$$

The results are tabulated in Table E4.4.

TABLE E4.4
Reaction Rates Calculated from the Data in Table E4.3

C_{Af} mol/liter	$1/(-r_A)_f$ (liter·hr)/mol
1.20	0.83
1.00	1.00
0.66	1.52
0.66	1.52
0.50	2.00
0.33	3.03

Next, $1/(-r_A)_f$ is plotted against C_{Af} and a smooth line is drawn through the experimental points (see Fig. E4.1). The estimated value of $1/(-r_A)_f$ at $C_A = 0.4$ is 2.5 so that $\tau = (C_{Ao} - C_{Af})[1/(-r_A)_f] = (1 - 0.4)(2.5) = 1.5$ hours

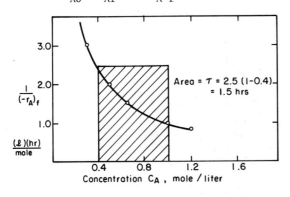

Figure E4.1. Determination of the space time
for Example 4.1.2-2.

We saw above that V/F is a natural term which occurs in the mass balance equation and defined it as the space time. Let us now define it more formally.

Space time is the time required to process one reactor volume of feed measured at specified conditions. Hence, it has the units of time. Although any arbitrary conditions of temperature, pressure, and phase (solid, liquid, or gas) may be used to measure the volume of feed, it is customary to use the actual conditions of the feed itself. We will use this convention.

Sometimes the term mean residence time is also used. This is the average time the fluid spends in the reactor before it leaves the system. When the reaction fluid has a constant density, then the mean residence time is equal to the space time. On the other hand, the mean residence time is not equal to space time if the fluid density changes. Only constant density situations will be considered herein.

Space velocity is the reciprocal of the space time and is the number of reactor volumes treated by the reactor in a unit time. Again the volume is measured at some specified conditions, which are usually the feed conditions. In biochemical engineering it is commonly called the dilution rate and has the units of reciprocal time.

$$D' = 1/\tau = F/V \tag{4.40}$$

In terms of dilution rate the CSTR equation (Eq. 4.39) becomes

$$D' = (-r_A)_f / (C_{Ao} - C_{Af}) \tag{4.41}$$

4.1.3 Plug-Flow Reactors

The steady-state mass balance equation on species A for a plug-flow reactor (PFR) as given by Eq. 2.32 is

$$-F\frac{dC_A}{dx} + r_A A_c = 0 \tag{2.32}$$

Since $A_c dx = dV$, Eq. 2.32 can be rearranged to yield

$$\int_0^V \frac{dV}{F} = \int_{C_{Af}}^{C_{Ao}} \frac{dC_A}{-r_A} \tag{4.42}$$

or

$$\tau = \int_{C_{Af}}^{C_{Ao}} \frac{dC_A}{-r_A} \tag{4.43}$$

where C_{Ao} and C_{Af} represent the influent and effluent reactant concentrations, respectively. A graphical representation of Eq. 4.43 is shown in Fig. 4.5. It is important to recognize that Eq. 4.43 is very similar to Eq. 2.27 for a batch reactor. Indeed, the right sides of both equations are identical. The only difference is that the left side is the reaction time for a batch reactor whereas it is the space time for a PFR. Consequently, the various batch reactor equations developed in

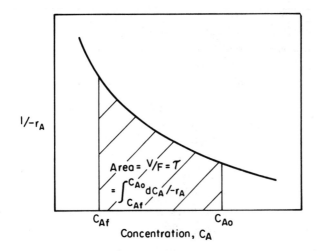

Figure 4.5. Graphical representation of the equation
for a constant density PFR (Eq. 4.43).

Section 4.1.1 can be used directly for PFR's by replacing the reaction time t by the
space time τ. For example, for the second-order aqueous reaction $2A \to C$, the PFR
equation can be written immediately from the batch reactor equation, Eq. 4.6,

$$k\tau = 1/C_{Af} - 1/C_{Ao} \tag{4.44}$$

Although variable volume reactions have not been considered herein, it should be
noted that the analogy between batch and plug-flow reactors does not apply for them.

As in the case of batch reactors the mass balance equation for PFR's, Eq. 4.43,
requires evaluation of an integral. This can be done exactly as it was done for a
batch reactor.

EXAMPLE 4.1.3-1

The third-order irreversible reaction $2A + B \to C$ is to be carried out in aqueous
phase using a PFR. The reaction rate is given by $-r_A = kC_A^2 C_B$ where $k = 5.35$
$(mol/liter)^{-2}(hr)^{-1}$.

(a) Determine the space time necessary to convert 90% of A in a feed containing
2 mol/liter of A and 1 mol/liter of B.

Solution:

Given information;

$$C_{Ao} = 2, \ C_{Bo} = 1, \ C_{Af} = 0.2,$$

$$-r_A = kC_A^2 C_B, \ k = 5.35$$

From the stoichiometric equation, $C_{Ao} - C_A = 2(C_{Bo} - C_B)$, so that

$$C_B = C_{Bo} - \frac{1}{2}(C_{Ao} - C_A) = 1 - \frac{1}{2}(2 - C_A) = C_A/2$$

Thus

$$C_{Bf} = 0.1$$

and

$$-r_A = 5.35 \; C_A^2 \; C_B = 5.35 \; C_A^2 \; C_A/2 = 2.675 \; C_A^3$$

Substitution of this rate expression into the PFR equation, Eq. 4.43, yields

$$\tau = \int_2^{0.2} dC_A/2.675C_A^3 = \frac{0.374}{-2}[C_A^{-2}] \Big|_{0.2}^{2}$$

$$= 0.187 \; [(1/0.2)^2 - (1/2)^2] = 4.63 \text{ hours}$$

(b) If the volumetric flow rate of the feed is 100 liters/hr, what should be the volume of the PFR in part a?

Solution:

   τ = 4.63 hours and τ = V/F.   Therefore

   V = τF = (4.63)(100) = 463 liters

(c) What volumetric flow rate can be handled by a PFR of 1000 liters while achieving the conversion in part a?

Solution:

   F = V/τ = 1000/4.63 = 216 liters/hr.

4.1.4 Size Comparisons of Reactors

BATCH VERSUS PLUG-FLOW REACTORS. We have shown above that when there is no volume change, as in constant-density reactions, the reaction time for a batch reactor, t, and the space time for a PFR, τ, are interchangeable. In other words, the fluid spends the same amount of time in both reactors. With a PFR the operation is continuous and the same space time is maintained throughout the run. With a batch reactor, on the other hand, the operation is not continuous and additional times are needed to charge the feed into and to discharge the product from the reactor. Therefore, in general, the productivity of a PFR is higher than that of a batch reactor of the same size. Nevertheless, batch reactors find wide applications in many industries, especially in handling relatively small-volume productions of very expensive materials because the same batch reactor can be used to produce a variety of products. Sequencing batch reactors have also been proposed for waste-water treatment.

CONTINUOUS STIRRED TANK VERSUS PLUG-FLOW REACTORS. For single reactions it is most convenient to use a graphical comparison of these reactors. For this purpose let us compare the graphical representation of the general CSTR equation, Fig. 4.4, with that of the general PFR equation, Fig. 4.5, as shown in Fig. 4.6. As long as

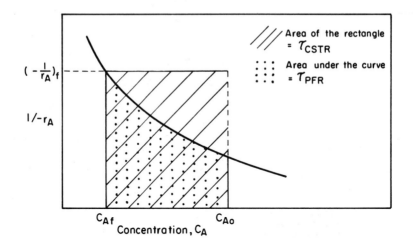

Figure 4.6. Graphical representation of CSTR and PFR sizes for
            a simple monotonic rate equation.

the curve of $-1/r_A$ is a monotonically decreasing function of concentration, C_A, the
area of the rectangle (τ_{CSTR}) will always be larger than the area under the curve
(τ_{PFR}). Therefore, it can be stated that as long as the reciprocal of the rate,
$-1/r_A$, is a monotonically decreasing function of the reactant concentration, C_A, the
size of a PFR required to reduce the reactant concentration to any level will be
smaller than that of a CSTR. Stated another way, if the volumes of the reactors are
the same, and the effluent concentration is to be kept the same, the PFR can process more fluid than the CSTR. This happens because in the CSTR the low reaction
rate associated with the effluent concentration prevails throughout the reactor
whereas in the PFR the reaction rate is fast at the inlet and decreases gradually
to that of the outlet.

When $-1/r_A$ is not a monotonically decreasing function of C_A, i.e., when it has
a local minimum or maximum, no general conclusion can be drawn and each case must be
treated carefully since the conclusion that can be drawn depends upon the influent
and effluent concentrations as well as upon the shape of the curve. This is illustrated in Fig. 4.7. Here, the size requirement for the PFR may be greater than,
equal to, or less than that of the CSTR, depending upon the desired effluent concentration, C_{Af}. For a low effluent concentration (Fig. 4.7c) the space time of a
PFR is less than that of a CSTR, while at a high effluent concentration (Fig. 4.7a)
the opposite is true. One can make a few generalizations from Fig. 4.7. If the
influent concentration, C_{Ao}, is equal to or less than the critical concentration at
which the curve shows the minimum, the required size for a PFR is smaller than that
for a CSTR, regardless of the desired effluent concentration. On the other hand,
if C_{Ao} is greater than the critical value, the three possibilities shown in Fig. 4.7
are possible. In this situation the relative sizes depend upon the desired effluent

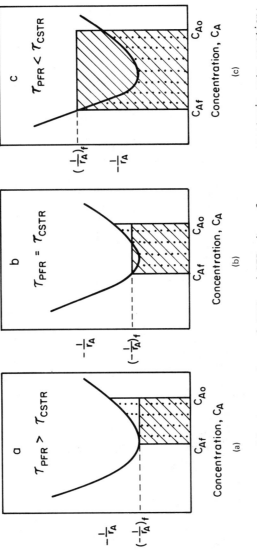

Figure 4.7.   Graphical comparison of CSTR and PFR sizes for a non-monotonic rate equation:
(a) $\tau_{PFR} > \tau_{CSTR}$;   (b) $\tau_{PFR} = \tau_{CSTR}$;   and   (c) $\tau_{PFR} < \tau_{CSTR}$.

concentration. Smaller CSTR's are required as long as C_{Af} is greater than or equal to the critical value. As the effluent concentration is decreased there will be a point at which the sizes are the same, and any further decrease in the desired effluent concentration beyond this point will result in smaller PFR's.

It is not necessary to resort to a graphical comparison. The size requirements can be compared analytically as long as the integral in Eq. 4.43 can be evaluated analytically. This is illustrated through an example.

EXAMPLE 4.1.4-1

For the first-order aqueous reaction A → C develop general space time expressions for a CSTR and a PFR and determine which reactor requires less volume to reduce the reactant concentration to any level.

Solution:

Given information;

$-r_A = kC_A$

From Eq. 4.1 and 4.39 we have

$$\tau_{PFR} = \int_{C_{Af}}^{C_{Ao}} dC_A/kC_A = \ln(C_{Ao}/C_{Af})/k$$

$$\tau_{CSTR} = (C_{Ao} - C_{Af})/kC_{Af} = (C_{Ao}/C_{Af} - 1)/k$$

Combining these yields

$$\tau_{PFR}/\tau_{CSTR} = \ln(C_{Ao}/C_{Af})/(C_{Ao}/C_{Af} - 1) < 1$$

Thus the plug-flow reactor will be smaller

4.1.5 Reactor Design for Autocatalytic Reactions

Autocatalytic reactions were discussed in Sections 3.2.1 and 3.3.4, where it was shown that microbial reactions fall under this category. Typical rate expressions for autocatalytic reactions are given by Eq. 3.33, and 3.97, and as shown in Fig. 3.1 a typical plot of the rate of consumption of the reactant versus the reactant concentration, shows a maximum. Therefore, a plot of $-1/r_A$ versus C_A will show a minimum and resemble those shown in Fig. 4.7. Consequently, the discussion in Section 4.1.4 should apply to autocatalytic reactions as well. In other words, the relative size requirement depends upon the influent and effluent concentrations as well as the shape of the curve of $1/-r_A$ versus C_A. Now, let us illustrate the unusual characteristics of autocatalytic reactions by comparing the size requirement analytically in an example.

EXAMPLE 4.1.5-1

A biological reactor is to be built to handle a reaction with the rate given by $-r_S = (1 - C_S)C_S/(1 + C_S)$. The units for concentration are mol/liter and those for the rate, mol/(liter·hr). Determine τ_{PFR} and τ_{CSTR} for the following two cases;

(a) $C_{So} = 0.4$ and $C_{Sf} = 0.1$ (b) $C_{So} = 0.8$ and $C_{Sf} = 0.4$.

Solution:

Given information;

$$-r_S = (1 - C_S)C_S/(1 + C_S)$$

The constant density equations, Eq. 4.39 and 4.43, yield

(a) $\tau_{CSTR} = (C_{So} - C_{Sf})/[(1 - C_{Sf})C_{Sf}/(1 + C_{Sf})]$

$\qquad = (0.4 - 0.1)/[(1 - 0.1)(0.1)/1.1] = 3.67$ hours

$$\tau_{PFR} = \int_{0.1}^{0.4} (1 + C_S)/[C_S(1 - C_S)] dC_S$$

$$= \int_{0.1}^{0.4} [1/C_S + 2/(1 - C_S)] dC_S = \ln(0.4/0.1) + 2\ln(0.9/0.6)$$

$\qquad = 2.197$ hours

Thus, $\tau_{PFR} < \tau_{CSTR}$

(b) $\tau_{CSTR} = (0.8 - 0.4)/[(1 - 0.4)(0.4)/(1 + 0.4)] = 2.33$ hours

$\qquad \tau_{PFR} = \ln(0.8/0.4) + 2\ln(0.6/0.2) = 2.89$ hours

Thus, $\tau_{PFR} > \tau_{CSTR}$

Thus, it can be seen that the relative sizes of the two reactors depend on the influent and effluent concentrations.

4.2 MULTIPLE REACTORS FOR SINGLE REACTIONS

For a given reaction it is not necessary to use only a single reactor to reduce the reactant concentration to a desired value; in fact, in some cases it may be advantageous to use a number of reactors in series. Therefore, let us consider the use of multiple reactors.

4.2.1 Multiple Batch Reactors

For a single reaction it is possible to use a number of batch reactors in parallel in place of one large reactor. With a large batch reactor it may be difficult to achieve perfect mixing or to meet the heating and cooling requirements so that the use of multiple batch reactors of smaller sizes may prove to be operationally de-

sirable. In this case, there is no added complexity. All that is needed is to
divide the total volume requirement into a number of smaller volume reactors.

4.2.2 Multiple Continuous Stirred Tank Reactors

Multiple CSTR's may be used in parallel or in series. For single reactions one basis
for comparing different arrangements is the effluent concentration; the arrangement
that gives the lowest effluent concentration is the best. Using this criterion,
there are a number of interesting questions that need to be answered. Among them
are (1) how should a feed be split among CSTR's in parallel, (2) how should CSTR's
of different sizes be sequenced, and (3) is a series arrangement better than an
equivalent parallel arrangement?

CSTR's IN PARALLEL. Consider a case in which a feed stream is treated by a
number of CSTR's in parallel. If the flow rate to each reactor is known, each
reactor can be treated independently, and therefore the material developed in
Section 4.1.2 applies directly. Usually, however, the feed distributions which result in the least amount of reactant in the effluent are not known *a priori* and must
be determined. For this purpose let us work with two CSTR's in parallel, as shown
in Fig. 4.8. The total amount of reactant is given by

$$\rho = C_{A1}F_1 + C_{A2}F_2 = C_{A1}F_1 + C_{A2}(F - F_1) \tag{4.45}$$

where C_{A1} and C_{A2} are given by the mass balance equations,

$$F_1(C_{Ao} - C_{A1}) + r_{A1}V_1 = 0 \tag{4.46}$$

and

$$(F - F_1)(C_{Ao} - C_{A2}) + r_{A2}V_2 = 0 \tag{4.47}$$

Since the objective is to minimize the total amount of reactant by properly selecting
F_1 (for given F), the problem is equivalent to taking the derivative of ρ with respect to F_1 and setting it to zero, i.e., $d\rho/dF_1 = 0$. It can be shown that as long

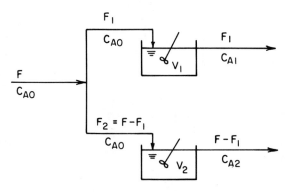

Figure 4.8. CSTR's in parallel.

as $-r_A$ is a monotonically increasing function of C_A the necessary and sufficient conditions are

$$\tau_1 = \tau_2 \tag{4.48}$$

or

$$F_1 = F \, V_1/(V_1 + V_2) \tag{4.49}$$

and

$$F_2 = F \, V_2/(V_1 + V_2) \tag{4.50}$$

Equation 4.48 states that when there are two CSTR's in parallel the feed should be split to give equal space times. This, of course, results in the same effluent reactant concentration from each reactor. Under this situation the two CSTR's in parallel are equivalent to a single CSTR, the volume of which is equal to the sum of the volumes of the two reactors. The same arguments can be applied to n CSTR's in parallel, leading to the general rule that the feed should be split among n CSTR's so that the space times are all equal. Then, the system is equivalent to one large CSTR whose volume is equal to the sum of the volumes of all reactors. It should be emphasized, however, that when the rate, $-r_A$, is not a monotonically increasing function of C_A, as in autocatalytic reactions, no generalization can be made and each case must be investigated individually.

EXAMPLE 4.2.2-1

Two CSTR's are to be used in parallel to reduce a reactant concentration from 0.1 mol/liter to 0.01 mol/liter. The reaction is second order with the rate given by $r_A = -C_A^2$. The reactor volumes are 5 and 10 liters, respectively. How should the flow rate be distributed and how much flow rate can the combined system handle?
Solution:

Since the rate is a monotonically increasing function of C_A, the optimal feed distribution is that which gives equal space time, $\tau_1 = \tau_2$.

$$\tau_1 = V_1/F_1 = (C_{Ao} - C_{Af})/(-r_A)_f = (0.1 - 0.01)/(0.01)^2 = 900 \text{ hours}$$

so that

$$F_1 = V_1/\tau_1 = 5/900 \text{ liter/hr}$$

$$F_2 = V_2/\tau_2 = V_2/\tau_1 = 10/900 \text{ liter/hr}$$

and the total feed rate

$$F = F_1 + F_2 = (5 + 10)/900 = 0.0166 \text{ liter/hr}$$

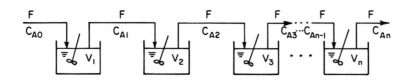

Figure 4.9. CSTR's in series.

CSTR's IN SERIES - ANALYTICAL APPROACH. Let us consider a series of n CSTR's for an aqueous reaction involving reactant A (see Fig. 4.9). From Eq. 4.39 the mass balance around each reactor gives

$$\tau_1 = V_1/F = (C_{Ao} - C_{A1})/(-r_A)_1 \tag{4.51}$$

$$\tau_i = V_i/F = (C_{Ai-1} - C_{Ai})/(-r_A)_i \qquad i = 2,3,4,\ldots,n-1 \tag{4.52}$$

$$\tau_n = V_n/F = (C_{An-1} - C_{Af})/(-r_A)_n \tag{4.53}$$

Equations 4.51 through 4.53 suggest that one can work with one reactor at a time starting from the first and moving down to the second, the third and so on to the last reactor. Let us take as an example, a first-order aqueous reaction with the rate given by $-r_A = kC_A$. From Eq. 4.51

$$\tau_1 = (C_{Ao} - C_{A1})/kC_{A1}$$

or

$$C_{A1} = C_{Ao}/(k\tau_1 + 1) \tag{4.54}$$

Likewise, Eq. 4.52 yields

$$C_{A2} = C_{A1}/(k\tau_2 + 1) \tag{4.55}$$

Substitution of Eq. 4.54 into Eq. 4.55 yields

$$C_{A2} = C_{Ao}/(k\tau_1 + 1)(k\tau_2 + 1) \tag{4.56}$$

Hence, it can be stated that

$$C_{Af} = C_{An} = C_{An-1}/(k\tau_n + 1)$$

$$= C_{Ao}/[(k\tau_1 + 1)(k\tau_2 + 1)(k\tau_3 + 1) \ldots (k\tau_n + 1)] \tag{4.57}$$

Equations 4.54 through 4.57 allow us to determine the concentration of A in any CSTR in terms of the feed concentration C_{Ao}. When the volumes of the CSTR's are identical, i.e., $V_1 = V_2 = V_3 = \ldots = V_n$, the space times are also identical so that Eq. 4.57 reduces to

$$C_{Af} = C_{Ao}/(k\tau + 1)^n \tag{4.58}$$

Equation 4.58 may be used to calculate the effluent concentration when the feed concentration, C_{Ao}, the space times, τ, and the number of CSTR's in series, n, are known. When it is desired to calculate the number of equally sized CSTR's (therefore, equal τ), required to reduce the reactant concentration of a first-order reaction from C_{Ao} to C_{Af}, one can solve Eq. 4.58 for n by taking the natural logarithm of both sides,

$$n = \ln(C_{Ao}/C_{Af})/\ln(k\tau + 1) \tag{4.59}$$

On the other hand, if one desires to use n reactors to reduce the concentration from C_{Ao} to C_{Af}, the space time required for each CSTR may be obtained by solving Eq. 4.58 for τ:

$$\tau = [(C_{Ao}/C_{Af})^{1/n} - 1]/k \tag{4.60}$$

EXAMPLE 4.2.2-2

For a first-order reaction with the rate given as $-r_A = kC_A$ (k = 16 hr^{-1}, C_A in g/l) it is desired to set up a series of CSTR's to reduce the concentration of a feed containing 10 g/l of A to less than 0.1 g/l. This system is to treat 10,000 liters/hr of the feed.

(a) Determine the number of reactors necessary to do the job if each has a volume of 1000 liters.

Solution:

Given information;

$C_{Af} < 0.1$ g/l, $C_{Ao} = 10$ g/l, $-r_A = 16 C_A$,

F = 10,000 liters/hr, $\tau = 1000/10,000 = 0.1$ hr.

Equation 4.59 gives

$n = \ln(10/0.1)/\ln[(16)(0.1) + 1] = 4.88$

Because we can't have partial reactors the number of reactors must be rounded up to the next integer, 5. Therefore, five 1000 liter CSTR's in series are required.

(b) If 6 CSTR's are to be used in series, what should be the volume of each reactor to meet the same effluent criterion?

Solution:

Equation 4.60 yields

$\tau = [(10/0.1)^{1/6} - 1]/16 = 0.0722$

Thus the reactor volume is given by

$V = \tau F = (0.0722)(10,000) = 722$ liters

(c) If a single CSTR is to be used, what should be its volume?

Solution:

Equation 4.39 gives

$$\tau = \frac{C_{Ao} - C_{Af}}{(-r_A)_f} = \frac{10 - 0.1}{16(0.1)} = 6.19$$

Thus the reactor volume required is

$$V = \tau F = (6.19)(10,000) = 61,900 \text{ liters}$$

Thus, a series of six 722 liter CSTR's would do the job of one 61,000 liter CSTR, a reduction of more than tenfold in total volume.

The total space time for a series of CSTR's is obtained by multiplying Eq. 4.60 by the number of CSTR's

$$n\tau = n[(C_{Ao}/C_{Af})^{1/n} - 1]/k \qquad (4.61)$$

In the limit as the number of equal sized CSTR's approaches infinity the right-hand side of Eq. 4.61 reduces to:

$$\lim_{n\to\infty} n\tau = \lim_{n\to\infty} \frac{(C_{Ao}/C_{Af})^{1/n} - 1}{k(1/n)} = \frac{1}{k} \ln(C_{Ao}/C_{Af}) \qquad (4.62)$$

Equation 4.62 is identical to the integrated plug-flow reactor equation for a first-order reaction. In other words, an infinite number of equally sized CSTR's in series is equivalent to a PFR of equal volume. Stated another way, a PFR may be broken up into an infinite number of infinitesimal sections, each of which may be considered as a CSTR. The implication of this statement is that a large number of CSTR's in series would mimic the performance of a single PFR, with volume equal to the sum of the individual CSTR volumes. In fact, the larger the number of CSTR's in series, the closer the performance is to that of a PFR. Since the use of a plug-flow reactor results in a lower effluent concentration than from a CSTR for a rate equation which is a monotonically increasing function of C_A, it can be concluded that the use of a large number of small CSTR's in series is better than that of a lesser number of large CSTR's of equivalent total volume. In practice, the greatest improvement in performance occurs in going from one CSTR to a two CSTR system, and the improvement becomes less significant as more and more CSTR's are added. For example, for the case considered in Example 4.2.2-2, six 722 liter CSTR's do the job of five 945 liter reactors, four 1351 liter reactors, three 2276 liter reactors, two 5625 liter reactors, or one 61,880 liter reactor.

The same approach may be applied to reactions of any order. For the second-order reaction, $-r_A = kC_A^2$, the above approach yields a recursive quadratic expression

$$C_{An} = [(1 + 4k\tau_n C_{An-1})^{1/2} - 1]/2k\tau_n \qquad (4.63)$$

which reduces to

$$C_{An} = [(1 + 4k\tau C_{An-1})^{1/2} - 1]/2k\tau \qquad (4.64)$$

when the space times are all equal, $\tau_1 = \tau_2 = \ldots \tau$.

For all practical purposes explicit expressions for the space time and the number of reactors in series, such as Eqs. 4.59 and 4.60, are limited to first-order reactions. For any other reactions it may be necessary to use an iterative approach to calculate the required space times of equal sized CSTR's, given the influent and effluent concentrations and the number of reactors. The simplest way to determine the number of CSTR's of known space time necessary to reduce the reactant concentration to a desired level is to solve Eq. 4.52 sequentially until the desired reduction in reactant concentration is achieved. For complicated rate expressions, however, it may be more convenient to use a graphical approach rather than an analytical one.

CSTR's IN SERIES - GRAPHICAL APPROACH. As was the case with a single CSTR it is not essential to have an analytical expression for the rate in order to determine the performance of a series of CSTR's. All that is needed is a graphical representation of the rate as a function of the reactant concentration. If it is desired to determine the reactant concentration in each CSTR in a series, the mass balance equation for each CSTR can be graphically represented on a plot of $(-r_A)$ versus C_A. Such a plot is given for an arbitrary rate in Fig. 4.10. The slope of the line PQ is obtained by dividing the ordinate QR by the abscissa PR, i.e., $(+r_A)_1/(C_{Ao} - C_{A1})$. This quantity can be recognized from Eq. 4.51 as $-1/\tau_1$. Similarly, the slope of the line RS should be $-1/\tau_2$, that of the line TU $-1/\tau_3$ and so on. Therefore, all that

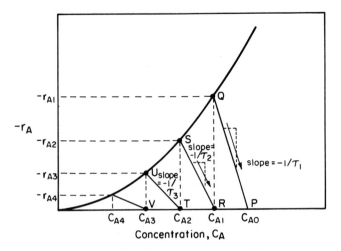

Figure 4.10.   Graphical representation of the performance of CSTR's in series.

is needed to determine the reactant concentration in each tank is to construct the graph as shown in Fig. 4.10. First, the feed concentration C_{Ao} is located on the abscissa (point P). Then a straight line of slope $-1/\tau_1$ is drawn through P and its intersection with the rate $(-r_A)$ curve (point Q), is located. The value of the abscissa corresponding to Q (point R) is obtained, and this is the reactant concentration in the first CSTR, C_{A1}. The same procedure is now repeated starting from point R. First, a straight line of slope $-1/\tau_2$ is drawn through R and the intersection of this line with the rate curve is labelled S. The value of abscissa corresponding to S is labelled as T, and this is the reactant concentration in the second CSTR, C_{A2}. It is obvious that for equal volume CSTR's these straight lines will all have the same slope, i.e.,

$$-1/\tau_1 = -1/\tau_2 = \ldots = -1/\tau_n \tag{4.65}$$

EXAMPLE 4.2.2-3

A plot of $(-r_A)$ versus C_A is given in Fig. E4.2. A feed containing 0.1 g/l of A is available at a rate flow of 100 liters/hr. It is desired to treat the feed so as to reduce the reactant concentration using three CSTR's in series, a 1 liter reactor followed by a 2 liter reactor followed by a 4 liter reactor. Determine the effluent concentration that would result from this arrangement.

Solution:

Given information;

$$-1/\tau_1 = -100, \ -1/\tau_2 = -50, \ -1/\tau_3 = -25 \ \mathrm{hr}^{-1}$$

$$C_{Ao} = 0.1 \ \mathrm{g/liter}$$

Following the procedure given above a straight line of slope equal to -100 $(= -1/\tau_1)$ is drawn through the point C_{Ao}, locating the intersection of the line with the curve

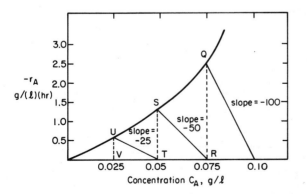

Figure E4.2. Graphical solution of Example 4.2.2-3.

(point Q), as shown in Fig. E4.2. After locating point R, another straight line of
slope equal to -50 (= -1/τ_2) is drawn through point R. The intersection of this line
with the curve -r_A is labelled S and the corresponding point T is located on the
abscissa. Finally, a line of slope equal to -25 (= -1/τ_3) is drawn through the
point T. Its intersection (point U), is located and the corresponding point on the
abscissa is labelled V. The numerical values are read off of the plot;
C_{A1} = 0.075 g/l, C_{A2} = 0.048 g/l, and C_{A3} = C_{Af} = 0.025 g/l. Thus, the effluent
concentration is 0.025 g/l.

CSTR's IN SERIES - SEQUENCING. Another important aspect of CSTR's in series is
the question of how one should arrange reactors of different sizes. Let us suppose
that there are two reactors of unequal sizes, one large and one small. The natural
question that arises is, would it make any difference whether the large one comes
first or last? Let us illustrate this through an example.

EXAMPLE 4.2.2-4
Two CSTR's are available, one having twice the volume of the other. Calculate the
effluent concentration that can be achieved by two arrangements, one in which the
large one is followed by the small one, and the other in which the small one is
placed ahead of the large one. Consider two cases of aqueous reactions, a first-
order (-r_A = kC_A) and a second-order (-r_A = kC_A^2) reaction. It may be assumed that
C_{Ao} = 1 mol/liter, k = 1 (1 hr^{-1} or 1 (liter)/(mol·liter)), V_L = 2V_S = 1000 liters
and F = 100 liters/hr.

(a) First-order reaction

  (Case I) The large one, V_L, followed by the small one, V_S.
  Equation 4.56 gives

$$C_{A2} = C_{Ao}/(k\tau_L + 1)(k\tau_S + 1)$$

$$= C_{Ao}/(2k\tau_S + 1)(k\tau_S + 1) \qquad (1)$$

  (Case II) The small one, V_S, followed by the large one, V_L.
  Equation 4.56 gives

$$C_{A2} = C_{Ao}/(k\tau_S + 1)(2k\tau_S + 1) \qquad (2)$$

Examination of equations (1) and (2) shows that for first-order reactions the
arrangements proposed are equivalent.

(b) Second-order reaction

  (Case I) The large reactor, V_L, followed by the small one, V_S.

Equations 4.56 and 4.57 give

$$(C_{Ao} - C_{A1})/kC_{A1}^2 = \tau_L \quad \text{or} \quad (1 - C_{A1}) = 10C_{A1}^2 \tag{3}$$

$$(CA_1 - C_{A2})/kC_{A2}^2 = \tau_S \quad \text{or} \quad (C_{A1} - C_{A2}) = 5C_{A2}^2 \tag{4}$$

On solving (4) and (3) we obtain

C_{A1} = 0.270 mol/liter

C_{A2} = 0.153 mol/liter

(Case II)   The small one, V_S, followed by the large one, V_L.

Equations 4.56 and 4.57 give

$$(1 - C_{A1}) = 5C_{A1}^2 \tag{5}$$

$$(C_{A1} - C_{A2}) = 10C_{A2}^2 \tag{6}$$

On solving (5) and (6) we obtain

C_{A1} = 0.358 mol/liter

C_{A2} = 0.146 mol/liter

The arrangement involving the small reactor followed by the large reactor gives a lower effluent concentration and therefore, is preferred.

As we have seen above the order in which two different size CSTR's are placed in series is immaterial for first-order reactions, whereas for second-order reactions the smaller reactor should be placed ahead of the larger one. As long as the rate equation is a monotonically increasing function of reactant concentration (n[th]-order reactions) some general statements can be made concerning a number of CSTR's in series:

(1)   for n < 1 the largest reactor should come first, followed by the second largest, the third largest and so on,

(2)   for n = 1 the order is immaterial (equivalent), and

(3)   for n > 1 the smallest reactor should come first, followed by the second smallest, the third smallest and so on.

These generalizations can be verified readily using the graphical approach discussed above. For example as shown in Fig. 4.11, for an nth-order reaction when n < 1 the rate curve is convex so that the arrangement involving a larger reactor followed by a smaller one (arrangement b) gives a lower effluent concentration as shown. The reader can construct a similar figure for n > 1 (rate curve concave) to show that a small reactor followed by a large one is best.

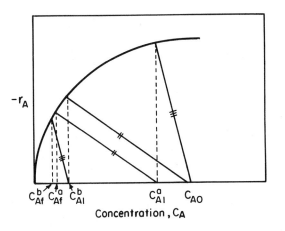

Figure 4.11. Graphical representation of the performance
of two different arrangements of two CSTR's
in series.

Another interesting question is whether the optimum condition of unequal size
reactors will be better than the same number of equal sized reactors in series. It
has been found that for n^{th}-order reactions the advantage of the optimum size
reactor system over the equal size reactor system is at most only a few percent [4].

PARALLEL VERSUS SERIES ARRANGEMENT OF CSTR's. Finally, one may ask the
question, given a number of reactors should these be used in a parallel or series
arrangement? It has been shown above that for a parallel arrangement the feed
should be split in such a manner as to make the space times all equal and that under
this situation the parallel arrangement of CSTR's is equivalent to one large CSTR
whose volume is equal to the sum of the volumes of all reactors in parallel. Hence,
the comparison between the parallel and series arrangements reduces to that between
one large reactor and a number of smaller CSTR's in series. Again, for reactions
whose rates are monotonically increasing functions of reactant concentration, we
have seen that a large number of small CSTR's in series gives a lower concentration
of reactant in the effluent than does one large CSTR. Therefore, it can be concluded that for reactions whose rate expressions are monotonically increasing
functions of reactant concentration, a series arrangement in proper sequence is always better than an equivalent parallel arrangement.

4.2.3 Multiple Plug-Flow Reactors

Plug-flow reactors may also be arranged in parallel or in series. As in the case of
multiple CSTR's one needs to consider the optimal arrangement and feed distribution.

PLUG-FLOW REACTORS IN PARALLEL. This situation is exactly the same as that of
CSTR's in parallel because when the flow rates and reactor volumes are known, each

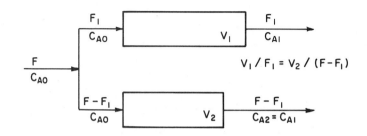

Figure 4.12. PFR's in parallel.

PFR may be treated independently of all others. Thus, the only problem associated with multiple PFR's in parallel is the determination of the feed distribution which minimizes the total amount of reactant in the effluent. Using the procedure outlined in Section 4.2.2 it can be shown that for reactions with monotonic rate expressions the optimal feed distributions require equal space times, i.e., $\tau_1 = \tau_2 = \ldots = \tau_n$. In this situation the whole system can be treated as one large PFR of volume equal to the total volume of the individual reactors. This is depicted in Fig. 4.12. When the rate expression is not monotonic, no general conclusion can be drawn and each case must be carefully investigated.

PLUG-FLOW REACTORS IN SERIES. In a series arrangement the PFR's are connected in series so that the effluent from the first reactor serves as the feed to the second, the effluent from the second serves as the feed to the third, and so on. This situation is equivalent to one large PFR of volume equal to the sum of individual reactors. This situation is illustrated in Fig. 4.13. Let the effluent concentrations of n number of PFR's in series be C_{A1}, C_{A2}, C_{A3}, $\ldots$, C_{An}, respectively, and the space times be τ_1, τ_2, τ_3, $\ldots$, τ_n, respectively. Then a single PFR with space time equal to the sum of the individual units, i.e., $\tau = \tau_1 + \tau_2 + \ldots + \tau_n$ is equivalent to n number of PFR's and reduces the reactant concentration from C_{Ao} to C_{An}. This conclusion is quite general and is independent of the shape of the rate expression. The sequence of the different size reactors is immaterial.

PARALLEL VERSUS SERIES ARRANGEMENT OF PLUG-FLOW REACTORS. For monotonic reaction rate expressions we have seen that n number of plug-flow reactors is

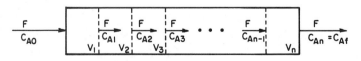

Figure 4.13. PFR's in series.

equivalent to one large PFR with volume equal to the sum of the individual units. For series arrangements the same conclusion was reached, regardless of the shape of the rate expressions. Therefore, it can be concluded that for reactions with mono- tonic rate expressions, a series arrangement of PFR's is equivalent to a parallel arrangement in which the feed is distributed to give equal space times in all reactors. Furthermore, under this situation, multiple PFR's in any parallel, series, or parallel-series network may be considered equivalent to a single large reactor.

EXAMPLE 4.2.3-1

An aqueous second-order reaction ($r_A = -kC_A^2$) is to be carried out in a reactor system consisting of four PFR's in two parallel branches. The first branch has a 100 liter reactor followed by a 50 liter reactor, while the second branch has a 25 liter reactor followed by a 50 liter one. Determine the reduction in reactant concen- tration that can be accomplished by the network of reactors. The total feed rate, F, is 1000 liters/hr, the feed concentration, C_{Ao}, is 1 mol/liter, and the rate constant, k, is 5 liters/(hr·mol).

Solution:

The first branch has two reactors in series; hence it may be considered as a single reactor of volume equal to the sum of the individual units, 100 + 50 = 150 liters. The second branch, likewise, may be considered as a single reactor of volume equal to 50 + 25 = 75 liters. We now have to determine the optimal feed split to these two reactors in parallel. As we have outlined above the best policy is to split the feed to provide the same space time in each parallel branch. Thus

$$\tau_1 = \tau_2$$

or

$$V_1/F_1 = V_2/(F - F_1)$$

The flow rate F_1 is obtained by solving for it:

$$F_1 = FV_1/(V_1 + V_2) = (1000)(150)/(150 + 75) = 666.7 \text{ liters/hr}$$

$$F_2 = F - F_1 = 1000 - 666.7 = 333.3 \text{ liters/hr}$$

The effluent concentration is obtained from

$$C_{A1} = 1/[1 + (5)(1)(150/666.7)] = 0.471 \text{ mol/liter} = C_{A2}$$

When the rate expression is not monotonic, as in autocatalytic reactions, it is not possible to generalize about the feed split for a parallel arrangement of PFR's or the equivalence of parallel and series arrangements. Therefore, in this situation, one must first formulate the problem of the optimal feed split for a parallel arrangement and obtain its solution. Then, one can compare the performance of this parallel arrangement with that of a series arrangement.

4.2.4 Combinations of CSTR's and PFR's

Both CSTR's and PFR's may be used in a single system, where they may be arranged either in parallel or in series. In a series arrangement they must be sequenced properly, while in a parallel arrangement the feed must be distributed properly. Ultimately, one must select the best arrangement when given a number of CSTR's and PFR's, consequently there must be a basis for making that selection. Although the ultimate criterion should be economic, in this section we again consider a simple criterion as the basis - minimum reactant in the effluent. As in previous sections we limit our discussion to those reactions whose rate expressions are monotonically increasing functions of reactant concentration.

CSTR AND PFR IN SERIES. In this arrangement the problem is proper sequencing of reactors of different types and sizes. It is instructive, therefore, to consider a simple situation involving a CSTR and a PFR for a first-order reaction. As illustrated in Fig. 4.14 there are two possibilities, a PFR followed by a CSTR, or a CSTR followed by a PFR. For the former the mass balance equations are

$$C_{Ap} = C_{Ao} \exp(-k\tau_p) \tag{4.66}$$

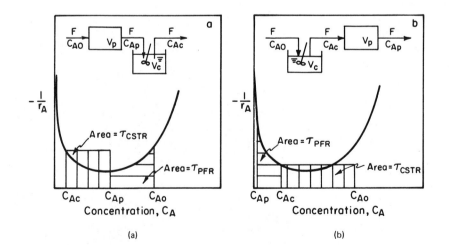

Figure 4.14.   Two arrangements involving a CSTR and a PFR:
               (a) a PFR followed by a CSTR;
               (b) a CSTR followed by a PFR.

and

$$C_{Ac} = C_{Ap}/(1 + k\tau_c) \qquad\qquad (4.67)$$

where the subscripts c and p are used to denote the CSTR and PFR, respectively. Substitution of Eq. 4.66 into Eq. 4.67 gives the effluent concentration for this arrangement

$$C_A = [C_{Ao}\exp(-k\tau_p)]/(1 + k\tau_c) \qquad\qquad (4.68)$$

For the arrangement involving a CSTR followed by a PFR the corresponding equations are

$$C_{Ac} = C_{Ao}/(1 + k\tau_c) \qquad\qquad (4.69)$$

and

$$C_{Ap} = C_{Ac}\exp(-k\tau_p) \qquad\qquad (4.70)$$

so that the effluent concentration, obtained from Eqs. 4.69 and 4.70, is

$$C_A = [C_{Ao}/(1 + k\tau_c)]\exp(-k\tau_p) \qquad\qquad (4.71)$$

Equations 4.68 and 4.71 are identical, implying that for first-order reactions it is immaterial whether the PFR is placed ahead of or behind the CSTR. This follows because in the mass balance equations for first-order reactions (Eqs. 4.66, 4.67, 4.69 and 4.70) the influent concentrations appear linearly.

Although the above approach may be used for other n^{th}-order reactions, non-linearity of the rate expressions ($n \neq 1$) makes it difficult to compare the resulting equations and one may have to resort to a numerical comparison. Using such techniques, it may be demonstrated that for $n > 1$, the PFR should be placed ahead of the CSTR, while the reverse order is preferred when $n < 1$. For n^{th}-order reactions we can state simple rules concerning the most effective use of a CSTR and a PFR in series:

(1) when $n = 1$, the ordering is immaterial.

(2) when $n > 1$, the PFR should be placed ahead of the CSTR.

(3) when $n < 1$, the CSTR should be placed ahead of the PFR.

When there is more than one CSTR or PFR, we can combine the above rules with those given in Sections 4.2.2 and 4.2.3:

(1) when $n = 1$, the ordering is immaterial.

(2) when $n > 1$, PFR's (any ordering) should be followed by the smallest CSTR, the second smallest CSTR, and so on to the largest CSTR.

(3) when $n < 1$, the largest CSTR, followed by the second largest CSTR, and so on to the smallest CSTR, should precede the PFR's (any ordering).

EXAMPLE 4.2.4-1

For a second-order reaction, $-r_A = C_A^2$, two CSTR's and one PFR in series are to be used. The reactor volumes are 1.0 and 2.0 liters, respectively for the CSTR's and the PFR is 3.0 liters. The feed is available at 1 liter/hr and the concentration C_{Ao} is 1 mol/liter. It is desired to investigate all possible series arrangements in order to pick the one which gives the lowest effluent concentration. (a) List all possible orderings by denoting the small CSTR, the large CSTR and the PFR by CS, CL, and P, respectively.

Solution:

Since there are three reactors six arrangements are possible.

(1) $P \rightarrow CS \rightarrow CL$ (2) $P \rightarrow CL \rightarrow CS$ (3) $CS \rightarrow P \rightarrow CL$

(4) $CL \rightarrow P \rightarrow CS$ (5) $CS \rightarrow CL \rightarrow P$ (6) $CL \rightarrow CS \rightarrow P$

(b) Eliminate some of the arrangements in (a) using the rules given in Section 4.2.2.

Solution:

Arrangements 2, 4, and 6 violate the rules for CSTR's in series. Since n = 2, the small CSTR should be placed ahead of the large CSTR. (c) Among the arrangements 1, 3, and 5 which one is expected to be the worst? Which is expected to be the best arrangement? Verify this numerically.

Solution:

Arrangement 5 is expected to be the worst of the three since the PFR is at the end instead of at the front.

Arrangement 1. The mass balance equations give

$$\int_{C_{A1}}^{C_{Ao}} \frac{dC_A}{-r_A} = \frac{1}{C_{A1}} - \frac{1}{C_{Ao}} = 3 \implies C_{A1} = C_{Ao}/(1 + 3C_{Ao}) = 1/(1 + 3) = 0.25$$

$$\frac{C_{A1} - C_{A2}}{C_{A2}^2} = 1 \implies C_{A2} = \frac{-1 + (1 + 4C_{A1})^{\frac{1}{2}}}{2} = 0.207$$

$$\frac{C_{A2} - C_{A3}}{C_{A3}^2} = 2 \implies C_{A3} = \frac{-1 + (1 + 8C_{A2})^{\frac{1}{2}}}{4} = 0.1575$$

Arrangement 3. The mass balance equations give

$$\frac{C_{Ao} - C_{A1}}{C_{A1}^2} = 1 \implies C_{A1} = \frac{-1 + (1 + 4C_{Ao})^{\frac{1}{2}}}{2} = 0.618$$

$$\frac{1}{C_{A2}} - \frac{1}{C_{A1}} = 3 \implies C_{A2} = C_{A1}/(1 + 3C_{A1}) = 0.2165$$

$$\frac{C_{A2} - C_{A3}}{C_{A3}^2} = 2 \Rightarrow C_{A3} = \frac{-1 + (1 + 8C_{A2})^{\frac{1}{2}}}{4} = 0.1632$$

Arrangement 5. The mass balance equations give

$$\frac{C_{Ao} - C_{A1}}{C_{A1}^2} = 1 \Rightarrow C_{A1} = \frac{-1 + (1 + 4C_{Ao})^{\frac{1}{2}}}{2} = 0.618$$

$$\frac{C_{A1} - C_{A2}}{C_{A2}^2} = 2 \Rightarrow C_{A2} = \frac{-1 + (1 + 8C_{A1})^{\frac{1}{2}}}{4} = 0.3595$$

$$\frac{1}{C_{A3}} - \frac{1}{C_{A2}} = 3 \Rightarrow C_{A3} = C_{A2}/(1 + 3C_{A2}) = 0.17296$$

Thus, the numerical values verify that arrangement 5 is the worst. The first arrangement is the best and the PFR should be placed at the front.

CSTR and PFR IN PARALLEL. Let us suppose that we have one CSTR and one PFR in a parallel network. How should the feed rate be distributed between the two reactors? For CSTR's in parallel or PFR's in parallel the feed should be split to give the same space time in each branch when the rate expression is monotonic. However, this is not the case when both a CSTR and a PFR are involved, as will be shown below. The rate of discharge of unreacted reactant in the effluent is given by

$$C_A F = C_{Ac} F_c + C_{Ap} (F - F_c) \tag{4.72}$$

where C_{Ac} and C_{Ap} stand for the reactant concentrations from the CSTR and PFR, respectively, F the total feed, and F_c the feed to the CSTR. For a first-order reaction the mass balance equations for the CSTR and PFR give

$$C_{Ac} = C_{Ao}/(1 + kV_c/F_c) \tag{4.73}$$

and

$$C_{Ap} = C_{Ao} \exp[-kV_p/(F - F_c)] \tag{4.74}$$

Substitution of Eqs. 4.73 and 4.74 into Eq. 4.72 and differentiation with respect to F_c give Eq. 4.75 when the derivative $(dC_A F/dF_c)$ is set equal to zero.

$$\frac{(1 + 2kV_c/F_c)}{(1 + kV_c/F_c)^2} = [1 + kV_p/(F - F_c)] \exp[-kV_p/(F - F_c)] \tag{4.75}$$

Equal space time is not a solution of Eq. 4.75. Equation 4.75 represents the necessary condition for the value of F_c to minimize the total amount of reactant in the

effluent and it may be solved iteratively for F_c, given the rate constant, k, the
reactor volumes, V_c and V_p, and the total feed rate, F. For example when $V_c = V_p = $
1 liter, k = 1 hr^{-1}, and F = 1 liter/hr, Eq. 4.75 yields F_c = 0.409 liter/hr. In
other words 40.9% of the feed should go to the CSTR and the remaining 59.1% to the
PFR. With this optimal feed distribution the effluent concentrations are 0.2093 and
0.1841 mol/liter for the CSTR and PFR, respectively and the concentration of reactant
in the mixed effluent is 0.1944 mol/liter. It is instructive to compare these values
with those that can be obtained under different operational modes. When the feed is
equally distributed (F_c = 0.5) to give the same space times in the two reactors, the
unreacted reactant leaving in the effluent is 0.2343 mol/liter, while a value of
0.2356 mol/liter is obtained when the feed is distributed to give the same effluent
concentrations in the two streams (F_c = 0.308). Thus, the optimal feed distribution
for a CSTR and a PFR in parallel is different from that which results in equal space
times or equal effluent concentrations.

The above approach may be taken for reactions of any order. However, the
algebra quickly becomes unwieldy and thus it can be better to look for a numerical
answer. First, a feed distribution is assumed, which allows one to calculate through
the mass balance equations the effluent concentrations from the CSTR and PFR. Then,
the rate at which the unreacted reactant leaves the system is calculated using
Eq. 4.72. This process is repeated for a large number of feed distributions until
one which gives the minimum value of C_A is obtained. This is illustrated below.

EXAMPLE 4.2.4-2

It is desired to use a CSTR (100 liters) and a PFR (100 liters) in parallel to treat
a stream containing a pollutant A, which reacts in accordance with a second-order
rate law, i.e., $-r_A = kC_A^2$ (k = 1.0 liter/(hr·mol), C_A in mol/liter). The stream
flow rate is 10 liters/hr and contains 0.1 mol/liter of A. Determine the feed
distribution which minimizes the concentration of pollutant in the combined effluent.
Solution:

The rate at which the pollutant is discharged is given by Eq. 4.72,

$$10C_A = C_{Ac}F_c + C_{Ap}(10 - F_c) \tag{1}$$

The CSTR and PFR mass balance equations are

$$(0.1 - C_{Ac})/C_{Ac}^2 = 100/F_c \tag{2}$$

and

$$100/(10 - F_c) = \int_{C_{Ap}}^{C_{Ao}} \frac{dC_A}{C_A^2} = \frac{1}{C_{Ap}} - \frac{1}{0.1} \tag{3}$$

Equations (2) and (3) may be solved for C_{Ac} and C_{Ap}, respectively and substituted into Eq. (1) to obtain

$$10C_A = F_c^2[(1 + 40/F_c)^{\frac{1}{2}} - 1]/200 + (10 - F_c)^2/(200 - 10F_c) \qquad (4)$$

Now, C_A is evaluated for a number of values of F_c and tabulated in Table E4.5

TABLE E4.5
Effect of Flow Split on Pollutant Concentration

F_c liters/hr	C_A mol/liter
3	0.04136
4	0.04103
5	0.04167
6	0.04327

Inspection of Table E4.5 suggests that the minimum value of C_A occurs when F_c is approximately 4 liters/hr. Further evaluation of C_A around an F_c value of 4.0 gives the concentrations listed in Table E4.6.

TABLE E4.6
Effect of Flow Split on Pollutant Concentration

F_c liters/hr	C_A mol/liter
3.8	0.0410201
3.83	0.041020019
3.84	0.041020016
3.85	0.0410201
3.9	0.041022

From which we conclude that the optimal value of F_c is 3.84 liters/hr and that the concentration of unreacted pollutant in the combined discharge is 0.04102 mol/liter.

PARALLEL VERSUS SERIES ARRANGEMENT. In Section 4.2.2 we concluded that a series arrangement of CSTR's is better than a parallel one for reactions whose rates increase with the reaction concentration, while in Section 4.2.3 both arrangements were found to be equivalent for PFR's. One would anticipate from these results that when both a CSTR and a PFR are involved, a series arrangement of proper sequence would be better than a parallel one, although this is difficult to prove. Numerical demonstrations are relatively simple, however, and therefore are given here. For a first-order reaction in a CSTR and a PFR of equal volume in series the effluent concentration is given by Eq. 4.71. Therefore, when a flow of 1.0 liter/hr is flowing to reactors with a volume of 1.0 liter each, and the first-order rate constant has a value of 1.0 hr^{-1}, the reactant concentration in the final effluent is 0.1839 mol/liter. For the same reaction the parallel arrangement with optimal feed split (F_c = 0.409 liter/hr) results in a final concentration of 0.2226 mol/liter. This demonstrates that for a first-order reaction the series arrangement is superior to the parallel one. For a second-order reaction, we saw in Example 4.2.4-2 that the parallel arrangement (F_c = 3.84 liters/hr and F_p = 6.16 liters/hr) gave a final reactant concentration of 0.04102 mol/liter. Let us now consider a series arrangement. Since this is a second-order reaction we need to place the PFR ahead of the CSTR. The mass balance equations are

$$\frac{100}{10} = \int_{C_{Ap}}^{0.1} \frac{dC_A}{C_A^2} = \frac{1}{C_{Ap}} - 10 \tag{4.76}$$

and

$$\frac{100}{10} = \frac{C_{Ap} - C_{Ac}}{C_{Ac}^2} \tag{4.77}$$

Solving Eq. 4.76 for C_{Ap} and substituting it into Eq. 4.77 yield

$$C_{Ac} = (\sqrt{3} - 1)/20 = 0.0366 \text{ mol/liter} \tag{4.78}$$

Thus the concentration of unreacted reactant is about 10% less than that resulting from the parallel arrangement. In general, for n^{th}-order reactions a series arrangement is superior to a parallel one.

4.2.5 Multiple Reactors for Reactions with Nonmonotonic Rates

We have already seen in Chapter 3 that autocatalytic reactions and biological reactions show rate expressions which first increase with the reactant concentration, reach the maximum, and then decrease with further increases in the reactant concentration.

As seen in Sections 4.1.4 and 4.1.5 the reactor selection and size requirements are quite different when the rate expression is not a monotonically increasing function of reactant concentration. Typical plots of $-1/r_A$ vs C_A are shown in Fig. 4.15, in which three distinct cases are shown. When the influent concentration, C_{Ao}, is less than or equal to the concentration at which $-1/r_A$ shows the minimum, C_{Am}, then the case is equivalent to those of monotonically increasing rate expressions. Thus, a PFR is always better than a CSTR and a number CSTR's in series is better than one large CSTR. On the other hand, when the influent as well as the effluent concentration are higher than C_{Am}, the opposite is true, i.e., one CSTR is better than a number of CSTR's in series, which in turn is better than a PFR.

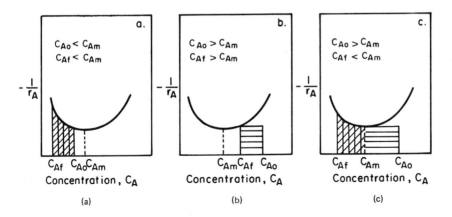

Figure 4.15. Reactor arrangements for reactions with nonmonotonic rate
equations: (a) a PFR; (b) a CSTR; (c) a CSTR followed by a PFR.

Finally, when the influent concentration is higher and the effluent concentration
is lower than C_{Am}, the best arrangement is a CSTR to reduce the concentration to
C_{Am} from C_{Ao} followed by a PFR to reduce the concentration to C_{Af} from C_{Am}. Any
other arrangement is inferior.

4.3 REACTORS FOR MULTIPLE REACTIONS

So far in this chapter we have considered the design of reactors for single reactions
in which only a few intermediates and products were involved. In multiple reaction
systems, on the other hand, many products and intermediates are involved, some of
which may be the desired result from the reactions and others of which may be undesirable. Thus, for design of reactors for multiple reactions there is another
aspect that plays an important role and that is product distribution. Suppose
reactant A can lead to B, which is the desired product, and at the same time also
lead to C, which is not desirable. In this situation the reactor should be designed
to give the maximum amount of B. Another case may be imagined, in which the reactant
A is transformed into B and B is in turn transformed into C, i.e., A → B → C. If C
were the desired product, the reaction should be carried out to completion so that
only the desired product, C would be present. On the other hand, if the intermediate
B were the desired product, it would be unwise to carry out the reaction to completion. In this section we will introduce the approach which must be used with problems
of this type. Generally it involves writing multiple mass balance equations which
may be solved simultaneously to quantitatively determine the size requirements for
the reactors needed to achieve a desired product distribution.

4.3.1 Parallel Reactions

The simplest parallel reactions are of the following types;

$$A \xrightarrow{k_1} D \qquad\qquad A + B \xrightarrow{k_1} D \qquad\qquad A + B + C \xrightarrow{k_1} D$$

$$A \xrightarrow{k_2} U \qquad\qquad A + B \xrightarrow{k_2} U \qquad\qquad A + C \xrightarrow{k_2} U \qquad (4.79)$$

$$\qquad (a) \qquad\qquad\qquad\qquad (b) \qquad\qquad\qquad\qquad (c)$$

where D and U stand for the desired and undesired products, respectively. The objective is to run the reaction so that as much as possible of the desired product is produced while minimizing the production of the undesired product. In other words, the ratio of the rates, r_D/r_U should be maximized by choosing the appropriate reactor type and reaction conditions. For example, consider Eq. (4.79b) with the corresponding rate expressions

$$A + B \xrightarrow{k_1} D \qquad\qquad r_D = k_1 c_A^a c_B^b \qquad\qquad\qquad (4.80)$$

$$A + B \xrightarrow{k_2} U \qquad\qquad r_U = k_2 c_A^c c_B^d \qquad\qquad\qquad (4.81)$$

where a, b, c, and d are assumed to be positive numbers. Therefore,

$$r_D/r_U = (k_1/k_2) c_A^{a-c} c_B^{b-d} \qquad\qquad\qquad (4.82)$$

We need to make this ratio as large as possible. Equation 4.82 suggests that C_A be kept as large as possible if a > c and as small as possible if a < c. Likewise, C_B should be kept as large as possible if b > d and as small as possible if b < d. When a = c, the ratio is independent of C_A as it is independent of C_B if b = d. One way of controlling reactant concentrations is to devise a contacting scheme. For example one reactant, say B, could be added slowly to a vessel containing a large concentration of A, thereby keeping C_B low and C_A high. Another way of keeping a reactant concentration high would be to use either a plug-flow reactor or a batch reactor, for the reactant concentration is usually low in a CSTR since it is maintained at a low effluent reactant concentration. It is also possible to make (k_1/k_2) large so as to make the ratio (r_D/r_U) large. Assuming that the rate constants are of the Arrhenius type

$$k_1/k_2 = k_{10}\exp(-E_1/RT)/k_{20}\exp(-E_2/RT) = (k_{10}/k_{20})\exp[-(E_1 - E_2)/RT] \qquad (4.83)$$

so that when $E_1 > E_2$, the reaction temperature should be made as large as possible, whereas when $E_1 < E_2$ it should be made as small as possible.

4.3.2 Series Reactions

Consider the first-order reactions in series, $A \xrightarrow{k_1} B \xrightarrow{k_2} C$, discussed in Section 3.2.3. Let us develop the mass balance equations for these reactions in a batch reactor (or a PFR) and a CSTR. For the batch reactor (or PFR) the independent mass balance equations are obtained using the rate expressions of Eqs. 3.65 and 3.66

$$dC_A/d\tau = -k_1 C_A, \qquad\qquad C_A(0) = C_{Ao} \qquad\qquad (4.84)$$

$$dC_B/d\tau = k_1 C_A - k_2 C_B, \qquad\qquad C_B(0) = C_{Bo} \qquad\qquad (4.85)$$

where τ is the space time for the PFR or reaction time for the batch reactor. Equation 4.84 is readily integrated to obtain

$$C_A = C_{Ao} \exp(-k_1 \tau) \qquad\qquad (4.86)$$

which is substituted into Eq. 4.85 to obtain

$$dC_B/d\tau + k_2 C_B = k_1 C_{Ao} \exp(-k_1 \tau) \qquad\qquad (4.87)$$

This is a first-order linear differential equation which may be integrated to yield

$$C_B = C_{Bo} \exp(-k_2 \tau) + C_{Ao} \frac{k_1}{k_2 - k_1} [\exp(-k_1 \tau) - \exp(-k_2 \tau)] \qquad\qquad (4.88)$$

Finally, the stoichiometric relationship gives

$$C_C = C_{Co} + (C_{Ao} - C_A) + (C_{Bo} - C_B) \qquad\qquad (4.89)$$

Typical curves showing C_A, C_B, and C_C as functions of τ are shown in Fig. 4.16 where

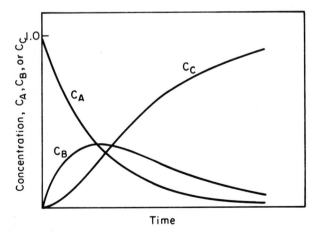

Figure 4.16.   Typical concentration profiles for two first-order
reactions in series: $A \rightarrow B \rightarrow C$.

it can be seen that if C is the desired product the reaction should be allowed to go to completion (large reaction time or space time). On the other hand, if the desired product is B, there is an optimum reaction or space time to obtain the highest concentration of B. This can be obtained by setting $dC_B/d\tau = 0$ in Eq. 4.87

$$k_2 C_{Bm} = k_1 C_{Ao} \exp(-k_1 \tau_m) \tag{4.90}$$

where the subscript m is used to denote the maximum condition. Substitution of Eq. 4.88 into Eq. 4.90 and solving for the optimal space or reaction time yields

$$\tau_m = \ln\{(k_2/k_1)[1 + (1 - k_2/k_1)(C_{Bo}/C_{Ao})]\}/(k_2 - k_1) \tag{4.91}$$

The optimum concentration is obtained by substituting Eq. 4.91 into Eq. 4.90

$$C_{Bm} = C_{Ao}(k_1/k_2)\{[1 + (1 - k_2/k_1)(C_{Bo}/C_{Ao})](k_2/k_1)\}^{-k_1/(k_2 - k_1)} \tag{4.92}$$

If B is not present initially, $C_{Bo} = 0$, and Eqs. 4.91 and 4.92 reduce to

$$\tau_m = \ln(k_2/k_1)/(k_2 - k_1) \tag{4.93}$$

and

$$C_{Bm} = C_{Ao}(k_1/k_2)^{k_2/(k_2 - k_1)} \tag{4.94}$$

For a CSTR the mass balance equations for A and B give

$$C_A = C_{Ao}/(1 + k_1 \tau) \tag{4.95}$$

and

$$C_B = C_{Bo}/(1 + k_2 \tau) + k_1 C_A \tau/(1 + k_2 \tau)$$

$$= C_{Bo}/(1 + k_2 \tau) + k_1 C_{Ao} \tau/(1 + k_1 \tau)(1 + k_2 \tau) \tag{4.96}$$

Again the space time giving the maximum C_B, C_{Bm}, is obtained by setting $dC_B/d\tau = 0$,

$$k_1 C_{Ao}(1 - k_1 k_2 \tau^2) - k_2(1 + k_1 \tau)^2 C_{Bo} = 0 \tag{4.97}$$

When $C_{Bo} = 0$, Eq. 4.97 reduces to

$$\tau_m = 1/(k_1 k_2)^{\frac{1}{2}} \tag{4.98}$$

and the corresponding C_{Bm} is obtained from Eq. 4.96

$$C_{Bm} = (k_1/k_2)^{\frac{1}{2}} C_{Ao}/[1 + (k_1/k_2)^{\frac{1}{2}}][1 + (k_2/k_1)^{\frac{1}{2}}] \tag{4.99}$$

The maximum product concentration C_{Bm} obtainable from a PFR (or a batch reactor) and a CSTR and the corresponding τ_m values are plotted in Fig. 4.17. It is apparent the PFR's give higher maximum intermediate concentrations, C_{Bm}, than the CSTR's, the difference being higher with increasing values of k_1/k_2. For instance C_{Bm}/C_{Ao} is

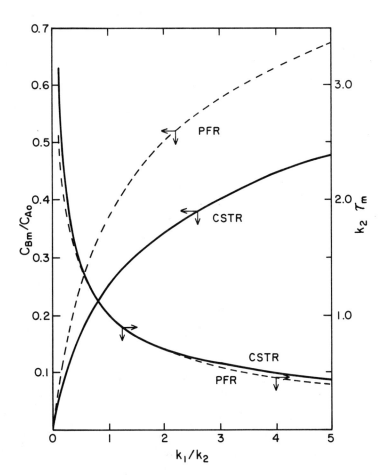

Figure 4.17. Typical comparison of the maximum intermediate
concentrations (C_B) in a PFR and a CSTR performing
a first-order series reaction:

$$A \xrightarrow{k_1} B \xrightarrow{k_2} C. \qquad C_{Bo} = 0$$

0.5 for the PFR and 0.34 for the CSTR when k_1/k_2 is 2. From the figure it may be
concluded that the temperature should be picked to give the highest k_1/k_2 and that
the reactions should be carried out in a PFR.

EXAMPLE 4.3.2-1

A first-order reaction in series, $A \xrightarrow{k_1} B \xrightarrow{k_2} C$, with $k_1 = k_2$ is to be carried out
in a PFR. What would be the optimum space time if the intermediate B is the desired
product? The feed contains only A, C_{Ao}. What would be the expected C_B?

Solution:

Equation 4.93 gives

$$\tau_m = \ln(k_2/k_1)/(k_2 - k_1) \qquad (1)$$

Since $k_2/k_1 = 1$, we cannot solve for τ_m directly. Instead we can let $k_2/k_1 = x$ and substitute it into (1)

$$\tau_m = \ln x/(k_1 x - k_1) = \frac{1}{k_1} \ln x/(x - 1)$$

We may then find τ_m as $x \to 1$. Using l' Hopital's rule

$$\lim_{x\to1} \tau_m = \lim_{x\to1} \frac{\ln x}{k_1(x - 1)} = \lim_{x\to1} \frac{\frac{1}{x}}{k_1} = \frac{1}{k_1}$$

The expected concentration is obtained from Eq. 4.94 using the same limiting process,

$$\lim_{x\to1} [C_{Bm} = C_{Ao}(k_1/k_2)^{\frac{k_2}{k_2 - k_1}}] = C_{Ao}\exp(-1) = 0.368\, C_{Ao}$$

Series reactions of mixed order higher than one may be analyzed by the same method. However, the above analytical approach often fails due to nonlinear rate expressions. In this situation one must resort to numerical integration of differential equations which arise from the differential mass balances for a PFR or a batch reactor. The procedure involved is the same, however. The mass balance equations are written for the reactant and the intermediate. The mass balance equation for the reactant A can be integrated analytically and substituted into the mass balance equation for the intermediate, which may have to be integrated numerically to obtain the concentration-vs-space time relationship for a PFR.

4.3.3 Parallel-Series Reactions

In certain multiple reactions one reactant is successively reacted to yield a series of intermediates, while another reactant reacts with one or more of the intermediates. These may be called parallel-series [3b] reactions since they are in series with respect to the first reactant and in parallel with respect to the second. For example, consider the production of ethylene glycol:

$$H_2O \xrightarrow{O(CH_2)_2} HOCH_2CH_2OH \xrightarrow{O(CH_2)_2} O=(CH_2CH_2OH)_2$$

$$(A) \xrightarrow{+(D)} (B) \xrightarrow{+(D)} (C) \qquad (4.100)$$

In this sequence water (A) is converted by the action of ethylene oxide (D) to
ethylene glycol (B) which in turn is converted to diethylene glycol (C), again by
the action of ethylene oxide (D). Hence the reactions are in series with respect
to A, B, and C, but are in parallel with respect to D.

$$A + D \xrightarrow{k_1} B$$

$$B + D \xrightarrow{k_2} C \qquad (4.101)$$

Such reactions are usually irreversible and bimolecular, hence, we shall consider
only those types. As with parallel and series reactions the desired product is
usually the intermediate, B, so that we want to maximize its concentration, C_B, by
proper selection of reactors and operating conditions.

BATCH REACTOR (or PFR). There are four species involved, A, B, C, and D. The
stoichiometric relationships are obtained using the method proposed in Section 2.4,

$$C_{Ao} - C_A + C_{Bo} - C_B + C_{Co} - C_C = 0 \qquad (4.102)$$

and

$$C_{Do} - C_D = C_{Ao} - C_A - (C_{Co} - C_C) \qquad (4.103)$$

Two mass balance equations are needed,

$$\frac{dC_A}{dt} = -k_1 C_A C_D \qquad C_A(0) = C_{Ao} \qquad (4.104)$$

and

$$\frac{dC_B}{dt} = k_1 C_A C_D - k_2 C_B C_D \qquad C_B(0) = C_{Bo} \qquad (4.105)$$

Equations 4.102 through 4.105 describe the system completely. We seek the maximum
C_B, C_{Bm}. Dividing Eq. 4.105 by Eq. 4.104 results in a first-order differential
equation

$$dC_B/dC_A = -1 + (k_2/k_1)C_B/C_A \qquad (4.106)$$

which is solved to obtain

$$C_B/C_{Ao} = [1/(1 - k_2/k_1) + C_{Bo}/C_{Ao}](C_A/(C_{Ao}))^{k_2/k_1} - (C_A/C_{Ao})/(1 - k_2/k_1) \qquad (4.107)$$

The necessary condition for maximum C_B, C_{Bm}, is $dC_B/dt = 0$ so that Eq. 4.105 yields

$$C_{Bm}/C_{Ao} = (k_1/k_2)C_A/C_{Ao} \qquad (4.108)$$

Substitution of Eq. 4.108 into Eq. 4.107 and simplification yield

$$C_A/C_{Ao} = [a + ab(1 - a)]^{1/(1 - a)} \qquad (4.109)$$

where

$$a = k_2/k_1$$

and

$$b = C_{Bo}/C_{Ao}$$

Substitution of Eq. 4.109 into Eq. 4.108 yields

$$C_{Bm}/C_{Ao} = [a^a + a^a b(1 - a)]^{1/(1 - a)} \qquad (4.110)$$

Equation 4.109 states that the reaction should be run until the concentration of species A reaches the level defined by the right hand side of it, at which point the maximum concentration of B is given by Eq. 4.110. It is not difficult to see that as the value of a decreases, the maximum value of C_B, C_{Bm}, increases and (C_A/C_{Ao}) decreases. In other words, to maximize C_{Bm}, a should be decreased by proper selection of reaction temperature (see Section 4.3.1) and the reaction should be run over a longer period as dictated by Eq. 4.109.

  CSTR. Mass balance equations on A and B yield

$$(C_{Ao} - C_A)/k_1 C_A C_D = \tau_c \qquad (4.111)$$

and

$$(C_{Bo} - C_B)/(k_2 C_B C_D - k_1 C_A C_D) = \tau_c \qquad (4.112)$$

Division of Eq. 4.112 by Eq. 4.111 and rearrangement yield

$$\frac{C_B}{C_{Ao}} = \frac{(C_A/C_{Ao})(1 - C_A/C_{Ao} + C_{Bo}/C_{Ao})}{C_A/C_{Ao} + (k_2 k_1)(1 - C_A/C_{Ao})} \qquad (4.113)$$

The necessary condition $dC_B/dC_A = 0$ yields

$$C_A/C_{Ao} = \{-a + [a(1 + b - ab)]^{\frac{1}{2}}\}/(1 - a) \qquad (4.114)$$

where $a = k_2/k_1$ and $b = C_{Bo}/C_{Ao}$. The optimal space time is defined by Eq. 4.114. The space time should be such that the value of C_A/C_{Ao} is given by the right side of Eq. 4.114. At this space time and C_A value the maximum C_B, C_{Bm}, is obtained by substituting Eq. 4.114 into Eq. 4.113. Once again, Eqs. 4.113 and 4.114 suggest that C_B may be maximized by proper selection of temperature to minimize a and by making the space time longer in accordance with Eq. 4.114.

EXAMPLE 4.3.3-1

Parallel-series reactions of Eq. 4.101 take place with the rate constants
k_1 = 2000exp(-5000/RT) and k_2 = 1000exp(-4000/RT) (T in °K). It is desired to
maximize the concentration of B. What type of reactor and temperature should be
used? Initially no B nor C is present. The range of operating temperatures is 25
to 60°C.

Solution:

   Given data: a = k_2/k_1 = 0.5exp(1000/RT)

                 b = 0

   Since a decreases with temperature and we note that one needs to make the
value of a as small as possible, we pick the highest allowable temperature;
60°C (333°K), at which a = (0.5)exp[1000/(1.987)(333)] = 2.266
We now evaluate each reactor separately.

   BATCH REACTOR (or PFR). Equation 4.109 gives C_A/C_{Ao} = 0.524 so that the
reaction should be stopped (or the space time so selected) to give 47.6% conversion
of A, at which point C_B reaches the maximum as given by Eq. 4.110

$$C_{Bm}/C_{Ao} = (2.266)^{2.266/(1 - 2.266)} = 0.231$$

   CSTR. Equation 4.114 gives C_A/C_{Ao} = 0.601 so that the space time should be
selected to give 39.9% conversion of A. Then, the maximum B concentration may be
calculated from Eq. 4.113, C_{Bm}/C_{Ao} = 0.159.

   From these we see that a PFR or batch reactor is preferred over a CSTR and
that the temperature should be made as high as possible.

   When C_{Bo} = 0, Eqs. 4.109 and 4.110 for the PFR reduce to

$$C_A/C_{Ao} = a^{1/(1 - a)} \tag{4.115}$$

and

$$C_{Bm}/C_{Ao} = a^{a/(1 - a)} \tag{4.116}$$

While Eqs. 4.113 and 4.114 for the CSTR reduce to

$$C_A/C_{Ao} = \sqrt{a}/(1 + \sqrt{a}) \tag{4.117}$$

and

$$C_{Bm}/C_{Ao} = 1/(1 + \sqrt{a})^2 \tag{4.118}$$

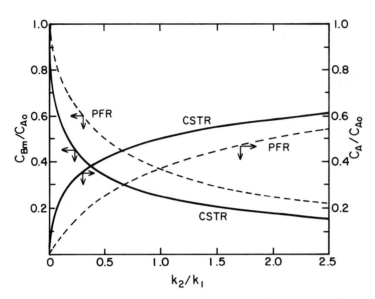

Figure 4.18. Typical comparison of the maximum intermediate
concentrations (C_B) in a PFR and a CSTR performing
an elementary parallel-series reaction:

$$A + D \xrightarrow{k_1} B; \quad B + D \xrightarrow{k_2} C. \qquad C_{Bo} = 0$$

The maximum values of C_B, C_{Bm}, and the corresponding values of C_A are given in
Fig. 4.18 for PFR's and CSTR's. Note that the PFR's give higher maximum concentrations.

The discussions presented for parallel and series reactions also apply to
parallel-series reactions. Therefore, we will not go into this but refer to
Sections 4.3.1 and 4.3.2.

4.4 REACTORS WITH EXTERNAL RECYCLE

In certain situations it is desirable or advantageous to recover the unreacted
reactants and return them to the reactor inlet. For example, in the synthesis of
ammonia from hydrogen and nitrogen the conversion is so low that the unreacted
hydrogen and nitrogen must be recovered and recycled. In activated sludge wastewater treatment plants the excess sludge (cells) generated is separated and recycled
back into the aerator to increase the rate of organics removal by increasing the
cell density. In trickling filter operations a portion of the effluent is recycled
back into the inlet to smooth out any hydraulic surges and to maintain sufficient
flow to keep all of the film wet at all times.

4.4.1 CSTR with Recycle

The schematic of a CSTR with recycle is shown in Fig. 4.19. The effluent from the CSTR is put through a separator and a portion of it is recycled back into the CSTR and the remainder is discharged as the effluent. The recycle flow rate is given by αF, where α is a positive constant and the concentration of A in the recycle is denoted by RC_A, where R is also a positive constant which is usually greater than or equal to 1. A value of $R > 1$ implies that the reactant A in the recycle is more concentrated than that in the CSTR, while $R = 1$ implies that the recycle concentration is the same as the CSTR concentration. The steady state mass balance equation around the CSTR is

$$FC_{Ao} + \alpha FRC_A - F(1 + \alpha)C_A + r_A V = 0 \qquad (4.119)$$

or

$$-r_A V = F(C_{Ao} - C_A) + \alpha F(R - 1)C_A \qquad (4.120)$$

The second term on the right hand side of Eq. 4.120 is due to recycle. Whenever the recycle stream is more concentrated than the CSTR effluent so that $R > 1$, the total rate of reaction, $-r_A V$, is higher than that which would result without recycle. On the other hand, when the composition of the recycle is identical to that of the CSTR so that $R = 1$, the recycle has no effect and Eq. 4.120 reduces to the equation without recycle, Eq. 2.20.

Equation 4.120 may be rewritten in terms of the feed to the reactor that would result by mixing the influent stream with the recycle stream. Referring to Fig. 4.19 we have

$$F(1 + \alpha)C_{AF} = FC_{Ao} + \alpha FRC_A \qquad (4.121)$$

or

$$C_{AF} = (C_{Ao} + \alpha RC_A)/(1 + \alpha) \qquad (4.122)$$

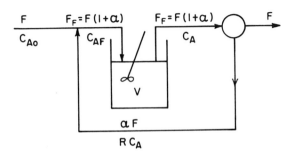

Figure 4.19. CSTR with recycle.

Then, the mass balance becomes

$$F(1 + \alpha)[C_{AF} - C_A] + r_A V = 0 \tag{4.123}$$

or

$$F_F(C_{AF} - C_{Af}) + (r_A)_f V = 0 \tag{4.124}$$

Equation 4.124 suggests that in terms of the feed to the reactor and its concentration, the mass balance is exactly the same as that without recycle. Therefore, the methods, analytical and graphical, developed in Sections 4.1 and 4.2 may be used also for the CSTR with recycle provided that we use the flow rate and the reactant concentration of the feed which results from combining the influent stream with the recycle stream.

4.4.2 PFR with Recycle

Consider a PFR with recycle of a part of the effluent stream back into the reactor as shown in Fig. 4.20. As in the case of CSTR the volumetric flow rate throughout the PFR is $F(1 + \alpha)$ and the inlet concentration is obtained by mixing the influent stream with the recycle stream,

$$C_{AF} = (C_{Ao} + \alpha R C_{Af})/(1 + \alpha) \tag{4.125}$$

The mass balance equation for a PFR without recycle, Eq. 4.42, must be modified to account for the fact that the volumetric flow rate and the inlet concentration are different, $F(1 + \alpha)$ and C_{AF}, respectively

$$V/F(1 + \alpha) = \int_{C_{Af}}^{C_{AF}} dC_A/(-r_A) \tag{4.126}$$

Equation 4.126 may be used in the same way as Eq. 4.42. Notice that unlike the case of a CSTR, even when R = 1 (the effluent being recycled without further concentration), Eq. 4.126 does not reduce to the equation of a PFR without recycle because the effluent being recycled has a different composition from that of the influent.

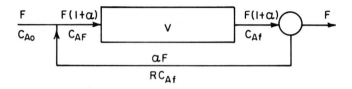

Figure 4.20. PFR with recycle.

EXAMPLE 4.4.2-1

Consider the third-order reaction that was treated in Example 4.1.3-1. We now allow a recycle of the effluent with the recycle ratio, α, set at 2. The volumetric flow rate is 100 liters/hr. Calculate the volume of PFR necessary to convert 90% of A in feed containing 2 mol/liter of A and 1 mol/liter of B.

Solution:

Given information;

$$C_{Ao} = 2, \ C_{Bo} = 1, \ C_{Af} = 0.2, \ \alpha = 2$$

$$k = 5.35, \ -r_A = 5.35C_A^2C_B$$

From the stoichiometric relationship

$$C_B = C_{Bo} - \tfrac{1}{2}(C_{Ao} - C_A) = 1 - \tfrac{1}{2}(2 - C_A) = C_A/2 \qquad (1)$$

Equation 4.122 gives

$$C_{AF} = [2 + (2)(1)(0.2)]/(1 + 2) = 0.8 \qquad (2)$$

The rate expression in terms of C_A is

$$-r_A = 5.35C_A^2C_B = 5.35C_A^2[1 - \tfrac{1}{2}(2 - C_A)] = 2.675C_A^3 \qquad (3)$$

Substitution of (2) and (3) into Eq. 4.126 gives

$$V = 100(1 + 1) \int_{0.2}^{0.8} dC_A/2.675C_A^3$$

$$= \left(\frac{200}{2.675}\right)\left(\frac{1}{2}\right)\left[\left(\frac{1}{0.2}\right)^2 - \left(\frac{1}{0.8}\right)^2\right] = 876.2 \text{ liters}$$

In this situation the recycle stream reduces the reactor feed concentration and since the rate is a monotonically increasing function of reactant concentration, it increases the necessary reactor volume.

As we have seen in Example 4.4.2-1 the recycle results in bigger reactor volumes for reactions whose rates are monotonically increasing functions of the reactant concentration. On the other hand, for reactions with nonmonotonic rate expressions it is possible to determine the recycle ratio which will result in the minimum reactor volume. To determine this optimum recycle ratio we need to differentiate the reactor volume, V, in Eq. 4.126 with respect to the recycle ratio, α, and to set $dV/d\alpha$ equal to zero. To do this we need to differentiate under an integral. The Leibnitz rule of differentiation [1b] allows us to do this. Given a function $g(x)$ defined as

$$g(x) = \int_{a(x)}^{b(x)} h(x,y)dy \tag{4.127}$$

the derivative is given by

$$dg(x)/dx = \int_{a(x)}^{b(x)} \frac{\partial h}{\partial x} dy + h[x,b(x)]\frac{db(x)}{dx} - h[x,a(x)]\frac{da(x)}{dx} \tag{4.128}$$

Applying this rule to Eq. 4.126 yields

$$dV/d\alpha = F(1 + \alpha)[(1/-r_A)_{C_{AF}}(dC_{AF}/d\alpha) - 0]$$

$$+ F\int_{C_{Af}}^{C_{AF}} dC_A/(-r_A) = 0 \tag{4.129}$$

or

$$[(1 + \alpha)/(-r_A)_{C_{AF}}]dC_{AF}/d\alpha = -\int_{C_{Af}}^{C_{AF}} dC_A/(-r_A) \tag{4.130}$$

where $dC_{AF}/d\alpha$ is obtained from Eq. 4.125,

$$dC_{AF}/d\alpha = (RC_{Af} - C_{Ao})/(1 + \alpha)^2 \tag{4.131}$$

To eliminate α in Eq. 4.130, we note from Eq. 4.125 that

$$C_{Ao} = (1 + \alpha)C_{AF} - \alpha RC_{Af} \tag{4.132}$$

or

$$RC_{Af} - C_{Ao} = RC_{Af} - [(1 + \alpha)C_{AF} - \alpha RC_{Af}] = (1 + \alpha)(RC_{Af} - C_{AF}) \tag{4.133}$$

Substitution of Eqs. 4.131 and 4.133 into Eq. 4.130 yields

$$\int_{C_{Af}}^{C_{AF}} dC_A/(-r_A) = (\frac{1}{-r_A})_{C_{AF}}(C_{AF} - C_{Af}) \tag{4.134}$$

Equation 4.134 sets the optimum recycle ratio by placing a constraint between C_{AF} and C_{Af}. When the rate expression is known analytically, then Eq. 4.134 can be readily used to determine the optimum value of C_{AF} provided that C_{Af} is given. Then this value of C_{AF} is used in Eq. 4.125 to obtain the corresponding recycle ratio. When the rate expression is available in graphical form, one can pick a value of C_{AF} and calculate the value of the integral. With the value of C_{AF} chosen, the right side of Eq. 4.134 can be readily evaluated. Obviously, with the first trial value of C_{AF} the right and left sides of Eq. 4.134 would not be the same so that a different value of C_{AF} must be picked and the possibility of its satisfying

Eq. 4.134 should be checked. This procedure should be repeated until the value of C_{AF} chosen satisfies Eq. 4.134. This value of C_{AF} may be substituted into Eq. 4.125 to obtain the optimal recycle ratio, α, and the corresponding space time from Eq. 4.126.

4.5 KEY POINTS

1. Regardless of the reaction order, the design equations for constant density batch reactors are obtained by substituting an appropriate rate equation into the constant density batch reactor mass balance equation, Eq. 2.27, and integrating between the initial concentration and the desired final concentration.

2. The integrated batch reactor mass balance equations for second- and mixed-order reversible reactions are identical in form.

3. When the mass balance equation for a batch reactor cannot be integrated analytically, the reciprocal of $-r_A$ should be plotted against C_A. The required reaction time is equal to the area under the curve between C_{Af} and C_{Ao}. It may be determined either by the trapezoidal method or by Simpson's rule.

4. The space time of a continuous flow reactor is the time required to process one reactor volume of feed measured at specified conditions, usually those of the feed. It is equal to the volume of reactor divided by the flow rate.

5. In terms of space time, the mass balance equation for a CSTR can be written as:

$$\tau = (C_{Ao} - C_{Af})/(-r_{Af}) \tag{4.39}$$

   Note that it is only necessary to know the reaction rate at one concentration (the effluent) to calculate the space time required to reduce the concentration from C_{Ao} to C_{Af}. Unlike the batch reactor equation, no integration is required.

6. The various batch reactor equations can be used directly for constant volume plug-flow reactors by replacing the reaction time t by the space time τ.

7. For constant density reactions, the reaction time for a batch reactor, t, and the space time for a plug-flow reactor, τ, are interchangeable.

8. As long as the rate, $-r_A$, is a monotonically increasing function of the reactant concentration, C_A, the feed should be split among n CSTR's in parallel or among n PFR's in parallel so that the space times are all equal. When $-r_A$ is not a monotonically increasing function of C_A no general conclusion can be drawn and each case must be examined individually.

9. As long as the rate, $-r_A$, is a monotonically increasing function of the reactant concentration, C_A, the feed should be split among n CSTR's in parallel or among n PFR's in parallel so that the space times are all equal. When $-r_A$ is not a monotonically increasing function of C_A no general conclusion can be drawn and each case must be examined individually.

10. The performance of n CSTR's in series for any order reaction can be determined by computing the performance of the first, then the second, etc., all the way down the chain.

11. An infinite number of equally sized CSTR's in series is equivalent to a PFR of equal volume.

12. When $-r_A$ is a monotonically increasing function of C_A, CSTR's in series are better than a single CSTR of equivalent total volume. The incremental improvement in performance declines as the number of reactors is increased.

13. As long as $-r_A$ is a monotonically increasing function of C_A, the following general statements can be made about CSTR's in series:
    a. If the reaction order is less than 1, the largest reactor should come first, followed by the second largest, etc.
    b. If the reaction order is 1.0, the sequencing of the reactors is immaterial.
    c. If the reaction order is greater than 1, the smallest reactor should come first, followed by the second smallest, etc.

14. For reactions whose rate expressions are monotonically increasing functions of reactant concentration, a series arrangement of CSTR's in proper sequence is always better than an equivalent parallel arrangement.

15. The sequence of different size PFR's in series is immaterial.

16. For monotonic reaction rate expressions multiple PFR's in any parallel, series, or parallel-series network may be considered equivalent to a single large reactor. If the rate expression is not monotonic, it is not possible to generalize.

17. Given an nth order reaction, the following rules concern the most effective use of CSTR's and PFR's in series:
    a. when $n = 1$, the ordering is immaterial.
    b. when $n > 1$, PFR's (any ordering) should be followed by the smallest CSTR, the second smallest CSTR, and so on to the largest CSTR.
    c. when $n < 1$, the largest CSTR, followed by the second largest CSTR, and so on to the smallest CSTR, should precede the PFR's (any ordering).

18. The optimal feed distribution for a CSTR and a PFR in parallel is different from that which results in equal space times or equal effluent concentrations. A unique distribution exists for each situation and thus no generalizations can be made.

19. For an nth order reaction, series arrangement of a CSTR and PFR is always superior to a parallel one.

20. In parallel reactions the ratio of rates, r_D/r_U, should be maximized by choosing the appropriate reactor type and reaction conditions, such as the reactant concentrations and temperature. The reactant concentrations dictate the choice of reactor and contacting scheme.

21. In series reactions the concentration of the desired product may be maximized by choosing the appropriate reactor type and reaction conditions, such as the reaction temperature and the space time. For series first-order reactions a PFR gives higher maximum concentrations and shorter space times than a CSTR.

22. The discussions presented for the parallel and series reactions also apply to the parallel-series reactions. The concentration of the desired product may be maximized by proper selection of reactor type and reaction conditions such as the temperature, the space time, and the reactant concentration. For bimolecular-type reactions the PFR gives higher maximum concentrations.

23. In terms of the feed to the reactor that would result by mixing the influent
     stream with the recycle stream the mass balance equations for the CSTR and PFR,
     Eqs. 4.124 and 4.126, are the same as those without recycle.

24. For those reactions with nonmonotonic rate expressions the optimum recycle
     ratio that minimizes the PFR volume is obtained through ordinary calculus
     using Leibnitz's rule and is given by Eq. 4.134.

4.6 STUDY QUESTIONS

The study questions are arranged in the same order as the text:

Section	Questions
4.1	1-19
4.2	20-30
4.3	31-34
4.4	35-37

1. The first-order reaction $A \overset{k}{\to} B$ with rate expression given by $-r_A = kC_A$
    ($k = 0.25$/hr, and C_A in mol/liter) is to be carried out in a constant-volume
    batch reactor. How long should the reaction be carried out to reduce the
    reactant concentration by 50%?

2. A reactant A decomposes in accordance with second-order kinetics. In a con-
    stant-volume batch reactor it took two hours to reduce the reactor concentra-
    tion to one-half. How long would it take to reduce it to one-fourth?

3. A reaction is suspected to follow first-order kinetics. The reaction was
    carried out in a constant-volume batch reactor and the reactant concentra-
    tion was monitored during the course of reaction. It took two hours to re-
    duce the reactant concentration to one-half of its initial value, while four,
    six and eight hours were required to reduce the concentration to one-fourth,
    one-eighth and one-sixteenth, respectively. Is the reaction first order? If
    yes, explain why and determine the rate constant. If no, explain your answer.

4. The reaction $A \to B$ is known to follow the rate expression given by
    $-r_A = kC_A/(K + C_A)$ where $k = 1.5$ mol/(liter·hr) and $K = 1.0$ mol/liter. How
    long would it take to reduce the reactant concentration from 2 mol/liter to
    1 mol/liter in a constant-volume batch reactor? How long to reduce it from
    1 mol/liter to 0.5 mol/liter?

5. Derive Eq. 4.6 from Eq. 4.3 and the condition $C_{Ao} = C_{Bo}$.

6. Derive Eq. 4.11 from Eq. 4.8 and the condition $C_{Ao} = C_{Co}$.

7. The reversible reaction $A \overset{\to}{\leftarrow} B$ was studied in a constant-volume batch reactor
    using various starting compositions and allowing the reaction to take place
    over a long time period. When the starting solution contained 1 mol/liter of
    A and 2 mol/liter of B, it remained at this composition, while a starting com-
    position of 3 mol/liter of A and 2 mol/liter of B gradually changed to contain
    1.67 mol/liter of A and 3.33 mol/liter of B. What final composition would you
    expect if the starting composition were 3 mol/liter of B and no A?

8. The first-order reversible liquid reaction $A \overset{\to}{\leftarrow} B$ takes place in a batch reactor
    with an initial charge of 1 mol/liter of A. At the end of 20.8 minutes
    $C_A = 0.75$ mol/liter whereas after a very long time the reaction fluid contains
    0.5 mol/liter each of A and B. Determine the reaction fluid composition at the
    end of 60 minutes.

9. A reaction involving one reactant has been studied in a constant volume batch reactor and the rates at various reactant concentrations are given in Table SQ4.1. Estimate the time required to reduce the reactant concentration from 1 mol/liter to 0.3 mol/liter using (a) the trapezoidal method and (b) Simpson's rule.

TABLE SQ4.1
Reaction Rates at Various Reactant Concentrations

$-r_A$ mol/(liter·min)	C_A mol/liter	$-r_A$ mol/(liter·min)	C_A mol/liter
3	1.3	0.295	0.6
1.82	1.1	0.171	0.5
1.37	1.0	0.0874	0.4
0.995	0.9	0.0369	0.3
0.700	0.8	0.0109	0.2
0.468	0.7	0.00137	0.1

10. For the reaction of Study Question 1 what space time is required to reduce the reactant concentration by 50% using a CSTR?

11. The forward and reverse reaction rate constants for the reversible elementary reaction $A + B \overset{\rightarrow}{\leftarrow} C + D$ are given as

    $k_1 = 2$ liters/(mol·hr)

    $k_{-1} = 1$ liter/(mol·hr)

    The reaction is to be carried out in a CSTR using a feed containing 1 mol/liter each of A and B. It is desired to reduce the concentration of A to 0.5 mol/liter. Determine the space time. If the feed is available at 1000 liters/hr what would be the reactor volume necessary to do this?

12. If the reaction in Study Question 2 is carried out in a CSTR what space time would be necessary to make the effluent concentration one-fourth of the feed concentration?

13. Determine the dilution rate necessary for a CSTR to reduce the reactant concentration in Study Question 2 from 2 to 1 mol/liter.

14. A second-order reaction is to be carried out in aqueous phase using a PFR. The rate expression is given by $r_A = -kC_A^2$ where $k = 1.53 \times 10^{-2}$ (mol/liter)$^{-1}$ (min)$^{-1}$. It is desired to reduce the reactant concentration from 1.0 mol/liter to 0.1 mol/liter. The feed is available at 1000 gal/hr. Determine the reactor volume necessary to achieve this task.

15. A bimolecular reaction involving two reactants A and B was carried out using a PFR. The aqueous feed stream contained 1.5 mol/liter of A and 0.5 mol/liter of B. When the space time was 0.5108 hr, the conversion of B was 50%. What would be the expected conversion if the space time were doubled? You may assume that $A + B \rightarrow$ Product and the reaction is elementary.

16. Repeat Study Question 11 using a PFR instead of a CSTR.

17. The rate expression for an aqueous reaction is given by $-r_A = k_1 C_A/(k_2 + C_A)$.

    When this reaction was carried out in a PFR using a feed containing 1.0 mol/liter of A, the conversions were 50 and 75% respectively at space times of 0.64 hour and 1.18 hours. What space time would be necessary for the PFR to achieve a 90% conversion? What would be the space time for a CSTR to achieve the same conversion?

18. For the reaction whose rates at various reactant concentrations are given in Table SQ4.1, estimate the PFR space time necessary to achieve 90% conversion of a feed containing 1 mol/liter of A. What would be the CSTR space time to do the same?

19. For the reaction whose rates at various reactant concentrations are given in Table SQ4.2, indicate the choice of reactor, PFR or CSTR, and the space time necessary for the following cases.

    (1) C_{Ao} = 0.8 mol/liter and C_{Af} = 0.4 mol/liter

    (2) C_{Ao} = 0.8 mol/liter and C_{Af} = 0.1 mol/liter

    and

    (3) C_{Ao} = 0.8 mol/liter and C_{Af} = 0.5 mol/liter.

    TABLE SQ4.2
    Reaction Rates at Various Reactant Concentrations

$-r_A$ mol/(liter·min)	C_A mol/liter	$-r_A$ mol/(liter·min)	C_A mol/liter
0.35	1.0	0.75	0.5
0.44	0.9	0.80	0.4
0.50	0.8	0.75	0.3
0.60	0.7	0.63	0.2
0.68	0.6	0.36	0.1

20. Two CSTR's are to be used in parallel for an aqueous second order reaction with the rate expression given by $-r_A = 1.3\ C_A^2$ where the rate constant is in units of (liters)/(mol·min) and C_A in mol/liter. The reactor volumes are 2 and 1 liter, respectively. The feed contains 0.9 mol/liter of A and is available at 10 liters/hr. It is desired to reduce C_A in the effluent as much as possible.

    How should the feed be distributed and what would be the rate of discharge of unreacted A in the effluent?

21. For the autocatalytic reaction whose rate is given by $-r_A = kC_A(C_{Ao} + C_{Bo} - C_A)$, where k = 0.15 liter/(mol·min) and the concentrations are given in units of mol/liter, determine the optimum feed distribution to a network of two CSTR's in parallel so as to minimize the rate of discharge of unreacted A in the effluent. The feed contains 1 mol/liter of A and B and is available at 100 liters/hr. The reactor volumes are 10 and 5 liters, respectively.

22. Three CSTR's of equal volume (2 liters each) are to be used in series. The reaction rate expression is given by $-r_A = k_1 C_A/(k_2 + C_A)$ where k_1 = 1.5 min^{-1}, k_2 = 0.5 mol/liter, and C_A is in units of mol/liter. The volumetric feed rate is 50 liters/min. Determine the effluent concentration.

23. Two CSTR's are to be used in series for the reaction whose rates are given in Table SQ4.1. The volumes are 10 and 20 liters. The feed concentration is 1 mol/liter and the feed rate is 10 liters/min. Determine the arrangement which gives the lowest concentration of A. What will that concentration be?

24. How many equal volume CSTR's in series are required to reduce the reactant concentration from 0.9 mol/liter to 0.1 mol/liter for the reaction whose rates are given in Table SQ4.1? The feed rate is 10 liters/min and the volume of each CSTR is 10 liters.

25. For the reaction whose rate is given by $-r_A = C_A/(1 + C_A)$ in which C_A is given in units of mol/liter and $-r_A$ in mol/(liter·hr), two CSTR's of volume equal to 10 and 20 liters, respectively, are to be used in series. Determine the optimum arrangement and the minimum effluent concentration that can be obtained. The feed contains 1.5 mol/liter of A and is available at 10 liters/hr.

26. Two CSTR's are available for the reaction whose rates are given in Table SQ4.2. The volumes are 50 and 100 liters, respectively. The feed contains 0.9 mol/liter of A and is available at 100 liters/min. Determine the arrangement which gives the minimum effluent concentration. What is the minimum effluent concentration?

27. Three PFR's are available for an aqueous reaction of first order. The reactor volumes are 10, 20 and 30 liters. Determine the arrangement which gives the minimum effluent concentration. Feed containing 15 mol/liter of A is available at 10 liters/hr. The rate constant $k = 1.3$ hr^{-1}. Determine also the effluent concentration.

28. Two CSTR's and two PFR's are available. The CSTR volumes are 10 and 20 liters while the PFR volumes are 5 and 10 liters. The reactors are to be used in series to handle an aqueous reaction whose rates are given by $-r_A = C_A/(2 + C_A)$, where C_A is in units of mol/liter and $-r_A$ in units of mol/(liter·hr). The feed rate is 10 liters/hr and the feed contains 1.2 mol/liter of A. Determine the arrangement which gives the minimum effluent concentration. What is the effluent concentration?

29. A CSTR and a PFR are to be used in parallel for the constant-density reaction whose rate expression is given in Study Question 28. Determine the feed distribution as well as the resulting minimum effluent concentration. The CSTR volume is 10 liters while that of the PFR is 5 liters. The feed contains 1.0 mol/liter of A and is available at 10 liters/hr.

30. Referring to the data given in Table SQ4.2 what arrangement of reactors (CSTR, PFR, or both) would give the minimum total space time? The feed contains 0.9 mol/liter of A and the desired effluent concentration is 0.2 mol/liter.

31. For the consecutive reaction $A \xrightarrow{k_1} B \xrightarrow{k_2} C$ each step is elementary with $k_1 = 2$ hr^{-1} and $k_2 = 1$ hr^{-1}. Determine the optimum space time for a PFR and the corresponding maximum C_B for a feed containing 0.8 mol/liter of A. What would be the corresponding concentrations of A and C?

32. Repeat Study Question 31 for a CSTR.

33. The parallel-series reactions of Eq. 4.101 are to be carried out in either a PFR or CSTR. The rate constants are equal, $k_1/k_2 = 1$. It is desired to maximize the concentration of B. What type of reactor should be used? Initially no B nor C is present. What concentration of B do you expect?

34. Derive Eq. 4.114 from Eq. 4.113.

35. Consider a second order reaction $A + 2B \rightarrow$ Product which takes place in aqueous phase. The rate expression is given by $-r_A = kC_AC_B$ where $k = 1.5$ (liters)/(mol·hr) and C_A and C_B are in units of mol/liter. This reaction is to be carried out in a CSTR. The unreacted B in the effluent is

instantaneously separated and fed back into the CSTR as shown in Fig. SQ4.1. Complete the mass balance and determine the unknown in the figure.

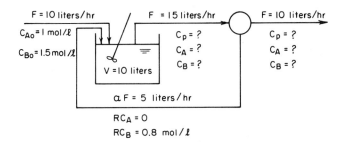

Figure SQ4.1. Reactor for Study Question 35.

36. Consider the third order reaction that was treated in Example 4.1.3-1. We now allow recycle of the effluent with the recycle ratio, α, set at 3. The volumetric flow rate is 100 liters/hr. Calculate the volume of PFR necessary to convert 95% of A in a feed containing 2 mol/liter of A and 1.2 mol/liter of B.

37. For the autocatalytic reaction whose rate equation is given by Eq. 3.33 where $k = 1.5$ liters/(mol·hr), determine the recycle rate which will minimize the PFR size and determine this size. The feed contains only A ($C_{Ao} = 1$ mol/liter) and is available at 10 liters/hr. The desired conversion is 98%.

REFERENCES AND FURTHER READING

1. C. R. Wylie, Jr., *Advanced Engineering Mathematics,* McGraw-Hill, Inc., New York, N.Y., a. 513-514; b. 591-592, 1951.

2. R. C. Fisher and A. D. Ziebur, *Calculus and Analytical Geometry,* Prentice-Hall, Inc., Englewood Cliffs, N.J., 229-230, 1961.

3. O. Levenspiel, *Chemical Reaction Engineering,* Second Edition, John Wiley & Sons, Inc., New York, N.Y., a. 100-101; b. 185-194, 1972.

4. S. Szepe and O. Levenspiel, *Industrial and Engineering Chemistry, Process Design and Development,* 3, 214, 1964.

Further Reading

Aris, R., *Elementary Chemical Reactor Analysis,* Prentice-Hall, Inc., Englewood Cliffs, N.J. 1974. See Chapters 7 through 10 for detailed design and control of various nonisothermal reactors.

Bamford, C. H. and Tipper, C. F. H. editors, *The Theory of Kinetics,* Vol. 2, Elsevier Publishing Co., Amsterdam, 1969. See Chapter 1 for batch reactors.

Capellos, C. and Bielski, B. H. J., *Kinetic Systems,* Wiley-Interscience, New York, N.Y., 1972. This gives many integrated mass balance equations for batch reactors.

Carberry, J. J., *Chemical and Catalytic Reaction Engineering,* McGraw-Hill Book Co., Inc., New York, N.Y. See Chapter 3.

Cooper, A. R. and Jeffreys, G. V., *Chemical Kinetics and Reactor Design,* Prentice-Hall Inc., Englewood Cliffs, N.J., 1973. See Chapters 4, 5 and 6.

Fogler, H. S., *The Elements of Chemical Kinetics and Reactor Calculations* (A Self-Paced Approach), Prentice-Hall Inc., Englewood Cliffs, N.J., 1974. See Chapter 4.

CHAPTER 5

ANALYSIS OF REACTION RATE DATA AND DETERMINATION OF REACTION RATE EXPRESSIONS

In previous chapters we assumed that the rate expressions were known and showed how they are used in appropriate mass balance equations for the design of various types of reactors. In this chapter we will discuss methods of obtaining and analyzing reaction rate data for the development of the needed rate expressions. In principle, any type of reactor can be used, but the simplest and most frequently used is a batch reactor operated at constant temperature and volume. In batch reactor experiments the reactor is charged with the reactant(s) and brought to the desired reaction conditions. Then, during the course of the reaction, samples are removed for analysis and the concentration is recorded against time. When more than one reactant is involved it is necessary to repeat the experiment with different initial compositions. When a CSTR is used to obtain rate data, feed is supplied continuously until a steady state is achieved. Then, the effluent concentration is recorded, and another steady-state run is initiated by changing the feed concentration and/or the feed rate. Thus a number of steady-state runs are required to obtain data relating reaction rate to concentration whereas a single unsteady-state run may be used to obtain the same information from a batch reactor. Plug-flow

reactors (PFR) are not normally used in environmental engineering for generating
reaction rate data from microbial slurries since it is simpler to use a batch
reactor. However a PFR is frequently used in the study of the kinetics of microorganisms growing on fixed films.

There are three methods available for analyzing rate data; algebraic, differential and integral. The algebraic method is used with data from a CSTR operating
at steady state because the reaction rate may be calculated algebraically from the
steady-state mass balance equation. The reaction rates thus obtained are then
correlated with the reactant concentrations to obtain a reaction rate expression.
Either the differential or the integral method must be used with data obtained from
a batch reactor because the data are in the form of concentration versus time and
do not provide a direct measure of the reaction rate as a function of concentration.
Therefore, the rate expression must be developed either by differentiating the
concentration-versus-time data to obtain rate-versus-concentration information which
may be correlated, or by fitting the integrated form of the hypothesized rate
expression to the concentration-versus-time data.

5.1 ALGEBRAIC METHOD

As mentioned in Section 2.3.1 the rate of consumption of reactant may be calculated
algebraically from the CSTR mass balance equation, Eq. 2.20. By changing the flow
rate, the reactor volume, or the feed concentration, the effluent concentration of
a CSTR can be varied experimentally and the corresponding rates calculated algebraically provided that sufficient time is allowed between runs to attain steady-
state conditions. The calculated rates must then be correlated with the effluent
reactant concentration to obtain the rate expression. The procedure may be summarized as follows for a reaction involving one reactant.

Step 1. Make a series of steady-state CSTR runs by varying the feed rate, F,
the feed concentration, C_{Ao}, or the reactor volume, V, and record the corresponding steady-state effluent concentrations as indicated in Table 5-1.

TABLE 5.1
Algebraic Analysis of CSTR Data

| Operational Conditions | | | Effluent | Rate from |
Conc.	Flow	Vol.	Conc.	Eq. 2.20
C_{Ao1}	F_1	V_1	C_{A1}	$-r_{A1}$
C_{Ao2}	F_2	V_2	C_{A2}	$-r_{A2}$
C_{Ao3}	F_3	V_3	C_{A3}	$-r_{A3}$
⋮	⋮	⋮	⋮	⋮
C_{Aon}	F_n	V_n	C_{An}	$-r_{An}$

Step 2. Calculate algebraically the rate of consumption of A, $-r_{Aj}$, for each run using the mass balance equation.

$$-r_{Aj} = (C_{Aoj} - C_{Aj})F_j/V_j \qquad\qquad (2.20)$$

Step 3. Correlate the rates thus calculated with the effluent concentrations to obtain the rate expression.

The key operation is step 3. Unless a proper functional form is selected for the relationship between the rate and the effluent concentration, the fit to the data may be very poor, thereby making the rate expression useless. Usually the experimenter will either assume a likely form for the rate expression or hypothesize a reaction mechanism and then develop the corresponding rate expression as was done in Chapter 3. Then, the proposed rate expression is tested graphically against the experimental data to see if the fit is satisfactory. If it is not, a new rate expression is proposed and the procedure repeated until a satisfactory fit is obtained. A satisfactory fit over the experimental ranges does not necessarily imply that the proposed rate expression truly reflects the intrinsic rate, but the broader the range of effluent concentrations covered in step 1, the greater the confidence one will have in the resulting rate expression.

Often the rate constants and reaction parameters appear linearly in a proposed rate expression so that it can be readily tested graphically. For example, if a second-order rate expression, $-r_A = kC_A^2$, is proposed, the rates of consumption of A calculated from the mass balance equation, $-r_{Aj}$, can be plotted against the squares of the effluent concentration, C_{Aj}^2. Then, if the plot results in a straight line passing through the origin, one may assume that the proposed rate expression is satisfactory, allowing determination of the rate constant, k, from the slope of the line. This is illustrated in Fig. 5.1. On the other hand, the rate constants and reaction parameters may appear nonlinearly in the proposed rate expression so that it may not be amenable to such simple graphical testing. For example, if a proposed

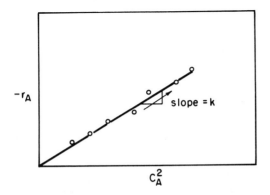

Figure 5.1. Test for a second-order reaction,
$-r_A = kC_A^2$.

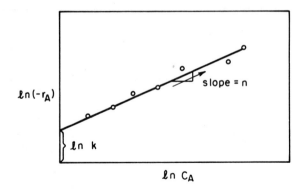

Figure 5.2.   Test for an nth-order reaction,
$$-r_A = kC_A^n.$$

rate expression is an nth-order reaction, $-r_A = kC_A^n$, in which n is unknown, it may
be necessary to manipulate the rate equation into a linear form suitable for graphical testing.   In this case we may take the natural logarithm of both sides to
obtain

$$\ln(-r_A) = \ln(k) + n \ln(C_A) \tag{5.1}$$

Thus $\ln(-r_A)$ should be plotted against $\ln(C_A)$. If the plot follows a straight line,
the proposed rate expression may be assumed to be satisfactory and the rate constant,
k, and the order of reaction, n, may be determined from the intercept and the slope
of the line, respectively. This is illustrated in Fig. 5.2. In other situations it
may not be possible to manipulate the rate equation into a linear form and therefore
one may have to rely upon a nonlinear least squares technique to estimate the parameters. Such situations arise infrequently but if they do, one may be able to check
various limiting behaviors of the proposed rate expression before relying on a nonlinear technique. Such partial analyses will be discussed later.

The procedure summarized above may be best illustrated through examples.

EXAMPLE 5.1-1
Given in Table E5.1 is a summary of the results from a series of steady-state runs
obtained when the feed rate to an 1.0 liter CSTR was varied. The feed concentration
was constant at 3.0 mol/liter of A. Develop a rate expression, r_A, consistent with
the data.

Solution: We follow the steps summarized above.

Step 1. This step has been given as a part of the problem statement.

TABLE E5.1
Data from a CSTR

F liters/hr	C_A mol/liter
0.05	0.284
0.1	0.395
0.2	0.530
0.4	0.740
0.8	0.990
1.6	1.27
3.2	1.61
6.4	2.01

TABLE E5.2
Data for Determination of the Rate Expression

C_A	$-r_A$	C_A^2	$k = -r_A/C_A^2$
0.284	0.136	0.0807	1.68
0.395	0.261	0.156	1.67
0.530	0.494	0.281	1.76
0.740	0.904	0.548	1.65
0.990	1.61	0.980	1.64
1.27	2.77	1.61	1.72
1.61	4.45	2.59	1.72
2.01	6.34	4.04	1.60

Step 2.   The rate corresponding to each effluent concentration is calculated using Eq. 2.20 and the results are summarized in Table E5.2.

$$-r_{Aj} = (C_{Aoj} - C_{Aj})F_j/V_j = (3 - C_{Aj})(F_j/1.0)$$

For the first run,

$$-r_{A1} = (3 - 0.284)(0.05) = 0.1358 \text{ mol}/(\text{liter} \cdot \text{hr})$$

Step 3.   We need to correlate the calculated rates with the effluent concentrations.   Comparing C_A and $-r_A$ we can see by inspection that they are not linearly correlated, and thus that the reaction is not first order.   As a first attempt, let us assume that the reaction is second order, $-r_A = kC_A^2$.   If this is a correct form $-r_A/C_A^2 = k$, i.e., for each run the rate, $-r_A$, divided by C_A^2 should yield the value of k.   Therefore, a consistent value of k from each run may be taken as satisfactory evidence for the proposed rate expression.   Calculated values of $-r_A/C_A^2$ are also shown in Table E5.2.   Since they are approximately constant (average = 1.68 $(\text{mol/liter})^{-1}(\text{hr})^{-1}$), we may accept the second-order rate expression as satisfactory.   Alternatively, we could have assumed an nth-order reaction, $-r_A = kC_A^n$, and plotted the data according to Eq. 5.1. If a straight line is obtained the proposed rate expression may be considered satisfactory and the values of n and k obtained from it. The reader should verify that indeed this method leads to the same results, i.e.,

$$-r_A = 1.7 \ C_A^2$$

EXAMPLE 5.1-2

A 1.0 liter CSTR was used to study an enzyme reaction, $S \overset{E}{\to} P$ and the experimental data in Table E5.3 were obtained with a feed containing 0.5 mol/liter of S and 1.2×10^{-4} mol/liter of enzyme E. Develop a rate expression for the consumption of S.

Solution:

Step 1. This has been given as a part of the problem statement.

Step 2. Calculate the rate using Eq. 2.20

$$-r_{Sj} = (C_{Soj} - C_{Sj})F_j/V_j = (0.5 - C_{Sj})(F_j/1.0)$$

The results are tabulated in Table E5.4.

TABLE E5.3
Data from a CSTR

| F | C_S |
liters/hr	mol/liter
0.1	0.047
0.2	0.089
0.4	0.16
0.6	0.21
0.8	0.25
1.0	0.27
2.0	0.37
3.0	0.40

TABLE E5.4
Data for Determination of the Rate Expression

C_S	$-r_S$	$1/C_S$	$-1/r_S$
0.047	0.0453	21.28	22.08
0.089	0.0822	11.24	12.17
0.16	0.136	6.250	7.353
0.21	0.174	4.762	5.747
0.25	0.20	4.000	5.000
0.28	0.22	3.704	4.545
0.37	0.29	2.703	3.846
0.40	0.30	2.500	3.333

Step 3. Either a form of the rate expression must be assumed or a mechanism must be hypothesized and the corresponding rate expression must be developed. If we postulate the mechanism, $E + S \rightleftarrows ES \rightarrow E + P$ as was done in Section 3.3.2 we obtain Eq. 3.81

$$-r_S = V_m C_S / (K_m + C_S) \qquad\qquad (3.81)$$

Therefore, the calculated rates, $-r_{Sj}$, should be tested to see if they can be correlated with the substrate concentration, C_{Sj}, in the form of Eq. 3.81. Note that the parameter K_m appears nonlinearly and it is not apparent how to manipulate Eq. 3.81 to put it into linear form so that the correlation between $-r_S$ and C_S can be determined. Several techniques for putting the equation into linear form have been used; the simplest is to take the reciprocal of each side.

$$-\frac{1}{r_S} = \frac{1}{V_m} + \frac{K_m}{V_m}\frac{1}{C_S}$$

which is of the form

$$y = a + bx$$

Therefore, if the proposed rate expression is valid, the plot of $-1/r_S$ versus $1/C_S$ should give a straight line with the slope equal to K_m/V_m and the intercept $1/V_m$. Therefore, we calculate $1/C_S$ and the corresponding $-1/r_S$ as shown in Table E5.4. The plot of the data is shown in Fig. E5.1. This is called a Lineweaver-Burke plot. The plot is a straight line and thus we can assume

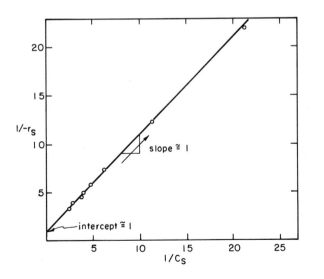

Figure E5.1. Test for enzyme reaction in Example 5.1-2.

that Eq. 3.81 is suitable for representing the data. Both the slope and the intercept are approximately equal to one, so that

$$a = 1/V_m = 1 \implies V_m = 1 \text{ mol}/(\text{liter}\cdot\text{hr})$$

$$b = K_m/V_m = 1 \implies K_m = 1 \text{ mol/liter}$$

It should be noted that Eq. 3.81 may be put into other linear forms as well, and two of these will be discussed in Chapter 10.

From the above examples we see that the procedure involves assuming a rate expression and testing it against experimental data. If the fit is satisfactory, the parameters (the rate constants, the order of reaction, etc.) can then be determined. Since a common procedure is to estimate the parameters from a linear plot it would be desirable to have a systematic method of fitting a straight line to a large number of points so as to obtain the best estimates of the slope and intercept. One method frequently used is the least squares fit, which results in the line which minimizes the sum of the squares of the differences between the experimental points and those predicted by the line. This technique may be used in connection with the algebraic, differential or integral methods of analysis.

So far we have limited the procedure to reactions involving only one reactant; however, the same approach may be used for reactions involving more than one reactant. For example, for reactions involving two reactants, A and B, one can vary C_{Ao} and C_{Bo} as well as the space time to obtain a series of steady-state effluent concentrations, C_A and C_B, from which the rates $-r_A$ and $-r_B$ may be calculated. The calculated rates must then be correlated with C_A and C_B by assuming a rate expression and checking it against the experimental data. The reaction parameters may be estimated by the least squares fit involving two independent variables. For complicated reactions involving many reactants it may help to consider the asymptotic behavior of the rate using various partial analysis methods. These will be covered later.

Finally, it should be mentioned that a CSTR may be operated under transient conditions and the resulting transient (unsteady-state) data may be used to develop the rate expression. For instance, under constant volume conditions, Eq. 2.17 may be written as

$$(C_{Ao} - C_{Af})/\tau + r_A = dC_{Af}/dt \qquad (5.2)$$

according to which

$$-r_A = (C_{Ao} - C_{Af})/\tau - dC_{Af}/dt \qquad (5.3)$$

so that the rate may be evaluated by knowing the space time, τ, the feed and effluent concentrations, C_{Ao} and C_{Af}, and the time derivative of the effluent concentration,

dC_{Af}/dt. The latter may be evaluated from either continuous or discrete recording of the effluent concentration, but in either case a derivative must be determined. Thus, transient CSTR data are very similar to batch reactor data because both require evaluation of derivatives. Hence, transient CSTR data may be analyzed by the same methods used for batch reactors, i.e., the differential or integral methods.

5.2 DIFFERENTIAL METHOD

The differential method of analysis of batch reactor data involves differentiation of the concentration profile to obtain the reaction rates. According to Eq. 2.25

$$r_A = dC_A/dt \qquad\qquad (2.25)$$

the time derivative of reactant concentration is equal to the rate of formation of A. Therefore, accurate experimental data of concentration versus time are needed so that the derivatives may be evaluated at various times, and therefore, at various reactant concentrations. In this manner, data of rate versus reactant concentration may be generated. The calculated rates may then be correlated with the reactant concentrations to obtain the desired rate expression just as they were in the algebraic method.

5.2.1 General Procedure

Let us summarize the procedure for a simple reaction involving one reactant.

Step 1. Make one or more batch runs and record the reactant concentrations as a function of time as shown in Table 5.2. Samples should be taken frequently to obtain accurate transient data.

Step 2. Calculate the rate of consumption of A, $-r_{Aj}$, at various reactant concentrations by evaluating the slopes of the concentration versus time curve.

$$-r_{Aj} = (dC_A/dt)_{C_{Aj}} \qquad\qquad (2.25)$$

TABLE 5.2
Differential Analysis of Batch Data

Time	Reactant Conc.	Rate
0	C_{Ao}	$-r_{Ao}$
t_1	C_{A1}	$-r_{A1}$
t_2	C_{A2}	$-r_{A2}$
t_3	C_{A3}	$-r_{A3}$
$\vdots$	$\vdots$	$\vdots$
t_n	C_{An}	$-r_{An}$

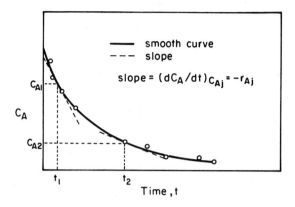

Figure 5.3.   Differentiation of concentration curve
              to find reaction rate.

Record the values as shown in Table 5.2 This is illustrated in Fig. 5.3.
Step 3. Correlate the rates thus calculated with the reactant concentrations
to obtain the rate expression.

In Step 2 differentiation of the reactant concentration versus time data is
required. This can be done analytically if the data are first fit to an analytical
function, for example, a polynomial function or a sum of exponential functions.
Computer programs are available which allow one to fit the reactant concentration
with a polynomial function, say

$$C_A = \sum_{i=1}^{n} a_i t^{i-1} \tag{5.4}$$

Then, the derivative may be evaluated analytically,

$$dC_A/dt = \sum_{i=1}^{n} (i-1) a_i t^{i-2} \tag{5.5}$$

The derivatives may be also evaluated graphically by drawing a smooth continuous
curve through the data points and then determining its slope at various reactant
concentrations. One way to determine the slope is through the use of a mirror. At
the point where the slope is to be determined a mirror is placed perpendicularly to
the paper so that the image of the curve is reflected in the mirror. A blank paper
is then used to cover up most of the curve in front of the mirror so that the small
region of the curve left visible between the paper and the mirror is practically a
straight line. The mirror is rotated until the curve segment in front of the mirror
forms a straight line with the image in the mirror, and a straight line is drawn
along the edge of the mirror. This line is the normal line. The slope may then be
determined by constructing a line perpendicular to the normal line.

Step 3 in the differential procedure is identical to that of the algebraic
method so that the comments made there apply here also. Thus the only difference
between the differential and algebraic methods is in the way the rates are obtained.

EXAMPLE 5.2.1-1

Given in Table E5.5 are batch reactor data for a simple reaction involving one
reactant. Using the differential method of analysis develop rate data which could
be used to evaluate the rate expression.

   Solution: We follow the step-by-step procedure summarized above.

   Step 1. This has been given as part of the problem statement.

   Step 2. The reactant concentrations are plotted against reaction times and a
smooth curve is drawn through them as shown in Fig. E5.2. The graphical procedure using a mirror is used to evaluate the slope of the curve at various points
and the results are summarized in Table E5.6. Note that it is not necessary to
evaluate the slope at the times at which the samples were taken. They can be
taken at any point.

TABLE E5.5
Data from a Batch Reactor

t min	C_A mol/liter
0	1
5	0.44
10	0.29
15	0.21
30	0.11
60	0.062
90	0.043
100	0.032

TABLE E5.6
Reaction Rate Data Obtained from Figure E5.2

C_A mol/liter	$-r_A$ mol/(liter·min)
1	0.26
0.5	0.063
0.3	0.023
0.2	0.010
0.1	0.0025
0.05	0.0006
0.04	0.0004
0.03	0.00023

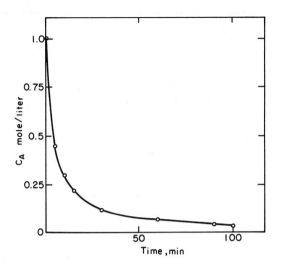

Figure E5.2. Concentration curve for Example 5.2.1-1.

EXAMPLE 5.2.1-2

Given in Table E5.7 is a summary of the data collected from a batch reactor containing two reactants, A and B. Develop a rate expression consistent with the experimental data.

Solution:

Step 1. This has been given as part of the problem statement

Step 2. The reactant concentrations, C_A and C_B, are plotted against time and smooth curves are drawn through the points as shown in Fig. E5.3. The slopes are evaluated graphically and tabulated in Table E5.8.

Step 3. Let us assume that $-r_A = k C_A^m C_B^n$ and try to estimate k, m and n simultaneously using the least squares fit. Taking the natural logarithm of the rate equation yields

$$\ln(-r_A) = \ln(k) + m \ln(C_A) + n \ln(C_B)$$

or

$$y = a + bx + cz$$

where a, b and c are constants to be determined by least squares techniques, yielding:

$$a = -4.627, \ b = 0.9900 \ \text{and} \ c = 1.0262$$

so that

$$k = 0.009787, \ m = 0.9900 \ \text{and} \ n = 1.0262$$

TABLE E5.7
Data from a Batch Reactor

t min	C_A mol/liter	C_B mol/liter
0	1	2
30	0.59	1.58
60	0.37	1.38
120	0.18	1.19
150	0.12	1.13
200	0.073	1.074
400	0.0092	1.009

TABLE E5.8
Reaction Rate Data Obtained from Figure E5.3

C_A	C_B	$-r_A = -dC_A/dt$
1	2	0.02
0.6	1.6	0.0095
0.5	1.49	0.0075
0.4	1.41	0.0055
0.3	1.3	0.0039
0.2	1.19	0.0024
0.1	1.1	0.0011
0.05	1.05	0.00053

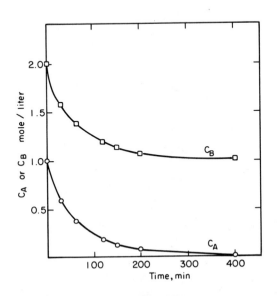

Figure E5.3. Smoothed concentration curves and
             graphical evaluation of slopes for
             Example 5.2.1-2.

and

$$-r_A = 0.009787 \; C_A^{0.99} C_B^{1.0262}$$

However, m and n are very close to unity and since for simple reactions we anticipate integer values for m and n, we proceed to fit the data with a bimolecular reaction rate, i.e., $-r_A = kC_AC_B$. In this case a straight line should result when $-r_A$ is plotted against C_AC_B, and Fig. E5.4 shows that it does, verifying that the postulated expression will work. The best estimate of the rate constant, k, may be obtained by using the least squares fit. Hence

k = 0.00998

so that

$$-r_A = 0.00998 \; C_AC_B$$

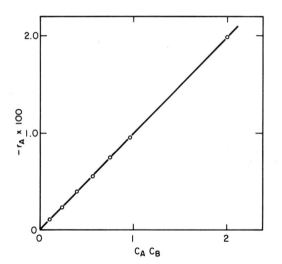

Figure E5.4.   Test for $-r_A = kC_AC_B$ in Example 5.2.1-2.

5.2.2 Specialized Techniques

In Chapter 3 we saw that the forms of the rate expressions for multiple reactions, reversible reactions and nonelementary reactions are complex; thus it may be difficult to test them by the differential method. One way to reduce the difficulty, however, is to perform experiments under special conditions so that parts of the proposed rate expression may be tested separately. In addition, it may also be possible to test the asymptotic behavior of the proposed rate expression. These approaches are generally referred to as the methods of isolation and two are commonly

used in conjunction with the differential method. They are the method of initial
rates and the method of excess.

METHOD OF INITIAL RATES. This method is frequently used for reversible
reactions and for reactions in which complications arise with time (for example,
an enzyme reaction in which the enzyme is unstable and deactivates with time.) Let
us illustrate it by means of a second-order reversible reaction,

$$A + B \underset{k_{-1}}{\overset{k_1}{\rightleftharpoons}} C + D \qquad\qquad (3.43)$$

with a postulated rate expression

$$r_A = -k_1 C_A C_B + k_{-1} C_C C_D \qquad\qquad (3.44)$$

If the reaction were run over a short time period with only A and B present (i.e.,
$C_{Co} = C_{Do} = 0$) and if the time derivative of the reactant concentration were taken
at $t = 0$, then the rate would be given by

$$(dC_A/dt)\Big|_{t=0} = r_A\Big|_{t=0} \cong -k_1 C_{Ao} C_{Bo} \qquad\qquad (5.6)$$

Thus the rate is a function only of C_A and C_B and not of C_C and C_D. In this manner
we can consider only the forward reaction and not the reverse reaction, i.e., the
forward reaction has been isolated. By performing a series of experiments in which
the initial concentrations of A and B are varied in the absence of C and D, the
resulting rates may be correlated with C_A and C_B thereby allowing the value of k_1
to be determined. The procedure can then be repeated for the reverse reaction by
working with only C and D (i.e., $C_{Ao} = C_{Bo} = 0$) so that the initial rate is

$$(dC_A/dt)\Big|_{t=0} = r_A\Big|_{t=0} \cong k_{-1} C_{Co} C_{Do} \qquad\qquad (5.7)$$

This allows the reverse reaction to be isolated and tested. Then, the complete rate
expression may be tested by running the reaction with all four species present and
comparing the observed results with the predicted.

METHOD OF EXCESS. When a reaction involves many reactants as in the case of
termolecular reactions

$$A + B + C \rightarrow \text{Products} \qquad\qquad (3.21)$$

with a rate expression

$$-r_A = k C_A C_B C_C \qquad\qquad (3.22)$$

it is possible to use a large excess of B and C (say $C_{Bo} = C_{Co} = 100\ C_{Ao}$) so that
the concentrations of B and C remain essentially unchanged (e.g., even at a 100%
conversion of A, C_B and C_C would vary only 1% so that $C_B = C_C \cong C_{Bo} = C_{Co}$). Under

these conditions, Eq. 3.22 may be written as

$$-r_A = (kC_{Bo}C_{Co})C_A = k'C_A \tag{5.8}$$

where $k' = kC_{Bo}C_{Co}$. Thus, the original reaction rate, which was third order, has been reduced to first order and the reaction has been isolated in terms of only A. Equation 5.8 is known as a pseudo-first-order reaction rate and k' as the pseudo-first-order reaction rate constant. The reaction rate may be estimated as a function of C_A using the differential technique and the value of k' determined in the usual way. A series of runs may then be made with different excess amounts of B and C (C_{Bo}, C_{Co}) and the corresponding k' determined for each run. The value of k' should be proportional to C_{Bo} and C_{Co} so that the rate constant k may be calculated using:

$$k = k'/(C_{Bo}C_{Co}) \tag{5.9}$$

After confirming that this technique is valid for the case in question, one should then make batch runs without an excess amount of any species and check the experimental rates with those predicted by the form of the rate expression, Eq. 3.22. Alternatively, one can repeat the above procedure in the presence of excess amounts of A and B or A and C. The method may be applied to rate expressions of any order, for example

$$-r_A = kC_A^a C_B^b C_C^c \tag{5.10}$$

although it is not restricted to power law forms of rate expressions.

Another form of rate expression which may be treated by the method of excess is

$$-r_A = kC_A^m/(K + C_A)^n \tag{5.11}$$

The rate expression given by Eq. 5.11 has two asymptotic behaviors. At high concentrations of A so that $C_A \gg K$, Eq. 5.11 reduces to

$$-r_A = kC_A^m/C_A^n = kC_A^{(m-n)} \tag{5.12}$$

so that the rate is $(m-n)^{th}$ order with respect to C_A, while for low values of A so that $C_A \ll K$ Eq. 5.11 reduces to

$$-r_A = kC_A^m/K^n = (k/K^n)C_A^m = k'C_A^m \tag{5.13}$$

so that the rate is m^{th} order. By running one or more batch runs with a very high initial concentration of A and terminating the reaction before C_A decreases by significant amounts it is possible to correlate calculated rates with C_A so that the values of k and (m - n) may be determined. Then by running one or more batch runs using very low values of C_A, it is possible to correlate the calculated rates with C_A giving the values of k/K^n and m. Therefore, all four parameter values, k, K, m and n, are completely determined. Finally the complete rate expression should be checked using a run in which the concentration of A is significant in comparison to the value of K.

The method of excess may be combined with the method of initial rates. For example, for the reaction of Eq. 3.43 one can start the reaction with initially no C nor D but with a large excess of B over A. Under these conditions $C_{Co} = C_{Do} = 0$ and $C_B = C_{Bo}$ so that a series of initial rate runs with varying C_{Ao} but with a fixed C_{Bo} measures the rate as a function of only A.

EXAMPLE 5.2.2-1

For a reversible reaction, $A + B \overset{\rightarrow}{\leftarrow} C + D$ a series of initial rate studies using excess amounts of reactants has been made. The initial rates were determined by taking the derivative of the concentration-versus-time data at $t = 0$. The results are summarized in Table E5.9. Determine the forward reaction rate expression consistent with the data.

Solution:

For Runs 1 through 4 the initial rate is a function only of C_A since C_{Bo} was held in excess and therefore almost constant. Furthermore, we see that the initial rate is proportional to C_{Ao}. Therefore

$$-r_A \Big|_{t=0} = k'C_A$$

$k' = 3.99$ when $C_{Bo} = 20$ (1)

For Runs 5 through 7 the same procedure yields

$k' = 0.250$ when $C_{Bo} = 5$ (2)

and for Runs 8 through 10

$k' = 0.0101$ when $C_{Bo} = 1$ (3)

Now we need to calculate the true rate constant from the pseudorate constant (k').

TABLE E5.9
Initial Rate Data

Run Number	C_A mol/liter	C_{Bo} mol/liter	$-r_A$ at $t = 0$ mol/(liter·hr)
1	1	20	3.95
2	0.5	20	2.05
3	0.1	20	0.40
4	0.05	20	0.195
5	0.5	5	0.120
6	0.1	5	0.025
7	0.05	5	0.0130
8	0.1	1	0.001
9	0.05	1	0.00049
10	0.07	1	0.00073

In the absence of any information we seek a power law form, i.e., $k' = k\, C_{Bo}^{b}$ where b is an integer or a simple fraction. It can be concluded by inspection that $k' = 0.01\, C_{Bo}^{2}$ so that the forward rate expression may be $-r_A = 0.01\, C_A C_B^2$. Additional experimental data would be needed to verify this.

Finally it should be mentioned again that the differential method of analysis is not restricted to batch reactors because transient CSTR data may also be analyzed by it. Equation 5.3 must be used to calculate the rate, which involves differentiation of the transient reactant concentrations with respect to time.

The differential method requires very accurate experimental data so that the time derivatives of the reactant concentrations may be taken reliably at various points. Therefore, when the data are scattered so badly that the derivatives cannot be taken reliably, one must seek an alternative method of analysis. That method is the integral method.

5.3 INTEGRAL METHOD

The integral method of analysis of batch reactor data requires as a first step the postulation of a particular form for the rate expression. This is substituted into the mass balance equation and integrated to obtain a theoretical equation for the concentration as a function of time which can then be manipulated into a linear form. Finally, the experimental data are plotted according to the linear form, and if a reasonably straight line can be drawn through them, it is assumed that the postulated rate expression is the proper one for modeling the experimental data. If, on the other hand, the experimental data deviate significantly from the predicted linear relationship, another rate expression must be postulated and the procedure repeated until a satisfactory expression is obtained. When a rate expression that can model the experimental data has been obtained, further verifications are necessary to ensure that it is indeed proper.

5.3.1 General Procedure

The procedure outlined above may be summarized as follows. We shall assume for convenience that the reaction involves only one reactant.

Step 1. Make one or more runs and record the reactant concentration as a function of time.

Step 2. Assume a rate expression, r_A, and substitute it into the batch reactor mass balance equation, Eq. 2.25. For example,

$$dC_A/dt = r_A = f_1(C_A) \tag{5.14}$$

Step 3. Analytically integrate Eq. 5.14 to obtain C_A as a function of time,

$$\int_{C_{Ao}}^{C_A} dC_A/f_1(C_A) = t \tag{5.15}$$

Step 4. Rearrange Eq. 5.15 into a linear form. Several examples will be
presented below.

Step 5. Plot the experimental data in the linear form and see if the plot
results in a straight line. If the fit is reasonable, then the assumed rate
expression $f_1(C_A)$ may be taken to be satisfactory. If the fit is poor or
needs to be improved further, another guess of r_A is made, say $f_2(C_A)$, and the
process repeated (Steps 3 through 5) until a good fit is obtained.

The integral method summarized above is easy to use when the rate expressions
are relatively simple so that the integration in Step 3 and the rearrangement in
linear form in Step 4 are straight forward. Therefore, we shall consider simple
elementary reactions first.

5.3.2 Applications of the Integral Method

The integral method of analysis is applied to some commonly occurring reactions
following the steps given above. It is assumed that the batch data have already
been obtained and hence we dispense with Step 1, unless the measurement needed
involves more than one species.

FIRST-ORDER REACTIONS: A → PRODUCTS.

Step 2. Assume $r_A = kC_A$, thus

$$dC_A/dt = -kC_A \qquad (5.16)$$

Step 3. Integrate Eq. 5.16

$$\ln(C_A/C_{Ao}) = -kt \qquad (5.17)$$

Step 4. Put Eq. 5.17 into a linear form

$$\ln(C_A) - \ln(C_{Ao}) = -kt \qquad (5.18)$$

A plot of $\ln(C_A)$ versus time will be linear.

Step 5. Plot the experimental data obtained in Step 1 in the form of $\ln(C_A)$
versus t as shown in Fig. 5.4. If the postulated rate expression is correct,

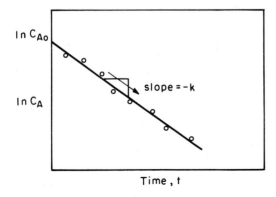

Figure 5.4. Test for an irreversible first-order reaction.

then a straight line will result and the slope will be -k, the first-order reaction rate constant, and the intercept $\ln(C_{Ao})$. The best estimate of k may be obtained by the least squares fit.

SECOND-ORDER REACTIONS: A + B → PRODUCTS.

Step 1. Make batch runs and record the reactant concentrations, C_A and C_B, as functions of time.

Step 2. Assume $r_A = -kC_A C_B$, thus

$$dC_A/dt = -kC_A C_B \qquad\qquad (5.19)$$

This equation cannot be integrated directly unless we can express C_B in terms of C_A. The stoichiometric relationship allows us to do this,

$$dC_A/dt = -kC_A(C_A - C_{Ao} + C_{Bo}) \qquad\qquad (5.20)$$

Step 3. Integrate Eq. 5.20 (see Eq. 4.3)

$$\ln[C_{Ao}(C_A - C_{Ao} + C_{Bo})/C_{Bo}C_A] = (C_{Bo} - C_{Ao})kt$$

or

$$\ln[(C_{Ao}/C_{Bo})(C_B/C_A)] = (C_{Bo} - C_{Ao})kt \qquad\qquad (5.21)$$

Step 4. Put Eq. 5.21 into a linear form by rewriting it

$$\ln(C_B/C_A) = \ln(C_{Bo}/C_{Ao}) + (C_{Bo} - C_{Ao})kt \qquad\qquad (5.22)$$

A plot of $\ln(C_B/C_A)$ versus time will be linear.

Step 5. Plot $\ln(C_B/C_A)$ versus time, as shown in Fig. 5.5. If the postulated rate expression is correct, then a straight line will result and the slope will be $(C_{Bo} - C_{Ao})k$ and the intercept $\ln(C_{Bo}/C_{Ao})$. A number of batch runs with various values of C_{Ao} and C_{Bo}, would further verify the validity of the assumed rate expression.

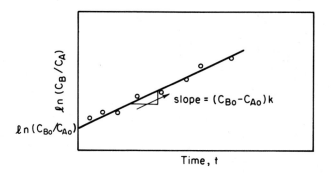

Figure 5.5. Test for a bimolecular reaction.

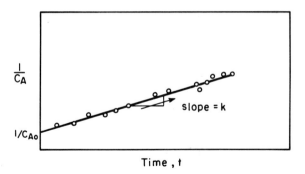

Figure 5.6.   Test for the reaction 2A → Product.

When $C_{Bo} = C_{Ao}$, the reaction is equivalent to A + A → Products and the integration step leads to

$$1/C_A = 1/C_{Ao} + kt \tag{5.23}$$

According to which $1/C_A$ versus t should be a straight line with the slope equal to k and the intercept $1/C_{Ao}$.   This is shown in Fig. 5.6.

ZERO-ORDER REACTIONS.

Step 2.   Assume $r_A = -k$, thus

$$dC_A/dt = -k \tag{5.24}$$

Step 3.   Integrate Eq. 5.24

$$C_A - C_{Ao} = -kt \tag{5.25}$$

Step 4.   Rearrange into a linear form

$$C_A = C_{Ao} - kt \tag{5.26}$$

A plot of C_A versus time will be linear

Step 5.   The plot of C_A versus t will result in a straight line with the slope equal to -k and the intercept C_{Ao}.   Note that the reaction time is restricted to less than C_{Ao}/k.   This is illustrated in Fig. 5.7.

N^{TH}*-ORDER REACTIONS.*

Step 2.   Assume $r_A = -kC_A^n$ $(n \neq 1)$, thus

$$dC_A/dt = -kC_A^n \tag{5.27}$$

Step 3.   Integrate Eq. 5.27

$$C_A^{(1-n)} - C_{Ao}^{(1-n)} = -(1 - n)kt \tag{5.28}$$

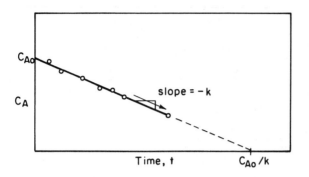

Figure 5.7. Test for a zero order reaction.

Step 4. Rearrange into a linear form. When the order of reaction, n, is known *a priori* or assumed,

$$C_A^{(1-n)} = C_{Ao}^{(1-n)} + (n - 1)kt \tag{5.29}$$

a plot of $C_A^{(1-n)}$ (with the assumed value of n) versus t will result in a straight line with the slope equal to $(n - 1)k$ and the intercept $C_{Ao}^{(1-n)}$ if the assumed rate expression with the specific value of n is correct. This is illustrated in Fig. 5.8.

If the assumed value of n is incorrect, the process must be repeated with a number of different values of n until a satisfactory fit is obtained. A better approach is to use a special technique, known as the method of half-life, which will be discussed later.

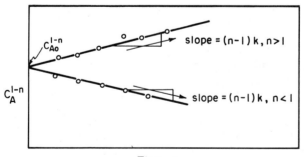

Figure 5.8. Test for an nth order reaction.

$$REVERSIBLE\ FIRST\text{-}ORDER\ REACTIONS: \quad A \underset{k_{-1}}{\overset{k_1}{\rightleftharpoons}} B$$

Step 1. Make batch runs with A and B until equilibrium is reached (i.e., the concentrations become constant) and record the concentrations as functions of time.

Step 2. Assume $r_A = -k_1 C_A + k_{-1} C_B = -(k_1 + k_{-1}) C_A + k_{-1}(C_{Ao} + C_{Bo})$,

thus

$$dC_A/dt = -(k_1 + k_{-1}) C_A + k_{-1}(C_{Ao} + C_{Bo}) \tag{5.30}$$

Step 3. Integrate Eq. 5.30 to obtain (see Eq. 4.15)

$$\ln\left(\frac{C_{Ao} - C_{Ae}}{C_A - C_{Ae}}\right) = (k_1 + k_{-1})t \tag{5.31}$$

where C_{Ae} is the equilibrium concentration of A which is determined experimentally in Step 1.

Step 4. Rearrange Eq. 5.31 to obtain a linear form

$$\ln(C_A - C_{Ae}) = \ln(C_{Ao} - C_{Ae}) - (k_1 + k_{-1})t \tag{5.32}$$

A plot of $\ln(C_A - C_{Ae})$ versus t will be linear.

Step 5. The plot of $\ln(C_A - C_{Ae})$ versus t will be a straight line with the slope equal to $-(k_1 + k_{-1})$ and an intercept of $\ln(C_{Ao} - C_{Ae})$. This is illustrated in Fig. 5.9. When $C_{Bo} = 0$

$$(C_{Ao} - C_{Ae})/C_{Ae} = k_1/k_{-1} \tag{3.41}$$

$$k_1/k_{-1} = (C_{Ao}/C_{Ae}) - 1 \tag{5.33}$$

and

$$slope = -(k_1 + k_{-1}) \tag{5.34}$$

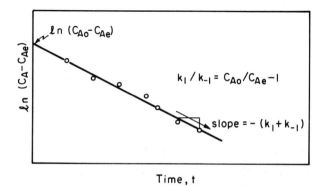

Figure 5.9. Test for a reversible first-order reaction, $A \rightleftharpoons B$.

then Eqs. 5.33 and 5.34 may be solved simultaneously to yield k_1 and k_{-1}. Finally, a run should be made with both A and B and the resulting concentration profiles should be tested with the model to confirm its validity.

PARALLEL FIRST-ORDER REACTIONS. $A \underset{k_2}{\overset{k_1}{\diagup}} \begin{smallmatrix} B \\ C \end{smallmatrix}$

Step 1. Make one or more runs and record C_A, C_B, and C_C as functions of time.

Step 2. Assume $r_B = k_1 C_A$ and $r_C = k_2 C_A$, thus

$$dC_A/dt = -(k_1 + k_2)C_A \tag{5.35}$$

Step 3. Integrate to obtain

$$\ln(C_{Ao}/C_A) = (k_1 + k_2)t \tag{5.36}$$

Step 4. Rearrange Eq. 5.36 into a linear form

$$\ln C_A = \ln C_{Ao} - (k_1 + k_2)t \tag{5.37}$$

A plot of $\ln C_A$ versus t will be linear.

Step 5. The plot of $\ln C_A$ versus time will result in a straight line with slope equal to $-(k_1 + k_2)$ if the assumed rates are correct. This is illustrated in Fig. 5.10. The individual rate constants are determined as follows:

$$r_B/r_C = (dC_B/dt)/(dC_C/dt) = k_1/k_2 \tag{5.38}$$

so that

$$dC_B = (k_1/k_2)dC_C$$

or

$$C_B - C_{Bo} = (k_1/k_2)(C_C - C_{Co}) \tag{5.40}$$

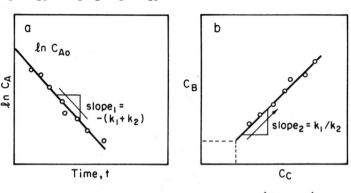

Figure 5.10. Test for parallel reactions, $A \overset{k_1}{\to} B$; $A \overset{k_2}{\to} C$.
(a) $\ln C_A$ versus time; (b) C_B versus C_C.

Therefore, the plot of C_B versus C_C will result in a straight line with slope equal to k_1/k_2, if the assumed rate expressions are correct. Thus we have $-(k_1 + k_2) = slope_1$ and $k_1/k_2 = slope_2$ which may be solved simultaneously to obtain k_1 and k_2.

AUTOCATALYTIC REACTIONS: A + B → 2B + C

Step 1.   Make batch runs and record C_A and C_B as functions of time.

Step 2.   Assume $r_A = -kC_A C_B = -kC_A(C_{Ao} + C_{Bo} - C_A)$, thus   (3.33)

$$dC_A/dt = -kC_A(C_{Ao} + C_{Bo} - C_A) \tag{5.41}$$

Step 3.   Integrate Eq. 5.41

$$\ln[C_{Ao}(C_{Ao} + C_{Bo} - C_A)/C_{Bo}C_A] = (C_{Ao} + C_{Bo})kt \tag{5.42}$$

Step 4.   Put Eq. 5.42 into a linear form

$$\ln[(C_{Ao} + C_{Bo} - C_A)/C_A] = -\ln(C_{Ao}/C_{Bo}) + (C_{Ao} + C_{Bo})kt \tag{5.43}$$

A plot of $\ln[(C_{Ao} + C_{Bo} - C_A)/C_A]$ or $\ln(C_B/C_A)$ versus t will be linear.

Step 5.   Plot the experimental data obtained in Step 1 in the form of $\ln(C_B/C_A)$ versus t, as shown in Fig. 5.11. If the postulated rate expression is correct, then a straight line will result and the slope will be equal to $(C_{Ao} + C_{Bo})k$.

ENZYME REACTIONS: $E + S \underset{k_{-1}}{\overset{k_1}{\underset{\leftarrow}{\rightarrow}}} ES \overset{k_2}{\rightarrow} E + P$

Step 1.   Make a batch run with a fixed amount of enzyme, C_{Eo}, and substrate, C_{So}. Record the concentration of substrate as a function of time. The enzyme is assumed stable so that $C_E = C_{Eo}$. Thus, the reaction time must be short enough to ensure $C_E \tilde{=} C_{Eo}$.

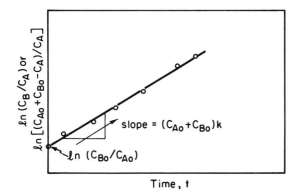

Figure 5.11.   Test for an autocatalytic reaction,
              A + B → 2B + C.

Step 2. Assume

$$-r_S = V_m C_S / (K_m + C_S) \qquad\qquad (3.81)$$

Then

$$dC_S/dt = -V_m C_S / (K_m + C_S) \qquad\qquad (5.44)$$

Step 3. Integrate Eq. 5.44, giving

$$\ln(C_{So}/C_S) + (C_{So} - C_S)/K_m = (V_m/K_m)t \qquad\qquad (5.45)$$

Step 4. Put Eq. 5.45 into a linear form

$$\ln(C_{So}/C_S)/(C_{So} - C_S) = -1/K_m + [V_m/K_m(C_{So} - C_S)]t \qquad\qquad (5.46)$$

A plot of $\ln (C_{So}/C_S)/(C_{So} - C_S)$ versus $t/(C_{So} - C_S)$ will be linear.

Step 5. Plot the experimental data in the form of $\ln (C_{So}/C_S)/(C_{So} - C_S)$ versus $t/(C_{So} - C_S)$ as shown in Fig. 5.12. If the postulated rate expression is correct, then a straight line will result and the slope will be V_m/K_m and the intercept $-1/K_m$. Since $V_m = k_2 C_{Eo}$, additional batch runs should be made by varying the enzyme concentration, C_{Eo}, to verify that V_m is proportional to C_{Eo} and thus determine the value of k_2.

5.3.3 Specialized Techniques

Sometimes specialized techniques can be employed in conjunction with the integral method of analysis, thereby simplifying the analyses which must be performed for complicated situations. Two of these are the fractional-life method and the method of excess.

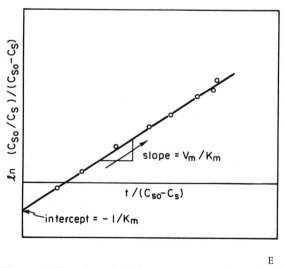

Figure 5.12. Test for the enzyme reaction, $S \xrightarrow{E} P$.

FRACTIONAL-LIFE METHOD. The fractional-life method makes use of the time, t_f, required for the concentration of a reactant in a batch reactor to drop to some arbitrary fraction, f, of its initial value. This time may then be plotted as a function of the initial reactant concentration, C_{Ao}, to estimate both the order of the reaction, n, and the reaction rate constant, k. Consider an nth order reaction. Substitution of $t = t_f$ and $C_A = fC_{Ao}$ into Eq. 5.29 and rearrangement yield:

$$t_f = [(f^{(1-n)} - 1)/(n - 1)k]C_{Ao}^{(1-n)} \tag{5.47}$$

Taking the natural log of both sides we obtain:

$$\ln(t_f) = \ln[(f^{(1-n)} - 1)/(n - 1)k] + (1 - n)\ln(C_{Ao}) \tag{5.48}$$

Thus a plot of $\ln(t_f)$ versus $\ln(C_{Ao})$ will yield a straight line with a slope equal to 1 - n and an intercept of $\ln[(f^{(1-n)} - 1)/(n - 1)k]$, thereby allowing both n and k to be determined from a single plot. Perhaps the most common fraction used is 0.5, i.e., f = 0.5, in which case the time is called the half-life. Figure 5.13 is an illustration of the plot obtained using it.

METHOD OF EXCESS. The use of this method with the differential method of analysis was discussed in detail earlier. Now let us look at its use with the integral method. As before, the basic idea is to supply all reactants except one in excess so that the reaction rate becomes a function of only that one reactant. This greatly simplifies the integration of the mass balance equation. For example, if the rate expression of Eq. 5.10 is substituted into the mass balance equation, Eq. 2.25, we obtain

$$dC_A/dt = -kC_A^a C_B^b C_C^c \tag{5.49}$$

If the normal integral technique were used, it would be necessary to express C_B and C_C in terms of C_A using stoichiometric relationships. Even then, it might be

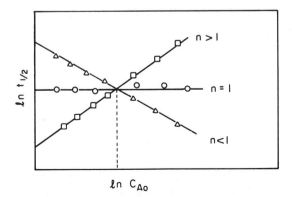

Figure 5.13.   Half-life method of determining
reaction order and rate constant.

difficult or impossible to integrate Eq. 5.49 unless the constants a, b and c are
simple integers. Now, let us suppose that we start the reaction with excess amounts
of B and C so that the amounts of them reacted during the course of the experiment
are negligibly small in comparison to the initial amounts, i.e., $C_B \cong C_{Bo}$ and
$C_C \cong C_{Co}$. Then, Eq. 5.49 reduces to a one-reactant reaction,

$$dC_A/dt = (-kC_{Bo}^b C_{Co}^c)C_A^a = -k'C_A^a \qquad\qquad (5.50)$$

which can be integrated readily. A series of runs should be made with different
values of C_{Bo} and C_{Co} so that the rate constant, k, can be recovered from one
pseudorate constant, k'. Batch runs should also be made using different combinations
of reactants in excess (i.e., C_A and C_B, as well as C_A and C_C) in order to test the
complete rate expression.

The integral method of analysis may be also applied to transient CSTR data.
For example, the unsteady-state mass balance equation for a first-order reaction in
a CSTR with space time τ is (see Eq. 5.2)

$$dC_A/dt + [(1/\tau) + k]C_A = (1/\tau)C_{Ao} \qquad\qquad (5.51)$$

Let us suppose that the reactor was initially filled with water so that $C_A(0) = 0$.
Then at time t = 0 the feed was changed to contain C_{Ao} and the value of C_A was
measured with time. Integration of Eq. 5.51 yields

$$C_A = \{(1/\tau)C_{Ao}/[(1/\tau) + k]\}\{1 - \exp[-[(1/\tau) + k]t]\} \qquad\qquad (5.52)$$

$$C_A = C_{A\infty}\{1 - \exp[-[(1/\tau) + k]t]\} \qquad\qquad (5.53)$$

or

$$\ln(1 - C_A/C_{A\infty}) = -[(1/\tau) + k]t \qquad\qquad (5.54)$$

in which $C_{A\infty}$ represents the ultimate steady-state value of C_A. According to
Eq. 5.54 a plot of $\ln(1 - C_A/C_{A\infty})$ versus time will give a straight line with the
slope equal to $-[(1/\tau) + k]$, if the reaction is first order. Since τ is known,
the rate constant k, may be determined from the slope of the line.

In summary, the integral method may be used to check for (or eliminate) simple
rate expressions or simple mechanisms. It also finds use when the experimental data
are scattered so that it is difficult to evaluate the rates reliably by differentiation. The differential method may be used to develop progressively a rate expression for a complex reaction, but it requires accurate experimental data that can
be differentiated reliably. In general, one would try the integral method first
because at worst it would allow some simple rate expressions to be eliminated. If
it is unsuccessful, the differential method should be tried. For complicated
reactions it may be necessary to use the specialized techniques to gain partial
asymptotic behaviors thereby allowing a complete rate expression to be built up

progressively. Once a reasonable fit is obtained, the proposed rate expression should be subjected to further experimental testing and if necessary, modifications should be made. This cyclic process must continue until the proposed rate expression has been verified thoroughly.

Finally it should be mentioned that even though all of the methods considered above used manipulations to force the parameters to appear linearly in the final form, we do not have to have linear parameters before we can evaluate the proposed model. There are methods of dealing with nonlinear parameters. However, a coverage of them is beyond the scope of this text and interested readers are encouraged to refer to literature in this area [1,2,3,4].

5.4 KEY POINTS

1. The algebraic method of analysis involves algebraic calculation of the rates of consumption of reactant from the steady-state CSTR mass balance equation. The calculated rates are correlated with the reactant concentrations to obtain the rate expression. The procedure involves assuming a rate expression and testing it against experimental data. If the fit is satisfactory, the parameters (the rate constants, the reaction order, etc.) are usually estimated by the method of least squares.

2. In the differential method of analysis of batch reactor data the reactant concentration profile is differentiated to obtain the reaction rates, which are then correlated with the reactant concentrations to obtain the rate expression.

3. For multiple reactions, reversible reactions and nonelementary reactions the method of initial rates may be used with the differential method to test the asymptotic behavior of a proposed rate expression. Only the initial rates are used so as to eliminate any complications that may arise with time.

4. The method of excess may be used with the differential method, in which a reaction may be carried out in the presence of large excess of one or more reactants so that their concentrations either remain practically constant or so large that the proposed rate expression reduces to a simpler form.

5. The differential method of analysis is not restricted to batch reactors. Transient CSTR data may be also analyzed by it using Eq. 5.3 to calculate the rates.

6. Very accurate experimental data are required for the differential method so that the time derivatives of the reactant concentrations may be taken reliably at various points. Otherwise, the integral method must be used.

7. The integral method of analysis of batch reactor data requires the postulation of a particular form for the rate expression which is substituted into the mass balance equation and integrated to obtain a theoretical equation for the concentration as a function of time. This equation is then manipulated into a linear form and the experimental data are plotted according to the linear form. If a reasonably straight line can be drawn through the points, the postulated rate expression is assumed to be the proper one for modeling the experimental data.

8. The fractional-life method makes use of the time, t_f, required for the concentration of a reactant in a batch reactor to drop to some arbitrary fraction of its initial value. This time may then be plotted as a function of the initial reactant concentration to estimate both the order of the reaction and the reaction rate constants.

9. The method of excess may be used with the integral method by supplying all reactants except one in excess so that the reaction rate becomes a function of only that one reactant, greatly simplifying the integration of the mass balance equation.

10. The integral method may be used to check for (or eliminate) simple rate expressions or simple mechanisms. It also finds use when the experimental data are scattered so that it is difficult to evaluate the rates reliably by differentiation.

11. The differential method may be used to develop progressively a rate expression for a complex reaction, but it requires accurate experimental data that can be differentiated reliably.

5.5 STUDY QUESTIONS

The study questions are arranged in the same order as the text:

Section	Questions
5.1	1-4
5.2	5-9
5.3	10-15

1. A homogeneous liquid phase reaction is studied in a CSTR. The results in Table SQ5.1 are obtained in steady-state runs with $C_{Ao} = 2.0$ mol/liter. Find the rate expression for this reaction, r_A.

TABLE SQ5.1
Data from a CSTR

τ min	C_A mol/liter
0.05	1.3
0.10	1.1
0.20	0.67
0.30	0.50
0.50	0.33
1.00	0.18
2.00	0.095

2. From the data given in Table SQ5.2 find a satisfactory rate expression for the liquid-phase reaction $A \rightarrow B$ taking place isothermally in a CSTR receiving feed with $C_{Ao} = 1.0$ mol/liter.

TABLE SQ5.2
Data from a CSTR

τ min	C_A mol/liter
0.5	0.73
1.0	0.62
5.0	0.35
10.0	0.27
20.0	0.20
50.0	0.13

3. An aqueous reaction is expected to follow the rate expression given by
    $r_A = -kC_A/(K + C_A)$. The experimental data are obtained using a one-liter CSTR
    receiving feed with $C_{Ao} = 1.0$ mol/liter and are given in Table SQ5.3. Show
    that the proposed rate expression can satisfactorily fit the data and determine
    the rate constants, k and K, by the least squares fit.

TABLE SQ5.3
Data from a CSTR

F liter/hr	C_A mol/liter
10	0.95
2	0.78
1	0.62
0.5	0.41
0.2	0.19
0.1	0.10

4. Find the rate expression for the aqueous reaction A + B → C from the results,
    given in Table SQ5.4, of steady-state runs made in a CSTR.

TABLE SQ5.4
Data from a CSTR

τ hr	C_{Ao} mol/liter	C_{Bo} mol/liter	C_A mol/liter	C_B mol/liter
0.1	1	2	0.83	1.83
0.5	1	2	0.53	1.53
1	1	2	0.39	1.39
5	1	2	0.13	1.13
0.2	1	1	0.84	0.84
1	1	1	0.59	0.59
5	1	1	0.34	0.34
10	1	1	0.26	0.26

5. The experimental data given in Table SQ5.5 are obtained at 30°C in a constant
    volume batch reactor. Find the rate expression which fits the data using the
    differential method.

TABLE SQ5.5
Data from a Batch Reactor

t min	C_A mol/liter
0	1.0
6	0.95
12	0.91
18	0.87
24	0.83
30	0.80
36	0.77
42	0.74
48	0.71

6. Table SQ5.6 gives the data obtained in a constant volume batch reactor for the aqueous decomposition of A. Develop the rate expression.

TABLE SQ5.6
Data from a Batch Reactor

t	C_A	$-dC_A/dt$
min	mol/liter	mol/(liter·min)
0	2.0	0.25
5	1.25	0.092
10	0.90	0.049
15	0.72	0.030
20	0.58	0.020
30	0.44	0.011
50	0.28	0.0048

7. A series of short-time runs using various substrate concentrations has been run in a batch reactor and the data are given in Table SQ5.7. The rate expression is suspected to be in the form of the Michaelis-Menten equation. Develop the rate expression consistent with the experimental data.

TABLE SQ5.7
Data from a Batch Reactor

t	C_A	t	C_A
min	mol/liter	min	mol/liter
0	2.000	0	0.500
5	1.991	5	0.494
10	1.982	10	0.488
15	1.973	15	0.483
0	1.000	0	0.1000
5	0.993	5	0.0982
10	0.985	10	0.0964
15	0.977	15	0.0942

8. A homogeneous reaction has been studied in a constant volume batch reactor and the results are given in Table SQ5.8. Determine whether these data can be reasonably fitted by the rate expression

$$-r_A = kC_A^a/(K + C_A)^b$$

If the fit is reasonable, evaluate the constants, a, b, k and K.

9. A homogeneous reaction A → B was studied in a CSTR under transient-state conditions. The reactor was receiving feed with C_{Ao} = 1.1 mol/liter at a flow rate of 1.0 liter/hr. The experimental data as well as the calculated time derivatives of reactant concentration are given in Table SQ5.9. Develop a rate expression consistent with the data.

10. A batch reactor was used to study a homogeneous reaction and the results are given in Table SQ5.10. Using the integral method, develop a rate expression consistent with the data.

TABLE SQ5.8
Data from a Batch Reactor for Study Question 8

t min	C_A mol/liter	$-r_A$ mol/(liter·min)
0	100	0.253
2	99.5	0.252
5	98.7	0.249
10	97.5	0.246
20	95.0	0.240
0	0.01	0.232×10^{-6}
2	0.00995	0.229×10^{-6}
5	0.00989	0.226×10^{-6}
10	0.00977	0.221×10^{-6}
20	0.00955	0.211×10^{-6}

TABLE SQ5.9
Transient-State Data from a CSTR for Study Question 9

t hr	C_A mol/liter	dC_A/dt mol/(liter·hr)
0	0	–
0.1	0.104	0.984
0.3	0.281	0.791
0.5	0.423	0.634
1.0	0.667	0.367
1.5	0.808	0.210
2.0	0.889	0.123
3.0	0.963	0.0405
4.0	0.987	0.0136
10.0	1.000	–

TABLE SQ5.10
Data from a Batch Reactor for Study Question 10

t min	C_A mol/liter
0	0.85
1	0.74
2	0.66
3	0.59
5	0.50
10	0.35
20	0.22
50	0.10

11. A reaction involving two reactants has been studied in a one liter batch
     reactor and the results are summarized in Table SQ5.11. Develop the rate
     expression.

12. A reversible reaction $A \underset{\leftarrow}{\rightarrow} B$ has been studied in a batch reactor and the time
     history is given in Table SQ5.12. Using the integral method of analysis
     develop a rate expression consistent with the data.

TABLE SQ5.11
Data from a Batch Reactor for Study Question 11

t min	C_A mol/liter	C_B mol/liter
0	1	2
1	0.98	1.98
5	0.90	1.90
10	0.83	1.82
20	0.69	1.68
30	0.59	1.60
60	0.38	1.39
100	0.23	1.24
200	0.07	1.07
∞	0	1.00

TABLE SQ5.12
Data from a Batch Reactor for Study Question 12

t hr	C_A mol/liter
0	1
0.1	0.90
0.2	0.81
0.3	0.74
0.5	0.66
1.0	0.54
3.0	0.50
10	0.50
20	0.50

13. Given in Table SQ5.13 are the experimental data for the decomposition of species
     A in a constant volume batch reactor. Determine the order of reaction and the
     rate constant.

TABLE SQ5.13
Data from a Batch Reactor

t min	C_A mol/liter
0	1
0.1	0.95
0.2	0.90
0.5	0.81
1.0	0.70
3.0	0.50
10.0	0.30
50.0	0.14
100.0	0.10

14. For reactions whose rate expressions are given by the Michaelis-Menten equation, $-r_A = kC_A/(K + C_A)$, develop the fractional life method including the equation corresponding to Eq. 5.48 and the type of plot necessary to determine the rate constants k and K.

15. Determine the order of reaction and the rate constant for a reaction, the characteristics of which are summarized in Table SQ5.14.

TABLE SQ5.14
Data from a Batch Reactor

Run #1		Run #2		Run #3		Run #4	
t	C_A	t	C_A	t	C_A	t	C_A
hr	mol/liter	hr	mol/liter	hr	mol/liter	hr	mol/liter
0	0.1	0	0.5	0	2.00	0	4.0
6.7	0.05	1.3	0.25	0.32	1.00	0.17	2.0
13.3	0.033	2.7	0.17	0.64	0.67	0.33	1.3
20.0	0.025	4.0	0.13	1.00	0.50	0.50	1.0

REFERENCES AND FURTHER READING

1. H. O. Hartley, "The modified Gauss-Newton method for the fitting of nonlinear regression functions by least squares," *Technometrics,* 3, 269-280, 1961.

2. D. L. Marquardt, "An algorithm for least squares estimation of nonlinear parameters," *Journal of the Society of Industrial and Applied Mathematics,* 2, 431-441, 1964.

3. V. Pereyra, "Iterative methods for solving nonlinear least squares problems," *Society of Industrial and Applied Mathematics Journal, Numerical Analysis,* 4, 27, 1967.

4. D. M. Himmelblau, *Process Analysis by Statistical Methods,* John Wiley & Sons, Inc., New York, N.Y., 176-200, 1970.

Further Reading

Bamford, C. H. and Tipper, C. F. H. editors, *The Theory of Kinetics,* Vol. 2, Elsevier Publishing Co., Amsterdam, 1969. See Chapter 1 for integration of complex reactions.

Benson, S. W., *The Foundations of Chemical Kinetics,* McGraw-Hill Book Co., Inc., New York, N.Y., 1960. See Chapters 2, 3, and 4 for simple kinetics, complex kinetics, and reaction orders and rate constants, respectively.

Capellos, C. and Bielski, B. H. J., *Kinetic Systems,* Wiley-Interscience, New York, N.Y., 1972. This gives many integrated equations for various reactions.

Carberry, J. J., *Chemical and Catalytic Reaction Engineering,* McGraw-Hill Book Co., Inc., New York, N.Y., 1976. See Chapter 2.

Fogler, H. S., *The Elements of Chemical Kinetics and Reactor Calculations (A Self-Paced Approach),* Prentice-Hall Inc., Englewood Cliffs, N.J., 1974. See Chapter 5.

Frost, A. A. and Pearson, R. G., *Kinetics and Mechanisms,* 2nd, John Wiley & Sons,
    Inc., New York, N.Y., 1961. See Chapter 8.

Levenspiel, O., *Chemical Reaction Engineering,* 2nd ed., John Wiley & Sons, Inc.,
    New York, N.Y., 1972. See Chapter 3.

Smith, J. M., *Chemical Engineering Kinetics,* 2nd ed., McGraw-Hill, Inc., New York,
    N.Y., 1970. See Chapter 2.

CHAPTER 6

REACTORS WITH NONIDEAL FLOW PATTERNS

In previous chapters we considered only reactors with idealized flow patterns, i.e., perfect CSTR's and PFR's. Real reactors often do not follow these idealized flow patterns because of imperfect mixing (which creates short circuiting and stagnant regions), channeling, and recycling of fluid. In these situations the deviation from ideality can be considerable and may vary widely depending upon reactor size. Even though in some cases it is possible to approximate the flow pattern as ideal, in other cases the use of an ideal flow pattern for design and scale-up may result in large error. Thus it is important that an engineer be able to evaluate the flow pattern in a reactor, and the major objective of this chapter is to provide the means for doing so. A secondary objective is to introduce the reader to the techniques for predicting the behavior of reactors with nonideal flow patterns.

6.1 FLOW OF TRACER THROUGH IDEAL REACTORS

The major technique for analyzing flow patterns involves introducing a tracer input into the fluid entering a reactor and observing the tracer output signal in the fluid leaving it. Any inert material that can be detected quantitatively and which does not disturb the flow pattern in the reactor can be used as a tracer. In addition, any type of input signal may be used although, in practice, step and impulse signals are used most frequently because they lead to simple analyses. A unit step function is defined by

$$S(t - T) = \{ \begin{matrix} 0 & t < T \\ 1 & t > T \end{matrix} \qquad\qquad (6.1)$$

and is shown in Fig. 6.1. Thus, $S(t - T)$ is constant for both $t < T$ and $t > T$ and has a unit step discontinuity at $t = T$; e.g., $S(t)$ is zero for all $t < 0$ and unity

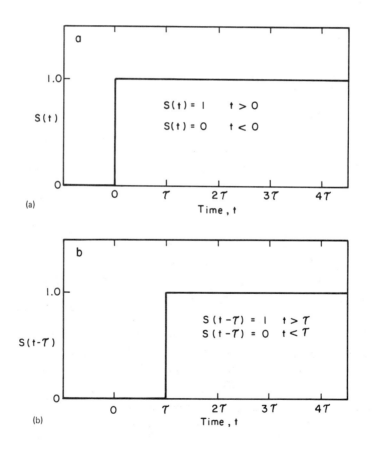

Figure 6.1. (a) A step function; (b) A delayed step function.

for all t > 0. The unit impulse function, or the Dirac delta function, is defined
by

$$\delta(t - T) = 0, \qquad t \neq T$$

$$\int_{-\infty}^{\infty} \delta(t - T)dt = 1 \qquad\qquad\qquad\qquad\qquad\qquad\qquad (6.2)$$

and is depicted in Fig. 6.2. This represents a mathematical idealization of a sudden
"jolt" of input to the system and it assumes that the "jolt" occurs instantaneously
in zero time. To understand this input better, think of a suddenly applied input of
constant magnitude which acts for a finite interval before ceasing suddenly, as shown
in Fig. 6.2b. Furthermore, let the magnitude and duration of the input be such that
their product is unity. Now, if we let the magnitude become very, very large and the
duration infinitesimally small, while keeping the product of duration and magnitude
equal to unity, the result is a unit impulse function.

These two functions belong to a class known as "generalized functions" which
possess properties useful in process analysis and mathematical solutions [1]. One
of these properties is that the unit impulse function is the derivative of the unit
step function [1].

$$\delta(t - T) = \frac{d}{dt}S(t - T) \qquad\qquad\qquad\qquad\qquad\qquad (6.3)$$

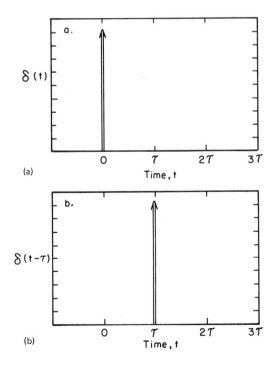

Figure 6.2.   (a) A Dirac delta function;
              (b) A delayed Dirac delta function.

An important consequence of this property is that the response of a linear system to a unit impulse input is the derivative of the response to a unit step input [2]. This allows us to recover the impulse response from a step response or vice versa. The expected responses of a CSTR or PFR to step and impulse inputs of tracers are developed below. Only constant density fluids under isothermal conditions are considered in this chapter.

6.1.1 Ideal PFR

The assumptions made in the definition of "plug flow" (Section 2.3.3) allow us to deduce the flow patterns through a PFR, which are quite simple as shown in Fig. 6.3. A constant flow of water is going through the reactor and a pulse of dye is added instantaneously (an impulse) to the feed stream as indicated in Fig. 6.3a. The response as measured by the concentration of dye in the fluid leaving the PFR will be that shown in Fig. 6.3b, i.e., after a delay of one space time, τ, the pulse will come out of the reactor in the same shape as it went in. The response of a PFR to a step change of the feed from clear water to a dye (with concentration of C_o) flowing at the same rate (Fig. 6.3c) is shown in Fig. 6.3d. Again, no dye will appear until one space time has passed, at which time the output will change instantaneously to the new condition, i.e., a step function with concentration of C_o. This response of a perfect PFR is not limited to an impulse or step input; no matter what the flow pattern in the feed, the effluent pattern will repeat it exactly after a time delay of one space time. In other words, a PFR acts as a time delay of one space time for unreactive substances. Mathematically, the mass balance equation for the tracer is obtained from Eq. 2.30

$$F \frac{\partial C}{\partial x} + A_c \frac{\partial C}{\partial t} = 0 \qquad (6.4)$$

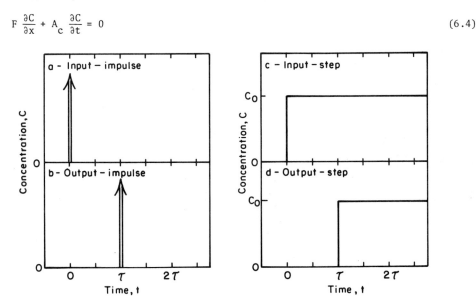

Figure 6.3.   Response of a perfect PFR to a Dirac delta (impulse) and a step input: (a) and (b) Impulse response; (c) and (d) Step response.

with the boundary and initial conditions given by

$$C (0,t) = C_o S(t) \text{ or } C_o \delta(t)$$

$$C (x,0) = 0 \tag{6.5}$$

The solution to these equations provides the flow patterns discussed above.

6.1.2 Ideal CSTR

The response of an ideal CSTR to a step or an impulse input of tracer is a little more complicated than the response of the ideal PFR. Therefore, we will write a mass balance on dye and solve it to obtain the flow patterns of an ideal CSTR. Consider a CSTR with volume, V, receiving a flow of tracer solution at constant flow rate, F. The unsteady-state mass balance equation for the nonreactive tracer is obtained from Eq. 2.17

$$FC_o - FC = VdC/dt \tag{6.6}$$

or

$$dC/dt + C/\tau = C_o/\tau \tag{6.7}$$

where C_o and C are the influent and effluent tracer concentrations. In the case of a step input the tracer concentration in the feed is zero initially, $C(0) = 0$, and then jumps to C_o and stays at that value. In response to this step input, the effluent tracer concentration slowly increases to C_o. Equation 6.7 is a first-order linear ordinary differential equation and hence can be integrated to yield:

$$C/C_o = 1 - \exp(-t/\tau) \tag{6.8}$$

The response represented by this equation is shown in Fig. 6.4b in which it is seen

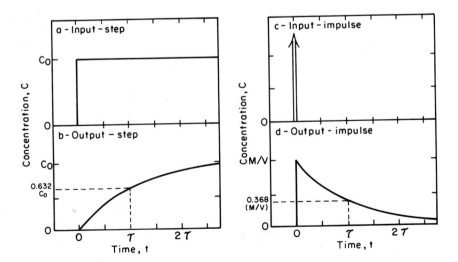

Figure 6.4.   Response of a perfect CSTR to a step and a Dirac delta (impulse) input: (a) and (b) Step response; (c) and (d) Impulse response.

that the effluent tracer concentration starts at zero and asymptotically approaches the ultimate value C_o. Equation 6.8 is a typical first-order system response to a step input and suggests that a plot of $\ln(1 - C/C_o)$ versus time for a CSTR should yield a straight line with slope equal to $-1/\tau$. This provides a simple means for checking for complete mixing. After one space time, that is at $t = \tau$, the effluent tracer concentration should be 63.2% of the ultimate value.

The response to an impulse input at $\tau = 0$ (Fig. 6.4c) is formulated by setting FC_o in Eq. 6.6 to be an impulse of mass M of the tracer introduced

$$M\delta(t) - FC = VdC/dt \tag{6.9}$$

The initial condition ($t = 0$) is zero. Equation 6.9 is a first-order linear ordinary differential equation whose solution is:

$$C = (M/V)\exp(-t/\tau) \tag{6.10}$$

The response given by Eq. 6.10 rises immediately to the concentration of M/V and decays exponentially as shown in Fig. 6.4d. A plot of $\ln(C)$ versus time should result in a straight line with the slope equal to $-1/\tau$. Furthermore, after one space time the effluent tracer concentration should be 36.8% of the initial tracer concentration (M/V). This also provides a simple method for determining whether a reactor approaches complete mixing. Finally, it should be noted that the impulse response, Eq. 6.10, may be obtained by differentiating the step response, Eq. 6.8, provided that the mass of tracer introduced, M, is equal to FC_o.

6.1.3 Ideal CSTR's in Series

As discussed in Section 4.2.2 it is sometimes advantageous to use a number of CSTR's in series rather than one large CSTR. Furthermore, a number of CSTR's in series can sometimes approximate one reactor with imperfect mixing. Therefore, it would be helpful to develop the expected flow patterns for a series of CSTR's. Consider N CSTR's of equal volume, V, ($V = V_T/N$) with flow F as shown in Fig. 4.9. At time zero the feed to the first tank is switched to one with a tracer concentration equal to C_o. The response from the first tank (given by Eq. 6.8) is the feed to the second tank. Hence, the following differential equation describes the transient tracer concentration in tank 2.

$$FC_o[1 - \exp(-t/\tau)] - FC_2 = V_2 dC_2/dt \tag{6.11}$$

or

$$dC_2/dt + C_2/\tau = (C_o/\tau)[1 - \exp(-t/\tau)] \qquad C_2(0) = 0 \tag{6.12}$$

where $\tau = V/F$ and C_2 is the tracer concentration leaving tank 2. Solving Eq. 6.12 yields:

$$C_2/C_o = 1 - (1 + t/\tau)\exp(-t/\tau) \tag{6.13}$$

By repeating this process for tanks $3, 4, \ldots, N$ we can obtain the following expression

$$C_N/C_0 = 1 - [1 + t/\tau + (t/\tau)^2/2! + \ldots + (t/\tau)^{N-1}/(N-1)!] \exp(-t/\tau) \qquad (6.14)$$

Tracer concentration profiles for various numbers of CSTR's in series are given in Fig. 6.5. There it can be seen that as the number of CSTR's in series is increased, the response due to a step input approaches that of a PFR. Indeed, the response of an infinite number of CSTR's is equivalent to that of a PFR. Thus, the step response resulting from a series of CSTR's lies somewhere between those of a single CSTR and a PFR. This implies that a real reactor which has the response of neither a CSTR nor a PFR may be simulated as N CSTR's in series. Hence, Eq. 6.14 may be used for two purposes; to predict the step response of N ideal CSTR's in series or to model the response of a real reactor with nonideal mixing.

The impulse response for N CSTR's in series may be obtained by differentiating Eq. 6.14. For equal volume reactors and the initial mass of tracer, M, introduced as an impulse

$$C_N = (M/V) [(t/\tau)^{N-1}/(N-1)!] \exp(-t/\tau) \qquad (6.15)$$

The impulse responses represented by Eq. 6.15 for various numbers of CSTR's in series

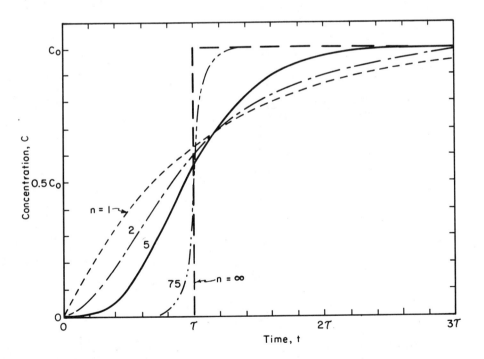

Figure 6.5. Responses of n CSTR's in series to a step input.

are shown in Fig. 6.6. There it can be seen that the addition of a second tank in series reduces the peak and draws out the tracer wave. However, as more tanks are added the peak concentration increases and moves toward the mean residence time.

6.2 NONIDEAL FLOW PATTERNS

Two types of ideal flow patterns have been considered as limiting cases; plug flow (in which each fluid element entering the vessel passes through the vessel without intermingling or mixing with other fluid elements) and perfectly mixed flow (in which the fluid elements down to a molecular scale are completely homogeneous). Actual process vessels show flow patterns that lie between these two extremes. Some fluid elements slip or move with different velocities causing channeling and dead space. For these phenomena to occur the fluid elements must not completely mix or intermingle but must be partially segregated as they move through the reactor. Sometimes the term "segregated flow" is used to denote the assumption that mixing or intermingling of fluid elements does not occur in the reactor. In this situation the effluent stream may be looked upon as being made up of fluid elements which have unique residence times.

There are many ways of estimating the degree of deviation of a flow pattern from an ideal one, but most rely upon experimental measurement of the residence time distribution (RTD), i.e., the distribution of times required by elements of fluid to pass through the reactor. Once the RTD has been determined, one can calculate the

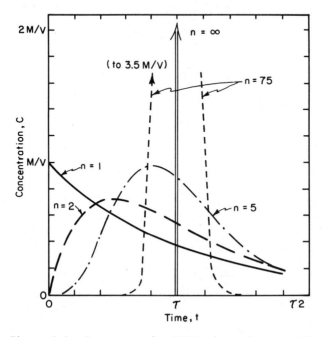

Figure 6.6. Responses of n CSTR's in series to a Dirac delta input.

expected conversion in the reactor by assuming the flow to be completely segregated.
Alternatively, the RTD can be used to estimate the axial diffusivity, thereby
allowing prediction of the conversion by superimposing axial mixing upon the plug
flow pattern. In yet another approach, the RTD is used to approximate the real
reactor by a network of ideal reactors (CSTR and PFR) and mixing zones (dead zone)
in series and/or parallel. The conversion is then calculated from this approximation. Therefore, let us consider the RTD analysis.

6.2.1 Residence Time Distribution

Much of the development of mixing and RTD concepts is due to Danckwerts [3] and
Zwietering [4]. We shall follow their notation and development here.

Residence time is the time taken by a molecule to pass through a reactor.
Fluid elements taking different routes require different lengths of time. Therefore,
the residence times of fluid elements in a reactor may take on different values. Let
the function, F(t), be the fraction of the elements in the effluent stream having
residence times less than t. With this definition it is apparent that $F(0) = 0$ and
$F(\infty) = 1$. In other words, none of the fluid can pass through the vessel in zero time
and all will come out eventually. This function, which is shown in Fig. 6.7, is
known as a cumulative distribution function or F curve. Another function is the
point distribution function, E(t), which is related to the cumulative distribution
function by:

$$E(t) = dF(t)/dt \qquad\qquad (6.16)$$

Thus, it follows that E(t)dt is the fraction of the effluent that has a residence
time between t and t + dt and thus it is the residence time distribution (RTD)
function of the fluid. The area under the RTD curve (also called the E curve),
between the limits 0 and ∞ is unity since the entire fraction of fluid must have

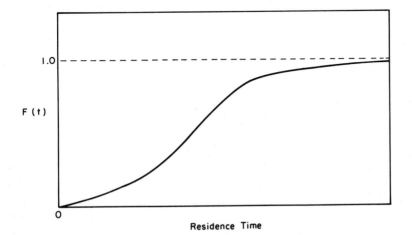

Figure 6.7. Cumulative residence time distribution function, F(t).

residence times between 0 and ∞,

$$\int_0^\infty E(t)\,dt = 1 \tag{6.17}$$

The fraction of fluid elements that has residence times less than t_1 is given by

$$F(t_1) = \int_0^{t_1} E(t)\,dt \tag{6.18}$$

while the fraction that has residence times greater than t_1 is

$$1 - F(t_1) = \int_{t_1}^\infty E(t)\,dt = 1 - \int_0^{t_1} E(t)\,dt \tag{6.19}$$

A typical E curve is shown in Fig. 6.8. Either the F or E curve may be used to account for nonideal flow.

Sometimes the space time of a reactor is not known and must be determined experimentally. Since a constant density is assumed for the fluid while it moves through the vessel and isothermal conditions are also assumed, the mean residence time may be expressed by Eq. 6.20.

$$\bar{t} = \tau = V/F = \int_0^1 t\,dF(t) \Big/ \int_0^1 dF(t) = \int_0^1 t\,dF = \int_0^\infty tE(t)\,dt \Big/ \int_0^\infty E(t)\,dt = \int_0^\infty tE(t)\,dt \tag{6.20}$$

6.2.2 Experimental Determination of RTD

The distribution functions for a given reactor and flow rate can be determined experimentally by introducing a tracer into the reactor input and observing the time response of the tracer concentration in the reactor effluent. The two most convenient types of tracer input to use are the step and impulse signals. A sinusoidal input may also be used, but the analysis is more complex [5].

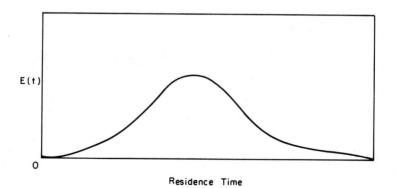

Figure 6.8. Point residence time distribution function, E(t).

STEP INPUT. Imagine a reactor subject to a step input of C_o in tracer concentration. The ratio of output to input concentration, C/C_o, would vary, depending on the extent of mixing, and might typically look like Fig. 6.9. Let us now try to relate this response to the cumulative distribution function. We can do this by dividing the effluent stream into two fractions, one of which has spent less time than t in the reactor, $F(t)$, and the other of which has spent more time than t, $1 - F(t)$. At time t, the fraction $F(t)$ will have concentration C_o, while the other fraction, $1 - F(t)$, will have no tracer in it. Therefore, the total flow of tracer in the effluent will be $F(t)C_oF + [1 - F(t)](0)F$ which must be equal to the mass flow of tracer at t, CF, that is

$$F(t)C_oF = CF \tag{6.21}$$

or

$$F(t) = C/C_o = C(t)/C_o \tag{6.22}$$

Equation 6.22 states that the cumulative distribution function, $F(t)$ is identical to the normalized tracer concentration response function due to a step input of C_o. Thus, Eq. 6.22 provides a convenient means of experimentally determining the F curve of a real vessel. All that is needed is a tracer whose concentration can be measured readily. Pure water should be fed continuously to the reactor and then the feed should be switched to one containing the tracer at concentration C_o. After the switch, the effluent concentration should be recorded until it reaches that of the feed. The effluent tracer concentration expressed as a fraction of the feed concentration, $C(t)/C_o$, is the cumulative distribution function $F(t)$. For example, for

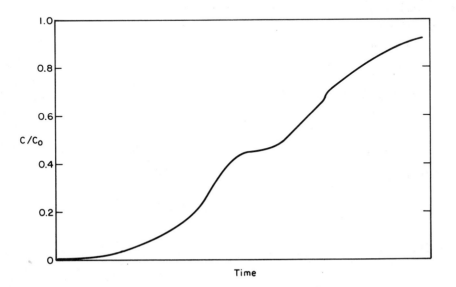

Figure 6.9. A typical step input response.

the ideal CSTR it is given by Eq. 6.8,

$$F(t)_{CSTR} = 1 - \exp(-t/\tau) \tag{6.23}$$

while for the PFR it is given by a delayed step function

$$F(t)_{PFR} = S(t - \tau) \tag{6.24}$$

For N CSTR's in series

$$F(t)_{CSTR's} = 1 - [1 + t/\tau + (t/\tau)^2/2! + \ldots + (t/\tau)^{N-1}/(N - 1)!]\exp(-t/\tau) \tag{6.25}$$

IMPULSE INPUT. A similar analysis can be used to establish that the point distribution function, $E(t)$, is identical to the normalized effluent tracer concentration curve resulting from an impulse input into the feed [6]. Instead of using this approach however, let us use an alternative one. We recall from an earlier section that $E(t) = dF(t)/dt$ (Eq. 6.16) and that the impulse input response is equal to the derivative of the step input response. Thus,

$$E(t) = dF(t)/dt = d[C(t)/C_o]_{step}/dt = [C(t)/C_o]_{impulse} \tag{6.26}$$

or

$$E(t) = C(t)_{impulse}/ \int_0^\infty C(t)_{impulse} \ dt \tag{6.27}$$

Equation 6.26 also suggests that the $F(t)$ function can be obtained from the impulse input response by integration, just as the $E(t)$ function can be obtained from the step input response by differentiation. Hence, either a step input or an impulse input may be used to obtain both $E(t)$ and $F(t)$. The choice between the two is primarily a matter of experimental convenience. However, it is generally wise to avoid differentiation of experimental data whenever possible. Thus the more commonly used technique is to obtain the $E(t)$ function from an impulse input, and then obtain the $F(t)$ function by integration if it is needed.

EXAMPLE 6.2.2-1

A tracer study using an impulse input was performed on a primary settling basin in a wastewater treatment plant. The concentration of tracer was measured in mg/1. The data are tabulated in Table E6.1. Tabulate and plot the $E(t)$ function. What is the average residence time?

Solution:

The $E(t)$ is evaluated by Eq. 6.27. The total amount of tracer introduced can be obtained by evaluating the integral with Simpson's rule, (Eq. 4.36):

TABLE E6.1
Impulse Response of a Primary Settler

t hr	C(t) mg/l	E(t)
0	0	0
0.3	6.0	0.172
0.6	22.0	0.630
0.9	25.5	0.730
1.2	22.5	0.645
1.5	16.5	0.473
1.8	11.0	0.315
2.1	7.2	0.206
2.4	4.4	0.126
2.7	2.1	0.060
3.0	0	0.00

$$\int_0^\infty C(t)_{impulse}\,dt = \frac{\Delta}{6}(C_0 + 4C_1 + 2C_2 + 4C_3 + 2C_4 + \ldots + 4C_{n-1} + C_n)$$

$$= \frac{\Delta}{6}(0 + (4)(6) + (2)(22) + (4)(25.5) + (2)(22.5) + (4)(16.5)$$

$$+ (2)(11) + (4)(7.2) + (2)(4.4) + (4)(2.1) + 0)$$

$$= \frac{0.6}{6}(349) = 34.9 \text{ mg/(liter·hr)}$$

Therefore $E(t) = C(t)_{impulse}/34.9$

The E(t) values are calculated and tabulated in Table E6.1. Figure E6.1 shows the
E(t) function. The average residence time is given by Eq. 6.20

$$\bar{t} = \int_0^\infty tE(t)\,dt / \int_0^\infty E(t)\,dt$$

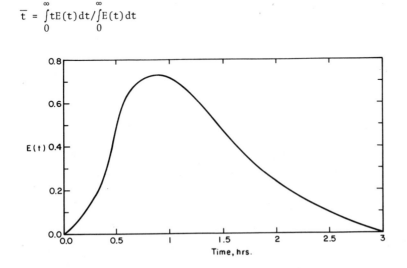

Figure E6.1.   Point residence time distribution function for
               Example 6.2.2-1.

which may be evaluated by Simpson's rule,

$$= \frac{\Delta}{6}(t_0 E_0 + 4t_1 E_1 + 2t_2 E_2 + \ldots + 4t_{n-1} E_{n-1} + t_n E_n)/$$

$$\frac{\Delta}{6}(E_0 + 4E_1 + 2E_2 + \ldots + 4E_{n-1} + E_n)$$

$$= 0.1[0 + (4)(0.3)(0.172) + (2)(0.6)(0.63) + (4)(0.90)(0.730) + (2)(1.2)(0.645)$$

$$+ (4)(1.5)(0.473) + (2)(1.8)(0.315) + (4)(2.1)(0.206) + (2)(2.4)(0.126)$$

$$+ (4)(2.7)(0.06)]/(0.1)[0 + (4)(0.172) + (2)(0.63) + (4)(0.73)$$

$$+ (2)(0.645) + (4)(0.473) + (2)(0.315) + (4)(0.206) + (2)(0.126)$$

$$+ (4)(0.06) + 0] = \frac{0.1(12.0936)}{0.1(9.996)} = 1.21 \text{ hrs.}$$

RTD FOR REACTORS IN SERIES. Let us now consider the RTD functions for a system consisting of one PFR and one CSTR. In the first system, made up of a CSTR followed by a PFR, the step input response of the CSTR is given by Eq. 6.8. This serves as the input to the PFR. Realization of the fact that a PFR merely acts to delay the input by a time equal to its space time, allows us to write immediately

$$C/C_0 = \{1 - \exp[-(t - \tau_{PFR})/\tau_{CSTR}]\}S(t - \tau_{PFR}) = F(t) \qquad (6.28)$$

For the reverse arrangement, the step input response of the PFR serves as the input for the CSTR, so that Eq. 6.7 can be written as

$$dC/dt + C/\tau_{CSTR} = C_0 S(t - \tau_{PFR})/\tau_{CSTR} \qquad (6.29)$$

which can be integrated to obtain

$$C/C_0 = \exp(-t/\tau_{CSTR})\int_0^t [\exp(t/\tau_{CSTR})]S(t - \tau_{PFR})dt/\tau_{CSTR} \qquad (6.30)$$

or

$$C/C_0 = \{1 - \exp[-(t - \tau_{PFR})/\tau_{CSTR}]\}S(t - \tau_{PFR}) = F(t) \qquad (6.31)$$

which is identical to Eq. 6.28. Thus, the RTD functions are identical for both systems. Yet, we know from Section 4.2.4 that the overall conversion depends on the actual reactor arrangement, with the exception of first-order reactions. Thus, this suggests that with the exception of first-order reaction systems the RTD function is not sufficient to predict the performance of actual reactors.

6.3 MODELS FOR REACTORS WITH NONIDEAL FLOW PATTERNS

We shall now consider four methods of estimating real reactor performance with RTD functions. These may be broadly classified into two types. The first type involves the use of the RTD function to determine the conversion by assuming the flow to be completely segregated. In the other type a network of ideal flow reactors is selected so that its theoretical RTD matches optimally the RTD for the actual reactor. The network is then assumed to represent the actual reactor and its performance is estimated by calculating the performance of the network of ideal flow reactors using the methods developed in Chapter 4. For example, the actual RTD function may be compared with those of various numbers of tanks in series and the one which most closely approximates the actual RTD is picked as the best. The performance of the actual reactor is then estimated by using these ideal CSTR's in series.

6.3.1 Actual RTD with Segregated Flow Model

As we discussed briefly in Section 6.2.2, the RTD functions are not unique, so that a number of different flow patterns can give the same one. However, for linear processes (first-order reactions) the conversion is independent of the actual flow pattern so that one may use any of them as long as the one chosen gives the same RTD function. The simplest pattern is that of segregated flow, which is equivalent to having an infinite number of parallel plug-flow reactors of different volumes and therefore of different residence times. Thus, in segregated flow the reactant concentration in each fluid element is determined solely by the residence time of that element as it moves through the reactor, and is given by the plug-flow reactor design equations developed in Chapter 4. Since the reactant concentration in each element depends on its residence time, the average concentration of the effluent may be obtained by averaging the concentrations of all fluid elements:

$$\overline{C}_A = \int_0^1 C_A dF = \int_0^\infty C_A E(t) dt \tag{6.32}$$

where E and F are the RTD functions and C_A is the reactant concentration in each fluid element. Because the performance of a perfect PFR is identical to that of a batch reactor at $t = \tau$, C_A for each element may be determined by the batch reactor equations of Section 4.1.1.

   REACTIONS WITH LINEAR RATE EXPRESSIONS. When the rate expression can be expressed as a linear function of reactant concentration, for example, $r_A = -kC_A$, $r_A = k_1 C_A - k_2$, etc. Eq. 6.32 can be used to predict correctly the performance of nonideal flow reactors. For example, consider an irreversible first-order reaction, as given by Eq. 4.2:

$$C_A = C_{Ao} \exp(-kt) \tag{4.2}$$

This may be substituted into Eq. 6.32 to obtain

$$\overline{C}_A = \int_0^\infty C_{Ao}[\exp(-kt)]E(t)dt = \int_0^1 C_{Ao}[\exp(-kt)]dF(t) \tag{6.33}$$

which can be evaluated either analytically or numerically depending upon the form in which $E(t)$ is available. Similar equations may be obtained for all reactions with linear rate expressions by substituting the integrated equation for C_A into Eq. 6.32. For a reversible first-order reaction, Eq. 4.16 is substituted into Eq. 6.32, yielding:

$$\overline{C}_A = \int_0^\infty \{C_{Ao} + (C_{Ao} - C_{Ae})\exp[-(k_1 + k_{-1})t]\}E(t)dt \tag{6.34}$$

For first-order consecutive reactions, $A \rightarrow B \rightarrow C$ Eq. 4.88 may be substituted into Eq. 6.32 giving:

$$\overline{C}_B = \frac{k_1 C_{Ao}}{k_1 - k_2} \int_0^\infty [\exp(-k_1 t) - \exp(-k_2 t)]E(t)dt \tag{6.35}$$

EXAMPLE 6.3.1-1

The effluent tracer concentration profile given in Table E6.2 represents the continuous response to a step input of a reactor which is to be used to carry out a first-order reaction with the rate expression

$$-r_A = kC_A \qquad k = 0.02 \text{ min}^{-1}$$

Determine the effluent reactant concentration in the real reactor when the influent concentration is 1000 mg/1.

TABLE E6.2
Step-Input Response for a Reactor

t min	C(t) mg/1	t min	C(t) mg/1
0	0	50	1.920
5	0.180	55	1.940
10	0.525	60	1.970
15	0.884	65	1.980
20	1.190	70	1.990
25	1.420	75	1.990
30	1.600	80	1.990
35	1.730	85	2.00
40	1.800	90	2.00
45	1.880		

Solution:

Since the step input response data are given we need to determine first the F(t) function, which is equal to $C(t)/C_o$. The feed concentration, C_o, is recognized as the ultimate value for the effluent tracer concentration. This value is 2.0 mg/l. Therefore, the cumulative distribution function is given by $F(t) = C(t)/2$. In Eq. 6.32 either E or F can be used. To avoid differentiation we use the F function, i.e.,

$$\overline{C}_A = \int_0^1 C_A dF \tag{1}$$

To evaluate this integral $C_A(t)$ must be plotted against F(t) and the area under the curve be evaluated. Equation 4.2 gives $C_A(t) = C_{Ao}\exp(-kt)$, while $F(t) = C(t)/2.0$ is available in Table E6.2. Thus, at a fixed time one can evaluate both $C_A(t)$ and F(t). These values are tabulated in Table E6.3 and plotted in Fig. E6.2. The

TABLE E6.3
Tabulation of $C_A(t)$ versus F(t)

t min	$C_A(t)$ mg/l	F(t)	t min	$C_A(t)$ mg/l	F(t)
0	1000	0	50	368	0.960
5	905	0.09	55	333	0.97
10	819	0.263	60	301	0.985
15	741	0.442	65	273	0.990
20	670	0.595	70	247	0.995
25	607	0.710	75	223	0.995
30	549	0.800	80	202	0.995
35	497	0.865	85	183	1.00
40	449	0.900	90	165	1.00
45	407	0.940			

integration is performed by using Simpson's rule, Eq. 4.36, and Fig. E6.2. We take $\Delta = 0.2$ and use the values of C_A at every 0.1 interval of F

$$\overline{C}_A = \frac{0.2}{6}[1000 + (4)(895) + (2)(840) + (4)(800) + (2)(756) + (4)(713)$$

$$+ (2)(665) + (4)(612) + (2)(500) + (4)(455) + 165] = 689.6 \text{ mg/l}$$

REACTIONS WITH NONLINEAR RATE EXPRESSIONS. When the rate expression is non-linear, the simple superpositional approach cannot be used. The RTD function is not sufficient because it is independent of point-to-point arrangement as we saw in Section 6.2.2. In general, the needed point-to-point information is difficult, if

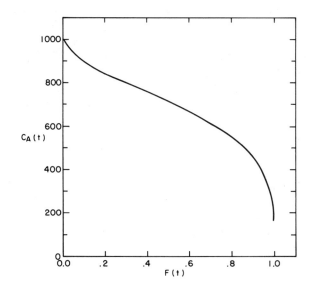

Figure E6.2. Plot of C_A versus F(t) for Example 6.3.1-1.

not impossible, to obtain. However, it can be shown [7a] that the superpositional
approach does give bounds to the conversion and thus it can provide useful information. For an nth order reaction, $-r_A = kC_A^n$, it gives the lower bound for n < 1
and the upper bound for n > 1. It should be noted that for zero-order reactions the
conversion is independent of flow pattern.

For reactors in which nonlinear reactions are to be carried out an alternate
approach is usually used. They are sometimes modeled by a dispersion term superimposed on a plug flow reactor, by a series of CSTR's with or without recycle, or by
a system of CSTR's and PFR's in series-parallel networks, etc., chosen to match the
model reactor tracer response with that of the real reactor. More sophisticated
approaches are available [7b,8] but their coverage is beyond the scope of this text.

6.3.2 CSTR's in Series Model

In this approach the real reactor is simulated by a series of equal volume CSTR's
for which the total volume is equal to that of the actual reactor. The only unknown
parameter is the number of CSTR's, n, which is to be picked so that the RTD function
of the model matches as closely as possible the RTD of the actual reactor. This can
be done by plotting either the F or E curve for the reactor and comparing it with the
curves in Figs. 6.5 or 6.6. Once the number of CSTR's is determined, the actual
reactor is replaced by the hypothetical model of n CSTR's in series for predicting
the performance. The performance of a series of equal volume CSTR's has been
covered in Chapter 4; it will not be repeated here.

EXAMPLE 6.3.2-1

Given in Table E6.4 are the response data from a reactor (τ = 30 min.) to which a step input of tracer was applied. Model this vessel with n tanks in series and determine the expected conversion for a second-order reaction, $-r_A = kC_A^2$, with k = 1.0 liter/(min·mol) and C_{Ao} = 1 mol/liter.

TABLE E6.4
Step-Input Tracer Response of a Reactor

t min	$C(t)/C_o$	t min	$C(t)/C_o$
0	0	45	0.88
5	0.006	50	0.96
10	0.017	55	0.98
15	0.084	60	0.99
20	0.220	65	0.99
25	0.380	70	0.99
30	0.550	75	0.99
35	0.700	80	0.99
40	0.800	90	1.00

Solution:

The step input response of Table E6.4 is plotted in Figure E6.3 and compared to those of ideal CSTR's in series in Fig. 6.5. The response of the vessel is almost identical to that of six equal volume CSTR's in series, therefore, we replace the actual reactor with 6 CSTR's. Equation 4.64 may be used to determine the effluent

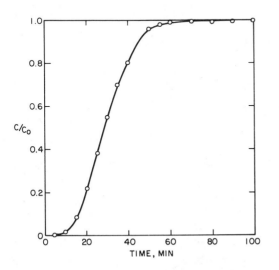

Figure E6.3. Plot of C/C_o versus time for Example 6.3.2-1.

concentration. Since the mean residence time for the actual reactor is 30 min, the equivalent mean residence time for each CSTR is 5 min. Solving Eq. 4.64 serially yields:

$$C_{A1} = [(1 + 4\tau k C_{Ao})^{\frac{1}{2}} - 1]/2\tau k = [(1 + 20)^{\frac{1}{2}} - 1]/10 = 0.358$$

$$C_{A2} = \{[1 + (4)(5)(0.358)]^{\frac{1}{2}} - 1\}/10 = 0.185$$

and

$$C_{A3} = \{[1 + (4)(5)(0.185)]^{\frac{1}{2}} - 1\}/10 = 0.117$$
$$\vdots$$
$$C_{A6} = \{[1 + (4)(5)(0.063)]^{\frac{1}{2}} - 1\}/10 = 0.0503$$

Thus the expected conversion is

$$x_A = (1.0 - 0.0503)(100) = 94.97\%$$

6.3.3 Axial Dispersion Model

In this model some degree of backmixing is superimposed upon a plug flow of fluid. The magnitude of the backmixing is assumed to be independent of the position in the reactor and is expressed by the "longitudinal" or "axial" dispersion coefficient, D_L, which is analogous to the coefficient of molecular diffusion in Fick's law of diffusion. Modeling the superimposed backmixing by axial dispersion requires writing an unsteady-state mass balance for tracer around an infinitesimal volume in the plug flow reactor. The molar flow rate of the tracer in and out of the control volume, as shown in Fig. 2.3, is made up of two terms, the bulk flow and axial dispersion (diffusion), so that Eq. 2.30 may be rewritten for the inert tracer as

$$- \frac{\partial}{\partial x} \left[F C_A - D_L A_c \frac{\partial C_A}{\partial x} \right] = A_c \frac{\partial C_A}{\partial t} \tag{6.36}$$

or

$$D_L \frac{\partial^2 C_A}{\partial x^2} - \frac{F}{A_c} \frac{\partial C_A}{\partial x} = \frac{\partial C_A}{\partial t} \tag{6.37}$$

This may be made dimensionless by substituting $z = x/L$ and $\theta = t/[L/(F/A_c)] = t/\tau$

$$\frac{D_L}{vL} \frac{\partial^2 C_A}{\partial z^2} - \frac{\partial C_A}{\partial z} = \frac{\partial C_A}{\partial \theta} \tag{6.38}$$

where v is F/A_c and (D_L/vL) is the dispersion number, which measures the extent of axial dispersion. This parameter is the inverse of the "Péclét" number, Pe. When the dispersion number is zero, there is no axial dispersion and therefore plug flow, whereas when it is infinitely large, complete backmixing exists and the reactor behaves as a CSTR.

As we have done previously for ideal vessels we need to solve Eq. 6.38 with appropriate initial and boundary conditions. For a step input the proper initial and boundary conditions [9] are

$$C_A(z,0) = 0 \qquad (6.39)$$

$$C_A(0,\theta) = C_o/v + (D_L/vL)\partial C_A/\partial z \qquad (6.40)$$

and

$$\frac{dC_A}{dz}(1,\theta) = 0 \qquad (6.41)$$

Equation 6.38 with the initial and boundary conditions given by Eqs. 6.39 and 6.41 is difficult to solve analytically so the solution must generally be obtained numerically. However, for small deviations from plug flow (i.e., when the dispersion number is small) an approximate analytical solution is available [10]. At the outlet, z = 1, the tracer concentration is given by

$$(C/C_o) = F(\theta) = \frac{1}{2}[1 - erf(\frac{1 - \theta}{2\ \theta^{0.5}(D_L/vL)^{0.5}})] \qquad (6.42)$$

where erf is the error function which is defined as

$$erf(x) = \frac{2}{\pi^{0.5}}\int_0^x exp(-y^2)\,dy \qquad (6.43)$$

and tabulated in any standard mathematical table. Equation 6.42 states that the $F(\theta)$ function for small values of the dispersion number is characterized by the mean residence time ($\tau = L/(F/A_c)$) and the dispersion number. For large values of (D_L/vL) the solution must be obtained numerically.

Figure 6.10 shows the step input tracer response of the axial dispersion model for various values of the dispersion number. Since the plot of F(t) versus t/τ may be characterized by the value of (D_L/vL), it would be convenient to have a way of matching the dispersion model with the actual RTD through proper selection of (D_L/vL). One way is to evaluate the derivative of the F(t) function at one mean residence time,

$$[dF(t)/dt]_{t=\tau} = [1/2\tau(\pi)^{0.5}](vL/D_L)^{0.5}$$

or

$$vL/D_L = 4\pi\tau^2\ [(dF/dt)_{t=\tau}]^2 \qquad (6.44)$$

Thus, the dispersion number can be approximated by determining the slope of the F curve at the time equal to one mean residence time (t = τ) and using Eq. 6.44. There are also several ways of determining the dispersion number using the E(t) curve. The reader is referred to the text by Levenspiel [7] where they are discussed in considerable detail.

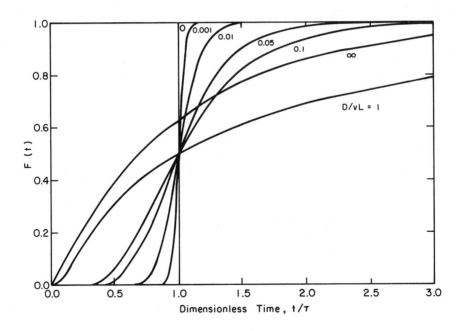

Figure 6.10.   Step input tracer response for the axial dispersion model.

EXAMPLE 6.3.3-1

Determine whether or not the dispersion model fits the RTD of the reactor in Example 6.3.2-1.

Solution:

The $F(t)$ function of Example 6.3.2-1 was plotted in Figure E6.3. Comparison of that curve with the ones in Fig. 6.10 show that a reasonably good fit can be made by a dispersion model with a dispersion number of 0.25. Thus, the reactor in Example 6.3.2-1 can be modeled by the axial dispersion method with the dispersion number of 0.25.

Having determined a dispersion number, the performance of the reactor can be predicted by the dispersion model with a reaction term. For any kinetics, the steady-state mass balance equation is obtained from Eq. 6.38 by dropping the unsteady-state term and incorporating a reaction term,

$$\frac{D_L}{vL} \frac{d^2C_A}{dz^2} - \frac{dC_A}{dz} + \tau r_A = 0 \tag{6.45}$$

The boundary conditions are

$$\frac{dC_A}{dz} = 0 \text{ at } z = 1 \tag{6.46}$$

and

$$C_A - \frac{D_L}{vL} \frac{dC_A}{dz} = C_{Ao} \text{ at } z = 0 \tag{6.47}$$

In general, a numerical technique must be used to solve Eq. 6.45 with its boundary conditions. However, for first-order reactions, $r_A = -kC_A$, it can be solved analytically to provide the solution at the exit $z = 1$ [9]:

$$C_A/C_{Ao} = \frac{4\alpha}{(1 + \alpha)^2 \exp[-Pe(1-\alpha)/2] - (1 - \alpha)^2 \exp[-Pe(1 + \alpha)/2]} \tag{6.48}$$

where

$$\alpha = (1 + 4k\tau/Pe)^{\frac{1}{2}} \tag{6.49}$$

and

$$Pe = vL/D_L \tag{6.50}$$

Equations 6.48 through 6.50 allow calculation of the effluent concentration from a PFR with axial dispersion, within which a first-order reaction is occurring. It can be shown that the effluent concentration predicted by Eq. 6.48 is always higher than that for a perfect PFR [11,12]. Thus nonideal flow lowers the conversion from first-order reactions in PFR's.

Another exception to the use of numerical methods for the solution of Eq. 6.45 is when the dispersion number is small. Under that circumstance the following expression has been developed [13]:

$$C_A - C_{Ap} = \frac{D_L}{vL} r_{Ae} \ln(\frac{r_{Ae}}{r_{Ai}}) \tag{6.51}$$

where C_A is the effluent concentration of the real reactor, C_{Ap} the effluent concentration from an ideal PFR, and r_{Ai} and r_{Ae} are the rates at the inlet and outlet of an ideal PFR with the same τ as the actual reactor. Equation 6.51 suggests that for reactions with rates which increase monotonically with reactant concentration, the axial mixing decreases the conversion, whereas for reactions with nonmonotonic rates it may increase it. For zero-order reactions the conversion is independent of local mixing.

EXAMPLE 6.3.3-2
A first-order reaction, with the rate constant equal to 1 min^{-1}, is to be carried out in the vessel whose RTD was given in Example 6.3.3-1. The space time is to be 1 min.

Determine the expected conversion resulting from this reactor and compare it with that of an ideal PFR.

Solution:

The axial dispersion number determined in Example 6.3.3-1 was 0.25 (= 1/Pe). Therefore, the effluent concentration of the reactor with axial dispersion is determined from Eq. 6.48 with $\alpha = [1 + (4)(1)(1)(0.25)]^{\frac{1}{2}} = 1.414$

$$C_A/C_{Ao} = (4)(1.414)/\{(1 + 1.414)^2 \exp[-4(0.414)/2] - (1 - 1.414)^2 \exp[-4(2.414)/2]\}$$

$$= 0.424$$

so that

$$x_A = 1 - 0.424 = 0.576$$

The ideal plug flow equation gives

$$C_A/C_{Ao} = \exp(-k\tau) = \exp(-1) = 0.368$$

or

$$x_A = 1 - 0.368 = 0.632$$

The axial dispersion reduced the conversion by 5.6%.

_____

6.3.4 Complex Network of Mixing Regions and Flow Modes

When the patterns of fluid flow through a reactor deviate greatly from plug or completely mixed flow due to channeling, bypassing, recirculation, dead space, etc., neither the dispersion model nor the tanks in series model can satisfactorily characterize them. In these situations the usual approach is to consider the real vessel to consist of a network of flow regions with various modes of flow between and around them. This technique was proposed by Cholette and Cloutier [14] and is based on the premise that just as adding different numbers of tanks in series changes the RTD so will adding different types of mixers in series and parallel. The main difference is that by adding different types of regions and flows, it is possible to reproduce almost any RTD. The usual flow regions include PFR's, CSTR's, PFR's with dispersion, and dead space while the modes of flow between and around them are bypassing, recirculation and cross flow. A few examples of networks of ideal mixers are given in Fig. 6.11. The problem is to find the volumes of the various regions and the flow rates of the various flow modes such that the RTD of the model network matches as closely as possible that of the actual vessel. The hypothetical network model is then used to predict the performance of the actual reactor. The model must fit as closely as possible the actual flow regions in the reactor or it will have no predictive value. This need for reality must always be balanced against the ability

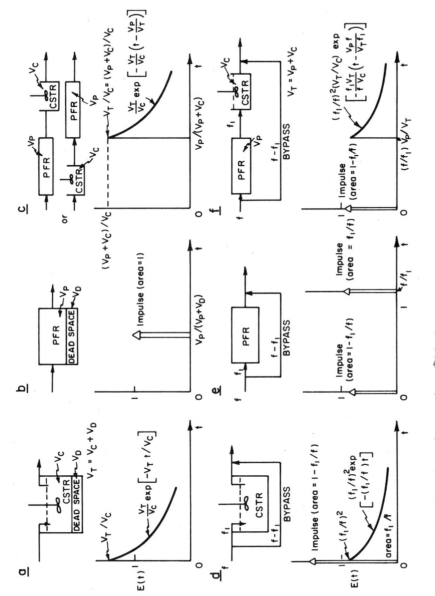

Figure 6.11. Examples of networks of ideal reactors.

to match the RTD, however. In other words, if a large number of regions and flow
modes is employed, the chance of matching the actual RTD is greater, but so is the
possibility of not matching the actual fluid regime required in the vessel. Therefore, in fitting an actual RTD one must aim for the simplest network which is consistent with the facts and whose regions are suggested by the actual vessel. This
will have the additional benefit of minimizing the mathematical complexity of the
model.

As shown in Fig. 6.11 the RTD functions of various arrangements are very different and distinctive. Indeed, these distinctive features may be used to diagnose
pathological flows in reactors [7b]. To be able to match the shapes of actual RTD
functions with a network of mixers requires familiarity with the distinctive features
of each mixing region and flow type. This is beyond the scope of this text, but
detailed discussions of network models are found elsewhere [7b,8]. Here, as an
illustration of the technique, we will consider the development of a network model
involving a parallel arrangement of a PFR and a CSTR with short circuiting [15]. The
arrangement is given in Fig. 6.12. A fraction of the feed (f_1) goes through the CSTR
of volume f_2V, while a fraction (f_3) goes through the PFR of volume $(1 - f_2)V$. The
remaining fraction of the feed $(1 - f_1 - f_3)$ by-passes. Let us assume that the
arrangement has been at a steady state with a feed whose tracer concentration is C_o.
At time zero, the feed is changed to that containing no tracer. The mass balance
for the CSTR is

$$-f_1 FC_c = f_2 VdC_c/dt \tag{6.52}$$

or

$$dC_c/dt + C_c(f_1/f_2)\tau = 0 \tag{6.53}$$

$$C_c = C_o \text{ at } t = 0 \tag{6.54}$$

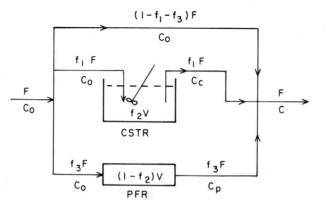

Figure 6.12. A parallel network model consisting of
a PFR and a CSTR with short circuiting.

which can be integrated to obtain

$$C_c = C_o \exp\{-t[(f_1/f_2)/\tau]\}$$ (6.55)

The PFR is a time delay device so that

$$C_p = C_o - C_o S[t - \tau(1 - f_2)/f_3]$$ (6.56)

where S[] stands for a delayed step function as defined in Eq. 6.1. The network effluent concentration is obtained from the tracer balance at the terminal mixing point,

$$FC = (1 - f_1 - f_3)(F)(0) + f_1 FC_c + f_3 FC_p$$ (6.57)

Substitution of Eqs. 6.55 and 6.56 into Eq. 6.57 yields

$$C/C_o = f_3 - f_3 S[t - \tau(1 - f_2)/f_3] + f_1 \exp\{-[(f_1/f_2)/\tau]t\}$$ (6.58)

Equation 6.58 can be used with data from an $E(t)$ curve to choose the values of f_1, f_2, and f_3 which will make the model RTD fit the actual one. The second term in the right hand side of Eq. 6.58 is zero for $t < \tau(1 - f_2)/f_3$ and is equal to f_3 for $t > \tau(1 - f_2)/f_3$. Therefore, for $t > \tau(1 - f_2)/f_3$ the plot of $\ln(C/C_o)$ versus t should give a straight line with slope of $-(f_1/f_2)/\tau$ while its extrapolated intercept gives f_1. Also it should show a discontinuity at $t = \tau(1 - f_2)/f_3$, the magnitude of which is equal to f_3. These features are so indicated in Fig. 6.13. Note that the intercept is equal to $(f_1 + f_3)$. There are three parameters (f_1, f_2, and f_3) which may be adjusted to match the RTD of the model as closely as possible with that of the actual RTD. The main features to consider in this situation are the intercept, $f_1 + f_3$, the extrapolated intercept f_2, the point of discontinuity, $t = \tau(1 - f_2)/f_3$, the slope of the line for $t > \tau(1 - f_2)/f_3$, $-f_1/f_2$, etc. Only three features are independent. Once the best values of f_1, f_2 and f_3 are selected, the performance of the real reactor may be predicted by the network.

6.4 KEY POINTS

1. The two most common input tracer signals for analyzing flow patterns through vessels are the step and impulse signals.

2. The unit impulse function is the derivative of the unit step function.

3. A perfect PFR acts as a time delay of one space time for unreactive substances. Thus the output pattern of a dye will be exactly like the input pattern and will appear one space time after initiation of the input pattern.

4. The response of a perfect CSTR to a step input of dye is an exponential increase in dye concentration which starts at zero and approaches the ultimate value asymptotically. The response to an impulse input is an exponential decay.

5. The response of a series of CSTR's to an input tracer signal lies somewhere between that of a single CSTR and a PFR. Thus a real reactor which has the response of neither a CSTR nor a PFR may be simulated as N CSTR's in series.

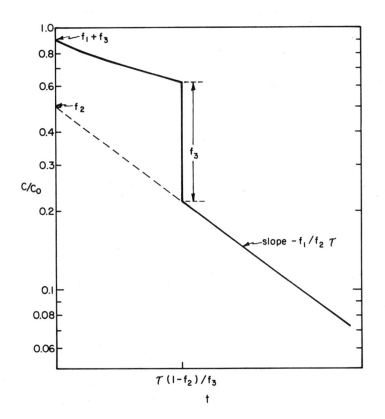

Figure 6.13.   A semilog plot of C/C_0 versus time for the
network of Figure 6.12.

6. A residence time distribution is the distribution of times required by elements
    of fluid to pass through a reactor.

7. The cumulative residence time distribution function is identical to the normal-
    ized tracer concentration response function due to a step input to the feed.

8. The point residence time distribution function is identical to the normalized
    tracer concentration response function due to an impulse input to the feed.

9. With the exception of first-order reaction systems, the RTD function is not
    sufficient to predict the performance of actual reactors. It is sufficient
    with first-order and other linear reaction rate expressions, however.

10. Superposition of the rate expression upon the RTD is called a segregated flow
    model for nonideal flow. It will only predict reactor response for reactions
    with linear rate expressions, however.

11. The reactor response for reactions with nonlinear rate expressions may be pre-
    dicted by a CSTR's in series model, an axial dispersion model, or a model based
    upon a network of CSTR's and PFR's.

12. When the CSTR's in series model is used to predict the response of a nonideal
     reactor, that response is assumed to be equivalent to the one obtained from n
     ideal CSTR's in series. The value of n is chosen so that the tracer response
     curve for the nonideal reactor matches as closely as possible the curve from
     the CSTR's in series.

13. The axial dispersion model uses the dispersion number to characterize the
     deviation of the tracer response curve from that of an ideal plug-flow reactor.
     The dispersion number may be evaluated by determining the slope of the F curve
     at one mean residence time.

14. When the patterns of flow through a reactor deviate greatly from plug or com-
     pletely mixed flow neither the dispersion model nor the tanks in series model
     can satisfactorily characterize them. They may often be modeled, however, as
     a network of flow regions with various modes of flow between and around them.

6.5 STUDY QUESTIONS

The study questions are arranged in the same order as the text:

Section	Questions
6.1	1-5
6.2	6
6.3	7-12

1. Explain in words the difference between a unit step function and a unit impulse
    function.

2. If a perfect PFR with space time τ receives an impulse of dye at time t = 0,
    what will be the output response pattern?

3. A perfect CSTR with a volume of 4.0 liters receives a constant flow of water at
    a rate of 2.0 liters/hr. At time zero, the flow is changed to water containing
    a dye at a concentration of 1000 mg/l. What will be the concentration of dye
    in the reactor after 1 hr; 2 hrs; 10 hrs?

4. A reactor with a volume of 4.0 liters receives a constant flow of water at a
    rate of 3.0 liters/hr. At time zero an impulse of dye is added to the reactor
    which instantaneously places a dye concentration of 1200 mg/l into it. Effluent
    samples taken 1,2,3,4 and 6 hrs after addition of the dye have concentrations
    of 567, 268, 126, 60 and 13 mg/l. Is the reactor a perfect CSTR?

5. Consider a chain of CSTR's, each with a volume of 1.0 liter, receiving a flow
    of water at a rate of 1.0 liter/hr. If at time zero the input stream was
    changed to a dye at a concentration of 1000 mg/l, what would be the concentration
    of dye in the effluent from the third reactor after 2 hrs?

6. A tracer study using an impulse input was performed on a reactor, and the data
    are tabulated in Table SQ6.1. Tabulate and plot the E(t) function. What is
    the average residence time? Also plot the F(t) function.

7. Explain why, in general, the RTD function does not provide sufficient infor-
    mation to allow prediction of the performance of an actual reactor.

8. The reactor in Study Question 6 is being used to carry out a first-order reaction
    which has the rate expression:

    $-r_A = kC_A \qquad k = 0.012 \text{ min}^{-1}$

TABLE SQ6.1
Tracer Concentrations in the Effluent from a Reactor Which Has Been
Subjected to an Impulse of Dye into the Influent Stream.

t hr	C mg/1	t hr	C mg/1	t hr	C mg/1	t hr	C mg/1
0.5	76	2.5	256	4.5	112	8.0	11
1.0	184	3.0	224	5.0	84	9.0	5
1.5	251	3.5	185	6.0	45	10.0	2
2.0	270	4.0	146	7.0	22		

Using the segregated flow model, determine the effluent concentration when
the influent concentration is 2000 mg/1.

9. If the reactor in Study Question 6 were to be simulated by a CSTR's in series
model, how many should be used?

10. Repeat Study Question 8 using the CSTR's in series model.

11. If the reactor in Study Question 6 were to be simulated by the axial dispersion
model, what dispersion number should be used?

12. Repeat Study Question 8 using the axial dispersion model.

REFERENCES AND FURTHER READING

1. T. P. G. Liverman, *Generalized Functions and Direct Operational Methods,*
Prentice-Hall Inc., Englewood Cliffs, N.J., 1964.

2. W. Kaplan, *Operational Methods for Linear Systems,* Addison-Wesley Publishing
Co., Inc., Reading, Massachusetts, 89-93, 1962.

3. P. V. Danckwerts, "Continuous flow systems-Distribution of residence times,"
Chemical Engineering Science, 2, 1-13, 1953.

4. T. N. Zwietering, "The degree of mixing in continuous flow systems," *Chemical
Engineering Science,* 11, 1-15, 1959.

5. H. Kramers and G. Alberda, "Frequency analysis of continuous flow systems,"
Chemical Engineering Science, 2, 173-181, 1953.

6. J. M. Smith, *Chemical Engineering Kinetics,* McGraw-Hill Book Company, New York,
N.Y., 249-251, 1970.

7. O. Levenspiel, *Chemical Reaction Engineering,* John Wiley & Sons, Inc., New
York, N.Y., a. 269, b. 253-314, 1972.

8. D. M. Himmelblau and K. B. Bischoff, *Process Analysis and Simulation: Deter-
ministic Systems,* John Wiley & Sons, Inc., New York, N.Y., 59-83, 1968.

9. J. F. Wehner and R. H. Wilhelm, "Boundary conditions of flow reactor," *Chemical
Engineering Science,* 6, 89-93, 1959.

10. O. Levenspiel and K. B. Bischoff, "Backmixing in the design of chemical
reactors," *Industrial and Engineering Chemistry,* 51, 1431-1434, 1959.

11. O. Levenspiel and K. B. Bischoff, "Reaction rate constant may modify the effects
of backmixing," *Industrial and Engineering Chemistry,* 53, 313-314, 1961.

12. I. Pasquon and M. Dente, "Heat and mass transfer in methanol synthesis - Optimum operating conditions of the reactor," *Journal of Catalysis,* 1, 508-520, 1962.

13. O. Levenspiel and K. B. Bischoff, "Patterns of flow in chemical process," *Advances in Chemical Engineering,* 4, 95-198, 1963.

14. A. Cholette and L. Cloutier, "Mixing efficiency determinations for continuous flow systems," *Canadian Journal of Chemical Engineering,* 37, 105-112, 1959.

15. A. R. Cooper and G. V. Jeffreys, *Chemical Kinetics and Reactor Design,* Prentice-Hall, Inc., Englewood Cliffs, N.J., 270-271, 1971.

Further Reading

Carberry, J. J., *Chemical and Catalytic Reaction Engineering,* McGraw-Hill, Inc., New York, N.Y., 1976. See Chapter 3 for detailed treatments, especially of recycle models and nonisothermal cases.

Cooper, A. R. and Jeffreys, G. V., *Chemical Kinetics and Reactor Design,* Prentice-Hall, Inc., Englewood Cliffs, N.J., 1971. See Chapter 7.

PART III

BIOCHEMICAL AND MICROBIOLOGICAL BACKGROUND

The models used to describe microbial activity in biochemical reactors are highly empirical, as they have arisen from macroscopic observations of the way in which cultures respond to environmental changes. This does not suggest that they are without theoretical foundation, however, since on the whole researchers have sought to build a mechanistic rationale for the models in order to understand their strengths and weaknesses. One of the dangers of employing simple mathematical models to describe complex biochemical systems is that some users may be unaware of the underlying biochemical phenomena and therefore attempt to extend their application beyond the valid regions. The purpose of this part of the book is to reduce that danger. While by no means exhaustive, the material in the next five chapters is meant to summarize the most important points needed for an understanding of what actually occurs in biochemical operations, and to demonstrate current attempts to describe them mathematically.

CHAPTER 7

THE ECOLOGY OF BIOCHEMICAL REACTORS

Life on earth depends on a complex network of elemental transformations. Until
recently those transformations were in reasonable balance, so that changes occurred
fairly gradually. As man developed an industrial society, however, he increased
the rate of input of elements into the system to the point that the balance was
broken. The result was pollution. Faced with the problem of correcting the imbalance, man has, logically, attempted to employ the natural biochemical transformations. Since the rates of those transformations could not be controlled in the

natural environment, he constructed artificial ones in which the concepts of reactor
engineering could be applied. The results were the biochemical operations which are
the subject of this book. Before those operations can be understood it is necessary
to comprehend the various interacting natural systems and the important role that
microorganisms play in them.

7.1 ROLE OF MICROORGANISMS IN NATURE

In order to organize knowledge, the world is generally subdivided into four major
sectors. Three of those sectors reflect the physical attributes of matter: the
atmosphere is the gaseous phase, the hydrosphere is the liquid phase, and the
lithosphere is the solid phase. Matter is continually being cycled from one phase
to another by chemical changes which are influenced by the organisms living within
the phases. For example, direct changes in the structures of chemicals are brought
about by the metabolism of the organisms. These can influence the changes indirectly,
on the other hand, by altering the pH or oxidation-reduction potential of the media
in which the chemical changes are occurring. The importance of organisms in the
movement of matter among the physical sectors is reflected in the designation of a
fourth sector, the biosphere, which incorporates all living matter regardless of the
physical sector in which it is found.

7.1.1 The Cycle of Matter

The maintenance of the biosphere is dependent upon the cyclic movement of all elements through the physical sectors, but particularly upon the cycles of carbon,
oxygen, nitrogen, and sulfur. A steady input of energy is required to drive the
cycles because all involve changes in the oxidation states of the elements. These
changes in oxidation state are quite important because they ensure that elements will
always be available in the proper form for any given group of organisms.

Organisms within the biosphere can be grouped into two major categories: the
primary producers and the consumers. The producers take the major elements in inorganic form and through the use of solar energy convert them into high energy organic forms. The resultant organic compounds then pass through the food chains of
the consumers, providing energy and matter for conversion into cellular material.
Before the elements can again be available to the producers they must be returned
to the inorganic state in a conversion called mineralization. One form of pollution
results when the input of materials to the mineralization step exceeds the rate at
which they can be processed in the natural environment without causing its deterioration. Biochemical operations seek to carry out mineralization in controlled environments where the desired reaction rates can be maintained.

The agents responsible for mineralization are microorganisms, primarily bacteria
and fungi. The large contribution made by them reflects their ubiquity in the biosphere, their high rates of growth and metabolism, and their collective ability to
degrade all naturally occurring organic compounds. This collective ability is the

result of the large number of diverse species of microorganisms existent in the biosphere rather than any unusual versatility of a few individual species. Indeed, a
single microbial species may be capable of mineralizing only a few organic compounds.
Thus, a biochemical operation, like the biosphere, contains a community of interacting populations of organisms.

Microorganisms also play a large role as producers because in the hydrosphere
unicellular algae are the primary agents responsible for the conversion of inorganic
compounds into organic ones. Their extremely small size provides a large amount of
surface area for photosynthetic activity. These organisms are only rarely encouraged in biochemical operations, of course, because their actions are counter to
the usual objective of mineralization.

7.1.2 The Cycles of Carbon and Oxygen

The movements of carbon and oxygen through the biosphere are intimately linked and
consequently their cycles are generally considered together. The carbon cycle is
shown in Fig. 7.1. The oxygen cycle is fairly simple and thus it is not shown
schematically.

The main reservoirs of oxidized carbon available for synthesis by the producers
are in the atmosphere and hydrosphere. In the atmosphere the carbon is in the form

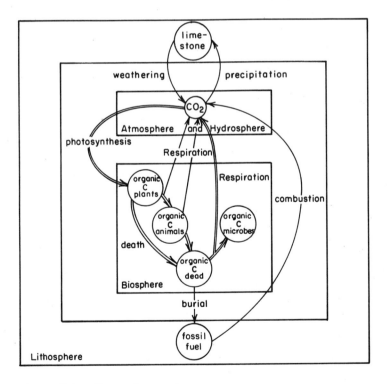

Figure 7.1. The carbon cycle.

of carbon dioxide and is directly available to terrestrial plants. In the hydrosphere dissolved CO_2 is in equilibrium with the carbonate system:

$$CO_2 + H_2O \rightleftharpoons H_2CO_3 \rightleftharpoons H^+ + HCO_3^- \rightleftharpoons 2H^+ + CO_3^= \tag{7.1}$$

Thus it is available to aquatic plants and algae in several forms. When the concentration of calcium carbonate in the water becomes too high, it precipitates leading
to limestone and dolomite. In the insoluble form the carbon is not available to the
biosphere. Weathering of the rock, aided by chemical conditions altered by metabolic activity, returns much of the carbon to the hydrosphere where it again becomes
available to the biosphere.

The prime force leading to the conversion of oxidized inorganic carbon to
reduced organic carbon is photosynthesis by plants and algae. In the process,
hydrogen from water is utilized with the evolution of molecular oxygen. Once the
carbon is in the organic form the process of mineralization begins to convert it
back to the inorganic form. The organic compounds move up the food chain with a
portion of the carbon being released as CO_2 by respiration at each level. Ultimately, the higher life forms die, and the organic compounds are then acted upon by
the ubiquitous microbes, releasing the carbon as CO_2 through respiration. Aerobic
respiration reactions utilize molecular oxygen as the terminal hydrogen acceptor
reducing it to water. Thus, the oxygen cycle may be depicted simply as

$$H_2O \underset{\substack{\text{respiration} \\ \text{combustion}}}{\overset{\text{photosynthesis}}{\rightleftharpoons}} O_2 \tag{7.2}$$

The conversion of inorganic carbon to organic carbon is referred to as carbon
fixation and amounts to approximately 2.8×10^{10} tons per year on a global scale.

In addition to the storage in inorganic form, sequestration of carbon may also
occur in the organic state. Some organic constituents of plants are highly resistant to microbial attack and tend to accumulate in nature after the more readily de-
composable compounds have been mineralized. Most of the organic matter of soil consists of such resistant residues, which are called humus. Peat bogs contain large
amounts of humus materials. Under proper conditions of temperature and pressure
such materials are converted to coal and other fossil fuel. The combustion of the
fuels returns the carbon to the inorganic state as CO_2.

The majority of the biochemical operations are aimed at the mineralization of
carbon. Only recently has attention been focused on the nitrogen cycle.

7.1.3 The Nitrogen Cycle

Most nitrogen is stored in the atmosphere in molecular form, N_2, which is chemically
inert and not suitable for most living forms. However, there are a few specialized
organisms which can convert it to organic form, i.e., "fix" it. Although nitrogen

fixation is very important in nature, with the exception of some experimental units, it plays no important role in biochemical operations.

Organic nitrogen containing compounds synthesized by plants and algae serve as the nitrogen source for animals. Animals excrete nitrogen containing compounds during their normal metabolism. As these compounds are mineralized by microorganisms, the nitrogen is released as ammonia. The same is true when a plant or animal dies and undergoes decomposition. This release of ammonia is called ammonification. The ammonia is available to plants and microorganisms for assimilation into new cell material, bringing about more organic nitrogen. A key feature about the transformation of organic nitrogen to ammonia and back is that there is no change in its oxidation state. Although not shown in Fig. 7.2, man has directly intervened in the nitrogen cycle by chemically converting molecular nitrogen to ammonia, which is then applied as fertilizer for crops.

A portion of the ammonia in the biosphere is oxidized first to nitrite (NO_2^-) and then to nitrate (NO_3^-) by two highly specialized kinds of bacteria. This conversion, nitrification, does not alter the availability of the nitrogen to plants and algae since they can assimilate nitrate as well as ammonia. Once the nitrate is in the cell it is reduced to ammonia for incorporation into organic compounds. The assimilation of nitrate is regulated closely by growth requirements; there is no significant excretion of nitrogen from plants. Nitrate is the principle nitrogenous material available in the soil, but it is very soluble, and thus is easily leached and carried by water to the ocean where it serves as the nitrogen source for algae. Some bacteria can assimilate nitrate but do so only when ammonia is absent or in very low concentration.

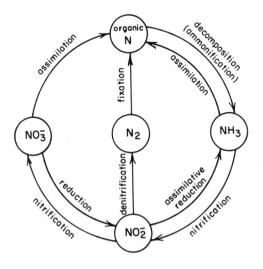

Figure 7.2. The nitrogen cycle.

Many different types of bacteria can use nitrate in place of oxygen as a final hydrogen acceptor in metabolism if oxygen is not present. This reduction employs a different biochemical mechanism called denitrification. The major gaseous end product is molecular nitrogen, although other less reduced forms such as N_2O, also may be observed. In this way nitrogen is removed from the biosphere and returned to the atmosphere.

Recently there has been much interest in biochemical operations employing the nitrogen cycle. Discharges of ammonia in wastewater treatment plant effluents can lead to nitrification with resultant oxygen depletion in the receiving stream. Consequently, nitrification is often carried out in a biochemical operation within the plant. Nitrate can be a nuisance to the receiving body in some cases, consequently biochemical operations also employ denitrification to return the nitrogen to the atmosphere.

7.1.4 The Sulfur Cycle

The final cycle of importance in the biosphere is the sulfur cycle depicted in Fig. 7.3. The assimilation of sulfate is similar to the assimilation of nitrate in two respects: (1) it must become reduced to be incorporated into organic compounds; and (2) only enough sulfur is assimilated to provide for the growth of the organisms; no reduced products are excreted. Most sulfur is present in the biosphere as sulfate. Sulfides are rapidly and spontaneously oxidized in the presence of oxygen, and elemental sulfur can be oxidized by bacteria to sulfate. Although sulfate has a valence of +6, most organic sulfur-containing compounds contain a thiol group containing sulfur with a valence of -2. This is why the reduction of sulfur is required.

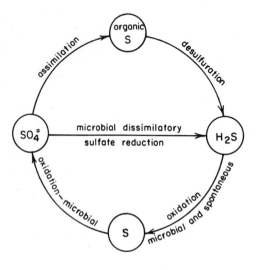

Figure 7.3. The sulfur cycle.

When sulfur containing organic compounds undergo decomposition the sulfur is liberated as H_2S with no valence change. Under anaerobic conditions the hydrogen sulfide can cause problems because of odor and toxicity. Under aerobic conditions, however, its rapid oxidation to sulfate alleviates those problems. If sulfate ions are present in an anaerobic environment, sulfate reducing bacteria will use them as terminal hydrogen acceptors, also resulting in the formation of hydrogen sulfide.

No biochemical operations employ the sulfur cycle directly, although there is often evidence of its presence, e.g., the familiar odor of septic sewage caused by the release of hydrogen sulfide. Furthermore, crown corrosion in sewers is the result of microbial oxidation of dissolved sulfides to sulfates, causing an acid condition.

7.2 CLASSIFICATIONS OF MICROORGANISMS

Before the advent of microscopy and the discovery of microorganisms, the classification of living things was relatively simple: they were either plants or animals. When microbes were first observed they were placed into one of the familiar categories, depending upon their general characteristics. As knowledge of them grew, however, it became apparent that many did not fit into either traditional class. This led to many arguments among scientists until eventually a third kingdom of life was established: the protists, which is distinguished from the plant and animal kingdoms by the relatively simple organization of its members. Many protists are unicellular, but even the multicellular ones lack the internal differentiation into separate cell types and tissue regions characteristic of plants and animals. The organisms generally recognized as protists are microscopic and include algae, bacteria, fungi, and protozoa. All are important in biochemical operations, although some to a greater extent than others.

7.2.1 General Characteristics of Protists

Protists are all small, but visible with the light microscope. Their organizational structure on the other hand, can only be discerned with the electron microscope. It was recognized quite early that the protists share common features with all other cells, i.e., that they contain a nucleus surrounded by cytoplasm, which is enclosed by a semipermeable membrane that separates the cell from the external environment. Only after the perfection of electron microscopy, however, was it found that protists actually fall into two categories, one having a more highly organized cell structure than the other. The more complex protists are called eucaryotes because of their resemblance to the cells of plants and animals ("eu" meaning true and "caryote", from the Greek karyon, meaning kernel or nucleus). The simpler protists are called procaryotes, suggesting a life form not showing a classical nucleus, and therefore being a more primitive, or prior, condition. The eucaryotic cell is the unit of structure in protozoa, fungi, and most groups of algae whereas the procaryotic cell is the unit of structure in bacteria and blue-green algae. Electron

microscopy has also done much to reveal the characteristics of a fourth group of
pseudoorganisms which are not visible through a light microscope: the viruses.
These are generally less than 0.1μm in diameter and do not possess a complete organizational structure that will allow them to reproduce outside of a living host.

   FINE STRUCTURE OF EUCARYOTIC CELLS. A schematic diagram of eucaryotic cells,
which are generally larger than 20μm in diameter, is shown in Fig. 7.4. While all
protists are separated from the environment by cellular membranes, the distinguishing
characteristic of eucaryotic cells is the presence of internal regions which are
bounded by unit membranes separate from the outer cell membrane. Among these internal regions are the nucleus, mitochondria, chloroplasts (in photosynthetic
organisms), lysosomes, and vacuoles.

   The nucleus contains the genetic material of the cell and is separated from the
cytoplasm by the nuclear membrane. The nucleus consists of a number of structural
subunits called chromosomes which are composed of deoxyribonucleic acid (DNA) linked
to a special protein, known as a histone. The replication of the individual chromosomes occurs in the resting nucleus before the onset of division. The nucleus
then divides by a complex process known as mitosis so that each new nucleus is
exactly like the original one. Cell division is triggered by and closely linked to
nuclear division so that each cell characteristically has a single nucleus. Thus
reproduction is usually asexual, although sexual reproduction may infrequently occur.

   Mitochondria and chloroplasts are the sites of energy generation. Cellular
respiration occurs in the mitochondria whereas photosynthesis occurs in chloroplasts.
Both of the organelles exhibit a fine structure characterized by inner membranes

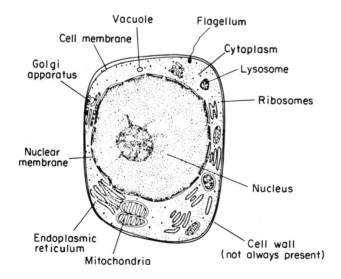

Figure 7.4. Diagram of a typical eucaryotic cell.
             (Adapted from R. Mitchell, Introduction
             to Environmental Microbiology).

along which are arrayed the enzymes which catalyze the reactions required for energy
transformation and generation. Chloroplasts and mitochondria are never assembled
from their constituent parts within the cell but always arise from the division of
a preexisting organelle. This implies that all the genetic material in these cells
is not confined to the nucleus and that the machinery for protein synthesis is likewise dispersed.

Vacuoles and lysosomes are involved with the ingestion and digestion of food.
Cells without rigid walls can envelop solid food (phagocytosis) and bring it into
the cell in a special compartment called a vacuole. Within the cell, those vacuoles
come into contact with lysosomes, which contain hydrolytic (digestive) enzymes. The
enzymes enter the vacuole where they act upon the food particle. Vacuoles are also
formed by pinocytosis, which is a form of cellular drinking analogous to phagocytosis. Specialized vacuoles, called contractile vacuoles, are involved in the
osmoregulation of eucaryotic cells without cell walls. Cells almost always exist in
a hypotonic environment so that osmotic pressure tends to force water into the cells.
Cells without walls would swell and burst if they did not contain some way of removing excess water. Contractile vacuoles serve that purpose by acting as cellular
pumps to discharge water.

The cytoplasm itself also has a fine structure, the most important parts of
which are the endoplasmic reticulum and the Golgi apparatus. The endoplasmic
reticulum is a folded membrane system which is continuous with the cell membrane.
Ribosomes are aligned on its surfaces and it serves as a site of protein synthesis.
The Golgi apparatus is a disc shaped body containing closely aligned membranes. It
appears to serve as a packaging structure within which vacuoles are formed. Lysosomes also appear to arise from the Golgi apparatus. The cytoplasm is frequently in
a state of motion, called cytoplasmic streaming which results in the movement of
organelles within the cell. It also is a means of locomotion in cells without cell
walls, by amoeboid movement.

Some eucaryotes are surrounded by rigid cell walls which provide a passive
means of osmoregulation. This cell wall prevents amoeboid movement so many of these
cells possess flagella, or whip-like appendages which provide a means of locomotion.
Flagella always arise from anchoring structures within the cytoplasm, known as basal
bodies. The external part of a flagellum consists of a circular system of micro-
tubules and is surrounded by an extension of the cell membrane.

FINE STRUCTURE OF PROCARYOTIC CELLS. The bacteria and blue-green algae are
usually smaller than 5μm in diameter and have much simpler structures than eucaryotes,
as can be seen from Fig. 7.5. These cells are bounded by a membrane similar to that
of eucaryotic cells, but there are only two major internal regions, the nucleus and
the cytoplasm, each relatively uniform in fine structure and not separated from one
another by a membrane.

The nucleus contains a single long molecule of DNA, unlinked to protein, and
nuclear division requires replication of that molecule followed by separation of the

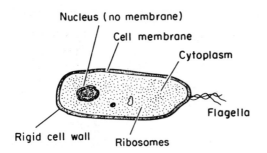

Figure 7.5.   Diagram of a typical procaryotic cell.
              (Adapted from R. Mitchell, Introduction
              to Environmental Microbiology).

daughter molecules. Nuclear and cell division are not closely geared with the result that nuclear division typically runs ahead of cell division during rapid growth.
Consequently, procaryotic cells often contain two or more groups of nuclear material
and only become uninucleate after growth has ceased and the organism has entered the
resting state.

The principal structural elements in the cytoplasmic region are the ribosomes,
which are smaller than those in eucaryotic cells. There is no endoplasmic reticulum
and thus these individual ribosomes serve as the sites of protein synthesis. Similarly, procaryotes contain no mitochondria or chloroplasts so that the enzymes for
respiration and photosynthesis are housed in the cell membrane or in foldings derived from it. The cell membrane serves a dual purpose, since it also acts in a
semipermeable capacity thereby regulating the flow of materials into and out of the
cell. The enzymes that mediate other parts of each metabolic process are located
throughout the cytoplasm.

Most procaryotes are surrounded by rigid cell walls which are of more complex
chemical structure than the walls of eucaryotes. Because of this wall, procaryotes
cannot undergo cytoplasmic streaming or form vacuoles for the intake of food. The
digestion of insoluble nutrients is extracellular, mediated by the excretion of
hydrolytic enzymes. The resultant small molecules then pass across the cell membrane by various mechanisms. The existence of the cell wall also negates the need
for an active osmoregulatory system such as contractile vacuoles.

The procaryotes usually move by the action of flagella, although in structure
they are simpler than those found in eucaryotic cells. Normally, a procaryotic
flagellum consists of a single helical fibril having dimensions similar to a single
microtubule in a eucaryotic flagellum. Furthermore, it protrudes through the cell
membrane, rather than being enclosed by an extension of it, and is anchored to a
much smaller basal body. The number and positioning of the flagella are often used
in identifying procaryotic cells.

FINE STRUCTURE OF VIRUSES. Viruses have a very simple chemical structure. In their extracellular phase, as virions, they consist only of a protein coat, called a capsid, surrounding a single kind of nucleic acid, either DNA or RNA (ribonucleic acid). With the exception of enzymes which aid in the penetration of the host cell they are devoid of enzyme activity. Consequently, they cannot be considered true cells, but rather represent a very simple life form which must take over a host cell to obtain the machinery for replication. When a virion attaches itself to a host cell it penetrates the membrane and injects its nucleic acid into the cell. In its intracellular phase, the virus exists as foreign nuclear material which takes over the regulation of the cell function and directs it towards the production of more virions. This usually results in the cell becoming filled to the bursting point with viruses. The cell membrane then breaks, spilling virions into the medium where each can infect other host cells.

7.2.2 Eucaryotes

The eucaryotic protists are generally divided into three major groups: eucaryotic algae, protozoa, and fungi. The eucaryotic algae are photosynthetic organisms containing chloroplasts. Some are unicellular while others are multicellular, although they possess no differentiation of cells and tissues. Protozoa are unicellular, nonphotosynthetic organisms whereas fungi are coenocytic (i.e., multinucleated), nonphotosynthetic organisms which grow in the form of a filamentous branched structure known as a mycelium.

EUCARYOTIC ALGAE. The primary classification of the algae is based upon four cellular properties: (1) the chemical nature of the cell wall (if present); (2) the organic storage product formed by the cell; (3) the nature of the photosynthetic pigments; and (4) the nature and arrangement of the flagella on motile cells. The arrangement of algae according to those characteristics is summarized in Table 7.1. The common cellular properties of each algal division suggest that its members are all representatives of a single evolutionary line. Thus for a particular cellular organization, evolution seems to have resulted in an increase in the complexity of the organism as a whole. Those algae which are completely enclosed by cell walls are osmotrophic and therefore dependent entirely upon dissolved substrates and nutrients. Others, however, ingest bacteria and other smaller microorganisms by phagotrophic means.

Algae are widely distributed in both fresh and salt waters. In fact, marine algae alone are equal in mass to all of the land plants and thus play an important role in the carbon-oxygen cycle. Their importance to the world's oxygen supply is one reason that ecologists are concerned about oil spills and other forms of ocean pollution.

The fact that many algae can grow in a completely inorganic medium both interests and dismays the engineer. It gives them the potential for removing inorganic

TABLE 7.1
MAJOR GROUPS OF EUCARYOTIC ALGAE

Group Name	Pigment System		Composition of Cell Wall	Nature of Reserve Matls.	Number & Type of Flagella	Range of Structure
	Chlorophylls	Other Special Pigments				
Green algae: division Chlorophyta	a + b	-	Cellulose	Starch	Generally two identical flagella per cell	Unicellular, coenocytic, filamentous; plant-like multicellular forms
Euglenids: division Euglenophyta	a + b	-	No wall	Paramylum and fats	One, two, or three flagella per cell	All unicellular
Dinoflagellates and related forms: division Pyrrophyta	a + c	Special carotenoids	Cellulose	Starch and oils	Two flagella, dissimilar in form and position on cell	Mostly unicellular, a few filamentous forms
Chrysophytes and diatoms: division Chrysophyta	a + c	Special carotenoids	Wall composed of two overlapping halves, often containing silica (some have no walls)	Leucosin and oils	Two flagella, arrangement variable	Unicellular, coenocytic, filamentous
Brown algae: division Phaeophyta	a + c	Special carotenoids	Cellulose and algin	Laminarin and fats	Two flagella, of unequal length	Plantlike multicellular forms
Red algae: division Rhodophyta	a	Phycobilins	Cellulose	Starch	No flagella	Unicellular; plantlike multicellular forms

Reproduced from Stanier, R. Y., Adelberg, E. A., and Ingraham, J.L., The Microbial World, 4th Edition, 1976.
Reprinted by permission of Prentice-Hall Inc., Englewood Cliffs, N.J.

nutrients from wastewaters, while simultaneously producing potentially useful cell
material. One of the problems associated with their use, however, stems from the
fact that during periods of darkness they can switch from photosynthetic to respiratory metabolism thereby removing dissolved oxygen from the medium. Naturally
occurring algae can be a nuisance, particularly when they proliferate rapidly
(bloom) as a result of inorganic nutrient enrichment of surface waters. Much of the
effort in the removal of these inorganics from wastewaters has been aimed at the
elimination of algal blooms.

PROTOZOA. Protozoa are broadly classified as nonphotosynthetic, typically
motile, unicellular, eucaryotic protists. Within the broad definition, four major
groups have been identified: Mastigophora, Sarcodina, Sporozoa, and Ciliata.

Mastigophora always possess flagella as the means of locomotion and are commonly
found as parasites in animals. Trypanosomes, for instance, live in the blood of
animals and cause many diseases, including African sleeping sickness. Other
Mastigophora are harmless inhabitants of the gut of vertebrates and invertebrates;
one, for example, lives in a symbiotic relationship with its host, the termite.

The familiar amoebas belong to the subdivision Sarcodina. Within this group
amoeboid locomotion is the most common, although some can produce flagella. Most
are free-living soil or water organisms which phagocytize smaller prey, such as
bacteria. They can often be seen grazing around the edges of activated sludge floc
particles. One particular Sarcodina, Entamoeba histolytica, is pathogenic, causing
amoebic dysentary in humans. It will grow only in the intestine, but can be trans-
mitted in cyst form by fecal contamination of food and water. The cysts are very
resistant to disinfection.

Sporozoa are a diverse group of osmophilic, parasitic protozoa which are either
immotile or show a gliding movement. The most important members of this class are
the coccidia, usually parasites of birds, and the plasmodia, which infect birds and
mammals, including humans. Although these organisms are of no importance in biochemical operations, they are significant to public health officials, being the
agents that cause malaria.

Ciliates are the largest and most varied group of protozoa possessing some of
the most structurally complex cells in the living world. At some stage in their life
cycle, all ciliates are motile by means of cilia, which are short, hair-like pro-
jections similar to flagella. Each cilium arises from a basal structure which is
interconnected with the others to form an elaborate, compound locomotion system.
The ciliates also have two kinds of nuclei: the micronucleus, which is concerned
only with inheritance and sexual reproduction, and the macronucleus, which is responsible for various aspects of cell growth and function. These organisms play an
important role in biochemical operations, where they feed upon bacteria and aid in
both bioflocculation and clarification.

FUNGI. The group of microorganisms called fungi includes more than 75,000
named species. Most are coenocytic (multinucleate) organisms which have a vegetative
structure known as a mycelium. The mycelium consists of a rigid, branching system
of tubes, fairly uniform in diameter. through which flows a multinucleate mass of
cytoplasm. A mycelium normally arises by the germination and outgrowth of a single
reproductive cell, or spore. Upon germination, the fungal spore puts out a long
filament, or hypha, which branches repeatedly as it elongates, to form the mycelium.
Growth is confined to the tips of the hyphae, but since the mycelium is capable of
almost indefinite growth, it frequently attains macroscopic dimensions. The habi-
tats of fungi are quite diverse, although most are aerobic. Many are aquatic although most are terrestrial, living in the soil or on dead plant matter, where they
play a large role in the carbon cycle. All fungi get their energy from the degradation of organic matter, and are capable of assimilating an extraordinarily wide
range of organic materials. In fact, they can often grow on the debris left after
bacterial degradation. Whereas this metabolic characteristic makes them valuable
in some biochemical operations, they become a nuisance in others because their
filamentous structure makes them difficult to remove in normal settling basins.

The classification of fungi is complicated, being based upon their morpho-
logical structure and their life cycle, so only a brief description will be given.
The phycomycetes are considered to be primitive fungi because they share two properties which separate them from other fungi: (1) Their asexual spores are formed
internally whereas the spores of other fungi are formed free at the tip of the
hyphae; (2) the mycelium shows no cross walls except in the region of specialized
cells. Such a mycelium is called a nonseptate mycelium. Many of the phycomycetes
are aquatic and are commonly referred to as water molds. Their spores are motile.
The terrestrial phycomycetes have nonmotile spores and are called zygomycetes.

Higher fungi have septate mycelia, i.e., they contain crosswalls between
nuclei. These organisms are still coenocytic, however, because the crosswalls contain pores through which the cytoplasm and nuclei can move freely. There are two
major groups, depending upon how the sexual spores are produced. Ascomycetes produce sexual spores within a sac, called an ascus. Basidiomycetes produce sexual
spores upon a club-shaped structure called a basidium. The most familiar examples
of the class Basidiomycetes are the mushrooms. There are a number of fungi which
do not produce sexual spores (or at least have not been observed to produce them).
These organisms are grouped under the classification Fungi Imperfecti. If they are
observed to produce sexual spores they are reclassified into one of the other
groups. Some of the imperfect fungi are human pathogens causing diseases of skin
membranes and lungs.

Yeasts are higher fungi that have lost the mycelial mode of growth and become
unicellular. Although budding is the predominant mode of multiplication, there are
a few that multiply by binary fission. Yeasts have many industrial uses, such as

alcoholic fermentations and single cell protein production. They are very common
in soil and water.

7.2.3 Procaryotes

The procaryotic protists exhibit much less structural variation then do the
eucaryotic ones, but no other major biological group is as metabolically diverse.
Other than the major difference in cellular organization, procaryotes differ from
eucaryotes in three other ways: (1) they are smaller; (2) their cell walls have a
unique chemical composition; and (3) they do not contain sterols or require them
for growth.

Procaryotes can be divided into two major groups: bacteria and blue-green
algae. The blue-green algae are easy to identify on physiological grounds. The
bacteria, on the other hand, share no common properties other than the fact that
they are procaryotic protists which are not blue-green algae. The bacteria may be
divided into three subgroups on the basis of structural and physiological characteristics. Myxobacteria have a thin, delicate cell wall and are motile by a
gliding movement. Spirochetes also have a thin and delicate cell wall, but they
have a unique structure and means of motility. The eubacteria are a large and
varied group, all having thick, rigid cell walls. The major characteristics of
these procaryotic groups are summarized in Table 7.2.

BLUE-GREEN ALGAE. The mechanism of photosynthesis in procaryotic and eucaryotic algae is identical although the photosynthetic pigments are different. Even
though the name implies that all are blue-green in color, species have been observed which are red, brown, brownish-purple, or almost black, depending upon the
pigment present. Many of these algae are unicellular and multiply by binary
fission. They are usually immotile. Others are multicellular, consisting of filaments of cells held together by a common outer wall. The filaments are not branched
and increase in length by repeated divisions of their component cells in one plane.
Reproduction occurs by liberation of short elements containing a few cells which
then divide to form a new filament. All of the filamentous forms are motile with
the entire filament being capable of gliding movement.

Blue-green algae have an exceptionally wide natural distribution, occurring
in soil, fresh water, and oceans in all parts of the world. Some of them can fix
atmospheric nitrogen. They are the only photosynthetic organisms to do so and
therefore have the simplest nutritional requirements of any organisms. When growth
of eucaryotic algae becomes limited by nitrogen, then nitrogen-fixing procaryotic
algae often predominate. This is a possible explanation of why blooms of blue-
green algae often occur late in the season under eutrophic conditions.

MYXOBACTERIA. The myxobacteria are small, rod-shaped unicellular organisms
that reproduce by transverse binary fission. They are all motile by gliding movement and are often grouped with the "gliding bacteria". One subgroup, the fruiting
myxobacteria, exhibits the most complex life cycle and behavioral pattern found in

TABLE 7.2
MAJOR GROUPS OF PROCARYOTIC PROTISTS

	Blue-Green Algae	Myxobacteria	Spirochetes	Eubacteria
Mechanism of cellular movement	Gliding or immotile	Gliding	Axial filament	Flagella or immotile
Cell Wall	Thick, rigid	Thin, flexible	Thin, flexible	Thick, rigid
Range of organismal structure				
Unicellular	+	+	+	+
Filamentous	+	-	-	+
Coenocytic (mycelial)	-	-	-	+
Shape of cells (unicellular forms)				
Rods	+	+	-	+
Spheres	+	-	-	+
Spirals	-	-	+	+
Resting cells (some members only of each group)	Akinetes	Microcysts	None	Endospores, cysts
Nutritional categories				
Photoautotrophic	+	-	-	+
Photoheterotrophic	-	-	-	+
Chemoautotrophic	+a	-	-	+
Chemoheterotrophic	+a	+	+	+

aFilamentous gliding organisms (e.g., Beggiatoa) which can be regarded for taxonomic purposes either as nonphoto-synthetic representatives of the blue-green algae or as a special group of filamentous, gliding bacteria.

Reproduced from Stanier, R. Y., Doudoroff, M. and Adelberg, E. A., The Microbial World, 3rd Edition, 1970.
Reprinted by permission of Prentice-Hall, Inc., Englewood Cliffs, N.J.

procaryotic organisms. These organisms obtain their nutrients by causing the lysis
of other organisms and are known to feed on a wide variety of bacteria, as well as
fungi, yeast, and algae. Most require amino acids for growth and with the exception
of some that use cellulose, cannot use carbohydrates. Members of the other subgroup,
the cytophaga, are long slendar rods which are widespread in soil and water. Some
cause diseases in fish.

   SPIROCHETES. The spirochetes are easily recognized by their characteristic
form. A cell is very long relative to its width, is heliocoidal in shape, and is
surrounded by a delicate, flexible wall. A unique axial filament is spirally wound
around the cell between the cell membrane and the flexible cell wall (or outer
sheath). The organisms are highly motile and can move through liquids by rapidly
rotating around the longitudinal axis or by lashing, bending, curling, and making
snakelike contortions. The mechanism of this motion is not yet understood. Free-
living spirochetes are inhabitants of mud and water where most appear to be anaerobic. Others are parasites of mollusks and vertebrates and include the agents responsible for several human diseases, including syphilis, yaws, relapsing fever,
and one type of infectious jaundice.

   EUBACTERIA. The rest of the procaryotes constitute a large and diverse group
of organisms known as the eubacteria. Table 7.3 is a listing of the major subgroups.
Since to describe each one would take more space than is possible in a text of this
type, we will discuss here only the major structural and developmental characteristics of eubacteria. Metabolic and energetic characteristics will be discussed in
Chapters 8 and 9.

TABLE 7.3
Major Groups of Eubacteria
_____

Photosynthetic bacteria
Gliding bacteria
Sheathed bacteria
Stalked, budding and appendaged bacteria
Gram-negative, aerobic, polarly flagellated heterotrophs
Sulfur- and iron-oxidizing chemoautotrophs
Nitrifying bacteria
Methane-oxidizing bacteria
Methane-producing bacteria
Rhizobium-Agrobacterium group
Azotobacter
Acetic acid bacteria
Enteric bacteria
Gram-negative cocci and coccobacilli
Lactic acid bacteria
Gram-positive cocci
Endospore-forming bacteria
Coryneform bacteria
Propionic acid bacteria
Mycobacterium
Actinomycetes
Mycoplasmas
_____

Adapted from Mitchell [4].

Reference to Table 7.2 reveals that unicellular eubacteria can be found with any of the three basic cellular shapes: rods, spheres, and spirals. All are characterized by their small size, with a minimum dimension of around 0.5µm. A spiral cell is called a spirillum and can be differentiated from a spirochete by the absence of an axial filament. In many eubacteria, separation of daughter cells occurs immediately after division so that they exist in a unicellular state. Others stay together, yielding cell aggregates. In rods and spirilla, division occurs at right angles to the long axis. Thus if these cells stay together, the result is a single-stranded chain of cells. It should be emphasized, however, that the organizational structure is not the same as the filamentous growth of blue-green algae in which the cells share a common outer wall. Rather, each cell has a separate wall and is connected to its neighbor only at the end. When a spherical cell (coccus) divides without separation of the daughter cells, more complex aggregates can result because spheres can divide in any number of planes. If all successive divisions occur in parallel planes, then a long chain of cocci will result, very much like a string of beads. On the other hand, if a coccus divides in one plane forming a pair, and if that pair then divides on a second plane at right angles to the first, the result will be flat plate with 4 cocci in it. With some organisms this has been observed to continue resulting in a square flat plate with 64 organisms in it. Finally, successive divisions can occur in three planes at right angles to each other, resulting in cubical packets of eight cells.

Table 7.2 also reveals that some eubacteria are motile by flagella while others are immotile, a difference which aids in identification. Furthermore, the placement of the flagella is an identifying characteristic. As a general rule, cocci and a large number of rods are immotile. In rods and spirilla, the two major types of flagellar attachment are polar and peritrichous. Peritrichous refers to the attachment of flagella at many points along the sides of the cell, but in polar flagellation the placement is limited to one or both ends of the cell. There are, in addition, two subgroups of polar flagellation. If the cell has only a single flagellum it is termed polar monotrichous flagellation. This occurs mostly in rods. Many spirilla have a large number of flagella at one or both poles, or, polar multitrichous flagellation.

Another characteristic listed in Table 7.2, which can be used to distinguish among the eubacteria is the organismal structure. By and large, the vast majority are unicellular, although they may grow in aggregates. True filamentous growth is rare, although it does occur. Generally, it occurs in a manner similar to that of blue-green algae, with about 10-30 cells per filament strand. One large group, the actinomycetes, exhibits mycelial growth similar to fungi. Many biochemical operations contain flocculent cultures of microorganisms. A floc particle is made up of many types of microorganisms growing as a closely associated group and thus is not a characteristic organismal structure.

A few eubacteria form resting cells. This characteristic is so unique that it serves as a guide to their identification. Some form spores much like eucaryotic cells. These are formed inside a vegetative cell, one per cell, and are referred to as endospores. These spores are often formed in response to adverse conditions and are very resistant to extremes of temperature, etc. They are released by cell lysis and can survive for long periods of time. When conditions are again conducive to growth they germinate, yielding a vegetative cell which then divides. Another type of resting cell formed by eubacteria is a cyst. It is formed by the shortening of a rod-shaped vegetative cell, accompanied by the secretion of an unusually thick cell wall. These resting cells are not significantly more heat resistant than the vegetative cell, although they can survive long periods of starvation.

One characteristic not listed in Table 7.2 that is commonly used to aid in the identification of eubacteria is the reaction to Gram stain. This technique, developed in 1884 by Christian Gram, subjects the organism to successive staining with crystal violet and iodine, followed by washing with an organic solvent such as alcohol or acetone. The cell protoplast is stained by the crystal violet and the iodine yielding a purple color. When certain cells are then washed with the solvent they are decolored. These are referred to as Gram-negative. The others are Gram-positive. Studies have shown that the composition and structure of the cell wall determines whether the protoplast will be decolored. Thus this rather simple test allows separation of the eubacteria into two broad groups based upon cell wall structure.

Much of the separation of eubacteria into individual species is based upon their metabolic and energetic characteristics. These characteristics largely determine the usefulness of the organisms in biochemical operations. They will be discussed in Chapters 8 and 9.

7.2.4 Viruses

Viruses were originally classified according to the type of cell which they infect, the means of transmission from cell to cell, and the disease symptoms produced. As more viruses were discovered, and more was learned about their structure and chemistry it became possible to establish more definitive criteria for classification. The first major division that can be made is one based upon the type of genetic material carried by the virion, DNA or RNA. Within each of those divisions, further subdivisions may be established based upon the structure of the capsid, or protein coat surrounding the nucleic acid. Still further subdivisions can be made according to the site of replication of the virus in the host cell, the size of the virus particle, and its susceptibility to various inactivating agents. All of the above are now considered to be the primary characteristics upon which classifications are made.

The range of virus hosts covers almost the whole living world. Viruses infecting warm-blooded animals are probably best known, for they are responsible for

diseases such as influenza, polio, smallpox, measles, mumps, and infectious hepa-
titis. Viruses also attack insects and plants, causing considerable economic loss
in the latter. Viruses have been found which attack fungi, but although they
probably exist, none have been isolated from eucaryotic algae and protozoa. Finally,
viruses which attack blue-green algae and bacteria have been isolated. The latter
are known as bacteriaphage and have been extensively studied because of the information that they provide about bacterial genetics. Phage no doubt play a role in
the microbial interactions existing in biochemical operations but no attempts have
yet been made to delineate them. The major interest in viruses by environmental
engineers has been in the removal and control of animal viruses in public water
supplies.

7.3 INTERACTIONS IN MICROBIAL COMMUNITIES

Although some industrially significant biochemical operations employ pure cultures
of microorganisms, very few wastewater treatment operations do. Almost all of them
run in open vessels receiving continuous innoculations from the surrounding environment, as well as from the wastewater itself. Consequently, these operations contain microbial communities of various degrees of diversity within which a multitude
of interactions are occurring.

7.3.1 General Nature of Microbial Interactions

Wastewater treatment operations contain communities of microorganisms made up of
populations of individual species. Within the community, changes can occur in the
populations present as the system responds to changes in the quantity and character
of the material entering in the feed stream, or to changes in the physical environment. If the nature of the waste being treated is such that it supports a broad
and diverse microbial community, then the community will adapt readily to changing
environmental conditions and the system will appear fairly stable from the macroscopic point of view. Under some circumstances the community may be restricted.
This leads to an unstable biochemical environment and a process which is difficult
to control. Generally, complex integrated communities with a large number of
diverse species are considered to be healthy ecosystems. The fact that they appear
stable from a macroscopic level should not be construed as indicating constancy at
the microscopic level, however. Indeed, the broad biochemical diversity responsible
for the macroscopic stability of the system is maintained by a complex series of
interactions between the microorganisms within it.

These interactions are categorized with regard to the effects that the organisms have on one another: neutral, benevolent, or antagonistic. Neutral interactions are those which have no observable effect, i.e., no real interaction. They
would be very difficult to confirm. Benevolent interactions are those in which one
organism is aided without any harm to the other. Antagonistic interactions involve
a detrimental effect upon one organism. Care must be used in categorizing

interactions because there is a tendancy to view the interaction as if it were the same under all conditions. This is not the case, because environmental and other factors may convert a benevolent relationship into an antagonistic one. The main reason for classifying interactions, however, is not to put each relationship into a specific niche, but rather to clarify in our minds the many possible ways that organisms affect one another, thereby helping us to understand a very complex situation.

The complexity of microbial ecosystems has been illustrated well by Brock [1a]. Consider a system consisting of 3 species, A, B and C. Let an arrow indicate that one organism is interacting with another. The first level of interaction would be between the individual species: $A \rightleftarrows B$, $A \rightleftarrows C$, $B \rightleftarrows C$. However, organism C may influence the interaction between A and B, etc., yielding:

$$
\begin{array}{ccc}
\boxed{\begin{array}{c} A \rightleftarrows B \\ \downarrow\uparrow \\ C \end{array}} \rightleftarrows &
\boxed{\begin{array}{c} A \rightleftarrows C \\ \downarrow\uparrow \\ B \end{array}} \rightleftarrows &
\boxed{\begin{array}{c} B \rightleftarrows C \\ \downarrow\uparrow \\ A \end{array}}
\end{array}
$$

Since each arrow can represent a neutral, benevolent, or antagonistic interaction, it can be seen that there are 81 possible interactions in the system, although all will never occur at one time. Add to this the effects of chemotaxis, which involves chemical signals between organisms, and the fact that most ecosystems contain more than three microbial species and the complexity of microbial communities becomes apparent. While it is doubtful that engineers will soon be able to formulate mathematically the effects of these interactions, a knowledge of their nature is helpful to a qualitative understanding of the systems.

7.3.2 Neutral Interactions

Only rarely would organisms be found which did not interact at all. About the only way that this could occur would be for two populations to be so different in environmental requirements that neither organism alters the qualities needed by the other. It seems likely that neutralism would be exhibited more frequently by extremely diverse organisms than by closely related ones, since related ones would probably compete for common resources.

7.3.3 Benevolent Interactions

The two most common types of benevolent interactions are commensalism and mutualism. Commensalism exists when one organism benefits while the other is unaffected. An example would be the use by one organism of material previously altered by another, such as in cellulose degradation. Cellulolytic microorganisms degrade the cellulose to yield glucose which is then used as a substrate by other microorganisms. Another form of commensalism would be the production of a growth factor or nutrient which cannot by synthesized by the second organism, but which is needed by it. The first organism excretes the material into the environment from which the second removes it.

Sometimes commensalism may involve an alteration of the physiological environment making it suitable for the growth of a second organism. For example, a facultative organism, which may live in either the presence or absence of oxygen, may grow and remove all of the oxygen from the microenvironment, thereby allowing growth of an obligate anaerobe. Finally, one organism may destroy or neutralize a toxic factor which would otherwise prevent growth of the second organism. Suppose an industrial waste contained phenol, which is toxic to many organisms. If the biochemical reactor contained a population capable of phenol degradation, then it would keep the concentration low enough to allow other populations to degrade the additional constituents of the waste. Many other examples of commensalism can be found in biochemical operations.

Mutualism is not as common as commensalism because it is an interaction in which both members benefit from the relationship. Each organism is deficient in its ability to produce a needed growth factor, but each excretes the growth factor needed by the other. Mutualistic interactions in the anaerobic degradation of cellulose are often cited, however, the most obvious example is the growth of algae and bacteria in oxidation ponds. Bacteria oxidize organic matter to produce carbon dioxide, which is then used by the algae, which produce oxygen for use by the bacteria. This example can also serve to illustrate how an interaction can change because during periods of darkness the algae require oxygen and then compete with the bacteria.

A third term which is sometimes used to denote a benevolent interaction is symbiosis. In that sense it is taken to mean an extremely close physical association and interchange of physiological functions. It has been suggested, however, that a broad definition to designate any type of coexistence is preferable to one denoting only beneficial associations [2].

7.3.4 Antagonistic Interactions

Antagonistic interactions range from simple competition to complex predator:prey relationships. Strict competition means that both organisms require some common resource which is present in limited amounts. Although one species may eventually predominate it arrives at that position without any direct attack on the other. This situation has been studied in laboratory reactors (CSTR) where it has been shown that the population capable of most rapid growth on the level of nutrient available will eventually cause the loss of the other organism from the system [3]. Pure competition occurs most frequently between closely related organisms with similar environmental requirements and is the interaction most easily expressed mathematically. In systems which are not completely mixed, regions of different concentrations allow different species to proliferate, thus decreasing competition. This often occurs in biochemical operations.

A number of antagonistic interactions can occur by alterations in the environment in a manner similar to (but opposite in effect to) benevolent interactions. For example, when facultative organisms consume all of the dissolved oxygen from the growth environment they prevent the growth of obligate aerobes. Similarly, the presence of algae and the resultant production of oxygen will prevent the growth of obligate anaerobes.

Several antagonistic interactions involve the release of chemicals which have a deleterious effect on some organisms. These are called amensalism. Antibiotics are organic chemicals which are produced by living organisms and which are able to kill or inhibit the growth of other living organisms. Although the controlled growth of antibiotics constitutes one of the largest industrial uses of biochemical operations, the importance of antibiotic production as a general controlling process in nature is still the subject of academic debate. Other chemicals of metabolic origin can be involved in antagonistic interactions although they are not antibiotics. For example, in alcoholic fermentations some organisms produce alcohol at high enough concentrations to prevent the growth of other microorganisms. Some produce organic acids, which can be directly inhibitory to other organisms, or can reduce the pH to levels where other organisms cannot grow. Finally, some organisms release lytic enzymes which attack the cell walls of other bacteria, causing them to break open. This gives the first group a definite advantage in any growth situation.

One very important form of an antagonistic interaction which is known to occur in biochemical operations is predation. Brock [1b] confines predation to phagotrophic organisms but Mitchell [4] takes the broader view that intermicrobial predation involves any microorganisms which prey on others. Many protozoa prey upon bacteria. Amoebas ingest them by phagocytosis, and thus fall within Brock's definition. Ciliates are known to play a role in biochemical operations where they prey upon bacteria. A form of predation is practised by bacteria of the genus *Bdellovibrio* which are tiny, highly motile, and obligately parasitic on other bacteria. They attach to the cell wall, lyse it, and consume the cytoplasm of the cell. Bacteria are not the only microorganisms subject to predation as algae, protozoa, and fungi are as well. Algae, for example, are often attacked by bacteria which excrete lytic enzymes and then consume the cytoplasm released when the cells lyse.

Bdellovibrio could also be classed under the antagonistic interaction known as parasitism because they can grow only by attacking other cells. Other examples of parasitism in microbial systems are viruses growing in bacteria, fungi on algae, and fungi on other fungi. The importance of parasitism in biochemical operations is not really known.

This brief description of the possible interactions between microorganisms serves to demonstrate the complex ecosystems at work in biochemical operations. Although the mathematical treatment of those operations does not account for those interactions, the astute engineer must be aware of them if he is to appreciate the limitations of his design models.

7.4 MICROBIAL ECOSYSTEMS IN BIOCHEMICAL OPERATIONS

An ecosystem is usually defined as the sum of the interacting elements (both bio-logical and environmental) in a limited universe. Consequently, each biochemical operation will develop a unique ecosystem governed by the physical design of the facility, the chemical nature of the wastewater going to it, and the biochemical changes wrought by the resident organisms on the system. The microbial community which develops in that ecosystem will be unique from the viewpoint of species diver-sity, being the result of physiological, genetic, and social adaptation. In fact, as changes occur in the physical and chemical environments, adaptation will operate to regulate the community structure. Thus it is impossible to generalize about the numbers and types of species which will be present. Nevertheless it would be in-structive to consider the general nature of community structures in biochemical operations and relate them to the environments in which the operations are carried out. The objective of such considerations is not just to list the organisms present, but rather to understand the role that each group plays in the operation.

7.4.1 Anaerobic Digestion

As indicated in Table 1.2, anaerobic digestion is usually carried out in a continuous stirred tank reactor in order to stabilize insoluble organic matter. The reactor commonly has a residence time of several days, receives a slurry of organic solids as a feed, and produces an effluent containing less organic matter as well as sub-stantial quantities of carbon dioxide and methane gas. The microbial community is relatively simple, consisting primarily of bacteria, although interesting and com-plex interactions are known to exist among them. Although fungi and protozoa have been observed in digester communities their importance is questionable [5].

Within the anaerobic digestion process, biodegradable organic compounds are hydrolyzed and degraded to yield a multitude of simpler compounds such as organic acids, carbon dioxide, and hydrogen gas which are then converted microbially to methane gas. Two broad classes of bacteria are involved in this transformation and live in a delicately balanced relationship. The first group, referred to as non-methanogenic bacteria convert the raw waste into organic acids, carbon dioxide and hydrogen gas. The second group, the methanogenic bacteria, carry out the production of methane gas. They are very substrate specific and are dependent on the first group for their supply of substrate; i.e., a commensal interaction. These organisms are very sensitive to changes in pH so that if the nonmethanogenic bacteria produce organic acids more rapidly than they can be removed, the pH drops, inhibiting the methanogenic bacteria and transforming the commensal interaction to amensalism. The delicate balance between these two bacterial groups is the biggest source of trouble in the process and often leads to difficulty in the operation of field-scale units.

Within the nonmethanogenic group of bacteria, fastidious obligate anaerobes are present in numbers one to two orders of magnitude greater than facultative bac-teria [6]. The predominant species are Gram-negative, non-spore-forming bacilli

which can produce acetic and butyric acids as well as carbon dioxide and hydrogen
gas. Although obligate anaerobes appear to have a competitive advantage over
facultative species, the latter group fills a small ecological niche which is not
yet well defined. It is possible, for example, that they play a role as scavengers
of any dissolved oxygen which enters the system, thereby maintaining the proper
growth environment for the obligate anaerobes. Other organisms which have been
shown to be of importance possess hydrolytic powers for carbohydrates, proteins, and
lipids, which are the major components in the feed. Although some attempts have
been made to establish the order of predominance in the nonmethanogenic community,
lab techniques have not been refined sufficiently for the results to be profit-
able [6]. Evidence does indicate, however, that complex nutritional requirements
exist for many of the organisms so that it is likely that commensalism and mutualism
play an important role within the nonmethanogenic group [3].

Studies on the nature of the methanogenic bacteria have been hampered by the
difficulties of working with very slow growing, obligate anaerobes. To date the
methane bacteria isolated from various environments tend to be unrelated morpho-
logically and range from minute cocci and larger sarcina to individual and chain-
forming bacilli [6]. The main features that they have in common are that they will
grow on only a limited number of substrates (methanol, formic acid, acetic acid,
CO_2, CO, and H_2) and that they produce methane gas. Mutualism is a common interaction within the group. For example, even though Methanobacterium omelianskii was
thought for many years to be a pure culture it is in reality a mutualistic interaction between two rod-shaped bacteria. The methane producing organism oxidizes
gaseous hydrogen with the subsequent reduction of CO_2 to methane. Its associate
oxidizes methanol to acetic acid and hydrogen gas. Since it is inhibited by hydrogen
gas, it is dependent upon the first organism to keep the concentration low. The
first organism, in turn, is dependent upon the second for its supply of hydrogen.

7.4.2 Activated Sludge

Activated sludge is the name used to describe aerobic biochemical operations which
use a flocculent microbial slurry to remove soluble and colloidal organic matter.
All employ gravity settling to separate the biomass from the treated effluent and
recycle to return the majority of it to the aeration chamber. As indicated in
Table 1.2 many reactor configurations are used, and to a degree, the one employed
influences the specific microbial community which developes. Nevertheless, there
are some generalities which can be made.

Species diversity in activated sludge is broader than that exhibited by the
community in an anaerobic digester. Whereas the latter contains procaryotic organisms almost exclusively, both procaryotes and eucaryotes play a significant role in
the former. Generally, the organisms in an activated sludge culture may be divided
into four major classes: floc-forming organisms, saprophytes, predators, and
nuisance organisms [7]. These are not distinct groups and in fact, any particular

organism may fit into more than one at a time or may change groups as the selective pressures within the community change.

Floc-forming organisms play a very important role in the process for without them the sludge could not be separated from the treated wastewater. Originally it was thought that the bacterium *Zooglea ramigera* was primarily responsible for floc formation but it has now been shown that a variety of bacteria are capable of flocculation [8]. Classification of organisms into the floc-forming group is complicated by the fact that protozoa and fungi can also cause bacteria to flocculate [7,8]. Nevertheless, this group is generally considered to be composed of bacteria. Flocculation is thought to be caused by natural polyelectrolytes, although their origin is uncertain.

The saprophytes are the organisms responsible for the degradation of organic matter. These are primarily bacteria and no doubt include most of the bacteria considered to be the floc formers. Nonflocculent bacteria are probably also present but are entrapped within the floc particles formed by the first group. The saprophytes can be subdivided into primary and secondary saprophytes, the primary ones being responsible for the degradation of the original substrates. No doubt the larger the number of substrates the more diverse the community will be because of less competition for the same substrates. The secondary saprophytes feed upon the metabolic products of the primary ones, indicating a high degree of commensalism within the community. Although most saprophytes are Gram-negative bacilli, there have been reports of fungi, yeast, and flagellated protozoa which compete directly with them for the original substrates [7].

The main predators in activated sludge communities are the protozoa which feed upon the bacteria. About 230 species have been reported to occur in activated sludge and they may constitute as much as 5% of the mass of biological solids in the system [8]. Ciliates are usually the dominant protozoa both numerically and from biomass estimations. All but one of them are known to feed on bacteria and the most important ones are either attached to or crawl over the surface of sludge flocs. On occasion both amoebae and flagellates may be seen in small numbers but they are not thought to play a major role in good settling, stable communities. It has been suggested that protozoa play a role in the formation of sludge flocs and contribute to the absence of dispersed bacteria in stable communities.

Nuisance organisms are those which interfere with the proper operation of the process when present in sufficient numbers. Most problems arise with respect to sludge settling and are the result of filamentous bacteria and fungi. If only a small percentage by weight of the community is made up of filamentous organisms, the effective specific gravity of the sludge flocs is reduced so much that the sludge is very difficult to separate by gravity settling. This leads to a situation known as bulking. The organisms that are often assigned responsibility for this situation are the bacterium *Sphaerotilus natans* and the fungus *Geotrichium*.

From the previous discussion it should be apparent that activated sludge represents a complex ecosystem within which many microbial interactions are occurring.
Pure competition is probably rare because of the diverse nature of the substrates
available but the more restricted the original substrate, the more severe the competition among the primary saprophytes. Even under that condition, however, there
will still be a diverse secondary population feeding on the many excretory products
of the primary saprophytes. An indication of the importance of commensalism and
mutualism in the community is the fact that only 8% of the bacteria isolated from
an activated sludge culture could grow without growth factors in a minimal medium [8]. Although predation by protozoa is an important interaction it has been
reported that neither phage nor *Bdellovibrio* play an important role in removing
bacteria from the sludge [3].

Selective mechanisms are also exerted upon the culture by environmental factors.
The importance of the nature of the incoming substrate has already been discussed.
Since this is an aerobic process the concentration of dissolved oxygen can also
play a role in the development of the community. After the sludge passes to the
sedimentation basin it may undergo a period without dissolved oxygen before being
returned to the aeration basin. The length of that period could well influence the
diversity of the community. Other factors of importance are pH and temperature.
Low pH's tend to favor the growth of filamentous fungi, while high temperature may
discourage the growth of protozoa. Finally, it has recently been discovered that
the hydraulic regime in the aeration basin plays an important role in determining
the relative numbers of filamentous and floc-forming bacteria [9]. This is one of
the first examples of how ecological studies can lead to better engineering design.
Hopefully, it will not be the last.

7.4.3 Trickling Filters

Trickling filters are biochemical reactors in which the biomass grows attached to a
solid surface over which the wastewater flows in thin sheets, providing the microbial
community with nutrients. Although the operation works best with soluble organic
matter, colloidal material is also removed by entrapment in the microbial film and
subsequent degradation.

Just as the activated sludge community is more diverse than the one in anaerobic digesters, so is the trickling filter community more diverse than that in activated sludge. In addition to a rich mixture of eucaryotic and procaryotic organisms,
trickling filters contain many higher life forms, notably nematodes, rotifers,
snails, sludge worms, and larvae of certain insects [10]. This more complex food
chain allows more complete oxidation of organic matter with the net result that
fewer excess organisms are produced. This has the beneficial effect of decreasing
the mass of solid material which must be disposed of.

The bacteria form the base of the food chain by acting upon the organic matter
in the wastewater. Soluble materials are taken up rapidly as the liquid passes by,

while colloidal-sized particles become entrapped in the rather slimy or gelatinous
layer built up by the bacteria. There they undergo attack by extracellular enzymes,
releasing small molecules which can be metabolized. The bacterial community is
composed of primary and secondary saprophytes, and though few studies have been done
to identify the specific bacteria present it seems likely that the populations would
be similar to those in activated sludge. For example, species of both *Sphaerotilus*
and *Zooglea* have been reported [10]. Most probably, the distribution of species
would change with depth through the filter.

Quite an extensive eucaryotic community is known to exist in trickling
filters [10]. Over 90 species of fungi have been reported and of these, more than
20 species were considered to be members of the permanent population. Their role
would be similar to that of the bacteria, i.e., saprophytic. Many protozoa have also
been found, with large communities of Sarcodina, Mastigophora, and Ciliata being
reported. Their roles are largely those of predators. During warm summer months
algae can flourish on the upper surfaces of the biomass. Usually green algae and
diatoms predominate.

Finally, trickling filters also contain a large metazoan community, consisting
of annelid worms, insect larvae, and snails. These feed on the microbial film and
are largely responsible for the phenomenon of sloughing in which the support medium
becomes bare.

Because of the very diverse nature of the community, the interactions within it
are extremely complex. In addition to the microbial interactions listed in the last
two sections, there is also the preying of the higher life forms upon the eucaryotes.
Few attempts have been made to relate· operational conditions to the ecology of the
system and we have even less understanding of this ecosystem than of anaerobic
digestion and activated sludge.

7.4.4 Oxidation Ponds

Oxidation ponds, also referred to as facultative stabilization ponds, are not shown
in Table 1.2 because they incorporate all objectives of wastewater treatment into a
single unit. They are large shallow ponds (25 ft.) with residence times between 45
and 75 days. If raw wastewater is introduced into them the solids settle to the
bottom where they undergo anaerobic decomposition; the colloidal and dissolved
organics are degraded by bacteria in the liquid; and inorganic nutrients are consumed by algae as they grow. Marias [11] has referred to stabilization ponds as the
first rung on the ladder of biochemical operations. While that is true in the sense
that engineers can exert little or no control over the operation once it is built,
it is misleading with regard to the ecological complexity of the system, for these
units are by far the most complex of all.

The ecology of the bottom layer is influenced by the anaerobic conditions therein and is similar to that of anaerobic digestion. The solid material is attacked

by nonmethanogenic bacteria. A portion of the soluble products is used by the methanogenic bacteria and the rest is released to the overlying liquid.

Within the overlying liquid, facultative bacteria degrade the soluble and colloidal organic matter, converting it to bacteria which drop to the bottom and undergo anaerobic decomposition. Conditions within the liquid zone evidently are not suitable for enteric bacteria because large die-offs have been reported [12]. In addition to facultative bacteria, over 100 species of filamentous fungi and 50 species of yeast have been found in these ponds [12], attesting to the complex nature of the ecosystem. If the pond is predominately aerobic, then protozoa can be found as well. Most of the oxygen used by the bacteria, fungi, and yeast comes from the algae which also reside in the liquid zone. The majority are eucaryotic algae, although blue-green ones can predominate at times. The type of eucaryotic algae present depends upon the mixing conditions in the pond [11]. If the pond is well mixed by wind and wave action, then immotile algae predominate and are pretty well distributed throughout the mixed zone. If the pond is stratified, on the other hand, then motile algae such as *Euglena* tend to predominate. In this case, only the upper reaches of the pond will be aerobic.

The interactions within the microbial community of an oxidation pond are obviously very complex. In addition to most of the ones found in anaerobic digestion and activated sludge, there is the mutualistic relationship between the bacteria and the algae. During hours of darkness, however, that relationship can switch to a competitive one as they both undergo respiration.

7.5 KEY POINTS

1. The maintenance of the biosphere is dependent upon the cyclic movement of all elements through the atmosphere, hydrosphere, and lithosphere, but particularly upon the cycles of carbon, oxygen, nitrogen, and sulfur.

2. Oxidized inorganic carbon (CO_2) is reduced to organic carbon by photosynthesis of plants and algae. The organic carbon, in turn, can be oxidized by a variety of microorganisms, converting it back to carbon dioxide.

3. Atmospheric nitrogen is converted into organic form by specialized organisms and thereafter is available for use by all other organisms. Eventually nitrogen is returned to the atmosphere through the action of other organisms, thereby completing the cycle.

4. Protists, which constitute the third kingdom of life, fall into two categories; the eucaryotes which have a complex internal structure, and the procaryotes which are simpler.

5. The major groups of eucaryotic protists are eucaryotic algae, protozoa, and fungi. Algae are photosynthetic organisms whereas protozoa and fungi are not.

6. There are two types of procaryotic protists: blue-green algae and bacteria. The types of bacteria are very diverse and share no common feature other than the fact that they are procaryotic protists which are not blue-green algae.

7. Microbial interactions are categorized with regard to the effects that the organisms have on one another: neutral, benevolent, or antagonistic.

8. Two common benevolent interactions are commensalism and mutualism. Commensalism exists when one organism benefits while the other is unaffected, whereas mutualism exists when both members benefit from the relationship.

9. Antagonistic interactions range from simple competition to complex predator:prey relationships.

10. Although the microbial community in an anaerobic digester is primarily bacterial, a number of complex microbial interactions occur.

11. The major microbial groups found in activated sludge are: floc-forming organisms, saprophytes, predators, and nuisance organisms.

12. The ecosystems of trickling filters are quite complex and contain many higher life forms in addition to a rich mixture of eucaryotic and procaryotic organisms.

13. The mutualistic interaction found between the bacteria and algae in an oxidation pond during the daylight hours switches to a competitive one at night.

7.6 STUDY QUESTIONS

The study questions are arranged in the same order as the text:

Section	Questions
7.1	1-2
7.2	3-6
7.3	7-8
7.4	9-10

1. Draw sketches depicting the carbon and oxygen cycles, showing how they are interrelated.

2. Draw a sketch depicting the nitrogen cycle, naming all steps within it.

3. Give the function of each of the following in eucaryotic cells: nucleus, mitochondria, chloroplasts, vacuoles, lysosomes, and cell wall.

4. Describe the function of each of the following in procaryotic cells: ribosomes, cell membrane, and flagella.

5. Discuss how eucaryotic algae, protozoa, and fungi interact with people and their attempts to control environmental pollution.

6. List and differentiate between the various types of procaryotic protists.

7. Give examples of commensalism and mutualism between microorganisms.

8. List some antagonistic interactions found among microorganisms.

9. List and give examples of three microbial interactions occurring in anaerobic digestion.

10. Describe the role played by each of the microbial groups in the activated sludge process.

REFERENCES AND FURTHER READING

1. T. D. Brock, *Principles of Microbial Ecology,* Prentice-Hall, Inc., Englewood
    Cliffs, N.J., a. 144-145; b. 142-144, 1966.

2. H. R. Bungay, III and M. L. Bungay, "Microbial interactions in continuous
    culture," *Advances in Applied Microbiology,* 10, 269-290, 1968.

3. J. L. Meers, "Growth of bacteria in mixed cultures," *Critical Reviews in
    Microbiology,* 2, 139-184, 1973.

4. R. Mitchell, "Ecological control of microbial imbalances," Chapter 11 in
    Water Pollution Microbiology, Edited by R. Mitchell, Wiley-Interscience, New
    York, N.Y., 273-288, 1972.

5. D. F. Toerian and W. H. J. Hattingh, "Anaerobic digestion - I - The microbiology
    of anaerobic digestion," *Water Research,* 3, 385-416, 1969.

6. E. J. Kirsch and R. M. Sykes, "Anaerobic digestion in biological waste treat-
    ment," *Progress in Industrial Microbiology,* 9, 155-237, 1971.

7. W. O. Pipes, "The ecological approach to the study of activated sludge,"
    Advances in Applied Microbiology, 8, 77-103, 1966.

8. E. B. Pike and C. R. Curds, "The microbial ecology of the activated sludge
    process" in *Microbial Aspects of Pollution,* Edited by G. Sykes and F. A.
    Skinner, Academic Press, London, 123-148, 1971.

9. J. Chudoba, *et al.,* "Control of activated sludge filamentous bulking - I -
    Effect of the hydraulic regime or degree of mixing in an aeration tank,"
    Water Research, 7, 1163-1182, 1973.

10. W. B. Cooke, "Trickling filter ecology," *Ecology,* 40, 273-291, 1959.

11. G. V. R. Marais, "Dynamic behavior of oxidation ponds," *Proceedings of the 2nd
    International Symposium for Waste Treatment Lagoons,* 14-46, June, 1970.

12. N. Porges and K. M. MacKenthun, "Waste stabilization ponds: Use, function,
    and biota," *Biotechnology and Bioengineering,* 5, 255-273, 1963.

Further Reading

Brock, T. D., *Biology of Microorganisms,* Second Edition, Prentice-Hall, Inc., Engle-
    wood Cliffs, N.J., 1974. An excellent general microbiology text. Chapter 16
    contains a particularly good discussion of the procaryotic groups. See
    Chapters 1, 2, 3, 9, 12, 14, 15, 16 and 17.

Doetsch, R. N. and Cook, T. M., *Introduction to Bacteria and their Ecobiology,*
    University Park Press, Baltimore, Md., 1973. Excellent photomicrographs de-
    picting the various types of bacteria. See Chapters 1, 2, 3 and 6.

Hobson, P. N. and Shaw, B. G., "The role of strict anaerobes in the digestion of
    organic matter," in *Microbial Aspects of Pollution,* Edited by G. Sykes and
    F. A. Skinner, Academic Press, London, 103-121, 1971.

Mitchell, R., *Introduction to Environmental Microbiology,* Prentice-Hall, Inc.,
    Englewood Cliffs, N.J., 1974. A survey of the interrelationships between
    microbiology, ecology, and pollution control. See Chapters 1, 2, 3, 4, 10,
    13 and 15.

Stanier, R. Y., *et al., The Microbial World,* Fourth Edition, Prentice-Hall, Inc.,
     Englewood Cliffs, N.J., 1976. A very complete microbiology text. Excellent
     photomicrographs. See Chapters 3, 4, 5, 12, 16, 17, 18, 19, 20, 21, 22, 23,
     24, 26, 27, 28 and 31.

CHAPTER 8

METABOLISM - THE KEY TO BIOCHEMICAL TRANSFORMATIONS

In the previous chapter the organisms responsible for the transformations occurring
in biochemical operations were discussed, as were their interactions within the
microbial communities. While such knowledge is important to an understanding of the
behavior of those communities, it does not tell us how the desired transformations
are carried out. To understand that we must examine the cells at the molecular
level. We must see how an organic substrate is broken down, how energy is derived,
and how the cell uses both energy and carbon to synthesize new cell material. In
other words, we must investigate how the metabolism of the cell functions.

8.1 COMPOSITION OF MICROBIAL CELLS

The first requirement of an investigation of metabolism is a knowledge of how the
cell is made, i.e., what elements are required and how they are arranged into the
cellular constituents.

8.1.1 Elementary Composition

The first attempts to analyze microbial cells concentrated on the elementary composition. Most contain about 50% carbon, 20% oxygen, 10-15% nitrogen, 8-10%
hydrogen, 1-3% phosphorus and 0.5-1.5% sulfur on a dry weight basis [1]. Although
variations in these percentages occur as growth conditions are changed [2], it is
common practice to assume a constant empirical formula when writing stoichiometric
equations for microbial growth. On an ash-free basis, the formula $C_5H_7O_2N$ has
found wide-spread use [3]. The ash left after incineration includes potassium,
calcium, magnesium, sodium, iron and trace amounts of manganese, cobalt, copper,
molybdenum, and zinc. Although these elements are known to play a role in metabolism,
their content varies rather widely, reflecting more the nature of the medium and the
growth conditions than actual variations in composition of the protoplasm.

8.1.2 Chemical Composition

Of more importance than the elemental composition is the chemical composition because the roles played by molecules within the cell are determined by their structure. Most microbial constituents fall into four categories: nucleic acids,
proteins, carbohydrates, and lipids. Some nucleic acids constitute the genetic
material which serves as a storehouse of all the information needed for the cell to
function. Others act to interpret that information, using it as a blueprint for the
formation of specialized proteins, called enzymes, which actually perform the cellular functions by catalyzing chemical reactions. Noncatalytic proteins serve a
structural role in the cell, as do lipids and carbohydrates. Proteins and lipids
occur together in the cell membrane, while cell walls contain lipids, proteins, and
carbohydrates. Carbohydrates are also found in nucleic acids as well as in energy
reserves within the cell. The relative weights of these classes of compounds vary
widely depending upon the growth conditions of the microbial culture [4]. As one

example, a typical rapidly growing culture of *E. coli* will contain approximately 60% protein, 15% lipid, 19% nucleic acid, and 6% carbohydrate.

NUCLEIC ACIDS. Cells contain two types of nucleic acid, deoxyribonucleic acid (DNA) and ribonucleic acid (RNA), both of which are polymers of nucleotides, consisting of a nitrogenous base, a sugar, and phosphate. The two differ structurally, however: RNA contains the pentose ribose (Rib), while DNA contains deoxyribose (dRib). They also have different functions: DNA is genetic material; RNA aids in the synthesis of protein and in the transfer of information from the DNA to the site of protein synthesis.

In DNA the bases are the purines adenine (Ade) and guanine (Gua), and the pyrimidines cytosine (Cyt) and thymine (Thy). The nucleotides are linked through the phosphates into a chain which can be represented as:

$$-\text{dRib} - \textcircled{P} - \text{dRib} - \textcircled{P} - \text{dRib} - \textcircled{P} - \text{dRib} - \textcircled{P} - \text{dRib} - \textcircled{P} -$$
$$\quad\quad | \quad\quad\quad\quad | \quad\quad\quad\quad | \quad\quad\quad\quad | \quad\quad\quad\quad |$$
$$\quad\text{Cyt} \quad\quad\quad \text{Ade} \quad\quad\quad \text{Thy} \quad\quad\quad \text{Ade} \quad\quad\quad \text{Gua}$$

It is this sequence of nucleotides which constitutes the genetic code. As we will see in Section 8.4.4 each sequence of three nucleotides specifies one bit of information which will be translated and used in the synthesis of proteins. DNA usually occurs as a double stranded molecule in which the individual strands are linked through hydrogen bonding between complementary bases. Because of bonding distances the purine adenine can pair only with the pyrimidine thymine and guanine can pair only with cytosine:

$$-\text{dRib} - \textcircled{P} - \text{dRib} - \textcircled{P} - \text{dRib} - \textcircled{P} - \text{dRib} - \textcircled{P} - \text{dRib} - \textcircled{P} -$$
$$\quad\quad | \quad\quad\quad\quad | \quad\quad\quad\quad | \quad\quad\quad\quad | \quad\quad\quad\quad |$$
$$\quad\text{Cyt} \quad\quad\quad \text{Ade} \quad\quad\quad \text{Thy} \quad\quad\quad \text{Ade} \quad\quad\quad \text{Gua}$$
$$\quad : \quad\quad\quad\quad : \quad\quad\quad\quad : \quad\quad\quad\quad : \quad\quad\quad\quad :$$
$$\quad\text{Gua} \quad\quad\quad \text{Thy} \quad\quad\quad \text{Ade} \quad\quad\quad \text{Thy} \quad\quad\quad \text{Cyt}$$
$$\quad\quad | \quad\quad\quad\quad | \quad\quad\quad\quad | \quad\quad\quad\quad | \quad\quad\quad\quad |$$
$$-\text{dRib} - \textcircled{P} - \text{dRib} - \textcircled{P} - \text{dRib} - \textcircled{P} - \text{dRib} - \textcircled{P} - \text{dRib} - \textcircled{P} -$$

It is this complementary pairing which allows new molecules of DNA to be formed with little chance for error to occur in the sequence of nucleotides. The normal configuration of the molecule is that of a double helix and the hydrogen bonding between the bases in the complementary strands stabilizes the structure. Because of the base pairing the amount of adenine is always equal to the amount of thymine (likewise with guanine and cytosine). The mole percentage of guanine - cytosine pairs in DNA is constant for a given microbial species and this fact is often used in classifying microorganisms.

As the name implies, ribonucleic acid contains the pentose ribose rather than deoxyribose. It contains the bases adenine, guanine, and cytosine as in DNA, but the second pyrimidine is uracil (Ura) instead of thymine. It, too, is arranged in a chain, although it is single stranded:

-Rib - (P) - Rib - (P) - Rib - (P) - Rib - (P) - Rib - (P) -
 | | | | |
 Gua   Ura   Ade   Ura   Cyt

There are three kinds of RNA, each with a separate function. Messenger RNA (mRNA)
is formed by complementary base pairing with a portion of one strand of the DNA in
a process called transcription. Again, the base pairing allows the nucleotide
sequences to be transcribed with a high degree of accuracy. The strand of RNA then
contains the information needed to allow synthesis of one or more protein molecules,
hence its name as "messenger". The mRNA moves to the ribosomes, which are made of
protein and a special RNA called ribosomal RNA (rRNA). The mRNA interacts with the
ribosome, thereby bringing about translation of the code carried by the mRNA and the
subsequent synthesis of new protein. The third type of RNA, transfer RNA (tRNA),
acts to bring the subunits of protein into place in the proper sequence to form the
protein coded for by the mRNA.

PROTEINS. Proteins perform diverse functions within the cell. Many are
catalytic and as such are called enzymes. The characteristic which distinguishes
enzymes from nonbiological catalysts is their high degree of specificity for a
particular substrate. This specificity results from both the nature of the individual monomers forming the active site and the conformation of the entire molecule.
Since each enzyme is very specific, a microbial cell must contain a large number of
them. Typically a cell can produce 1000 or so, which are synthesized as needed,
under the direction of the genetic information carried in the DNA. Other proteins
have a structural function in cell walls and membranes. While this function is also
related to the three-dimensional configuration of the molecule, there is no catalytic
activity involved. Finally, proteins aid in the movement of microbes. For example,
bacterial flagella consist wholly of protein.

The monomers of proteins are amino acids, which have the general formula:

$$NH_2 - CH - \overset{\displaystyle O}{\overset{\displaystyle \|}{C}} - OH$$
$$|$$
$$R$$

Each amino acid has a separate side chain, R, and 20 different ones commonly occur
in proteins. The amino acids are joined by peptide bonds into a chain:

$$NH_2 - CH - \overset{O}{\overset{\|}{C}} - NH - CH - \overset{O}{\overset{\|}{C}} - NH - CH - \overset{O}{\overset{\|}{C}} - etc.$$
$$\quad\quad | \quad\quad\quad\quad | \quad\quad\quad\quad |$$
$$\quad\quad R_a \quad\quad\quad R_b \quad\quad\quad R_c$$

A chain consisting of only a few amino acids is called a peptide chain, or polypeptide. The sequence of amino acids is called the primary structure and determines

the function of the protein. It is this sequence which is coded for by the DNA.

The chains do not usually exist in a straight extended form, but instead, may coil into a spiral with eleven amino acids for each three turns. This α-helix is called the secondary structure and is stabilized principally by hydrogen bonds between NH and O = C groups in the chain. Many proteins consist of several polypeptide chains joined together through disulfide bridges. The amino acid cysteine contains a sulfhydryl group in its side chain:

$$
\begin{array}{c}
\qquad\qquad O \\
\qquad\qquad || \\
NH_2 - CH - C - OH \\
\quad\ |\ \\
\quad\ CH_2 \\
\quad\ |\ \\
\quad\ SH
\end{array}
$$

When two chains come together the sulfhydryl groups may be oxidized to form a disulfide bridge (- S - S -), thereby linking them. Two cysteine monomers in a single chain can likewise join together to form an intrachain bridge. Other bonds can be formed by side group interactions thus forcing the entire molecule to assume a specific, three-dimensional shape, called the tertiary structure of the protein which helps to form the site at which catalysis occurs. Bonding restrictions within the site limit the molecules which can fit into it, thereby imparting to enzymes the high degree of chemical specificity for which they are noted. Both the secondary and tertiary structures of a protein are determined solely by the primary structure, which was in turn coded for by the DNA.

CARBOHYDRATES. Carbohydrates are simple organic compounds containing only carbon, hydrogen and oxygen, usually in the ratio CH_2O. Unlike proteins and nucleic acids they contain no nitrogen, phosphorus, or sulfur. Many carbohydrates exist within the cell as monomers, usually as intermediate compounds in metabolic pathways. Others are found as constituent parts of polymers. Some of those are subunits of complex compounds such as nucleic acids, while others appear in polymers which are predominantly carbohydrate, i.e., polysaccharides. The simplest polysaccharides contain only one type of subunit and are sometimes found as storage products in cells grown in carbohydrate-rich media. When the energy supply in the medium is exhausted, the storage products are then utilized. Other polysaccharides are found outside of the cells where they form capsules or slime layers which serve a protective role. If more than one carbohydrate monomer is present in a polymer, the polymer is called a heteropolysaccharide. These are often found as capsules, and are also associated with the cell wall of Gram-positive bacteria (as teichoic acids). The last type of carbohydrate-containing material is a complex polysaccharide, which also includes lipids or amino acids. These are often found in cell walls. A compound called peptidoglycan, consisting of two sugar derivatives and a small group of amino acids forms a rigid layer in the walls of both Gram-positive and Gram-negative bacteria, although it is much thicker in Gram-positive ones. In addition, the walls of

Gram-negative bacteria contain a layer composed of lipopolysaccharide and protein which usually lies outside of the peptidoglycan layer.

LIPIDS. Most lipids of importance in microorganisms are found in association with other polymers. Gram-negative bacteria have a relatively high lipid content in their walls whereas Gram-positive bacteria do not. This lipid is usually linked to protein or polysaccharide. Although the function of the lipopolysaccharide layer is not clearly defined, the lipoprotein layer has some of the properties of a semi-permeable membrane. The most important semipermeable membrane, the cell membrane, is composed mostly of phospholipids.

Many species of bacteria produce poly-β-hydroxybutyrate which serves as a lipid-like storage material and can be utilized during periods of starvation. It occurs in crystalline form and appears to be the same in all species studied.

8.2 NUTRITION OF MICROORGANISMS

Studies of the elementary composition of microbial cells have revealed those elements which are important to growth. Investigations of the chemical composition have determined how those elements are incorporated into cell material. The study of nutrition reveals the sources of the elements.

8.2.1 Carbon

Photosynthetic organisms, which obtain their energy from light, and chemoautotrophic organisms, which obtain their energy from the oxidation of inorganic compounds, use carbon dioxide as the primary source of cellular carbon. Since the carbon dioxide must be reduced, a considerable amount of energy must be expended. Microorganisms do not take in carbon dioxide from the air, but rather from the liquid surrounding them.

Heterotrophic organisms obtain their cellular carbon from organic compounds present in the environment. They must also obtain energy from the oxidation of the organic compounds. Consequently, a large portion of the carbon will be excreted from the cell, either as carbon dioxide or as organic waste products, and thus will be lost for purposes of synthesis. Many microorganisms can use a single organic compound as a carbon and energy source, synthesizing all other compounds required for growth. Some organisms, on the other hand, must be provided with specific organic compounds which they are incapable of synthesizing. Compounds which must be supplied are called growth factors and the requirements for them are highly specific. In biochemical operations using mixed microbial communities it is likely that many of the organisms require growth factors which are being supplied by their neighbors.

8.2.2 Nitrogen

Nitrogen is required for the synthesis of proteins and nucleic acids. If insufficient nitrogen is present in the medium the culture will not be able to remove all of the organic matter because it will not be able to continue to synthesize cell material

after the nitrogen is gone. This is particularly important in wastewater treatment
and the engineer must insure that a wastewater being subjected to biological treatment always contains ample nitrogen. The quantity of nitrogen required can be determined from the stoichiometry of microbial growth. This will be discussed in
Section 9.4.3.

Nitrogen can be obtained by microorganisms in either inorganic or organic
forms. The most common inorganic forms are as ammonium or nitrate ions. Ammonium
ion can be assimilated directly by the cell through amination of a keto acid to form
glutamic acid. The amino group may then be transferred to other keto acids by
transamination, thereby forming other amino acids. Once the amino acids are formed
the nitrogen can be easily incorporated into proteins and nucleic acids. When
nitrate is supplied as the only nitrogen source, it must first be reduced to ammonia
through an energy requiring step called assimilatory nitrate reduction before the
bacterium can use it. It is then incorporated in the same manner as described above.
Most photosynthetic and some nonphotosynthetic organisms can assimilate nitrogen in
the oxidized state although it must eventually be reduced to ammonia before it can
be incorporated into amino acids. Organic nitrogen, as in amino acids, purines,
pyrimidines, etc., can be incorporated directly.

8.2.3 Sulfur

Sulfur is required for the synthesis of proteins, and in concept its metabolism is
similar to that of nitrogen. Most organisms can use inorganic sulfate which is
reduced to sulfide by energy requiring steps. It then reacts with the amino acid
serine to form the sulfur containing amino acid cysteine. Other organic sulfur
compounds can be synthesized from the reduced sulfur of cysteine. Since the cell
contains so few sulfur containing compounds the requirement for sulfur is much less
than the requirement for nitrogen. Usually the sulfate concentration in wastewaters
is sufficient to provide the sulfur needed by biochemical operations.

8.2.4 Phosphorus

Phosphorus is required for the synthesis of nucleic acids and phospholipids. It also
plays a very important role in energy transfer. Insufficient phosphorus will limit
growth and prevent the complete removal of the organic substrate in a manner similar
to insufficient nitrogen. Engineers must see that the amount of phosphorus available in a wastewater undergoing biological treatment is sufficient. This, too,
can be estimated by stoichiometry, as discussed in Section 9.4.3. Most, and perhaps
all, microorganisms are able to utilize inorganic orthophosphate. If the organic
matter being degraded contains phosphorus it can be incorporated by the growing cells
through the mediation of enzymes called phosphatases.

8.2.5 Oxygen

Oxygen is a basic constituent of organic compounds and consequently is found in all
cell material. When an organism is obtaining its carbon and energy from the degradation of organic compounds a large portion of the oxygen in the cell material

originates in the organic substrate. The rest of it is provided by water.

The role of molecular oxygen in microbial growth is as the terminal oxidizing agent in aerobic respiration. If oxygen is the only terminal electron acceptor that the cell can use, the organism is called obligately aerobic. Obligately anaerobic organisms are those which obtain energy only from reactions that do not involve the utilization of molecular oxygen. In some cases molecular oxygen is toxic. Most of these use fermentative metabolism, i.e., an organic compound serves as the terminal electron acceptor. Facultative anaerobes are organisms which may grow either in the presence or absence of molecular oxygen. Most shift from fermentation to respiration when oxygen is present. Many organisms found in aerobic biochemical operations fall into this category. Finally, microaerophilic organisms are obligate aerobes which grow best at very low oxygen concentrations.

8.2.6 Mineral Nutrition

As discussed in Section 8.1.1 the ash content of microorganisms contains many minerals which play important roles in microbial metabolism. These can generally be divided into two groups on the basis of the cellular requirements for them. The requirements for the micronutrients are so low that it will seldom, if ever, be necessary to add them to wastewaters undergoing biological treatment. Although the need is rare, it may sometimes be necessary to add a macronutrient, particularly when wastewaters are chemically treated by a process which removes cations before being treated biologically.

MACRONUTRIENT MINERALS. Potassium is a universally required mineral because a variety of enzymes are activated by it. Magnesium is also required to activate enzymes, especially those involving phosphate transfer. In addition, it is thought to help stabilize ribosomes, cell membranes, and nucleic acids. Calcium is involved in the stability of the cell wall, as well as in the heat stability of bacterial spores. Iron is required by virtually all organisms since it is essential as an electron carrier in oxidation - reduction reactions. The last macronutrient is sodium, but it is not required by all microorganisms. Its role is not clear.

MICRONUTRIENT MINERALS. Cobalt is needed for the formation of vitamin B-12 which is required for single carbon transfer reactions. Zinc plays a structural role in certain enzymes by helping to hold together protein subunits in the proper configuration for enzyme activity. Molybdenum is present in certain compounds involved in assimilatory nitrate reduction and thus must be present if only nitrate nitrogen is available to the cell. Copper plays a role in certain oxidation-reduction reactions, while manganese can sometimes substitute for magnesium.

8.2.7 Nutritional Categories

The simplest useful nutritional classification of microorganisms takes into account two factors: the nature of the energy source and the principal carbon source. When considering the latter, the requirements for growth factors are ignored because they do not normally contribute materially to the carbon requirement. Considering the first factor, organisms which use light as an energy source are called phototrophic whereas those which use chemical energy are called chemotrophic. Similarly, organisms which use carbon dioxide as the principal carbon source are called autotrophic while those which use organic compounds are called heterotrophic. Combining these factors leads to four basic nutritional categories. Organisms which use light energy and receive their carbon from carbon dioxide are called photoautotrophic. This group includes the algae and certain photosynthetic bacteria. Photoheterotrophic organisms use light for energy but derive their carbon from organic compounds. The main members of this group are the purple, nonsulfur bacteria. All chemoautotrophic organisms are bacteria. They use carbon dioxide as their carbon source and gain their energy from the oxidation of reduced inorganic compounds, such as ammonia, sulfide, or hydrogen. The ammonia oxidizing bacteria are important in biochemical operations. By far the most important group, however, is the chemoheterotrophic organisms which obtain both their carbon and their energy from organic compounds. This group, which includes most bacteria as well as all protozoa and fungi, may be subdivided on the basis of the chemical manner in which the energy yielding organic substrate is degraded. Respiratory organisms couple the oxidation of the substrate with the reduction of an externally supplied inorganic oxidizing agent. If the inorganic agent is oxygen the activity is called aerobic respiration, whereas if it is not, the activity is called anaerobic respiration. Fermentative organisms, on the other hand, couple with internally supplied organic oxidizing agents.

8.3 THE CHEMICAL NATURE OF METABOLISM

No matter what the nutritional category of an organism, all of its activities are brought about by chemical reactions which obey the normal laws of chemistry and thermodynamics. Microorganisms have developed specialized techniques for harnessing those reactions, however, and in this section we will investigate them.

8.3.1 Thermodynamic Properties of Living Systems

The second law of thermodynamics states that not all of the energy released by oxidation reactions will be available to do work because part of it will be lost as an increase in entropy. Thermodynamics, however, can only tell us the maximal amount of energy available to a cell; it cannot tell us the extent to which microbial growth will occur because that depends upon the efficiency of energy utilization by the particular organism involved. Some energy will always be lost as heat and the amount is an indication of the thermodynamic inefficiency in the transfer of energy from the energy-yielding to the energy-requiring reactions. As a general rule, however, most microorganisms are fairly efficient so that they must be in very high concentration before the total amount of heat lost is sufficient to affect the temperature

of their environment. The reason for this level of efficiency is that energy-
yielding and energy-requiring reactions are closely coupled through chemical
mediators.

8.3.2 Energetic Coupling through ATP

The most important compound coupling energy release to energy utilization is adenosine triphosphate, ATP:

$$
\begin{array}{ccccccc}
& O & & O & & O & \\
& \parallel & & \parallel & & \parallel & \\
HO - & P & \sim O - & P & \sim O - & P & - \text{Adenosine} \\
& | & & | & & | & \\
& OH & & OH & & OH &
\end{array}
$$

In ATP the two bonds designated by the wavy lines ($\sim$) are energy rich. This means
that if they were broken by chemical hydrolysis the reaction would be exergonic with
the liberation of a substantial amount of energy as heat (8 kcal/mole for each bond).
In biological systems the energy is not liberated as heat but is used instead to
drive reactions which require energy (endergonic) through coupled chemical reactions
catalyzed by enzymes. Such reactions exist at a number of stages in the biosynthetic
process. First, they are required to drive the entry of many nutrients into the cell
by a process known as active transport. Second, they are used to convert those
nutrients into intermediary metabolites of low molecular weight, i.e., amino acids,
sugar phosphates, nucleotides, fatty acids, etc. Finally, the energy of ATP is used
to achieve polymerization of those intermediates into the various macromolecules of
the cell: proteins, polysaccharides, nucleic acids, and lipids. Many of these
reactions release the terminal phosphate group as inorganic phosphate, leaving adenosine diphosphate, ADP. In other cases, particularly in the synthesis of macromolecules, one or both of the high energy phosphate groups are transferred to an acceptor
molecule, with part of the bond energy being preserved in the newly formed molecule.
That molecule can then react directly with the growing macromolecule, using the
energy liberated by the subsequent removal of the phosphate group(s) to make the
required chemical bond. If both of the high energy phosphate groups are lost from
ATP during this process, the resulting molecule is called adenosine monophosphate,
AMP.

ATP can be regenerated from both ADP and AMP. The first step in the regeneration of ATP from AMP is the reaction of one mole of AMP with one mole of ATP to
yield two moles of ADP. There are then several types of reactions which can regenerate ATP from ADP. One type involves direct formation of ATP through coupling
with the removal of high energy phosphate bonds from organic molecules within the
cell. Another type occurs through the use of the energy available in thiol esters
to add an inorganic phosphate to ADP. Both of these are referred to as substrate-
level phosphorylation because the ATP is formed by reactions which directly involve
organic substrates. They are the primary means of ATP formation in organisms which
carry out fermentations. Much more ATP can be formed by reactions associated with

the terminal electron transfer chain, which will be discussed in Section 9.1.3.
These reactions can only be carried out by organisms growing by respiration. Because they can form more ATP than organisms growing by fermentations, respiratory
organisms gain more energy from their substrates.

8.3.3 Oxidations and Reductions of Organic Compounds

An organic substrate is not directly oxidized to carbon dioxide and water in a
single chemical step because there is no energy conserving mechanism that could trap
so large an amount of energy. The loss of energy as heat would be extremely high.
Consequently, biological oxidation occurs in small steps. Basically, oxidation requires the transfer of an electron from the substance being oxidized to some acceptor
molecule which will subsequently be reduced. In most biological systems, each step
in the oxidation of the substrate involves the removal of two electrons(e^-) and the
simultaneous loss of two protons ($2H^+$). The combination of the two losses is equivalent to the molecule having lost two hydrogen atoms (H_2) and consequently the
reaction is referred to as a dehydrogenation. The electrons and protons are not
just released into the cell but are instead transferred to an acceptor molecule.
The acceptor molecule will not accept the protons until it has accepted the electrons,
and thus it is referred to as an electron acceptor. Since the net result of accepting an electron and a proton is the same as accepting a hydrogen atom, such
acceptors are also called hydrogen acceptors.

8.3.4 Role of Pyridine Nucleotides

The major hydrogen acceptors in microbial cells are two carrier molecules known as
pyridine nucleotides. They are nicotinamide adenine dinucleotide, NAD, and
nicotinamide adenine dinucleotide phosphate, NADP. Both contain the vitamin
nicotinamide which undergoes reversible oxidation and reduction:

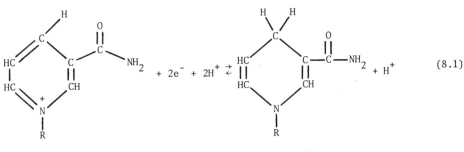

NAD or NADP   NADH or NADPH

The reaction can be symbolized as:

$$NAD^+ + 2H \rightleftarrows NADH + H^+ \tag{8.2}$$

or

$$NADP^+ + 2H \rightleftarrows NADPH + H^+ \tag{8.3}$$

It can be seen that when NAD and NADP are reduced, one of the protons combines with the molecule while the other remains in solution as a hydrogen ion, although it will play a part in any subsequent oxidation. Thus, the reduced form of NAD^+ is written as $NADH + H^+$, although often, for convenience, the oxidized and reduced forms are symbolized as NAD and $NADH_2$, respectively. Most enzymatic reactions require either NAD or NADP preferentially. This presents no difficulties, however, because a reversible reaction occurs which allows interconversion of the two:

$$NADPH_2 + NAD \overset{\rightarrow}{\leftarrow} NADP + NADH_2 \tag{8.4}$$

The reduced forms of the pyridine nucleotides are said to contain reducing power and consequently they can serve as a source of electrons and protons required to carry out biosynthetic reactions. The requirement for reducing power reflects the difference between the oxidation state of a nutrient and that of the final product of biosynthesis. Most reducing power is required for carbon since it is the major element in the cell. Chemoheterotrophic organisms have a smaller requirement for reducing power than do autotrophs because the latter take in carbon as carbon dioxide, which is highly oxidized, whereas carbon in cell material has an oxidation state similar to carbohydrate. Reducing power may also be required to convert nitrate nitrogen to ammonia, or sulfate to sulfide.

Reduced NAD and NADP also carry electrons to the terminal electron transport chain thereby returning the pyridine nucleotides to the oxidized form. The electrons are eventually transferred to the terminal hydrogen acceptor, and in the process ATP is formed. This provides the energy in a chemical form which can be used by the cell.

This brief description of metabolism has indicated that two forms of coupling are involved, linking degradation with biosynthesis: energetic coupling through ATP and reductive coupling through NAD and NADP. It was also mentioned that oxidations and reductions occur in small steps, so that the cell can efficiently manipulate the energy released or required. Consequently, it is not surprising to learn that chemical reactions do not occur as random events in the cell, but are organized into interconnected metabolic pathways, or routes, which are sequences of enzymatically catalyzed reactions wherein the products of one reaction become the reactants in the subsequent reaction.

8.4 METABOLIC ROUTES

There are five major types of metabolic routes: energy producing, carbon incorporating, amphibolic, anabolic, and anaplerotic. Within chemoheterotrophic organisms the first two are combined and called catabolic routes. Autotrophic organisms use separate and distinct pathways for each function. Amphibolic routes play a dual role in the metabolism of chemoheterotrophs, with their products functioning both as intermediates in energy release from the organic substrate and as components of larger molecules formed by biosynthesis. In autotrophic organisms the amphibolic routes act only to supply components for synthesis. Anabolic pathways are the routes wherein the more complex molecules of the cell are synthesized from the intermediates of the amphibolic routes. As intermediates are removed for use there must be some way to replenish them and this function is performed by anaplerotic routes. Each of these will be discussed in this section.

8.4.1 Routes for Energy Production and Carbon Incorporation

CHEMOHETEROTROPHIC ORGANISMS. The suffix "-troph" means nourishing, and thus the name chemoheterotroph signifies that these organisms obtain their nourishment from a diverse group of organic compounds. Because such oxidations must be carried out in small steps, the organic molecules are changed in size and structure as they proceed through the pathways, becoming both smaller and more highly oxidized. These smaller molecules serve as intermediates for incorporation into the biosynthetic processes of the cell, so that these degradative pathways serve a dual purpose, supplying both energy and carbon to the organism. Consequently, a portion of the carbon atoms entering the cell in the organic substrate will be given off either as carbon dioxide or as metabolic end products such as ethanol, while the rest are incorporated into new cell material. The fate of the carbon atoms depends upon the type of organism and the nature of the terminal hydrogen acceptor.

Figure 8.1 depicts the general flow of organic compounds into heterotrophic organisms and shows that substrate degradation can be considered to occur in three phases [5]. Phases I and II constitute the catabolic routes. These lead to phase III, which contains amphibolic and anaplerotic routes. This section will consider only the catabolic routes. The others will be discussed in subsequent sections, although it should be recognized that some aerobic organisms derive a large portion of their energy from the amphibolic routes.

Phase I constitutes a preparatory stage of catabolism because it is here that large molecules are broken down into their small constituent units. These reactions are normally carried out extracellularly to reduce the size of the molecule sufficiently to allow its transport into the cell and usually utilize single enzymatically catalyzed reactions rather than complete pathways. Some common examples of phase I catabolism are listed in Fig. 8.1. Proteins are broken down into amino acids; large carbohydrates are degraded into their constituent di- or mono-saccharides; fats are broken down into glycerol and fatty acids. Less than 1% of the total energy available in the macromolecules is released, and this is not generally available to the cell for useful work. It is usually lost as heat.

During phase II metabolism the small molecules produced in phase I are incompletely degraded, liberating about one-third of their total energy to the cell. In the process a number of different products are formed which can enter the amphibolic pathways of phase III and eventually be incorporated into new organic compounds by the biosynthetic processes of the organism. It is significant that in spite of the

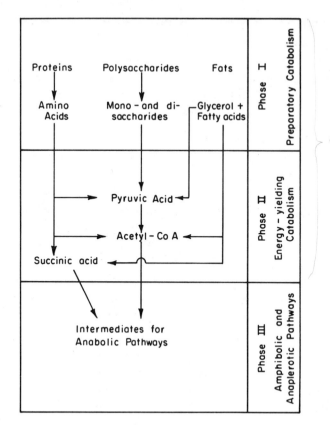

Figure 8.1.   The phases of degradative metabolism
              in microorganisms.

large number of starting compounds for phase II catabolism, there are relatively few
end products. This reduces the number of metabolic routes required by the organism,
thereby minimizing the number of enzymes which must be synthesized and contributing
to greater biochemical efficiency. For example, pyruvate is a very common product
of phase II catabolism, resulting from the degradation of carbohydrates, glycerol,
and a number of amino acids. Succinate is another common end product of amino acid
and fatty acid degradation, as is acetyl-Coenzyme A. There are a number of other
such products, although for illustrative purposes these are the only ones shown in
Fig. 8.1.

      Just because there are relatively few products from phase II catabolism, one
should not draw the conclusion that the number of metabolic pathways is small. Quite
to the contrary, the number of pathways is large. The nature of a given pathway depends upon both the type of compound being degraded and the organism performing the
degradation. Generally, however, the diversity caused by the nature of the substrate
is much larger than that caused by the organism because many organisms contain
similar pathways.

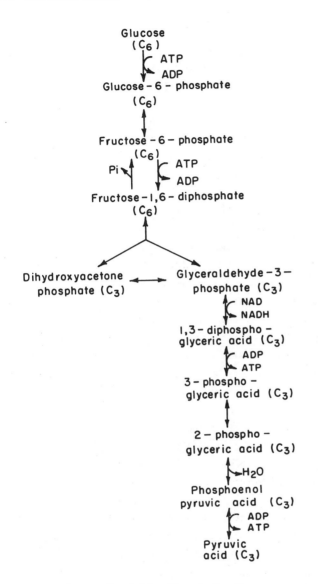

Figure 8.2. The Embden-Meyerhof pathway.

Figure 8.2 depicts the Embden-Meyerhof pathway, a very common pathway for the
degradation of glucose, which will be described briefly to illustrate the general
nature of the reactions in catabolic routes. After entering the cell the glucose is
phosphorylated, which requires ATP and yields ADP plus glucose-6-phosphate. This
initial phosphorylation activates the molecule (i.e., raises its energy level) for
the subsequent reactions. An isomerization and another phosphorylation lead to
fructose-1,6-diphosphate which is split into two three-carbon molecules, glyceral-
dehyde-3-phosphate and dihydroxyacetone phosphate. Up to this point there has been

no oxidation, although two high energy phosphate bonds from ATP have been used. The
first oxidation occurs in the conversion of glyceraldehyde-3-phosphate to 1,3-
diphosphoglyceric acid. Coupled with this oxidation is the conversion of inorganic
phosphate into organic phosphate. This synthesis of a new high energy phosphate
bond utilizes energy which would otherwise be lost as heat and is an example of substrate level phosphorylation. In the next step one of the phosphate bonds is returned to ADP forming new ATP. Since there are two moles of triose formed per mole
of glucose utilized, two moles of ATP are generated per mole of glucose, thereby re-
covering the energy initially invested in the pathway. As the reaction sequence
proceeds, more ATP is made with the result that by the time pyruvate is formed there
has been a net generation of two moles of ATP per mole of glucose utilized. In addition, two moles of NADH$_2$ have been formed, which supply reducing power to the cell.

The fate of pyruvate depends upon both the organism and the environment. If
the organism is in an aerobic environment or is capable of carrying out anaerobic respiration in an anaerobic one, most of the pyruvate will pass on to phase III where
more energy will be released and intermediates will be formed for use in the anabolic
routes. If the organism is a fermentative one in an anaerobic environment, only
enough pyruvate will enter phase III to meet the requirements of the anabolic pathways for intermediates, and little energy will be produced. Instead, most of the
pyruvate will pass to further phase II reactions which oxidize the excess NADH$_2$ produced in the Embden-Meyerhof pathway, thereby producing reduced organic end products
such as ethanol.

As indicated earlier, the Embden-Meyerhof pathway is just one example of the
many pathways that exist in phase II catabolism. All, however, produce some energy
for the cell and yield an end product which can enter the amphibolic pathways. The
nature and fate of the end products, however, will depend upon the environment in
which the organisms are grown.

AUTOTROPHIC ORGANISMS. The term autotroph means self-nourishing and probably
derives from the fact that these organisms appear to survive without any external
sustenance. Actually they obtain their energy either from sunlight (photoautotrophs)
or from the oxidation of inorganic compounds such as elemental sulfur, ammonia, or
ferrous iron (chemoautotrophs). Both obtain their carbon from the fixation of carbon
dioxide, and thus, unlike the heterotrophs, carbon incorporation is quite distinct
from energy production. Let us consider examples of the pathways by which each is
accomplished.

Energy Production. The pathway for energy production in algae (which are photoautotrophic) is shown in Fig. 8.3 where it can be seen that both ATP and NADPH$_2$ are
produced. When a quantum of light at 400-500nm strikes a molecule of the pigment
chlorophyll, the chlorophyll undergoes a change in properties, called excitation,
which captures the energy of the light. As a result, an electron is driven off so
that the chlorophyll molecule becomes positively charged. This excitation state is

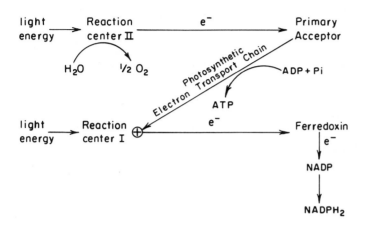

Figure 8.3. The coupling of two light reactions providing
energy for eucaryotic photoautotrophs.

carried through the photosynthetic unit to the chlorophyll of reaction center II
where charge separation occurs, resulting in the transfer of an electron from the
hydroxyl radical of water to the positive charge in the chlorophyll unit, leading to
the formation of oxygen. As a result of these reactions an electron has been liberated which migrates through an electron-transport system until it reaches cytochrome
f which becomes reduced. During this passage, one high energy phosphate bond in ATP
is synthesized. The reduced cytochrome f is oxidized at reaction center I where the
chlorophyll is sensitive to light at 700nm, and the electron is eventually transferred to NADP, forming $NADPH_2$. It is also possible that a second mole of ATP is
formed during this transfer. This whole scheme is called noncyclic photophosphoryl-
ation and is found in all plants and algae.

Photosynthetic bacteria possess a separate scheme, called cyclic photophos-
phorylation, which produces no oxygen and no $NADPH_2$. These organisms must obtain
their reducing power from constituents in the environment.

Chemoautotrophs obtain their energy from the oxidation of inorganic compounds.
Of the many known chemoautotrophs the ones of most importance to wastewater treatment
belong to the genera *Nitrosomonas* and *Nitrobacter* because they are responsible for
the oxidation of ammonia and nitrite nitrogen, respectively, a process called nitrification. The pathway by which *Nitrosomonas* oxidizes ammonia to nitrite is still
largely speculative, but has been hypothesized to be [6]:

$$NH_4^+ + energy \rightarrow (NH_2) \rightarrow NH_2OH \rightarrow (NHOH) \rightarrow (NOH) \rightarrow NO \rightarrow NO_2^- \qquad (8.5)$$

The compounds in parentheses are thought to be free radical-like intermediates which
are bound to the enzymes. It has been suggested that a preliminary activation of
ammonia is required to initiate the pathway. This probably requires a compound such

as ATP. Energy is derived from the oxidation of hydroxylamine (NH_2OH) and electron flow is presumably from hydroxylamine through an electron transport chain to oxygen. It is during this electron flow that ATP is generated by oxidative phosphorylation. In *Nitrobacter* the energy-yielding process appears to consist of a single step oxidation of nitrite to nitrate [6]:

$$NO_2^- + H_2O \rightarrow NO_3^- + 2H^+ + 2e^- \tag{8.6}$$

Again the electrons pass to an electron transport chain where ATP is generated.

A unique feature of most chemoautotrophs is that no NAD or NADP participates in the oxidation of the energy source. Consequently, the question arises as to where they obtain the $NADH_2$ or $NADPH_2$ required for carbon fixation and biosynthesis. Apparently these organisms use a process called reversed electron transport. Electrons normally flow from $NADH_2$ to oxygen through the electron transport chain, thereby generating ATP. In reversed electron transport the opposite occurs so that part of the ATP generated by the cell must be utilized to form $NADH_2$.

Carbon Fixation. Autotrophic organisms can obtain all of their carbon from carbon dioxide fixation by the Calvin cycle which utilizes ATP and $NADPH_2$ to reduce the carbon to the oxidation level of carbohydrate. All autotrophs possess this pathway, however chemoautotrophic bacteria use $NADH_2$ for the reductions rather than $NADPH_2$ [6]. The CO_2 is trapped by reaction with the five-carbon compound ribulose diphosphate (RuDP) to give two moles of 3-phosphoglyceric acid (PGA):

$$\tag{8.7}$$

This has not changed the oxidation level of the carbon atom, however, so that two subsequent reactions are needed to reduce it to the level of carbohydrate. First, ATP reacts with PGA to form 1,3-diphosphoglyceric acid, thereby activating the carboxyl group:

$$\tag{8.8}$$

In the next step, $NADPH_2$ is used in the actual reduction, resulting in the formation
of glyceraldehyde-3-phosphate (GAP) and the release of inorganic phosphate:

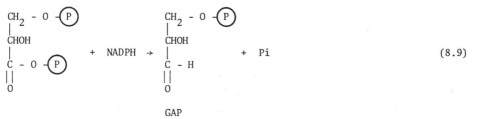

$$(8.9)$$

GAP

The new carbon atom is now at the oxidative level of a carbohydrate. Because only
one carbon has been incorporated, the process must be repeated two more times to
achieve the net formation of one mole of GAP. In addition, each time a mole of
carbon is incorporated, a mole of ribulose diphosphate is required, so that if the
process is to continue, there must be some way of regenerating the RuDP. It is the
mechanism of regeneration which gives this pathway the cyclic nature shown in
Fig. 8.4. Three moles of RuDP react with 3 moles of CO_2 and produce 6 moles of GAP,
consuming 6 moles of ATP and $NADPH_2$ in the process. Because GAP is a three-carbon
compound, and three carbon atoms have been incorporated, this is equivalent to producing one new mole of GAP. Five other moles of GAP are thus available for use in
the regeneration cycle. A mole of GAP then reacts with a 3-carbon compound formed
from another GAP to form a 6-carbon compound which reacts with a third mole of GAP
to give a 5-carbon and a 4-carbon compound. The 5-carbon compound undergoes transformation to ribulose phosphate which is then phosphorylated with ATP to regenerate
one mole of RuDP. The 4-carbon compound reacts with another 3-carbon compound
arising from GAP to give a 7-carbon compound which in turn reacts with the last GAP
to give two 5-carbon compounds which are each converted to ribulose phosphate. Each

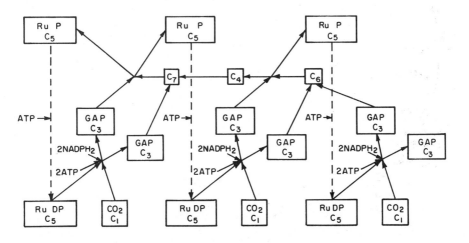

Figure 8.4. Simplified diagram of the Calvin cycle for the
incorporation of carbon by obligate autotrophs.

of the ribulose phosphates is then phosphorylated with ATP to regenerate the final
two moles of RuDP. In this way the cycle is prepared to begin again.

In summary, the production of one mole of glyceraldehyde-3-phosphate requires
nine moles of ATP and six moles of $NADPH_2$, all of which must be provided either by
light sensitive reactions or by the oxidation of inorganic compounds.

8.4.2 Amphibolic Routes

Returning to Fig. 8.1 it can be seen that the end products of phase II catabolism in
chemoheterotrophic organisms enter phase III where they serve as the starting points
of the amphibolic pathways. The amphibolic (from the Greek amphi, meaning both)
pathways serve two very important roles in heterotrophs, the production of energy
and the provision of intermediates for biosynthesis. Perhaps the most important
amphibolic pathway is the tricarboxylic acid (TCA) cycle, and a brief discussion of
it will serve to illustrate the general characteristics of such routes.

Figure 8.5 depicts the TCA cycle, which operates essentially as a mechanism
for the complete oxidation of acetate, resulting in the formation of two moles of

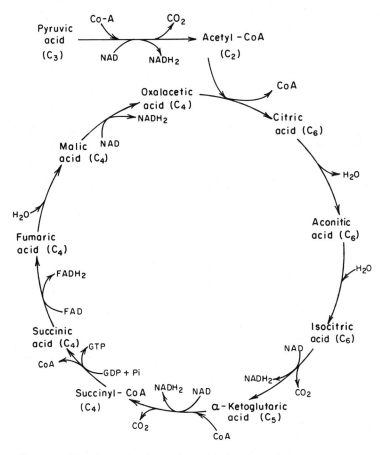

Figure 8.5. The tricarboxylic acid (TCA) cycle.

CO_2 for each mole of acetate entering. As shown in Fig. 8.1, any compound which can be dissimilated to yield acetate directly, or pyruvate (which is convertible to acetate), can be fed into the cycle and oxidized. To enter the pathway the acetate must be activated through the formation of a high energy bond with coenzyme A (CoA). For example, the pyruvate formed by the Embden-Meyerhof pathway can be decarboxylated to form acetyl-CoA and $NADH_2$. The 2-carbon acetyl group of acetyl-CoA combines with the 4-carbon compound oxalacetic acid to form citric acid, a 6-carbon compound. The energy of the acetyl-CoA bond is used to drive this reaction. Dehydration, decarboxylation, and oxidation reactions follow until ultimately oxalacetic acid is regenerated. Three oxidations are coupled to the reduction of NAD, leading to $NADH_2$. The NAD must be regenerated by the transfer of the hydrogen to some suitable acceptor compound. If it is not, the oxidation of the substrate will cease. Organisms growing by respiration processes can oxidize the $NADH_2$ through an electron transport chain, thereby resulting in the formation of ATP. Substrate level phosphorylation occurs as succinyl-CoA is converted to succinic acid, with the resultant formation of GTP, which can be used to form more ATP. Consequently, the TCA cycle serves as an important source of reducing power ($NADH_2$) and of energy (ATP).

Acetyl-CoA is not the only form in which compounds can enter the pathway. For example, the degradation of amino acids in phase II catabolism can lead to succinic acid and to α-ketoglutaric acid, both of which are intermediates in the cycle.

Because the TCA cycle is an amphibolic pathway it also serves an important function in supplying the cell with intermediates to be used in anabolic pathways. Both α-ketoglutaric acid and oxalacetic acid are important starting compounds for the synthesis of amino acids, so a portion of the molecules formed leave the cycle for that purpose.

Autotrophic organisms have a need for these same intermediates and the TCA cycle provides them. It does not produce energy, however, due to a slight alteration which prevents the pathway from functioning in a cyclic manner. The destruction of the cyclic operation is the result of the absence of the enzyme responsible for the oxidation of α-ketoglutaric acid. The result is a branched pathway as shown in Fig. 8.6. The reactions responsible for the incorporation of acetyl-CoA and the subsequent formation of the important intermediate α-ketoglutaric acid function in the usual cyclic sense. The remainder of the reactions operate in a reverse manner to furnish 4-carbon dicarboxylic acids such as oxalacetic acid and succinic acid, which are important in biosynthetic pathways.

The pathway even operates in this noncyclic manner in some heterotrophic organisms. When an organism is growing in a fermentative mode the opportunity for regeneration of NAD from $NADH_2$ is greatly reduced. Consequently, the normal TCA cycle would not function for want of NAD. However, the cell still needs the intermediates formed by the TCA cycle enzymes. These, therefore, are supplied in the same way that they are in autotrophic organisms.

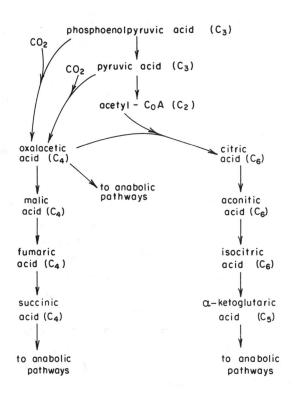

Figure 8.6. The amphibolic nature of TCA cycle reactions
             in obligate autotrophs. *(Adapted from Stanier, et al.).*

8.4.3 Anaplerotic Routes

Consider for a moment the growth of a culture of aerobic, heterotrophic organisms on
the 6-carbon sugar, glucose. This compound would pass through phase II metabolism
yielding pyruvate and subsequently acetyl-CoA which would enter the TCA cycle of
phase III. Because glucose is the sole source of carbon, the organisms must make
all of the many compounds needed to synthesize new cell material. This means that a
large amount of the α-ketoglutaric acid, succinic acid, and oxalacetic acid will be
removed from the TCA cycle to supply the biosynthetic needs. Examination of Fig. 8.5
reveals that if acetyl-CoA is the only compound entering the cycle while CO_2 and the
various intermediates are being removed from it, the cycle will soon be depleted of
intermediates and will cease functioning. This is prevented by anaplerotic (meaning
replenishing) pathways which act to resupply the cycle with dicarboxylic acids.

   There are a number of anaplerotic routes, some consisting of only single enzymes,
which act to provide the cell with 4-carbon dicarboxylic acids. One important one,
which enables microorganisms to grow on 2-carbon compounds, is the glyoxylate shunt
which bypasses the decarboxylation steps of the TCA cycle. The shunt consists of
two enzymes, the first of which catalyzes the cleavage of citric acid into succinic

acid and glyoxylic acid. The succinic acid reenters the TCA cycle while the glyoxylic acid reacts with acetyl-CoA to form malic acid, which can also enter the cycle, providing needed 4-carbon intermediates.

8.4.4 Anabolic Routes

In Section 8.1.2 it was seen that the majority of the organic material in microbial cells consists of four classes of relatively large molecules (macromolecules): nucleic acids, proteins, polysaccharides, and lipids. Furthermore, it was seen that these macromolecules are composed of a relatively small number of simple compounds arranged in systematic order. For example, only five bases are needed to synthesize the nucleic acids, and 20 amino acids to make the proteins. In all, no more than 70 different precursor monomers are needed for the biosynthesis of all four classes of compounds. In addition, the cell contains a number of organic constituents which serve other roles, such as coenzymes and carriers. Nevertheless, the total number of relatively low molecular weight organic compounds that must be synthesized by any microbe is probably only between 100 and 200. All of these are synthesized from the intermediates of the amphibolic routes, by the pathways referred to as the anabolic routes.

   BIOSYNTHESIS OF MONOMERS. There are several features common to the anabolic routes for the synthesis of monomers. Although the overall effect of the anabolic routes is opposite to that of the catabolic routes, the two show little or no similarity in their component enzymatic steps. In fact, the anabolic pathway for the synthesis of a particular compound generally starts from amphibolic intermediates quite different from those produced through the catabolism of that compound. There is, however, a link between anabolism and catabolism because the latter must provide the energy and reducing power needed by the former. Breakage of the high energy phosphate bond of ATP provides the energy needed to join molecules in biosynthetic reactions. Reducing power is generally provided by $NADPH_2$ which is formed by the reduction of NADP through coupled oxidation of the $NADH_2$ arising from catabolism. Some biosynthetic reactions can use $NADH_2$ directly, as well. The number of main anabolic pathways is considerably smaller than the number of products formed because most biosynthetic pathways are branched. This means that several products can be formed by routes diverging from a single primary biosynthetic reaction sequence, thereby providing economy for the cell by reducing the number of enzymes required.

   Despite the branched nature of anabolic pathways, the number of routes is far in excess of the space available for discussion herein. Consequently, only one such branched pathway will be presented to illustrate the general nature of anabolic routes for the synthesis of monomers.

   Figure 8.7 presents the main reactions found in the pathway for the aspartate family of amino acids. This pathway starts from oxalacetic acid of the TCA cycle and yields five amino acids (aspartic acid, methionine, threonine, isoleucine, and lysine) as well as one amide (glutamine). Aspartic acid arises directly from

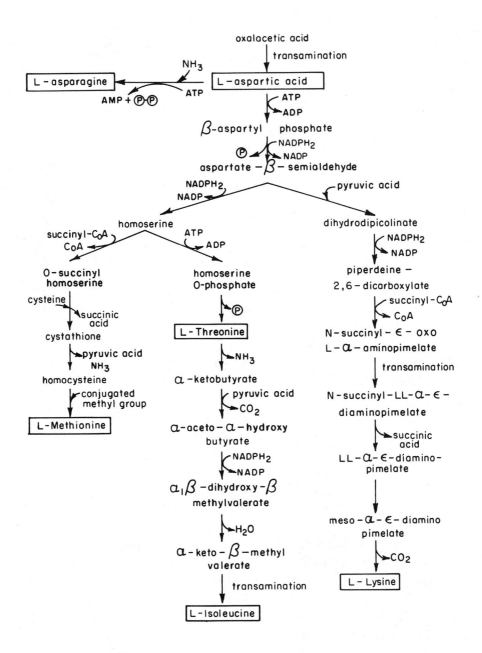

Figure 8.7. Pathway for the synthesis of the aspartate family of amino acids.

oxalacetic acid by transamination, which is a reaction whereby the amino group of one amino acid is transferred to an α-keto acid, resulting in a new amino acid and another α-keto acid. The usual donor molecule is glutamic acid. Asparagine comes from aspartic acid via reaction with ammonia. The reaction is an energy requiring one, consequently ATP is consumed. The four remaining amino acids are all formed from the β-semialdehyde of aspartic acid, which is synthesized in two reactions requiring both energy and reducing power. Aspartate-β-semialdehyde is the point at which the pathway for lysine biosynthesis diverges from the main sequence. Note that succinyl-CoA is required in the lysine pathway and that it comes from the TCA cycle. Later in the sequence, succinic acid is released and returns to the TCA cycle. This illustrates well the role of amphibolic pathways as sources and sinks of intermediate compounds. Homoserine is a second branch point in the main pathway, with one branch leading to methionine and the other to threonine and isoleucine. The reduction of aspartate-β-semialdehyde to form homoserine is shown as requiring $NADPH_2$, although $NADH_2$ may be used as well. Note that in both branches amphibolic intermediates are required, and that transamination again plays a role in the formation of an amino acid.

   Although the individual steps are different, other biosynthetic pathways share the general features of the pathway in Fig. 8.7. Those interested in the details of other pathways should consult the books listed at the end of the chapter. Let us now turn our attention to the way in which the end products of these pathways are used for assembly of the macromolecules of the cell.

   ASSEMBLY OF MACROMOLECULES.

   General Principles. In Section 8.1.2 the general characteristics of proteins, nucleic acids, and polysaccharides were discussed. Polysaccharides were shown to be composed of repeating subunits of carbohydrate, and as such are referred to as repetitive polymers. They may be subdivided into two major classes, depending upon whether the subunits are all the same (homopolymers) or chemically similar but nonidentical (heteropolymer). Generally the number of different subunits in heteropolymer polysaccharides are relatively few, e.g., three or four. Polysaccharides occur as both simple chains and as branched chains. Several examples follow:

   chain homopolymer: -A-A-A-A-A-A-A-A-A-A-

   chain repetitive heteropolymer: -A-B-C-D-A-B-C-D-A-B-C-D-

   branched chain homopolymer: -A-A-A-A-A-A-A-A-A-A-A
                                                                |
                                          -A-A-A-A-A-A-A-A-A-A-A-
                                                                     |
                                             -A-A-A-A-A-A-A-A-A-A-A-

   branched chain repetitive heteropolymer: A-B-C-D
                                                                  |
                                                         A-B-C-D
                                                                  |
                                                         A-B-C-D

In contrast to carbohydrates, proteins and nucleic acids are always linear, nonrepetitive heteropolymers. Proteins are the chemically more complex of the two because they are composed of up to 20 different subunits -- the amino acids (A-T):

A-F-G-K-B-S-J-A-A-H-M-T-K-B-O-N-H-C-E-R-S-S-

It will be recalled that the sequence of amino acids is the primary structure of the molecule and that it determines the role the protein will play within the cell. These molecules can be quite large, with most containing over 100 amino acids. Nucleic acids vary widely in size, depending upon the specific type, but all are made of four subunits -- the nucleotides (A-D):

A-D-C-D-A-B-C-B-B-A-D-C-A-D-B-C-A-A-A-D-C-B-B-C-

The sequence of these subunits is particularly important because it is involved in the storage and translation of genetic information.

The fact that polysaccharides are repetitive polymers, whereas proteins and nucleic acids are not, has a profound effect upon the mechanisms by which they are made. Repetitive polymers can be made by the successive action of only a small number of enzymes. For example, the branched chain repetitive heteropolymer shown earlier could be made by the repetitive action of four:

$$A + B \rightarrow A\text{-}B \qquad\qquad\qquad (8.10)$$

$$A\text{-}B + C \rightarrow A\text{-}B\text{-}C \qquad\qquad\qquad (8.11)$$

$$A\text{-}B\text{-}C + D \rightarrow A\text{-}B\text{-}C\text{-}D \qquad\qquad\qquad (8.12)$$

$$A\text{-}B\text{-}C\text{-}D + A\text{-}B\text{-}C\text{-}D \rightarrow A\text{-}B\text{-}C\text{-}D$$
$$\qquad\qquad\qquad\qquad\qquad |$$
$$\qquad\qquad\qquad\qquad A\text{-}B\text{-}C\text{-}D \qquad\qquad\qquad (8.13)$$

This mechanism would not be feasible for nucleic acids or proteins, however, because an astronomically large number of different enzymes would be required to attach the subunits in the proper order. On the other hand, if the subunits were arranged in order on some sort of template, relatively few enzymes would be required to join them. This is exactly what happens, and no matter whether the product is DNA, RNA, or protein, the template is always a nucleic acid.

The subunits of all biopolymers are linked together by anhydride bonds which can be broken easily by mild hydrolysis, suggesting that bond formation is thermodynamically unfavorable and requires energy. Consequently, polymerization is always preceded by activation of the monomers in which they are attached to carrier molecules. Such activation requires the expenditure of at least one mole of ATP (or equivalent) per mole of monomer and converts the monomers to anhydrides attached to the carrier molecules by energy rich bonds. As a consequence, each monomer can be readily incorporated into the growing polymer chain because its transfer is thermodynamically favorable.

Nucleoside monophosphates are the monomers of nucleic acids. They are all activated by conversion to the corresponding triphosphates, which are similar to ATP in energy content. For the synthesis of RNA these are adenosine triphosphate (ATP), guanosine triphosphate (GTP), uridine triphosphate (UTP), and cytidine triphosphate (CTP). For DNA, the corresponding deoxynucleotide triphosphates are used (dATP, dGTP, dCTP) with thymidine triphosphate (dTTP) taking the place of UTP.

Monosaccharides occur in the cell primarily as phosphate esters which are activated by reaction with one of the ribonucleotide triphosphates (ATP, UTP, GTP, or CTP). These reactions always lead to the formation of inorganic pyrophosphates as shown by the reaction of glucose-6-phosphate with ATP:

$$\text{glucose-6-phosphate} + \text{ATP} \rightarrow \text{ADP-glucose} + \text{P-P} \tag{8.14}$$

Amino acids are activated in two steps, both catalyzed by specific enzymes. In the first, the amino acid reacts with ATP to yield a molecule of amino acyl-AMP:

$$\begin{array}{c}\quad O \\ \quad \| \\ NH_2\text{-CH-C-OH} \\ \quad | \\ \quad R\end{array} + \begin{array}{c}O\ O\ O \\ \|\ \|\ \| \\ \text{adenosine-P-P-P-OH} \\ |\ \ |\ \ | \\ OH\ OH\ OH\end{array} \rightarrow \begin{array}{c}O\quad\ O \\ \|\quad\ | \\ NH_2\text{-CH-C-O-P-adenosine} \\ \quad |\qquad | \\ \quad R\qquad OH\end{array} + \textcircled{P} - \textcircled{P} \tag{8.15}$$

In the second, the amino acyl group is transferred to a molecule of transfer RNA (tRNA) specific for the amino acid being activated:

$$\text{amino acyl-AMP} + \text{tRNA} \rightarrow \text{amino acyl-tRNA} + \text{AMP} \tag{8.16}$$

The tRNA acts as the carrier molecule and plays an active role in protein synthesis.

Synthesis of Polysaccharides. Polysaccharide synthesis is relatively simple. Once the sugar molecules are activated they are added stepwise to an existing chain (primer) by a single specific enzyme which catalyzes each successive addition:

$$\text{ADP-G} + \text{G-G-G-G-G-G} \rightarrow \text{G-G-G-G-G-G-G} + \text{ADP} \tag{8.17}$$

The formation of branched homopolymers is catalyzed by a "branching" enzyme that cleaves small fragments from the end of the linear polysaccharide chain and reinserts them at another point:

$$\begin{array}{c}\qquad\qquad\ G \\ \qquad\qquad\ \backslash \\ \qquad\qquad\ G \\ \qquad\qquad\ \backslash \\ \overset{\frown}{\text{G-G-G-G-G-G}} \rightarrow \text{G-G-G-G-G}\end{array} \tag{8.18}$$

Specific chain elongation and branching enzymes are required for each type of polysaccharide.

Biosynthesis of DNA. As discussed in Section 8.1.2 the sequence of deoxy-nucleotides in DNA constitutes the genetic code of a cell and contains all of the information needed for it to function. In order for the cell to reproduce generation after generation without any alteration of its genetic makeup, the mechanism of DNA replication must produce exact copies. The fact that DNA occurs as a double strand, and the ability of adenine to pair with thymine and of guanine to pair with cytosine forms the basis whereby the exact copy may be made. If a complementary copy (i.e., one that contains adenine in place of thymine, etc.) is made of each of the original strands and the resulting four strands are divided into two pairs of one old and one new, then the original pair will have been duplicated:

```
-dG-dT-dA-dT-dC-dG-          new
-dC-dA-dT-dA-dG-dC-          old
 :   :   :   :   :   :
-dG-dT-dA-dT-dC-dG-          old
-dC-dA-dT-dA-dG-dC-          new
```

Thus, it is seen that DNA serves as its own template. The synthesis begins at one end by separation of the two strands and proceeds sequentially to the other end of the original DNA chain:

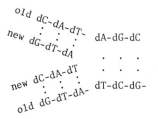

Since the activated precursors are the deoxyribonucleotide triphosphates (dATP, dCTP, dGTP, dTTP), condensation occurs with the elimination of inorganic pyrophosphate. The reaction is catalyzed by the enzyme DNA polymerase:

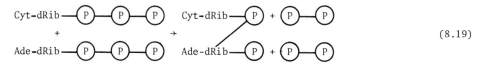

$$(8.19)$$

Biosynthesis of RNA. There must also be an accurate method of obtaining the proper nucleotide sequence in RNA, particularly for messenger RNA (mRNA) which must carry the genetic code from the DNA to the site of protein synthesis. This is accomplished in a manner quite similar to that of DNA replication, with DNA serving as the template. The major differences are that the precursor molecules are the ribonucleotide triphosphates ATP, GTP, CTP, and UTP rather than the deoxyribonucleotide triphosphates, and that base pairing occurs with only one strand of the DNA. The enzyme involved is RNA polymerase:

```
-dC-dA-dT-dA-dG-dC-
 :   :   :   :   :   :
-dG-dT-dA-dT-dC-dG-          DNA template
- C- A- U- A- G- C-          RNA product
```

Biosynthesis of Protein. Each enzyme within the cell is characterized by a unique and specific primary structure, and this is what differentiates one enzyme from another. It has been stated that the sequence of deoxynucleotides in DNA constitutes the genetic code of the cell, which means that the information carried within the DNA specifies the primary structures of all protein molecules. Consequently, if a cell needs to synthesize β-aspartokinase, the enzyme responsible for the phosphorylation of aspartic acid in Fig. 8.7, it must retrieve the information from the DNA and use it to organize amino acids in the proper sequence. DNA, however, does not participate directly in the process of protein synthesis. Rather, the structural gene on the DNA specifying the amino acid sequence for the needed protein is transcribed into a strand of messenger RNA (mRNA) which does participate directly in protein synthesis. Because mRNA is a complementary copy of DNA it carries as much information in its nucleotide sequence as did the original structural gene. As you might surmise, there is a unique mRNA for each kind of protein which is being made. The mRNA interacts with a ribosome to form a site for the assembly of amino acids into protein.

The next problem is how a linear polymer composed of four subunits (mRNA) can specify the sequence of subunits in a polymer containing twenty (protein). This is accomplished by means of a triplex code, i.e., each sequence of three nucleotides (called a codon) specifies one amino acid. Since there are 64 possible triplet sequences of four subunits, one or more distinct codons exist for each amino acid. No codon ever specifies more than one amino acid, however.

How is the sequence of codons in the mRNA translated into the amino acid sequence of the protein? When activation of the amino acids occurs each amino acyl group is attached to a molecule of transfer RNA (tRNA) specific for it, there being at least one tRNA corresponding to each amino acid. Each tRNA molecule contains at a specific site within its structure a sequence of three nucleotides, the anticodon, that recognizes the complementary sequence on the mRNA coding for the amino acid being carried by the tRNA. Base pairing can thus occur between the codon and the anticodon, and it is the specific association between the two that insures that the amino acids are joined in the proper sequence. Each tRNA molecule with its associated activating enzyme is the real translator of the genetic code because it is the combination of the two that matches a particular amino acid with its codon. In other words, an amino acid cannot recognize its place in the growing polypeptide chain without being first combined specifically with a tRNA molecule carrying the proper triplet code.

Now, how is the actual protein synthesis performed? This is where the ribosomes enter the picture because one of their functions is to orient the mRNA and tRNA into appropriate juxtaposition. Consider Fig. 8.8. First a complex is formed with the ribosome attached to the beginning of the mRNA. The tRNA carrying the first amino acid (AA_1) is bound to the ribosome at site P (peptide site), with its anticodon

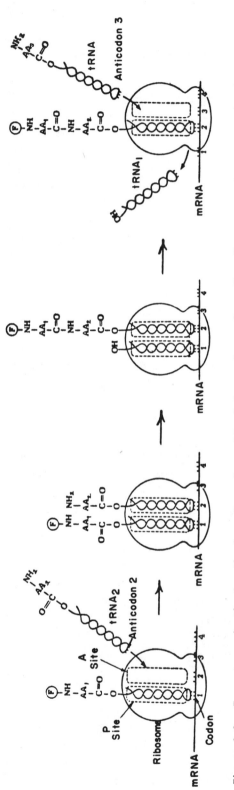

Figure 8.8. Four stages in the lengthening of a polypeptide chain on the surface of a ribosome. (Adapted from Stanier, et al.).

paired with the first codon in the mRNA. A second tRNA whose anticodon complements the second codon in the mRNA also is bound to the ribosome, but at site A (acceptor site). A series of enzymatic reactions then takes place so that the carboxyl group of AA_1 is joined to the amino group of AA_2. AA_2 remains attached to its tRNA. The ribosome now moves the distance of one codon along the mRNA molecule and the first tRNA molecule is discharged. The second tRNA, which contains the peptide chain moves to site P, thereby opening up site A for acceptance of the tRNA whose anticodon is complementary to codon 3. The whole series of reactions is now repeated: AA_1-AA_2 is transferred to the amino group of AA_3, forming AA_1-AA_2-AA_3 which remains attached to tRNA #3; the empty tRNA is discharged, and the one bearing the peptide chain moves to site P. The process is repeated until the ribosome reaches one of the chain-terminating codons, at which point the polypeptide chain is released from the tRNA-ribosome-mRNA complex. When the ribosome reaches the end of the mRNA molecule, it detaches to be used again. When the last ribosome reaches the end of the mRNA strand, the mRNA is broken down into its constituent nucleotides which can be used again to synthesize new mRNA.

8.5 REGULATION OF METABOLISM

8.5.1 General Principles

The preceding discussion of metabolic routes has given some indication of the biochemical complexity of microbial cells. Obviously, if a cell is to function in a reasonably efficient manner its metabolic routes cannot all act in an independent fashion but must, instead, be coordinated in some way. Indeed, the highly sophisticated control system that has evolved in microbial cells constitutes one of the most fascinating areas for study in modern biology.

Consider for a moment the types of controls that would be advantageous to an organism. Suppose that a culture was growing in minimal media on a single carbon and energy source, e.g., glucose. It would be necessary for each cell to synthesize all of the compounds required for it to function and grow. If an amino acid was added to the medium it would be wasteful of both carbon and energy for the cell to go on making that amino acid. Consequently, control mechanisms exist whereby the activities of one or more enzymes in the pathway forming the amino acid are inhibited, thereby reducing or stopping its production. Such feedback inhibitions are called fine controls because they act rapidly to prevent existing enzymes from wasting energy and carbon. Furthermore, if the amino acid continued to be available exogenously, it would be wasteful for the cell to continue making the enzymes in the pathway when they were not being utilized. Consequently, a mechanism (called repression) exists which controls the rates of synthesis of the enzymes. Repression is a coarse control because it is much more slowly acting, both in its imposition and its removal. Although the addition of the amino acid rapidly prevents synthesis of new enzymes, the existing ones are not rapidly destroyed, but are instead merely diluted in the biomass by growth of cells which do not contain the enzyme. Similarly, if the amino acid was removed from the medium, two or more generations might elapse before the amount of enzymes rose to the original level.

Similar control mechanisms exist for catabolic enzymes. Some microorganisms have the ability to utilize more than a hundred different organic compounds as their sole source of carbon and energy, and the use of any one of them might require the activity of many enzymes. Thus, such a cell must possess the ability to synthesize several hundred catabolic enzymes. It would be inefficient for the organism to make all of them if the substrates upon which they act were not present. Enzyme induction is the coarse control mechanism which allows the synthesis of many catabolic enzymes only when their substrates are present. Alternatively suppose two substrates, A and B, are present in the medium. Both can be used by the culture but A can be used much more rapidly. If A can be used rapidly enough to supply all of the biosynthetic needs of the cells, there would be no need to degrade B because its degradative intermediates would just add to the intracellular pool sizes without affecting the rate of growth. The coarse control mechanism known as catabolite repression prevents the synthesis of the enzymes needed to degrade B as long as A is present. Finally, feedback inhibition exists within degradative pathways to balance the rate of metabolite flow into amphibolic pathways with the flow rate of those metabolites out into anabolic pathways. This is a fine control mechanism.

Let us now look in more detail at examples of these control mechanisms.

8.5.2 Regulation of Anabolism

A good picture of the regulation of anabolism is provided by the pathway for the production of the aspartate family of amino acids, which was shown in Fig. 8.7. In Fig. 8.9 the feedback loops for the control mechanisms are superimposed upon a simplified version of the pathway. Let us first consider the enzymes subject to fine control by feedback inhibition of their activities. Examination of the figure shows that all enzymes subject to end product inhibition are either at the start of an entire route, where compounds enter from the amphibolic pathway, or at branch points leading to a specific product. For example, enzyme L is the first committed step in the production of isoleucine, and it is subject to feedback inhibition by that compound. Lysine has a similar effect upon enzyme Q, as does methionine upon enzyme E. These are all cases of simple feedback inhibition because only one compound exerts the control. The control of step D is more complicated, because there are two forms of the enzyme catalyzing it, D_1 and D_2, called isoenzymes. This situation is a result of the multiplicity of products ultimately arising from their action. Enzyme D_1 is subject to feedback inhibition by threonine but D_2 is not at all susceptible to feedback inhibition. Thus the production of homoserine cannot be stopped completely by inhibition. An even more complicated situation exists in the case of enzyme B,

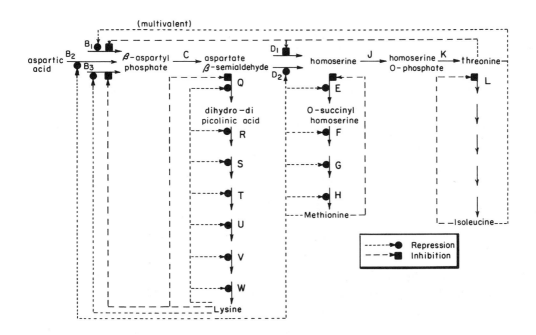

Figure 8.9. Simplified diagram of the pathway for synthesis of the aspartate
family of amino acids showing metabolic controls. Each solid
arrow represents a reaction catalyzed by one enzyme. The dashed
arrows show feedback inhibition while the dotted arrows represent
repression.

aspartokinase, which has three forms. B_1 is inhibited by threonine and B_3 is inhibited by lysine, but B_2 is not subject to feedback inhibition. All three enzymes
produce β-aspartyl phosphate which enters a single pool for use in the downstream
portions of the pathway, consequently, this type of control provides fine tuning of
the rate of metabolite flow from aspartate into the remainder of the pathway in response to the needs for the various end products.

Not all organisms possess multiple forms of the same enzyme, consequently other
methods of control have evolved. For example, some organisms possess only a single
form of aspartokinase which is subject to multivalent feedback inhibition. Its
activity is little affected by either lysine or threonine individually, but in the
presence of both, a severe inhibition occurs. At least six different types of feedback inhibition patterns have been discovered.

The enzymes of the aspartate family are also subject to control by repression
as shown in Fig. 8.9. Repression acts to decrease the rate of synthesis of an
enzyme in relation to the rates of synthesis of other enzymes within the cell. Under
severe repression, synthesis will stop completely. Repression acts at the level of
transcription, i.e., it prevents the production of the mRNA coding for the enzyme.
The less mRNA there is within the cell, the slower the rate of enzyme synthesis.

Enzyme B_3 is subject to repression by lysine, and even though B_2 was not susceptible to inhibition, its synthesis may be repressed by methionine. The synthesis of enzyme B_1 is controlled by both threonine and isoleucine in what is called multivalent repression. The concept is similar to multivalent feedback inhibition. Methionine represses the synthesis of enzyme D_2, whose activity was not subject to control by feedback inhibition, whereas D_1, which could be inhibited, is not subject to repression. Thus it can be seen that the fine and coarse controls work in concert. Finally all of the enzymes Q through W are repressed by lysine while E through H are repressed by methionine. This is termed cumulative repression and is often indicative of the fact that the codes for the enzymes are all adjacent in the DNA.

All other biosynthetic pathways are subject to similar types of control schemes. Consideration of this one, however, should be sufficient to demonstrate the complex, yet efficient, control system that has developed within microbial cells.

8.5.3 Regulation of Catabolism

Anabolic enzymes are normally present within the cell, and their synthesis will only be stopped when they are no longer needed. With the exception of the major common pathways of degradation (for example, the Embden-Meyerhof pathway) catabolic enzymes are normally not present in the cell and will only be synthesized when they are needed. Teleologically, it is easy to imagine why this might be so: the cell "knows" what compounds it normally must synthesize to make new cells, but it doesn't "know" what carbon and energy sources will be available to it. Those enzymes which are normally present in large amount are called constitutive, while those which are not are called inducible. The quantity of an inducible enzyme will increase in response to the presence of the inducer, usually the substrate upon which the enzyme acts. Induction, therefore, is defined as an increase in the rate of synthesis of an enzyme with respect to the rates of synthesis of other enzymes within the cell, in response to the presence of the inducer. Induction operates at the level of transcription and acts to allow the production of mRNA coding for the needed enzyme. The amount of mRNA produced, and consequently the rate of enzyme synthesis, is responsive to the concentration of inducer in the medium. A lag is exhibited between the time of introduction of the inducer and the time at which the enzyme level is sufficiently high to allow observation of the removal of the substrate from the medium. Likewise, upon exhaustion of the inducer, the cells will continue to contain the inducible enzyme until it is lost by dilution due to new growth. Consequently, this is a coarse control mechanism.

The classical example of an inducible enzyme system is that for the degradation of lactose as shown in Fig. 8.10. Lactose itself acts as the inducer for two enzymes, galactoside permease which is responsible for the transport of lactose into the cell, and β-galactosidase which splits lactose into its constituent monosaccharides, glucose and galactose. This situation, in which one compound causes two or more enzymes to be synthesized at about the same rate is called coordinate induction. The glucose

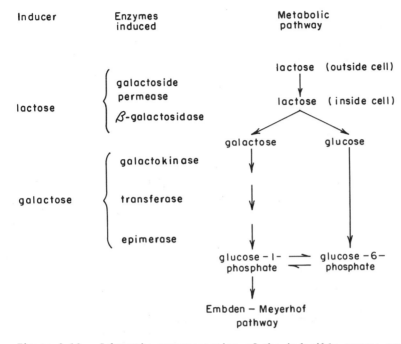

Figure 8.10. Schematic representation of the inducible enzyme system for the degradation of lactose by *E. coli*. *(Adapted from Stanier, et al.)*.

enters directly into the constitutive Embden-Meyerhof pathway. The galactose, on the other hand, induces the synthesis of a sequence of three enzymes which prepare it for entry into the Embden-Meyerhof pathway. This situation, in which the inducer for enzymes is itself the product of prior inducible enzymes is called sequential induction. Both coordinate and sequential induction are common occurences in microbial cells.

Catabolite repression is a mechanism whereby the synthesis of an inducible enzyme may be slowed down or stopped in spite of the presence of the inducer in the medium. As an example, if a cell is allowed to grow at maximal rate on glucose and lactose, the degradation of glucose alone, which can occur by constitutive enzymes, is sufficient to saturate all of the biosynthetic pathways. As a result the levels of ATP within the cell would be high and nothing would be gained by adding the degradation of lactose to the flow of carbon and energy. Because of the high ATP levels, the amount of cyclic 3'-5'-AMP in the cell will be reduced, thereby preventing the inducer from allowing the transcription of mRNA needed for the synthesis of the inducible enzymes. Under conditions of restricted growth, however, when the levels of ATP are lower, it is possible for the inducible enzymes to be synthesized and for both glucose and lactose to be used concurrently. Glucose is not the only compound that can exert catabolite repression because the key to repression is the ability of the carbon source to be degraded rapidly enough to saturate the biosynthetic pathways.

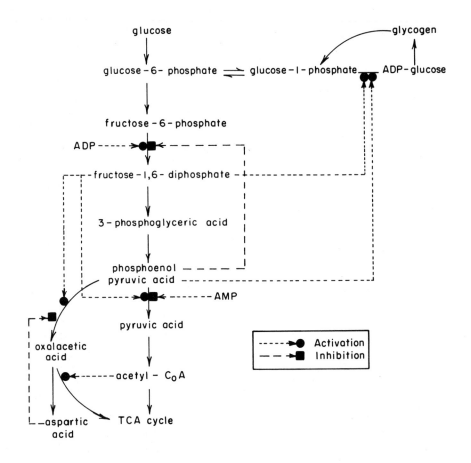

Figure 8.11.   Metabolic control of the Embden-Meyerhof pathway and of glycogen
               synthesis in *E. coli*. Each solid arrow represents a reaction
               catalyzed by one enzyme. The dashed arrows show feedback
               inhibition while the dotted arrows represent activation.
               (From R. Y. Stanier, et al., The Microbial World, Fourth Edition,
               265, 1976. Reprinted by permission of Prentice-Hall, Inc.,
               Englewood Cliffs, N.J.).

Catabolic enzymes are also subject to fine controls, particularly constitutive
ones which are not controlled by induction or repression. As an example, consider
the Embden-Meyerhof pathway. In *E. coli* its regulation is coupled with control of
the synthesis of glycogen, which that organism can use as a reserve food material.
As shown in Fig. 8.11 some enzymes of this route are subject to activation as well
as to inhibition. When activation occurs the rate of activity of an enzyme is increased in the presence of an effector molecule, rather than being decreased. Note
how high levels of fructose-1,6-diphosphate and phosphoenol pyruvic acid act to pro-
mote the conversion of glucose to glycogen both by activating enzymes for the formation of glycogen and by inhibiting enzymes for the degradation of glucose. On the

other hand, ADP and AMP, which would be in high concentration if the level of ATP was low, encourage the degradation of glucose with the resultant production of ATP.

8.5.4 Significance of Regulation to Biochemical Operations

Although most of the work to delineate the various metabolic control mechanisms has been performed with pure cultures, a considerable effort has gone into showing that the same controls are at work in the mixed microbial communities used in wastewater treatment [7]. Superimposed upon these controls, however, is the phenomenon of population dynamics.

From a practical point of view, induction is probably the most important control because whenever a new component is introduced into a wastewater treatment system the requirement for induction will create a time lag before the component is efficiently removed. Often this will be followed by a period in which the relative proportions of the microbial populations within the community shift, but the primary response is a physiological one. Catabolite repression is of secondary importance in slurry reactors because at the growth rates employed the various organic constituents within the waste will be removed concurrently. It is likely that in many fixed-film reactors, however, the bacteria near the influent end will be growing rapidly enough to cause catabolite repression to exist, thereby causing sequential substrate removal. Although this could have a profound effect upon the kinetics of the process, our knowledge of controls is still too limited to allow the phenomenon to be modeled mathematically for design purposes. Nevertheless, awareness of the consequences of these control mechanisms is important to the wastewater treatment engineer.

8.6 KEY POINTS

1. A commonly used empirical formula for ash-free cells is $C_5H_7O_2N$.

2. The relative weights of nucleic acids, proteins, carbohydrates, and lipids found in a microbial culture depend upon its growth conditions.

3. Cells contain two types of nucleic acids, DNA and RNA. DNA is genetic material; RNA aids in the synthesis of protein and in the transfer of information from the DNA to the site of protein synthesis.

4. The chemical nature of a protein is determined by the sequence of amino acids within it, which is coded for by the DNA.

5. Carbohydrates are simple organic compounds, containing only carbon, hydrogen, and oxygen, which play a number of roles in the cell.

6. Autotrophic organisms obtain their cellular carbon from carbon dioxide whereas heterotrophic organisms obtain theirs from organic compounds in their environment.

7. Nitrogen and phosphorus are required for microbial growth and can be obtained in either organic or inorganic forms.

8. The oxygen in cell material comes either from water or from organic compounds being degraded by the cell. Molecular oxygen is used as a terminal electron acceptor during aerobic growth.

9.   The four major nutritional categories of microorganisms are: photoautotrophic, chemoautotrophic, photoheterotrophic, and chemoheterotrophic.

10. Adenosine triphosphate, ATP, is the medium of energy exchange within microbial cells.

11. Biological oxidations occur in small steps with the electrons being transferred with the aid of carrier molecules known as pyridine nucleotides.

12. The reduced forms of the pyridine nucleotides can serve as a source of electrons and protons required to carry out biosynthetic reactions or they can carry electrons to the terminal electron transport chain.

13. During phase I catabolism large molecules are broken down into their small constituent parts so that they may be taken into the cell, but little energy is released. During phase II catabolism the small molecules are incompletely degraded, liberating about one-third of their total energy to the cell.

14. Unlike heterotrophs, the pathways for energy production and carbon incorporation are distinctly different in autotrophs.

15. Autotrophic organisms can obtain all of their carbon from carbon dioxide fixation by the Calvin cycle. The generation of 1 mole of glyceraldehyde-3-phosphate, a 3-carbon compound, requires 9 moles of ATP and 6 moles of $NADPH_2$ which must be supplied either by light sensitive reactions or by the oxidation of inorganic compounds.

16. Amphibolic pathways serve two important roles in heterotrophic organisms, the production of energy and the provision of intermediates for biosynthesis.

17. Anaplerotic routes act to resupply amphibolic pathways with dicarboxylic acids.

18. Anabolic pathways are those in which biosynthesis occurs. The energy needed to perform synthesis reactions is obtained from ATP and the reducing power is provided by $NADPH_2$.

19. Before being incorporated in a growing biopolymer, each subunit must be activated by attachment to a carrier molecule. This activation requires the expenditure of at least one mole of ATP (or equivalent) per mole of monomer.

20. Protein synthesis occurs on ribosomes with the aid of messenger and transfer RNA.

21. Cells possess fine control mechanisms which regulate the activity of enzymes and coarse control mechanisms which regulate the quantity of enzymes present.

22. A constitutive enzyme is one which is normally present in large amounts within the cell whereas an inducible enzyme is one that is not.

8.7 STUDY QUESTIONS

The study questions are arranged in the same order as the text.

Section	Questions
8.1	1-3
8.2	4
8.3	5-6
8.4	7-12
8.5	13-14

1. Explain why new molecules of DNA or molecules of mRNA can be formed with little chance for an error to occur in the sequence of nucleotides.

2. Draw the chemical structure for a peptide chain.

3. Describe the different roles played by carbohydrates in a cell.

4. Describe the major characteristics of the four nutritional categories of microorganisms.

5. Explain how ATP serves as a medium of energy transfer.

6. Describe the roles played by NAD and NADP in electron transfer within the cell.

7. Describe where ATP is formed in the Embden-Meyerhof pathway.

8. Explain why the production of one mole of glyceraldehyde-3-phosphate by the Calvin cycle requires 9 moles of ATP and six moles of $NADPH_2$.

9. Describe the points in the TCA cycle at which reducing power and energy are produced.

10. Explain how DNA replication occurs and how it guarantees that an exact copy will be made.

11. Explain how the subunits in mRNA code for the subunits in the resultant protein.

12. Explain how protein synthesis occurs.

13. Differentiate between the roles of inhibition and repression in a microbial cell, and explain how they enhance its efficiency.

14. Define induction and repression with reference to catabolic enzymes.

REFERENCES AND FURTHER READING

1. S. E. Luria, "The bacterial protoplasm: Composition and organization," in *The Bacteria*, Vol 1, edited by I. C. Gunsalus and R. Y. Stanier, Academic Press, New York, N.Y., 1-34, 1960.

2. F. C. Neidhardt, "Effects of environment on the composition of bacterial cells," *Annual Review of Microbiology*, 17, 61-86, 1963.

3. N. Porges, *et al.*, "Principles of biological oxidation," in *Biological Treatment of Sewage and Industrial Wastes, Vol. 1, Aerobic Oxidation*, edited by J. McCabe and W. W. Eckenfelder, Jr., Reinhold Publishing Corp., New York, N.Y., 35-48, 1956.

4. D. Herbert, "The chemical composition of microorganisms as a function of their environment," in *Microbial Reaction to Environment*, 11th Symposium of the Society for General Microbiology, Cambridge University Press, Cambridge, England, 391-416, 1961.

5. H. L. Kornberg, "The co-ordination of metabolic routes," in *Function and Structure in Microorganisms*, 15th Symposium of the Society for General Microbiology, Cambridge University Press, Cambridge, England, 8-31, 1965.

6. R. N. Doetsch and T. M. Cook, *Introduction to Bacteria and Their Ecobiology,* University Park Press, Baltimore, Md., a. 257-259; b. 268-270, 1973.

7. A. F. Gaudy Jr. and E. T. Gaudy, "Mixed microbial populations," *Advances in Biochemical Engineering, 2,* 97-143, 1972.

Further Reading

Brock, T. D., *Biology of Microorganisms,* Second Edition, Prentice-Hall, Inc., Englewood Cliffs, N.J., 1974. An excellent general microbiology text. See Chapters 4, 5, 6 and 7.

Cohen, G., *The Regulation of Cell Metabolism,* Holt, Rinehart and Winston, Inc., New York, N.Y., 1968. An advanced reference book detailing all aspects of metabolic control.

Lamanna, C., *et al., Basic Bacteriology - Its Biological and Chemical Background,* Fourth Edition, The Williams and Wilkins Company, Baltimore, Md., 1973. An advanced and detailed treatise of bacterial biochemistry. An excellent reference book. See Chapters 10, 12, 13 and 14.

Mandelstam, J. and McQuillen, K., editors, *Biochemistry of Bacterial Growth,* Second Edition, John Wiley & Sons, New York, N.Y., 1973. A detailed, yet very readable account of bacterial biochemistry. Chapter 8, dealing with coordination of metabolism is particularly good. See Chapters 3, 4, 5, 6 and 8.

Sokatch, J. R., *Bacterial Physiology and Metabolism,* Academic Press, London and New York, 1969. An excellent reference book on bacterial biochemistry. See Chapters 3, 6, 7, 8, 10, 11, 12, 13, 14, 15, 16, 17, 18 and 19.

Stanier, R. Y., *et al., The Microbial World,* Fourth Edition, Prentice-Hall, Inc., Englewood Cliffs, N.J., 1976. A very complete microbiology text. See Chapters 6, 7 and 8.

Thimann, K. V., *The Life of Bacteria,* Second Edition, The Macmillan Company, New York, N.Y., 1963. A comprehensive bacteriology text. See Chapters 8, 11, 12, 13, 14, 15, 16, 17, 20, 21, 22 and 23.

Watson, J. D., *Molecular Biology of the Gene,* Second Edition, W. A. Benjamin, Inc., New York, N.Y., 1970. A very readable and enjoyable treatise on molecular biology. See Chapters 8-14.

CHAPTER 9

ENERGETICS AND BIOCHEMICAL EFFICIENCY

In Chapter 8 we investigated the basic metabolic routes whereby microorganisms derive carbon and energy, and synthesize new cell material. We also looked briefly at
the control mechanisms responsible for the coordination of those routes. These
intracellular factors, however, are strongly influenced by the environment in which
the organism is grown, particularly as it affects the fate of the electrons removed
from the substrate during the biooxidation process. As was mentioned briefly in the
last chapter, the fate of those electrons governs the amount of useful energy that can
be obtained by the culture from the substrate, and thereby the efficiency of growth.

The efficiency of growth is of vital concern to the engineer designing a biochemical operation because it affects many factors: the efficiency with which materials can be removed by the process; the amount of excess biomass produced by the
culture which must be disposed of; the quantities of nutrients such as nitrogen and
phosphorus which must be supplied; and the amount of terminal hydrogen acceptor
(e.g., oxygen or nitrate ion) which must be provided. Consequently this chapter will
investigate the factors influencing biochemical efficiency and will demonstrate how
stoichiometry can be used to predict the needs of a culture once its efficiency has
been established.

9.1 CONSERVATION OF ENERGY

The medium of energy exchange within the cell is adenosine triphosphate, ATP.
Consequently, any discussion of the conservation of energy is really a discussion of
the generation of ATP. ATP is formed by two types of phosphorylation reactions:
substrate-level and oxidative. During substrate-level phosphorylation ATP is formed
directly by coupled reactions within a catabolic or amphibolic pathway. This is the
mechanism by which ATP is formed in the Embden-Meyerhof pathway (Fig. 8.2) and GTP
(which can transfer its energy for the formation of ATP) within the TCA cycle
(Fig. 8.5). Only small amounts of ATP can be generated in this way. Much larger
amounts can be generated by oxidative phosphorylation which occurs as electrons removed during oxidation of the substrate are passed along an electron transport chain
toward the terminal hydrogen acceptor. The ultimate fate of those electrons, and
consequently the amount of ATP that can be generated, depends upon both the organism
and its growth environment.

9.1.1 Fate of Electrons in Microbial Metabolism

When electrons and protons are removed from an organic compound during biological
oxidation they are received by an electron transfer coenzyme such as nicotinamide
adenine dinucleotide, NAD (see Section 8.3.4), which is reduced to $NADH_2$. Because
there is a limited quantity of coenzyme within the cell, the reduced form must be
reoxidized or the oxidation of the substrate will cease for lack of a place to put
the hydrogen atoms removed. There are two ways by which the reduced coenzyme may be
oxidized. Fermentation occurs when the hydrogen acceptor is an organic compound
which has been generated within the cell by metabolism of the original substrate.
Respiration occurs when the acceptor is an inorganic compound from the medium. Respiration may be further subdivided with respect to the inorganic acceptor molecule.
If it is molecular oxygen the respiration is referred to as aerobic. If it is some
other inorganic ion, such as nitrate, sulfate, or carbonate, the respiration is
called anaerobic.

The nature of carbon flow within the organism is also influenced by the type of terminal hydrogen acceptor. The organic compounds which act as the hydrogen acceptor in fermentations are metabolites derived from the degradation of the original substrate, so that the substrate ultimately gives rise to a mixture of end products, some of which are more oxidized than it and others of which are more reduced. Each oxidation of an organic molecule must be balanced by the reduction of another, consequently, the original substrate must be neither highly oxidized nor highly reduced. This limits the kinds of compounds that can be degraded by fermentation. In respiration, because the terminal hydrogen acceptor is an inorganic molecule from the medium, there is no internal balance of organic oxidations and reductions required. Consequently, respiratory metabolism usually permits complete oxidation of the organic substrate to carbon dioxide and water. Figure 9.1 summarizes the basic features of fermentations and respirations.

The oxidation state of a compound determines the maximum amount of energy available from it, and the more reduced it is, the more energy it contains. The objective of energy metabolism is to conserve as much of that energy as possible in a form available to the cell; for example, by the formation of ATP. The energy content of an organic compound can be quantified by measuring the amount of heat released when it is oxidized completely to carbon dioxide. For example, oxidation of one mole of ethanol to carbon dioxide releases 316 kcal, whereas oxidation of acetic acid, which is in a more oxidized state to begin with, releases only 269 kcal/mol. As the most completely oxidized form of carbon, carbon dioxide can never serve as an energy source. The maximum energy available from biological oxidation of a substrate is the difference between its energy content and the energy content of the end products derived from it. It is thus easy to see why respiratory metabolism makes more energy available to the cell than does fermentation. Consider the biodegradation of glucose. During aerobic respiration it can be oxidized completely to carbon dioxide, so that

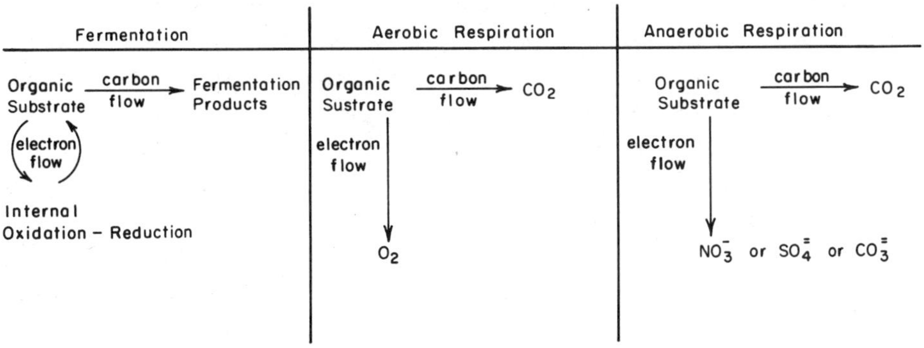

Figure 9.1. Summary of electron and carbon flow in heterotrophic metabolism.

the energy available is 686 kcal/mol. If it is fermented to ethanol, however, the energy content of the two moles of ethanol formed is 632 kcal, which means that only 54 kcal are available to the cell from each mole of glucose metabolized. Since ATP has a free energy of around 8 kcal/mol, it is apparent that much less ATP can be formed per mole of glucose degraded by fermentation.

9.1.2 ATP Generation in Fermentation

All ATP generation during fermentation is by substrate-level phosphorylations. Referring to Fig. 8.2 it can be seen that ATP formation occurs at two places in the Embden-Meyerhof pathway: it is coupled with the oxidation of glyceraldehyde-3-phosphate and with the conversion of phosphoenol-pyruvate to pyruvic acid. As a result of these reactions four moles of ATP are formed per mole of glucose degraded. Since two moles of ATP are required to initiate biodegradation, there is thus a net production of only two moles.

The oxidations and reductions are not balanced in Fig. 8.2 because two moles of $NADH_2$ are formed per mole of glucose. Part of that $NADH_2$ will be used to form $NADPH_2$ which will then be used as reducing power for biosynthesis. The metabolism becomes fermentation when the $NADH_2$ in excess of that required for reducing power is re-oxidized by an organic molecule, thereby forming an organic end product which is released to the medium. For example, pyruvic acid could be reduced to lactic acid:

$$
\begin{array}{ccccccc}
\text{COOH} & & & & \text{COOH} & & \\
| & & & & | & & \\
\text{C} = \text{O} & + & NADH_2 & \rightarrow & \text{CHOH} & + & \text{NAD} \\
| & & & & | & & \\
\text{CH}_3 & & & & \text{CH}_3 & &
\end{array}
\qquad (9.1)
$$

pyruvic acid   lactic acid

Pyruvic acid can also enter an amphibolic pathway, such as the TCA cycle, where it can be converted to intermediary metabolites for the cell, and/or be oxidized to provide energy. In order for the catabolic function of the amphibolic pathway to be significant, however, a large supply of oxidized coenzyme (e.g., NAD, FAD, etc.) must be available. Since the quantity of coenzyme is limited by its regeneration by pyruvic acid and other organic acceptors, the generation of energy by amphibolic pathways is of little importance during fermentation. Thus for this mode of metabolism, amphibolic pathways act primarily in an anabolic capacity to supply intermediary metabolites. The reactions proceeding from glucose to pyruvic acid occur in a wide variety of microorganisms, while the reactions for the fermentative oxidation of the reduced coenzyme are quite diverse, although all rely upon some product derived from pyruvic acid. The products are often characteristic of a particular group of organisms and some are listed in Table 9.1.

TABLE 9.1
Types of Fermentations of Various Microorganisms

Type of Fermentation	Products	Organisms
Alcoholic	Ethanol, CO_2	Yeast
Lactic acid	Lactic acid	*Streptococcus, Lactobacillus*
Mixed acid	Lactic acid, Acetic acid, Ethanol, CO_2, H_2	*Escherichia, Salmonella*
Butanediol	Butanediol, Ethanol, Lactic acid, Acetic acid, CO_2, H_2	*Aerobacter, Serratia*
Butyric acid	Butyric acid, Acetic acid, CO_2, H_2	*Clostridium butyricum*
Acetone-butanol	Acetone, Butanol, Ethanol	*Clostridium acetobutylicum*
Propionic acid	Propionic acid	*Propionibacterium*

Many other compounds may be degraded by fermentations, including other carbohydrates, organic acids, alcohols, amino acids, and purines. Various pathways are
used, but in all cases ATP is formed only by substrate-level phosphorylations and
most of the energy remains in the products. The products are equally diverse, but
may be classified into several groups: steam volatile acids (formic, acetic, and
propionic acids); nonvolatile acids (lactic and succinic acids); volatile neutral
compounds (ethanol, acetone, and isoproponal); nonvolatile neutral compounds
(glycerol and 2,3-butanediol); and gases (hydrogen and CO_2). The array of fermentation end products is characteristic for a particular organism, although the relative amounts of each are influenced strongly by environmental conditions, such as pH.
No matter what the character of the fermentation products however, two facts always
apply: fermentations occur only in the absence of an external hydrogen acceptor,
and the quantity of ATP produced per mole of substrate is always low.

9.1.3   ATP Generation in Respiration

The feature that distinguishes respiration from fermentation is that substrate
electrons are ultimately passed to an inorganic compound rather than to an organic
one. This has two significant effects upon the energetics of the cell. First, because the end product of the catabolic pathway is not required for reoxidation of

the coenzyme it is free to enter an amphibolic pathway where it can be oxidized completely to carbon dioxide. This generates much more reduced coenzyme. Second, ATP is produced by oxidative phosphorylation during passage of the electrons from the reduced coenzyme to the final acceptor, so that much more ATP can be formed. This is the result of the presence in the cell of a special set of carrier enzymes that form the respiratory electron transport chain.

ELECTRON TRANSPORT CHAIN. Electron transport chains (ETC) are found in both eucaryotes and procaryotes and all share common features. All are highly organized and are localized within membranes. All contain flavoproteins and cytochromes which accept electrons from a donor such as $NADH_2$ and pass them in discrete steps to a terminal acceptor. All conserve some of the energy released by coupling the electron transfer to the synthesis of ATP. The ETC in eucaryotic organisms is located within the mitochondria and most of our knowledge about the chain has come from studies of them. The characteristics of the ETC's in various eucaryotes have been found to be remarkably uniform. Within procaryotic organisms the ETC is associated with the cytoplasmic membrane, although little is known about its actual arrangement. The nature of the ETC's found in different bacteria is quite varied and the components of the chain often differ markedly from those characteristic of mitochondrial systems. Because none of the bacterial systems have yet been really well characterized, most of the discussion herein will use the mitochondrial system as a model.

As shown in Fig. 9.2, an electron transport chain consists of flavins, quinones, and cytochromes which are alternately reduced and oxidized as electrons are passed between them in transit to the terminal acceptor. Flavins are proteins bound to a derivative of riboflavin which consists of an isoalloxazine ring which is reduced as it accepts electrons and is oxidized as it loses them. Two flavins are known, flavin mononucleotide (FMN) and flavin adenine dinucleotide (FAD), although the latter is more common. In the mitochondrial ETC the major quinone is ubiquinone although a diversity of other quinones occur in bacteria. The cytochromes are iron-containing porphyrin rings attached to proteins. They undergo oxidation and reduction through loss or gain of an electron by the iron atom at the center of the ring. Several cytochromes are known which differ in their oxidation-reduction potential, and

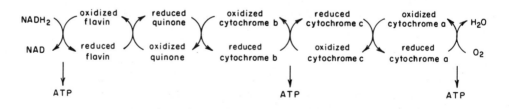

Figure 9.2. Simplified schematic diagram of a typical electron transport chain showing suspected sites of oxidative phosphorylation.

these are designated by letters such as a, b, c, etc. A cytochrome can transfer
electrons to another that has a higher oxidation-reduction potential. A typical
ETC contains three cytochromes as shown in the figure.

   OXIDATIVE PHOSPHORYLATION. ATP formation within the electron transport chain
is called oxidative phosphorylation. Although the mechanism is not well understood,
it is quite different from substrate-level phosphorylation. In mitochondria the reoxidation of one molecule of $NADH_2$, which involves the passage of two electrons
through the chain, results in the formation of three molecules of ATP when the terminal acceptor is oxygen. One is thought to be formed in association with the reoxidation of $NADH_2$ by the flavin, another in electron transfer from the flavin to
cytochrome c, and the third in the oxidation of cytochrome a by oxygen, as mediated
by the enzyme cytochrome oxidase. In bacteria the amount of ATP formed per electron
pair transferred has not been established with certainty, although it appears to be
lower. Referring to Fig. 8.5 it can be seen that a large amount of $NADH_2$ is formed
during oxidation of pyruvic acid by the TCA cycle. Because up to three moles of ATP
are formed per mole of $NADH_2$ reoxidized by the ETC, the importance of respiration as
an energy yielding mechanism is apparent.

   EFFECTS OF TERMINAL HYDROGEN ACCEPTOR ON EFFICIENCY OF OXIDATIVE PHOSPHORY-
LATION. During aerobic respiration, molecular oxygen serves as the terminal hydrogen acceptor, thereby forming water. If no oxygen is present another inorganic
compound can serve as the electron acceptor (anaerobic respiration). From the standpoint of biochemical operations, the reduction of nitrate ion is very important because it forms the basis for denitrification, the process by which nitrogen is removed from a wastewater in the gaseous state. It is thought that N_2 formation from
nitrate follows a multistep reaction sequence:

$$NO_3^- \rightarrow NO_2^- \rightarrow NO \rightarrow N_2O \rightarrow N_2 \qquad\qquad (9.2)$$

Nitrate is reduced to nitrite by the enzyme nitrate reductase utilizing electrons
from cytochrome b. The nitrite is further reduced to the gases N_2O and N_2, but the
biochemistry of these reactions is not well understood. It is thought, however,
that NO is an intermediate which is formed by the activity of cytochrome oxidase
which transfers electrons from cytochrome c to the nitrite ion. Referring to Fig.
9.2 it can be seen that if electrons pass to nitrate from cytochrome b and to nitrite from cytochrome c then cytochrome a is bypassed, which acts to shorten the
electron transport chain. Consequently, a maximum of only two moles of ATP would be
formed per mole of $NADH_2$ reoxidized, rather than the three possible from aerobic
respiration. Therefore, growth by anaerobic respiration of nitrate is less efficient than growth by aerobic respiration.

   The reduction of nitrate is performed by some facultative bacteria when molecular oxygen is not present. As such it is an alternative mode of respiratory energy
yielding metabolism and the organic substrate is ultimately oxidized to carbon

dioxide. This is not the case with the reduction of sulfate, however, which is carried out by only a small group of organisms which are obligate anaerobes. Sulfate reduction is thus not an alternative to aerobic respiration and is of little significance to biochemical operations. Of more significance is the reduction of carbon dioxide to yield methane. Like sulfate reduction it occurs in strictly anaerobic bacteria and does not utilize a typical aerobic electron transport chain. The best understood mechanism for the synthesis of methane is the reduction of carbon dioxide with molecular hydrogen which is thought to consist of two simple coupled reactions involving the enzyme hydrogenase and ferredoxin. Other mechanisms of methane formation are thought to exist and this topic is still undergoing active research.

9.1.4 Effect of Environmental Conditions on Efficiency of ATP Production

From the preceding discussion it is apparent that the environment in which cells are grown will influence the efficiency of ATP production. Consider the case of a culture of hypothetical facultative organisms capable of performing aerobic respiration, anaerobic respiration, or fermentation. If they are placed in an environment rich in oxygen they will degrade the substrate by aerobic respiration thereby forming a large amount of ATP. If they are placed in an environment devoid of oxygen, but rich in nitrate ion, they will utilize anaerobic respiration to produce a sizeable quantity of ATP, although less than would be formed during aerobic respiration. Finally, if they are placed in an environment devoid of both oxygen and nitrate they will attack the substrate by fermentative pathways which will result in even less ATP. The flow of electrons in these situations is summarized in Fig. 9.3. Let us now estimate quantitatively the differences that would result from the three cases for the degradation of glucose.

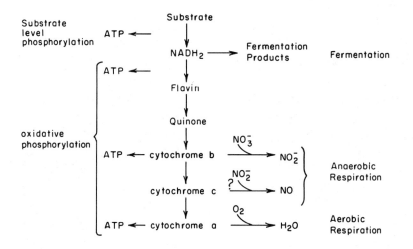

Figure 9.3.   Generalized diagram of possible routes of electron flow in microorganisms. Not all alternatives can be used by every organism.

Consider first the case of fermentation of glucose by the Embden-Meyerhof pathway with ethanol as the end product. Referring to Fig. 8.2 it can be seen that two moles of ATP are formed per mole of glucose attacked. The change in free energy for the conversion of one mole of glucose to two moles of ethanol is about 54 kcal as determined in Section 9.1.1. Consequently the cell would expend 27 kcal of energy per mole of ATP formed.

Reference to Fig. 8.5 shows that when the two moles of pyruvic acid formed by the Embden-Meyerhof pathway are degraded by the TCA cycle, two more moles of ATP are formed (from GTP) by substrate-level phosphorylation. In addition, eight moles of $NADH_2$ are formed which will be added to the two formed in the Embden-Meyerhof pathway, for a total of ten which must be reoxidized by the electron transport chain. In addition, two moles of $FADH_2$ are formed.

If aerobic respiration is possible, three moles of ATP will be formed from the oxidation of each mole of $NADH_2$ and two moles from each mole of $FADH_2$. This results in 34 moles of ATP which can be formed by oxidative phosphorylation. When added to the four from substrate-level phosphorylation, a total of 38 moles of ATP can be formed per mole of glucose - 19 times as much as from fermentation! This has been achieved at the expense of the total oxidation of glucose to carbon dioxide, which is a change in free energy of 686 kcal/mole. Consequently, the cell expended 18 kcal of energy per mole of ATP formed. Thus, not only is more ATP formed during aerobic respiration, but it is formed more efficiently than by fermentation.

If the culture is growing by anaerobic respiration with nitrate as the terminal hydrogen acceptor, it will produce the same amount of $NADH_2$ and $FADH_2$ as it did in aerobic respiration. However, when a mole of $NADH_2$ is reoxidized, only two moles of ATP will be formed, for a total of 20. Furthermore, only one mole of ATP will be formed per mole of $FADH_2$ which is oxidized, for a total of two. Thus, oxidative phosphorylation will result in only 22 moles of ATP per mole of glucose. Substrate-level phosphorylation will still yield four ATP's however, so that the total ATP production will be 26 moles per mole of glucose. This is much more than could be achieved by fermentation, but somewhat less than the result from aerobic respiration. Since the glucose is oxidized to carbon dioxide, the free energy change is still 686 kcal/mole so that the energy expended per mole of ATP formed is 26.4 kcal. This is almost the same as with fermentation.

The amount of cell material that can be formed from a substrate is proportional to the amount of ATP that can be generated. Consequently, growth by aerobic respiration would result in 19 times more cell mass than from fermentation, while anaerobic respiration would result in 13 times as much. However, most of the energy originally available in the glucose was not used during the fermentation. It is still in the medium in the form of end products, although they are not available to the culture unless there is a change of environment.

9.1.5 Other Factors Affecting Efficiency of ATP Generation

In addition to the growth environment there are other factors which influence the
generation of ATP; for example, the nature of the substrate being degraded. If the
substrate is already fairly highly oxidized, only a small amount of $NADH_2$ will be
generated, and consequently little ATP will be formed. On the other hand, if the
substrate is highly reduced, a large amount of $NADH_2$ will be formed, leading to
large quantities of ATP. Thus more growth would be expected from the degradation of
a mole of reduced substrate than from a mole of an oxidized one. Even if the two
substrates were at about the same oxidation state, however, different amounts of
ATP may be formed by their degradation if substantially different pathways are
employed. Consequently, biochemical efficiency also influences ATP production.

The effects of biochemical efficiency are also reflected by the fact that the
nature of the organism performing the degradation can also influence ATP production.
During aerobic metabolism the differences in biochemical efficiency will be relatively small because most of the ATP arises from oxidative phosphorylation. During
fermentative growth, however, the differences can be substantial because of the
importance of substrate-level phosphorylation. As a case in point, consider the
fermentation of glucose by the yeast *Saccharomyces cerevisiae* and the bacterium
Pseudomonas linderi [1]. Both produce ethanol and carbon dioxide as the end products,
but the yeast uses the Embden-Meyerhof pathway in which two moles of ATP are generated per mole of glucose fermented, whereas *P. linderi* ferments glucose via the
Entner-Doudoroff pathway, which results in only one mole of ATP. While this is
admittedly an extreme case, other examples have been noted [2].

As we have seen, many factors can influence the efficiency of energy conservation within a cell. The amount of growth that will result from that energy depends upon how efficiently it is utilized.

9.2 UTILIZATION OF ENERGY

There are a multitude of ways in which a cell can expend energy, but for the purposes
of discussion these will all be grouped under two major headings, as shown in
Fig. 9.4 [3]. Energy for synthesis includes that required to form the bonds made in
anabolic pathways and to reduce the carbon source to the proper oxidation-reduction
level for incorporation into cell material. Energy for maintenance represents the
energy required to keep the cell functioning. All of the energy to meet these two
requirements comes from the energy-yielding metabolism of the cell and the central
role of ATP in this exchange is noted in Fig. 9.4 by the cyclic nature of the
energy carrier.

9.2.1 Energy for Synthesis

Energy is required for the synthesis reactions of anabolic pathways as can be seen
by referring back to Fig. 8.7 where the pathway leading from oxalacetic acid to the
aspartate family of amino acids was given. Using the route to methionine as an

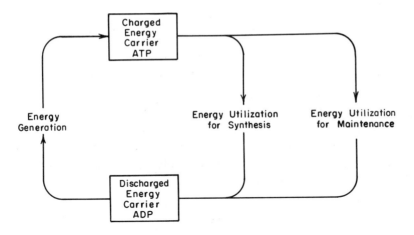

Figure 9.4.   Schematic diagram of energy flow in microorganisms.
(Adapted from McCarty [3].).

example, one mole of ATP is required directly, but more importantly, two moles of
$NADPH_2$ are used as reducing power. Each of those represents electrons which will
not be transferred to the electron transport chain, and thus will not be available
for the generation of ATP. Consequently, the use of two moles of $NADPH_2$ is equivalent to the use of six moles of ATP in energy expenditure. Furthermore, succinyl-
CoA is used to activate homoserine for the addition of cysteine, with succinic acid
ultimately being released. Referring to the TCA cycle, Fig. 8.5, it can be seen
that this bypasses the synthesis of GTP; thus the succinyl-CoA is equivalent to the
use of another ATP. In sum, the equivalent of eight moles of ATP is required to
synthesize one mole of methionine from oxalacetic acid. Looking at other biosynthetic reactions in a similar manner gives an indication of the large amount of
energy that must be expended by the cell to synthesize all of the small molecules
needed for growth.

Considerations like those above lead to the conclusion that much more energy
would be required for a culture to grow in a minimal medium in which only a single
organic compound is supplied as both carbon and energy source than in a complex
medium in which all required amino and nucleic acids are supplied. Actually, however, such a conclusion is false [2]. The energy needed to synthesize the amino
acids amounts to only about 10 percent of the total energy needed to synthesize new
cell material because most of the energy is used to synthesize the polymers of the
cells, such as the enzymes, nucleic materials, membranes, etc. Since these are too
large to be transported into the cell they must be formed inside even when the
monomers are provided in the medium. Consequently, while the complexity of the
growth medium has some effect on the energy required for synthesis, it is not large.

Of more importance are the oxidation state and size of the carbon source. The
oxidation state of cell material is roughly the same as that of carbohydrate [2].

Consequently if the carbon source is more oxidized than that, energy must be expended
to reduce it to the proper level. If the carbon source is more reduced, it will be
oxidized to the proper level during normal biodegradation and thus no extra energy
will be required. As a general rule, a carbon source at an oxidation state higher
than carbohydrate will require more energy to be incorporated into cell material
than will one at a lower oxidation state. Pyruvic acid occupies a unique position
in metabolism because it lies at the end of many catabolic pathways and at the beginning of many anabolic and amphibolic ones. As such it provides carbon atoms in
a form which can be incorporated easily into other molecules. Indeed three-carbon
fragments play an important role in the biosynthesis of many compounds. If the
carbon source contains more than three carbon atoms it can be broken down to size
without the expenditure of large amounts of energy. If it contains less than three
carbon atoms, however, energy must be expended to provide three-carbon fragments for
incorporation. Consequently, substrates containing few carbon atoms require more
energy for synthesis than do large ones.

Carbon dioxide, which is used by autotrophic organisms as their chief carbon
source, is an extreme example of the factors just discussed, being a one-carbon
compound in which the carbon is in the highest possible oxidation state. Consequently, the energy for synthesis for autotrophic growth is very much higher than
for heterotrophic growth. As a result, the amount of cell material that can be
formed per unit of available charged energy carrier is quite low.

9.2.2 Energy for Maintenance

Energy for maintenance is the term applied to the energy needed to maintain the
status quo in a microbial culture, i.e., to keep cells functioning even in the absence of growth. Duclaux in 1898 was probably the first microbiologist to distinguish between energy for synthesis and energy for maintenance [4]. Nevertheless it
is only recently that the concept of energy for maintenance has been established for
bacteria and that was as a result of the development of techniques allowing the
continuous culture of microorganisms.

Cellular processes, whether mechanical or chemical, require energy for their
performance, and unless a supply is available these essential processes will cease
and the cell will become disorganized and die. Mechanical processes include motility,
osmotic regulation, molecular transport, and in the case of eucaryotic organisms,
cytoplasmic streaming. While it might be argued that motility could be dispensed
with in some organisms, this argument would not hold for all because some require
motility to find food. Osmotic regulation is quite important in all cells, even in
those protected by a rigid cell wall, and pump mechanisms, such as contractile
vacuoles, exist in cells to counteract the normal tendency of osmotic pressure to
force water into them. Cell membranes are permeable to many small molecules, such
as amino acids, and because of the high concentrations within the cell these tend to
diffuse into the medium. Active transport mechanisms operate to bring such molecules

into the cell against the concentration gradient. Cytoplasmic streaming and the movement of materials within eucaryotes is often required for their proper functioning. For example, phagocytosis is used by many for the ingestion of food. These all require energy.

Although mechanical factors contribute to the energy requirement for maintenance they are probably secondary to the chemical ones in importance. Microbial cells represent chemical organization and many of the components within them have higher free energies than the original substrates from which they were formed. In general, because of this organization, energy must be available to counteract the normal tendency toward disorder (i.e., overcome entropy). The chemical processes contributing to the energy requirement for maintenance are those involved in resynthesis of structures such as the cell wall, flagella, the cell membrane, and the catabolic apparatus. Within *E. coli* for example, energy for the resynthesis of proteins and nucleic acids appears to be a major portion of the maintenance requirement [5].

There is still some question about whether the maintenance energy requirement is dependent upon the rate at which the organisms are growing. Mandelstam [6] has reported that the rate of turnover of macromolecules (protein and RNA) is inversely proportional to the growth rate of the organisms, which suggests that the maintenance energy requirement would vary in a similar fashion. Experiments with carbon limited continuous cultures, on the other hand, suggest that the requirement is constant for any given culture condition (pH, temperature, etc.) and independent of the specific growth rate of the organisms [4]. Because most biochemical operations are carbon limited continuous cultures, engineers generally consider the maintenance energy requirement to be constant.

Assuming that the maintenance energy requirement is constant, what energy sources are used to supply it? The answer to that question depends upon the conditions of growth of the organisms. If an exogenous energy supply is available a portion of it will be used to meet the maintenance requirement and the remainder will be used for growth. As the rate of supply of energy is decreased, less and less will be available for new growth. When the point is reached at which the rate of energy supply just balances the rate at which energy must be used for maintenance, no net growth will occur because all energy will be used to maintain the *status quo*. If the rate of energy supply is reduced still further, the difference between the supply and the maintenance requirement will be met by the degradation of energy sources available within the cell, i.e., by endogenous metabolism. This will cause a decline in the mass of the culture. Finally, if no external energy source is available all of the maintenance requirements must be met by endogenous metabolism. When the point is reached at which endogenous supplies are not sufficient, the cells deteriorate and die.

The nature of the materials serving as substrates for endogenous metabolism depends both upon the species of the organism and the conditions under which the culture has been grown. For example, when *E. coli* is grown rapidly in a glucose-mineral

salts medium it stores glycogen. If those cells are then placed in an environment
devoid of exogenous substrate they will utilize the glycogen as an endogenous energy
source. Amino acids and proteins show little net catabolism until the glycogen is
gone. When grown in tryptone medium, on the other hand, *E. coli* accumulates little
glycogen. As a result, endogenous metabolism utilizes nitrogenous compounds immediately. Other organisms use still other compounds, including RNA and the lipid, poly-
β-hydroxybutyrate.

One question that has intrigued microbiologists concerns the route of energy
flow when sufficient exogenous substrate is available to supply the maintenance requirements of the culture. Does endogenous metabolism continue under those circumstances so that the energy of maintenance is used to continually resupply the materials being degraded? Or, alternatively, does endogenous metabolism cease so that the
exogenous energy source goes directly into maintenance functions? The evidence is
still not conclusive in either direction. Although such a question is of fundamental
scientific significance, it has little bearing in the macroscopic energy balances
used by engineers to mathematically model biochemical operations. Consequently, because of the simpler forms taken by the models, most engineers assume that endogenous
metabolism continues at all times.

9.3 YIELD

Biochemical operations can be used for wastewater treatment because they transform
soluble materials, which are difficult to remove, into insoluble materials, (e.g.,
cells), which can be removed relatively easily by physical means and gases (e.g.,
CO_2), which are released to the atmosphere. The amount of insoluble material formed
during this transformation is of considerable significance to the design enegineer
because it represents a process byproduct which must be disposed of. Consequently,
the engineer must have some means of predicting the quantity of cell material that
will be formed during the course of biochemical treatment. This has led to the
concept of yield.

9.3.1 Concept of Yield

Yield is defined as the amount of biomass formed per unit amount of substrate removed
from the medium. The rather vague term "amount" is used purposely at this time because there are a number of specific definitions which can be attached to it, all of
which lead to different numerical values of the yield. These will be the topic of
Section 9.3.2. The word substrate is taken to mean the growth limiting nutrient.
For example, if the organic carbon source is the growth limiting nutrient the yield
should be based on the amount of it removed. However, yields can be computed on the
basis of other materials as well.

The choice of the word yield to express the concept of a quantitative relationship between the consumption of substrate and the production of biomass reflects the
scientific background of biochemical engineering. Many early experiments on

microbial growth were directed toward the production of cell material. Consequently they were concerned with maximizing their output, i.e., the amount of product formed per unit of input involved, or the yield. This term, with its emphasis upon production, has become accepted by wastewater treatment engineers as well, in spite of their desire to minimize production. This minimization is not a trivial problem because a large portion of the cost of wastewater treatment is involved with the disposal of the excess biomass, or sludge.

The amount of microbial growth occurring during the degradation of a substrate is a result of the relative efficiencies of energy conservation and energy utilization. Because these efficiencies are strongly affected by the nature of the substrate and the culture utilizing it, it was expected that they would influence the yield, as indeed they do. However, it was also found that even under constant environmental conditions the yield values observed for a specific substrate and organism were not consistent. The reason for this variability was not recognized, however, until the significance of the energy of maintenance was recognized.

Consider a culture growing at a constant rate. The rate of energy generation will be fixed in proportion to the growth rate, as will the rate of energy utilization for synthesis. These two rates will not be equal, however, because of the need for energy for maintenance. In fact, the rate of energy generation must equal the sum of the rates of energy utilization for maintenance and for synthesis:

$$r_{eg} = -r_{eus} - r_{eum} \qquad (9.3)$$

where

r_{eg} = rate of energy generation
$-r_{eus}$ = rate of energy utilization for synthesis
$-r_{eum}$ = rate of energy utilization for maintenance

The rate of cell growth, r_G, will be proportional to the rate of energy utilization for synthesis and the proportionality constant will be the efficiency of that energy utilization, E_{eus}:

$$r_G = E_{eus} \, (-r_{eus}) \qquad (9.4)$$

The rate of substrate utilization, $-r_s$, will be proportional to the rate of energy generation and the proportionality constant will be the efficiency of energy generation, E_{eg}:

$$-r_s = r_{eg}/E_{eg} \qquad (9.5)$$

The yield, Y, is defined as the amount of cell material formed per unit amount of substrate removed from the medium. Over a fixed time period at steady state this is equal to the ratio of the rates of cell growth and substrate removal:

$$Y = r_G/-r_s \qquad (9.6)$$

which implies that:

$$Y = (E_{eus}E_{eg})(-r_{eus}/r_{eg}) \tag{9.7}$$

However, the rate of energy utilization for synthesis is the difference between the
rate of energy generation and the rate of energy utilization for maintenance
(Eq. 9.3), so that the equation for the yield can be expressed as:

$$Y = (E_{eus}E_{eg})[\frac{r_{eg} - (-r_{eum})}{r_{eg}}] \tag{9.8}$$

Now, we saw earlier that $-r_{eum}$ is considered to be constant. However, r_{eg} depends
upon the growth conditions, particularly the growth rate. When cells are allowed to
grow in unrestricted growth r_{eg} will be maximal with the result that $-r_{eum}$ is
negligible in comparison to it so that the yield is maximal. If, on the other hand,
the rate of growth is restricted to some low value by restricting the rate of energy
supply, r_{eg} will be small so that $-r_{eum}$ becomes a significant percentage of it,
thereby decreasing the yield. In fact, if the rate of energy supply becomes so low
that $-r_{eum}$ exceeds r_{eg}, the yield would become negative! This is what happens
during endogenous metabolism when the cell mass decreases.

Obviously, if the yield is subject to so much variation it is of little value
as a predictive tool for design purposes. Consequently, a new parameter, the true
growth yield, Y_g, was defined as the amount of cell material formed per unit of
substrate utilized in the absence of maintenance energy requirements [4]. As such,
it cannot be measured directly, but is instead obtained by extrapolation of observed
yield data. Techniques for performing that extrapolation will be discussed in
Chapter 12. During periods of rapid, unrestricted growth, however, the observed
yield approaches Y_g and consequently measurements made under those circumstances are
often used as approximations of it.

The advantage in the use of the true growth yield is that it is not affected
by the growth rate under which the culture is grown. Furthermore, in Section 9.2.2
it was stated that most engineers model energy flow by considering endogenous metabolism to occur at all times. The use of a parameter such as Y_g facilitates the
formulation of such a model.

It should not be assumed that Y_g is a universal constant having one value for
all substrates and all organisms because it is influenced by the efficiencies of both
energy generation and energy utilization for synthesis as can be seen by letting r_{eum}
approach zero in Eq. 9.8. Section 9.3.3 will discuss the factors that influence Y_g,
but first we must consider the various ways of expressing it.

9.3.2 Methods of Expressing Y_g

In the definitions of yield presented in the previous section the unspecific term
"amount" was used. The most convenient measure of the amount of cells is the dry
mass, generally expressed in grams. The amount of substrate consumed can be

expressed in four ways: as the equivalent amount of ATP which could be formed from
it; as the equivalent amount of total energy removed from the medium; as the number
of electrons initially available in the substrate for transfer to the terminal
hydrogen acceptor or for incorporation into cell material; and as the reduction in
the amount of organic matter present, with the concentration expressed as COD or
some other similar measurement (see Chapter 11). Each one of these methods leads to
a different expression for the yield, thus let us look at each in more detail.

   YIELDS BASED UPON ATP FORMED. A yield value based upon the amount of ATP which
could be formed from the degradation of the substrate has a certain appeal. From the
discussion of the first part of this chapter it is known that the mass of cell material which can be formed from a substrate depends upon two factors: the efficiency of energy generation (i.e., the amount of ATP formed), and the efficiency of
energy utilization. By expressing the yield as the mass of cells formed per unit
of ATP generated from the substrate, one of these factors is removed and thus the
variability of Y_g is likely to be reduced. Indeed, experiments initiated by Bauchop
and Elsden [1] and extended by others have shown that a large number of organisms
produce 10.5 grams of cells per mole of ATP formed. Consequently, many researchers
considered that to be a "universal" constant. Recent research, however, has suggested that this value is low because it fails to account for the maintenance energy
requirements of the cells [7]. Furthermore, there is reason to question the validity
of considering the yield based upon ATP to be a constant.

   In order for the mass of cells formed per mole of ATP to be constant, it would
be necessary for the efficiency of cell production to be the same for all organisms
and all substrates. However, from the information in Section 9.2.1 we know that
this is not the case. The energy required to form the cells (and hence the mass of
cells formed per unit of available energy) will depend upon the size and oxidation
state of the substrate molecule as well as the complexity of the medium. While it
is possible that a constant value of Y_g exists for the rather narrow range of substrate and media employed in microbiological research, it is unlikely that this will
be the case for the broad range of conditions encountered in wastewater treatment.
Consequently, even if the amount of ATP formed from a substrate were known, it would
still be necessary to experimentally determine the amount of cell material that
could be formed.

   From a practical standpoint an even more serious objection to the use of Y_g
based upon ATP is the uncertainty associated with estimates of the amount of ATP
formed from various substrates under different conditions. Knowledge of the amount
of ATP which can be formed anaerobically from glucose and other simple sugars is
relatively good. However, there is still a great deal of doubt about the amount of
ATP formed by oxidative phosphorylation in bacteria, so that it is difficult to predict the amount of ATP that will be formed under aerobic conditions. Add to this
the fact that most wastewaters contain a mixture of substrates of variable chemical
complexity and it becomes apparent that it would be almost impossible to predict

the amount of ATP likely to be formed. Consequently, yields based upon ATP production find little engineering application.

   YIELDS BASED UPON ENERGY TAKEN FROM MEDIUM. For the growth of bacteria in any
culture medium, the energy consumed (e_c) from the medium must be the sum of that
incorporated biosynthetically into cell material (e_a) and that expended by catabolism
(e_d):

$$e_c = e_a + e_d \qquad\qquad\qquad (9.9)$$

The yield based upon the energy taken from the medium is defined as the grams of
cells formed per kcal of total energy consumed, e_c. It is estimated by dividing the
mass of cells by the sum of e_a and e_d. A good approximation of e_a is provided by
the heat of combustion of dry cells. For bacteria, a value of 5.41 kcal/gram of
ash-free cells has been reported as the average from 15 different species, with a
coefficient of variation of only 3.5 percent [8]. Other work [9], however, has
indicated that e_a depends upon the growth conditions so it is unlikely that a constant value can be assumed. The value of e_d may be approximated as the difference
between the heat of combustion of the original substrate and the end products from
its degradation. Determination of heat of combustion values on a routine basis is
rather tedious, particularly for dilute aqueous samples such as wastewater. Consequently, the main appeal of a yield based upon the energy from the medium would be
if it were constant. Payne [10] reported an average value of 0.130 grams dry
weight/kcal expended for nine different organisms growing anaerobically on a number
of different substrates, with a range from 0.092 to 0.164. For aerobic growth the
average for eight organisms was 0.116 with a range from 0.075 to 0.143. Consequently, it can be seen that this yield is not constant. Indeed, we would not expect it to be because it does not take into account the efficiencies of energy
generation and utilization. Therefore, its utility is limited, although the energy
balance upon which it is based is an important concept and does find considerable
use.

   YIELDS BASED ON ELECTRONS AVAILABLE IN SUBSTRATES. As long as its empirical
formula is known, any substrate may be characterized by its constituent number of
available electrons. The yield per available electron may be determined by dividing
the mass of cells formed per mole of substrate consumed by the number of electrons
available per mole. Like the energy yield in the previous section, this one does
not account for the efficiencies of energy generation and utilization, and hence
should not be expected to be constant. In fact, values from 19 different species of
organisms have been reported to range from 1.09 to 4.96 grams of cells per available
electron [10]. The advantage to this type of yield is that it is easily determined
as long as the formula for the substrate is known, and thus it is useful for experimental work. It is impossible to measure directly, however, if the chemical nature

of the substrate is unknown, although it is closely related to another yield which
is commonly used in the design of biochemical operations.

   YIELDS BASED ON COD CONSUMED. The COD test is used by wastewater treatment
engineers to measure the concentration of a waste of unknown chemical composition.
It is based upon the complete chemical oxidation of the organic compounds to carbon
dioxide and water, and the results are expressed on a mass basis in terms of the
amount of oxygen that would be required if it were the electron acceptor (i.e.,
mg of oxygen required per liter of sample). Yields based upon this parameter can be
determined easily because all that is required is that the COD of the growth medium
be determined initially and at the end of growth, after the cells are removed. The
difference between the two is the amount of organic matter removed by the organisms,
expressed as COD. When the mass of cells formed is divided by the mass of COD removed the result is the yield expressed as the grams of cells formed per gram of COD
consumed.

   The yield based on COD can be converted easily to the yield per available
electron. One mole of oxygen has a mass of 32 grams, and four electrons are required to reduce one mole. Thus the yield per gram of COD can be converted to the
yield per electron by multiplying by eight.

   Because the yield per gram of COD can be determined so easily, it finds frequent use in engineering. It is obvious from the previous section that this yield
will not be constant for all substrates and all organisms, but this detracts little
from its usefulness as it is just one of several parameters which must be determined
experimentally during wastewater treatability studies. These will be discussed in
Chapter 12. Let us now consider some of the factors that influence Y_g, expressed as
grams of cells formed per gram of COD removed.

9.3.3 Factors Affecting Y_g

By this time it should be clear that the nature of the substrate will influence Y_g.
Nevertheless to give some indication of the effects that substrate can have,
Table 9.2 summarizes data from one species, *Aerobacter aerogenes,* which was grown in
unrestricted batch growth in minimal media [11]. The values were not reported on a
COD basis, but were converted to it for this table.

   The previous discussion has also made it clear that the species of organism will
also affect Y_g, although the effect will not be as great as the effect of substrate.
Table 9.3 presents values for various species growing aerobically on glucose in minimal media. The data were compiled by Payne [10] from a number of different published
reports, consequently some of the variation may be due to differences in experimental
conditions rather than to the species. As with Table 9.2, the values were not
originally reported on a COD basis, but were converted to it.

   The growth environment, including media complexity, type of terminal hydrogen
acceptor, pH, and temperature, will affect Y_g. For example, cells grown in complex
media will exhibit slightly higher values than will cells grown in minimal media.

TABLE 9.2
Y_g Values for Aerobic Growth of *Aerobacter aerogenes* on Various Substrates

Substrate	Y_g [a] (grams cells/gram COD removed)
Sorbitol	0.47
Mannitol	0.46
Sucrose	0.45
Citrate	0.43
L-Arabinose	0.41
Fructose	0.40
Maltose	0.39
Galactose	0.38
Glucose	0.38
Galactonate	0.38
Glycerol	0.37
Mannose	0.36
Gluconate	0.35
Ribitol	0.35
Galacturonate	0.35
Glucolronate	0.35
Dihydroxyacetone	0.33
Ribose	0.33
Xylose	0.33

[a] Values were calculated from data in Hadjipetrou *et al.* [11].

TABLE 9.3
Y_g Values for Aerobic Growth of Various Organisms on Glucose in Minimal Media

Organism	Y_g [a] (gram cells/gram COD removed)
Aerobacter aerogenes	0.38
Aerobacter cloacae	0.41
Arthrobacter globiformis	0.49
Bacterium HR	0.38
Candida utilis	0.48
Escherichia coli	0.48
Pseudomonas aeruginosa	0.37
Pseudomonas fluorescens	0.36

[a] Values were calculated from data in Payne [10].

Growth by aerobic respiration will produce higher yields than will growth by anaerobic respiration using nitrate, because more ATP is produced by the former. The yield from fermentations will depend upon the reduced end products. When they are all soluble compounds the true growth yield based upon the amount of COD removed from the medium is not much different from that obtained with aerobic cultures [12,13]. However, when methane is produced, so that most of the reduced end product is lost from the system as a gas, then the COD removed from solution is actually much higher than the COD utilized by the bacteria. Under these circumstances the yield per unit of COD removed is about an order of magnitude lower than for aerobic growth. The pH of the medium has long been known to affect microbial growth, but no quantitative studies of the influence upon Y_g have been published. It is likely, however, because of its effect upon transport, that pH will also affect Y_g, with a maximum around 7. Temperature also affects the observed yield, but much of that can be attributed to the changing requirements for maintenance energy as the temperature is changed. Recent research, however, indicates that temperature also influences the true growth yield, Y_g [14,15,16,17]. Figure 9.5 summarizes the data from those reports. Although the significance of temperature is apparent, more research is needed before generalizations can be made. Nevertheless, it is important for engineers to recognize that temperature has an effect. A final factor which might influence Y_g is the composition of the microbial community. When the culture is heterogeneous, the waste products from one species might serve as growth factors for another, thereby converting a seemingly minimal medium into a complex one. Consequently it might be anticipated that the yields from mixed cultures would be higher than those from pure cultures growing on the same medium. A comparison of the two types revealed this to be the case [18].

9.3.4 Constancy of Y_g in Biochemical Operations

Biochemical operations are used to treat wastewaters containing mixtures of substrates by using heterogeneous cultures. Thus it is apparent that Y_g will depend

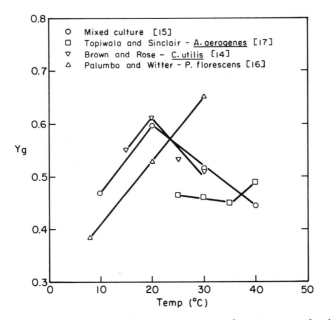

Figure 9.5. Effect of temperature on the true growth yield Y_g. (*From R.E. Muck and C.P.L. Grady, Jr., "Temperature Effects on Microbial Growth in CSTR's," Journal of the Environmental Engineering Division, ASCE, Vol. 100, No. EE5, Proceedings Paper #10848, 1147-1163, October 1974. Reprinted by permission of the American Society of Civil Engineers.*)

upon both the character of the wastewater and the particular culture that develops
on it. It is important that this variability be recognized by engineers designing
biochemical operations, because then the numbers will be treated with the proper
care. As will be seen in Chapter 10, similar conclusions can be reached about the
kinetic parameters associated with biochemical operations. This means that designers
must utilize considerable judgement and allow for uncertainty. This situation does
not prevent certain generalities from being made, however. For example, examination
of a large number of yield values indicates that Y_g will generally lie within the
range of 0.40-0.60 grams of cells per gram of COD removed for aerobic heterotrophs
utilizing carbohydrates [19]. Similar values would be expected for other substrates
of approximately the same oxidation state. The range for autotrophs will be much
lower, however, e.g., 0.05-0.29 grams of cells per gram of nitrogen oxidized [20].
Although ranges such as these provide the engineer with an idea of the magnitude to
be expected, designs should be based upon estimates of Y_g obtained from laboratory-
and pilot-scale studies of the particular waste to be treated.

9.4 STOICHIOMETRY OF MICROBIAL GROWTH

Stoichiometry provides a system for keeping track of the reactants that participate
in a chemical reaction and the products that are produced from them. When applied
to a biochemical reaction it reveals the amount of cell material formed, terminal
hydrogen acceptor used, and nutrients required for microbial growth from a given
carbon and energy source; all valuable information to the designer of a wastewater
treatment system.

9.4.1 The "Pseudostoichiometry" of Biological Systems

Consider a closed flask containing a medium consisting of water, glucose, ammonia,
inorganic phosphate, and trace elements. The atmosphere above the medium is charged
with oxygen in an amount sufficient to allow complete oxidation of the glucose to
carbon dioxide and water. These materials are the reactants. Now, to the flask add
a minute quantity of microorganisms to serve as a catalyst and continually agitate
the flask to facilitate transfer of the oxygen to the liquid. The reaction that
will then proceed is one in which carbon from the glucose is combined with nitrogen,
phosphorus, oxygen, and hydrogen to produce new microbial cells. Energy for the
synthesis is derived from the simultaneous biologically catalyzed oxidation of a
portion of the glucose, with the resultant release of carbon dioxide and water. If
we examine the flask at the moment that the last amount of glucose is removed we
would conclude that the products of the reaction are microorganisms, carbon dioxide,
and water. Determination of the quantities of these materials would allow us to
calculate the coefficients required to write a stoichiometric equation for the
reaction, and if our measurements were correct we could be reasonably confident
about our equation. Or could we? Are the products really stable products?

Now suppose we set the flask aside for some time before remeasuring the concentration of bacteria. Would it be the same? No, because the microorganisms would have undergone endogenous metabolism to satisfy their requirements for maintenance energy. In addition, if predator organisms were present in the system they would have reacted to the product organisms in the same way that the original biocatalyst organisms reacted to the glucose. Theoretically, this process could continue until all of the carbon from the glucose ended up as carbon dioxide, and the only organisms left in the system were equal in mass to those added originally as the biocatalyst.

This demonstrates that the only stable end point for a biological system is when it ceases to exist or function as a biological system. The only true stoichiometric equation that can be written for the biological oxidation of organic matter is a trivial one, because in an oxidizing environment all of the carbon will go to carbon dioxide. This is not to say that meaningful stoichiometric equations cannot be written. They can be, but we must choose the point at which to strike the balance. For most biochemical operations, this will be the point at which the yield is equal to the observed yield expected in the actual operation. On occasion, however, it may be of interest to write the equation so that the calculated yield is equal to Y_g, the true growth yield. The important point to grasp, however, is the variable nature of the end point of biological reactions, and consequently the "pseudostoichiometric" nature of the equations.

9.4.2 The Equation for Microbial Growth

GENERALIZED EQUATION. In generalized terms, the equation for microbial growth can be written as:

Carbon source + Energy source + Terminal hydrogen acceptor + Nutrients

$\underrightarrow{\text{microorganisms}}$ Microorganisms + End-products   (9.10)

It would be desirable to be able to write a quantitative equation of the same form for any specific situation, no matter what the carbon source, energy source, or hydrogen acceptor (we will exclude photosynthetic reactions since they greatly complicate the issue, and are only rarely used in biochemical operations). McCarty [21] has devised a technique, using the concept of half reactions, whereby this may be done.

All nonphotosynthetic microbially mediated reactions consist of two components, one for synthesis and one for energy. The carbon in the synthesis component ends up in cell material, whereas the carbon in the energy component ends up as carbon dioxide. Microbially mediated reactions are also oxidation-reduction reactions and thus involve the transfer of electrons from a donor, which is generally the substrate being oxidized, to an acceptor. Consequently, McCarty [21] has written three types of half reactions, one for cell material (R_c), one for the electron donor (R_d), and one for the electron acceptor (R_a). To facilitate their combination, these are all written on an electron equivalent basis, with the electrons on the right side of the reaction. The overall stoichiometric equation (R) is just the sum of the half reactions:

$$R = R_d - f_e R_a - f_s R_c \qquad (9.11)$$

The minus terms mean that the half reactions R_a and R_c must be inverted before adding so that the right side of the equation becomes the left side. The term f_e represents the fraction of the electron donor which is used for energy and f_s represents the portion used for synthesis. As such they quantify the end point of the reaction. In other words, they are related to the observed yield. In order for Eq. 9.11 to balance properly:

$$f_e + f_s = 1.0 \qquad (9.12)$$

Before the half reactions can be written, however, empirical formulas must be available for cell material and the various possible organic electron donors.

EMPIRICAL FORMULA FOR CELL MATERIAL. Various empirical formulas have been proposed to represent the organic composition of microbial cells. One of the oldest and most widely accepted in the field of wastewater treatment is $C_5H_7O_2N$ [22]. Other formulas consisting of the same elements have also been used, but they all result in about the same COD when it is calculated from a balanced stoichiometric equation representing complete oxidation of the carbon [3]. Another formula has been proposed which includes phosphorus, $C_{60}H_{87}O_{23}N_{12}P$ [23]. While awareness of the need for phosphorus is essential, it is not really necessary to include phosphorus in the empirical formula because the amount required generally equals about one-fifth of the nitrogen requirement. Consequently, it can be calculated even when the simpler empirical formula is used.

All empirical formulas for cell material seek to represent in a simple way material composed of a highly complex and integrated mixture of organic molecules. Furthermore, because the relative quantities of those molecules change as the growth conditions of the culture change, it would be a purely fortuitous circumstance if a single formulation for cell composition applied to all cases. An estimate of the constancy of overall elemental composition can be obtained by measuring the COD and heat of combustion of microbial cells grown under various conditions. If those factors are relatively constant then that would imply that the ratios of the elements C, H, O and N are relatively constant in spite of the known variability in the actual chemical composition. Investigations of that sort have indicated that the elemental composition is indeed a function of the growth conditions, although current data are not sufficient to establish how it varies [9]. Thus while an empirical formula can be written for cell material the general applicability of it is doubtful and one should view with caution equations which claim to depict "the biochemical reaction".

Even though the variability of the chemical composition of cell material makes the adoption of a single empirical formula for all conditions impossible, the concepts stated in Eq. 9.10 are still valid, and many important relationships can be demonstrated through its use. Consequently, for illustrative purposes, the formula $C_5H_7O_2N$ will be assumed herein.

CHEMICAL COMPOSITION OF ELECTRON DONOR. In a laboratory or experimental situation the exact composition of the electron donor is often known. For example, if glucose were the energy source, its empirical formula $C_6H_{12}O_6$ would be used. If a synthetic medium contained glucose, acetic acid, and glutamic acid, then the empirical formula and half reaction for each could be written separately. These could then be combined to get R_d by multiplying each half reaction by the fraction of its particular electron donor in the medium (on an electron equivalent basis) and adding them together.

An actual wastewater presents a more difficult situation because the chemical composition of the electron donor is seldom known. One approach would be to analyze the waste for C, H, O and N, and then construct the empirical formula from the results. A half reaction could then be written for that particular formula [21]. As an example, the empirical formula for the organic matter in domestic wastewater has been estimated to be $C_{10}H_{19}O_3N$. Alternatively, if the COD, organic nitrogen, organic carbon, and volatile solids content of the waste are known they can be used to generate the half reaction [21]. Finally, if the wastewater contains predominately carbohydrate, protein, and lipid, knowledge of their relative concentrations can be used to write the equation for microbial growth because each can be represented by a generalized empirical formula: i.e., CH_2O, $C_{16}H_{24}O_5N_4$, and $C_8H_{16}O$ respectively. The half reactions for each are multiplied by the fraction of the component in the waste and then added together to get R_d.

COMPOSITION OF ELECTRON ACCEPTOR. The composition of the electron acceptor depends upon the environment in which the microbes are growing. If it is aerobic, the acceptor will be oxygen. If an anaerobic fermentation is taking place the acceptor will depend upon the particular fermentation. For example, in the lactic acid fermentation (Section 9.1.2) pyruvic acid is the acceptor. During methane production carbon dioxide can be considered the electron acceptor. Finally, nitrate can serve as the electron acceptor during anaerobic respiration. Half reactions can be written for each of these.

DETERMINATION OF f_s. Once the nature of the electron donor, electron acceptor, and the chemical composition of cell material have been identified, the fraction of energy used for synthesis, f_s, must be determined before the balanced stoichiometric equation can be written. This fraction is related to the observed yield of the culture which must be measured in the laboratory for the particular growth conditions under consideration. The units of f_s are equivalents of cells formed per equivalent of electron donor consumed, whereas the units of Y are grams of cells formed per gram of COD removed. As discussed in Section 9.3.2 Y can be converted to grams of cells formed per equivalent by multiplying by 8. An equivalent of bacterial cells depends upon the empirical formula used. If $C_5H_7O_2N$ is used, one mole of ash-free cells weighs 113 grams and one equivalent is 1/20 of a mole, or 5.65 grams of volatile (organic) microbial solids. Thus:

$$f_s = \frac{(Y \frac{\text{grams volatile solids formed}}{\text{gram COD removed}})(8 \frac{\text{grams COD}}{\text{equivalent}})}{5.65 \frac{\text{grams volatile solids}}{\text{equivalent}}} = 1.42Y \qquad (9.13)$$

To use Eq. 9.13 Y must be expressed on a volatile solids basis. Often, however, Y is expressed on a total solids basis, as is the case in this book. The two may be interconverted through use of the fact that microbial cells are around 85% volatile. The observed yield, Y, is some fraction of the true growth yield, Y_g, which depends upon the growth conditions imposed upon the culture. In Chapter 12 it will be shown how Y and Y_g are related, and an equation (12.18) will be given which allows prediction of Y (and consequently f_s) from Y_g for a specific reaction environment. In the mean time, for illustrating the techniques of writing stoichiometric equations, it will be assumed that Y is known.

HALF REACTIONS. Table 9.4 contains a list of common half reactions that can be used in constructing stoichiometric equations for microbial growth. Reactions 1 and 2 represent R_c for the formation of cell material. Both are based on the empirical formula $C_5H_7O_2N$, but one uses ammonia nitrogen as the nitrogen source whereas the other uses nitrate. Reactions 3 through 6 are half reactions R_a for the electron acceptors oxygen, nitrate, sulfate, and carbon dioxide. Reactions 7 through 17 are half reactions R_d for organic electron donors. The first of these represents the general composition of domestic wastewater, while the next three are for wastes composed primarily of proteins, carbohydrates, and lipids, respectively. Reactions 11 through 17 are for specific organic compounds of interest in some biochemical operations. The last eight reactions represent possible autotrophic electron donors. Reactions 19 and 20 are for nitrification.

USE OF HALF REACTIONS. The best way to illustrate the use of the half reactions is by an example.

EXAMPLE 9.4.2-1

Write the stoichiometric equation for aerobic microbial growth on a carbohydrate using ammonia nitrogen as the nitrogen source, under conditions such that the observed yield is 0.59 grams of cells formed per gram of COD removed.

To do this we must make use of Eq. 9.11

$$R = R_d - f_e R_a - f_s R_c \qquad (9.11)$$

$$f_s = 1.42\ Y \qquad (9.13)$$

Y in Eq. 9.13 is expressed as grams of volatile solids formed per gram of COD removed. Since cells are 85% volatile,

$$Y = (0.85)(0.59) = 0.50$$

Thus

$$f_s = (1.42)(0.50) = 0.71$$

$$f_e = 1.00 - 0.71 = 0.29 \qquad (9.12)$$

Therefore

$$R = R_d - 0.29 R_a - 0.71 R_c$$

The electron donor is carbohydrate and the acceptor is oxygen. Thus from Table 9.4:

$$R_d = 1/4\ CH_2O + 1/4\ H_2O = 1/4\ CO_2 + H^+ + e^-$$

$$R_a = 1/2\ H_2O = 1/4\ O_2 + H^+ + e^-$$

Since ammonia is the nitrogen source, R_c is:

$$R_c = 1/20\ C_5H_7O_2N + 9/20\ H_2O = 1/5\ CO_2 + 1/20\ HCO_3^- + 1/20\ NH_4^+ + H^+ + e^-$$

Applying Eq. 9.11 gives:

$$R_d = 0.25\ CH_2O + 0.25\ H_2O = 0.25\ CO_2 + H^+ + e^-$$

$$-0.29\ R_a = 0.0725\ O_2 + 0.29\ H^+ + 0.29\ e^- = 0.145\ H_2O$$

$$-0.71\ R_c = 0.142\ CO_2 + 0.0355\ HCO_3^- + 0.0355\ NH_4^+ + 0.71\ H^+ + 0.71\ e^-$$

$$= 0.0355\ C_5H_7O_2N + 0.3195\ H_2O$$

$$R = 0.25\ CH_2O + 0.0725\ O_2 + 0.0355\ NH_4^+ + 0.0355\ HCO_3^-$$

$$= 0.0355\ C_5H_7O_2N + 0.108\ CO_2 + 0.2145\ H_2O$$

This can be normalized to one mole of carbohydrate by dividing through by 0.25

$$CH_2O + 0.29\ O_2 + 0.142\ NH_4^+ + 0.142\ HCO_3^-$$

$$= 0.142\ C_5H_7O_2N + 0.432\ CO_2 + 0.858\ H_2O$$

TABLE 9.4
Oxidation Half Reactions

Reaction Number	Half Reactions

Reactions for Bacterial Cell Synthesis (R_c)

Ammonia as Nitrogen Source:

1. $\frac{1}{20} C_5H_7O_2N + \frac{9}{20} H_2O \qquad = \frac{1}{5} CO_2 + \frac{1}{20} HCO_3^- + \frac{1}{20} NH_4^+ + H^+ + e^-$

Nitrate as Nitrogen Source:

2. $\frac{1}{28} C_5H_7O_2N + \frac{11}{28} H_2O \qquad = \frac{1}{28} NO_3^- + \frac{5}{27} CO_2 + \frac{29}{28} H^+ + e^-$

Reactions for Electron Acceptors (R_a)

Oxygen:

3. $\frac{1}{2} H_2O \qquad = \frac{1}{4} O_2 + H^+ + e^-$

Nitrate:

4. $\frac{1}{10} N_2 + \frac{3}{5} H_2O \qquad = \frac{1}{5} NO_3^- + \frac{6}{5} H^+ + e^-$

Sulfate:

5. $\frac{1}{16} H_2S + \frac{1}{16} HS^- + \frac{1}{2} H_2O \qquad = \frac{1}{8} SO_4^= + \frac{19}{16} H^+ + e^-$

Carbon Dioxide (Methane Fermentation):

6. $\frac{1}{8} CH_4 + \frac{1}{4} H_2O \qquad = \frac{1}{8} CO_2 + H^+ + e^-$

Reactions for Electron Donors (R_d)

Organic Donors (Heterotrophic Reactions)

Domestic Wastewater:

7. $\frac{1}{50} C_{10}H_{19}O_3N + \frac{9}{25} H_2O \qquad = \frac{9}{50} CO_2 + \frac{1}{50} NH_4^+ + \frac{1}{50} HCO_3^- + H^+ + e^-$

Protein (Amino acids, proteins, nitrogenous organics):

8. $\frac{1}{66} C_{16}H_{24}O_5N_4 + \frac{27}{66} H_2O \qquad = \frac{8}{33} CO_2 + \frac{2}{33} NH_4^+ + \frac{31}{33} H^+ + e^-$

Carbohydrate (Cellulose, Starch, Sugars):

9. $\frac{1}{4} CH_2O + \frac{1}{4} H_2O \qquad = \frac{1}{4} CO_2 + H^+ + e^-$

Grease (Fats and Oils):

10. $\frac{1}{46} C_8H_{16}O + \frac{15}{46} H_2O \qquad = \frac{4}{23} CO_2 + H^+ + e^-$

Acetate:

11. $\frac{1}{8} CH_3COO^- + \frac{3}{8} H_2O \qquad = \frac{1}{8} CO_2 + \frac{1}{8} HCO_3^- + H^+ + e^-$

TABLE 9.4
Continued

Reaction Number	Half Reactions

Propionate:

12. $\frac{1}{14} CH_3CH_2COO^- + \frac{5}{14} H_2O$ $= \frac{1}{7} CO_2 + \frac{1}{14} HCO_3^- + H^+ + e^-$

Benzoate:

13. $\frac{1}{30} C_6H_5COO^- + \frac{13}{20} H_2O$ $= \frac{1}{5} CO_2 + \frac{1}{30} HCO_3^- + H^+ + e^-$

Ethanol:

14. $\frac{1}{12} CH_3CH_2OH + \frac{1}{4} H_2O$ $= \frac{1}{6} CO_2 + H^+ + e^-$

Lactate:

15. $\frac{1}{12} CH_3CHOHCOO^- + \frac{1}{3} H_2O$ $= \frac{1}{6} CO_2 + \frac{1}{12} HCO_3^- + H^+ + e^-$

Pyruvate:

16. $\frac{1}{10} CH_3COCOO^- + \frac{2}{5} H_2O$ $= \frac{1}{5} CO_2 + \frac{1}{10} HCO_3^- + H^+ + e^-$

Methanol:

17. $\frac{1}{6} CH_3OH + \frac{1}{6} H_2O$ $= \frac{1}{6} CO_2 + H^+ + e^-$

Inorganic Donors (Autotrophic Reactions):

18. Fe^{++} $= Fe^{+3} + e^-$

19. $\frac{1}{8} NH_4^+ + \frac{3}{8} H_2O$ $= \frac{1}{8} NO_3^- + \frac{5}{4} H^+ + e^-$

20. $\frac{1}{6} NH_4^+ + \frac{1}{3} H_2O$ $= \frac{1}{6} NO_2^- + \frac{4}{3} H^+ + e^-$

21. $\frac{1}{6} S + \frac{2}{3} H_2O$ $= \frac{1}{6} SO_4^= + \frac{4}{3} H^+ + e^-$

22. $\frac{1}{16} H_2S + \frac{1}{16} HS^- + \frac{1}{2} H_2O$ $= \frac{1}{8} SO_4^= + \frac{19}{16} H^+ + e^-$

23. $\frac{1}{8} S_2O_3^= + \frac{5}{8} H_2O$ $= \frac{1}{4} SO_4^= + \frac{5}{4} H^+ + e^-$

24. $\frac{1}{2} H_2$ $= H^+ + e^-$

25. $\frac{1}{2} SO_3^= + \frac{1}{2} H_2O$ $= \frac{1}{2} SO_4^= + H^+ + e^-$

(Adapted from McCarty [21].)

9.4.3 Use of the Equation for Microbial Growth

After the equation for microbial growth has been obtained it may be used to calculate the amount of terminal hydrogen acceptor required, as well as the amounts of nitrogen and phosphorus which must be supplied by using the concepts of Section 2.1. Furthermore, by making use of the fact that Y (and consequently f_s) is a function of the growth conditions imposed upon the microorganisms, the equation may be used to show how those needs change as the growth conditions are changed [24].

DETERMINATION OF THE QUANTITY OF TERMINAL HYDROGEN ACCEPTOR NEEDED. The amount of terminal hydrogen acceptor needed can be determined directly from the balanced stoichiometric equation. Referring to the equation derived in Example 9.4.2-1, it can be seen that 0.29 moles of oxygen are needed for each mole of carbohydrate removed under the growth conditions used in the example. Since the molecular weights of oxygen and carbohydrate are 32 and 30, respectively, 0.31 grams of oxygen [(0.29)(32/30)] will be required for every gram of carbohydrate removed. This requirement would change, however, if the growth conditions were changed. For example, suppose the growth conditions were such that the observed yield was 0.40. Then f_s would be 0.568 and the stoichiometric equation would be:

$$CH_2O + 0.432\ O_2 + 0.1136\ NH_4^+ + 0.1136\ HCO_3^-$$

$$= 0.1136\ C_5H_7O_2N + 0.4544\ CO_2 + 0.8864\ H_2O \qquad (9.14)$$

This causes the oxygen requirement to increase to 0.432 moles per mole of carbohydrate, or 0.461 grams of oxygen per gram of carbohydrate removed. Note that the requirement for oxygen increased when the observed yield decreased.

If nitrate were serving as the terminal hydrogen acceptor under conditions of anaerobic respiration the amount needed could be calculated from the stoichiometric equation obtained when half reaction No. 4 was used in place of No. 3 as R_a in Eq. 9.11.

DETERMINATION OF QUANTITY OF NUTRIENT NEEDED. The amount of nitrogen required can also be calculated from the stoichiometric equation. From the equation in Example 9.4.2-1 it can be seen that 0.142 moles of ammonia nitrogen are needed per mole of carbohydrate used. One gram mole of nitrogen has a mass of 14, consequently, 0.066 grams of nitrogen [(0.142)(14/30)] are needed for each gram of carbohydrate removed. Like the oxygen requirement, this requirement will also change if the growth conditions are changed.

The phosphorus requirement can be estimated as one-fifth of the nitrogen requirement on a mass basis.

The provision of sufficient nutrients is essential if efficient biological wastewater treatment is to be achieved because without them the microorganisms will not be able to perform their synthesis reactions. The result will be an uncoupling of the energy generating and energy utilizing reactions. One mechanism for this is the production of low energy storage products which often results in a gelatinous mass in the culture which interferes with proper liquid-solid separation. Although the nutrients needed in greatest quantity are nitrogen and phosphorus, many other elements are needed by the organisms, as noted in Section 8.2.6. These are not normally included in the stoichiometric equation because of the complicating effect that they would have. Wastewaters usually contain ample quantities of them but their presence should not be taken for granted because severe problems can result if they are not present in sufficient quantities [25,26].

The requirements for nutrients have been well documented in the literature, with the work of Sawyer [27] serving as the cornerstone. Sawyer recognized the dependency of the nutrient requirement upon the growth conditions of the organisms, although his values have been reduced by others to ratios which are widely, and not always judiciously, applied. For example, figures such as 1 gram of nitrogen per 30 grams of COD removed and 1 gram of phosphorus per 140 grams of COD removed are encountered often. The danger in such fixed ratios is that they might either underestimate or overestimate the nutrient requirement. The former situation would result in poor treatment efficiency while the latter would be uneconomical. Such ratios provide good guidelines for rough estimates, but the exact requirements can only be determined through either an experimental program or field tests once the installation is completed.

9.5 KEY POINTS

1. Adenosine triphosphate, ATP, is the medium of energy exchange within a cell. It may be formed by substrate-level and oxidative phosphorylation.

2. Metabolism for energy conservation may be divided into three categories depending upon the character of the terminal hydrogen acceptor. They are fermentation, anaerobic respiration, and aerobic respiration. Respiratory metabolism makes more energy available to the cell than does fermentation.

3. ATP generation during fermentation is by substrate-level phosphorylation so that the quantity of ATP produced per mole of substrate used is low.

4. Electron transport chains are highly organized sequences of flavoproteins and cytochromes which accept electrons from a reduced coenzyme and pass them in discrete steps to a terminal acceptor. The generation of ATP is associated with some of the steps through coupled oxidative phosphorylation reactions. The amount of ATP formed is influenced by the nature of the terminal acceptor.

5. The amount of ATP that can be formed as a result of the oxidation of a substrate is proportional to the oxidation state of the carbon in the substrate because that state determines the energy available from oxidation of the substrate.

6. Energy for synthesis includes that required to form the bonds made in anabolic pathways and to reduce the carbon source to the proper oxidation-reduction level for incorporation into cell material. Consequently, the energy for synthesis associated with highly oxidized substrates is greater than that associated with highly reduced ones, and substrates containing few carbon atoms require more energy for synthesis than do those with many.

7. Energy for maintenance is the term applied to the energy needed to keep the cells in a culture functioning in the absence of growth. If an exogenous supply of energy is not available, maintenance needs will be met by endogenous metabolism.

8. The yield, defined as the amount of biomass formed per unit amount of substrate removed, is a measure of the relative efficiencies of energy conservation and utilization. Because of the need for maintenance energy the yield depends upon the rate at which the culture is growing. The true growth yield, which is the yield in the absence of maintenance energy requirements, is independent of the growth rate of the culture.

9. The yield is the "amount" of cells formed per unit "amount" of substrate used. The amount of cells is generally measured on a dry mass basis. The amount of substrate used may be expressed in terms of (a) the quantity of ATP which may be formed from it; (b) the kcal of total energy consumed; (c) the number of electrons originally available in the substrate; and (d) the amount of chemical oxygen demand (COD) consumed. Environmental engineers usually use the last one.

10. The true growth yield is influenced by the nature of the substrate, the species of organisms present, the complexity of the growth medium, the type of terminal hydrogen acceptor available, the pH of the medium, the temperature, and the composition of the microbial community. Consequently, Y_g must be determined experimentally for each particular situation.

11. The only true stoichiometric equation which can be written for the aerobic biological oxidation of organic matter is a trivial one in which all of the substrate carbon goes to carbon dioxide. A "pseudostoichiometric" equation may be written, however, by specifying the end point of the reaction, which is determined by the yield.

12. Before the equation for microbial growth can be written the following factors must be known: the empirical formula for cell material, the chemical composition of the electron donor, the chemical composition of the electron acceptor, and the yield.

13. The half-reaction method of McCarty provides a simple technique whereby the equation for microbial growth may be written for a large number of situations.

14. The equation for microbial growth may be used to estimate the amounts of terminal hydrogen acceptor and nutrients required for the degradation of a given amount of substrate. Those quantities will change as the yield of the culture changes. The phosphorus requirement may be estimated as one-fifth of the nitrogen requirement.

9.6 STUDY QUESTIONS

The study questions are arranged in the same order as the text.

Section	Questions
9.1	1-5
9.2	6-7
9.3	8-10
9.4	11-14

1. Differentiate between fermentation, anaerobic respiration, and aerobic respiration. How do they compare with regard to efficiency of ATP generation?

2. Explain why respiratory metabolism makes more energy available to the cell then does fermentation.

3. Draw a sketch of a typical mitochondrial electron transport chain and indicate on it the sites of oxidative phosphorylation.

4. By way of an example, explain why the nature of the terminal hydrogen acceptor affects the efficiency of ATP production.

5. Explain on theoretical grounds why you might expect more ATP to be formed from the aerobic metabolism of acetic acid (CH_3COOH) than from oxalic acid ($(COOH)_2$).

6. Explain on theoretical grounds why it would take more energy to synthesize cell material from oxalic acid than from glucose.

7. Discuss the possible sources of maintenance energy for a cell. Explain why and how the sources might change as the growth conditions of the culture are changed.

8. Explain why the yield changes as the growth rate changes.

9. Compare the advantages and disadvantages of the different ways of expressing the yield and explain why environmental engineers usually express it on the basis of the amount of oxygen demand consumed.

10. Explain why each of the following factors influences the true growth yield:
    (a) the nature of the substrate.
    (b) the species of organism present.
    (c) the complexity of the growth media.
    (d) the type of terminal hydrogen acceptor available.
    (e) the pH of the media.
    (f) the temperature.
    (g) the composition of the microbial community.

11. Explain why the only true stoichiometric equation for aerobic biological oxidation of organic matter is a trivial one.

12. Using the half-reaction technique, write the equation for microbial growth for each of the following situations:
    (a) aerobic growth on domestic wastewater with nitrate as the nitrogen source. The yield is 0.50 grams of cells formed per gram of COD removed.
    (b) growth on a carbohydrate with nitrate as the terminal hydrogen acceptor and ammonia as the nitrogen source. The yield is 0.40 grams of cells formed per gram of COD removed.
    (c) fermentation of glucose to ethanol with ammonia as the nitrogen source. The yield is 0.35 grams of cells formed per gram of COD removed from the medium.

13. An aerobic culture is growing on glucose with ammonia as the nitrogen source. By means of the equation for microbial growth determine the grams of nitrogen and oxygen which must be provided per gram of COD removed for each of the following situations:
    (a) Y = 0.55 grams of cells formed/gram COD removed
    (b) Y = 0.40 grams of cells formed/gram COD removed
    (c) Y = 0.25 grams of cells formed/gram COD removed

14. Estimate the grams of phosphorus required per gram of COD removed for each
     situation in Study Question 13.

REFERENCES AND FURTHER READING

1. T. Bauchop and S. R. Elsden, "The growth of microorganisms in relation to their
    energy supply," *Journal of General Microbiology, 23,* 457-469, 1960.

2. J. C. Senez, "Some considerations on the energetics of bacterial growth."
    Bacteriological Reviews, 26, 95-107, 1962.

3. P. L. McCarty, "Thermodynamics of biological synthesis and growth," *Proceedings
    of the Second International Conference on Water Pollution Research,* Pergamon
    Press, New York, N.Y., 169-199, 1965.

4. S. J. Pirt, "Maintenance energy of bacteria in growing cultures," *Proceedings
    of the Royal Society (London). Series B, 163,* 224-231, 1965.

5. A. G. Marr, *et al.* "The maintenance requirement of *Escherichia coli,*" *Annals
    of the New York Academy of Science, 102,* 536-548, 1963.

6. J. Mandelstam, "The intracellular turnover of protein and nucleic acids and
    its role in biochemical differentiation," *Bacteriological Reviews, 24,* 289-308,
    1960.

7. A. H. Stouthamer, and C. Bettenhausen, "Utilization of energy for growth and
    maintenance in continuous and batch cultures of microorganisms," *Biochimica et
    Biophysica, Acta, 301,* 53-70, 1973.

8. G. J. Prochazka, *et al.* "Calorific contents of microorganisms," *Biotechnology
    and Bioengineering, 15,* 1007-1010, 1973.

9. C. P. L. Grady Jr., *et al.* "Effects of growth conditions on the oxygen
    equivalence of microbial cells," *Biotechnology and Bioengineering, 17,* 859-872,
    1975.

10. W. J. Payne, "Energy yield and growth of heterotrophs," *Annual Review of
    Microbiology, 24,* 17-52, 1970.

11. L. P. Hadjipetrou, *et al.* "Relation between energy production and growth of
    Aerobacter aerogenes," *Journal of General Microbiology, 36,* 139-150, 1964.

12. D. A. Rollag, "The effect of anaerobic pretreatment on an activated sludge
    system," Thesis presented to Purdue University, at West Lafayette, Ind, in
    1975, in partial fulfillment of the requirements for the degree of Doctor
    of Philosophy.

13. J. F. Andrews and E. A. Pearson, "Kinetics and characteristics of volatile
    acid production in anaerobic fermentation processes," *International Journal
    for Air and Water Pollution Research, 9,* 439-461, 1965.

14. C. M. Brown and A. H. Rose, "Effects of temperature on composition and cell
    volume of *Candida utilis,*" *Journal of Bacteriology, 97,* 261-272, 1969.

15. R. E. Muck and C. P. L. Grady Jr., "Temperature effects on microbial growth
    in CSTR's," *Journal of the Environmental Engineering Division, ASCE, 100,*
    1147-1163, 1974.

16. S. A. Palumbo and L. D. Witter, "Influence of temperature on glucose utilization by *Pseudomonas fluorescens*," *Applied Microbiology*, 18, 137-141, 1969.

17. H. Topiwala and C. G. Sinclair, "Temperature relationships in continuous culture," *Biotechnology and Bioengineering*, 13, 795-813, 1971.

18. L. J. Hettling, *et al.* "Kinetics of the steady state bacterial culture - II - Variation in synthesis," *Proceedings of the 19th Industrial Waste Conference,* Purdue University Engineering Extension Series No. 117, 687-715, 1964.

19. M. Ramanathan and A. F. Gaudy Jr., "Studies on sludge yield in aerobic systems," *Proceedings of the 26th Industrial Waste Conference,* Purdue University Engineering Extension Series No. 140, 665-675, 1971.

20. A. W. Lawrence and P. L. McCarty, "Unified basis for biological treatment design and operation," *Journal of the Sanitary Engineering Division, ASCE,* 96, 757-778, 1970.

21. P. L. McCarty, "Stoichiometry of biological reactions," *Progress in Water Technology,* 7, 157-172, 1975.

22. S. R. Hoover and N. Porges, "Assimilation of dairy wastes by activated sludge - II - The equations of synthesis and rate of oxygen utilization," *Sewage and Industrial Wastes,* 24, 306-312, 1952.

23. P. L. McCarty, "Phosphorus and nitrogen removal by biological systems," *Proceedings of the Wastewater Reclamation and Reuse Workshop,* Lake Tahoe, Calif., 226, June 1970.

24. J. H. Sherrard, "Kinetics and stoichiometry of completely mixed activated sludge," *Journal of the Water Pollution Control Federation,* 49, 1968-1975, 1977.

25. J. L. Carter and R. E. McKinney, "Effects of iron on activated sludge treatment," *Journal of the Environmental Engineering Division, ASCE,* 99, 135-152, 1973.

26. D. K. Wood and G. Tchobanoglous, "Trace elements in biological waste treatment," *Journal of the Water Pollution Control Federation,* 47, 1933-1945, 1975.

27. C. N. Sawyer, "Bacterial nutrition and synthesis" in *Biological Treatment of Sewage and Industrial Wastes, Vol 1, Aerobic Oxidation,* edited by J. McCabe and W. W. Eckenfelder, Jr., Reinhold Publishing Corp., New York, N.Y., 3-17, 1956.

Further Reading

Brock, T. D., *Biology of Microorganisms,* Second Edition, Prentice-Hall, Inc., Englewood Cliffs, N.J., 1974. An excellent general microbiology text. See Chapter 4.

Dawes, E. A. and Ribbons, D. W., "Some aspects of the endogenous metabolism of bacteria," *Bacteriological Reviews,* 28, 126-149, 1964. Reviews the evidence for the existence of endogenous metabolism and discusses the materials most likely to serve as endogenous reserves. Also discusses the need for energy of maintenance.

Doetsch, R. N. and Cook, T. M., *Introduction to Bacteria and their Ecobiology,* University Park Press, Baltimore, Md. 1973. Chapter 5 provides a very good discussion of anaerobic respiration.

Gaudy, A. F. Jr. and Gaudy, E. T., "Biological concepts for design and operation of the activated sludge process," *Environmental Protection Agency Water Pollution Research Series,* Report No. 17090 FQJ 09/71, Sept, 1971. Provides a good discussion of the stoichiometry of microbial growth and of yield. See Chapters 3 and 4.

Lamanna, C., *et al., Basic Bacteriology - Its Biological and Chemical Background,* Fourth Edition, The Williams and Wilkins Company, Baltimore, Md., 1973. Contains an excellent discussion of endogenous metabolism and maintenance energy. See Chapter 12.

McCarty, P. L., "Energetics of organic matter degradation," Chapter 5 in *Water Pollution Microbiology,* edited by R. Mitchell, John Wiley and Sons, Inc., New York, N.Y., 91-118, 1972. Reviews the thermodynamics of bioenergetics.

Stanier, R. Y., *et al. The Microbial World,* Fourth Edition, Prentice-Hall, Inc., Englewood Cliffs, N.J., 1976. A very complete microbiology text. See Chapters 6, 9 and 10.

Stebbing, N., "Feedback control mechanisms in microorganisms and efficiency of growth," *Sub-Cellular Biochemistry,* 2, 169-182, 1973. Discusses how changes in efficiency of energy generation and utilization could be responsible for variation in yield.

CHAPTER 10

KINETICS OF BIOCHEMICAL SYSTEMS

Research into the kinetics of microbial growth has proceeded down two paths during the last 25 years. One path has been fundamental in scope and has been followed by experimenters using pure microbial cultures. The other has been more applied and has been followed by researchers working with mixed microbial cultures. Both paths have their starting points in the early work of Monod [1] and while they have often diverged, on occasion they have converged leading to concepts that can be applied to all biochemical operations. This chapter will introduce those concepts in mathematical terms so that they can be used in Part IV to develop models for biochemical operations.

The rates at which microorganisms carry out their various functions depend upon the interrelationships between the rates of energy supply and energy utilization, which are determined by the rates of the enzymes in the pathways. Although much fundamental work has been performed by biochemists on the kinetics of enzymatic reactions and the effects of metabolic controls within the cells, the extreme complexity of the many interacting reactions has prevented the mathematical integration of that knowledge into a mechanistic model for microbial growth. Consequently, most research into the kinetics of microbial growth has taken a macroscopic viewpoint and equations have been developed from experimental observations of growing cultures. This does not mean, however, that enzyme kinetics is unimportant. Indeed, many of the techniques used to evaluate microbial growth kinetics had their foundations in enzyme kinetics. Consequently, the first section of this chapter will deal with that subject.

10.1 ENZYME KINETICS

Although no single theory has been able to provide a detailed account of the mechanisms by which enzymes catalyze specific reactions, the one which has gained widest acceptance postulates that substrates bind to specific regions on the enzymes. These "active sites" contain amino acid side chains or other functional groups which may serve as acids or bases or as acceptors or donors of electrons thereby facilitating the reactions [2]. Thus, these functional groups are involved directly in the reaction mechanism through the formation of enzyme-substrate complexes. The existence of such complexes was proposed in 1902 by Henri [3] who postulated that an enzyme forms a complex with the substrate which is subsequently broken down to yield the product and release the free enzyme. The first generally acceptable mathematical formulation for the kinetics of enzyme reactions was put forward in 1913 by Michaelis and Menten [4] and has subsequently served as the foundation upon which most current formulas are based. Thus, it would be instructive to review the concepts and assumptions involved in their formulation for a single substrate reaction.

10.1.1 Michaelis-Menten Kinetics

The basic assumption is that enzymatic catalysis occurs through a series of elementary reactions involving the enzyme-substrate complex. Using the approach taken in Section 3.3.2, where the reaction rate expressions for catalytic reaction were presented, the result is:

$$S + E \underset{k_{-1}}{\overset{k_1}{\rightleftharpoons}} ES \tag{10.1}$$

and

$$ES \overset{k_2}{\longrightarrow} E + P \tag{10.2}$$

where S, E, ES and P represent the substrate, free enzyme, enzyme-substrate complex, and product, respectively. Henri [3] and Michaelis and Menten [4] assumed that the reaction in Eq. 10.1 is in equilibrium so that

$$k_1 C_E C_S = k_{-1} C_{ES} \tag{10.3}$$

or

$$C_{ES} = C_E C_S / K_m^* \tag{10.4}$$

where

$$K_m^* = k_{-1}/k_1 \tag{10.5}$$

The parameter K_m^* is called the dissociation constant and is the reciprocal of the equilibrium constant. Again, following the approach of Section 3.3.2 a mass balance on enzyme gives

$$C_E + C_{ES} = C_{Eo} \tag{10.6}$$

in which C_{Eo} and C_E represent the initial and free enzyme concentrations. The reaction rate is given by

$$r_p = k_2 C_{ES} \tag{10.7}$$

Since the concentration of enzyme-substrate complex, C_{ES}, is not readily measurable it must be replaced by the concentrations of readily measurable species, such as the substrate and the initial enzyme present. Combining Eqs. 10.4 and 10.6 allows this to be done:

$$C_{ES} = C_{Eo} C_S / (K_m^* + C_S) \tag{10.8}$$

Thus,

$$r_p = k_2 C_{Eo} C_S / (K_m^* + C_S) \tag{10.9}$$

or

$$r_p = V_m C_S / (K_m^* + C_S)$$ (10.10)

where

$$V_m = k_2 C_{Eo}$$ (10.11)

Although Eq. 10.10, which relates the rate of an enzymatically catalyzed reaction to the concentration of substrate, is consistent with experimental evidence, its derivation was not rigorous. The problem lies in the assumption that the reaction in Eq. 10.1 is in equilibrium, because the appearance of ES in both Eq. 10.1 and Eq. 10.2 prevents that from being true. Briggs and Haldane [5] recognized this and invoked the pseudosteady-state hypothesis in order to perform a rigorous derivation. Their derivation was presented in Section 3.3.2 and led to Eq. 3.81

$$r_p = -r_s = V_m C_S / (K_m + C_S)$$ (3.81)

where

$$K_m = (k_{-1} + k_2)/k_1$$ (10.12)

Comparison of Eq. 3.81 with Eq. 10.10 shows that they are of identical functional form and differ only by the constants K_m^* and K_m as defined by Eqs. 10.5 and 10.12, respectively. In recognition of the pioneering work of Michaelis and Menten, K_m is called the "Michaelis-Menten constant" and consequently equations of the form of 10.10 and 3.81 are commonly called Michaelis-Menten equations. Following this practice, we will write the Michaelis-Menten equation as

$$v = -r_s = V_m C_S / (K_m + C_S)$$ (10.13)

where v is used to denote the "velocity." A typical plot of this equation is given in Fig. 10.1 where it can be seen that the reaction rate is approximately

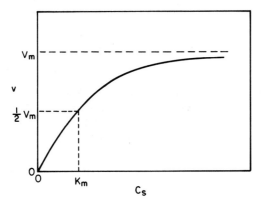

Figure 10.1. Plot of the "velocity" of an enzymatic
              reaction as a function of reactant con-
              centration as depicted by the Michaelis-
              Menten equation (Eq. 10.13).

first order with respect to substrate at low concentrations and zero order at
high concentrations.

When one wishes to use Eq. 10.13 to express the effect of substrate concentration upon the rate of a reaction, numerical values must be obtained for the two
parameters V_m and K_m. The usual approach is to perform a series of batch experiments
using different substrate concentrations and to measure the reaction rate before the
substrate concentration can change significantly (initial rates). In addition, since
V_m is equal to $k_2 C_{Eo}$ it is necessary to confirm experimentally that the rate is proportional to the enzyme concentration. This may be done by using various enzyme
concentrations.

Once data are available relating v to C_S (for constant C_{Eo}) several different
techniques may be used to estimate V_m and K_m. The most obvious one is to just use
a curve like that in Fig. 10.1. Examination of it reveals that V_m can be estimated
by measuring the reaction rate at high substrate concentration. Furthermore, rearrangement of Eq. 10.13 reveals that when the substrate concentration is numerically
equivalent to K_m, v is one-half of V_m. Thus, in Fig. 10.1, the abscissa value corresponding to an ordinate value of $V_m/2$ is equal to K_m. Even though this technique for
obtaining V_m and K_m is simple in concept, it is seldom used because it is inaccurate,
thereby making it difficult to obtain reliable values for the parameters. More reliable estimates can be obtained by transforming the Michaelis-Menten equation into
a linear form.

One common transformation is obtained by taking the reciprocal of both sides of
Eq. 10.13 and rearranging it:

$$\frac{1}{v} = \frac{1}{V_m} + \frac{K_m}{V_m} \frac{1}{C_S} \tag{10.14}$$

Equation 10.14 is called the Lineweaver-Burk equation, and a plot of it is shown in
Fig. 10.2. There it can be seen that V_m may be obtained from the intercept and K_m
from the slope.

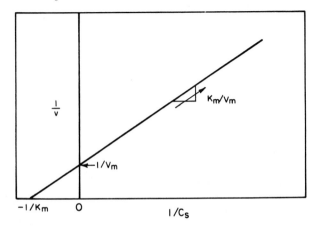

Figure 10.2.   Lineweaver-Burk linearization of the
               Michaelis-Menten equation.

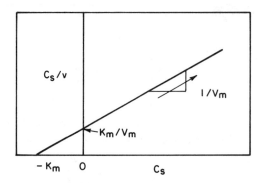

Figure 10.3. Hanes linearization of the
              Michaelis-Menten equation.

Another useful transformation is obtained by multiplying both sides of the Lineweaver-Burk equation by C_S:

$$\frac{C_S}{v} = \frac{C_S}{V_m} + \frac{K_m}{V_m} \qquad\qquad (10.15)$$

This equation was originally suggested by Hanes [6] and yields a straight line when C_S/v is plotted as a function of C_S as shown in Fig. 10.3. In this case V_m is obtained from the slope and K_m from the intercept.

Finally, Eq. 10.13 may also be linearized by multiplying both sides by $(K_m + C_S)$ and dividing both sides by C_S to give:

$$v = V_m - K_m v/C_S \qquad\qquad (10.16)$$

Thus a plot of v versus v/C_S will give a straight line with slope equal to $-K_m$ and intercept V_m. Such a plot is called a Hofstee plot and is shown in Fig. 10.4.

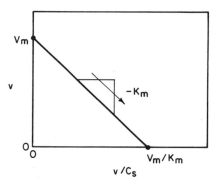

Figure 10.4. Hofstee linearization of the
              Michaelis-Menten equation.

Several points should be noted about each of these methods. In the Lineweaver-Burk plot the most accurately known values (around V_m) will tend to cluster near the origin, while those which are least accurately known will be far from the origin and thus will have the most influence upon the slope. This tends to make the Lineweaver-Burk plot the least reliable of the three [7]. When the data are plotted by the Hanes equation, the points near V_m are spread out so that the slope, $1/V_m$, may be determined accurately. Often, however, the intercept appears near the origin, which makes an accurate determination of K_m difficult. In spite of that, this technique can be quite reliable, especially when a least squares fit is used. Finally, the Hofstee equation contains the dependent variable, v, in both coordinates, which makes the use of the least squares fitting technique invalid. This technique does spread the data out, however, so that reasonable fits may be obtained by eye.

In addition to the technique described above, it is also possible to determine the parameters in the Michaelis-Menten equation by the integral method using data taken from a single batch reactor over an extended period of time. For a constant density batch reaction the mass balance equation is:

$$\frac{dC_S}{dt} = r_s = \frac{-V_m C_S}{K_m + C_S} \tag{10.17}$$

or

$$\frac{K_m}{V_m}\frac{dC_S}{C_S} + \frac{1}{V_m}dC_S = -dt \tag{10.18}$$

Integrating between the limits of C_{So} and C_S we obtain:

$$\frac{K_m}{V_m}\ln(C_{So}/C_S) + \frac{1}{V_m}(C_{So} - C_S) = t \tag{10.19}$$

which can be rearranged to give

$$\frac{1}{t}\ln(C_{So}/C_S) = \frac{V_m}{K_m} - \frac{C_{So} - C_S}{K_m t} \tag{10.20}$$

Thus a plot of $\frac{1}{t}\ln(C_{So}/C_S)$ versus $(C_{So} - C_S)/t$ will give a straight line with a slope of $-1/K_m$ and an intercept of V_m/K_m. This is shown in Fig. 10.5. Like the Hofstee plot, the dependent variable (C_S) appears in both axes so that the least squares technique cannot be employed.

The Michaelis-Menten equation was derived for a single substrate reaction involving only one enzyme-substrate complex. Another common mechanism is for the enzyme-substrate complex to be transformed into an enzyme-product complex before the product is released. Even in this case, however, the equation expressing the effect of the substrate concentration upon the reaction rate is identical to Eq. 10.13, although V_m and K_m are defined by more complex expressions. Consequently the

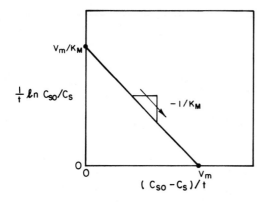

Figure 10.5. Application of the integral method
to batch data from an enzymatic
reaction for determination of the
parameters in the Michaelis-Menten
equation.

Michaelis-Menten equation is often suitable for expressing the rates of single substrate reactions which involve reaction mechanisms that are more complex than the
simple situation for which it was derived. This helps to explain its wide usage.

Not all enzymatic reactions involve only one substrate, and thus the Michaelis-
Menten equation is not universally applicable. Consequently much work has gone into
the determination of the rate expressions for a number of types of reactions. Space
does not permit us to discuss them, however, so the interested reader is urged to
consult the references under Further Reading.

10.1.2 Enzyme Inhibition

An enzyme inhibitor is a compound which acts to reduce the rate of an enzymatically
catalyzed reaction by binding with either the free enzyme and/or with the enzyme-
substrate complex. Because such compounds can have a marked effect upon the rate of
a reaction, a brief review of the three most common reaction rate expressions would
be helpful. In addition, under some circumstances, either the substrate or the
product can inhibit the reaction if its concentration gets too high. These are
called substrate inhibition and product inhibition, respectively. They too can be
important. Thus, let us look at these five types of inhibition.

COMPETITIVE INHIBITION. An inhibitor which is classed as competitive tries to
bind at the same active site as the substrate and therefore competes with the substrate for a spot on the enzyme. The mechanism of this type of inhibition can be
represented by:

$$E + S \xrightleftharpoons[k_{-1}]{k_1} ES \xrightarrow{k_2} E + P$$

$$E + I \xrightleftharpoons[k_{-3}]{k_3} EI \tag{10.21}$$

When the enzyme-inhibitor complex is formed the substrate cannot react with the enzyme and thus no product is formed. If the procedure developed in Section 3.3.2 is applied to this situation, the result is:

$$r_p = V_m C_S / [K_m(1 + C_I/K_I) + C_S] \tag{10.22}$$

where

$$K_I = k_{-3}/k_3 \tag{10.23}$$

and V_m and K_m are defined by Eqs. 10.11 and 10.12, respectively. If we let K_m' be defined as

$$K_m' = K_m(1 + C_I/K_I) \tag{10.24}$$

Eq. 10.22 reduces in form to the Michaelis-Menten equation. Examination of Eq. 10.24 shows that the effect of a competitive inhibitor is to increase the Michaelis-Menten constant, thereby requiring that a higher substrate concentration be present to attain a rate equal to that attained in the absence of the inhibitor. The presence of the inhibitor has no effect upon V_m, however, because if the concentration of substrate is very high it will compete effectively for the sites on the enzymes. These effects are illustrated in Fig. 10.6a.

UNCOMPETITIVE INHIBITION. An uncompetitive inhibitor binds with the enzyme-substrate complex to form an inactive enzyme-substrate-inhibitor complex which cannot undergo further reaction to yield the product. This mechanism can be represented by:

$$E + S \xrightleftharpoons[k_{-1}]{k_1} ES \xrightarrow{k_2} E + P$$

$$ES + I \xrightleftharpoons[k_{-3}]{k_3} ESI \tag{10.25}$$

Application of the usual procedure leads to:

$$r_p = V_m C_S / [K_m + C_S(1 + C_I/K_I)] \tag{10.26}$$

If we define V_m' and K_m'' as:

$$V_m' = V_m/(1 + C_I/K_I) \tag{10.27}$$

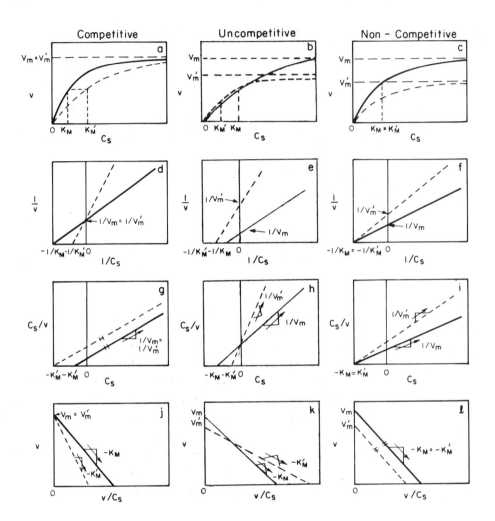

Figure 10.6. Typical plots for identifying the types of enzyme inhibition.
The solid curves represent the uninhibited case whereas the
dashed curves are the inhibited case. Parts a, b, and c are
Michaelis-Menten plots of enzyme velocity versus substrate
concentration for competitive, uncompetitive, and noncompetitive
inhibition, respectively; Parts d, e, and f are Lineweaver-Burk
plots of the same data; Parts g, h, and i are Hanes plots; and
Parts j, k, and l are Hofstee plots.

and

$$K_m'' = K_m/(1 + C_I/K_I) \tag{10.28}$$

Eq. 10.26 reduces in form to the Michaelis-Menten equation. Examination of
Eqs. 10.27 and 10.28 shows that the effect of an uncompetitive inhibitor is to reduce both V_m and K_m. A high substrate concentration will not overcome the effect of
the inhibitor because it binds with the enzyme-substrate complex rather than with
the free enzyme. An increase in substrate concentration serves only to increase the
concentration of enzyme-substrate complex with which the inhibitor can bind. The effect of this type of inhibitor is shown in Fig. 10.6b.

NONCOMPETITIVE INHIBITION. A noncompetitive inhibitor can combine with both
the free enzyme and the enzyme-substrate complex so that the reaction mechanism may
be represented by

$$E + S \underset{k_{-1}}{\overset{k_1}{\rightleftharpoons}} ES \xrightarrow{k_2} E + P$$

$$E + I \underset{k_{-3}}{\overset{k_3}{\rightleftharpoons}} EI \tag{10.29}$$

$$ES + I \underset{k_{-4}}{\overset{k_4}{\rightleftharpoons}} ESI$$

The usual approach can be used to develop the rate expression:

$$r_p = V_m C_S/(K_m + C_S)(1 + C_I/K_I) \tag{10.30}$$

If V_m' is defined by Eq. 10.27, then this equation also reduces to the Michaelis-
Menten equation. Thus it can be seen that the effect of a noncompetitive inhibitor
is to reduce V_m without affecting K_m. This is illustrated in Fig. 10.6c.

The easiest way to distinguish between the three types of reversible inhibitors
is through the use of one of the linearized plots discussed earlier. Experiments
should be run in which the effect of substrate concentration upon the reaction rate
is measured in the presence and absence of the inhibitor thereby yielding two linear
plots. The relative positions of the two lines will be characteristic of the type
of inhibition. Figure 10.6 illustrates how each type will appear for each of the
three linearized forms of the Michaelis-Menten equation. After the type has been
identified the value of K_I can be estimated by using the value of K_m' or V_m' obtained
from the graph in conjunction with the values of K_m and V_m obtained from the
uninhibited data.

SUBSTRATE INHIBITION. When their concentrations are very high, some substrates will bind with the enzyme-substrate complex as well as with the free enzyme. When this occurs, an enzyme-substrate-substrate complex is formed which cannot undergo further reaction to yield the product. This mechanism may be represented by:

$$E + S \underset{k_{-1}}{\overset{k_1}{\rightleftharpoons}} ES \overset{k_2}{\longrightarrow} E + P$$

$$ES + S \underset{k_{-3}}{\overset{k_3}{\rightleftharpoons}} SES \tag{10.31}$$

Application of the usual approach leads to:

$$r_p = k_2 C_{Eo} C_S / (K_s + C_S + C_S^2/K_s') \tag{10.32}$$

where K_s and K_s' are the dissociation constants for ES and SES, respectively. A typical plot of Eq. 10.32 is shown in Fig. 10.7. The substrate concentration at which the rate is maximum is obtained by differentiating Eq. 10.32 with respect to C_S and settling it equal to zero. The value of V_m is given by:

$$V_m = k_2 C_{Eo} / [1 + 2(K_s/K_s')^{0.5}] \tag{10.33}$$

The substrate concentration at which the maximum rate occurs is equal to $(K_s K_s')^{0.5}$.

PRODUCT INHIBITION. Occasionally, the product of an enzymatically catalyzed reaction may act to inhibit the reaction forming it. The simplest mechanism by which this could occur would be for the product to bind with the enzyme-substrate

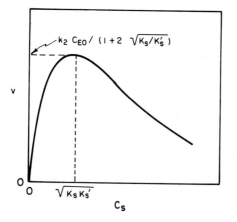

Figure 10.7.   Plot of Eq. 10.32 representing
               substrate inhibition of an
               enzymatic reaction.

complex, forming an unreactive enzyme-substrate-product complex. This may be represented by:

$$E + S \underset{k_{-1}}{\overset{k_1}{\rightleftharpoons}} ES \overset{k_2}{\longrightarrow} E + P$$

$$ES + P \underset{k_{-3}}{\overset{k_3}{\rightleftharpoons}} ESP$$

$$(10.34)$$

Application of the usual approach leads to:

$$r_p = V_m C_S / [K_m + C_S(1 + C_p/K_p)] \tag{10.35}$$

where

$$K_p = k_{-3}/k_3 \tag{10.36}$$

Comparison of Eq. 10.35 with Eq. 10.26 shows that product inhibition of this type would tend to behave like uncompetitive inhibition although the changing value of C_p would tend to complicate the results. Other mechanisms could be hypothesized which would lead to alternative rate equations but they need not be presented here.

Although the study of enzyme kinetics is an important field, microbial cells contain so many enzymes which interact in such complex ways (see Sections 8.4 and 8.5) that it is currently impossible to write mechanistic models expressing cellular processes in fundamental terms. Nevertheless, the basic concepts delineated by enzyme kinetics have strongly influenced the development of empirical models characterizing cell growth. Let us therefore turn our attention to the whole cell and the kinetic expression of its activities.

10.2 EVENTS OCCURRING IN MICROBIAL CULTURES

The growth of a microbial culture is a complex phenomenon composed of a number of simultaneously occurring events. The relative magnitudes of the respective rates determine what the net effect is upon the culture. The primary events are the utilization of substrate and the concurrent growth of organisms. These two events are closely related because it is only through the utilization of substrate that energy and carbon are made available for cell growth. This interrelationship was discussed in Section 9.3.1 and was expressed mathematically by the yield which was defined as the ratio of the rate of growth to the rate of substrate utilization. The cells must also use energy for maintenance, as was discussed in Section 9.2.2. If no exogenous energy source is present, maintenance energy will be provided by endogenous energy reserves and the cell mass in the culture will decline, or decay. If the culture is a mixed microbial community, predator organisms will prey upon those lower in the food chain. From the macroscopic point of view this also leads to a decrease in the total mass of the culture. Finally, there will always be some death of organisms, i.e., they will lose their ability to use substrate and to divide. The result of this is to make only a portion of the biomass active, a consequence which can be quite important in some biochemical operations. Even after death, however, organisms continue to play a role in the ecosystem. They may lyse, releasing soluble materials to the environment where other organisms can use them as substrates. Alternatively, they may be ingested by predators, thereby reducing their mass in the culture. Again, from a macroscopic point of view, the net effect of these is a decline in cell mass.

For purposes of kinetic modeling, the events occurring in microbial cultures can be grouped into three categories. Because cell growth and substrate utilization are proportional to each other they can be considered together. The use of exogenous energy for maintenance requirements, the decay of cell mass due to endogenous metabolism, and the consumption of bacteria by predators all have the same effect, i.e., to reduce the cell mass. Consequently, they can be grouped together into one category, normally referred to simply as decay. Finally, all events leading to death can be considered together. In the remainder of this chapter these events will be considered in detail.

10.3   CELL GROWTH AND SUBSTRATE UTILIZATION

10.3.1   First-Order Nature of Cell Growth

Bacteria divide by binary fission and consequently if they are grown rapidly in a batch reactor the number (or mass) of viable cells will increase in an exponential fashion. Therefore the reaction rate for bacterial growth can be expressed as a first-order equation:

$$r_{GX_V} = \mu X_V \qquad\qquad (10.37)$$

where r_{GX_V} is the rate of production of viable bacteria [mg/(liter·hr)], X_V is the concentration of viable bacteria (mg/liter), and μ is the specific growth rate constant (hr^{-1}). The growth rate constant is referred to as a specific rate constant because it defines the rate of cell growth in terms of the concentration of cells present, i.e., the mass of viable cells formed per unit mass of viable cells present.

In Section 9.3.1 the true growth yield, Y_g, was defined as the ratio of the rate of cell growth in the absence of maintenance energy requirements to the rate of substrate removal. This can be expressed mathematically as

$$Y_g = \frac{r_{GX_V}}{-r_s} \qquad\qquad (10.38)$$

where $-r_s$ is the rate of substrate disappearance [mg/(liter·hr)]. Combining Eqs. 10.37 and 10.38 yields:

$$-r_s = (\mu/Y_g)X_v \qquad (10.39)$$

Consequently it can be seen that the rate of substrate removal is first order with respect to the concentration of viable cells. The term in parentheses is given a new symbol and called the specific rate of substrate removal:

$$q = \mu/Y_g \qquad (10.40)$$

Thus Eq. 10.39 may be rewritten:

$$-r_s = qX_v \qquad (10.41).$$

10.3.2 Effect of Substrate Concentration on the Specific Growth Rate Constant
Originally, exponential growth of bacteria was considered to be possible only when all nutrients, including the substrate, were present in high concentration. In the early 1940's, however, it was found that bacteria could grow exponentially even when one nutrient was present in only limited amount [1]. Furthermore, the value of the specific growth rate constant, μ, depended upon the concentration of that particular nutrient which was present in least supply in proportion to the amount needed for growth. This growth limiting nutrient could be the organic carbon source, nitrogen, or any other factor needed by the organisms for growth. Since that time the generality of this observation has been substantiated often so that it can now be considered a basic concept of microbial kinetics [8].

NATURE OF THE RELATIONSHIP. The effect of the concentration of growth limiting nutrient upon the specific growth rate constant can be seen most easily by considering a hypothetical experiment [9]. Prepare a series of identical bacterial growth flasks by adding to each dilution water containing excess concentrations of nitrogen, phosphorus, and all other inorganic nutrients. Next add a different amount of organic substrate to each flask and inoculate each with a small amount of cells acclimated to the particular substrate. Then follow the changes in cell concentration with time. The resulting curves will be similar to those depicted in Fig. 10.8. The most obvious difference between the curves is that the ones from flasks with higher initial substrate concentrations have greater slopes. The slopes do not increase indefinitely, however, but approach a maximum value. Thus, as the substrate concentration is increased the curves become superimposable. The data plotted in the upper portion of Fig. 10.8 may be quantified by plotting on semilog coordinates. The exponential growth phase corresponds to the linear portion and μ may be calculated from the time required to double the cell mass (t_d) during that phase, as indicated in the figure. It is apparent from Fig. 10.8 that μ is a

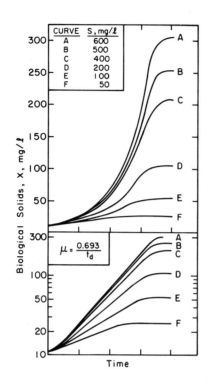

Figure 10.8.   Idealized curves showing effects
              of initial substrate concentration
              on the rate and total amount of
              microbial growth in batch reactors.
              (Adapted from Gaudy and Gaudy [9].)

function of the initial substrate concentration in the flasks. This relationship
can be seen clearly in Fig. 10.9, where it is apparent that μ increases as the substrate concentration is increased and that it approaches some maximum value which
will be called μ_m.

Although this example has been in terms of the organic substrate, similar
relationships have been found for nitrogen, phosphorus, and other inorganic
nutrients when all other constituents were present in excess. The shape of the
curve, however, will depend upon the particular nutrient that is limiting growth,
although μ_m will be the same for a given culture growing under fixed environmental
conditions. For example, if nitrogen or phosphorus were the growth limiting
nutrient, the curve might approach μ_m more rapidly than indicated in Fig. 10.9.

MATHEMATICAL EXPRESSION OF THE μ:S RELATIONSHIP. The question of the best
mathematical formula to express the relationship shown in Fig. 10.9 has been the
subject of much debate. No one yet knows enough about the mechanisms of microbial

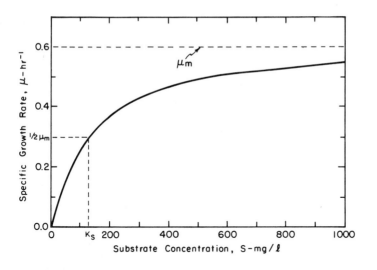

Figure 10.9.   Typical plot of the relationship between the
               specific growth rate constant, μ, and the
               substrate concentration, S.

growth to propose a mechanistic equation which will characterize growth exactly.
Instead, experimenters have observed the effects of various factors upon growth and
have then attempted to fit empirical mathematical equations to their observations.
Consequently, all equations that have been proposed are curve fits and the only
valid arguments for the use of one over another are goodness of fit, mathematical
utility, and broad acceptance.

   The equation with historical precedence and greatest acceptance is the one
proposed by Monod [1]. Having performed experiments similar to the hypothetical one
above, he observed similar results. The work was later extended and refined by
workers using continuous cultures of single bacterial species growing on defined
media and the conclusion was reached that the curve could be approximated adequately
by the equation for a rectangular hyperbola [8]. Consequently, Monod proposed the
equation:

$$\mu = \frac{\mu_m S}{K_s + S} \tag{10.42}$$

where S is the concentration of growth limiting substrate and K_s is the saturation
constant. K_s determines how rapidly the curve approaches μ_m and is defined as the
substrate concentration at which μ is equal to half of μ_m.

   Because of the similarity of Eq. 10.42 to the Michaelis-Menten equation in
enzyme kinetics (Eq. 10.13), many people have erroneously concluded that Monod proposed the equation on theoretical grounds. While the Michaelis-Menten equation can
be derived from a consideration of the rates of chemical reactions catalyzed by

enzymes, and thus has a theoretical basis, the Monod equation is strictly empirical. The shape depicted in Fig. 10.9 is too general in nature to be used as mechanistic evidence.

Equation 10.42 has been found to fit the data from many pure cultures growing on single substrates and has been used extensively in the derivation of equations describing the continuous cultivation of microorganisms. It has not been blindly accepted, however, and other workers have proposed alternative equations which fit their data better [10,11,12].

EFFECTS OF PARAMETERS ON SHAPE OF μ:S CURVE. The effects of μ_m and K_s upon the shape of the μ:S curve can be visualized most easily through graphs. Figure 10.10 shows the effect of changing μ_m while maintaining a constant K_s value. There it can be seen that higher values of μ_m will result in more rapid specific growth rates for any substrate concentration. Figure 10.11 shows what happens when K_s is varied while holding μ_m constant. When K_s is very small, the curve rises steeply and breaks to the right very sharply. In this case μ is only sensitive to small changes in the substrate concentration at very low values. A high value of K_s causes the μ:S curve to approach its asymptote much more slowly, and therefore μ is sensitive to changes in S over a much broader range. A system giving a more gradual curve, such as A in Fig. 10.10 or E in Fig. 10.11, tends to be more stable in continuous culture.

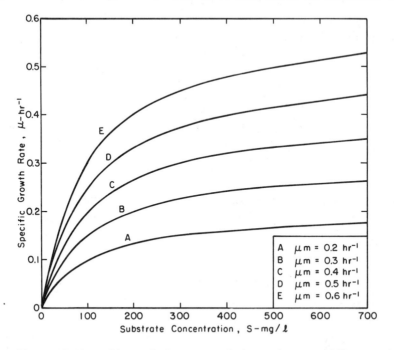

Figure 10.10.   Effect of the value of the maximum specific growth
                rate constant, μ_m, on the curvature of the Monod
                equation (Eq. 10.42). The value of K_s used was
                100 mg/1.

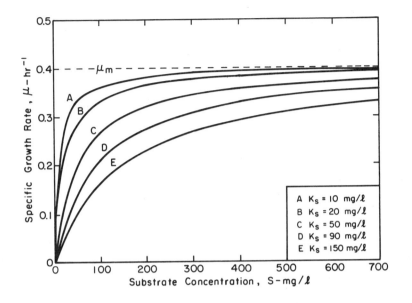

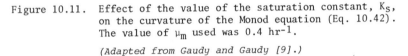

Figure 10.11. Effect of the value of the saturation constant, K_S, on the curvature of the Monod equation (Eq. 10.42). The value of μ_m used was 0.4 hr^{-1}.

(Adapted from Gaudy and Gaudy [9].)

FIRST-ORDER APPROXIMATION. If K_S is very, very large the curvature in the μ:S curve is reduced so that the region of low substrate concentration approaches linearity. The reason for this can be seen by examining Eq. 10.42. When K_S is large with respect to the substrate concentration the S term in the denominator becomes unimportant and the equation may be approximated as a first-order equation:

$$\mu \cong \frac{\mu_m S}{K_S} \qquad\qquad (10.43)$$

Although this equation is often easier to use than the Monod equation, care should be exercised in its use because serious error can result if S is not small with respect to K_S.

Activated sludge systems are often characterized by low values of S and high values of K_S. Consequently, Eq. 10.43 is often used with such systems. Garrett and Sawyer [13] were the first to propose such an equation because they had observed that the specific growth rate constant for bacteria was directly proportional to the substrate concentration at low values and independent of it at high ones. They recognized that these two conditions were specialized cases of the Monod equation; nevertheless, others who adopted their first-order equation incorrectly considered it to be an alternative equation.

10.3.3 Effect of Substrate Concentration on the Specific Substrate Removal Rate
While most of the researchers in the field of microbial kinetics were concentrating
on the specific growth rate constant others chose to investigate the effects of substrate concentration on the rate at which each organism removed substrate [14,15,16].
Biologists [16] called the rate a "specific" rate, in analogy to the specific growth
rate, while engineers [14] often called it a "unit" rate. Nevertheless, the concept
was the same.

    ANALOGY TO μ. In deriving Eq. 10.41 a new term, q, was defined. It is called
the specific rate of substrate removal and is related to the specific growth rate
constant through the true growth yield as given by Eq. 10.40. Because μ is a
function of the concentration of growth limiting substrate, it is apparent that q
is also:

$$q = \left(\frac{\mu_m S}{K_s + S}\right)\left(\frac{1}{Y_g}\right) \tag{10.44}$$

Equation 10.44 is often expressed in terms of the maximum specific substrate removal
rate, q_m:

$$q = \frac{q_m S}{K_s + S} \tag{10.45}$$

where

$$q_m = \mu_m/Y_g \tag{10.46}$$

Equation 10.45 is empirical and is accepted on the same pragmatic grounds as the
Monod equation. Some researchers, [14], while recognizing that a hyperbolic type
relationship exists between q and S prefer to leave it in a graphical form rather
than tie it to a specific equation.

    FIRST-ORDER APPROXIMATION. Just as was the case with the Monod equation,
Eq. 10.45 can be approximated by a first-order equation when the saturation constant
is large in relation to the substrate concentration:

$$q \cong \frac{q_m S}{K_s} \tag{10.47}$$

Eckenfelder [17] has called the ratio of q_m over K_s the mean reaction rate
coefficient:

$$k_e = q_m/K_s \tag{10.48}$$

The validity of the approximation expressed by Eq. 10.47 depends upon S being small
with respect to K_s. Just as with the use of the first-order approximation for μ,
care should be exercised when using Eq. 10.47.

10.3.4 Complications Caused by the Use of Heterogeneous Cultures with Heterogeneous Substrates

The Monod equation was developed from experiments using pure cultures of bacteria growing on single organic compounds. When the growth of microorganisms in biochemical operations for wastewater treatment is being considered, however, two complicating factors enter the picture. The first is that wastewaters do not contain single organic compounds. Rather, they contain a mixture of compounds whose concentration must be measured by a nonspecific test such as COD. Consequently, the concentration of biodegradable COD is usually considered to be the substrate. The second factor is that biochemical operations contain complex microbial communities rather than single species. These communities are usually in a continuous state of flux with constant changes in the relative magnitudes of the species present. This can have a drastic effect upon the observed kinetics so that the growth "constants" are seldom constant.

Many investigators have studied the relationship between μ and S in mixed microbial populations in order to ascertain whether it can be represented by an equation like that of Monod [9,15,18,19,20]. It is generally agreed that it can. Several authors have pointed out the effects that the culture conditions have upon the character of the microbial community that develops, and have stated that the kinetic parameters (μ_m and K_s) obtained from continuous culture studies are in reality average values resulting from many predominant species [9,21,22]. Consequently it has been recommended that μ_m and K_s be characterized by ranges of values rather than single ones, just as was recommended for Y_g. It can be concluded, however, that the Monod equation represents a reasonable model with which to describe the kinetics of microbial growth in wastewater treatment systems, and consequently it is widely used.

10.3.5 Relationship of Growth Limiting Nutrient to Treatment Objective

One result of the dependence of specific growth rate upon the concentration of growth limiting nutrient is that the growth limiting nutrient must be the material that the biochemical operation is designed to remove. If an operation is to be designed to remove soluble organic matter, then predictions must be made of the rates of removal of that organic matter. If the concentration of organic matter is not the rate limiting factor then it would be impossible to write the rate expressions in terms that would yield useful information concerning the size of the reactor needed to remove it.

Three essential nutrients must be provided when a microbial culture removes organic matter by growth: nitrogen, phosphorus, and oxygen. The stoichiometry of the system determines the total quantity of each which will be needed for cell growth. The concentration in the influent to the reactor, however, must be in excess of the stoichiometric amount. If only the stoichiometric amount were present, microbial growth would remove it all, leaving only negligible concentrations in solution. The consequence is that nitrogen, phosphorus, or oxygen could become the rate limiting material. If that occurred, the S in Eq. 10.42 would be the concentration of one of them rather than the concentration of organic matter, making the equation invalid for the prediction of the rates of removal of organic matter.

For the ranges of specific growth rate found in most biochemical operations removing soluble organic matter, the concentrations of oxygen, nitrogen, and phosphorus required in excess of the stoichiometric amount to ensure that carbon is growth limiting can be estimated. Normally oxygen is continually supplied by gas transfer, and a concentration of 1 mg/liter dissolved in the medium will generally ensure fully aerobic metabolism of dispersed cultures with growth rates unaffected by the oxygen concentration [23]. Goel and Gaudy [24] determined that K_s for ammonia nitrogen during normal heterotrophic growth lies between 1.5 and 4.0 mg/liter as N. Using 0.50 hr^{-1} as a representative value of μ_m it can be shown that if the influent ammonia nitrogen concentration exceeds the stoichiometric amount by 1 mg/liter as N, nitrogen will not be rate limiting at the growth rates normally employed in wastewater treatment. Although some work has been done on kinetic limitation of heterotrophs by phosphorus, the results are not as clear as those with nitrogen. Attempts to measure the limiting phosphorus concentration in both pure and mixed microbial cultures have found that it was too low to detect with the techniques available [25]. Consequently, if the concentrations of phosphorus in the influent exceed the stoichiometric amount by 0.5 mg/liter as P, phosphorus should not be rate limiting. In some biochemical operations the organisms pass through a growth cycle, and nutrients will be taken up during one phase, and then released during another. To prevent nutrient limitation during the phase of nutrient uptake the amounts presented above should be in excess of the maximum quantity removed, not the net amount as determined by the final effluent.

If a biochemical operation is being designed to bring about nitrification, then the kinetic parameters must be determined with ammonia and nitrite nitrogen as the growth limiting nutrients and every effort must be made to ensure that no other kinetic limitation occurs. Similar statements can be made about all other biochemical operations. The importance of this concept should never be forgotten.

10.3.6 Representative Parameter Values

The values of the parameters μ_m and K_s are very dependent upon the organism and substrate employed. If a given species of organism is grown on each of several substrates under fixed environmental conditions the values of μ_m and K_s observed will depend upon the substrates. Likewise, if several pure cultures are fed the same substrate under otherwise identical environmental conditions, the values of μ_m and K_s will depend upon the species of organisms. Generally, however, substrates which are difficult to degrade will be characterized by low values of μ_m and high values of K_s. Conversely, substrates which are easy to degrade will have high values of μ_m and low values of K_s.

Given the organism dependency of μ_m and K_s, one would anticipate that these parameters would be characterized by considerable variability in mixed cultures, even when grown on single substrates. This is exactly the case. For example, when 22 experiments were performed using heterogeneous cultures of sewage origin grown on glucose, μ_m varied from 0.31 to 0.77 hr^{-1} and K_s varied from 11 to 181 mg/liter as COD in spite of relatively constant environmental conditions [9]. This emphasizes the fact that there are no single values which can be used as the kinetic parameters for a heterogeneous culture. Other examples of μ_m and K_s for mixed cultures have been reported in the literature and some are summarized in Table 10.1 to give an indication of the types of values often found.

Several researchers have reported parameter values for domestic sewage and these are listed in Table 10.2. Part of the differences in the μ_m values were caused by the way in which the cultures were grown prior to the tests to determine them. Chiu *et al.* [22] and Ghosh and Pohland [21] have reported that continuous cultures grown slowly had predominantly slow growing organisms (i.e., low values of μ_m) while those grown rapidly had mostly fast growing organisms. The culture giving a μ_m value of 0.55 hr^{-1} was grown rapidly prior to testing whereas that giving a value of 0.16 hr^{-1} was grown slowly. Thus it is logical that the μ_m

TABLE 10.1
Typical Values of the Monod Parameters for Mixed Cultures

Substrate	μ_m hr.$^{-1}$	K_s mg/liter	Basis for K_s	Reference
Glucose	0.31 - 0.77	11 - 181	COD	9
Glucose	0.69	26	COD	22
Glucose	0.18	-	-	13
Glucose	-	133	BOD5	62
Glucose	0.36	103	COD	26
Glucose	0.38 - 0.49	11 - 29	weight	27
Glucose	0.35	8	weight	63
Lactose	0.20 - 0.53	33 - 55	weight	27
Skim Milk	0.10	100	BOD5	64
Skim Milk	0.12	110	COD	65
Sucrose	0.28 - 0.55	6 - 17	weight	27
Glutamic acid	0.59 - 0.78	47 - 95	weight	27
Serine	0.43 - 0.54	30 - 50	weight	27
Peptone	0.26	109	BOD5	13
Peptone	-	296	BOD5	62
Acetic acid	0.29 - 0.36	41 - 47	weight	27
Propionic acid	0.37 - 0.38	6 - 17	weight	27
Poultry waste	3.0	500	BOD5	28
Soybean waste	0.50	355	BOD5	28
Textile waste	0.29	86	BOD5	28

TABLE 10.2
Typical Values of the Monod Parameters for
Mixed Cultures Growing on Domestic Sewage

μ_m hr.$^{-1}$	K_s mg/liter	Basis for K_s	Reference
0.40	60	COD	27
0.46	55	COD	27
0.16	22	COD	66
0.55	120	BOD$_5$	28

values are different. This emphasizes the importance of growth conditions upon the apparent kinetics of mixed microbial cultures. If the parameter values determined during lab studies are to be representative of those anticipated in a field-scale treatment plant, then the growth conditions utilized in the lab study must be similar to those to be used in the treatment plant.

The examples of the kinetic parameters presented in the preceding paragraphs all relate to the removal of soluble biodegradable organic matter by aerobic cultures. The Monod model may be used to define the rates of substrate removal in anaerobic cultures as well. In general, μ_m is an order of magnitude lower and K_s is an order of magnitude higher than the values reported in Table 10.1 [15].

The mean reaction rate coefficient, k_e, has found considerable use in industrial waste treatment practice. Table 10.3 is a summary of values collected from the literature [29]. Three things should be noted about the values. First, most are much lower than the values that might be calculated from μ_m and K_s for simple compounds in Table 10-1. This is a reflection of both the slow conditions under which most cultures were grown and the difficulty of degradation of many wastewaters. Second, the range of values found in the table extends over an order of magnitude, indicating that some wastes may be degraded more than ten times faster than others. Third, compared to many industrial wastes, domestic sewage is fairly easily degraded. Consequently, plants designed to handle domestic wastewaters may not perform effectively if they subsequently receive industrial wastes.

All of the examples presented in Tables 10.1 through 10.3 have been for heterotrophic growth with an organic compound as the growth limiting substance. Equation 10.42 is applicable to autotrophic growth as well, and Table 10.4 summarizes some parameter values which have been reported in the literature for nitrification [15]. Two things are worth noting. First, μ_m is one or two orders of magnitude less than for aerobic heterotrophic growth. Consequently, systems with very low growth rates are required for nitrification. Second, K_s is extremely low, implying that the growth rate is independent of the substrate concentration except at very low values. Both of these factors give nitrification reactors unique characteristics which will be discussed in Chapter 20.

TABLE 10.3
Typical Values of the Mean Reaction Rate Coefficient for Wastewaters[a]

Wastewater Type	k_e liter/mg·hr	Coefficient Basis	Reference
Ammonia base, semi-chemical	4.6×10^{-4}	BOD_5	67
Brewery	2.2×10^{-4}	BOD_5	67
Chemical industry	$1.4 \times 10^{-4} - 2.0 \times 10^{-4}$	BOD_5	49
Coke plant, ammonia liquor	11.0×10^{-4}	COD	68
Domestic sewage	$10.8 \times 10^{-4} - 13.7 \times 10^{-4}$	BOD_5	49
Organic chemical	$0.50 \times 10^{-4} - 0.73 \times 10^{-4}$	BOD_5	67
Petrochemical	$2.4 \times 10^{-4} - 2.8 \times 10^{-4}$	BOD_5	49
Pharmaceutical	$2.1 \times 10^{-4} - 5.7 \times 10^{-4}$	BOD_5	49
Phenolic	0.92×10^{-4}	BOD_5	67
Pulp and Paper	4.17×10^{-4}	BOD_5	69
Refinery	$3.5 \times 10^{-4} - 10.0 \times 10^{-4}$	BOD_5	67, 69
Rendering	15.0×10^{-4}	BOD_5	49
Tetraethyl lead	7.1×10^{-4}	BOD_5	67
Textile	1.5×10^{-4}	BOD_5	70
Thiosulfate	$1.1 \times 10^{-4} - 2.1 \times 10^{-4}$	COD	71
Vegetable oil	3.1×10^{-4}	BOD_5	67

[a] *(Adapted from [29]).*

TABLE 10.4
Typical Values of the Monod Parameters for Mixed Cultures
Oxidizing Ammonia and Nitrite Nitrogen at 20°C[a]

Substrate	μ_m hr.$^{-1}$	K_s mg/liter N	Reference
NH_3-N	0.022	3.6	56
NH_3-N	0.014	1.0	72
NH_3-N	0.027	0.6	55
NO_2-N	0.017	1.1	56
NO_2-N	0.006	2.1	72
NO_2-N	0.035	1.9	55

[a] *(Adapted from [15]).*

10.4 MICROBIAL DEATH AND VIABILITY

For many years microbiologists have studied the death rates of bacteria which have
been subjected to stress (adverse pH, pressure, high temperature, low temperature,
radiation, etc.). On the other hand, only recently have biologists begun to study
spontaneous death of microorganisms [30,31]. In a sense this is rather curious because environmental engineers have long recognized that only a portion of the biomass present in biochemical operations is actually alive and contributing to the
activity therein [32].

10.4.1 The Concepts of Natural Death and Viability

The viability of a microbial culture is defined as the number of live or viable
cells present divided by the total number [30,33]:

$$v = \frac{X_v}{X_v + X_d} \tag{10.49}$$

where X_d represents dead cells. In most engineering studies X_v and X_d are taken to
represent the mass of viable and dead cells, respectively. As currently defined,
a viable cell is one that will divide and form a colony on solid growth media, such
as an agar plate [30]. A dead cell would be one which has lost that ability. Under
some circumstances it is possible to obtain microbes having many of the characteristics of normal, living organisms, but which are unable to divide. Such organisms
are committed to death and are called moribund. Moribund organisms can contribute
to substrate removal; however, because of the extreme difficulty of detecting them
we will ignore their existence. Generally, substrate removal is attributed only to
the active, viable cells in the culture.

The exact nature of natural death is not yet understood, although many factors
are known to influence it. Natural death is undoubtedly related to the requirement
for maintenance energy and the associated endogenous decay that develops when that
energy cannot be supplied from external sources. Eventually the point is reached
where all storage polymers are depleted and degradation of the essential components
begins. Although starved, nongrowing bacteria show no evidence for involvement of
degradative damage to DNA at death, slight damage to the structure of the genetic
material could cause a loss of viability [30]. In addition, evidence indicates that
autodegradation of RNA is very important to cell death.

10.4.2 Evidence for Natural Death in Growing Systems

Weddle and Jenkins [34] summarized the historical work which established that large
portions of the microbial cultures used in wastewater treatment are nonviable.
Much of it is indirect evidence involving comparisons of substrate removal rates
and enzyme activities. They went one step further, however, and showed that the
viability of their culture was between 10 and 20 percent when it was grown under
conditions similar to the activated sludge process. Furthermore, the viability was

a function of the growth condition and could be made to approach 100 percent.
Walker and Davis [35] have substantiated these findings.

Evidence from the pure culture literature supports the findings of Weddle and
Jenkins. Postgate and Hunter [36], using a pure culture of *Aerobacter aerogenes,*
found viabilities on the order of 40 to 50 percent when the organisms were grown at
rates similar to those found in activated sludge. Tempest *et al.* [37] also obtained
reduced viabilities when organisms were grown continuously at very low rates, implying that death occurs continuously even in the presence of growth. In addition,
they observed gross distortions in the morphology of their culture which changed
normally rod shaped organisms into long filamentous type strands. This indicates
that many activated sludges may contain deformed organisms which could contribute
to the low viabilities observed.

10.4.3 Expression for the Rate of Natural Death

Very little work has been done on the mathematical representation of natural death,
primarily because of the paucity of fundamental data upon which to base such
equations. McKinney [38] was one of the first engineers to attempt to model the
generation of an inactive microbial mass in biochemical operations. He assumed that
the active organisms would be undergoing endogenous metabolism and that some cells
would die and lyse as a result. A portion of the cell residue was considered to be
nondegradable and thus would build up in the medium. This cell residue was called
endogenous mass and is cell mass which would not contribute to substrate removal.
As such, it is similar in concept to the dead cells defined earlier. Others have
taken a similar approach [39,40]. None, however, seek to express the viability of
the population or relate it to a rate constant for the generation of dead cells.

Westberg [41] tried a more sophisticated approach by hypothesizing a specific
death rate constant which was inversely propotional to the substrate concentration.
Unfortunately, no data were presented to verify the concept.

The only model for bacterial death which has been checked against experimental
data is the one proposed by Sinclair and Topiwala [33] for death in pure cultures
of *A. aerogenes.* Although the form of the equation used has no mechanistic justification it gave good experimental correlation and consequently can be considered
as an empirical model. The equation says simply that the death rate of viable
bacteria, r_{DX_v}, is directly proportional to the concentration of viable bacteria
in the medium.

$$r_{DX_v} = -\gamma X_v \tag{10.50}$$

where γ is a first-order death rate coefficient, which is assumed to be constant for
a given environmental condition. The rate of generation of dead bacteria, r_{GX_d}, is
just equal to the rate of death of the live bacteria.

$$r_{GX_d} = -r_{DX_v} \tag{10.51}$$

This model has been applied to activated sludge systems and has been able to predict many observations from that biochemical operation [42]. Consequently, Eq. 10.50 will be employed herein.

10.4.4 Representative Parameter Values

Little data is available in the literature relating to the value of γ. Grady and Roper [42] summarized all that were available in 1974 and these are presented in Table 10.5. The approximate nature of the rate equation can be seen from the fact that γ depends upon how rapidly the culture is being grown. Because of the type of data available it was not possible to calculate γ values for actual activated sludge systems [42]. From consideration of the data in Table 10.5, and the normal growth rate of activated sludge, it is probable that γ will be less than 0.005 hr^{-1} for such systems.

10.5 MICROBIAL DECAY

10.5.1 The Concept of Decay

In Section 9.2.2 the requirement of microorganisms for energy of maintenance, which is thought to be needed at a constant rate per cell, was discussed. When equations are written for microbial growth in biochemical reactors consideration must be given to this requirement and two alternative approaches are possible.

In the first approach all maintenance requirements are considered to be supplied by degradation of endogenous reserves. All exogenous energy is used for synthesis and the route of energy flow is assumed to be sequential, i.e., energy flows from the medium into the cell for synthesis of protoplasm and endogenous reserves, and then from the endogenous reserves to meet the maintenance requirement. In the second approach energy flow is considered to be concurrent, i.e., a portion of the exogenous energy source is used for maintenance and the remainder is used for synthesis.

TABLE 10.5
Typical Values of the Specific Death Rate Coefficient[a]

Organism	Substrate	Growth Rate Range, hr^{-1}	γ hr^{-1}
A. aerogenes	Glucose	> 0.05	0.0065
A. aerogenes	Glycerol	> 0.05	0.0088
A. aerogenes	Glycerol	< 0.05	0.0055

[a] (Data extracted from [42]).

Conceptually, the idea of concurrent energy flow is more attractive. Even though the evidence concerning the route of energy flow for maintenance purposes is inconclusive, the existence of metabolic control mechanisms suggests that an organism would not synthesize material while simultaneously degrading it. From a modeling point of view, however, there are difficulties associated with the concept of concurrent flow. For example, endogenous metabolism is known to occur under some conditions. Consequently the model would have to include a term for it and provide a way for it to increase in importance as the rate of supply of energy fell below the needs for maintenance, such as after exhaustion of the exogenous energy supply in a batch or plug-flow reactor. Such variable rate expressions can be complicated.

Mathematically, the sequential concept is much easier to use. All exogenous substrate is assumed to result in cell material with the amount being formed per unit of substrate removed determined by the true growth yield (the amount of cell material formed in the absence of maintenance requirements, Section 9.3.1). A portion of that cell mass is then degraded to supply maintenance energy. As a result the net yield is less than Y_g. Because the cells are assumed to undergo endogenous metabolism at all times the same rate equation is applicable to all types of reactors. The end result of equations based upon this concept is the same as the result of equations based upon the concurrent energy flow concept. These are simpler, however.

It is common in many biochemical operations for the microorganisms to be without exogenous substrate for extended periods of time. Consequently, many environmental engineers [15,17,18,38,39] have adopted the sequential energy flow concept as a means of accounting for maintenance requirements and endogenous metabolism in mathematical models for biochemical operations.

The term representing the loss of cell mass is usually called microbial decay. This name is a result of the fact that the loss of mass in a heterogenous culture is due to more than just endogenous metabolism. For example, predators remove bacteria, resulting in the growth of more predators. The efficiency of growth of predators upon bacteria, however, is similar to that of bacteria upon the original substrate so that even though new predators are being formed a reduction in microbial mass results. Although some of the more sophisticated models attempt to express mathematically the interactions within a multi-level food chain most design equations for biochemical operations just reflect the loss of mass empirically through the decay term. Another factor contributing to the reduction in mass is cell lysis. As some organisms die, they begin to leak their internal contents into the medium. Only the remaining cell wall is large enough to be measured as insoluble cell mass in the medium, consequently each time a cell lyses there is a reduction in mass. This, too, is accounted for in the cell decay term.

Thus it can be seen that decay is actually a composite term, incorporating the effects of many factors. It is strictly empirical. However, its use can be justified on the basis of the satisfactory job that it does in predicting the changes in cell mass that will occur as the growth conditions on a biochemical reactor are changed.

10.5.2 Rate Expressions for Decay

When equations are written to express the loss of biomass by decay it is generally
assumed that decay occurs in a first-order manner with a constant coefficient. It
is reasonable that the rate of decay would depend upon the types of cells (viable
or dead) undergoing decay:

$$r_{dX_v} = -b_v X_v \tag{10.52}$$

$$r_{dX_d} = -b_d X_d \tag{10.53}$$

where b_v and b_d are the specific decay rate constants for viable and dead cells, respectively. Viable cells would tend to be lost by all of the mechanisms discussed
in the previous section whereas dead cells would be lost primarily by predation and
lysis. Consequently, b_v is likely to be larger than b_d.

Most work dealing with the specific decay rate has not considered the viability
of the culture but has instead measured an overall specific rate of decay, b, of the
combined mass of viable and dead cells. Because b_d is likely to be smaller than b_v,
one would expect the magnitude of b to decrease as the viability of the culture
decreased [42]. This is exactly what happens [39]. In spite of this evidence concerning the variable nature of b, however, it is common practice to assume that it
will be constant for a given culture. In other words, it is assumed that the
values of the rate constants are equal:

$$b_v = b_d = b \tag{10.54}$$

Because only values of b have been reported in the literature and because of the
scarcity of data on the individual rate constants, it will be assumed herein that
Eq. 10.54 is valid. Such an assumption will not detract materially from the predictive ability of the models developed in Part IV of this book because of the approximate nature of the decay concept to begin with.

10.5.3 Representative Parameter Values

Table 10.6 contains values of b for pure and mixed cultures which have been grown
on simple media in the laboratory. Looking first at the pure cultures it can be
seen that the species of organism influences the value of b. The carbon source
upon which the culture was grown also has some effect, probably because of the influence it will have upon the type of endogenous reserves in the cells. Of even
more interest, however, is the difference between the values observed in pure and
mixed cultures. The values obtained with mixed cultures were often an order of
magnitude lower than those obtained with pure cultures. Examination of the references from which the values were taken reveals that the mixed cultures exhibiting
low values of b were grown at much slower rates than the pure cultures. Furthermore, when mixed cultures are grown at rates similar to those used for the pure
cultures the values observed are similar to those of the pure cultures.

TABLE 10.6
Typical Values of the Specific Decay Rate Coefficient
for Cultures Grown on Simple Substrates

Organism	Substrate	b hr^{-1}	Reference
A. aerogenes	Glucose	0.080	43
A. aerogenes	Glycerol	0.042	44
A. cloacae	Glucose	0.041	44
E. coli	Glucose	0.028	45
P. fluorescens	Glucose	0.12	46
P. fluorescens	Glucose	0.11	47
Mixed culture	Glucose	0.019	22
Mixed culture	Glucose	0.023 (30°C)	26
Mixed culture	Glucose	0.0036	73
Mixed culture	Sucrose and Peptone	0.0018	48
Mixed culture	Peptone	0.0033	48
Mixed culture	Skim milk	0.0019	64

Table 10.7 presents values of b observed for mixed cultures growing on various
wastewaters. Most wastewater cultures are grown relatively slowly and the values
in the table are similar in magnitude to those of the slowly growing mixed cultures
in Table 10.6. Barnard [49] has discussed the factors influencing b and has concluded that for wastewater treatment systems many values lie around 0.002-0.003 hr^{-1}.
Examination of Table 10.7 reveals that this is true for domestic sewage but that for
other wastes the nature of the wastewater has an effect.

TABLE 10.7
Typical Values of the Specific Decay Rate
Coefficient for Cultures Grown on Wastewaters[a]

Wastewater Type	b hr^{-1}	Reference
Domestic sewage	0.0020 - 0.0029	15
Poultry Processing	0.030	28
Pulp and Paper	0.0083	69
Pulp and Paper	0.0015	74
Refinery	0.010	69
Shrimp Processing	0.067	75
Soybean	0.006	28
Textile	0.030 - 0.050	76
Textile	0.0014	70
Thiosulfate	0.00042 - 0.00083	71
Vegetable and Fruit Processing	0.0012 - 0.0079	74
Whey, Cottage Cheese Manufacturing	0.0023	77

[a] *(Adapted from [29]).*

10.6 PHYSICAL FACTORS AFFECTING KINETICS

It has long been recognized that temperature and pH affect the rate at which microbial cultures grow. It is only recently, however, that these effects have been
quantified through use of the kinetic parameters. In this section we will see how
pH and temperature influence the values of μ_m, K_s, γ and b.

10.6.1 Temperature

Temperature can exert an effect upon biological systems in two ways: by influencing
the rates of enzymatically catalyzed reactions and by affecting the rate of diffusion of substrate to the cell. The importance of both has not always been recognized and this has led to some confusion in the quantification of temperature effects. Generally, temperature effects found in the lab are more pronounced than
those in the field. This is due in part to the fact that full-scale reactors are
much more apt to be diffusion controlled. In this section we will consider only
the effects upon reaction rates. Diffusional effects will be considered in Part V
where individual reactor types are discussed.

METHODS OF EXPRESSING TEMPERATURE EFFECTS. There are three techniques in
common use to quantify the effects of temperature upon biochemical operations.
Arrhenius [50] was the first to quantify the effects of temperature upon a biological reaction when in 1889 he suggested an equation to describe the effects of
temperature upon the enzymatic hydrolysis of sugar. The equation proposed was:

$$k = A \exp(-u/RT) \tag{10.55}$$

where k is the temperature dependent reaction rate coefficient, A is a constant,
u is the temperature characteristic, R is the gas constant, and T is the absolute
temperature. The constant A and the Arrhenius temperature characteristic u can be
determined from a plot of ln k versus 1/T:

$$\ln(k) = -\frac{u}{R}\left(\frac{1}{T}\right) + \ln(A) \tag{10.56}$$

A positive value of u indicates that k increases as the temperature increases.
Many biological reactions have been shown to fit this expression, although generally
over a limited temperature range.

The second technique for quantifying temperature effects was proposed by
Phelps [51]:

$$k_1 = k_2 \theta^{(T_1 - T_2)} \tag{10.57}$$

where k_1 and k_2 are the reaction rate coefficients at temperatures T_1 and T_2, respectively, and θ is the thermal coefficient. For most biochemical operations T_2
is chosen as 20°C. The thermal coefficient, θ, is determined from a plot of
ln(k) versus T.

Recently, Novak [52] has proposed a third equation:

$$k_1 = k_0 \exp[C(T_1 - T_0)] \tag{10.58}$$

where k_1 is the reaction rate at temperature T_1, k_o is the reaction rate at the reference temperature T_o, and C is the temperature coefficient. This equation reduces to Eq. 10.57 when C is set equal to the $\ln(\theta)$. The value of C is determined from a plot of $\ln(k)$ versus T.

A comparison of the three equations reveals that theoretically the same data could not fit all three. In reality, however, because of the relatively small temperature range associated with biochemical operations, and the normal scatter in microbial data, it is usually possible to obtain a satisfactory fit with any of them. Because the Arrhenius equation is commonly used in the microbiological literaure it will be used in this chapter where the consideration of temperature effects is limited to reaction rates.

EFFECTS OF TEMPERATURE UPON KINETIC PARAMETERS. Because of its traditional role in characterizing batch microbial growth many studies have reported the effects of temperature upon μ_m. It is widely accepted that μ_m increases as T is increased until some maximum value is reached. After that point μ_m decreases, due in part to the inactivation of enzymes. Equations like 10.55 through 10.58 are limited to the range in which μ_m increases with increasing T. Organisms are grouped into three categories depending upon the temperature range over which such an equation can be applied. Of chief concern in biochemical operations are mesophilic organisms which grow well over the range of 10-35°C. The two other groups, psychrophilic and thermophilic have ranges on either side and find use under special conditions.

Table 10.8 is a summary of the Arrhenius characteristics, u, obtained from the literature [26]. They are characterized by a large range, particularly those obtained from batch culture. The data obtained from continuous culture may be more indicative of what would be expected in biochemical operations. Although no trends are clear, it appears that u depends upon the microbial species. Care should be taken when extrapolating this data to mixed cultures because as the cultures are grown at different temperatures different predominant populations will develop [53]. The values in Table 10.8 do, nevertheless, provide a good idea of the types of values that can be expected.

Relatively few studies have been performed in which the effects of temperature upon the saturation constant, K_s, were measured. Among those that have been performed there is no consensus concerning the nature of the effect. For example, one study [39] reported a decrease in K_s when the temperature was increased. This was performed with a pure culture degrading glucose. Another study [52] reported an increase in K_s with increasing temperature for mixed cultures degrading sewage, glucose, a synthetic waste, and fatty acids, all aerobically. A third study [26], done with a mixed culture degrading glucose, reported that K_s decreased slightly as the temperature was raised from 10°C to 20°C, and then rose significantly as the temperature was raised to 30°C. A fourth study [54], done with yeast, reported an increase in K_s with increasing T. Thus, at this time, it appears that K_s is likely to increase with

338

10/ KINETICS OF BIOCHEMICAL SYSTEMS

TABLE 10.8
Typical Values of the Arrhenius Temperature
Characteristic for the Maximum Specific Growth Rate[a]

Temperature Range, °C	Organism	Growth Media	u cal/mol	Reference
18-39.4	E. coli	Glucose	15,100[b]	78
10-30	P. aeruginosa	Glucose	28,300[b]	79
0-20	a psychrophil	Glucose	17,000[b]	79
20-30	a psychrophil	Glucose	10,000[b]	79
0-15	Pseudomonas No. 92	Lactose	34,800[b]	80
0-30	Pseudomonas No. 69	Lactose	20,500[b]	80
10-15	E. coli	Glucose	40,000[b]	81
15-30	E. coli	Glucose	18,000[b]	81
8-20	Gibberella fujikuroi	Glucose	10,900	82
20-28	Gibberella fujikuroi	Glucose	5,100	82
23-37	A. aerogenes	Glucose	10,900	83
22-37	E. coli	Glycerol; N-limited	13,300	84
20-30	P. fluorescens	Glucose	16,600	46
25-40	A. aerogenes	Glucose	8,400	43
10-30	Mixed cultures	Glucose	12,500	26

[a] (Adapted from [26]).

[b] Temperature characteristic, in calories per mole, calculated from data presented in article.

increasing temperature for aerobic growth of heterotrophs. Similar results have been obtained with autotrophs [55,56]. For anaerobic cultures the effect appears to be the opposite, as two papers [15,57] have reported that K_s decreased as the temperature was increased.

Because both μ_m and K_s are functions of temperature, it is important that both be considered when attempts are made to predict the effects of temperature upon the efficiency of biochemical operations. A recent study has shown that much of the confusion in the literature has been caused by the failure of researchers to recognize that point [58].

The microbial decay coefficient, b, is also influenced by temperature. Table 10.9 is a summary of the values found for the Arrhenius temperature characteristic, u, with rapidly growing pure and mixed cultures. All values of u are positive, indicating that the rate of decay increases as the temperature increases. This is in agreement with the general concept of the specific decay constant as a measure of maintenance energy requirements and endogenous metabolism. For these fairly rapidly growing cultures the effect of temperature upon decay appears to be closely correlated to the effect upon growth. For each of the three cultures for which values of u were available for both μ_m and b the ratio of u for b to u for μ_m was 1.1.

TABLE 10.9
Typical Values of the Arrhenius Temperature
Characteristic for the Specific Decay Rate Coefficient[a]

Temperature Range, °C	Organism	Growth Media	u cal/mol	Reference
20-30	P. fluorescens	Glucose	18,200	46
8-30	P. fluorescens	Glucose	8,430	47
25-40	A. aerogenes	Glucose	9,000	43
10-40	Mixed culture	Glucose	13,900	26

[a] (Adapted from [26]).

In Section 10.5.3 it was stated that the magnitude of the microbial specific
decay coefficient, b, depends upon the rate at which the culture has been grown.
The rate of growth appears to affect the temperature characteristic as well. Two
reports [59,60] indicate that temperature has little effect upon b for slowly growing
cultures. In addition, the effect of temperature upon b could not be characterized
by the Arrhenrius equation. Thus care should be exercised in extrapolating data
from rapidly growing cultures to slowly growing ones.

No work has been reported on the effects of temperature upon the death rate
coefficient, γ.

10.6.2 pH

The pH of the medium has long been known to affect μ_m. Most cultures display a
region in which μ_m is relatively insensitive to pH, with a sharp decrease in μ_m for
changes in pH on either side of that region. The band of insensitivity is relatively
small, generally covering a range of 2 or 3 pH units. For most bacteria the region
is centered around pH 7, although some, particularly anaerobic acid-forming bacteria,
exhibit a lower value. Many fungi are also characterized by a low optimum pH.
Surprisingly, little quantitative information is available on the effects of pH upon
μ_m, and no simple models are available.

Very little work has been done on the effects of pH upon the saturation constant. One theoretical study developed equations for the effects of pH upon K_s
based upon the influence that pH has upon the transport of material across the cell
membrane [61]. Their equation predicted that K_s would be relatively insensitive to
pH over a range of 2 to 3 units, but then would increase rapidly as the pH was moved
out of that range.

When the effects of pH are considered, care must be taken to ensure that the
proper observations are made. For example, some organic compounds can be ionized,
but only the nonionized form is available to the cell as substrate. Changing the pH
of the medium would change the concentration of the available substrate even though
the total concentration of the compound hadn't changed. If experiments were run in
which μ was measured as a function of the concentration of compound in the medium,

changing the pH would have an apparent effect upon μ_m and K_s. On the other hand, if only the nonionized compound had been used as an indication of the substrate concentration, changing the pH might have no effect upon the parameters. Thus it can be seen that the effects of pH are more complicated than those of temperature.

No reports have been found in which the effects of pH on b and γ were measured. Nevertheless, we would expect the effects on b to be similar to those on μ_m. The effects on γ would probably be just the opposite, i.e., over some range around neutrality γ should be relatively independent of pH, whereas it should increase sharply outside of that range.

10.7 KEY POINTS

1. The effect of the substrate concentration, C_S, upon the rate of an enzymatically catalyzed reaction, v, can be expressed by the Michaelis-Menten equation:

$$v = \frac{V_m C_S}{K_m + C_S} \tag{10.13}$$

in which V_m is the maximum possible rate and K_m is the Michaelis-Menten constant. K_m is numerically equivalent to the substrate concentration at which v is equal to $V_m/2$.

2. The values of V_m and K_m may be estimated by plotting data depicting the effects of C_S upon v according to a linearized form of Eq. 10.13.

3. Reversible inhibitors of enzymatically catalyzed reactions may often be classified as competitive (effect is upon K_m), uncompetitive (effect is upon both V_m and K_m), or noncompetitive (effect is upon V_m). The substrate and the product of the reaction may also act as inhibitors.

4. Four events occur simultaneously in a microbial culture: growth, substrate utilization, decay, and death.

5. The rates of cell growth and substrate removal are both first order with respect to the concentration of viable cells in the culture. They are also proportional to each other, with the proportionality constant being the true growth yield.

6. The growth limiting substrate is that particular nutrient which is present in least supply in proportion to the amount needed for growth. Both the specific growth rate and the specific substrate removal rates are functions of the concentration of growth limiting substrate.

7. As the concentration of growth limiting substrate, S, is increased, the specific growth rate, μ, will increase until a maximum value, μ_m is attained. The relationship between μ and S can be expressed by the equation for a rectangular hyperbola:

$$\mu = \frac{\mu_m S}{K_s + S} \tag{10.42}$$

K_S is called the saturation constant and is defined as the substrate concentration at which μ is equal to $\mu_m/2$. Equation 10.42 is commonly called the Monod equation.

8. When K_S is much, much larger than S, Eq. 10.42 may be replaced with a first-order approximation:

$$\mu \cong \frac{\mu_m S}{K_S} \qquad\qquad (10.43)$$

9. The Monod equation is a reasonable model with which to describe the kinetics of microbial growth in wastewater treatment systems. Because of the heterogeneous nature of the culture and the substrate, however, μ_m and K_S should be characterized by ranges of values rather than single ones.

10. The growth limiting nutrient in a biochemical operation should be the one the operation is designed to remove. Consequently, if the objective of the operation is the removal of soluble organic matter, nitrogen, phosphorus, and oxygen should be supplied in excess of the stoichiometric amounts.

11. The values of the various kinetic parameters representing growth and substrate removal are influenced strongly by both the types of organisms and types of substrates in the system.

12. Although the exact causes are uncertain, it is known that some of the bacteria in a growing culture will lose the ability to use substrate and reproduce. Consequently the viability of the culture will be less than 100%.

13. For modeling purposes the rate of natural death in a microbial culture is taken to be first order with respect to the number or mass of viable cells in the culture.

14. Decay is a composite term which accounts for a reduction in yield caused by predation, cell lysis, and the need for maintenance energy.

15. For modeling purposes the rate of decay in a microbial culture is taken to be first order with respect to the masses of viable and dead cells present. For simplicity the specific decay rates for the two types of cells are assumed to be the same.

16. Because of the approximate nature of the equation for microbial decay, the numerical value of the specific decay rate, b, appears to be influenced by the rate at which the culture is grown.

17. Within a relatively narrow physiological range, the maximum specific growth rate constant, μ_m, increases as the temperature is increased, in a manner that can be described by the Arrhenius equation. For rapidly growing cultures the effect of temperature upon decay appears to be closely correlated to the effect upon growth. No conclusions can be drawn about the effects of temperature upon the saturation constant, K_S, or the specific death rate, γ.

18. pH can affect the various kinetic parameters describing microbial cultures. Generally there will be a band 2 to 3 pH units wide within which the values of the parameters are relatively insensitive to pH.

10.8 STUDY QUESTIONS

The study questions are arranged in the same order as the text.

Section	Questions
10.1	1-3
10.2	none
10.3	4-12
10.4	13
10.5	14
10.6	15-16

1. Initial rates for various substrate concentrations are given in Table SQ10.1 for an enzyme-catalyzed reaction. Estimate V_m and K_m by the three methods covered in Section 10.1.1.

TABLE SQ10.1
Effect of Substrate Concentration on Initial Velocity

c_S mol/liter	v mol/(liter·min)
1×10^{-3}	4.9×10^{-6}
1×10^{-4}	4.3×10^{-6}
5×10^{-5}	3.8×10^{-6}
1.5×10^{-5}	2.5×10^{-6}
1×10^{-5}	2.0×10^{-6}
5×10^{-6}	1.3×10^{-6}

2. A certain enzyme-catalyzed reaction is known to follow Michaelis-Menten kinetics with $V_m = 5.2 \times 10^{-5}$ mol/(liter·min) and $K_m = 1.3 \times 10^{-4}$ mol/liter. What would be the substrate concentration at the end of one hour if the reaction is carried out in a batch reactor with the initial substrate concentration of 2×10^{-4} mol/liter? What would be the expected substrate concentration at the end of one hour if the amount of enzyme added initially is doubled and everything else remains the same?

3. A certain compound is suspected of being an inhibitor of an enzyme-catalyzed reaction when present at a concentration of 1×10^{-4} mol/liter. The experimental data are given in Table SQ10.2. Determine the type of inhibition and evaluate the kinetic parameters.

TABLE SQ10.2
Effects of Inhibitor on Initial Velocity at Various Substrate Concentrations

c_S mol/liter	v mol/(liter·min)	
	With Inhibitor	Without Inhibitor
2×10^{-3}	12.5×10^{-6}	13.5×10^{-6}
2×10^{-4}	5.0×10^{-6}	7.5×10^{-6}
2×10^{-5}	7.1×10^{-7}	1.4×10^{-6}
1×10^{-5}	3.7×10^{-7}	7.1×10^{-7}
2×10^{-6}	7.5×10^{-8}	1.5×10^{-7}

4. What is meant by the terms specific growth rate and specific substrate removal rate?

5. Describe an experiment to show that the specific growth rate of bacteria is a function of the concentration of growth limiting substrate.

6. The specific growth rate of a bacterial culture can be characterized by the Monod equation. The culture is being grown in a reactor in which the specific growth rate can be held constant at a value of 0.15 hr^{-1}. Determine the substrate concentration for each of the following situations:

   (a) $\mu_m = 0.60$ hr^{-1}, $K_s = 200$ mg/liter

   (b) $\mu_m = 0.60$ hr^{-1}, $K_s = 50$ mg/liter

   (c) $\mu_m = 0.25$ hr^{-1}, $K_s = 200$ mg/liter

   (d) $\mu_m = 0.25$ hr^{-1}, $K_s = 50$ mg/liter

7. Using the information in the study question above, determine in which (if any) of the situations the use of the first-order approximation would be justified.

8. A culture with a true growth yield of 0.50 mg cells formed/mg substrate removed is growing with a specific growth rate of 0.10 hr^{-1}. Determine the specific rate of substrate removal in the culture.

9. The growth kinetics of a culture with a true growth yield of 0.45 mg cells formed/mg substrate removed can be characterized by the Monod equation with $\mu_m = 0.60$ hr^{-1} and $K_s = 150$ mg/l. Calculate the mean reaction rate coefficient, k_e.

10. Explain why μ_m and K_s are not likely to be constant for a wastewater treatment situation.

11. Explain why the growth limiting nutrient in an operation should be the one the operation was designed to remove.

12. Explain why the growth environment used in lab studies for the determination of kinetic parameters should be similar to that anticipated in the final treatment facility.

13. Define what is meant by "microbial death" and state evidence suggesting that it occurs in growing microbial systems.

14. Explain the concept of decay and how it is expressed mathematically.

15. Three techniques are often used to describe the effects of temperature upon microbial cultures. Describe each of them and tell how you would plot data to determine the values of the parameters in the equations.

16. The data in Table SQ10.3 were collected for the specific decay rate, b, as a function of temperature. Determine if each of the three temperature equations adequately fits the data, and if so, determine the temperature parameters.

TABLE SQ10.3
Effect of Temperature on the Specific Decay Rate

T °C	b hr^{-1}
10	0.0037
20	0.0095
30	0.0229
40	0.0372

REFERENCES AND FURTHER READING

1. J. Monod, "The growth of bacterial cultures," *Annual Review of Microbiology,* 3, 371-394, 1949.

2. A. White, *et al. Principles of Biochemistry,* Fourth Edition, McGraw-Hill Book Co., New York, 260, 1968.

3. V. C. R. Henri, "Lois ge'ne'rales de l'action des diastases," *Academy of Science, Paris,* 135, 916, 1902.

4. L. Michaelis and M. L. Menten, "Die Kinetik der Invertinwirkung," *Biochemische Zeitschrift,* 49, 333-369, 1913.

5. G. E. Briggs and J. B. S. Haldane, "A Note on the kinetics of enzyme action," *Biochemical Journal,* 19, 338-339, 1925.

6. C. S. Hanes, "Studies on plant amylases. I. The effect of starch concentration upon the velocity of hydrolysis by the amylase of germinated barley," *Biochemical Journal,* 26, 1406-1421, 1932.

7. J. E. Dowd and D. S. Riggs, "A comparison of estimates of Michaelis-Menten kinetic constants from various linear transformations," *Journal of Biological Chemistry,* 240, 863-869, 1965.

8. Z. Fencl, "Theoretical analysis of continuous culture systems," in *Theoretical and Methodological Basis of Continuous Culture of Microorganisms,* edited by I. Malek and Z. Fencl, Academic Press, New York, N.Y., 67-153, 1966.

9. A. F. Gaudy Jr. and E. T. Gaudy, "Biological concepts for design and operation of the activated sludge process," *Environmental Protection Agency Water Pollution Research Series, Report #17090 FQJ 09/71,* Sept. 1971.

10. H. Moser, "The dynamics of bacterial populations maintained in the chemostat," *Carnegie Institute of Washington Publication, No. 614,* 1958.

11. K. L. Schulze and R. S. Lipe, "Relationship between substrate concentration, growth rate, and respiration in *Escherichia coli* in continuous culture," *Archiv für Mikrobiologie,* 48, 1-20, 1964.

12. E. O. Powell, "The growth rate of microorganisms as a function of substrate concentration" in *Microbial Physiology and Continuous Culture,* edited by E. O. Powell, *et al.,* Her Majesty's Stationery Office, London, 34-55, 1967.

13. M. T. Garrett and C. N. Sawyer, "Kinetics of removal of soluble BOD by activated sludge," *Proceedings of the 7th Industrial Waste Conference*, Purdue University Engineering Extension Series No. 79, 51-77, 1952.

14. J. C. McLellan and A. W. Busch, "Hydraulic and process aspects of reactor design - I - Basic concepts in steady state analysis," *Proceedings of the 22nd Industrial Waste Conference*, Purdue University Engineering Extension Series No. 129, 537-552, 1967.

15. A. W. Lawrence and P. L. McCarty, "Unified basis for biological treatment design and operation," *Journal of the Sanitary Engineering Division, ASCE*, 96, 757-778, 1970.

16. N. van Uden, "Kinetics of nutrient-limited growth," *Annual Review of Microbiology*, 23, 473-486, 1969.

17. W. W. Eckenfelder, Jr., *Industrial Water Pollution Control*, McGraw-Hill Book Co., New York, N.Y., 1966.

18. D. W. Eckhoff and D. Jenkins, "Activated sludge systems, kinetics of the steady and transient states," Report No. 67-12 of the Sanitary Engineering Research Laboratory, University of California, Berkeley, 1967.

19. J. F. Andrews, "Kinetic models of biological waste treatment," *Biotechnology and Bioengineering Symposium No. 2*, 5-33, 1971.

20. S. Y. Chiu *et al.* "Kinetic model identification in mixed populations using continuous culture data," *Biotechnology and Bioengineering*, 14, 207-231, 1972.

21. S. Ghosh and F. G. Pohland, "Population dynamics in continuous cultures of heterogeneous microbial populations," *Developments in Industrial Microbiology*, 12, 295-311, 1971.

22. Chiu, S. Y. *et al.* "Kinetic behavior of mixed populations of activated sludge," *Biotechnology and Bioengineering*, 14, 179-199, 1972.

23. D. E. F. Harrison and S. J. Pirt, "The influence of dissolved oxygen concentration on the respiration and glucose metabolism of *Klebsiella aerogenes* during growth," *Journal of General Microbiology*, 46, 193-211, 1967.

24. K. C. Goel and A. F. Gaudy Jr., "Studies on the relationship between specific growth rate and concentration of nitrogen source for heterogeneous microbial population of sewage origin," *Biotechnology and Bioengineering*, 11, 67-78, 1969.

25. D. J. Schaezler, *et al.* "Kinetic and stoichiometric limitations of phosphate in pure and mixed bacterial cultures," *Proceedings of the 24th Industrial Waste Conference*, Purdue University Engineering Extension Series No. 135, 507-533, 1969.

26. R. E. Muck and C. P. L. Grady Jr., "Temperature effects on microbial growth in CSTR's," *Journal of the Environmental Engineering Division, ASCE*, 100, 1147-1163, 1974.

27. K. M. Peil and A. F. Gaudy Jr., "Kinetic constants for aerobic growth of microbial populations selected with various single compounds and with municipal wastes as substrates," *Applied Microbiology*, 21, 253-256, 1971.

28. W. L. Jorden, *et al.* "Evaluating treatability of selected industrial wastes," *Proceedings of the 26th Industrial Waste Conference,* Purdue University Engineering Extension Series No. 140, 514-529, 1971.

29. M. D. Mynhier and C. P. L. Grady Jr. "Design graphs for activated sludge process," *Journal of the Environmental Engineering Division, ASCE,* 101, 829-846, 1975.

30. J. R. Postgate, "Viability measurements and the survival of microbes under minimum stress," *Advances in Microbial Physiology,* 1, 1-23, 1967.

31. J. R. Postgate, "The viability of very slow-growing populations: A model for the natural ecosystem," *Bulletin from the Ecological Research Committee,* 17, 287-292, 1973.

32. W. R. Wooldrige and A. F. B. Standfast, "The biochemical oxygen demand of sewage," *Biochemical Journal,* 27, 183, 1933.

33. C. G. Sinclair and H. H. Topiwala, "Model for continuous culture which considers the viability concept," *Biotechnology and Bioengineering,* 12, 1069-1079, 1970.

34. C. L. Weddle and D. Jenkins, "The viability and activity of activated sludge," *Water Research,* 5, 621-640, 1971.

35. I. Walker and M. Davies, "The relationship between viability and respiration rate in the activated sludge process," *Water Research,* 11, 575-578, 1977.

36. J. R. Postgate and J. R. Hunter, "The survival of starved bacteria," *Journal of General Microbiology,* 29, 233-263, 1962.

37. D. W. Tempest, *et al.* "Studies on the growth of *Aerobacter aerogenes* at low dilution rates in a chemostat," in *Microbial Physiology and Continuous Culture,* edited by E. O. Powell *et al.,* Her Majesty's Stationery Office, London, 240-253, 1967.

38. R. E. McKinney, "Mathematics of complete mixing activated sludge," *Journal of the Sanitary Engineering Division, ASCE,* 88, No. SA3, 87-113, 1962.

39. B. L. Goodman and A. J. Englande, Jr., "A unified model of the activated sludge process," *Journal of the Water Pollution Control Federation,* 46, 312-332, 1974.

40. D. R. Christensen and P. L. McCarty, "Multi-process biological treatment model," *Journal of the Water Pollution Control Federation,* 47, 2652-2664, 1975.

41. N. Westberg, "A study of the activated sludge process as a bacterial growth process," *Water Research,* 1, 795-804, 1967.

42. C. P. L. Grady Jr. and R. E. Roper Jr., "A model for the bio-oxidation process which incorporates the viability concept," *Water Research,* 8, 471-483, 1974.

43. H. H. Topiwala and C. G. Sinclair, "Temperature relationship in continuous culture," *Biotechnology and Bioengineering,* 13, 795-813, 1971.

44. S. J. Pirt, "The maintenance energy of bacteria in growing cultures," *Proceedings of the Royal Society, (London), Series B,* 163, 224-231, 1965.

45. A. G. Marr, *et al.* "The maintenance requirement of *Escherichia coli,*" *Annals of the New York Academy of Science,* 102, 536-548, 1963.

46. R. H. Mennett and T. O. M. Nakayama, "Influence of temperature on substrate and energy conversion in *Pseudomonas fluorescens*," *Applied Microbiology*, 22, 772-776, 1971.

47. S. A. Palumbo and L. D. Witter, "Influence of temperature on glucose utilization by *Pseudomonas fluorescens*," *Applied Microbiology*, 18, 137-141, 1969.

48. J. H. Sherrard, "Control of cell yield and growth rate in the completely mixed activated sludge process," Thesis presented to the University of California, at Davis, Calif., in 1971, in partial fulfillment of the requirements for the degree of Doctor of Philosophy.

49. J. L. Barnard, "Discussion of the paper - 'A consolidated approach to activated sludge process design,'" *Progress in Water Technology*, 7, 73-90, 1975.

50. S. Arrhenius, "Über die Reaktionsgeschwindigkeit bei der Inversion von Rohrzucker durch Sauren," *Zeitschrift für Physikalische Chemic*, 4, 226-248, 1889.

51. E. B. Phelps, *Stream Sanitation*, John Wiley and Sons, Inc., New York, N.Y., 71-75, 1944.

52. J. T. Novak, "Temperature-substrate interactions in biological treatment," *Journal of the Water Pollution Control Federation*, 46, 1984-1994, 1974.

53. A. H. Benedict and D. A. Carlson, "Temperature acclimation in aerobic biooxidation systems," *Journal of the Water Pollution Control Federation*, 45, 10-24, 1973.

54. R. C. Jones and J. S. Hough, "The effect of temperature on the metabolism of baker's yeast growing in continuous culture," *Journal of General Microbiology*, 60, 107-116, 1970.

55. G. Knowles, *et al.* "Determination of kinetic constants for nitrifying bacteria in mixed culture, with the aid of an electronic computer," *Journal of General Microbiology*, 38, 263-278, 1965.

56. F. E. Stratton and P. L. McCarty, "Prediction of nitrification effects on the dissolved oxygen balance of streams," *Environmental Science and Technology*, 1, 405-410, 1967.

57. A. W. Lawrence, "Application of process kinetics to design of anaerobic processes," *Advances in Chemistry Series, No. 105*, 163-189, 1971.

58. A. H. Benedict and D. A. Carlson, "Rational assessment of the Streeter-Phelps temperature coefficient," *Journal of the Water Pollution Control Federation*, 46, 1792-1799, 1974.

59. C. W. Randall, "Discussion of 'Temperature effects on microbial growth in CSTR's,'" *Journal of the Environmental Engineering Division, ASCE*, 101, 458-459, 1975.

60. C. W. Randall, *et al.* "Temperature effects on aerobic digestion kinetics," *Journal of the Environmental Engineering Division, ASCE*, 101, 795-811, 1975.

61. L. A. Muzychenko, *et al.* "The influence of environmental factors on the kinetics of a biosynthetic process," *Pure and Applied Chemistry*, 36, 339-355, 1973.

62. H. B. Tench and A. V. Morton, "The application of enzyme kinetics to activated sludge research," *Journal of the Institute of Sewage Purification,* 478-486, 1962.

63. W. E. Gates and J. T. Marlar, "Graphical analysis of batch culture data using the Monod expressions," *Journal of the Water Pollution Control Federation,* 40, R469-R476, 1968.

64. A. L. Gram, "Reaction kinetics of aerobic biological processes," *S.E.R.L. Series 90, Report No. 2,* Sanitary Engineering Research Laboratory, University of California, Berkeley, May, 1956.

65. W. F. Milbury, "A development and evaluation of a theoretical model describing the effects of hydraulic regime in continuous microbial systems," Thesis presented to Northwestern University, Evanston, Ill, in 1964 in partial fulfillment of the requirements for the degree of Doctor of Philosophy.

66. P. Benedek and I. Horvath, "A practical approach to activated sludge kinetics," *Water Research,* 1, 663-682, 1967.

67. W. W. Eckenfelder, Commentary on 'A consolidated approach to activated sludge process design,'" *Progress in Water Technology,* 7, 35-40, 1975.

68. C. E. Adams, "Treatment of a high strength phenolic and ammonia wastestream by a single and multi-stage activated sludge process," *Proceedings of the 29th Industrial Waste Conference,* Purdue University Engineering Extension Series No. 145, 617-630, 1974.

69. R. A. Kormanik, "Design of two-stage aerated lagoons," *Journal of the Water Pollution Control Federation,* 44, 451-458, 1972.

70. J. L. Mahloch *et al.* "Treatability studies and design considerations for a dyeing operation," *Proceedings of the 29th Industrial Waste Conference,* Purdue University Engineering Extension Series No. 145, 44-50, 1974.

71. W. C. Kreye, *et al.* "Kinetic parameters and operating problems in the biological oxidation of high thiosulfate industrial wastewaters," *Proceedings of the 29th Industrial Waste Conference,* Purdue University Engineering Extension Series No. 145, 410-419, 1974.

72. A. L. Downing, *et al.* "Nitrification in the activated sludge process," *Journal of the Institute of Sewage Purification,* 130-158, 1964.

73. V. T. Stack and R. A. Conway, "Design data for completely mixed activated sludge treatment," *Sewage and Industrial Wastes,* 31, 1181-1190, 1959.

74. G. E. Gray, *et al.* "Biological treatment of vegetable processing wastes," *Proceedings of the 28th Industrial Waste Conference,* Purdue University Engineering Extension Series No. 142, 548-557, 1973.

75. C. R. Horn and F. G. Pohland, "Characterization and treatability of selected shellfish processing wastes," *Proceedings of the 28th Industrial Waste Conference,* Purdue University Engineering Extension Series No. 142, 819-831, 1973.

76. W. R. Domey, "Design parameters and performance of biological systems for textile plant effluents," *Proceedings of the 28th Industrial Waste Conference,* Purdue University Engineering Extension Series No. 142, 438-446, 1973.

77. T. P. Quirk and J. Hellman, "Activated sludge and trickling filtration treatment of whey effluents," *Journal of the Water Pollution Control Federation,* 44, 2277-2293, 1972.

78. F. H. Johnson and I. Lewin, "The growth rate of *Escherichia coli* in relation to temperature, quinine, and coenzyme," *Journal of Cellular and Comparative Physiology,* 28, 47-75, 1946.

79. A. D. Brown, "Some general properties of a phychrophilic pseudomonad: The effects of temperature on some of these properties and the utilization of glucose by this organism and *Pseudomonas aeruginosa,*" *Journal of General Microbiology,* 17, 640-648, 1957.

80. V. W. Greene and J. J. Jezeski, "Influence of temperature on the development of several psychrophilic bacteria of dairy origin," *Applied Microbiology,* 2, 110-117, 1954.

81. H. Ng, *et al.* "Damage and derepression in *Escherichia coli* resulting from growth at low temperatures," *Journal of Bacteriology,* 84, 331-339, 1962.

82. A. Borrow *et al.* "The effect of varied temperature on the kinetics of metabolism of *Gibberella fujikuroi* in stirred culture," *Canadian Journal of Microbiology,* 10, 445-466, 1964.

83. J. C. Senez, "Some considerations of the energetics of bacterial growth," *Bacteriological Reviews,* 26, 95-107, 1962.

84. D. Y. Ryu and R. I. Mateles, "Transient response of continuous cultures to changes in temperature," *Biotechnology and Bioengineering,* 10, 385-397, 1968.

Further Reading

Aiba, S., *et al., Biochemical Engineering,* Second Edition, Academic Press, Inc., New York, 1973. See Chapter 4.

Bailey, J. E. and Ollis, D. F., *Biochemical Engineering Fundamentals,* McGraw-Hill Book Company, New York, 1977. See Chapters 3 and 7.

Busch, A. W., *Aerobic Biological Treatment of Waste Waters, Principles and Practice,* Oligodynamics Press, Houston, 1971.

Cleland, W. W., "Steady state kinetics" in *The Enzymes,* Third Edition, P. D. Boyer (ed.), Academic Press, New York, 1-65, 1970.

Dawson, P. S. S. (editor) *Microbial Growth,* Dowden, Hutchinson and Ross, Inc. Stroudsburg, PA., 1974. A compilation of benchmark papers.

Dixon, M. and Webb, E. C., *Enzymes,* Second Edition, Academic Press, Inc., New York, 1964. See Chapters 4 and 7.

Gaudy, A. F. Jr. and Gaudy, E. T., "Microbiology of waste waters," *Annual Review of Microbiology,* 20, 319-336, 1966.

Gaudy, A. F. Jr. and Gaudy, E. T., "Mixed microbial populations," *Advances in Biochemical Engineering,* 2, 97-143, 1972.

Laidler, K. J., *The Chemical Kinetics of Enzyme Action,* Second Edition, Oxford Press, New York, 1973.

Lamanna, C., *et al.*, *Basic Bacteriology - Its Biological and Chemical Background,* Fourth Edition, The Williams and Wilkins Company, Baltimore, Md., 1973. See Chapters 7, 8 and 9.

Malek, I. and Fencl, Z. (editors), *Theoretical and Methodological Basis of Continuous Culture of Microorganisms,* Academic Press, New York, 1966.

Plowman, K., *Enzyme Kinetics,* McGraw-Hill, Inc., New York, 1972.

Tyteca, D., *et al.*, "Mathematical modeling and economic optimization of wastewater treatment plants," *CRC Critical Reviews in Environmental Control,* 8, 1-89, 1977.

White, A., *et al.*, *Principles of Biochemistry,* Fourth Edition, McGraw-Hill Book Co., New York, 1968. See Chapters 10, 11 and 12.

CHAPTER 11

MEASUREMENT OF PURIFICATION

In the previous chapter, rate equations were developed for the major events occurring in biochemical operations. The rates of microbial growth and substrate utilization were both related to S, the concentration of growth limiting nutrient in the system. As pointed out earlier, the material which acts as the growth limiting nutrient must be that which the operation is designed to remove. The question naturally arises, therefore, as to how the concentration of that material is to be measured. This question must be answered before the equations of Chapter 10 can be applied to the various operations described in Chapter 1.

11.1 MEASUREMENT OF INSOLUBLE MATERIAL

The efficiencies of unit operations designed to remove insoluble materials are easily assessed by standard physical measurements, such as the concentration of suspended solids. Various other tests allow differentiation between organic and inorganic solids as well as determination of the size distributions or the chemical composition of inorganic materials. The availability of such a large number of

351

tests makes it possible to measure removal efficiencies of specific components as
well as of total suspended solids.

11.2 MEASUREMENT OF SOLUBLE MATERIAL

With the exception of deionization, most unit operations involving soluble inorganic
materials are designed to remove specific substances (or at most, specific classes
of substances) and their efficiencies can be determined by measuring the concentrations of those specific materials. For example, the progress of nitrification
can be followed by measuring the loss of ammonia nitrogen and the increase in
nitrate nitrogen.

     Things are less well defined where the removal of soluble organics is concerned.
The problem arises from the fact that wastewaters contain hundreds of different
organic compounds, each in very low concentration, so that the use of specific tests
is not practical. Occasionally, an operation will be performed to remove a particular compound, such as phenol or cyanide, and in that case a specific test can be
used, although this is the exception, rather than the rule.

     An alternative to measuring the concentrations of individual species is the
determination of the concentrations of classes of compounds, as was suggested in
Section 9.4.2. For example, if an industrial waste consists primarily of carbohydrates the anthrone test could be used to measure their concentration [1]. Numerous
tests are available which could be used for determinations such as proteins, lipids,
alcohols, etc. [1]. The results could then be used to develop stoichiometric
equations for microbial growth.

     A test which is less specific, but which is finding increasing application, is
the measurement of total organic carbon (TOC) [2]. The proponents of this test consider one of its advantages to be that it measures the element which causes many of
the problems in lakes and streams - organic carbon. Critics, on the other hand,
point out that it does not differentiate between compounds which are subject to
biodegradation and those which are not. This is an important issue with respect to
biochemical operations because obviously such operations can only act upon compounds
which are biodegradable. Thus in order for TOC to be effective as a measure of the
performance of biochemical operations there must be some way of determining the
biodegradable TOC. Such a technique exists, and results in a measure of the total
biologically available carbon [3]. Nevertheless, the test still has not found wide
spread use because of another serious deficiency. It tells nothing about the oxidation state of the organic compounds. As seen in Chapter 9 the energy available in
an organic compound is proportional to its oxidation state. A highly reduced compound will provide more energy and require more terminal hydrogen acceptor than will
an equal mass of a highly oxidized compound. Consequently, a highly reduced compound
has a higher pollutional potential than does a highly oxidized one.

The oxidation state of a compound can be expressed in terms of the electrons available for transfer. As discussed in Section 9.3.2 the number of electrons can be translated directly into the amount of oxygen required to oxidize the compound to carbon dioxide and water, with 8 grams of oxygen being required per electron. This quantity of oxygen is referred to as oxygen demand and is an historical measure of the pollutional strength of a wastewater. Since most accepted measures of the concentrations of soluble organic compounds utilize the concept of oxygen demand we will look at it in more detail.

11.3 THE CONCEPT OF OXYGEN DEMAND

With the exception of the criterion of clearness most public images of water quality in a lake or stream relate to the amount of oxygen present. Fish and all other beneficial aquatic organisms require an aerobic environment. An anaerobic environment, on the other hand, favors the growth of organisms which produce the noxious gases associated with pollution. The effects of dissolved oxygen (DO) upon the quality of the water resource are so profound that most of the natural and technological uses to which water is put are seriously impaired if that oxygen supply is depleted. The prime reason for the depletion of DO is its utilization in aerobic metabolism by microorganisms feeding upon organic matter introduced into streams by wastewater discharges. The problem is aggravated by the facultative nature of many of those microorganisms, allowing them to continue to function even at negligible DO levels, thereby killing the higher organisms which are obligately aerobic. Consequently, as the Gaudys [4] have stated: "A conceptual principle has been established which recommends that the 'pollutional load' should be assessed, not by measuring its amount (i.e., amount of metabolizable organic matter in the waste) but by estimating the magnitude of its effect (i.e., the amount of oxygen that will be used because of the presence of the organic matter). Such a principle is defensible in an engineering sense, because it goes directly to the heart of the matter, potential depletion of the DO resource in the receiving stream, and because it measures a colligative effect of that organic matter in a wastewater which is readily available as organic carbon source to microorganisms without requiring determination of either the total amount or type of that organic matter."

11.3.1 Strength of the Concept

The strength of the concept of oxygen demand is in its dual character: it seeks to measure the amount of pollutant present in terms of the oxygen demand that would be exerted by the biodegradable material in it. Although the focus is upon biodegradable materials, because they are the only ones influenced by biochemical operations and are the only ones that have an effect upon the oxygen inventory of the receiving stream, one should not imply that biologically refractory materials are not important (consider the case of PCB's). It is the engineer's responsibility to

assess the total organic content of a wastewater, as well as the biodegradable
content, and then decide whether a biochemical operation is suited to the job. The
second important point of the concept is that it measures directly the amount of
oxygen required and thereby automatically compensates for the oxidation states of
the compounds involved. The number of electrons available can only be determined if
the chemical formula of the compound is known, a situation which is rare in wastewater treatment. The oxygen demand of a wastewater, on the other hand, may be determined quite easily using the COD test. Because that test does not differentiate between biodegradable and refractory organics, however, its use must be coupled with
some means of making that distinction. How this is done will be the subject of
Section 11.4.

11.3.2 Use of the Concept

The concept of oxygen demand is valid; the difficulty in its application arises from
the multiplicity of uses to which it is put; for example, measurement of operational
efficiency, plant design, and assessment of the assimilative capacity of a stream.

Because biochemical operations are only effective against biologically degradable organic compounds the only logical criterion against which to judge their performance is the removal of those compounds. Thus, any test used in the assessment
of operational efficiency must measure as closely as possible the total quantity of
biodegradable organic matter present. If an oxygen demand test is used it must be
based upon the amount of oxygen required to oxidize all of the biodegradable organic
matter completely to carbon dioxide and water.

The designs of many biochemical operations are governed primarily by the concentrations of soluble biodegradable organic matter. The concept of oxygen demand may
be used to measure that concentration, as long as the total quantity of material is
measured. The rates of removal may also be determined in terms of the time required
to reduce the oxygen demand of the material as long as total measurements are used.
In other words, the oxygen demand may be used in place of the actual quantity of
material as long as the measurement is as complete as possible so that the oxygen
demand is truly a measure of the amount of material.

An important aspect of the design of aerobic biochemical operations is the
amount of oxygen required. In Section 9.4.3 the use of a stoichiometric equation for
the determination of the amount of oxygen needed was discussed. It will be recalled
that the stoichiometric equation was constructed from half reactions. This was
possible because all of the electrons available in the original substrate must end up
either in the cell material formed or attached to the terminal hydrogen acceptor.
This concept was stated in Eq. 9.9 in terms of an energy balance. When oxygen demand
is used as the measure of the electrons available (and thus of the energy content)
this balance leads to a very simple way of calculating the oxygen requirement. The
amount of oxygen required is just equal to the change in the oxygen demand of the
substrate minus the oxygen demand of the cell material formed. This balance must

hold at any end point chosen for writing the stoichiometric equation, but the advantageous thing is that it is not even necessary to write the stoichiometric equation since an oxygen demand test can be used directly.

The third use to which the concept of oxygen demand could be put is the determination of the capacity of a stream to assimilate a waste without depleting the oxygen resource. Just as with plant design, knowledge of both quantities of material and their rates of degradation are required. Again, it is logical to express the quantity of material discharged in terms of its total biodegradable oxygen demand.

11.4 TESTS FOR THE MEASUREMENT OF BIOLOGICALLY DEGRADABLE ORGANIC MATTER

Two tests are currently used to measure the concentration of biologically degradable organic matter in terms of an oxygen demand. The traditional test, termed the biochemical oxygen demand (BOD) test, has been around in various forms since 1911 [5]. The newer test, called the total biological oxygen demand (T_bOD) test was developed during the 1950's and 1960's [6,7].

11.4.1 The Biochemical Oxygen Demand Test

The BOD test seeks to measure the quantity of biologically degradable organic matter in a wastewater in terms of the amount of oxygen required by microorganisms to oxidize it to carbon dioxide and water. This is done by measuring the change in DO concentration in sealed bottles over a period of several days. The change in DO concentration is commonly called the exertion of BOD.

EXERTION OF BOD. The sequence of events occurring during the course of BOD exertion can be delineated most easily through a hypothetical experiment. Suppose a sample of soluble organic matter is added, along with inorganic nutrients and dissolved oxygen, to BOD bottles in sufficient quantity to give a COD of S_o. The bottles are then innoculated with a mixed culture of microorganisms, sealed, and incubated at 20°C. Periodically over several days bottles are opened and analyzed for DO concentration, soluble COD concentration (S), and the concentration of suspended solids due to microorganisms (X). The DO values are subtracted from the initial value to obtain the oxygen uptake, or BOD exerted (y). Figure 11.1 is typical of the results that would be obtained if y, S, and X were all plotted as functions of time.

Several significant observations can be made from Fig. 11.1. The most striking one is that removal of the soluble organic matter occurs rapidly and that the point of complete removal of the biodegradable material corresponds to the maximum concentration of biomass within the system. This point also corresponds to the pause or plateau in the oxygen uptake curve. The plateau is not always present, in which case oxygen uptake follows a curve similar to the dashed one above that region. In most cases where the plateau occurs, it serves as a marker for the termination of substrate removal [8], and its significance as such was first recognized by Busch [9].

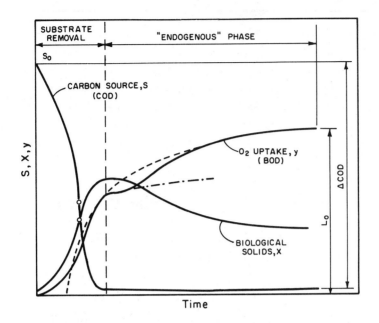

Figure 11.1 Generalized plot of substrate concentration,
             biological solids concentration, and oxygen
             utilization during exertion of biochemical
             oxygen demand. *(Adapted from Gaudy and Gaudy [4]).*

The plateau generally occurs within two days and thus it can be seen that the
majority of time spent in the BOD test is devoted to the destruction of the biomass
formed rather than to the removal of the original substrate.

   Because BOD bottles are seeded with heterogeneous microbial cultures, protozoa
and other predator organisms will be present. Studies on the role of protozoa
showed that the bacterial population reaches its peak at the point of soluble substrate removal [10,11]. The high bacterial population represents a food source for
the protozoa, which go through their own autocatalytic growth curve. Thus the declining biological solids curve is in reality a composite curve made up of the combined effects of bacterial loss and protozoan growth. The oxygen uptake responsible
for the dark line after the plateau is due to oxygen consumption by the protozoa
feeding upon the bacteria. If no protozoa were present the only cause of oxygen uptake after the plateau would be the endogenous metabolism of the bacteria. This is
presented by the dashed-dot curve. The distinctness of the plateau is due in part
to the relative numbers of bacteria and protozoa in the initial inoculum. If the
protozoan population is large the plateau will be indistinct or nonexistent.

   The preceding description has ignored the effects of autotrophic metabolism and
the oxygen demand associated with the conversion of ammonia nitrogen to nitrate because our primary concern is the determination of the oxygen demand of the organic

matter. The occurrence of nitrification can be prevented, and should be, when
measurements of the oxygen demand of biodegradable organic matter are required.

   CALCULATION OF BOD. Traditionally, the value used as the BOD of a wastewater
is the oxygen uptake after five days, and it is designated as BOD_5. From Fig. 11.1
it is apparent that BOD_5 is a relative value, dependent upon both the rate of oxygen
utilization and the total quantity of organic matter initially present. The BOD_5
value has no absolute meaning unless it is related to the total quantity of biodegradable organic matter because it is only one point on a curve which is approaching the oxygen demand of that material. To report the oxygen demand of the
biodegradable organic matter in a wastewater as BOD_5 is analogous to placing a
thermometer into a water bath and then reporting the temperature as the reading on
the thermometer at some arbitrary time before stability has occurred.

   The usual method of relating BOD_5 to the total oxygen demand of the biodegradable organic matter is to assume that the exertion of BOD follows a first-order
curve with the rate of oxygen utilization being proportional to the difference between the amount of oxygen used (y) and the ultimate BOD (L_o). This results in the
following equation:

$$y = L_o[1 - \exp(-K_1 t)] \tag{11.1}$$

where K_1 is a first-order rate constant. The value of y when t is equal to five
days is the BOD_5. The dashed curve in Fig. 11.1 follows Eq. 11.1 after correction
for a lag period. From Eq. 11.1 it can be seen that if K_1 is known, then the ultimate BOD may be calculated from the BOD_5. The utility of the BOD_5 test as a measure
of the total quantity of biodegradable organic matter present in a wastewater therefore depends upon two factors: (1) how closely the BOD curve can be approximated
by a first-order equation, and (2) how closely L_o approximates the total oxygen
demand of the biodegradable organic matter present.

   The validity of the assumption of a first-order rate curve depends upon the use
to which the BOD test is to be put. If the data are to be used to estimate the total
oxygen demand of the wastewater, the first-order approximation introduces little
error. If it is to be used to predict the course of BOD exertion in a stream, on
the other hand, the first-order curve can lead to serious errors.

   In order for the total oxygen demand of biodegradable organic matter to be
measured through the recording of oxygen uptake values in a biological growth system,
it would be necessary to measure oxygen uptake until all of the carbon in the original organic matter had been given off as carbon dioxide. Consider a carbon balance
on the system shown in Fig. 11.1. Initially, all of the system carbon is tied up in
either the substrate (S_o) or the cells (X_o). As growth occurs, part of the substrate is used for energy and the carbon is respired as carbon dioxide. The remainder is incorporated into new cell material. Consequently, at the plateau there
is little soluble carbon left; most of it has been converted into cell material

while the remainder has been respired. As oxygen utilization continues during the
"endogenous" phase, more carbon is respired as carbon dioxide as the bacteria
utilize endogenous reserves for maintenance energy and the protozoa use the bacteria
for food. Again, part of the carbon is incorporated into cell material (this time
protozoan) and the remainder is respired. Eventually protozoan growth will stop
and the system will enter a condition of true endogenous metabolism. At that time
the rate of evolution of the original carbon as carbon dioxide will be very slow
and may approach zero while an appreciable fraction of the carbon is still tied up
in cell material. While it is theoretically possible that all of the carbon could
eventually be given off as carbon dioxide, thereby returning the cell concentration
to X_o, the likelihood of this occurrence in a BOD bottle is remote. Consequently,
L_o may not be even a reasonable approximation of the total oxygen demand of the
biodegradable substrate.

   $UTILITY\ OF\ BOD$. To be useful as an indicator of the quantity of organic
matter present it is necessary for a test to measure the total quantity. BOD_5 is
not a measure of the total amount and thus its utility for design and operation of
biochemical operations is limited. In theory, the ultimate BOD (L_o) is the parameter that should be used, since it attempts to measure the total quantity. In
practice, however, it does not do this. Consequently, the BOD test, in spite of its
historical place in the field of environmental engineering, is not particularly
well suited to the design and operation of biochemical operations.

   In addition to its use as a measure of the quantity of organic matter, many
workers have attempted to use the BOD test to measure the rate at which biodegradation will occur. The rate of BOD exertion in a bottle bears no resemblance to
the rate of waste utilization in a treatment plant. The reaction environments are
totally different and BOD bottle rates should never be used for design purposes.
At most, rate data from the BOD test can give an engineer an estimate of the relative ease of degradation of several wastes.

   The use of BOD rate data for the estimation of stream assimilative capacity is
more acceptable, although it too suffers from weaknesses. Errors associated with
its use stem from two sources: (1) the representation of the oxygen uptake curve
as first order, and (2) differences between the reaction environments in a stream
and in the BOD bottle. The importance of both of these factors and some alternative
approaches to the problem have been discussed in a series of papers [12,13,14].

11.4.2 The Total Biological Oxygen Demand Test

The T_bOD test attempts to overcome the objections to the BOD test as a measure of
the quantity of organic matter present in a wastewater by making use of some of the
concepts illustrated by Fig. 11.1. In that figure it was seen that soluble substrate removal occurs fairly rapidly so that all soluble biodegradable organic
matter is removed from solution in a relatively short time period. The actual time
required is governed by the relative magnitudes of S_o and X_o. The larger X_o is

with respect to S_o, the more rapidly it will occur. Consequently, if S_o could be measured easily in terms of oxygen demand, and if the residual organic matter left after all biodegradable material has been removed could also be measured, then the difference between the two would be equal to the oxygen demand that was biodegradable, i.e., the T_bOD [4,15].

It will be recalled that the COD test offers a rapid and convenient means for measuring the oxygen demand of organic material in water. If S_o were measured by the COD test, the value would contain both biodegradable and nonbiodegradable material. However, if the concentration of residual organic matter were also measured by the COD test it would contain only nonbiodegradable material (provided a filtered sample was used so that it contained no microbial cells). Consequently, the difference between the two values would be the COD of only the biodegradable material. Since the COD test oxidizes most compounds to carbon dioxide and water, this change in COD is actually equal to the total oxygen demand of the soluble biodegradable organic matter, i.e., the T_bOD.

Hiser and Busch [7] developed this concept into a short-term (8 hour) measure of the T_bOD of a soluble waste. This was done through the use of a heterogeneous mass culture of acclimated organisms so that the substrate removal occurred within a short time period. Mullis and Schroeder [16] have extended the test to include insoluble organics as well. However, examination of their technique reveals that they assume 100% biodegradability of the insoluble organic matter. A more reasonable assumption would be that the ratio of biodegradable to nonbiodegradable organics in the solid phase is the same as that in the liquid phase, thereby allowing calculation of the insoluble T_bOD through comparison with the easily measured soluble T_bOD. Alternatively, it is possible to estimate the fraction of the COD due to insoluble material which is resistant to biodegradation. This can be done with a long term batch experiment and will be discussed in Section 12.4. No matter how the T_bOD is estimated, however, to avoid confusion it is always wise to specify whether a T_bOD value is for soluble, insoluble, or mixed materials.

This parameter meets all of the requirements for a suitable test that were set forth in the preceding sections of this chapter. In addition, its utility in the design and operation of biochemical operations has been demonstrated [3]. Consequently, henceforth in this book, all waste concentrations will be expressed as T_bOD. By so doing all of the uses of the oxygen demand concept discussed in Section 11.3.2 may be realized.

11.5 KEY POINTS

1. The measurement of the concentration of total organic carbon (TOC) can be used to assess the amount of soluble organic matter in a wastewater, but it will tell nothing about the pollutional potential of that organic matter.

2. The concept of oxygen demand may be used to measure the amount of pollutant in a wastewater in terms of the oxygen demand that would be exerted by the biodegradable material in it.

3. As long as the oxygen demand test measures the amount of oxygen required to oxidize all of the biodegradable organic matter completely to carbon dioxide and water, it may be used to assess the effectiveness of biochemical operations, to predict the amount of oxygen required in aerobic biochemical operations, and to estimate the assimilative capacity of streams.

4. Because it does not measure the total quantity of biodegradable organic matter present in a wastewater the BOD test is not particularly well suited to the design and operation of biochemical processes.

5. The total biological oxygen demand (T_bOD) test measures the total quantity of soluble biodegradable organic matter in a wastewater by determining the change in chemical oxygen demand (COD) during the course of biodegradation.

11.6 STUDY QUESTIONS

The study questions are arranged in the same order as the text.

Section	Questions
11.2	1
11.3	2,3
11.4	4-7

1. Explain why the TOC test tells nothing about the pollutional potential of an organic compound.

2. Explain why the dual character of oxygen demand makes it a valuable tool for measuring the presence of an organic pollutant in a water.

3. List and discuss the three major uses of an oxygen demand test.

4. Describe the "exertion of BOD" in a sealed environment, discussing all events that are known to occur.

5. Explain why the BOD test is not particularly well suited to the design and operation of biochemical processes.

6. Describe how to determine the T_bOD of a wastewater containing only soluble organic matter.

7. The initial soluble COD in a batch experiment was 365 mg/l and the residual soluble COD after growth had stopped was 85 mg/l. What is the T_bOD of the wastewater?

REFERENCES AND FURTHER READING

1. M. Ramanathan, et al., Selected Analytical Methods for Research in Water Pollution Control, Publication M-2 of the Center for Water Research in Engineering, Oklahoma State University, Stillwater, Oklahoma, 1968.

2. L. L. Hiser, "A new approach to controlling biological processes," Environmental Science and Technology, 4, 648-651, 1970.

3. A. W. Busch, *Aerobic Biological Treatment of Waste Waters, Principles and Practice,* Oligodynamics Press, Houston, Texas, 151-173, 1971.

4. A. F. Gaudy Jr. and E. T. Gaudy, "Biological conepts for design and operation of the activated sludge process," *Environmental Protection Agency Water Pollution Research Series,* Report No. 17090, FQJ, 09/71, 9-18, Sept, 1971.

5. W. J. O'Brien and J. W. Clark, "The historical development of the biochemical oxygen demand test," Bulletin No. 20, Engineering Experiment Station, New Mexico State University, Las Cruces, N.M., 1962.

6. J. M. Symons, *et al.,* "A procedure for determination of the biological treat-ability of industrial wastes," *Journal of the Water Pollution Control Federation,* 32, 841-852, 1960.

7. L. L. Hiser and A. W. Busch, "An 8-hour biological oxygen demand test using mass culture aeration and COD," *Journal of the Water Pollution Control Federation,* 36, 505-516, 1964.

8. E. D. Schroeder, "Importance of the BOD plateau," *Water Research,* 2, 803-809, 1968.

9. A. W. Busch, "BOD progression in soluble substrates," *Sewage and Industrial Wastes,* 30, 1336-1349, 1958.

10. M. N. Bhatla and A. F. Gaudy Jr., "Role of protozoa in the diphasic exertion of BOD," *Journal of the Sanitary Engineering Division, ASCE,* 91, #SA3, 63-87, 1965.

11. R. P. Canale and F. Y. Young, "Oxygen utilization in bacterial-protozoan community," *Journal of the Environmental Engineering Division, ASCE,* 100, 171-185, 1974.

12. W. P. Isaacs and A. F. Gaudy Jr., "Comparison of BOD exertion in a simulated stream and in standard BOD bottles," *Proceedings of the 22nd Industrial Waste Conference,* Purdue University Engineering Extension Series No. 129, 165-182, 1967.

13. E. M. Jennelle and A. F. Gaudy Jr., "Studies on the kinetics and mechanisms of BOD exertion in dilute systems," *Biotechnology and Bioengineering,* 12, 519-539, 1970.

14. K. M. Peil and A. F. Gaudy Jr., "A rational approach for predicting the dis-solved oxygen profile in receiving waters," *Biotechnology and Bioengineering,* 17, 69-84, 1975.

15. A. W. Busch, *et al.,* "Short-term total oxygen demand test," *Journal of the Water Pollution Control Federation,* 34, 354-362, 1962.

16. M. K. Mullis and E. D. Schroeder, "A rapid biochemical oxygen demand test suitable for operational control," *Journal of the Water Pollution Control Federation,* 43, 209-215, 1971.

Further Reading

American Public Health Association, *Standard Methods for the Examination of Water and Wastewater,* 14th Edition, APHA, New York, N.Y., 1975. A detailed labora-tory manual outlining all procedures used in the measurement of wastewaters.

Gaudy, A. F. Jr. "Biochemical oxygen demand," in *Water Pollution Microbiology,*
    edited by R. Mitchell, John Wiley and Sons, Inc., New York, N.Y., 305-332,
    1972. An excellent review of the BOD test and its uses.

PART IV

THEORY: MODELING OF IDEAL BIOCHEMICAL REACTORS

The primary function of a mathematical model is to reduce a complex system to the minimum terms essential for its description so that those terms may be manipulated, thereby helping us to understand how the system will respond under a variety of conditions. Generally, mathematical models do not describe a system completely, but if the terms employed are chosen with care, the model response will be qualitatively similar to the real system. In Part III we considered in detail the major events occurring during microbial growth; now the mathematical descriptions of those events will be combined with the concepts presented in Part II in order to develop models for microbial growth in a number of reactor configurations. Chapter 12 considers a simple CSTR, Chapter 13 a CSTR with cell recycle, while Chapter 14 covers fixed-film reactors. It will be assumed that the reactors are ideal, with respect to both fluid flow and the response of the microbial culture. In other words, we will investigate how the cultures would respond if the mathematical models were absolutely correct. In Part V any significant deviations from ideality will be discussed and incorporated into the application of the models to design.

CHAPTER 12

CONTINUOUS STIRRED TANK REACTORS

By returning to Chapter 1 and studying Fig. 1.1 and Table 1.2, it can be seen that
the single continuous stirred tank reactor (CSTR) is the simplest reactor configuration used in biochemical operations, finding application in aerated lagoons,
anaerobic contact, aerobic digestion, and anaerobic digestion. Multiple CSTR's in
series introduce another degree of complexity. Their use has largely been confined
to aerated lagoons although there has been much interest in their application to
anaerobic digestion. These two arrangements have also found considerable use in
microbiological research (where they are referred to as chemostats because they
provide a constant chemical environment) and in the fermentation industry. Consequently much of our knowledge about microbial growth has come from them.

Because of their simplicity, CSTR's provide an ideal system in which to study the modeling of microbial growth. In this chapter we will develop models describing the growth of microbial cultures in single and multiple CSTR's and use them to gain an understanding of how such systems behave. For simplicity, the models will be confined to aerobic heterotrophic organisms in an environment that contains ample nitrogen, phosphorus, and oxygen so that the soluble organic substrate (expressed as T_bOD) is the growth limiting material. With minor modifications, however, the same models can be applied to anaerobic heterotrophic growth and autotrophic growth. In Part V the models will be modified as needed to allow their use in the design of named biochemical operations.

12.1 SINGLE CSTR

A schematic diagram of a single CSTR is shown in Fig. 12.1. A reactor with volume V receives a flow at rate F containing soluble substrate at concentration S_o and suspended solids at concentration M_o. The suspended solids are made up of microorganisms at concentration X_o, biodegradable suspended solids at concentration Z_{bo}, and inert suspended solids at concentration Z_{io}:

$$M_o = X_o + Z_{bo} + Z_{io} \qquad\qquad (12.1)$$

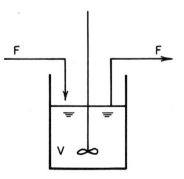

Influent	Reactor and Effluent
S_o - Soluble, biodegradable substrate, mg/l	S - Soluble, biodegradable substrate, mg/l
X_o - cells, mg/l	X - total cells, mg/l
Z_{io}- inert solids, mg/l	X_v- viable cells, mg/l
Z_{bo}- biodegradable solids, mg/l	X_d- dead cells, mg/l
Z_o -total (non-cell) solids, mg/l	Z_i -inert solids, mg/l
M_o - total suspended solids	M_o- total suspended solids

Figure 12.1. Schematic of a CSTR.

All inputs to the reactor are assumed to be constant, as are pH, temperature, and other environmental conditions. Within the reactor microorganisms utilize the substrate thereby growing to concentration X while reducing the concentration of soluble substrate to S. Because the reactor is completely mixed the concentrations of materials in the effluent are the same as those in the reactor.

12.1.1 Basic Model--Soluble Substrate with Monod Kinetics

In order to facilitate an understanding of how microorganisms behave when growing in a CSTR the simplest case should be considered first. The basic model upon which most research has been done considers only a soluble input, i.e., M_o equal to zero. This has the effect of making microorganisms the only solids in the reactor:

$$M = X \tag{12.2}$$

CONCENTRATION OF SOLUBLE SUBSTRATE. Equation 10.42 (the Monod model) stated that the specific growth rate of microorganisms is related to the concentration of growth limiting substrate in the medium surrounding them:

$$\mu = \frac{\mu_m S}{K_s + S} \tag{10.42}$$

This suggests that if it were possible to control the specific growth rate of a culture, the substrate concentration would also be controlled:

$$S = \frac{\mu K_s}{\mu_m - \mu} \tag{12.3}$$

A CSTR allows such control, as can be seen from a mass balance on viable organisms within the reactor.

$$0 - FX_V + r_{GX_V} V + r_{DX_V} V + r_{dX_V} V = 0 \tag{12.4}$$

The input term is zero because M_o is zero. Viable organisms are flowing out at concentration X_V, however, so there is an output term. There are three reaction terms: growth, death, and decay, the rates for which were given in Chapter 10.

$$r_{GX_V} = \mu X_V \tag{10.37}$$

$$r_{DX_V} = -\gamma X_V \tag{10.50}$$

$$r_{dX_V} = -bX_V \tag{10.52}$$

Death and decay cause a loss of organisms and thus have negative signs. Finally, because we are considering only steady-state models in this book, the accumulation term will be zero. Equation 12.4 states simply that the organisms must be growing as rapidly as they are being lost by washout, death, and decay. Recalling the definition of space time from Chapter 4:

$$\tau = V/F \tag{4.38}$$

substitution may be made into Eq. 12.4, which after simplification yields:

$$\mu = \frac{1}{\tau} + \gamma + b \tag{12.5}$$

Equation 12.5 suggests that since γ and b are fixed for a given microbial population, the growth rate may be controlled by manipulating τ. As pointed out in Chapter 4, this can be done by manipulating either the flow rate with a given size reactor, or the reactor size for a given flow rate.

The effect of τ on the concentration of soluble substrate in a reactor may be found by substituting Eq. 12.5 into Eq. 12.3:

$$S = \frac{K_s(1/\tau + \gamma + b)}{\mu_m - (1/\tau + \gamma + b)} \tag{12.6}$$

Equation 12.6 states that the concentration of soluble substrate in a CSTR is controlled solely by τ. Furthermore, because the influent substrate concentration does not appear in it, it implies that S is independent of S_o. This has been shown to be the case for a pure culture growing on a single substrate [1]. However, this is not the case when mixed cultures are employed and a general test such as T_bOD is used as a measure of the substrate concentration [1,2]. In that situation much of the effluent T_bOD is composed of metabolic intermediates with the effect that for a fixed τ, S is directly proportional to S_o [3]. For steady-state modeling, this dependence can be accounted for by considering K_s to be a function of S_o. That will not be done explicitly herein, however, although the implicit dependence should be kept in mind.

Examination of Eq. 12.5 reveals that as τ becomes very large (so that $1/\tau \to 0$), the specific growth rate of the organisms must equal the sum of the specific rates of death and decay. This means that in a single CSTR substrate must always be present to drive the required growth reaction. Consequently, there is a minimum substrate concentration that can be approached in a CSTR and it is given by the limit of Eq. 12.6 as τ approaches infinity:

$$S_{min} = \frac{K_s(\gamma + b)}{\mu_m - (\gamma + b)} \tag{12.7}$$

Equation 12.7 shows that S_{min} depends upon the kinetic parameters, and thereby upon both the nature of the substrate undergoing degradation and the type of culture degrading it. If a CSTR is being considered for the treatment of a waste to a desired effluent concentration, that concentration should be compared with S_{min}. If it is less than S_{min}, some other treatment operation must be used because a single CSTR cannot do the job.

The maximum rate at which bacteria can grow upon a given substrate depends upon the influent substrate concentration:

$$\hat{\mu} = \frac{\mu_m S_o}{K_s + S_o} \tag{12.8}$$

Consequently, the minimum space time, τ_{min}, for a given situation can be calculated by setting μ in Eq. 12.5 equal to $\hat{\mu}$ in Eq. 12.8 and rearranging:

$$\tau_{min} = \frac{K_s + S_o}{S_o(\mu_m - \gamma - b) - K_s(\gamma + b)} \tag{12.9}$$

The minimum space time is also called the point of washout because at shorter space times all organisms are washed out of the reactor and no growth will occur. At washout no organisms are produced and no substrate is utilized, i.e., the process has failed.

EXAMPLE 12.1.1-1

A culture of microorganisms is being grown in a CSTR with a volume of 8 liters. The medium contains lactose as the sole carbon source and all inorganic nutrients are present in excess. When the concentration of lactose is 1000 mg/liter as T_bOD the kinetic parameters have the values shown in Table E12.1.

TABLE E12.1
Kinetic Parameters for a Culture Growing on
Lactose at S_o = 1000 mg/liter T_bOD

Symbol	Value
μ_m	0.80 hr^{-1}
K_s	165.0 mg/liter
Y_g	0.55 mg cells/mg T_bOD
b	0.01 hr^{-1}
γ	0.005 hr^{-1}

(a) At what flow rate would washout of the culture occur?
Using Eq. 12.9 determine τ_{min}:

$$\tau_{min} = \frac{165 + 1000}{1000(0.80 - 0.01 - 0.005) - 165(0.01 + 0.005)}$$

$$\tau_{min} = 1.5 \text{ hrs.}$$

Consequently, washout will occur when

$$F = 8 \text{ liters}/1.5 \text{ hrs} = 5.33 \text{ liters/hr}$$

(b)   What will be the concentration of soluble T_bOD in the effluent when the flow
      is 5.5 liters/hr?

      A flow of 5.5 liters/hr will cause the reactor to wash out, consequently the
      concentration of T_bOD in the effluent will be the same as in the influent,
      namely 1000 mg/liter.

(c)   What will be the concentration of soluble T_bOD in the effluent when the flow
      is 0.125 liters/hr?

      τ = 8 liters/0.125 liters/hr = 64 hours

      Using Eq. 12.6 to determine S:

      $$S = \frac{165(1/64 + 0.005 + 0.01)}{0.80 - (1/64 + 0.005 + 0.01)}$$

      S = 6.6 mg/liter T_bOD

CONCENTRATION OF SUSPENDED SOLIDS.   For a soluble substrate the only suspended
solids in the reactor will be microorganisms, as given by Eq. 12.2.   Furthermore,
the microbial population will be composed of two components, viable cells, X_v, and
dead cells, X_d:

$$X = X_v + X_d \tag{12.10}$$

The concentration of viable organisms that will develop as a result of substrate
utilization can be determined from a mass balance on substrate:

$$FS_o - FS + r_sV = 0 \tag{12.11}$$

where $-r_s$ is the rate of substrate utilization as given by Eq. 10.39:

$$-r_s = (\mu/Y_g)X_v \tag{10.39}$$

Substitution of Eq. 10.39 into Eq. 12.11 and simplification yield:

$$X_v = \frac{Y_g(S_o - S)}{\mu\tau} \tag{12.12}$$

However, Eq. 12.5 may be substituted for μ, yielding

$$X_v = \frac{Y_g(S_o - S)}{1 + \gamma\tau + b\tau} \tag{12.13}$$

Equation 12.6 showed that the effluent substrate concentration depended upon τ.
Consequently, the viable cell concentration depends only upon the influent substrate
concentration, the space time of the reactor, and the kinetic parameters characteristic of the substrate.   In a CSTR receiving a fixed concentration of a given substrate the viable cell concentration therefore depends upon τ and is not a variable

that can be altered by the operator. This is an important concept that should be remembered.

The concentration of dead organisms must be obtained by a mass balance on them. The input is zero but dead cells are flowing out at concentration X_d. There are two reaction terms: generation and decay. Thus:

$$0 - FX_d + r_{GX_d} V + r_{dX_d} V = 0 \qquad (12.14)$$

Recalling the equations for the rate constants from Chapter 10:

$$r_{GX_d} = -r_{DX_v} = \gamma X_v \qquad (10.51)$$

$$-r_{dX_d} = bX_d \qquad (10.53)$$

substitution and simplification yield:

$$X_d = \frac{\gamma X_v \tau}{1 + b\tau} \qquad (12.15)$$

The total cell concentration can be obtained by substituting Eq. 12.13 and 12.15 into Eq. 12.10 and simplifying:

$$X = \frac{Y_g(S_o - S)}{1 + b\tau} \qquad (12.16)$$

Equation 12.16 indicates that the total cell concentration is independent of the specific death rate of the culture. This is a result of the assumption made in Eq. 10.54 that the decay rate of dead cells is the same as the decay rate of viable cells. There is some evidence that the assumption is not correct [4]. However, as discussed in Chapter 10 quantitative data relative to the values of the two decay rates are scarce. Equation 12.16 is widely used and will be sufficient for our purposes.

In Chapter 9 it was stated that the observed yield changed as the growth conditions in the reactor changed. The nature of this variation can be seen by rearrangement of Eq. 12.16. The observed yield was defined as the mass of cells formed per mass of COD removed, or in the symbols used here:

$$Y = \frac{X}{S_o - S} \qquad (12.17)$$

Dividing both sides of Eq. 12.16 by $(S_o - S)$ and substitution of Eq. 12.17, reveals that:

$$Y = \frac{Y_g}{1 + b\tau} \qquad (12.18)$$

Thus it can be seen that the observed yield, Y, depends upon the space time of the reactor, and that the larger τ is, the smaller the yield. This is the direct result of decay within the culture.

The viability of the culture may be determined through use of Eq. 10.49

$$v = \frac{X_v}{X_v + X_d} \qquad\qquad (10.49)$$

The effect of τ upon the viability may be seen by substituting Eq. 12.15 into Eq. 10.49 and simplifying:

$$v = \frac{1 + b\tau}{1 + b\tau + \gamma\tau} \qquad\qquad (12.19)$$

This states that large values of τ will result in low viabilities.

EXAMPLE 12.1.1-2

Continue with the problem begun in Example 12.1.1-1.

(a) What will be the cell concentration when the flow is 0.125 liters/hr?
     From part c of example 12.1.1-1

$\tau = 64$ hrs

$S = 6.6$ mg/liter T_bOD

Using Eq. 12.16 we can determine X:

$$X = \frac{(0.55)(1000 - 6.6)}{1 + (0.01)(64)}$$

$X = 333$ mg/liter

(b) What is the viability of the culture?
     Eq. 12.19 may be used:

$$v = \frac{1 + (0.01)(64)}{1 + (0.01)(64) + (0.005)(64)}$$

$v = 0.84 = 84\%$

OXYGEN REQUIREMENT. Equation 9.9 stated that the energy consumed by a microbial culture (i.e., removed from the medium) must equal the energy actually expended plus the energy incorporated into cell material. The energy in a compound is related to the electrons available for transfer, which can in turn be related to the amount of oxygen required to completely oxidize the compound, or the oxygen demand. Stated in terms of oxygen demand, therefore, Eq. 9.9 can be restated as "the amount of oxygen demand removed from the medium must equal the oxygen actually used by the culture plus the oxygen demand of the cells formed." This balance was the basis of the T_bOD test discussed in Chapter 11, and as discussed in Section 11.4.2, it can provide a means for computing the steady-state oxygen requirement of a microbial culture.

The steady-state oxygen requirement can be calculated by a mass balance on oxygen demand. The input term is just the flow times the influent substrate concentration, expressed as T_bOD. The output term contains two components: the unused soluble substrate and the cells leaving the reactor. The latter term may be expressed as the product of the cells leaving and the oxygen demand per unit mass of cells, β. The coefficient β depends upon the empirical formula of cell material, and if $C_5H_7O_2N$ is used, has a value of 1.41 mg O_2/mg of ash-free cells, or 1.20 mg O_2/mg dry weight (because bacteria have an ash content of around 15%). Alternatively β could be measured directly, as will be discussed in Section 12.4. The reaction term is just the rate of oxygen utilization, $-r_o$ [in mg/(liter·hr)] times the reactor volume and carries a negative sign because it removes oxygen demand. Assuming steady-state conditions, the balance is:

$$FS_o - FS - F\beta X + (-r_o V) = 0 \qquad (12.20)$$

Substitution of Eq. 12.16 for X and rearrangement gives the equation for the rate of oxygen utilization in mg/(liter·hr):

$$r_o = \frac{(S_o - S)(1 + b\tau - \beta Y_g)}{\tau(1 + b\tau)} \qquad (12.21)$$

Often the mass rate of oxygen utilization, mg/hr, is required in the design of a biochemical operation. Letting RO stand for the mass rate, and recognizing that

$$RO = r_o V \qquad (12.22)$$

Eq. 12.21 may be altered to read:

$$RO = \frac{F(S_o - S)(1 + b\tau - \beta Y_g)}{1 + b\tau} \qquad (12.23)$$

Because S is a function of τ it can be seen that for a fixed influent condition the only variable affecting the rate of oxygen utilization is τ, the space time of the reactor.

EXAMPLE 12.1.1-3

Continue with the problem begun in Example 12.1.1-1.

(a) What is the oxygen requirement of the reactor when τ = 64 hours, if β = 1.20?

From part c of Example 12.1.1-1, S = 6.6 mg/liter.

Using Eq. 12.23, the oxygen requirement is:

$$RO = \frac{(0.125)(1000 - 6.6)\ [1 + (0.01)(64) - (1.20)(0.55)]}{1 + (0.01)(64)}$$

RO = 74.2 mg/hr

(b)   What percentage of the influent oxygen demand is satisfied in the reactor?

Mass flow rate of oxygen demand in = FS_o

$$= (0.125)(1000)$$

$$= 125 \text{ mg/hr}$$

% oxygen demand satisfied = $(\frac{74.2}{125})100 = 59\%$

12.1.2 First-Order Approximation

As discussed in Section 10.3.2, occasions arise in which the steady-state substrate concentration is much less than the saturation constant so that the Monod equation may be simplified into a first-order equation:

$$\mu = \frac{\mu_m}{K_s} S \tag{10.43}$$

The use of Eq. 10.43 does not alter any of the mass balances nor does it alter the fact that μ is controlled by τ as expressed in Eq. 12.5. All that it does is to simplify the relationship between S, the effluent substrate concentration, and τ. Substitution of Eq. 10.43 into Eq. 12.5 and simplification give:

$$S = \frac{K_s}{\mu_m} (1/\tau + \gamma + b) \tag{12.24}$$

Many workers prefer to use the mean reaction rate coefficient, k_e, when applying the first-order approximation. Recalling Eqs. 10.48 and 10.46 it can be seen that:

$$k_e = (\frac{\mu_m}{K_s}) \frac{1}{Y_g} \tag{12.25}$$

Substitution of Eq. 12.25 into 12.24 gives the equation for the effluent substrate concentration in terms of k_e:

$$S = \frac{1/\tau + \gamma + b}{Y_g k_e} \tag{12.26}$$

Comparison of Eqs. 12.24 and 12.26 with Eq. 12.6 shows the effect that the first-order approximation has. It should be emphasized again that care must be exercised in the application of the first-order approximation to ensure that the basic assumption (i.e., $S \ll K_s$) is valid.

None of the other equations are affected by the first-order assumption, except through the effect on S.

EXAMPLE 12.1.2-1

Continue with the problem begun in Example 12.1.1-1.

(a) Calculate the mean reaction rate coefficient.

Use Eq. 12.25

$$k_e = \frac{(0.80)}{(165)(0.55)}$$

k_e = 0.0088 liter/mg·hr

(b) How much error would result from use of the first-order approximation to calculate the substrate concentration when τ = 64 hrs?

Use Eq. 12.26 to estimate S

$$S_{est} = \frac{(1/64) + 0.005 + 0.01}{(0.55)(0.0088)}$$

S_{est} = 6.3 mg/liter

From part c of Example 12.1.1-1, S is actually 6.6 mg/liter, thus

$$Error = (\frac{6.6 - 6.3}{6.6})100 = 4.5\%$$

Thus the first-order approximation would be satisfactory.

(c) What error would result if the first-order approximation was used when

τ = 10 hours?

Use Eq. 12.26 to estimate S

$$S_{est} = \frac{(1/10) + 0.005 + 0.01}{(0.55)(0.0088)}$$

S_{est} = 23.8 mg/liter

Use Eq. 12.6 to find the actual S

$$S = \frac{165 (1/10 + 0.005 + 0.01)}{0.80 - (1/10 + 0.005 + 0.01)}$$

S = 27.7 mg/liter

$$Error = (\frac{27.7 - 23.8}{27.7})100 = 14.1\%$$

In this case, the first-order approximation is not satisfactory.

12.1.3 Modifications of the Basic Model

The influents to most wastewater treatment plants contain suspended solids, therefore if the models are to reflect reality they must be capable of taking that fact
into account. As shown in Fig. 12.1 there are three types of suspended solids which
are of interest, because each will affect the reactor in a different way. These are
inert solids, microorganisms, and biodegradable solids. Traditionally, environmental
engineers have measured two forms of suspended solids, volatile and nonvolatile.
Unfortunately these subdivisions are not particularly useful for model development.
For example, McKinney [5] states that around 40 percent of the volatile suspended
solids in domestic sewage are nonbiodegradable (inert) in wastewater treatment
systems. It makes no difference whether inert solids are volatile or nonvolatile
because they both will affect the reactor in the same way. Conversely, it does make
a difference whether volatile solids are inert or biodegradable because each will
affect the reactor in different ways.

   INERT SOLIDS. Nothing happens to inert solids in a CSTR. Consequently there
is no reaction term in the material balance, which leads to the conclusion that the
concentration in the reactor will be equal to that in the influent:

$$Z_i = Z_{io} \tag{12.27}$$

The only effect this has upon the system is to increase the concentration of
suspended solids:

$$M = X + Z_i \tag{12.28}$$

In a situation like this, in which the suspended solids in the reactor contain more
than one component, M is often referred to as mixed liquor suspended solids (MLSS).

EXAMPLE 12.1.3-1

What would be the concentration of suspended solids in the CSTR of Example 12.1.1-2
if the influent contained 50 mg/liter of inert solids?

   Nothing will happen to the inert solids so the concentration in the reactor
will also be 50 mg/liter. The total concentration of suspended solids is given by
Eq. 12.28

   M = 333 + 50

   M = 383 mg/liter

   MICROBIAL SOLIDS. The effect of microbial solids in the influent can be seen by
a mass balance on viable organisms:

$$FX_{vo} - FX_v + r_{GX_v} V + r_{DX_v} V + r_{dX_v} V = 0 \tag{12.29}$$

where x_{vo} is the concentration in the influent.

Substitution of the appropriate rate equations into Eq. 12.29 and rearrangement yield:

$$\mu = (1 - \frac{X_{vo}}{X_v}) \frac{1}{\tau} + \gamma + b \tag{12.30}$$

Equation 12.30 shows that the viable organisms in the influent reduce the specific growth rate by an amount related to their contribution to the total viable biomass. This means that if two reactors have the same τ and the same influent substrate concentration, but one has viable cells in the influent, then it will produce an effluent with a lower substrate concentration.

The equation for the effluent substrate concentration may be found by substituting Eq. 12.30 into Eq. 12.3 in a manner analogous to that used with the basic model. The result however will contain X_v, which is unknown. Consequently we must first find the equation for X_v. This is done by a mass balance on substrate and leads to Eq. 12.12. When Eq. 12.30 is substituted into Eq. 12.12 the result is:

$$X_v = \frac{X_{vo} + Y_g(S_o - S)}{1 + b\tau + \gamma\tau} \tag{12.31}$$

When Eq. 12.31 is substituted into Eq. 12.30, and the resulting equation is substituted into Eq. 12.3, the final result is a quadratic equation for the substrate concentration:

$$[\mu_m - (1/\tau + \gamma + b)]S^2 - [\mu_m(X_{vo}/Y_g + S_o) + (K_s - S_o)(1/\tau + \gamma + b)]S$$

$$+ S_o K_s(1/\tau + \gamma + b) = 0 \tag{12.32}$$

The easiest way to use the equations is to calculate the substrate concentration first, and then use that result to calculate the viable cell concentration.

The concentration of dead cells can be obtained from a mass balance on them, considering their concentration in the influent, X_{do}. The resulting equation is:

$$X_d = \frac{X_{do} + \gamma X_v \tau}{1 + b\tau} \tag{12.33}$$

The equation for the total cell concentration may be obtained by substituting Eq. 12.31 and 12.33 into Eq. 12.10 and simplifying:

$$X = \frac{X_o + Y_g(S_o - S)}{1 + b\tau} \tag{12.34}$$

where

$$X_o = X_{vo} + X_{do} \tag{12.35}$$

Comparison of Eq. 12.34 with Eq. 12.16 shows that they differ only by the term X_o in the numerator. One assumption implicit in these equations is that once the cells enter the reactor they will become indistinguishable from cells arising from growth within the reactor and thus will undergo death and decay at the same specific rates.

The equation for the viability comes from the substitution of Eq. 12.31 and 12.34 into Eq. 10.49:

$$v = (\frac{1 + b\tau}{1 + b\tau + \gamma\tau}) [\frac{X_{vo} + Y_g(S_o - S)}{X_o + Y_g(S_o - S)}] \qquad (12.36)$$

The first term on the right hand side will be recognized as the viability in a CSTR receiving only soluble substrate. If the cells entering were 100% viable, then Eq. 12.36 would reduce to Eq. 12.17. Usually they will not be, however, so that the viability in the reactor will be less than it would be in a comparable CSTR not receiving biomass in the influent.

The oxygen requirement may be determined by a mass balance on oxygen demanding material, as was done in Eq. 12.18. The main difference is that the cells entering contribute an oxygen demand to the reactor:

$$FS_o + F\beta_o X_o - FS - F\beta X + (-r_o V) = 0 \qquad (12.37)$$

β_o is the oxygen demand coefficient of influent cells and may be the same as β. Substitution of Eq. 12.34 for X, and rearrangement yield:

$$RO = \frac{F\{(S_o - S)(1 + b\tau - \beta Y_g) + X_o[(\beta_o - \beta) + \beta_o b\tau]\}}{1 + b\tau} \qquad (12.38)$$

Inspection of Eq. 12.38 shows that it simplifies to Eq. 12.21 when X_o is equal to zero.

EXAMPLE 12.1.3-2

Continue with the problem begun in Example 12.1.1-1, only add 50 mg/liter of cells with a viability of 90% to the influent.

(a) What will be the concentration of soluble T_bOD in the effluent when the flow is 5.5 liters/hr?

$\tau = (8 \text{ liters})/(5.5 \text{ liters/hr}) = 1.45 \text{ hrs}$

$X_{vo} = (50)(0.90) = 45 \text{ mg/liter}$

Use Equation 12.32 to find S

$[0.80 - (1/1.45 + 0.005 + 0.01)]S^2 - [(0.80)(45/0.55 + 1000)$

$\qquad + (165 - 1000)(1/1.45 + 0.005 + 0.01)]S$

$\qquad + (1000)(165)(1/1.45 + 0.005 + 0.01) = 0$

$0.095\ S^2 - 277\ S + 116268 = 0$

Using the quadratic equation to find S, gives

$S = 508$ mg/liter

(b) What is the viable cell concentration?

Using Eq. 12.31,

$$X_v = \frac{45 + (0.55)(1000 - 508)}{1 + (0.01)(1.45) + (0.005)(1.45)}$$

$X_v = 309$ mg/liter

(c) What is the dead cell concentration?

From Eq. 12.33

$$X_d = \frac{5 + (0.005)(309)(1.45)}{1 + (0.01)(1.45)}$$

$X_d = 7.1$ mg/liter

(d) Assuming that $\beta_o = \beta = 1.20$, what percent of the influent oxygen demand is satisfied?

Influent oxygen demand $= (S_o + \beta_o X_o)F$

$$\qquad\qquad\qquad\qquad = [1000 + (1.20)(50)]5.5$$

$$\qquad\qquad\qquad\qquad = 5830 \text{ mg/hr}$$

Find the oxygen requirement from Eq. 12.38

$$RO = \frac{(5.5)\{(1000 - 508)[1 + (0.01)(1.45) - (1.20)(0.55)] + (50)(1.20)(0.01)(1.45)\}}{1 + (0.01)(1.45)}$$

$RO = 950$ mg/hr

Percent satisfied $= (\frac{950}{5830})\ 100 = 16.3\%$

(e) What is the specific growth rate of the organisms in the reactor?

Use Eq. 12.30

$\mu = (1 - 45/309)(1/1.45) + 0.005 + 0.01$

$\mu = 0.604\ hr^{-1}$

(f) What space time would give this same specific growth rate if $X_o = 0$?
     Use Eq. 12.5

     $0.604 = 1/\tau + 0.005 + 0.01$

     $\tau = 1.70$ hrs

On some occasions CSTR's are used to reduce the amount of biomass produced in another biochemical operation. In that case the concentration of cells in the influent, X_o, is large, but the concentration of soluble substrate, S_o, is very small. If S_o can be considered to be approximately equal to S (i.e., both near zero) then the equations for concentration of biomass in the CSTR, the viability of that biomass, and the oxygen consumed by it can all be simplified. The concentration of biomass will be:

$$X = \frac{X_o}{1 + b\tau} \tag{12.39}$$

The viability will be

$$v = \left(\frac{1 + b\tau}{1 + b\tau + \gamma\tau}\right)\left(\frac{X_{vo}}{X_o}\right) \tag{12.40}$$

The oxygen consumption rate will be:

$$RO = \frac{F\{X_o[(\beta_o - \beta) + \beta_o b\tau]\}}{1 + b\tau} \tag{12.41}$$

EXAMPLE 12.1.3-3

A CSTR is being used to degrade microbial cells. A stream containing cells at a concentration of 600 mg/liter is flowing at a rate of 1 liter/hr into a CSTR with a volume of 120 liters. The oxygen demand coefficient for the cells, β_o is 1.30 mg T_bOD/mg cells, and the decay rate, b, is 0.02 hr^{-1}.

(a) What is the cell concentration in the reactor?
     Use Eq. 12.39

     $$X = \frac{600}{1 + (0.02)(120)}$$

     $X = 176$ mg/liter

(b) What will be the oxygen consumption rate if the oxygen demand coefficient of the cells in the reactor, β, is 1.20?

Use Eq. 12.41

$$RO = \frac{(1.0)\{600[(1.30 - 1.20) + (1.30)(0.02)(120)]\}}{1 + (0.02)(120)}$$

RO = 568 mg/hr

BIODEGRADABLE SOLIDS. The soluble substrate model has been studied in considerable detail because it is amenable to investigation under well defined and controlled experimental conditions. Study of the kinetics of degradation of solid substrates is much more complex because of our inability to differentiate experimentally
between undegraded substrate and the microorganisms which have grown as a result of
biodegradation. Solid substrates must be rendered soluble before they can be ingested, thus bacteria must utilize enzymes on their surfaces or excreted into the
medium to carry out phase I catabolism (Section 8.4.1). As a result, the concentration of soluble materials will be greatest around the solid substrate particles
so that the bacteria will tend to grow either attached to the particles or in aggregates around them. This intimate association between reactant and product makes
verification of theoretical relationships very difficult. As a consequence no
generally accepted models are available, although the subject is under study [6].
Therefore, the usual approach is one of approximation.

Two situations arise in which solid substrates are introduced in CSTR's: when
they form the majority of the substrate so that the primary objective is their
destruction, and when their contribution is small in comparison to the soluble
substrate so that the primary objective is the destruction of the soluble material.

When the solid material constitutes the majority of the substrate, the usual
assumption is that destruction occurs in a first-order fashion with respect to the
concentration of biodegradable solid material [7,8]. No distinction is made between
the biodegradable substrate and the microbial culture. Under such an assumption the
equation describing the concentration of biodegradable solids (substrate + cells) is
the same as Eq. 12.39, in which X_o and X are taken to mean the concentrations of
biodegradable solids in the influent, M_{bo} and reactor, M_b, respectively, rather than
just the cell material. In this case the rate constant, b, is a lumped parameter
which incorporates the effects of all of the various events occurring in the reactor.
The rate of oxygen uptake will be given by Eq. 12.41, in which β_o and β represent
the amount of oxygen required to oxidize completely a unit mass of influent or
reactor solids, respectively. Although this model is admittedly crude it finds
considerable use in the field, as will be seen in Chapter 17.

In the second situation, in which the primary objective is the destruction of
soluble organic matter, the major role of the solid substrate is to increase the
effective substrate concentration entering the reactor. Several assumptions are

then made so that the equations derived for the soluble substrate model may be applied with only minor modifications. The first assumption is that the microbial
growth rate is controlled by the concentration of soluble organic matter present
within the reactor. The second assumption is that no solid substrate actually
exists within the reactor but that it acts to increase the cell concentration above
that which would be formed from the soluble substrate alone. This is equivalent to
assuming the existence of an imaginary transformation reactor preceding the CSTR
which converts all of the biodegradable solids to soluble substrate with conservation of T_bOD. These assumptions are required by the fact that any solid substrate
will become enmeshed in the biomass and will thus be analytically indistinguishable
from it. Although they are not stated in exactly the same terms, these assumptions
are commonly made in the modeling of biochemical operations [5,9].

The only difference between the equations derived earlier for the soluble substrate case, and the ones applicable to this case is in notation. Because the concentration of soluble substrate in the reactor is not affected by the presence of
the solid substrate, Eq. 12.6 is unchanged, as are Eqs. 12.7, 12.8, and 12.9. In
determining the minimum space time from Eq. 12.9 the concentration of soluble substrate in the influent, S_o, should be used because only the soluble substrate determines the specific growth rate of the organisms. The concentration of microorganisms
will be given by Eq. 12.16 but the symbol will be changed to X' to indicate the
"effective cell concentration," i.e., the combination of cells and undegraded solid
substrate. Likewise, the effective influent substrate concentration, S_o', must be
used, as well as the effective yield, Y_g'. The techniques for determining S_o' and Y_g'
will be discussed in Section 12.4. Finally, the oxygen requirement may be calculated from Eq. 12.21 using S_o' and Y_g'. The value of β, the cell oxygen demand coefficient, should be a measured rather than an assumed value as it is likely to differ
considerably from that of pure microbial cells. In fact, it is likely that β will
depend upon τ, thereby reflecting the changing contribution of undegraded solid
substrate to the effective cell concentration, X', as τ is changed. Relatively
little error will be introduced by this technique for reactors with long space times.
As the space time is reduced, however, the assumptions become weaker. Consequently
these equations should only be used at space times considerably larger than τ_{min}.

COMBINATION OF SOLIDS. The effects of inert solids may be combined with the
effects of either of the other types quite easily, since all that it affects is the
total suspended solids concentration. Whenever the influent contains microbial
solids and solid substrate, but little soluble substrate, Eqs. 12.39 and 12.41 may
be used with X_o and X representing the total of the two solids (i.e., M_{bo} and M_b)
rather than just cells. If the influent contains microbial solids, solid substrate,
and soluble substrate the situation will be more difficult because of the problems
associated with determination of X_o, the concentration of cells in the influent.

There is no currently accepted technique for handling that situation. Fortunately, most wastewaters can be grouped into one of the other categories with little risk of serious error.

12.1.4 Performance of CSTR as Predicted by Models

The major utility of the models as presented herein is to give us an appreciation for the way in which microbial CSTR's behave under a variety of circumstances. The only operational variable which can be manipulated in a CSTR is the space time, τ. Let us, therefore, first look at how τ affects the performance of a reactor. We will then turn our attention to how changes in the various parameters affect performance. The effects are most easily seen by examining graphs of the equations. Table 12.1 lists the base values of the parameters and variables used to obtain many of the graphs in this section.

EFFECTS OF τ. Figure 12.2 is a typical plot showing how the space time of a CSTR affects the concentrations of soluble substrate and total cells (dead plus viable) in it. In addition, the viability is shown which allows calculation of either the live or dead cell concentration from the total cell concentration. Looking first at the soluble substrate curve it can be seen that when τ is less than 2.4 hours, no growth occurs and no substrate is removed. That space time corresponds to τ_{min} as given by Eq. 12.9. For space times below τ_{min}, washout of the culture occurs so that the reactor serves no purpose. As τ is increased, growth becomes established and substrate removal occurs. For the kinetic parameters employed, soluble substrate removal occurs very rapidly, with reduction of the concentration from 500 to 50 mg/liter with a space time of only 6.3 hours. Further increases in

TABLE 12.1
Values of Kinetic Parameters and Variables Used
to Plot Figures 12.2, 12.3, and 12.4

Symbol	Value
μ_m	0.5 hr^{-1}
K_s	100 mg/liter T_bOD
Y_g	0.5 mg cells/mg T_bOD
b	0.004 hr^{-1}
γ	0.004 hr^{-1}
β	1.20 mg O_2/mg cells
F	1.0 liter/hr
S_o	500 mg/liter T_bOD
X_o	0
Z_o	0

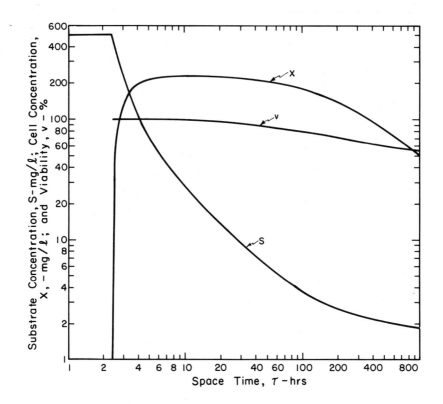

Figure 12.2.   Effect of space time on the performance of a single CSTR.
Parameters are listed in Table 12.1.

τ bring about smaller incremental changes in S, although they do make the reactor
more stable. For example, compare the differences in S resulting from a one hour
change in τ at space times of 4 and 40 hours. At short space times, the cell concentration curve reflects the changes in substrate concentration. Under this condition the cells are growing rapidly and the death and decay rates are small in
comparison to the growth rate so that the observed yield approaches the true growth
yield as shown by Eq. 12.18. This makes the cell concentration proportional to the
substrate removed during this period. As the space time is increased beyond 12
hours the relative importance of death and decay start to increase, with the result
that both the viability and the total cell concentration start to decline. Since
the viable cell concentration is the product of the viability and the total cell
concentration, it declines even more rapidly than the total cell concentration.
The decline in the total cell concentration at long space times reflects the decline
in the observed yield as indicated by Eq. 12.18, since substrate removal is essentially complete throughout the period.

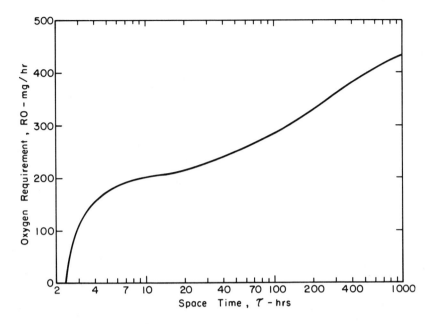

Figure 12.3. Effect of space time on the oxygen requirement of a CSTR.
Parameters are listed in Table 12.1.

One of the implications of Fig. 12.2 is that larger values of τ bring about
more complete destruction of the original organic substrate. At low space times
much of the carbon from the substrate is left in the cell material formed but as τ
is increased more and more of the carbon is released as carbon dioxide as the cells
undergo decay. This may be seen clearly in Fig. 12.3, which depicts the oxygen requirement as calculated with Eq. 12.23. The rate at which oxygen demand is entering
the CSTR is 500 mg/hr (S_oF), so that it is easy to see that as τ is increased more
of this demand is met. Comparing Figs. 12.2 and 12.3 it can be seen that when τ is
10 hours the effluent substrate concentration is 28 mg/liter but that only 196 mg/hr
of oxygen are used because the majority of the substrate removed has been converted
to new cells, which are present at a concentration of 230 mg/liter. Thus, if the
cells could be removed from the liquid, the mass flow rate of oxygen demand would be
reduced from 500 mg/hr to 28 mg/hr at the expense of only 196 mg/hr of oxygen. However, 230 mg/hr of cells would have to be disposed of. If τ was increased to 250
hours there would be only a slight increase in the soluble substrate removal, but
there would be a considerable change in the oxygen used and in the cells to be disposed of. In that case the soluble substrate concentration is 2.5 mg/liter, so that
the mass of substrate removed per hour has been increased from 472 mg/hr to 497.5, a
slight improvement. The oxygen requirement, on the other hand, is now 348 mg/hr,
almost double the previous amount, while the cells to be disposed of are only
125 mg/hr.

A CSTR can be said to accomplish two tasks, removal of soluble substrate and stabilization of a waste. The meaning of removal of soluble substrate is apparent. Stabilization refers to conversion of the carbon in the original waste to a stable end product such as carbon dioxide. It is sometimes also called mineralization. The removal of soluble substrate may be assessed easily by measuring the concentration of substrate remaining. Stabilization may be assessed by comparing the oxygen requirement to the oxygen demand entering the reactor. If the oxygen requirement is 40% of the influent oxygen demand, for example, then the efficiency of stabilization is 40%. The efficiencies of each of those tasks are shown in Fig. 12.4. There it can be seen that a CSTR can achieve a very high efficiency of substrate removal at relatively low space times. It is relatively poor as a stabilization device, however, requiring large values of τ to achieve stabilizations in excess of 50%. This emphasizes that biochemical operations are conversion processes. There is a rapid conversion of soluble substrate carbon to cell material, with only a small percentage of the oxygen demand of the substrate being met, i.e., the amount needed to supply the energy for synthesis. As τ is increased, cell decay becomes more important and is what converts the carbon which had been incorporated into the cell material into carbon dioxide, and therefore is what brings about stabilization. In practice, the efficiency of a process utilizing a CSTR can only approach the efficiency of soluble substrate removal if the cells can be removed from the liquid before discharge. Otherwise, because the cells in suspension represent oxygen demand, the efficiency of the process will be limited to the efficiency of stabilization.

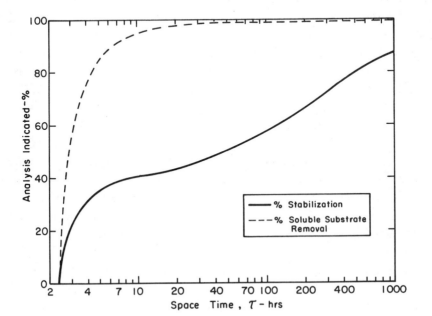

Figure 12.4. Effect of space time on percent stabilization and
percent substrate removal in a CSTR. Parameters
are listed in Table 12.1.

Now let us consider what happens when 50 mg/liter of microorganisms with a
viability of 80% are added to the influent stream. Plots of Eqs. 12.32, 12.34, and
12.36 are shown in Fig. 12.5, where the curves from Fig. 12.2 are reproduced to
allow comparison. The most striking effect of having cells in the influent is to
prevent washout. As long as cells are in the influent, there will be cells in the
reactor, no matter how small τ gets. Instead of causing washout in which the cell
concentration approaches zero, the effect of reducing τ is to make the cell concentration in the reactor approach that in the influent. Because there are always
cells in the reactor, there will always be substrate removal and consequently the
curve for S does not have the discontinuity characteristic of a CSTR receiving no
cells. Instead S approaches S_o asymptotically as the space time becomes small.
Under that condition the organisms are growing and removing substrate at a very
rapid specific rate but the time for reaction is not large enough to bring about a
large removal. The viability curve is also somewhat different in that the viability
reaches a maximum for intermediate values of τ, but then decreases on either side.
The decline in viability at long space times is the result of the increased importance of the death rate in proportion to the growth rate, just as it was for the

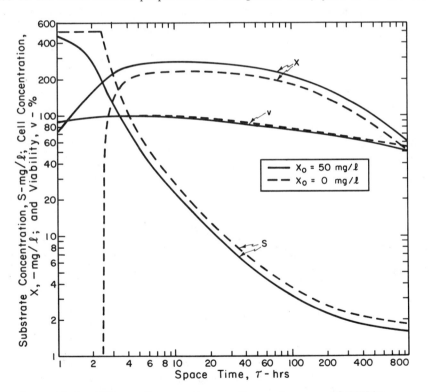

Figure 12.5. Effect of space time on the performance of CSTR's
with and without microbial cells in the influent.
Parameters are listed in Table 12.1.

totally soluble substrate case. At small values of τ, however, the decline in
viability is caused by the increased contribution of X_o to the total cell mass. As
the percentage of influent cells in the culture approaches 100 percent the viability
will approach that of X_o. In this case the viability of the influent cells was 80%,
so that the viability in the reactor approaches 80% as τ is decreased. Figure 12.6
compares the oxygen requirement for a CSTR with cells in the influent to one without
them. At low values of τ there is much more oxygen consumption in the reactor receiving cells because of the growth and substrate removal occurring. At intermediate values of τ there is only a small difference between the curves and this is
caused by the slightly greater substrate removal in the reactor receiving cells.
However, the curves diverge again as τ becomes large. This is because of the oxygen
demand of the cells in the influent. This demand is not exerted, however, until τ
becomes large because the mechanism for its exertion is decay, and as we saw in the
previous figures, decay does not become important until τ becomes large.

Another situation which was considered in Section 12.1.3 was the introduction
of biodegradable solids in the influent. Curves will not be presented for this case
because they are very similar to those in Figs. 12.2 and 12.3. The only significant
difference would be in the suspended solids curve. The actual cell concentration
would approach zero as τ approached τ_{min}, however the effective cell concentration,
X' would approach a concentration equal to that of the biodegradable solids in the
influent.

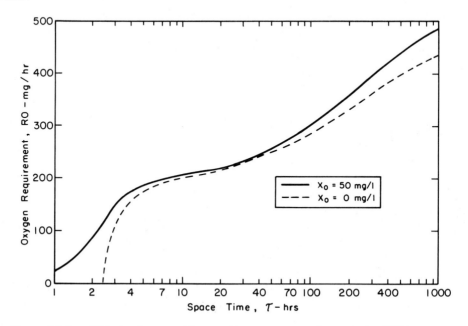

Figure 12.6. Effect of space time on the oxygen requirement in CSTR's
             with and without microbial cells in the influent.
             Parameters are listed in Table 12.1.

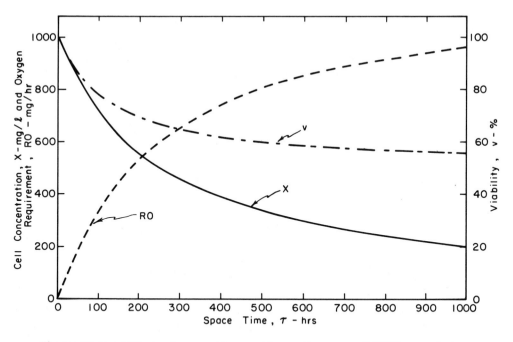

Figure 12.7.   Effect of space time on the performance of CSTR's receiving
only microbial cells. Parameters are listed in Table 12.2.

One other case that should be considered here is the situation when the influent organic matter consists almost wholly of microorganisms, with little or no
biodegradable soluble substrate present. Figure 12.7 presents plots of Eqs. 12.39,
12.40 and 12.41 for the situation listed in Table 12.2. It was assumed that β and
β_o, the oxygen demand coefficients of cells in the reactor and in the influent,
respectively, are equal. Examination of Fig. 12.7 shows that for the value of b

TABLE 12.2
Values of Kinetic Parameters and Variables Used
to Plot Figures 12.7, 12.14, 12.15 and 12.16

Symbol	Value
b	$0.004 \ hr^{-1}$
γ	$0.004 \ hr^{-1}$
β	$1.20 \ mg \ O_2/mg \ cells$
β_σ	$1.20 \ mg \ O_2/mg \ cells$
S_o	0.0
X_o	$1000 \ mg/liter$
Z_o	0
F	$1.0 \ liter/hr$
v	$100 \ \%$

used to generate the curves, CSTR's with short space times have little effect upon
the concentration of microorganisms. As τ is increased, however, the cell concentration decreases in a manner that is characteristic of first-order rate expressions.
Since all of the oxygen requirement comes from the decay of the cells, its curve is
just the reverse of the cell curve. The viability of the culture declines as τ is
increased, as would be anticipated from the previous examples.

EFFECTS OF KINETIC PARAMETERS. The space time of a CSTR is the only variable
under the control of the designer or operator; however, it is not the only factor
affecting the performance of such a reactor. Examination of the equation for the
performance of a CSTR shows that each of the kinetic parameters μ_m, K_s, Y_g, b and γ
will influence it as well. The primary effect of μ_m is on the substrate concentration. A higher value of μ_m allows the organisms to grow faster at a given substrate concentration, with the effect of giving a lower reactor substrate concentration for any given value of τ. The maximum specific growth rate constant also
exerts a strong influence on τ_{min}, so that organisms with high μ_m's can grow in
CSTR's with short space times. The effect of μ_m on the cell concentration is
strongest at short space times where the effect upon S is strongest. It has almost
no effect at large values of τ, however. The effects of changes in K_s are similar
to those of changes in μ_m, except that the magnitudes are reversed, e.g., decreasing
the value of K_s acts to decrease S and τ_{min}. In contrast to μ_m and K_s which affected primarily the substrate concentration, the decay rate constant, b, exerts
its main influence upon the cell concentration and oxygen requirement. A high decay
rate means that the operation will be more efficient at oxidizing the carbon in the
original substrate to carbon dioxide, consequently the cell concentration will be
low and the oxygen requirement high. The effect will be especially pronounced at
long space times. Changes in the true growth yield will also primarily affect the
cell concentration and the oxygen requirement. High yields will result in more
cells but the culture will require less oxygen because less of the substrate will
have been oxidized.

Seldom does a situation occur in which only one parameter changes. Usually all
will change. For example, as discussed in Chapter 9, temperature can affect all of
them and the response of the system will depend upon how the changes interact. An
example of this is shown in Fig. 12.8 which is a plot of the cell and substrate
concentrations for the situations listed in Table 12.3. The dashed curves represent
a low temperature situation and the dashed-dot curves a high one. The solid curves
are the same as in Fig. 12.2 and are presented for comparison. It can be seen that
an increase in temperature might give a greater efficiency of soluble substrate removal at any value of τ. It might produce more cells at low space times due to
greater substrate removal, whereas it might produce less at long space times because
of the increased importance of decay. The effects of a low temperature are just the
opposite. The curves in Fig. 12.8 serve to emphasize the point that one must learn

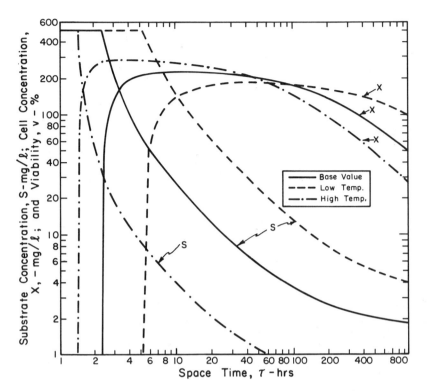

Figure 12.8.   Effect of changing all parameters on the performance
of a CSTR.   Parameters are listed in Table 12.3.

TABLE 12.3
Values of Kinetic Parameters and Variables
Used to Plot Figure 12.8

	Value[a]	
Symbol	Low Temp.	High Temp.
μ_m	0.30 hr^{-1}	0.70 hr^{-1}
K_s	300 mg/liter T_bOD	20 mg/liter T_bOD
Y_g	0.4 mg cells/mg T_bOD	0.6 mg cells/mg T_bOD
b	0.001 hr^{-1}	0.01 hr^{-1}
γ	0.002 hr^{-1}	0.008 hr^{-1}
β	1.20 mg O_2/mg cells	1.20 mg O_2/mg cells
F	1.0 liter/hr	1.0 liter/hr
S_o	500 mg/liter T_bOD	500 mg/liter T_bOD
X_o	0	0
Z_o	0	0

[a] Parameters for base case are same as Table 12.1.

to think in terms of the group of parameter values which characterize a particular substrate and culture rather than thinking of individual parameters. A low value of Y_g, for example, may not mean a low cell concentration if the culture also has a low decay rate, as evidenced by the low temperature curves.

12.2 CSTR's IN SERIES

Schematic diagrams of CSTR's arranged in series are shown in Fig. 12.9. Figure 12.9a illustrates a simple chain in which all flow enters the first reactor and proceeds sequentially through the others. The diagram shows all stages of equal size, but that is not necessary, nor in some cases desirable. Figure 12.9b illustrates a chain with multiple substrate addition points. It, too, can have reactors of various size within it, thus, it is apparent that there are a large number of possible combinations. Optimization techniques may be used to find the minimum total reactor volume required to achieve a given effluent quality from either type of chain.

12.2.1 Conceptual Problems in the Use of Simple Steady-State Models

The usual approach in modeling CSTR's in series is to apply the equations that were discussed in Section 12.1 and to assume constant kinetic parameters throughout the

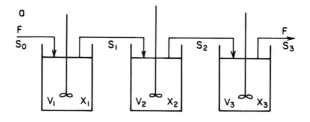

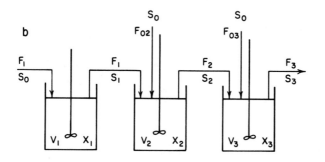

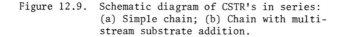

Figure 12.9.   Schematic diagram of CSTR's in series:
(a) Simple chain; (b) Chain with multi-stream substrate addition.

chain. It should be recognized, however, that certain conceptual problems arise when this is done. Models have been proposed which overcome some of these problems [10,11,12] but their level of sophistication is beyond the scope of this text. Nevertheless the relatively simple models of Section 12.1 have found considerable use in the theoretical literature [13,14] although their validity in multi-stage systems has not been verified experimentally to the extent that it has for single-stage systems.

The major difficulty associated with the models is that they were developed from data collected on microorganisms kept at steady state in a constant physical and chemical environment. When organisms flow through a chain of reactors, on the other hand, they are subjected to a different environment in each reactor, which can cause their physiological states to change, thereby changing their growth characteristics. In addition it can change the relative sizes of the different microbial populations within the community. Another factor is the possibility of sequential substrate removal. If the cells in the first reactor are growing very rapidly, metabolic controls may prevent degradation of some of the components of the substrate, whereas others may be almost completely removed. The net effect could be to make the composition of the substrate entering the second reactor quite different from that entering the first. The steady-state models also assumed a constant specific rate for death and decay. It is entirely possible, however, that the rate of decay could increase as substrate is completely depleted in the latter stages of a multi-stage system. Furthermore, it is logical to expect the rate of death to increase as the cells become depleted of reserve materials due to decay. All of these factors could have the effect of changing the values of the kinetic parameters as the flow proceeds downstream.

Most of these deficiencies are more important for a large number of tanks in series than for a small number. As will be seen later, however, there is little to be gained in practice by using a long series. Thus for most applications (i.e., 2 or 3 tanks) the models of Section 12.1 will give a reasonable qualitative description of what will occur. Consequently those models will be adopted here. Emphasis will be placed upon the soluble substrate model and the case in which only microorganisms are in the feed.

12.2.2 A Simple Chain of CSTR's

The simple chain of CSTR's is shown in Fig. 12.9a. The simplest form contains only two stages and as the number of stages increases the system increasingly resembles a plug-flow reactor. Emphasis here will be placed upon only a few reactors in the sequence.

EQUATIONS FOR SYSTEM PERFORMANCE. First consider the soluble substrate model based upon the Monod equation (Eq. 10.42). The first-order approximation (Eq. 10.43) should not be employed for this system because when several reactors are used the concentration of substrate in some of the reactors can exceed the saturation constant, K_s, thus invalidating the assumption of the first-order approximation. The first-stage reactor behaves exactly like the CSTR discussed in Section 12.1.1. Consequently, the concentration of soluble substrate leaving stage one is given by Eq. 12.6, the concentration of viable cells is given by Eq. 12.13, the concentration of dead cells is given by Eq. 12.15, the total cell concentration is given by Eq. 12.16, the viability by Eq. 12.19, and the oxygen requirement by Eq. 12.23. Furthermore, washout will occur anytime τ_1 is less than the value given by Eq. 12.9.

The equations for the second, and subsequent, reactors may be derived in a manner similar to that used for a single CSTR receiving both soluble substrate and cells in the influent. Application of a mass balance on viable cells in the nth-stage leads to an equation similar to Eq. 12.30:

$$\mu_n = (1 - \frac{X_{v(n-1)}}{X_{v(n)}}) \frac{1}{\tau_n} + \gamma + b \qquad (12.42)$$

Thus the specific growth rate of the nth stage is always less than that of the preceding stage if they have the same space time. The equation for the soluble substrate concentration will be a quadratic equation similar to Eq. 12.32:

$$[\mu_m - (1/\tau_n + \gamma + b)]S_n^2 - [\mu_m(X_{v(n-1)}/Y_g + S_{n-1})$$
$$+ (K_s - S_{n-1})(1/\tau_n + \gamma + b)]S_n$$
$$+ S_{n-1}K_s(1/\tau_n + \gamma + b) = 0 \qquad (12.43)$$

Likewise, the equation for the viable cell concentration is similar to Eq. 12.31:

$$X_{v(n)} = \frac{X_{v(n-1)} + Y_g(S_{n-1} - S_n)}{1 + b\tau_n + \gamma\tau_n} \qquad (12.44)$$

The dead and total cell concentrations are given by equations similar to Eqs. 12.33 and 12.39 respectively:

$$X_{d(n)} = \frac{X_{d(n-1)} + \gamma X_{v(n)}\tau_n}{1 + b\tau_n} \qquad (12.45)$$

$$X_n = \frac{X_{n-1} + Y_g(S_{n-1} - S_n)}{1 + b\tau_n} \qquad (12.46)$$

The viability in the nth stage is given by a generalization of Eq. 12.36:

$$v_n = (\frac{1 + b\tau_n}{1 + b\tau_n + \gamma\tau_n})(\frac{X_{v(n-1)} + Y_g(S_{n-1} - S_n)}{X_{v(n)} + Y_g(S_{n-1} - S_n)}) \qquad (12.47)$$

Finally the oxygen requirement for the nth reactor is given by an equation similar to Eq. 12.38:

$$RO_n = \frac{F\{(S_{n-1} - S_n)(1 + b\tau_n - \beta_n Y_g) + X_{n-1}[(\beta_{n-1} - \beta_n) + \beta_{n-1}b\tau_n]\}}{1 + b\tau_n} \tag{12.48}$$

The subscript on the oxygen demand coefficient of the cells, β, is to indicate that the term may change from stage to stage due to changes in the chemical composition of the cells. The total amount of oxygen required by the system is the sum of the requirements for each reactor. If β is constant throughout the chain, the equation reduces to:

$$RO = F[(S_o - S_n) - \beta X_n] \tag{12.49}$$

where

$$\beta_{n-1} = \beta_n = \beta$$

EXAMPLE 12.2.2-1

Return to the problem of Example 12.1.1-1, but substitute two 4 liter CSTR's in series for the single 8 liter CSTR.

(a) Determine the substrate concentration remaining in the effluent from the second reactor if the flow rate is 1.0 liter/hr.

Use Eq. 12.43 to do this. Examination of that equation shows that both S_1 and and $X_{v(1)}$ are needed. Thus, determine S_1 from Eq. 12.6, using $\tau_1 = 4/1 = 4$ hrs

$$S_1 = \frac{165(1/4 + 0.005 + 0.01)}{0.80 - (1/4 + 0.005 + 0.01)}$$

$S_1 = 81.7$ mg/liter

Use Eq. 12.13 to find $X_{v(1)}$

$$X_{v(1)} = \frac{0.55(1000 - 81.7)}{1 + (0.005)(4) + (0.01)(4)}$$

$X_{v(1)} = 476$ mg/liter

Now use Eq. 12.43 to find S_2

$[0.80 - (1/4 + 0.005 + 0.01)]S_2^2 - [0.80(476/0.55 + 81.7)$

$\quad + (165 - 81.7)(1/4 + 0.005 + 0.01)]S_2$

$\quad + (81.7)(165)(1/4 + 0.005 + 0.01) = 0$

$0.535S_2^2 - 780S_2 + 3572 = 0$

$\quad S_2 = 4.6$ mg/liter

(b)   What is the total cell concentration in each reactor?

Use Eq. 12.16 to find the cell concentration in the first reactor:

$$X_1 = \frac{0.55(1000 - 81.7)}{1 + (0.01)(4)}$$

X_1 = 486 mg/liter

Use Eq. 12.46 to find the cell concentration in the second reactor:

$$X_2 = \frac{486 + 0.55(81.7 - 4.6)}{1 + (0.01)(4)}$$

X_2 = 508 mg/liter

(c)   What percent of the overall oxygen requirement must be supplied to reactor 1?

Use Eq. 12.23 for reactor 1

$$RO_1 = \frac{(1.0)(1000 - 81.7)[1 + (0.01)(4) - (1.2)(0.55)]}{1 + (0.01)(4)}$$

RO_1 = 336 mg/hr

Use Eq. 12.48 for reactor 2

$$RO_2 = \frac{1.0\{(81.7 - 4.6)[1 + (0.01)(4) - (1.2)(0.55)] + 486[(1.2)(0.01)(4)]\}}{1 + (0.01)(4)}$$

RO_2 = 50.6 mg/hr

RO = 336 + 50.6 = 386.6

% to No. 1 = $\frac{336}{386.6}$(100) = 87%

The equations for reactors in series can be applied to an influent stream containing cells and soluble substrate by simply using Eqs. 12.30 through 12.38 to characterize the first stage and then using the equations above for the subsequent stages. In the event that the influent stream contains only cells with a negligible amount of soluble substrate, the equations can be simplifed considerably. The concentration of biomass in the nth reactor will be

$$X_n = \frac{X_{n-1}}{1 + b\tau_n} \tag{12.50}$$

The viability in the nth reactor will be

$$v_n = (\frac{1 + b\tau_n}{1 + b\tau_n + \gamma\tau_n})(\frac{X_{v(n-1)}}{X_{n-1}}) \tag{12.51}$$

The oxygen requirement for the nth reactor will be:

$$RO_n = \frac{X_{n-1}[(\beta_{n-1} - \beta_n) + \beta_{n-1}b\tau_n]}{1 + b\tau_n} \tag{12.52}$$

The total oxygen requirement is just the sum of that for each of the other reactors. All of the above equations apply to the general case of unequal size reactors. The easiest way to use the equations is to calculate stepwise down the chain, using the output from one reactor as the input to the subsequent one.

EXAMPLE 12.2.2-2

(a) If the single CSTR in Example 12.1.3-3 is replaced with two equal sized reactors in series with a total volume of 120 liters, what will be the concentration of cells in the second reactor?

Use Eq. 12.39 to find the concentration in the first reactor:

$$X_1 = \frac{600}{1 + (0.02)(60)}$$

$$X = 273 \text{ mg/liter}$$

Use Eq. 12.50 to find the concentration in the second reactor:

$$X_2 = \frac{273}{1 + (0.02)(60)}$$

$$X_2 = 124 \text{ mg/liter}$$

(b) What fraction of the cells in the second reactor are viable, if the viability of the cells entering the first reactor is 0.95, and γ is 0.005 hr^{-1}? First we must find the viability in the first reactor using Eq. 12.40:

$$v_1 = \left(\frac{1 + (0.02)(60)}{1 + (0.02)(60) + (0.005)(60)}\right)(0.95)$$

$$= 0.836$$

Then we must use Eq. 12.51 to find v_2:

$$v_2 = \left(\frac{1 + (0.02)(60)}{1 + (0.02)(60) + (0.005)(60)}\right)(0.836)$$

$$= 0.736$$

EXAMPLES OF SYSTEM PERFORMANCE. Figure 12.10 shows the effects of both chain length and total space time on the performance of a chain of equal sized reactors receiving a soluble substrate. All curves were calculated using the parameters shown in Table 12.1 with the exception of β, which was 1.41. Since all reactors in a chain are the same size, the space time of a single reactor, is given by:

$$\tau_n = \frac{\tau_T}{n} \tag{12.53}$$

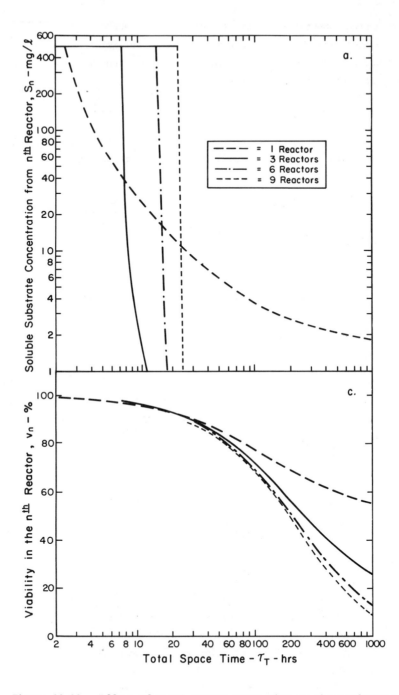

Figure 12.10.   Effect of total system space time on the performance
of a chain of CSTR's:
(a) Soluble substrate concentration in last reactor;
(c) Viability in last reactor.

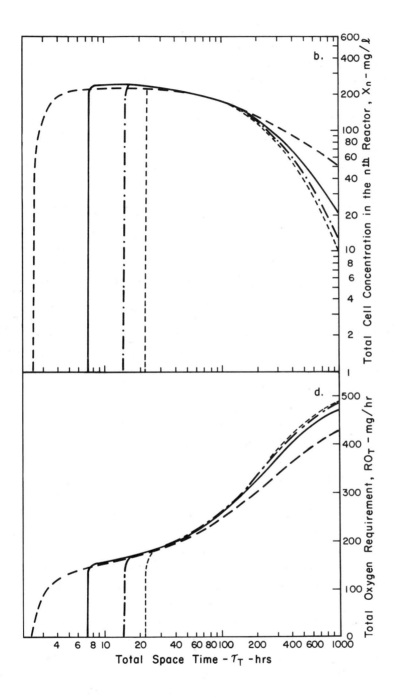

Figure 12.10.   (Continued)
(b) Total cell concentration in last reactor;
(d) Oxygen requirement for the entire chain.

where τ_T is the space time in the entire chain. Since the first reactor behaves as a single-stage reactor, it will wash out any time its space time is less than that given by Eq. 12.9. Since all reactors have the same space time, if the first washes out, so will the second, and so on down the chain until the entire chain fails. For the parameters used, a single stage reactor will wash out at a space time of 2.4 hrs. If two reactors are used, washout will occur when τ_1 is 2.4 hrs, i.e., when τ_T is 4.8 hours. Likewise, if 3 reactors are used, τ_T must be greater than 7.2 hours to prevent washout. Thus, the larger the number of reactors in a chain, the greater the total volume must be to prevent washout. It can be seen from Fig. 12.10a that for a desired effluent substrate concentration, for example, 5 mg/l, the total volume diminishes and then increases as the number of reactors is increased. The increase is due to the washout phenomenon. As the number of reactors is increased the total volume needed to prevent washout is larger than the volume needed to achieve the desired effluent substrate concentration in a smaller chain. Thus in practice, there is little to be gained from a long chain of reactors so that chains longer than three are seldom employed.

A second property of a chain of reactors is that it is capable of achieving a lower substrate concentration than a single reactor of equal total volume. This is a result of the nature of the substrate removal expression. If a single reactor is used, the reaction rate throughout the entire volume is equal to that associated with the effluent, i.e., a very low value. In the chain, however, earlier stages have higher substrate concentrations, and thereby have higher removal rates. Since the overall removal rate is higher in the chain than in a single tank of equal volume, a lower substrate concentration will result. The property of reactors in series that makes them of most value for wastewater treatment is that there is no lower limit on the substrate concentration attainable. As given by Eq. 12.7, in a single CSTR the minimum substrate concentration will be attained when growth just balances death and decay. In any reactor after the first stage, however, growth is not needed to balance death and decay since organisms are being added to the reactor in the feed. Thus it is possible for the substrate concentration to approach zero. The steepness of the substrate concentration curve in a long chain often has the effect of preventing use of the minimum sized chain needed to achieve a desired effluent concentration. This is because of possible instability and washout. For example, the system space time needed to achieve an effluent concentration of 2 mg/l in a 9 reactor chain is 24 hours, yet τ_{min} is only 22 hours. It would be unwise to use a τ_T so close to τ_{min}, and consequently additional space time must be provided just to make the total τ larger than τ_{min}. This is another reason why long chains are seldom used.

Figure 12.10b shows that greater decay occurs in a series of reactors, but that extremely long space times are required to achieve it. Note that a considerable advantage is gained in going from 1 to 3 reactors, but that little additional benefit is gained by adding more than 3 reactors in series. Figure 12.10c shows that there is also a loss of viability when reactors in series are used, consequently a larger percentage of the population will not be contributing anything to the system. As shown in Fig. 12.10d slightly more oxygen is required because more cell decay occurs. More important, however, is the distribution of oxygen utilization. In a single tank, oxygen is used at a steady rate throughout. When the same volume is divided into a number of tanks, however, the uptake rate of oxygen is more rapid in the early stages where substrate utilization occurs and slower in the latter where endogenous metabolism is more important. This is depicted in Fig. 12.11, which shows the distribution for a total space time of 100 hours. The total amount of oxygen transferred to each reactor system, however, is essentially the same, as was shown in Fig. 12.10d.

It can be seen from Figs. 12.10 and 12.11 that there are two distinct disadvantages associated with the use of equal size reactors in series. One is that the critical space time is increased in direct proportion to the number of reactors in

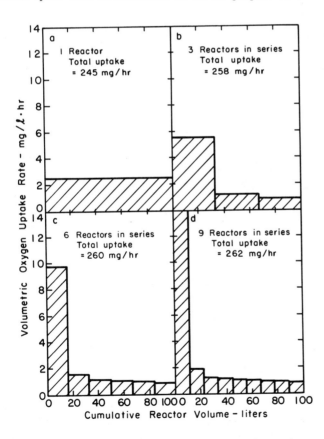

Figure 12.11.   Effect of number of tanks in series on
the oxygen uptake pattern in a chain of
CSTR's. Total space time = 100 hours.

the chain. This has the effect of making the system unstable at short space times. The second disadvantage is that very high oxygen uptake rates are exhibited in the early stages. This means that very high oxygen transfer rates are required in them. Both of these can be overcome by using larger reactors in the earlier stages of the chain.

Figure 12.12 shows the effects upon the effluent substrate concentration of changing the sizes of the reactors within a three reactor chain. As before, all curves were calculated using the parameters in Table 12.1, except for β which was 1.41. The total cell concentration and oxygen requirement are essentially the same as in the equal sized reactor chain. Figure 12.12 illustrates that shifting the majority of the reactor volume toward the front of the chain reduces the space time at which washout occurs without sacrificing the chief advantage of the chain, i.e., the ability to achieve substrate and cell concentrations lower than a single-stage reactor. In addition, since the majority of the oxygen requirement is distributed evenly throughout the first reactor, the oxygen uptake rate is much closer to that

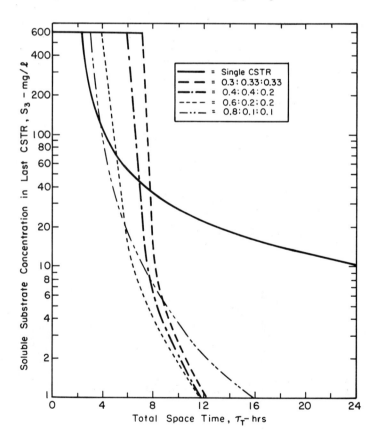

Figure 12.12. Effect of volume distribution on soluble substrate
concentration in the last CSTR of a three-reactor chain.

of a simple CSTR, as shown in Fig. 12.13. The most effective distribution of reactor volume depends upon the kinetic parameters describing the waste being treated and must be determined for each situation. Generally, however, distinct advantages can be obtained by putting the majority of the volume in the first reactor.

The advantages of using reactors in series are not as great when the influent contains only microbial cells. Figure 12.14 shows the effects of the total space time on the cell concentration in the effluent from the last reactor of chains containing 2, 3, and 6 equal volume reactors. For comparison, the curve for a single reactor is reproduced from Fig. 12.7. The parameters listed in Table 12.2 were used to construct the curves. From the figure it can be seen that little is gained from the use of long chains and that 2 reactors in series have a larger incremental advantage than do larger chains. These curves are in agreement with what one could expect from a first-order reaction as discussed in Chapter 4. Figure 12.15 shows the effects of the number of tanks in series on the oxygen uptake pattern. Comparison with Fig. 12.11 shows that the first tank does not have the high requirement characteristic of a chain receiving soluble substrate. This is because most of that high uptake was due to soluble substrate removal; since no soluble substrate is present, all oxygen uptake is due to endogenous metabolism, which declines slowly as the cell mass declines.

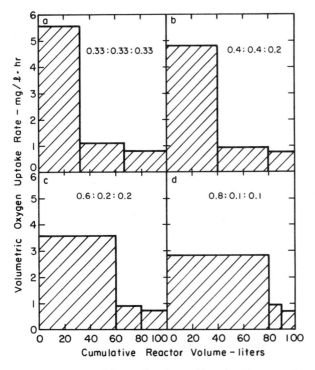

Figure 12.13.   Effect of volume distribution on the oxygen uptake pattern in a three-CSTR chain. Total space time = 100 hrs. The uptake rate for a single CSTR of equal space time is 2.45 mg/(liter·hr).

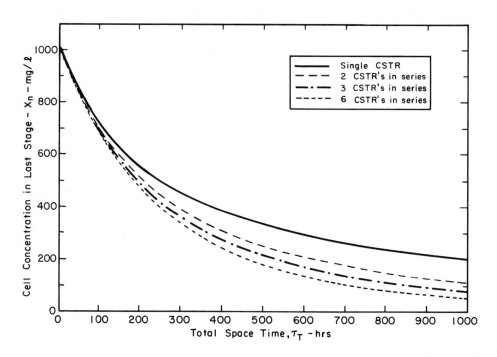

Figure 12.14.   Effect of total space time on performance of CSTR's receiving
               only microbial cells.   Parameters are listed in Table 12.2.

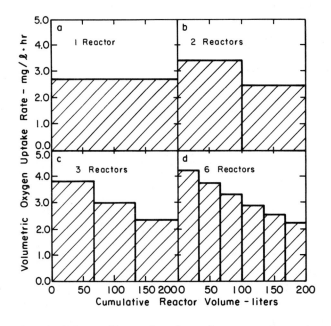

Figure 12.15.   Effect of number of reactors in series on the oxygen
               uptake pattern in a chain of CSTR's receiving only
               microbial cells.   Total space time = 100 hours.

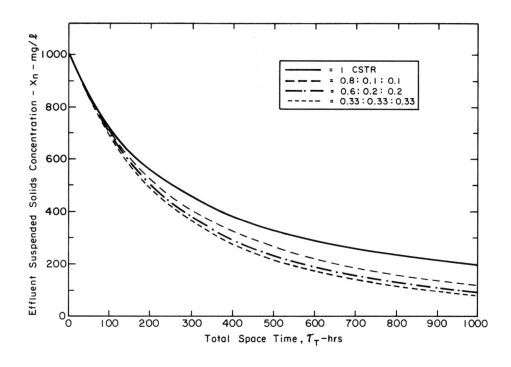

Figure 12.16. Effect of volume distribution on the performance of a three-
reactor chain of CSTR's receiving only microbial cells.
Parameters are listed in Table 12.2.

Figure 12.16 shows the effects upon the effluent cell concentrations of changing
the volume distribution within a three reactor chain receiving only microbial cells
in the influent. The curves were calculated with the parameters given in Table 12.2,
and the curve for a single-stage reactor is reproduced for comparison. Examination
of the figure shows that an equal distribution of reactor volumes is the most effective and that any deviation from that distribution results in a decrease in cell
degradation. This is characteristic of a first-order reaction which is how cell
decay has been modeled. Another characteristic of a first-order reaction is that a
volume distribution of A:B:C gives the same performance as a distribution of C:B:A.
Since reactors receiving cells cannot wash out, the effects depicted in Fig. 12.12
for a chain receiving soluble substrate are not manifested in this case. This, together with the fact that the oxygen requirements in the individual reactors of the
chain are not drastically different, implies that there is little reason to use anything other than an equal volume distribution in this case.

12.2.3 Chain of CSTR's with Multiple Feed Points
A chain of CSTR's with multiple feed points is shown in Fig. 12.9b. The volumes of
the reactors may be the same or different, and the feed may either all be the same
or originate from different sources. When the composition of the feed to each

reactor is different, the kinetic constants for each reactor will be different, thereby complicating the modeling. For that situation, the second and subsequent reactors should be modeled as reactors receiving soluble substrate plus cells, and the computations should be made stepwise down the chain. No special equations are required then. Here we will consider only the case of a single type of substrate at one concentration.

EQUATIONS FOR SYSTEM PERFORMANCE. Consider a two reactor chain consisting of reactors with volumes V_1 and V_2. The first reactor receives feed at rate F_1. The effluent from reactor 1 flows to the second, carrying cells at concentration X_1. Reactor 2 also receives flow directly from the source at rate F_{02}. The combined flow leaves the reactor at rate F_2, where

$$F_2 = F_1 + F_{02} \qquad (12.54)$$

The space time for reactor 2, τ_2, is defined in terms of the total flow through the reactor, F_2.

$$\tau_2 = V_2/F_2 \qquad (12.55)$$

It is also convenient mathematically to define two artificial space times, one with respect to the flow from reactor 1:

$$\tau_{12} = V_2/F_1 \qquad (12.56)$$

and the other with respect to the flow entering the reactor directly:

$$\tau_{02} = V_2/F_{02} \qquad (12.57)$$

These can be related by the expression

$$\frac{1}{\tau_2} = \frac{1}{\tau_{12}} + \frac{1}{\tau_{02}} \qquad (12.58)$$

The substrate concentration in the raw feed is S_o, and will be considered the same for all reactors here, although equations can be derived easily for reactors receiving different concentrations of the same type waste.

All equations describing system performance will be derived using the Monod model (Eq. 10.42). Because of the likelihood of high substrate concentrations in some stages, the first-order model should not be used. The first-stage reactor is described by the equations presented in Section 12.1.1. Although the equations will only be presented for a two reactor chain similar ones can be derived easily for three or more.

The equation for the specific growth rate in the second reactor may be obtained by applying a steady-state mass balance to viable cells; which reduces to:

$$\mu_2 = \frac{1}{\tau_2} - \frac{1}{\tau_{12}} \frac{X_{v1}}{X_{v2}} + b + \gamma \qquad (12.59)$$

This differs from the equation for a simple chain by the fact that both the total flow and the flow from the preceding reactor influence the growth rate, but reduces to Eq. 12.42 when F_{02} is equal to zero. Whereas in a single stream system the specific growth rate of the second stage must always be less than that of the first, in a multistream system this is no longer true. In fact by judicious selections of F_1 and F_{02} it is possible for μ_2 to approach μ_m without danger of washout in the system. This characteristic is used by microbiologists who wish to maintain a rapidly growing culture for a long period. It has not been exploited by wastewater treatment engineers although it could be employed advantageously when pretreatment is desired.

The equation for viable cell concentration in the second stage may be obtained by substituting Eq. 12.59 into a mass balance on substrate, and rearranging:

$$X_{v2} = \frac{(X_{v1}/\tau_{12}) + Y_g[(S_1/\tau_{12}) + (S_o/\tau_{02}) - (S_2/\tau_2)]}{1/\tau_2 + b + \gamma} \tag{12.60}$$

When this is substituted into Eq. 12.59, and it, in turn, is substituted into the Monod equation, the result is a quadratic equation for the substrate concentration:

$$[\mu_m - (1/\tau_2 + b + \gamma)]S_2^2 - \{(\mu_m X_{v1}/Y_g)(\tau_2/\tau_{12}) + (1/\tau_2 + b + \gamma) K_s$$

$$+ [\mu_m - (1/\tau_2 + b + \gamma)][S_1(\tau_2/\tau_{12}) + S_o(\tau_2/\tau_{02})]\} S_2$$

$$+ K_s(1/\tau_2 + b + \gamma)[S_1(\tau_2/\tau_{12}) + S_o(\tau_2/\tau_{02})] = 0 \tag{12.61}$$

This may be solved directly for S_2 and the result used to calculate the viable cell concentration from Eq. 12.60. Similar equations may be derived for three or more stages.

The concentration of dead cells may be obtained from a mass balance on dead cells; which can be rearranged to give:

$$X_{d2} = \frac{X_{d1}/\tau_{12} + \gamma X_{v2}}{1/\tau_2 + b} \tag{12.62}$$

The total cell concentration equation can be obtained by combining Eqs. 12.60 and 12.62:

$$X_2 = \frac{X_1/\tau_{12} + Y_g[S_1/\tau_{12} + S_o/\tau_{02} - S_2/\tau_2]}{1/\tau_2 + b} \tag{12.63}$$

The oxygen requirement within the second reactor can be calculated from a balance on oxygen-demanding material:

$$RO_2 = F_1 S_1 + F_{02} S_o - F_2 S_2 + \beta_1 F_1 X_1 - \beta_2 F_2 X_2 \tag{12.64}$$

The total oxygen requirement will be the sum of that in each reactor.

PERFORMANCE OF A MULTISTREAM CHAIN. Herbert [14] has discussed in detail the properties of a multistream chain as revealed through mathematical models. Although his model did not incorporate the death and decay concepts his conclusions are applicable here because of the high growth rates normally employed in multistream systems. Figure 12.17 depicts what will occur as the feed to the first of two equivolume reactors is held constant while the feed to the second is increased. If no new feed enters the second reactor the space time will be the same as the first, the specific growth rate will be very low, almost all substrate will be removed and the cell concentration will be essentially the same as in the first stage. As the flow to the second stage is increased, however, the specific growth rate in it increases until it reaches μ_m. The substrate concentration must also increase to maintain the higher growth rate. This reduction in the amount of substrate removed in turn causes the cell concentration to decrease. Once μ_m is reached the cells can grow no faster, so less substrate is removed as the flow rate is increased further. In spite of the very fast flow rate, however, washout cannot occur due to the continuous innoculation of the second-stage reactor.

Herbert [14] also investigated the effects of changes in the first stage upon the performance of the second stage. He was able to show that the second stage is

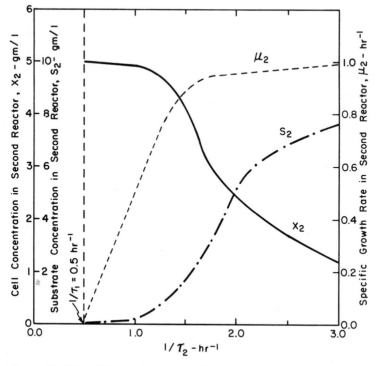

Figure 12.17. Effect of varying the inflow to the second stage
of a two-stage, multistream CSTR system. The
space time of the first stage remains fixed at
2.0 hours. *(Adapted from Herbert, [14]).*

relatively insensitive to changes in the space time of the first stage as long as
the washout point is not approached. Thus, the system is very stable, which makes it
a good research tool for microbiological studies. It also makes it more attractive
than a single stage system for industrial fermentations because the second stage can
be selected to maximize production of a desired product without fear of washout.

12.2.4 Determination of Optimal Input and Volume Distributions

From the previous discussion it is apparent that a number of different flow patterns
could be used for any given application employing CSTR's. The question then arises
as to the best configuration. There are two situations which could be encountered
in seeking the best configuration. One would be to seek the minimum effluent substrate concentration for a given total reactor volume. The other would be to seek
the minimum total volume to achieve a desired effluent substrate concentration.

Grieves and Kao [15] have investigated the effects of input and volume distributions upon the effluent substrate concentration leaving a three reactor chain of
fixed total volume. The characteristics of the system and the kinetic parameters
employed are given in Table 12.4. The results showed that the feed distribution
which produced the lowest effluent substrate concentration depended upon the volume
distribution. For example, if three equivolume tanks were used, the best feed policy
was to add all of it to the first reactor. This same conclusion was reached whenever
the majority of the system volume was housed in the first reactor (i.e., $V_1:V_2:V_3$
of 0.6:0.2:0.2 and 0.8:0.1:0.1). On the other hand, if a large middle tank was used
(0.2:0.6:0.2), or if the volume was shifted toward the rear (0.2:0.4:0.4), then the
best feed policy was to put 70% of the raw feed to the first unit and 30% to the
second. This reduction in the feed quantity to the first stage was required to prevent washout of the first stage as its volume was reduced. The optimum amount fed
to the first stage would therefore depend upon its volume. In general, however, unless the feed to the first tank must be reduced to prevent washout, the optimal feed
policy will be to place all feed into the first tank.

TABLE 12.4
Parameters Used by Grieves and Kao [15]

Symbol	Value
μ_m	0.20 hr^{-1}
K_s	100 mg/liter COD
Y_g	0.5 mg cells/mg COD
b	0.002 hr^{-1}
γ	0.000 hr^{-1}
F	1 liter/hr
S_o	5000 mg/liter COD
X_o	0
Z_o	0
V_T	20 liters

When all feed is to the first tank the optimum volume distribution will depend upon the total volume. This can be seen by examining Fig. 12.12. For that example, if τ_T = 12 hrs, the best volume distribution of the four considered is 0.4:0.4:0.2. On the other hand, if τ_T = 10 hrs, the best distribution is 0.6:0.2:0.2. For the situation studied by Grieves and Kao (τ_T = 20 hrs.), the best volume distribution when all feed entered the first tank was 0.6:0.2:0.2, and this combination also gave better results than any other feed and volume distribution.

Very often CSTR's must be used to reduce the concentration of soluble substrate to a specific value. The question thus arises as to what volume and input distribution will achieve that concentration with the minimum total volume. Generally, the optimum input distribution is to put it all into the first reactor. The best volume distribution will then depend upon the desired effluent concentration as shown by Fig. 12.12. There it can be seen that if the desired effluent concentration is 10 mg/l the best distribution is 0.6:0.2:0.2 whereas if it is 1 mg/l the volume distribution resulting in the least total volume is 0.4:0.4:0.2. Because the optimum volume distribution depends upon the soluble substrate concentration desired, no generalizations can be made. Nevertheless, for any given situation the optimum distribution can be found relatively easily by applying conventional optimization techniques.

Another way in which CSTR's are sometimes used is to obtain a desired degree of stabilization of a waste containing only a soluble substrate. Since the degree of stabilization depends upon the amount of cell material discharged in the effluent, the reactor chain would have to be chosen to achieve a desired cell concentration in the final reactor. This usually requires relatively long space times, so that the soluble substrate concentration will be negligible. The best feed policy in this case is also to put all feed into the first reactor. The second and third reactors will act primarily to allow microbial decay, and will thus be similar to reactors receiving only cells. Since cell decay follows a first-order rate expression, the optimal policy is to distribute the volume evenly among the stages. This policy will result in the minimum volume for a given degree of stabilization or in the best degree of stabilization for a given volume. This same policy applies to a chain of reactors receiving only cells in the influent.

12.3 VERIFICATION OF MODELS

The majority of work on model verification has been performed on the simple soluble substrate model. Such a model served as the basis for the development of the continuous culture of microorganisms and its application to the fermentation industry as well as to the study of microbial physiology. Herbert and his colleagues [13,16,17] were instrumental in the verification of the theory for pure microbial cultures so that today the studies performed using the theory are too numerous to mention. Several studies are of particular importance however, because they represent refinements of the basic model as originally presented. Into this category fall the work of Pirt [18], who established the need for a term to represent the maintenance energy requirement, and Sinclair and Topiwala [19], who initiated the use of a bacterial death rate. Generally though, verification of the trends expressed by the model has come through the work of many researchers as evidenced by the numerous symposia which are listed at the end of the chapter.

Within the field of wastewater treatment the validity of the trends predicted by the model has been established through the efforts of many researchers. Those efforts were organized and summarized by Lawrence and McCarty [20] into a unified concept which put biochemical operation design on a rational basis. Many complications arise when models based upon pure culture systems are applied to wastewater treatment; nevertheless, our body of knowledge is now expanding to the point where the general trends predicted by the model no longer are in doubt. For example, the work of Chiu *et al.* [21] demonstrated good correlations between predicted and observed results with a model of this general form even though significant shifts in microbial predominance occurred [22]. Furthermore, the work of Jorden *et al.* [23], Horn and Pohland [24], and Kormanik [25] have demonstrated the applicability of the general equations to actual wastewaters, while Eckenfelder [9] has made wide use of the first-order approximations in design.

Gaudy and Gaudy [26] have made a valuable point concerning the verification of kinetic equations. While discussing the empirical nature of the models they state:

> Experimentally, a relatively steady-state develops (in a CSTR), and these equations can be used to model it and to predict levels of X and S; the steady-state does not inevitably develop because of these equations or because of any theory on which they are based. The fact that a process can be described quantitatively with formulas of predictive value is not necessarily ascribable to correctness of theory or hypothesis. The equations are best looked upon as an outcome of enlightened empiricism. They are important equations because they are basic to description of the behavior in X and S of completely mixed, continuous culture reactors.

This should always be remembered.

12.4 TECHNIQUES FOR ASSESSING KINETIC PARAMETERS

The main use of the models presented in the preceding sections is in design. Samples of waste are collected in the field and returned to the lab where treatability studies are performed. If a CSTR is being considered as the treatment method, bench-scale reactors will be run at a number of space times and samples will be collected and analyzed to determine the values of the various system variables. That data must then be used to estimate the values of μ_m, K_s, Y_g, b and γ so that the models may be used to determine the value of τ required to meet the performance objectives.

Many of the equations describing the performance of a CSTR may be reduced to linear form, consequently graphical techniques are usually used to determine the parameters. Because some of the parameters appear in more than one equation it is

necessary to determine them in a sequential manner. This, coupled with the fact
that two or more parameters will be determined in a single plot, means that the
estimated values are interrelated, e.g., an error in the estimate of K_s will effect
the value of μ_m obtained, etc. This should be recognized and accounted for when
parameters are determined and used. Less emphasis should be placed upon individual
parameters than upon the entire set determined for a particular wastewater. Because
of this interrelationship attempts have been made to determine all parameters simultaneously using the computer [21]. However, to allow parameter estimation by those
without access to a computer, only the graphical procedure will be presented here.

12.4.1 Soluble Substrate Situation

The only organic matter that will be influenced by biochemical operations is that
which is biodegradable. Often, however, the COD test is the most convenient way of
measuring the concentration of organic matter in the waste. Since it measures both
biodegradable and nonbiodegradable organic matter there must be some way to estimate
the concentration of nonbiodegradable material so that it may be subtracted from the
COD to determine S_o and S expressed as T_bOD.

The concentration of nonbiodegradable COD is estimated by application of the
fact that the specific growth rate (μ) and the specific rate of substrate removal (q)
will be zero when the concentration of biodegradable substrate is zero. If a plot
is made of q as a function of COD the curve will pass through the origin if all of
the COD is biodegradable. If a portion of the COD is nonbiodegradable, the curve
will have a positive abscissa intercept, equal in magnitude to the nonbiodegradable
COD. This is shown in Fig. 12.18.

The specific rate of substrate removal in a CSTR is given by the combination
of Eqs. 10.41 and 12.11:

$$q = \frac{F(S_o - S)}{VX_v} \tag{12.65}$$

The specific rate may be calculated using COD values because the nonbiodegradable
COD will appear in both the influent and effluent, and thus will cancel out. The
specific rate of substrate removal should be based upon the viable cell concentration
because only then will q be proportional to μ. If viability data is not available,
however, q must be estimated by the total cell concentration although this may
introduce some error into the estimation of the nonbiodegradable COD. The working
equation used to estimate q under this circumstance is:

$$q' = \frac{F(C_o - C)}{VX} \tag{12.66}$$

where C_o and C refer to the soluble COD in the influent and in the reactor.

Figure 12.18a illustrates how the plot will look if data are taken at relatively
short space times so that a large range of reactor substrate concentrations is
covered. There will be certain conditions under which the effect of substrate

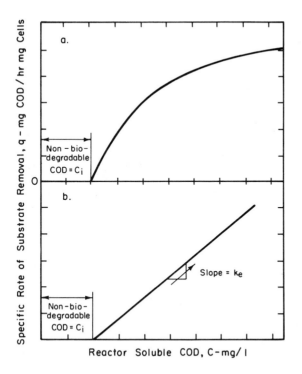

Figure 12.18.   Determination of the nonbiodegradable COD
from CSTR data: (a) Large range of substrate
concentrations; (b) Low substrate concentrations where the first-order approximation
is valid.

concentration on the specific rate of removal will be first order so that only the
lower portion of Fig. 12.18a will be applicable. This means that the plot may be
treated as a straight line as shown in Fig. 12.18b. A linear least squares analysis
may then be used to aid in the estimation of the nonbiodegradable COD, although care
should be exercised to ensure that the first-order approximation is valid.

Failure to correct for the nonbiodegradable COD, C_i, will make it impossible to
estimate some of the parameters, and thus this determination should always be made.
The assumption will be made in the rest of this discussion that S_o and S refer only
to the T_bOD, where:

$$S_o = C_o - C_i \tag{12.67}$$

$$S = C - C_i \tag{12.68}$$

It has been assumed in this book that the decay rates of live and dead cells are
the same. Techniques are available for estimating the rates for each classification
individually [4]; however, since more work needs to be done to verify the validity
of the two decay rates this discussion will be limited to the case where they are

equal. The procedure described for estimating Y_g and b is applicable even when the microbial specific death rate is assumed to be zero.

The true growth yield and decay rate constant are found from a linearization of Eq. 12.16 [23]:

$$\left(\frac{S_o - S}{X}\right) = \left(\frac{b}{Y_g}\right)\tau + \frac{1}{Y_g} \tag{12.69}$$

When a plot is made of $(S_o - S)/X$ as a function of the space time in the CSTR the result is a straight line with slope b/Y_g and ordinate intercept of $1/Y_g$, as shown in Fig. 12.19. Note that Y_g is estimated at the point where τ is zero, i.e., when the cells would be growing infinitely fast. This meets the definition of Y_g since no decay would be occurring under those circumstances. The space time is the independent variable and a linear least squares line of best fit may be used to estimate the slope and intercept.

The specific bacterial death rate, γ, may be determined by putting Eq. 12.19 in double reciprocal form:

$$\frac{1}{v} = 1 + \gamma \left(\frac{\tau}{1 + b\tau}\right) \tag{12.70}$$

The reciprocal of the viability is plotted versus $\tau/(1 + b\tau)$. The line is drawn through an ordinate intercept of 1 and the resulting slope is γ, the specific death rate. This is illustrated in Fig. 12.20. Any error in the estimate of the decay rate, b, will carry over into the estimate of γ.

After b and γ have been determined, μ_m and K_s may be found from a linearized form of Eq. 12.6. Because the Monod equation is similar in form to the Michaelis-Menten equation for enzymatically catalyzed reactions any of the techniques

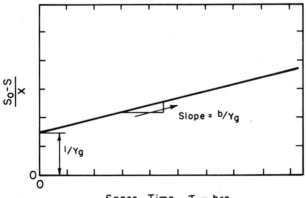

Figure 12.19. Determination of Y_g and b.

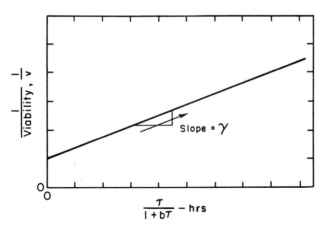

Figure 12.20.   Determination of γ.

discussed in Chapter 10 may be used to linearize Eq. 12.6.   If the line is to be
drawn by eye the Hanes plot is preferable.

$$\frac{S}{1/\tau + b + \gamma} = (\frac{1}{\mu_m})S + \frac{K_s}{\mu_m} \tag{12.71}$$

This is shown in Fig. 12.21.   The scales can be chosen to give good estimates of μ_m
and K_s.   The least squares technique cannot be used to find the line of best fit,
however, because both axes contain terms which are subject to error (i.e., S).

If it is desired to use the least squares line of best fit, then the Hofstee
plot should be used:

$$\frac{(1/\tau + b + \gamma)}{S} = \frac{\mu_m}{K_s} - (\frac{1}{K_s})(1/\tau + b + \gamma) \tag{12.72}$$

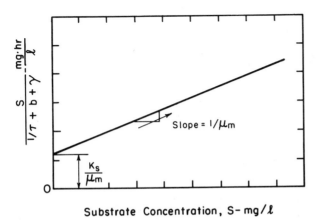

Figure 12.21.   Hanes plot for the determination of μ_m and K_s.

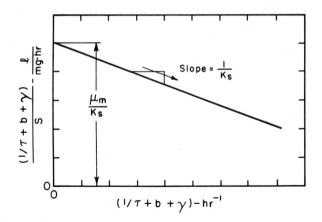

Figure 12.22. Hofstee plot for the determination of μ_m and K_s.

This is drawn as shown in Fig. 12.22. The use of the reciprocal of S may amplify
the error in S and make it difficult to get a line by eye. Since the independent
variable $(1/\tau + \gamma + b)$ appears in each axis there will be some degree of inevitable
correlation.

In Chapter 10 it was noted that the double reciprocal or Lineweaver-Burk method
of plotting Eq. 12.6 will give a deceptively good fit, even with unreliable points.
Thus it should be avoided [27].

No matter what technique is used, it will not be possible to estimate μ_m and
K_s if S includes nonbiodegradable COD, because the plots will not yield straight
lines, or will give lines which have slopes or intercepts of the wrong sign. This
emphasizes the importance of using T_bOD as a measure of S.

The oxygen demand coefficient of the cell material, β, must be known before the
oxygen requirement for the system can be determined using Eq. 12.22. This can be
measured easily during laboratory studies by determining the COD of the unfiltered
effluent T, as well as the COD of the filtrate, C. The difference between the two,
divided by the cell concentration in the reactor gives a direct measure of β. There
is no mathematical expression to describe β as a function of τ, so an experimental
plot should be prepared for use in design.

In those cases where K_s is very high, or where all S values are very low, it is
sometimes difficult to get accurate estimates of μ_m and K_s. In that case the first-
order approximation is usually used. The first-order rate constant is k_e, the mean
reaction rate coefficient, and it relates the specific rate of substrate removal, q,
to the biodegradable COD, S.

$$q = k_e S \qquad\qquad (12.73)$$

Thus k_e may be estimated as the slope of a plot of q versus S, as shown in
Fig. 12.18b.

EXAMPLE 12.4.1-1

A culture of microorganisms was grown in a CSTR on a complex soluble medium with a COD of 1000 mg/l, and the data in Table E12.2 were collected. Using the data determine μ_m, K_s, Y_g, b and γ for the culture.

TABLE E12.2
Data Collected from a CSTR

τ hrs.	Soluble COD mg/liter	Suspended solids-mg/liter	Viability %
4	347	285	98.5
8	106	378	97.1
12	70	382	95.8
16	55	377	94.6
32	36	345	90.8
64	28	289	85.5

(a) First find C_i, the nondegradable COD so that the substrate concentration may be expressed as T_bOD. This requires computation of specific substrate removal rates by Eq. 12.65. The values are then plotted as a function of the soluble COD(C) as shown in Fig. E12.1. The nonbiodegradable COD is 16 mg/l. This should be subtracted from all COD values to get the concentrations in terms of T_bOD. Therefore S_o = 984 mg/l, and S is as given in Table E12.3.

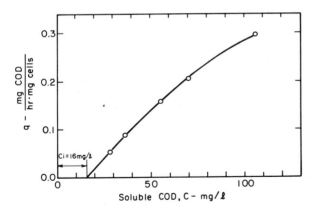

Figure E12.1. Determination of C_i.

TABLE E12.3
Biodegradable Substrate Concentration in a CSTR

τ hr	S mg/liter T_bOD
4	331
8	90
12	54
16	39
32	20
64	12

(b)   Next we must find Y_g and b from a plot of Eq. 12.69. Thus values of $(S_o - S)/X$ were calculated and plotted as shown in Fig. E12.2. There it can be seen that b = 0.008 hr^{-1} and Y_g = 0.45 mg cells/mg T_bOD utilized.

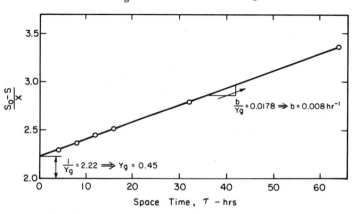

Figure E12.2. Determination of b and Y_g.

(c)   The value of γ is determined from a plot of Eq. 12.70, as shown in Fig. E12.3.

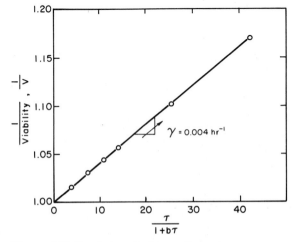

Figure E12.3. Determination of γ.

γ is determined from the slope to be 0.004 hr^{-1}.

(d)   Now that γ and b are known μ_m and K_s may be determined from a Hanes plot, as given in Fig. E12.4.   The data points which are most important to the determination of μ_m are those with the highest values of S because the cells were growing the closest to μ_m for those values.   Also, because of the term $1/\tau + \gamma + b$ in the denominator of the ordinate, very small errors in the estimation of C_i can result in large errors in the ordinate at low values of S. Consequently more weight should be given to the points corresponding to higher values of S when drawing the line.   The values of μ_m and K_s are 0.395 hr^{-1} and 168 mg/l respectively.

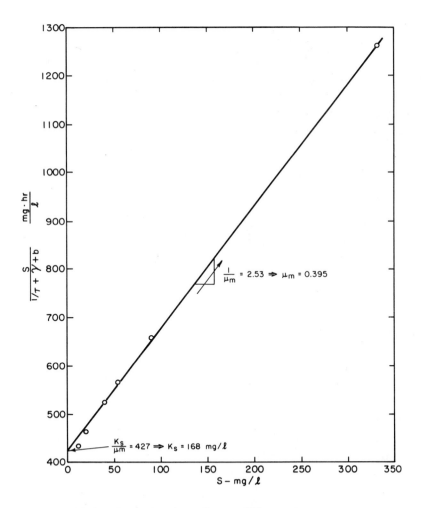

Figure E12.4.   Determination of μ_m and K_s.

12.4.2 General Situation--Soluble Substrate Plus Inert and Biodegradable Solids

Many wastewaters contain suspended solids. Because the rates of degradation of solid substrates are low it is preferable to remove as many of the solids as possible prior to any biochemical operation which is being designed primarily for the removal of soluble organic matter. Unfortunately, clarification cannot economically achieve 100% efficiency, so that allowances must be made for the effects of suspended solids when biochemical operations are designed. The procedure outlined below is intended to correct the soluble substrate equations for the presence of solids. In Chapter 17 consideration will be given to those aerobic biochemical operations intended for the destruction of insoluble organic matter and anaerobic operations will be considered in Chapter 21.

The values of μ_m and K_s (or k_e) depend only upon the soluble T_bOD and thus they will be determined from measurements of S as was done in the soluble substrate case. The major effects of the solid substrate will be upon the estimation of Y_g, b and β. As was discussed in Section 12.1.3 the usual approach is to consider the suspended solids to be composed of only inert solids and biomass, even though biodegradable solids are known to be present. This has the effect of making Y_g, b and β pseudo-constants which incorporate the effects of the influent solids. Because long space times are needed for degradation of solid substrates, only data from reactors with large values of τ should be used for parameter estimation.

In order to characterize this type of waste the following data should be collected during operation of a lab- or pilot-scale CSTR:

C_o - Soluble COD concentration of the influent waste.

C - Soluble COD concentration in the reactor.

T_o - Total COD concentration of the influent waste.

T - Total COD concentration in the reactor.

Z_o - Total suspended solids concentration in the influent waste.

M - Suspended solids concentration in the reactor.

v - Viability of the microbial solids.

V - Reactor volume under aeration.

F - Influent flow rate.

In addition, long term batch experiments should be performed to determine the following:

f - Fraction of influent solids which is inert.

f' - Fraction of the particulate COD in the influent which is inert.

If primary sedimentation is to be used in the full scale plant then all of the tests described should be run on settled waste.

The influent suspended solids are generally composed of a mixture of inert and biodegradable solids. Since a portion of the organic or volatile solids in the waste may be nonbiodegradable, biodegradation must be allowed to occur so that the nondegradable material may be detected as that remaining. Consequently, a long

term experiment is required so that time will be adequate for decay of the biomass
resulting from the degradation of the solid substrate. The first step is to separate the solids from the wastewater by centrifugation or membrane filtration. The
separation conditions should be sufficient to remove the colloidal material larger
than 0.45 μ. The solids should then be resuspended in an inorganic nutrient solution
to the original concentration, Z_o. This procedure minimizes the influence of the
original soluble substrate and the biomass which would result from its degradation.
The resuspended solids should then be blended or macerated to reduce the particle
size as much as possible. This will increase the surface area available for biological attack and thereby reduce the time required for complete degradation of the
biodegradable material to be achieved. It should be emphasized that this treatment
makes the test invalid for rate studies. The purpose of the test is to determine
the fraction of the material which can be degraded biologically. Next the blended
slurry should be placed into an aerated batch reactor and seeded with a small
inoculum of acclimated bacteria. It is important that the original inoculum be
small so that the solids added will have a negligible effect upon the concentration
of solids in the reactor. The slurry should be aerated until the solids concentration reaches a constant value. That may take a period of several weeks. In order
to avoid erroneous results a strict accounting should be kept of all samples removed
from the reactor so that any water lost by evaporation can be replaced with distilled
water prior to sampling.

The fraction of inert solids in the influent waste, f, may be estimated from
the change in suspended solids concentration over the course of the batch experiment.
The assumption is made that the final stable suspended solids concentration is due
to inert solids.
Consequently:

$$f = \frac{\text{Final stable suspended solids concentration}}{\text{Initial suspended solids concentration}} \tag{12.74}$$

The assumption is then made that f is constant so that

$$Z_{io} = fZ_o \tag{12.75}$$

For safety, several determinations of f should be made over the course of the
treatability studies and an average value used.

In order to determine the effective influent substrate concentration, S_o' and
the oxygen demand coefficient of the MLSS, it is important that the fraction of the
particulate COD which is due to inert solids, f', be determined. This, too, will be
assumed to remain constant so several determinations should be made concurrently
with the measurements of f. Since the soluble fraction of the waste was removed in
the preparation of the sample for the batch test, all COD would be due to the solid
or particulate fraction. Consequently, during the batch test measurements should be
made of the COD. The assumption is made that the final stable COD is due to the

inert solids. Consequently,

$$f' = \frac{\text{Final stable COD concentration}}{\text{Initial COD concentration}} \tag{12.76}$$

The soluble, nonbiodegradable COD, C_i, may be determined exactly as described in Section 12.4.1 for a totally soluble waste. The influent soluble T_bOD, S_o, and the effluent soluble T_bOD, S, may be found from Eqs. 12.67 and 12.68, respectively.

The solids in the CSTR are composed of inert solids, biomass, and undegraded solid substrate. Since it is very difficult to distinguish experimentally between the biomass and the undegraded solid substrate these will be grouped together into "effective biomass", X'. This requires that the influent substrate concentration, decay rate, and true growth yield in Eq. 12.16 all be replaced with "effective values":

$$X' = \frac{Y_g'(S_o' - S)}{1 + b'\tau} \tag{12.16'}$$

The value of X' may be found by subtracting the inert solids in the reactor from the total suspended solids concentration, M, as shown in Eq. 12.77:

$$X' = M - fZ_o \tag{12.77}$$

The effective substrate concentration, S_o', is the sum of the soluble and particulate biodegradable COD's. However, during the CSTR experiments, measurements are only made of the total and soluble COD's in the influent. Consequently, use also must be made of f' and C_i to calculate S_o'. The COD in the waste due to particulate matter, P_o, is the difference between the total and soluble COD's.

$$P_o = T_o - C_o \tag{12.78}$$

The effective COD in the influent is that remaining after subtracting the soluble and particulate inert COD values from the total COD. Noting that the particulate inert COD is given by $f'P_o$, the effective COD is given by

$$S_o' = T_o - C_i - f'P_o \tag{12.79}$$

Substitution of Eq. 12.78 for P_o gives

$$S_o' = T_o - C_i - f'(T_o - C_o) \tag{12.80}$$

The values of b' and Y_g' can be found from a linearized form of Eq. 12.16', after substitution of Eqs. 12.75 and 12.78 for X' and S_o' respectively:

$$\frac{T_o - C_i - f'(T_o - C_o) - S}{M - fZ_o} = \frac{b'}{Y_g'}\tau + \frac{1}{Y_g'} \tag{12.81}$$

The plot is similar to that in Fig. 12.19 and the parameters are determined from the slope and intercept as indicated.

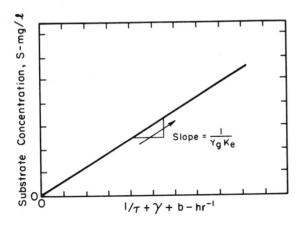

Figure 12.23. Plot for the determination of k_e.

The specific bacterial death rate, γ, may be determined exactly as it was in Section 12.4.1.

The kinetic parameters for the Monod equation may be determined after b' and γ are known by using the procedure outlined in Section 12.4.1.

The simplest way to determine the mean reaction rate coefficient, k_e, is through the use of Eq. 12.26, as shown in Fig. 12.23. Determination of k_e using the unit rate of substrate removal requires that only the viable cell concentration be used. The presence of the solid substrate makes this more complex, and consequently, the use of Eq. 12.26 is recommended.

The value of β', the oxygen demand coefficient of the effective cell concentration, X', may be found by COD measurements on the CSTR during the experiments. The effective cell concentration is given by Eq. 12.77. The total COD in the reactor, T, is made up of three components: the COD due to effective cells; the COD due to soluble material, C; and the COD due to inert solids, $f'P_o$. However, P_o was given by Eq. 12.78. Thus:

$$\beta' = \frac{T - C - f'(T_o - C_o)}{M - fZ_o} \tag{12.82}$$

As in Section 12.4.1 a plot may be made of β' versus τ for use in the design calculations.

EXAMPLE 12.4.2-1

A CSTR contains a microbial culture that is growing on a medium which contains soluble organic matter, solid organic matter, and inorganic solids. The concentration of suspended solids in the influent is 120 mg/liter and 60% of it is

biodegradable. The total COD of the medium is 640 mg/liter of which 520 is soluble.
A long term batch experiment has indicated that 70% of the COD due to particulate
matter is biodegradable. Samples taken from the CSTR reveal that the soluble COD
is 100 mg/liter whereas the total (unfiltered) COD is 470 mg/liter. Furthermore
the concentration of suspended solids is 300 mg/liter. What is the value of β',
the oxygen demand coefficient of the effective cell material?

Use Eq. 12.82. The values to be substituted are

T_o = 640 mg/liter of COD

C_o = 520 mg/liter of COD

Z_o = 120 mg/liter of suspended solids

T = 470 mg/liter of COD

C = 100 mg/liter of COD

M = 300 mg/liter of suspended solids

f = 1.00 - 0 60 = 0.40

f' = 1.00 - 0.70 = 0.30

$$\beta' = \frac{470 - 100 - 0.30(640 - 520)}{300 - (0.40)(120)}$$

β' = 1.33 mg T_bOD/mg effective cell material

12.5 KEY POINTS

1. Because the specific growth rate (μ) of microorganisms is related to the concentration of growth limiting substrate (S) surrounding them, if μ can be controlled, then S can be controlled. In a CSTR, μ is controlled by the space
   time (τ); thus, so is S.

2. The concentration of cells in a CSTR depends upon the space time of the reactor
   and the amount of substrate removed. The observed yield and the viability
   depend only upon τ and decrease as τ is increased.

3. The oxygen requirement for a CSTR may be determined from an energy balance on
   the reactor. Stated in terms of oxygen demand, the balance is: "the amount
   of oxygen demand removed from the medium must equal the oxygen actually used
   by the culture plus the oxygen demand of the cells formed."

4. When the steady-state concentration is much less than K_S, the first-order
   approximation of the Monod equation may be used. The only equation which this
   affects, however, is the one for the substrate concentration in the reactor.

5. Nothing happens to inert solids in a CSTR so their concentration in the
   reactor is the same as in the influent.

6. The effect of microbial solids in the influent to a CSTR is to reduce the
   specific growth rate by an amount related to their contribution to the total
   viable biomass in the reactor. This requires alteration of the equations by
   which the concentrations of the various constituents are calculated.

7. The modeling approach taken when the influent contains biodegradable solids depends upon the relative amount of soluble substrate mixed with them.

8. The performance of a CSTR receiving soluble substrate is depicted by Fig. 12.2. At low space times no substrate removal and growth occur because the organisms can't grow rapidly enough. As τ is increased, the substrate concentration declines in response to the decline in the specific growth rate. The cell concentration in the reactor at small space times is low because substrate removal is incomplete. At long space times the cell concentration declines because of the importance of cell decay. The viability declines as τ is increased because the declining growth rate increases the relative importance of the death rate.

9. As the space time in a CSTR is increased, the amount of oxygen required increases because more of the waste is stabilized (i.e., oxidized to CO_2).

10. The most important effect of having microbial cells in the influent to a CSTR is to prevent washout, thereby allowing some substrate removal even at very small space times.

11. Changes in μ_m and K_s are reflected primarily in the substrate concentration whereas changes in Y_g and b have their largest effect upon the cell concentration.

12. Because the simple mathematical models of this chapter were developed from data on microbial growth in a constant chemical environment, care should be exercised when they are applied to tanks in series.

13. The equations describing the performance of the second and subsequent reactors in a multi-CSTR chain are essentially the same as those for a single CSTR receiving cells in the influent.

14. A chain of equivolume CSTR's has the following properties:
    (a) The larger the number of CSTR's in a chain, the greater the total volume must be to prevent washout, consequently chains containing more than three reactors are seldom used.
    (b) A chain of reactors is capable of achieving a lower substrate concentration than a single reactor whose volume is equal to the total volume of the chain.
    (c) Greater cell decay will occur in a chain than in a single CSTR.
    (d) Most of the oxygen required must be supplied to the first reactor.

15. Using a large reactor as the first reactor in a chain of CSTR's removing soluble substrate reduces the system space time at which washout occurs and makes it easier to transfer the oxygen to the first reactor, without detracting materially from the advantages associated with a chain.

16. By proper distribution of the feed to a two reactor chain it is possible to hold the specific growth rate of the cells in the second reactor near μ_m without causing washout.

17. For the removal of soluble substrate in a chain of CSTR's the optimum input distribution is generally to put all feed into the first reactor (unless it causes washout). No generalization can be made about the volume distribution leading to the minimum total volume since it will depend upon the desired effluent substrate concentration.

18. Verification of the models has come from the work of many people. It should be recognized, however, that the models are empirical so that their predictive value is not necessarily ascribable to correctness of theory or hypothesis.

19. The parameters μ_m, K_s, Y_g, b and γ may all be determined from linear plots of the data collected from running a CSTR at different values of τ.

20. When the influent to a CSTR contains solids, the fraction of those solids which is inert (f) and the fraction of the particulate COD (contributed by those solids) which is inert (f') must be determined before the kinetic parameters can be estimated.

21. When the influent to a CSTR contains biodegradable solids, μ_m and K_s are still determined from soluble T_bOD data. Y_g, b and β all become pseudoparameters, however, which incorporate the effects of the influent biodegradable solids.

12.6 STUDY QUESTIONS

The study questions are arranged in the same order as the text.

Section	Questions
12.1	1-13
12.2	14-20
12.3	None
12.4	21-24

1. Explain why the concentration of soluble substrate in a CSTR is controlled by the space time of the reactor.

2. Explain why there will always be soluble substrate present in a CSTR, regardless of how large τ is made.

3. Explain why there is a minimum space time below which microorganisms cannot grow in a CSTR.

4. Explain why the oxygen requirement in an aerobic CSTR may be calculated from a mass balance on "oxygen demand."

5. A feed containing a soluble substrate with a T_bOD of 1000 mg/l is flowing at a rate of 100 liters/hr into an aerobic CSTR which contains a mixed population of microorganisms. The parameters characterizing the culture are given in Table SQ12.1.

TABLE SQ12.1
Kinetic Parameters for a Microbial Culture

Parameter	Value
μ_m	$0.40\ hr^{-1}$
K_s	$120\ mg/l\ T_bOD$
b	$0.004\ hr^{-1}$
γ	$0.003\ hr^{-1}$
Y_g	$0.45\ mg\ cells/mg\ T_bOD$
β	$1.20\ mg\ O_2/mg\ cells$

(a) If τ = 100 hours, determine the following:
    (1) The concentration of soluble T_bOD in the effluent.
    (2) The total cell concentration in the reactor and the effluent.
    (3) The viability of the culture.
    (4) The mg/hr of oxygen which must be supplied to the reactor.

(b) Would it be possible to achieve an effluent soluble T_bOD of 1.5 mg/l
    with this substrate?

(c) Find τ_{min} for this feed.

(d) What size reactor would be required to reduce the soluble substrate con-
    centration to 20 mg/l? What concentration of viable cells would be
    present in the reactor?

(e) If the first-order approximation was used to predict the soluble substrate
    concentration in the reactor of part d, how much error would there be?

6. Explain how the presence of inert solids in the influent affects the perfor-
    mance of a CSTR.

7. Explain the effect that the presence of microbial solids in the influent to a
    CSTR has upon the specific growth rate of cells in the reactor.

8. Rework Study Question 5a and 5d by adding 100 mg/l of cells with a viability
    of 70% to the influent.

9. A CSTR is being used to degrade microbial cells. A stream containing cells at
    a concentration of 800 mg/l is flowing at a rate of 2 liters/hr into a CSTR
    with a volume of 480 liters. The oxygen demand coefficient for the cells, β_o,
    is 1.40 mg T_bOD/mg cells, and the specific decay rate, b, is 0.007 hr^{-1}.

(a) What is the cell concentration in the reactor?

(b) What is the oxygen consumption rate in mg/hr if β of the cells in the
    reactor is 1.20?

10. Explain how Eqs. 12.39 and 12.41 are modified for use in the situation where
    the influent contains primarily insoluble organic matter. What does b repre-
    sent in this situation?

11. Describe the approach used to model the situation in which the influent to a
    CSTR contains both soluble and insoluble substrate.

12. Describe and explain the effects of τ upon the performance of a CSTR.

13. Contrast the relative effectiveness of a CSTR as a remover of soluble substrate
    and a stabilizer of organic matter.

14. Explain some potential problems associated with the application of the models
    to CSTR's in series.

15. Given the feed and culture described in the problem statement of Study Question
    5, determine the following if the flow is directed to two 2000 liter CSTR's
    in series:

(a) The concentrations of cells and substrate in the effluent from the second
    reactor.

(b) The fraction of the oxygen requirement which must be supplied to each
    reactor.

16. Rework Study Question 9 using two CSTR's in series, each with a volume of 250 liters.

17. Explain why a chain of CSTR's can achieve a lower substrate concentration than a single CSTR with a volume equal to the total volume of the chain.

18. Explain why most of the oxygen required by a chain of CSTR's removing soluble substrate must be supplied to the first reactor.

19. Explain why it is often advantageous to make the first reactor in a chain of CSTR's larger than the others.

20. Rework Study Question 9 using two CSTR's in series, each with a volume of 240 liters, but put 40% of the flow directly into the first reactor and 60% directly into the second.

21. Describe how to determine the concentration of nonbiodegradable COD in a CSTR.

22. Describe how to use CSTR data to determine μ_m, K_s, Y_g, b, γ and β.

23. A culture of microorganisms was grown in a CSTR on a complex soluble substrate with a COD of 870 mg/l, and the data given in Table SQ12.2 were collected. Using the data, determine μ_m, K_s, Y_g, b and γ for the culture.

TABLE SQ12.2
Data from a CSTR at Steady State

τ hrs	Soluble COD mg/l	Suspended Solids mg/l	Viability %
4	828	19	98.9
8	162	317	97.8
12	121	324	96.9
16	106	321	96.0
32	88	291	93.1
64	82	240	89.1

24. Describe how to determine the fraction of inert solids (f) in a wastewater and the fraction of the COD due to influent solids which is inert (f').

REFERENCES AND FURTHER READING

1. C. P. L. Grady Jr., et al. "Effects of growth rate and influent substrate concentration on effluent quality from chemostats containing bacteria in pure and mixed culture," *Biotechnology and Bioengineering*, 14, 391-410, 1972.

2. C. P. L. Grady Jr. and D. R. Williams, "Effects of influent substrate concentration on the kinetics of natural microbial populations in continuous culture," *Water Research*, 9, 171-180, 1975.

3. G. T. Daigger and C. P. L. Grady Jr., "A model for the bio-oxidation process based on product formation concepts," *Water Research*, 11, 1049-1057, 1977.

4. C. P. L. Grady Jr. and R. E. Roper, Jr., "A model for the bio-oxidation process which incorporates the viability concept," *Water Research*, 8, 471-483, 1974.

5. R. E. McKinney and R. J. Ooten, "Concepts of complete mixing activated sludge," *Transactions of the 19th Sanitary Engineering Conference*, University of Kansas, 32-59, 1969.

6. S. Takahashi, *et al*, "Metabolism of suspended matter in activated sludge treat-
    ment," in *Advances in Water Pollution Research*, Vol 4, edited by S. H. Jenkins,
    Pergamon Press, London, 341, 1969.

7. R. M. Stein, *et al*, "A practical model of aerobic sludge stabilization,"
    *Proceedings of the Third Annual Environmental Engineering and Science
    Conference*, University of Louisville, Kentucky, 781-795, 1973.

8. C. W. Randall, *et al.*, "Aerobic digestion of trickling filter humus,"
    *Proceedings of the Fourth Annual Environmental Engineering and Science
    Conference*, University of Louisville, Kentucky, 557-577, 1974.

9. W. W. Eckenfelder, Jr., *Industrial Water Pollution Control*, McGraw-Hill,
    New York, 1966.

10. H. M. Tsuchiya, *et al.*, "Dynamics of microbial cell populations," *Advances in
    Chemical Engineering*, $\underline{6}$, 125-206, 1966.

11. D. Ramkrishna, *et al.*, "Dynamics of microbial propagation: Models considering
    endogenous metabolism," *Journal of General and Applied Microbiology*, $\underline{12}$,
    311-327, 1966.

12. D. Ramkrishna, *et al.*, "Dynamics of microbial propogation: Models considering
    inhibitors and variable cell composition," *Biotechnology and Bioengineering*,
    $\underline{9}$, 129-170, 1967.

13. D. Herbert, "A theoretical analysis of continuous culture systems," *Continuous
    Culture of Microorganisms*, Society of Chemical Industry, London, Monograph #12,
    21-53, 1960.

14. D. Herbert, "Multi-stage continuous culture" in *Continuous Culture of Micro-
    organisms*, Proceedings of the 2nd International Symposium on Continuous Culture,
    Edited by I. Malek, *et al.*, Academic Press, New York, 23-44, 1964.

15. R. B. Grieves, and R. Kao. "Input and volume distributions for continuous
    culture," *Biotechnology and Bioengineering*, $\underline{10}$, 497-510, 1968.

16. D. Herbert, *et al.*, The continuous culture of bacteria; A theoretical and
    experimental study," *Journal of General Microbiology*, $\underline{14}$, 601-622, 1956.

17. E. O. Powell, "The growth rate of microorganisms as a function of substrate
    concentration," in *Microbiol Physiology and Continuous Culture*, edited by
    E. O. Powell, *et al.*, Her Majesty's Stationery Office, London, 34-55, 1967.

18. S. J. Pirt, "The maintenance energy of bacteria in growing cultures,"
    Proceedings of the Royal Society, Series B, $\underline{163}$, 224-231, 1965.

19. C. G. Sinclair, and H. H. Topiwala, "Model for continuous culture which con-
    siders the viability concept," *Biotechnology and Bioengineering*, $\underline{12}$, 1069-1079,
    1970.

20. A. W. Lawrence and P. L. McCarty, "Unified basis for biological treatment
    design and operation," *Journal of the Sanitary Engineering Division, ASCE*,
    $\underline{96}$, 757-778, 1970.

21. S. Y. Chiu, *et al.*, "Kinetic model identification in mixed populations using
    continuous culture data," *Biotechnology and Bioengineering*, $\underline{14}$, 207-231, 1972.

22. S. Y. Chiu, *et al.*, "Kinetic behavior of mixed populations of activated sludge,"
    Biotechnology and Bioengineering, $\underline{14}$, 179-199, 1972.

23. W. L. Jorden, *et al.*, "Evaluating treatability of selected industrial wastes," *Proceedings of the 26th Industrial Waste Conference,* Purdue University Engineering Extension Series No. 140, 514-529, 1971.

24. C. R. Horn and F. G. Pohland, "Characterization and treatability of selected shellfish processing wastes," *Proceedings of the 28th Industrial Waste Conference,* Purdue University Engineering Extension Series No. 142, 819-831, 1973.

25. R. A. Kormanik, "Design of two-stage aerated lagoons," *Journal of the Water Pollution Control Federation,* 44, 451-458, 1972.

26. A. F. Gaudy Jr. and E. T. Gaudy, "Biological concepts for design and operation of the activated sludge process," *Environmental Protection Agency Water Pollution Research Series,* Report #17090 FQJ 09/71, Sept. 1971.

27. J. E. Dowd and D. S. Riggs, "A comparison of estimates of Michaelis-Menten kinetic constants from various linear transformations," *The Journal of Biological Chemistry,* 240, 863-869, 1965.

Further Reading

Aiba, S., *et al.*, *Biochemical Engineering,* 2nd Edition, Academic Press, New York, N.Y., 1973. A text dealing with the application of biochemical engineering to the fermentation industry. Chapter 5 deals with the modeling of biochemical reactors.

Andrews, J. F., "Kinetic models of biological waste treatment," *Biotechnology and Bioengineering Symposium No. 2,* 5-33, 1971.

Blanch, H. W. and Dunn, I. J., "Modelling and simulation in biochemical engineering," *Advances in Biochemical Engineering,* 3, 127-165, 1974.

Dean, A. C. R., *et al.* (editors), *Environmental Control of Cell Synthesis and Function,* Academic Press, London, 1972. Proceedings of the 4th International Symposium on the Continuous Culture of Microorganisms. Microorganisms.

Gaudy, A. F. Jr. and Gaudy, E. T., "Mixed microbial populations," *Advances in Biochemical Engineering,* 2, 97-143, 1972.

Herbert, D., "Some principles of continuous culture," *Recent Progress in Microbiology,* 7, 381-396, 1958. Presents the fundamental theory of microbial growth in CSTR's.

Lawrence, A. W., "Modeling and simulation of slurry biological reactors," in *Mathematical Modeling of Water Pollution Control Processes,* Edited by T. M. Keinath and M. Wanielista, Ann Arbor Science, Ann Arbor, Mich., 1975.

Malek, I. (editor), *Continuous Cultivation of Microorganisms,* Publishing House of the Czechoslovic Academy of Science, Prague, 1958. Proceedings of the 1st International Symposium on the Continuous Culture of Microorganisms. Contains papers outlining the basic theory of microbial growth in a CSTR.

Malek, I. and Beran, K., "Continuous cultivation of microorganisms, a review," *Folia Microbiologica,* 7, 388-411, 1962. Reviews papers published between Oct. 1960 and June 1962.

Malek, I., *et al.*, (editors), *Continuous Cultivation of Microorganisms,* Academic Press, New York, N.Y., 1964. Proceedings of the 2nd International Symposium on the Continuous Culture of Microorganisms. Contains papers dealing with the models for a chain of CSTR's.

Malek, I., *et al.*, (editors), *Continuous Cultivation of Microorganisms,* Academic Press, New York, N.Y., 1969. Proceedings of the 4th International Symposium on the Continuous Culture of Microorganisms.

Malek, I. and Fencl, Z., "Continuous cultivation of microorganisms, a review," *Folia Microbiologica,* 6, 192-209, 1961. Reviews papers published between Oct. 1959 and Oct. 1960.

Malek, I. and Fencl, Z., (editors), *Theoretical and Methodological Basis of Continuous Culture of Microorganisms,* Academic Press, New York, N.Y., 1966. Chapter 3 presents a theoretical analysis of both single and multistage CSTR's.

Malek, I. and Hospodka, J., "Continuous cultivation of microorganisms, a review," *Folia Microbiologica,* 5, 120-139, 1960. Reviews papers published up to Oct. 1959.

Malek, I. and Ricica, Jr., "Continuous cultivation of microorganisms, a review," *Folia Microbiologica,*

   a. 9, 321-344, 1964. Review papers published between July 1962 and Dec. 1963.

   b. 10, 302-323, 1965. Reviews papers published in 1964.

   c. 11, 479-535, 1966. Reviews papers published in 1965.

   d. 13, 46-96, 1968. Reviews papers published in 1966.

   e. 14, 254-277, 1969. Reviews papers published in 1967.

   f. 15, 129-149, 1970. Reviews papers published in 1968.

   g. 15, 377-416, 1970. Reviews papers published in 1969.

Moser, H., "The dynamics of bacterial populations maintained in the chemostat," Carnegie Institute of Washington, Publication No. 614, 1958.

Painter, P. R. and Marr, A. G. "Mathematics of microbial populations," *Annual Review of Microbiology,* 22, 519-548, 1968.

Powell, E. O. *et al.* (editors). *Microbial Physiology and Continuous Culture,* Her Majesty's Stationery Office, London, 1967. Contains papers which verify the theoretical models.

Ricica, J., "Multistage systems," in *"Methods in Microbiology,* Vol 2, edited by J. R. Norris and D. W. Ribbons, Academic Press, New York, N.Y., 329-348, 1970. Deals with the theory of CSTR's in series.

Ricica, J., "Continuous cultivation of microorganisms, a review," *Folia Microbiologica,*

   a. 16, 389-415, 1971. Reviews papers published in 1970.

   b. 17, 398-432, 1972. Reviews papers published in 1971.

   c. 18, 418-448, 1973. Reviews papers published in 1972.

   d. 19, 397-436, 1974. Reviews papers published in 1973.

Tempest, D. W., "The continuous cultivation of microorganisms: 1, Theory of the
    chemostat" in *Methods in Microbiology,* Vol 2, edited by J. R. Norris and D. W.
    Ribbons, Academic Press, New York, N.Y., 259-276, 1970. Deals with the theory
    of a single CSTR.

van Uden, N., "Kinetics of nutrient-limited growth," *Annual Review of Microbiology,*
    23, 473-486, 1969.

CHAPTER 13

CONTINUOUS STIRRED TANK REACTORS WITH CELL RECYCLE

Although the simple continuous stirred tank reactor finds many applications in wastewater treatment it suffers from two inherent disadvantages. First, relatively long
space times are required to achieve high levels of soluble substrate removal and
extraordinarily long ones are needed for waste stabilization; second, the reactor
space time governs its performance, so that if anything happens to alter τ the substrate removal efficiency will also be altered. Furthermore, since τ is the only
operational variable influencing performance, if the kinetic characteristics of the

substrate change, then the substrate removal efficiency will also change unless something is done to alter τ. In other words, no effective control of effluent quality
is possible, and consequently the use of a CSTR is restricted to those circumstances
in which control is relatively unimportant. Fortunately, a simple modification of
the CSTR enhances its controllability and allows reactors with relatively short
space times to achieve high degrees of waste stabilization. That modification involves the separation of the microbial cells from the effluent liquid with subsequent
recycle of a portion of them to the reactor.

Examination of Table 1.2 reveals that CSTR's employing cell recycle play an
important role in wastewater treatment. Thus in this chapter we will investigate
models describing them in order to gain an understanding of the important effect
that such a simple modification has upon system performance. Just as in Chapter 12,
the models will be confined to the growth of aerobic heterotrophic organisms in an
environment in which soluble organic substrate is the growth limiting material,
although with minor modifications they can be applied to the growth of autotrophs
and anaerobic heterotrophs as well. In Part V the models will be modified as needed
to allow their use in the design of named biochemical operations.

13.1 SINGLE CSTR

A schematic diagram of a single CSTR with cell recycle is shown in Fig. 13.1.
Because the reactor is completely mixed the concentrations of materials leaving it
are the same as those in it but upon reaching the settler the suspended solids
settle to the bottom so that the concentrations in the final effluent (X_e, etc.)
are much less than those in the reactor. Upon reaching the bottom of the settler
the solids are removed for recycle to the reactor. Solids must also be wasted from
the system, and two alternative methods are shown in the diagram. If they are
wasted from the settler underflow their concentration will be the same as that of
the solids in the recycle flow and the flow entering the settler will be $F + F_r$.
If they are wasted directly from the reactor, their concentration will be the same
as that of the solids in the reactor and the flow entering the settler will be
$F - F_w + F_r$. System performance is independent of the method by which solids are
wasted although the operational simplicity of the system depends very much upon
it [1].

In addition to the assumptions listed in Chapter 12, three others will be made
in the development of the models. The first is that all soluble substrate removal
occurs in the reactor and none in the settler. The concentration of substrate
entering the settler will be low enough to make the specific rate of substrate
utilization very low, and the rapid separation of cells from the liquid will make
their concentration in the bulk of the liquid low. Consequently, the rate of substrate removal in the settler will be negligible. One result of this assumption
is that the soluble substrate concentrations in the final effluent, sludge recycle,

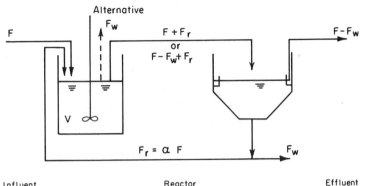

Influent

S_o – Soluble, biodegradable
  substrate, mg/l
X_o – cells, mg/l
Z_{io} – inert solids, mg/l
Z_{bo} – biodegradable solids, mg/l
Z_o – total (non-cell) solids, mg l
M_o – total suspended solids, mg/l

Reactor

S – Soluble, biodegradable
  substrate, mg/l
X – total cells, mg/l
X_v – viable cells, mg/l
X_d – dead cells, mg/l
Z_i – inert solids, mg l
M – total suspended solids, mg/l

Effluent

S_e – Soluble, biodegradable
  substrate, mg/l
X_e – total cells, mg/l
X_{ve} – viable cells, mg/l
X_{de} – dead cells, mg/l
Z_{ie} – inert solids, mg l
M_e – total suspended solids, mg/l

Recycle

S_r – Soluble, biodegradable
  substrate, mg/l
X_r – total cells, mg/l
X_{vr} – viable cells, mg/l
X_{dr} – dead cells, mg/l
Z_{ir} – inert solids, mg/l
M_r – total suspended solids, mg/l

Wastage

S_w – Soluble, biodegradable
  substrate, mg/l
X_w – total cells, mg/l
X_{vw} – viable cells, mg/l
X_{dw} – dead cells, mg/l
Z_{iw} – inert solids, mg/l
M_w – total suspended solids, mg/l

Figure 13.1. Schematic of a CSTR with cell recycle.

and wastage streams are all the same as in the reactor:

$$S_e = S_r = S_w = S \qquad (13.1)$$

The second assumption is that all biological activity stops in the settler so that
no further decay and death occur. Obviously there will be some activity because the
cells cannot cease metabolizing instantly. However, any that is present will act to
lower the concentration of dissolved oxygen in the biomass, which will in turn
probably act to reduce the rate of activity to a negligible level. The exact
effects of dissolved oxygen tension upon microbial activity are not completely
clear [2], however, so although the assumption appears reasonable it is open to some
question. Nevertheless, the qualitative predictions of the model should still be
valid. The third assumption is that the mass of organisms carried in the settler
and recycle stream is small compared to the mass in the reactor. Even though this
assumption has been the topic of recent debate [3] the effects of its violation will
be small because in cases where it is not true a corresponding change in the definition of the main operational variable can be made. It is reasonable for most systems
at steady state, however, especially if a suction type sludge pickup system is used
to remove the sludge rapidly from the bottom of the settler.

13.1.1 Basic Model--Soluble Substrate with Monod Kinetics

The simplest case should be investigated first in order to facilitate understanding. Therefore, only soluble substrate will be considered, making M_o equal to zero so that only microorganisms are present in the reactor and Eq. 12.2 still applies:

$$M = X \tag{12.2}$$

CONCENTRATION OF SOLUBLE SUBSTRATE AND CONCEPT OF MEAN CELL RESIDENCE TIME. The concentration of soluble substrate within the reactor is governed by the specific growth rate constant of the microorganisms as stated in Eq. 12.3:

$$S = \frac{\mu \, K_s}{\mu_m - \mu} \tag{12.3}$$

Thus, the problem of determining the concentration of soluble substrate becomes one of stating μ in terms of a manipulable parameter. First consider the situation in which cells are wasted from the bottom of the settler and perform a mass balance on viable cells within the reactor [4,5]. The input of cells in the feed is zero because M_o is zero, but viable cells are entering the reactor at concentration X_{vr} via the recycle flow, F_r. Viable organisms are flowing out at concentration X_v and rate $F + F_r$. Three reactions, growth, death, and decay, occur just as they did in the simple CSTR, so that when the mass balance equation is written and substitution is made for the reaction terms as in Eq. 12.4, the result is:

$$F_r X_{vr} - (F + F_r)X_v + \mu X_v V - bX_v V - \gamma X_v V = 0 \tag{13.2}$$

The recycle flow is proportional to the influent flow and the proportionality factor is called the recycle ratio:

$$\alpha = F_r/F \tag{13.3}$$

Rearrangement of Eq. 13.2 and substitution of the definition of space time, τ:

$$\tau = V/F \tag{4.38}$$

lead to an equation similar to Eq. 12.5:

$$\mu = G(1/\tau) + \dot{\gamma} + b \tag{13.4}$$

where

$$G = 1 + \alpha - \alpha(X_{vr}/X_v) \tag{13.5}$$

This equation is quite important because it shows that μ may be controlled independently of τ through manipulation of the term G. Thus, the addition of cell recycle to a CSTR has made it possible to adjust the specific growth rate, μ, thereby determining reactor performance, while maintaining a constant space time. Alternatively, it allows maintenance of a constant specific growth rate even when flow conditions alter τ. In other words, recycle places the reactor under operational control thereby overcoming one of the disadvantages associated with a simple CSTR. The other disadvantage of the simple CSTR was that long space times were often required to achieve high degrees of substrate removal. Examination of Eq. 13.4 suggests, however, that if G could be made small by proper manipulation of α, low specific growth rates could be obtained even though τ was small. Thus the use of cell recycle also overcomes the other disadvantage of a simple CSTR.

The use of the term G as expressed by Eq. 13.5 can be misleading because the ratio (X_{vr}/X_v) is related to α and cannot be chosen independently of it. A more realistic picture can be obtained by performing a mass balance on viable cells in the settler. In accordance with the assumptions listed at the start of this chapter, there will be no reaction in the settler, and at steady state the accumulation term will be zero, so that the balance simply states that input equals output. Furthermore, in order to simplify the equations, the concentrations of solids escaping in the effluent will be assumed to be zero, making the only output by the underflow. Thus the balance is:

$$(F + \alpha F)X_v - \alpha F X_{vr} - F_w X_{vw} = 0 \tag{13.6}$$

Recognizing that X_{vr} and X_{vw} are the same because they come out of the settler at the same point, Eq. 13.6 may be rearranged to give:

$$1 + \alpha - \alpha(X_{vr}/X_v) = F_w X_{vr}/FX_v \tag{13.7}$$

or

$$G = F_w X_{vr}/FX_v \tag{13.8}$$

In other words, the value of G is actually set by the rate at which cells are lost from the system $(F_w X_{vr})$ and not by the recycle ratio. Consequently, Eq. 13.4 states exactly the same thing that Eq. 12.5 stated, i.e., that the cells in a reactor system must grow at a rate which balances their loss by washout, death, and decay.

At this point it would be convenient to introduce a new variable. Recall that the definition of the space time is the volume of fluid contained in a reactor divided by the volume lost per unit time. Now, if both the numerator and denominator of Eq. 4.38 are multiplied by X_v the result is

$$\tau = VX_v/FX_v \tag{13.9}$$

Since in a CSTR the concentration of viable cells in the reactor is the same as the concentration in the effluent, Eq. 13.9 states that τ is equal to the mass of viable cells in the reactor divided by the mass of viable cells lost from the reactor per unit time, i.e., τ is equal to the residence time of the viable bacteria within the reactor. Thus τ determines μ in a simple CSTR because it determines the residence time of the bacteria within it. Using this same concept, we can now define a new residence time, called the mean cell residence (MCRT), θ_c, [6,7]:

$$\theta_c \equiv \frac{\text{Mass of viable bacteria in a reactor}}{\text{Mass of viable bacteria lost per unit time}} \qquad (13.10)$$

Assuming that the mass of bacteria contained in the settler is negligible with respect to the mass in the reactor, and that all loss of bacteria is by wastage the MCRT for a CSTR with cell recycle and wastage from the settler underflow is:

$$\theta_c = VX_v/F_w X_{vr} \qquad (13.11)$$

Examination of Eq. 13.8 reveals that if the top and bottom of the right side are both multiplied by V and substitutions of Eqs. 13.11 and 4.38 are made, the result is:

$$G = \tau/\theta_c \qquad (13.12)$$

Finally, when Eq. 13.12 is substituted into Eq. 13.4 we find that

$$\mu = 1/\theta_c + \gamma + b \qquad (13.13)$$

Comparison of this equation with Eq. 12.15 for a simple CSTR shows that they are of the same form, with the only difference being the substitution of θ_c for τ. Thus the specific growth rate of the organisms in a CSTR with recycle is independent of the space time of the reactor and dependent only upon the mean residence time of the cells within it. Furthermore, since the MCRT may be controlled through manipulation of the wastage rate, no matter what happens to the flow through the reactor it is possible to control the MCRT at a desired value.

Equation 13.11 states that the MCRT may be controlled simply by manipulating the wastage flow rate, F_w. However the manner in which F_w is manipulated depends upon the recycle ratio, α, because it determines the ratio X_{vr}/X_v as shown by Eq. 13.7. The relationship is easier to see if Eq. 13.11 is substituted into Eq. 13.7, yielding:

$$\frac{F_w}{F} = \frac{\tau\alpha}{\theta_c(1 + \alpha) - \tau} \qquad (13.14)$$

Letting F_w/F be defined as w:

$$w \equiv \frac{F_w}{F} \qquad (13.15)$$

we see that

$$w = \frac{\tau\alpha}{\theta_c(1 + \alpha) - \tau} \qquad (13.16)$$

which says that the wastage flow rate, expressed as a fraction of the inflow rate, depends upon the space time of the reactor, the recycle ratio, and the desired mean

cell residence time. Thus once the desired MCRT is known, the wastage flow required
to maintain it may be determined from Eq. 13.16 after α and τ are known. Equation
13.16 also tells us that the wastage fraction must be changed whenever the recycle
ratio, α, is changed if the MCRT is to be maintained at a constant value. This
follows from the fact that the sludge is wasted from the settler underflow and is
thus at the same concentration as the recycle flow. Because a mass balance must be
maintained on the settler, the ratio X_{vr}/X_v will change whenever α is changed and
it is this change in the concentration of solids being wasted that necessitates the
change in w.

The necessity for changing w each time α is changed led to the use of the alternative sludge wastage scheme shown in Fig. 13.1 [1]. In order to see how this scheme
simplifies things, let us perform a mass balance on viable cells in the reactor, and
derive a new equation for μ. The mass balance is similar to Eq. 13.2:

$$F_r X_{vr} - F_w X_v - (F - F_w + F_r)X_v + \mu X_v V - b X_v V - \gamma X_v V = 0 \qquad (13.17)$$

and reduces to it:

$$F_r X_{vr} - (F + F_r)X_v + \mu X_v V - b X_v V - \gamma X_v V = 0 \qquad (13.2)$$

Therefore the mass balance on viable cells still leads to Eqs. 13.4 and 13.5. Now
perform a mass balance on the settler:

$$(F - F_w + \alpha F)X_v - \alpha F X_{vr} = 0 \qquad (13.18)$$

which may be rearranged to give:

$$1 + \alpha - \alpha(X_{vr}/X_v) = F_w/F \qquad (13.19)$$

or

$$G = F_w/F \qquad (13.20)$$

Thus the concentration of cells in the recycle flow does not enter into the expression for G. Evoking the definition of MCRT, and noting that cells are wasted at
concentration X_v, we see that:

$$\theta_c = V X_v / F_w X_v \qquad (13.21)$$

or

$$\theta_c = \frac{V}{F_w} \qquad (13.22)$$

Substitution of Eq. 13.22 into Eq. 13.20 and rearrangement still lead to Eq. 13.12,
however:

$$G = \tau/\theta_c \qquad (13.12)$$

Consequently, Eq. 13.13 still defines the specific growth rate:

$$\mu = 1/\theta_c + \gamma + b \tag{13.13}$$

The advantage of the alternative system of sludge wastage can be seen by substituting Eq. 13.12 into Eq. 13.20:

$$\frac{F_w}{F} = \frac{\tau}{\theta_c} \tag{13.23}$$

or

$$w = \tau/\theta_c \tag{13.24}$$

Thus the wastage fraction is independent of the recycle ratio and need be changed only when τ changes. Actually, as shown by Eq. 13.22, the MCRT can be held constant by maintaining a constant wastage flow, F_w, no matter what happens to τ. This considerably simplifies the operation of a CSTR with cell recycle. Equation 13.22 also implies that no knowledge of the cell concentration is needed to determine the MCRT in a system wasting sludge from the reactor. This is another advantage of that wastage scheme, since cell concentrations are needed to determine the MCRT when wastage is from the settler underflow (see Eq. 13.11). Although Eq. 13.11 suggests that knowledge of the viable cell concentration is needed to calculate the MCRT when wastage is from the settler underflow the total cell concentration may be used instead. This follows from the fact that the viable cells will be dispersed evenly throughout the suspended solids in the reactor so that when the cells reach the settler they will settle without regard to whether they are live or dead. Consequently:

$$\frac{X_{vr}}{X_v} = \frac{X_r}{X} \tag{13.25}$$

so that Eq. 13.11 becomes

$$\theta_c = VX/F_w X_r \tag{13.26}$$

Having determined that both sludge wastage schemes lead to the same expression for the specific growth rate of the cells and that the MCRT is the operational parameter of primary importance, we can continue with the derivation of the equations in terms of the MCRT. The effect of MCRT on the concentration of soluble substrate may be found by substituting Eq. 13.13 into Eq. 12.3:

$$S = \frac{K_s(1/\theta_c + \gamma + b)}{\mu_m - (1/\theta_c + \gamma + b)} \tag{13.27}$$

Equation 13.27 is analogous to Eq. 12.6 for a simple CSTR, the only difference being the substitution of θ_c for τ. Thus the concentration of soluble substrate in a CSTR with cell recycle is dependent only upon the MCRT, θ_c, and is independent of the space time, τ. Equation 13.27 also implies that S is independent of S_o. However, as pointed out in Chapter 12 this is not true for a simple CSTR when mixed cultures are employed and a general test such as T_bOD is used as a measure of the substrate concentration; nor is it true for a CSTR with recycle [8]. Thus the same precautions suggested in Chapter 12 should be taken here.

Examination of Eq. 13.13 reveals that as the MCRT becomes very large (so that $1/\theta_c \to 0$), the specific growth rate of the organisms must equal the sum of the specific rates of death and decay, just as was necessary when τ became very large in a CSTR. Furthermore, the limit of Eq. 13.27 as θ_c approaches ∞ is the same as Eq. 12.7, so that the minimum attainable substrate concentration in the two reactors is the same:

$$S_{min} = \frac{K_s(\gamma + b)}{\mu_m - (\gamma + b)} \qquad (12.7)$$

Because substrate must always be present to drive the growth reaction, the presence of recycle has no effect upon S_{min}. Consequently, if a desired effluent substrate concentration cannot be attained in a simple CSTR, the addition of recycle will not alter that fact.

The maximum specific rate at which bacteria can grow in the reactor is still given by Eq. 12.8:

$$\hat{\mu} = \frac{\mu_m S_o}{K_s + S_o} \qquad (12.8)$$

Thus the minimum MCRT, θ_{cmin}, can be calculated by setting μ in Eq. 13.13 equal to $\hat{\mu}$ in Eq. 12.8 and rearranging:

$$\theta_{cmin} = \frac{K_s + S_o}{S_o(\mu_m - \gamma - b) - K_s(\gamma + b)} \qquad (13.28)$$

Washout will occur for values of θ_c less than θ_{cmin}. In theory there is no minimum space time, τ, for a CSTR with cell recycle because as long as the cells can be separated from the effluent and returned to the reactor so that the MCRT is greater than θ_{cmin}, growth can be maintained. In practice however, τ should not be made less than θ_{cmin}. If τ were less than θ_{cmin} and something happened to the cell recycle so that the reactor behaved as a simple CSTR, then the MCRT would be the same as τ, washout would occur, and the process would fail. Furthermore, if τ were less than θ_{cmin}, the process would be difficult to restart once failure had occurred. Thus to be safe, τ should be kept larger than θ_{cmin}.

EXAMPLE 13.1.1-1

A culture of microorganisms is being grown in a CSTR with cell recycle, with a volume of 8 liters. The medium contains lactose as the sole carbon source and all inorganic nutrients are present in excess. When the concentration of lactose is 1000 mg/liter as $T_b OD$ the kinetic parameters have the values shown in Table E13.1.

TABLE E13.1
Kinetic Parameters for a Culture Growing on
Lactose at S_o = 1000 mg/liter $T_b OD$

Symbol	Value
μ_m	0.80 hr^{-1}
K_s	165.0 mg/liter
Y_g	0.55 mg cells/mg $T_b OD$
b	0.01 hr^{-1}
γ	0.005 hr^{-1}

(a) What is the maximum permissible flow rate through the reactor?

When the flow is maximum, τ will be minimum. τ_{min} should not be less than θ_{cmin}. Thus calculate θ_{cmin} using Eq. 13.28:

$$\theta_{cmin} = \frac{165 + 1000}{1000(0.8 - 0.01 - 0.005) - 165(0.01 + 0.005)}$$

θ_{cmin} = 1.49 hrs.

Therefore,

τ_{min} = 1.49 hrs

and the maximum permissible flow is

F = 8 liters/1.49 hrs = 5.37 liters/hr

(b) The flow through the reactor is 1.0 liter/hr and sludge is wasted directly from the reactor at a rate of 0.1 liter/hr. What is the concentration of soluble $T_b OD$ in the effluent?

First we must calculate θ_c using Eq. 13.22:

θ_c = 8 liters/(0.1 liter/hr)

θ_c = 80 hrs

Then we must use Eq. 13.27 to find S:

$$S = \frac{165(1/80 + 0.005 + 0.01)}{0.80 - (1/80 + 0.005 + 0.01)}$$

S = 5.9 mg/liter $T_b OD$

(c) The flow through the reactor is 1.0 liter/hr and sludge is wasted from the settler underflow at a rate of 0.1 liters/hr. The recycle rate is such that X/X_r is 0.5. What is the concentration of soluble $T_b OD$ in the effluent?

First we must calculate θ_c using Eq. 13.26:

$$\theta_c = \frac{(8 \text{ liters})(0.5)}{(0.1 \text{ liters/hr})}$$

$$\theta_c = 40 \text{ hrs}$$

Then we must use Eq. 13.27 to find:

$$S = \frac{165(1/40 + 0.005 + 0.01)}{0.80 - (1/40 + 0.005 + 0.01)}$$

$$S = 8.7 \text{ mg/liter}$$

Thus even though the volume of wastage flow was the same as in part b, the fact that the concentration of cells in the wastage was higher in part c caused a larger mass to be wasted per unit time, thereby decreasing θ_c and increasing S.

CONCENTRATION OF SUSPENDED SOLIDS. For a soluble substrate the only suspended solids in the reactor will be microorganisms, which can be divided into two groups, viable and dead cells. This will be true both in the reactor:

$$X = X_v + X_d \tag{12.10}$$

and in the settler underflow:

$$X_r = X_{vr} + X_{dr} \tag{13.29}$$

Furthermore the fraction of viable cells in each location will be the same, as given by Eq. 13.25.

The only source of viable organisms in the system is from growth due to substrate utilization and the presence of recycle does not alter that fact. Therefore, just as in Chapter 12 the equation for viable cells is determined from a mass balance on substrate:

$$FS_o + \alpha FS - (F + \alpha F)S + r_s V = 0 \tag{13.30}$$

where $-r_s$ is again the rate of substrate utilization as given by Eq. 10.39

$$-r_s = (\mu/Y_g)X_v \tag{10.39}$$

Because the term αFS cancels out of Eq. 13.30, substitution of 10.39 into 13.30 and simplification yield the same result as in Chapter 12, i.e.:

$$X_v = \frac{Y_g(S_o - S)}{\mu\tau} \tag{12.12}$$

In this case, however, the equation for the specific growth rate constant, μ, is different, so that substitution of Eq. 13.13 into 12.12 yields:

$$X_v = \frac{Y_g(S_o - S)}{[(1/\theta_c) + \gamma + b]\,\tau}$$ (13.31)

This states that the concentration of viable cells depends upon both θ_c and τ. Furthermore, rearrangement of Eq. 13.31 shows that for a fixed MCRT (which determines S), the product $X_v\tau$ is a constant:

$$X_v\tau = \frac{Y_g(S_o - S)}{1/\theta_c + \gamma + b}$$ (13.32)

Thus a reactor with a short space time will have a higher concentration of viable cells in it than one with a long space time. In a simple CSTR the viable cell concentration was fixed by τ and could not be altered by the operator as long as τ was constant. This is not true when recycle is employed, however. Examination of Eqs. 13.31 and 13.32 shows that for a fixed τ the viable cell concentration may be changed by changing θ_c.

The concentration of dead microorganisms can be obtained from a mass balance on them. Regardless of the method of cell wastage, the equation contains terms for inflow from the recycle, outflow, and reaction by generation and decay:

$$\alpha F X_{dr} - (F + \alpha F)X_d + \gamma X_v V - b X_d V = 0$$ (13.33)

When this is rearranged it gives:

$$1 + \alpha - \alpha(X_{dr}/X_d) = (\gamma X_v \tau / X_d) - b\tau$$ (13.34)

However, because the fraction of live cells in the reactor and in the settler underflow are the same, a similar statement can be made about the dead cells, so that

$$\frac{X_{dr}}{X_d} = \frac{X_{vr}}{X_v}$$ (13.35)

Substitution of Eq. 13.35 into Eq. 13.34 and use of the definition of G lead to:

$$G = (\gamma X_v \tau / X_d) - b\tau$$ (13.36)

Now, no matter from where cells are wasted, Eq. 13.12 is true, so that substitution of it into 13.36 gives:

$$X_d = \frac{\gamma X_v}{1/\theta_c + b}$$ (13.37)

This is similar to Eq. 12.15.

The total cell concentration can be obtained by summing Eqs. 13.31 and 13.37, and then rearranging:

$$X = (\frac{\theta_c}{\tau}) \frac{Y_g(S_o - S)}{1 + b\theta_c} \tag{13.38}$$

If there is no recycle so that θ_c and τ are the same, this equation reduces to Eq. 12.16. Consequently the comments made about that equation are applicable here as well. In addition, this equation states that the $X\tau$ product is fixed for any given value of θ_c. Thus as far as theory is concerned, once the MCRT has been chosen to give a desired effluent substrate concentration any combination of reactor size and cell concentration may be used as long as it gives the proper $X\tau$ product. There are practical limits, of course, and these will be discussed in Chapter 16.

In Chapter 12, Eq. 12.18 showed how the observed yield depended upon τ in a simple CSTR. Using a similar approach here, it can be shown that

$$Y = \frac{Y_g}{1 + b\theta_c} \tag{13.39}$$

Thus the observed yield in a CSTR with recycle is independent of τ and dependent only upon the MCRT. As the MCRT is increased the opportunity for cell decay increases and the yield thus decreases.

The culture viability may be determined by substituting Eqs. 13.31 and 13.38 into the definition of viability as given by Eq. 10.13. The result is:

$$v = \frac{1 + b\theta_c}{1 + b\theta_c + \gamma\theta_c} \tag{13.40}$$

which reverts to Eq. 12.19 when there is no recycle so that θ_c and τ are equal.

The required recycle ratio depends primarily upon the degree of concentration of biological solids that is desired in the final settler, and to a lesser degree upon the ratio of τ to θ_c. Since all cells that enter the settler must leave it, the equation relating α to the other variables may be obtained from a mass balance on cells in the settler. When cell wastage is from the settler underflow the balance leads to an equation analogous to 13.7:

$$1 + \alpha - \alpha(X_r/X) = F_w X_r/FX \tag{13.41}$$

Rearrangement gives

$$\alpha = \frac{1 - (F_w/F)(X_r/X)}{(X_r/X) - 1} \tag{13.42}$$

and substitution of Eqs. 13.8, 13.12, and 13.25 yields:

$$\alpha = \frac{1 - \tau/\theta_c}{(X_r/X) - 1} \tag{13.43}$$

Actually, this equation is applicable no matter which method of cell wastage is employed. (The derivation of the same equation for the case of cell wastage from the reactor is left as an exercise for the reader). The wastage rate, of course, will depend upon the location of the wastage as well as upon α, τ, and θ_c. For wastage from the settler underflow:

$$F_w = F[\frac{\tau\alpha}{\theta_c(1 + \alpha) - \tau}] \tag{13.14}$$

For wastage from the reactor:

$$F_w = F(\tau/\theta_c) \tag{13.23}$$

Equation 13.43 simply gives the recycle ratio required to satisfy the mass balance on cells around the settler. It says nothing about whether the settler can attain the degree of concentration (X_r/X) desired. For any given inflow to the settler the maximum possible degree of concentration will depend upon the settler surface area and the thickening properties of the solids [9]. Thus, in practice the choice of a recycle ratio must be made only after consideration has been given to the thickening characteristics of the cells. Several authors have shown that the volume of the CSTR and the surface area of the final settler are interrelated through the cell concentration, X, and the recycle ratio, α. Consequently, it is possible to arrive at a combination of CSTR and settler sizes that results in a least cost design [10,11,12,13,14]. This will be considered again briefly in Chapter 16. For the remainder of this chapter, however, we will assume that any α can be chosen and that no limits exist on X_r/X.

EXAMPLE 13.1.1-2
Continue with the problem begun in Example 13.1.1-1.
(a) What is the total cell concentration in the reactor when the inflow is 1.0 liter/hr and cells are wasted directly from the reactor at a rate of 0.1 liter/hr?

From Example 13.1.1-1b, θ_c = 80 hrs, S = 5.9 mg/liter, and τ = 8.0 hrs. Thus, using Eq. 13.38:

$$X = \frac{(0.55)(1000 - 5.9)}{[(1/80) + 0.01]\ 8}$$

X = 3038 mg/liter

(b) What recycle ratio will be required if it is desired to concentrate the cells by a factor of 2.0 for recycle (i.e., X_r/X = 2.0)?

Use Eq. 13.43:

$$\alpha = \frac{1 - (8.0/80)}{2.0 - 1}$$

α = 0.9

(c) What is the total cell concentration in the reactor when the inflow is 1.0 liter/hr and sludge is wasted from the settler underflow at a rate of 0.1 liter/hr? The recycle rate is such that $X_r/X = 2.0$.

From Example 13.1.1-1c, θ_c = 40 hrs, S = 8.7 mg/liter and τ = 8 hrs. Using Eq. 13.38:

$$X = \frac{(0.55)(1000 - 8.7)}{[(1/40) + 0.01]\ 8}$$

$$X = 1947 \text{ mg/liter}$$

(d) What recycle ratio is required in part c?

Use Eq. 13.43:

$$\alpha = \frac{1 - (8/40)}{2.0 - 1}$$

$$\alpha = 0.8$$

(e) What is the observed yield in part c?

Use Eq. 13.39:

$$Y = \frac{0.55}{1 + (0.01)(40)}$$

$$Y = 0.39 \text{ mg cells/mg } T_b OD \text{ removed.}$$

OXYGEN REQUIREMENT. The oxygen requirement for the reactor may be determined from an oxygen demand balance in a manner analogous to that in Section 12.1.1. Assuming steady-state conditions:

$$FS_o + \alpha FS + \alpha F\beta X_r - (F + \alpha F)S - (F + \alpha F)\beta X + (-r_o V) = 0 \tag{13.44}$$

Rearrangement and simplification yield:

$$r_o \tau = S_o - S - [1 + \alpha - \alpha(X_r/X)]\beta X \tag{13.45}$$

However, the bracketed term is G (Eq. 13.5) which is equal to τ/θ_c (Eq. 13.12). When those substitutions are made, along with Eq. 13.38 for X, the result is:

$$r_o = \frac{(S_o - S)(1 + b\theta_c - \beta Y_g)}{\tau(1 + b\theta_c)} \tag{13.46}$$

This says that the rate in mg/(liter·hr) depends upon both τ and θ_c. However, using the definition of τ and letting RO stand for the mass rate of oxygen uptake as was done in Eq. 12.22, it can be seen that RO depends only on θ_c:

$$RO = \frac{F(S_o - S)(1 + b\theta_c - \beta Y_g)}{1 + b\theta_c} \tag{13.47}$$

This is analogous to Eq. 12.23. As the MCRT is increased, the mass of oxygen consumed per unit time increases, due to the increased importance of decay. If for a given MCRT, τ is increased, then the same mass would be consumed from a larger volume, thereby making the rate in mg/(liter·hr) decrease. This is what Eq. 13.46 says.

EXAMPLE 13.1.1-3

Continue with the problem begun in Example 13.1.1-1. How many mg/hr of oxygen would be needed for operation under the conditions listed in Example 13.1.1-1b, if β = 1.20 mg O_2/mg cells?

From that example, θ_c = 80 hrs, S = 5.9 mg/liter and F = 1.0 liter/hr. Using Eq. 13.47:

$$RO = \frac{1.0(1000 - 5.9)[1 + (0.01)(80) - (1.2)(0.55)]}{1 + (0.01)(80)}$$

RO = 629 mg/hr

EXCESS MICROORGANISM PRODUCTION RATE. One of the primary reasons for using a CSTR with cell recycle is to provide an effluent of high quality; that is, one that is not only low in soluble substrate but also low in suspended solids. In order to accomplish the latter the cells are separated from the effluent and a concentrated slurry is recycled to the reactor. However, as the equations of this chapter have shown, it is necessary to continually waste some of the cells if a steady state is to be achieved. The disposal of those cells constitutes a major expense in the treatment of a wastewater, and thus it would be convenient to be able to relate the mass which must be wasted to the other parameters describing the process. The mass wastage rate is just the wastage flow multiplied by the concentration of cells in the wastage stream, and it has been given the symbol P_x [6]. Thus, the equation for P_x depends upon the point of sludge wastage:

From the settler underflow

$$P_x = F_w X_r \tag{13.48}$$

From the reactor

$$P_x = F_w X \tag{13.49}$$

However, by returning to the definitions of mean cell residence time given by Eqs. 13.26 and 13.22 it can be seen that P_x is the same no matter where wastage is from:

$$P_x = VX/\theta_c \tag{13.50}$$

Furthermore, substitution of Eq. 13.38 for X gives a more general equation for P_x:

$$P_x = \frac{FY_g(S_o - S)}{1 + b\theta_c} \tag{13.51}$$

Since S depends only upon the MCRT, it can be seen that P_x depends only upon the MCRT and will decrease as the MCRT increases. This is due to the increased importance of decay at long mean cell residence times.

EXAMPLE 13.1.1-4

How many mg/hr of cells would have to be disposed of in Example 13.1.1-1b?

Using Eq. 13.51:

$$P_x = \frac{(1.0)(0.55)(1000 - 5.9)}{1 + (0.01)(80)}$$

$$P_x = 304 \text{ mg/hr}$$

RELATIONSHIP OF MCRT TO PROCESS LOADING FACTOR. Before the introduction of MCRT as the basic independent variable in the design and control of a CSTR with cell recycle, most designers used the process loading factor. The process loading factor, U, is defined as the mass of substrate used over a finite time divided by the mass of organisms contained in the reactor [15].

$$U \equiv \frac{F(S_o - S)}{XV} \tag{13.52}$$

That U is related to θ_c may be shown in the following manner. Rearrangement of the mass balance on soluble substrate given in Eq. 13.30 shows that

$$-r_s = \frac{F(S_o - S)}{V} \tag{13.53}$$

However, in Eq. 10.41 r_s was defined as

$$-r_s = qX_v \tag{10.41}$$

which implies that

$$q = \frac{F(S_o - S)}{VX_v} \tag{13.54}$$

The term q was called the specific substrate removal rate and is the mass of substrate used per unit time divided by the mass of viable bacteria in the reactor.

Furthermore Eq. 10.40 stated that

$$q = \mu/Y_g \qquad (10.40)$$

Substitution of Eq. 13.13 for μ gives:

$$qY_g = 1/\theta_c + \gamma + b \qquad (13.55)$$

Now, from the definition of viability we know that

$$X_v = vX \qquad (13.56)$$

Substitution of Eq. 13.56 into Eq. 13.54 and use of Eq. 13.52 tells us that

$$q = U/v \qquad (13.57)$$

Thus,

$$UY_g = v(1/\theta_c + \gamma + b) \qquad (13.58)$$

However, Eq. 13.40 expressed the effects of θ_c upon the viability, so when it is substituted into Eq. 13.58 and the result is simplified, we see that

$$UY_g = 1/\theta_c + b \qquad (13.59)$$

Thus the process loading factor is inversely proportional to the MCRT and both may be used in process design and operation. There will be instances in which the use of the MCRT is easier, however, such as when the reactor contains solids other than bacteria and these will be discussed in Section 13.1.3.

EXAMPLE 13.1.1-5

Continue with the problem begun in Example 13.1.1-1. Calculate the process loading factor on the reactor when it is operated under the conditions listed in Example 13.1.1-1b.

From Example 13.1.1-1b we know that θ_c = 80 hrs. From Table E13.1 we know that Y_g = 0.55 and b = 0.01. Thus, using Eq. 13.59

$$U = \frac{(1/80) + 0.01}{0.55}$$

U = 0.041 mg T_bOD/(hr·mg cells)

The process loading factor could also be calculated from the definition (Eq. 13.52). From the problem statement and Example 13.1.1-1b we know that F = 1.0 liter/hr, S_o = 1000 mg/liter T_bOD, S = 5.9 mg/liter T_bOD, and V = 8.0 liters. Furthermore, from Example 13.1.1-2a we know that X = 3038 mg/liter cells.

Thus

$$U = \frac{1.0(1000 - 5.9)}{(3038)(8.0)}$$

$$U = 0.041 \text{ mg T}_b\text{OD/(hr}\cdot\text{mg cells)}$$

13.1.2 First-Order Approximation

As discussed in Section 12.1.2, occasions often arise in which the steady-state substrate concentration is much less than the saturation constant so that the Monod equation may be simplified into a first-order equation:

$$\mu = \frac{\mu_m S}{K_s} \tag{10.43}$$

This situation often arises when CSTR's with recycle are being used, thereby making it difficult to evaluate μ_m and K_s separately. In that case, the mean reaction rate coefficient, k_e, is used. That parameter was given in Chapter 12 as

$$k_e = (\frac{\mu_m}{K_s}) \frac{1}{Y_g} \tag{12.25}$$

so that

$$\mu = Y_g k_e S \tag{13.60}$$

As pointed out in Chapter 12, the use of Eq. 10.43 (or 13.60) does not alter any of the mass balances nor does it alter the fact that μ is controlled by the MCRT as expressed in Eq. 13.13. All that it does is to simplify the relationship between S, the reactor substrate concentration, and the MCRT. Substitution of Eq. 13.60 into Eq. 13.13 and simplification yield:

$$S = \frac{1/\theta_c + \gamma + b}{Y_g k_e} \tag{13.61}$$

This equation is analogous to Eq. 12.26. None of the other equations are affected by the first-order approximation, except through the effect on S. Once more, however, it should be emphasized that Eq. 13.61 will only be valid when the first-order approximation can be justified and thus care should be exercised in its use.

13.1.3 Modifications of the Basic Model

The necessity for considering the effects of influent suspended solids was discussed in Section 12.1.3, and those comments also apply here. In fact, consideration of the effects of influent solids is even more important for a CSTR with recycle, because the recycle magnifies them. Consequently we will again look at what happens when the three types of solids (inert solids, microorganisms, and biodegradable solids) are present in the influent.

INERT SOLIDS. Because of the recycle, the concentration of inert solids in the reactor will be higher than in the feed, as can be seen by a mass balance on inert solids:

$$FZ_{io} + \alpha FZ_{ir} - F(1 + \alpha)Z_i = 0 \tag{13.62}$$

Rearranging terms gives:

$$Z_i = \frac{Z_{io}}{1 + \alpha - \alpha(Z_{ir}/Z_i)} \tag{13.63}$$

If the inert solids are entrapped in the biological solids, then the entire mass will settle together so that:

$$\frac{Z_{ir}}{Z_i} = \frac{X_{vr}}{X_v} \tag{13.64}$$

Substitution of Eq. 13.64 into Eq. 13.63 reveals that the denominator of 13.63 is G (Eq. 13.5). Thus:

$$Z_i = Z_{io}/G \tag{13.65}$$

Furthermore, regardless of the point from which solids are wasted, G is equal to the ratio of the space time to the MCRT (Eq. 13.12) so that the concentration of inert solids in the reactor depends upon both:

$$Z_i = Z_{io}(\theta_c/\tau) \tag{13.66}$$

The mixture of inert and microbial solids in a CSTR with recycle is called the mixed liquor suspended solids (MLSS), M:

$$M = X + Z_i \tag{12.28}$$

Substitution of Eqs. 13.38 and 13.66 into 12.28 shows the effect of the MCRT and the space time upon M:

$$M = (\frac{\theta_c}{\tau}) [\frac{Y_g(S_o - S)}{1 + b\theta_c} + Z_{io}] \tag{13.67}$$

Looking at the bracketed term it can be seen that the contribution of cells to the MLSS decreases as the MCRT is increased whereas the contribution of inert solids does not. Thus the percentage of cells in the MLSS decreases as the MCRT is increased. In addition, Eq. 13.40 showed that the viability of the cells decreases as the MCRT is increased. Consequently, less of the MLSS actually contribute to substrate removal as the MCRT is increased.

The mass rate at which solids must be disposed of will also be increased by the presence of the inert solids. Letting P_m represent the "production rate" of excess

MLSS, the equation representing it can be found by substituting M for X in Eq. 13.50, yielding:

$$P_m = F[\frac{Y_g(S_o - S)}{1 + b\theta_c} + Z_{io}] \tag{13.68}$$

Since nothing happens in the biochemical operation to reduce the amount of inert solids present, the mass of solids to be disposed of will just be increased by the mass rate of input of inert solids.

EXAMPLE 13.1.3-1

Continue with the problem begun in Example 13.1.1-1.

(a) What will be the MLSS concentration when the reactor is operated under the conditions listed in Example 13.1.1-1b if the influent contains 100 mg/liter of inert solids?

From example 13.1.1-1b and the problem statement we know that θ_c = 80.0 hrs, τ = 8 hrs, S_o = 1000 mg/liter, S = 5.9 mg/liter, Y_g = 0.55, and b = 0.01 hr^{-1}. Using Eq. 13.67

$$M = (\frac{80}{8})[\frac{0.55(1000 - 5.9)}{1 + (0.01)(80)} + 100]$$

$$M = 4038 \text{ mg/liter}$$

(b) What fraction of the MLSS is made up of cells?

From Example 13.1.1-2a we know that X = 3038 mg/liter. Thus

$$\text{fraction of cells} = \frac{3038}{4038} = 0.75$$

(c) What fraction of the MLSS is made up of viable cells?

Using Eq. 13.31 and the fact that γ = 0.005 hr^{-1} we can find the viable cell concentration:

$$X_v = \frac{0.55(1000 - 5.9)}{[(1/80) + 0.005 + 0.01] \ 8}$$

$$X_v = 2485 \text{ mg/liter}$$

Thus:

$$\text{fraction of viable cells} = \frac{2485}{4038} = 0.62$$

Thus only 62% of the MLSS is contributing to substrate removal.

(d) What is the excess sludge production rate?

Using Eq. 13.68

$$P_m = 1.0 \ [\frac{(0.55)(1000 - 5.9)}{1.0 + (0.01)(80)} + 100]$$

$$P_m = 404 \text{ mg/hr}$$

Comparison of this with P_x in Example 13.1.1-4 shows that when 100 mg/hr of inert solids were added to the influent, the amount of solids to be disposed of increased by 100 mg/hr. Since they were inert they were not affected by the reactor.

It is common in practice to define the process loading factor in terms of the MLSS concentration rather than the cell concentration:

$$U_M = \frac{F(S_o - S)}{VM} \tag{13.69}$$

Nevertheless, Eq. 13.69 cannot be related simply to the specific growth rate of the microorganisms because it contains solids which are not biological in origin and thus U_M cannot be easily related to S. This problem does not exist when the MCRT is used as the operational variable, however. If solids are wasted directly from the reactor, the equation defining θ_c does not even contain a term for the solids concentration (see Eq. 13.22). If solids are wasted from the settler underflow, the MCRT may be computed from MLSS data because the solids generally settle homogeneously so that

$$\frac{M}{M_r} = \frac{X_v}{X_{vr}} \tag{13.70}$$

Substitution of Eq. 13.70 into Eq. 13.11 yields:

$$\theta_c = VM/F_w M_r \tag{13.71}$$

Consequently the use of the MCRT allows operational control of the reactor regardless of the types of solids present. This is a significant advantage.

MICROBIAL SOLIDS. The effects of microbial solids on a CSTR with recycle are very similar to those on a simple CSTR. By employing a mass balance on viable organisms in a reactor receiving them at concentration X_{vo} it can be seen that:

$$\mu = [1 + \alpha - \alpha(X_{vr}/X_v) - X_{vo}/X_v] \, 1/\tau + \gamma + b \tag{13.72}$$

Furthermore, substitution of Eqs. 13.5 and 13.12 into 13.72 yields:

$$\mu = (\tau/\theta_c - X_{vo}/X_v)1/\tau + \gamma + b \tag{13.73}$$

which simplifies to Eq. 12.30 when the MCRT is equal to the space time. Thus, just as in a simple CSTR, microbial solids in the influent act to reduce the specific growth rate below the value that would be attained in a similar reactor receiving no cells. A mass balance on soluble substrate leads to Eq. 12.12 and substitution of Eq. 13.73 for μ gives:

$$X_v = (\frac{\theta_c}{\tau}) \, [\frac{X_{vo} + Y_g(S_o - S)}{1 + \gamma\theta_c + b\theta_c}] \tag{13.74}$$

Comparison of Eq. 13.74 with Eq. 13.31 reveals that the viable cell concentration will be higher than in a reactor receiving no solids. Substitution of Eq. 13.74 into 13.73, with subsequent substitution of the resulting equation into 12.3 yields a quadratic equation for the substrate concentration:

$$[\mu_m - (1/\theta_c + \gamma + b)]S^2$$
$$- [\mu_m(X_{vo}/Y_g + S_o) + (K_s - S_o)(1/\theta_c + \gamma + b)]S$$
$$+ S_o K_s(1/\theta_c + \gamma + b) = 0 \tag{13.75}$$

The similarity of this equation to Eq. 12.32 is apparent. The concentration of dead cells can be obtained from a mass balance on them, considering their concentration in the influent, X_{do}. The resulting equation is:

$$X_d = \frac{X_{do}(\theta_c/\tau) + \gamma\, X_v \theta_c}{1 + b\theta_c} \tag{13.76}$$

The equation for the total cell concentration may be obtained by summing Eqs. 13.74 and 13.76 and simplifying:

$$X = (\frac{\theta_c}{\tau})\, [\frac{X_o + Y_g(S_o - S)}{1 + b\theta_c}] \tag{13.77}$$

where

$$X_o = X_{vo} + X_{do} \tag{12.35}$$

Thus the total cell concentration will be increased by the presence of cells in the influent. It should be noted, however, that the presence of cells in the influent will not increase the MLSS concentration as much as the presence of an equal amount of inert solids. This is because the microbial cells are subject to decay. The equation for the viability of the cells in the reactor comes from substitution of Eqs. 13.74 and 13.77 into Eq. 10.13:

$$v = (\frac{1 + b\theta_c}{1 + b\theta_c + \gamma\theta_c})\, (\frac{X_{vo} + Y_g(S_o - S)}{X_o + Y_g(S_o - S)}) \tag{13.78}$$

This equation is similar to Eq. 12.36 and thus has the same implication, i.e., that the viability of the cells in a reactor receiving cells in the influent will be less than the viability of those in a reactor at the same MCRT which is not receiving cells. The final equation of importance is the one describing the oxygen require-ment, which can be obtained by performing a mass balance on oxygen demand in the reactor. When the symbol β_o is used to represent the oxygen demand coefficient of the influent cells, the resulting equation is:

$$RO = \frac{F\{(S_o - S)(1 + b\theta_c - \beta Y_g) + X_o[(\beta_o - \beta) + \beta_o b\theta_c]\}}{1 + b\theta_c} \tag{13.79}$$

which is analogous to Eq. 12.38. It also simplifies to Eq. 13.47 when X_o is zero.
The excess microorganism production rate may be found by substitution of Eq. 13.77
into Eq. 13.50:

$$P_x = F[\frac{X_o + Y_g(S_o - S)}{1 + b\theta_c}] \qquad\qquad (13.80)$$

Since decay of the microbial cells will occur in the reactor, the increase in excess
microorganisms to be disposed of will be less than their mass input to the reactor.

EXAMPLE 13.1.3-2

Continue with the problem begun in Example 13.1.1-1, only add 50 mg/liter of cells
with a viability of 90% to the influent. The reactor is operated under the conditions
listed in Example 13.1.1-1b, i.e., F = 1.0 liter/hr, with sludge wasted from the
reactor at a rate of 0.1 liter/hr.

(a) What is the concentration of soluble substrate?

The presence of cells in the influent has no effect upon θ_c, so from
Example 13.1.1-1b it is 80 hrs. X_{vo} is 45 mg/liter. Equation 13.75 must be used
to find S

$$[0.80 - (1/80 + 0.005 + 0.01)]S^2$$

$$- [0.80(45/0.55 + 1000) + (165 - 1000)(1/80 + 0.005 + 0.01)]S$$

$$+ (1000)(165)(1/80 + 0.005 + 0.01) = 0$$

which gives

$$S = 5.42 \text{ mg/liter}$$

(b) What is the total cell concentration?

Using Eq. 13.77

$$X = \frac{50 + (0.55)(1000 - 5.42)}{[(1/80) + 0.01]\ 8}$$

$$X = 3316 \text{ mg/liter}$$

Comparing this answer to the one in Example 13.1.1-2a shows that the addition of
50 mg/liter of cells to the influent raised the concentration of cells in the
reactor by 278 mg/liter.

(c) What is the excess microorganism production rate?

Using Eq. 13.80

$$P_x = 1.0 \ [\frac{50 + (0.55)(1000 - 5.42)}{1.0 + (0.01)(80)}]$$

$$P_x = 331.7 \text{ mg/hr}$$

Comparison of this with the value of P_x in Example 13.1.1-4 shows that when 50 mg/hr of cells were added to the influent, the amount to be disposed of increased by 27.7 mg/hr. The rest were destroyed by decay in the reactor.

Like the simple CSTR, a CSTR with recycle is sometimes used to reduce the amount of biomass produced in another biochemical operation prior to its disposal. Because in that case X_o is large and S_o is very small, the equations for the cell concentration, viability, oxygen requirement and excess cell production may be simplified. The results are:

$$X = (\frac{\theta_c}{\tau})\ [\frac{X_o}{1 + b\theta_c}] \tag{13.81}$$

$$v = (\frac{1 + b\theta_c}{1 + b\theta_c + \gamma\theta_c})\ (\frac{X_{vo}}{X_o}) \tag{13.82}$$

$$RO = F\{\frac{X_o[(\beta_o - \beta) + \beta_o b\theta_c]}{1 + b\theta_c}\} \tag{13.83}$$

$$P_x = F[\frac{X_o}{1 + b\theta_c}] \tag{13.84}$$

BIODEGRADABLE SOLIDS. The methods of incorporating the effects of biodegradable solids were covered in detail in Section 12.1.3, and the material presented there is also applicable to a CSTR with cell recycle. Consequently a detailed discussion will not be presented here, although three points should be emphasized. First, biodegradable solids frequently accompany soluble substrates. Therefore, although the primary objective of a biochemical reactor will be the removal of the soluble substrate, provisions must be made for the solid substrate which may be with it. Thus the equations for soluble substrate will be used herein, with appropriate changes in notation as discussed in Section 12.1.3. The second point is that the presence of undegraded solid substrate in the MLSS makes it difficult to relate the process loading factor, U_M, to the specific growth rate of the microorganisms. The use of the MCRT as the main process descriptive variable avoids that problem, however. The final point is that the errors associated with the use of the techniques of Section 12.1.3 will be smaller as the MCRT is made larger, just as they decreased for larger values of τ in a CSTR. Because CSTR's with recycle are normally designed and operated with large MCRT's, the errors will thus be small.

COMBINATION OF SOLIDS. This situation is just a combination of the others and thus can be handled by combining the equations in a manner similar to that discussed in Section 12.1.3.

13.1.4 Performance of CSTR with Recycle as Predicted by the Models

The major value of the models presented in this chapter is that they can help us understand the way in which microbial CSTR's with recycle behave under a variety of circumstances. We saw during the derivation of the equations that the most important operational variable was the MCRT. Consequently, in this section we will see through graphs of the equations how the MCRT affects reactor performance, and note those circumstances in which τ and/or α have effects separate from those caused by changes in the MCRT. The graphs were generated by computer simulation using the typical parameter values listed in Table 12.1. Therefore, they may be compared directly with the graphs from Chapter 12.

EFFECTS OF MCRT. Figure 13.2 is a typical plot showing how the MCRT of a CSTR with recycle affects the concentrations of soluble substrate and total cells in it. In addition, the viability is shown, which allows calculation of either the live or dead cell concentrations from the total. The MCRT affects the soluble substrate concentration and viability in exactly the same manner that τ influenced them in a

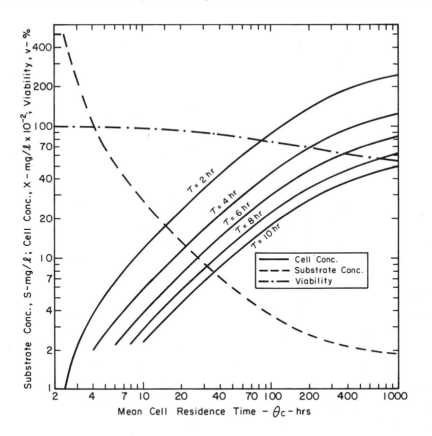

Figure 13.2. Effect of MCRT on the performance of CSTR's with recycle, with various space times. All values of α. Parameters are listed in Table 12.1.

simple CSTR. Consequently those curves are identical with the ones shown in
Fig. 12.2 and the discussion of Section 12.1.4 is applicable here, as long as one
remembers to substitute θ_c for τ. The significant point to note, however, is that
in this case both S and v are independent of τ. Thus the same effluent can be
achieved with a τ of 2 hours as with a τ of 10 hours, as long as both reactors have
the same MCRT. It is this fact which provides the operational and design flexibility
inherent in CSTR's with recycle. The cell concentration, on the other hand, is
affected by τ, and is, in fact, inversely proportional to it, as shown by Eq. 13.38.
For example, at θ_c = 100 hours, a cell concentration of 8800 mg/liter would be
required in a reactor with a two hour space time, whereas only 2200 mg/liter would
be needed if τ were 8 hours. In other words, for a fixed influent condition, a
fixed mass of cells will result for each MCRT. Equation 12.16 said exactly the same
thing for a simple CSTR -- that each specific growth rate had a fixed mass of cells
associated with it. By separating those cells from the main body of fluid flow, we
are able to house them in a smaller container, and thus must have a higher concentration. Another way of looking at it is that the removal of a certain mass of
substrate at a fixed specific rate will always require the same mass of cells. Only
the volume in which they are contained determines their concentration.

Computation of θ_{cmin} by Eq. 13.28 reveals that it is 2.4 hours, thus the use of
a reactor with a space time of 2 hours would not be wise. The curve for a τ of 2
hours was presented, however, to illustrate that a CSTR with recycle can operate
with a τ less than θ_{cmin}, as long as the MCRT is greater than θ_{cmin}. The MCRT cannot
be less than τ, however. Thus the poorest performance that can be achieved by a
CSTR with recycle is that associated with a simple CSTR with equal space time. While
this information can help a designer to choose the value of τ for a facility, the
main factor determining the choice of τ is the economics of the entire system; i.e.,
the reactor, settler, pumps, etc. Although systems analysis is beyond the scope of
this text, the practical factors which should be considered in choosing τ will be
discussed in Chapter 16. The main point to arise from this discussion, however, is
that τ is a variable which can be chosen freely without affecting the soluble
substrate concentration.

An increase in the MCRT results in a decrease in the specific growth rate and
specific substrate removal rate of the culture. Consequently, an increase in the
MCRT requires an increase in the mass of organisms present if the flow rate and
reactor volume are constant. This is reflected in Fig. 13.2 by the slopes of the
cell concentration curves, which are all positive. Cell decay, however, still
increases in importance as the MCRT is increased, just as it did in a CSTR as the
space time was increased. The increased importance of decay at larger MCRT's is
reflected by the fact that the slopes decrease as the MCRT increases.

The increased importance of cell decay at larger MCRT's can also be seen by examining the curves for the excess microorganism production rate (P_x) and the oxygen requirement (RO) in Fig. 13.3. The curve for P_x reflects the net synthesis in the system, i.e., the mass of cells formed less those lost by decay. For this particular example, the maximum production occurs at an MCRT around 13 hours because most soluble substrate has been removed and converted to cells whereas decay is not yet a significant factor. As the MCRT is increased, little additional substrate removal can occur so that as decay becomes more important the mass of cells to be disposed of declines. This decrease in cells is achieved at the expense of more oxygen usage, however. The oxygen requirement curve is identical to the one in Fig. 12.3 since the MCRT affects the oxygen requirement in a CSTR with recycle in the same way that τ affected it in a simple CSTR. In this case, however, the mass of oxygen required per unit time is independent of τ.

The main objective when using a CSTR with cell recycle is the removal of soluble substrate, and thus the efficiency of the biological reactor should be judged on that basis. The effects of MCRT upon efficiency are the same as the effects of τ upon it in a CSTR and such a curve was shown in Fig. 12.4. Examination of that figure shows that a CSTR with recycle is capable of achieving high efficiencies of substrate removal at relatively low MCRT's. The overall process efficiency, on the other hand, also depends upon the concentration of suspended

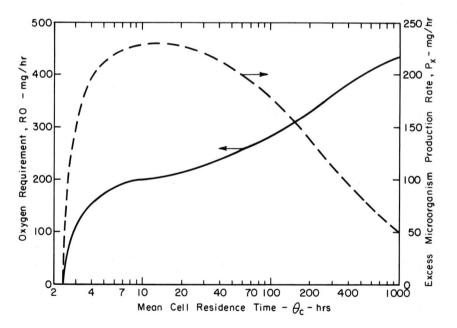

Figure 13.3.   Effect of MCRT on oxygen requirement and excess
              microorganism production rate in a single CSTR
              with cell recycle. All values of α.
              Parameters are listed in Table 12.1.

solids in the effluent. In many cases that concentration exceeds considerably the
concentration of soluble substrate, and therefore system performance may be governed
more by settler efficiency than by the ability of the microbial mass to remove
soluble substrate. When judgements are made concerning the performance of a CSTR
with cell recycle, distinction should be made between soluble and solid material in
the effluent so that action may be taken to correct the operation responsible for
poor performance. Too often the biochemical operation is condemned for what in
reality is failure of the physical operation of sedimentation. This is not to imply,
however, that the two operations do not interact, because they do. For example, the
MCRT of the biochemical operation can have an important effect upon the manner in
which the organisms settle, and this acts as a constraint upon the biochemical
reactor which will be discussed in Chapter 16. Nevertheless, the performance of
the CSTR should be judged only on the basis of its ability to remove soluble
substrate.

Examination of the equations describing performance of a CSTR with recycle,
as well as examination of Figs. 13.2 and 13.3 shows that the recycle ratio, α, has
no effect upon system performance. Consequently the recycle ratio may be chosen on
the basis of economics by considering the effect that it has upon the size of the
final settler and on the operational costs of the facility [14]. If cells are wasted
directly from the reactor, α will have no effect upon w, the waste cell flow rate
expressed as a fraction of the inflow rate, F, as shown in Eq. 13.24 and Fig. 13.4.

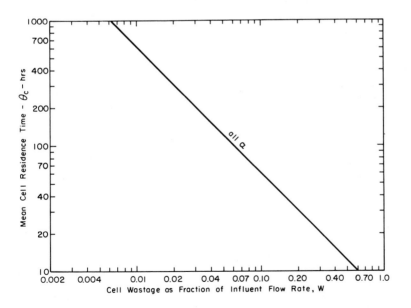

Figure 13.4.   Effects of wastage flow rate and recycle ratio
               on the MCRT of a single CSTR with cell recycle.
               Wastage directly from the CSTR. V = 6.0 liters,
               F = 1.0 liter/hr. Parameters are in Table 12.1.

For each value of τ, a fixed value of w will result in a single MCRT, regardless of α because the concentration of cells in the wastage flow is independent of α. If cells are wasted from the settler underflow, on the other hand, α will affect the value of w required to achieve a desired MCRT, as shown in Eq. 13.16 and Fig. 13.5. For a fixed MCRT, the mass of cells to be wasted, P_x, is constant. Since P_x is the product of the waste cell concentration and the wastage flow rate, it follows that if α affects the concentration it will also affect the flow rate. Thus the effect of α on w shown in Fig. 13.5 is a result of the effect of α on X_r. Comparison of Figs. 13.4 and 13.5 makes it evident why it is easier to operate a system in which sludge is wasted directly from the reactor, although the volume which must be wasted is larger.

One factor which must be considered when deciding upon the recycle ratio, regardless of the point of sludge wastage, is the resulting return sludge concentration, X_r. If X_r is not practical then another recycle ratio must be chosen. Figure 13.6 shows the relationship between α and X_r for one situation. In that case it may be impractical to use a recycle ratio less than 0.3 because it is usually difficult to settle biological sludges to concentrations greater than 12,000 mg/liter. These and other practical factors will be considered in more detail in Chapter 16. The important points to remember now are that α will determine X_r regardless of the point from which sludge is wasted but that α will only affect the MCRT when sludge is wasted from the settler underflow.

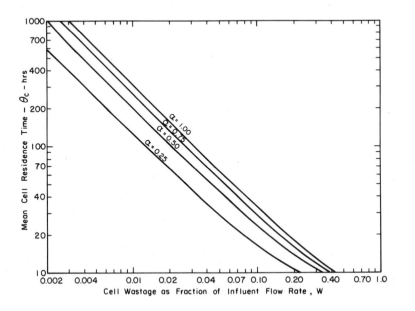

Figure 13.5.   Effects of wastage flow rate and recycle ratio
              on the MCRT of a single CSTR with cell recycle.
              Wastage from settler underflow. V = 6.0 liters,
              F = 1.0 liter/hr. Parameters are listed in
              Table 12.1.

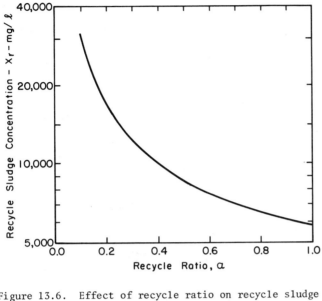

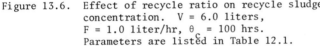

Figure 13.6.   Effect of recycle ratio on recycle sludge
concentration. V = 6.0 liters,
F = 1.0 liter/hr, θ_c = 100 hrs.
Parameters are listed in Table 12.1.

The addition of inert solids to the influent will affect only the concentration of suspended solids in the reactor. As was shown by Eq. 13.66 the presence
of recycle causes the concentration of inert solids in the reactor to be higher than
it is in the feed, and the degree of build-up will depend upon the ratio of the MCRT
to the space time. Furthermore, that fact that inert solids are not subject to
decay, whereas biological solids are, leads to an increase in the fraction of inert
solids in the MLSS when the MCRT is increased as illustrated in Fig. 13.7. For that
example, inert solids contribute more to the MLSS concentration than do the cells
when the MCRT is greater than 360 hours. This means that much of the expense of
solids recycle may be associated with the pumping of solids which contribute nothing
to the process. The potential contribution of inert solids to the total MLSS concentration emphasizes the importance of suspended solids removal prior to biochemical
operations which utilize cell recycle.

It was seen in Chapter 12 that the main effect of the presence of microbial
cells in the influent to a single CSTR was to prevent washout of the reactor. That
would also be the main effect in a recycle reactor. Furthermore, since Eqs. 13.75,
13.78 and 13.79 are the same as Eqs. 12.30, 12.34, and 12.36 respectively (when θ_c
is substituted for τ) the plots of S, v, and RO as functions of the MCRT in a recycle
reactor would be identical with the plots of S, v, and RO as functions of τ shown in
Figs. 12.5 and 12.6. The effect of influent microbial solids on the cell concentration in the CSTR with recycle is simply to raise it for all values of the MCRT,

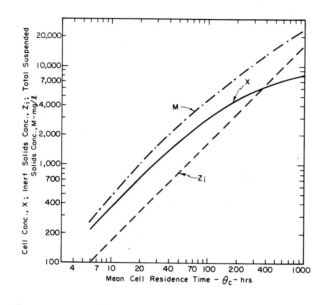

Figure 13.7. Effect of MCRT on the suspended solids
              concentration in a CSTR with recycle
              with a space time of 6 hours, receiving
              100 mg/liter of inerts in the feed.

although the curves will still be shaped like the ones in Fig. 13.2. Because of the
similarity to curves which have already been presented no new figures will be given
for this case. Likewise, no curves will be presented for the situation in which the
influent contains biodegradable solids because they would be very similar to those
given in Figs. 13.2 and 13.3. The only exception will be the suspended solids curve,
which will be higher.

EFFECTS OF KINETIC PARAMETERS. A detailed discussion of the effects of kinetic
parameters upon the performance of a CSTR was presented in Chapter 12. In general,
all of the comments presented there are applicable to this situation, as long as one
recognizes that the effects will depend upon the MCRT instead of upon the space time.
Consequently, no further discussion of the effects of kinetic parameters will be
presented at this time.

13.2 CSTR's IN SERIES

The use of CSTR's in series with recycle of cells around the entire chain is fairly
common in wastewater treatment, consequently we will investigate in this section the
performance of such systems. In Section 12.2.1 it was noted that certain conceptual
problems exist when simple steady-state models are applied to reactors in series,
and thus the reader should review that section if his memory of them is vague.
Nevertheless, the qualitative predictions of the model are reasonable, particularly

when the chain contains no more than four reactors and the MCRT within the chain is well removed from θ_{cmin}. Consequently we will use the models to see how various chain configurations will perform when the feed contains only soluble substrate. Application of the other cases is left as an exercise for the reader.

13.2.1 Generalized System and Model

Figure 13.8 contains a schematic diagram of the generalized system which contains four tanks because that is a common number in practice. The nomenclature within the reactors and the various streams is the same as in Fig. 13.1. The concentrations within the reactors are differentiated by numerical subscripts, e.g., X_i refers to the total cell concentration in reactor i where i = 1-4. In the most general situation feed may enter any or all of the reactors and recycle may be returned to any or all of them. The fraction of feed entering tank i is designated by ϕ_i and

$$\sum_{i=1}^{4} \phi_i = 1.0 \tag{13.85}$$

Likewise, the fraction of the recycle flow entering tank i is designated by ρ_i and

$$\sum_{i=1}^{4} \rho_i = 1.0 \tag{13.86}$$

The reactor volumes are designated by V_i, and they need not be the same although that is the only situation considered herein. The symbol V_T denotes the total reactor volume,

$$V_T = \sum_{i=1}^{4} V_i \tag{13.87}$$

and the total space time, τ_T, is defined as

$$\tau_T \equiv V_T/F \tag{13.88}$$

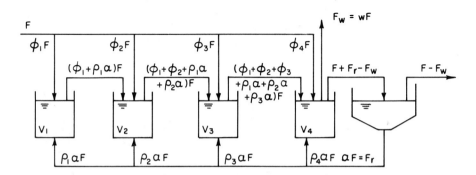

Figure 13.8. Schematic diagram of CSTR's in series with cell recycle.

Because of the recycle around the chain it is impossible to develop explicit equations describing the system performance. Instead, mass balances must be written for substrate, viable cells, and dead cells in each reactor and in the settler, yielding 14 nonlinear equations which must be solved simultaneously on the computer using any one of a number of library subroutines. After the concentrations of substrate, viable cells and dead cells in each tank have been determined, the total cell concentration and oxygen requirement in each tank may be calculated. The mean cell residence time for cells within the system is defined by:

$$\theta_c \equiv \frac{V_1 X_1 + V_2 X_2 + V_3 X_3 + V_4 X_4}{F_w X_4} \tag{13.89}$$

Thus, after the cell concentrations are known the mean cell residence time may be calculated so that it can be used as a system variable. Sludge could be wasted from the settler underflow rather than from reactor 4 as shown, as long as appropriate changes are made in Eq. 13.89. For simplicity of calculation, however, only wastage from tank 4 will be used herein.

The mass balance equations follow. Substitution has been made for the various rate constants from Chapter 10.

SUBSTRATE. The balance on substrate in the settler shows that under the assumptions made at the start of this chapter, the concentration in the recycle stream is just S_4. This substitution has been made in the other mass balances, thereby reducing the number of equations by one.

Tank 1:

$$\phi_1 FS_o + \rho_1 \alpha FS_4 - (\phi_1 + \rho_1 \alpha) FS_1 - (\frac{\mu_m S_1}{K_s + S_1}) \frac{X_{v1}}{Y_g} V_1 = 0 \tag{13.90}$$

Tank 2:

$$\phi_2 FS_o + \rho_2 \alpha FS_4 + (\phi_1 + \rho_1 \alpha) FS_1$$
$$- [\phi_1 + \phi_2 + (\rho_1 + \rho_2)\alpha] FS_2 - (\frac{\mu_m S_2}{K_s + S_2}) \frac{X_{v2}}{Y_g} V_2 = 0 \tag{13.91}$$

Tank 3:

$$\phi_3 FS_o + \rho_3 \alpha FS_4 + [\phi_1 + \phi_2 + (\rho_1 + \rho_2)\alpha] FS_2$$
$$- [\phi_1 + \phi_2 + \phi_3 + (\rho_1 + \rho_2 + \rho_3)\alpha] FS_3 - (\frac{\mu_m S_3}{K_s + S_3}) \frac{X_{v3}}{Y_g} V_3 = 0 \tag{13.92}$$

Tank 4:

$$\phi_4 FS_o + \rho_4 \alpha FS_4 + [\phi_1 + \phi_2 + \phi_3 + (\rho_1 + \rho_2 + \rho_3)\alpha] FS_3$$
$$- (1 + \alpha) FS_4 - (\frac{\mu_m S_4}{K_s + S_4}) \frac{X_{v4}}{Y_g} V_4 = 0 \tag{13.93}$$

VIABLE CELLS.

Tank 1:

$$\rho_1 \alpha F X_{vr} - (\phi_1 + \rho_1 \alpha) F X_{v1} + (\frac{\mu_m S_1}{K_s + S_1}) X_{v1} \, V_1 - \gamma X_{v1} V_1 - b X_{v1} V_1 = 0 \qquad (13.94)$$

Tank 2:

$$\rho_2 \alpha F X_{vr} + (\phi_1 + \rho_1 \alpha) F X_{v1} - [\phi_1 + \phi_2 + (\rho_1 + \rho_2) \alpha] F X_{v2}$$

$$+ (\frac{\mu_m S_2}{K_s + S_2}) X_{v2} V_2 - \gamma X_{v2} V_2 - b X_{v2} V_2 = 0 \qquad (13.95)$$

Tank 3:

$$\rho_3 \alpha F X_{vr} + [\phi_1 + \phi_2 + (\rho_1 + \rho_2) \alpha] F X_{v2} - [\phi_1 + \phi_2 + \phi_3 + (\rho_1 + \rho_2 + \rho_3) \alpha] F X_{v3}$$

$$+ (\frac{\mu_m S_3}{K_s + S_3}) X_{v3} V_3 - \gamma X_{v3} V_3 - b X_{v3} V_3 = 0 \qquad (13.96)$$

Tank 4:

$$\rho_4 \alpha F X_{vr} + [\phi_1 + \phi_2 + \phi_3 + (\rho_1 + \rho_2 + \rho_3) \alpha] F X_{v3} - (1 + \alpha) F X_{v4}$$

$$+ (\frac{\mu_m S_4}{K_s + S_4}) X_{v4} V_4 - \gamma X_{v4} V_4 - b X_{v4} V_4 = 0 \qquad (13.97)$$

Settler.

$$(1 + \alpha - w) F X_{v4} - \alpha F X_{vr} = 0 \qquad (13.98)$$

DEAD CELLS.

Tank 1:

$$\rho_1 \alpha F X_{dr} - (\phi_1 + \rho_1 \alpha) F X_{d1} + \gamma X_{v1} V_1 - b X_{d1} V_1 = 0 \qquad (13.99)$$

Tank 2:

$$\rho_2 \alpha F X_{dr} + (\phi_1 + \rho_1 \alpha) F X_{d1} - [\phi_1 + \phi_2 + (\rho_1 + \rho_2) \alpha] F X_{d2}$$

$$+ \gamma X_{v2} V_2 - b X_{d2} V_2 = 0 \qquad (13.100)$$

Tank 3:

$$\rho_3 \alpha F X_{dr} + [\phi_1 + \phi_2 + (\rho_1 + \rho_2) \alpha] F X_{d2} - [\phi_1 + \phi_2 + \phi_3 + (\rho_1 + \rho_2 + \rho_3) \alpha] F X_{d3}$$

$$+ \gamma X_{v3} V_3 - b X_{d3} V_3 = 0 \qquad (13.101)$$

Tank 4:

$$\rho_4 \alpha FX_{dr} + [\phi_1 + \phi_2 + \phi_3 + (\rho_1 + \rho_2 + \rho_3)\alpha]FX_{d3} - (1 + \alpha)FX_{d4}$$

$$+ \gamma X_{v4}V_4 - bX_{d4}V_4 = 0 \tag{13.102}$$

Settler.

$$(1 + \alpha - w)FX_{d4} - \alpha FX_{dr} = 0 \tag{13.103}$$

OXYGEN REQUIREMENT. After the concentrations of substrate and cells in each tank are determined, the oxygen requirement in each tank may be calculated using the following equations which are simply mass balances on oxygen demand.

$$RO_1 = F[\phi_1 S_o + \rho_1 \alpha(S_4 + \beta X_r) - (\phi_1 + \rho_1 \alpha)(S_1 + \beta X_1)] \tag{13.104}$$

$$RO_2 = F\{\phi_2 S_o + \rho_2 \alpha(S_4 + \beta X_r) + (\phi_1 + \rho_1 \alpha)(S_1 + \beta X_1)$$

$$- [\phi_1 + \phi_2 + (\rho_1 + \rho_2)\alpha](S_2 + \beta X_2)\} \tag{13.105}$$

$$RO_3 = F\{\phi_3 S_o + \rho_3 \alpha(S_4 + \beta X_r) + [\phi_1 + \phi_2 + (\rho_1 + \rho_2)\alpha](S_2 + \beta X_2)$$

$$- [\phi_1 + \phi_2 + \phi_3 + (\rho_1 + \rho_2 + \rho_3)\alpha](S_3 + \beta X_3)\} \tag{13.106}$$

$$RO_4 = F\{\phi_4 S_o + \rho_4 \alpha(S_4 + \beta X_r) + [\phi_1 + \phi_2 + \phi_3 + (\rho_1 + \rho_2 + \rho_3)\alpha](S_3 + \beta X_3)$$

$$- (1 + \alpha)(S_4 + \beta X_4)\} \tag{13.107}$$

The total oxygen requirement is just the sum of the individual tank requirements:

$$RO_T = \sum_{i=1}^{4} RO_i \tag{13.108}$$

EXCESS SLUDGE PRODUCTION. The excess sludge production rate is given by a modified form of Eq. 13.49, since wastage is from tank 4:

$$P_x = F_w X_4 \tag{13.109}$$

The system equations can be solved for any distributions of reactor volume, influent flow, and recycle flow. There are four cases of particular interest, however, all of which will use equal reactor volumes. In the first case the system corresponds to a simple chain, with all feed and all recycle going to reactor 1. In case two, the feed and recycle are distributed evenly among the four tanks. In case three, the feed is distributed evenly among the four tanks and the recycle is added to tank 1, whereas in case four, all recycle is returned to tank 1 and all feed enters tank 3. In order to obtain an understanding of how the flow scheme in each case affects the performance of the reactor system, the equations were solved using the parameters listed in Table 12.1, allowing comparison of the response with the response of a single reactor.

13.2.2 Chain with All Feed and Recycle to Tank 1

Examination of Fig. 13.8 reveals that a simple chain of CSTR's results when ρ_1 and ϕ_1 are both set equal to 1.0, making the other values of ρ and ϕ all zero. The performance of a multi-tank chain can be expressed in terms of the MCRT (Eq. 13.89), although it should be recognized that the MCRT is no longer related to the specific growth rate of the microorganisms in the way that it was for a single CSTR (Eq. 13.13). This is because the specific growth rate is different in each reactor of the chain. Nevertheless the MCRT does represent a net average specific growth rate for the system because at steady state the mass of cells wasted from the system must equal the net amount grown, (i.e., the growth minus the decay). Thus if two reactor systems have the same MCRT they will have the same net formation of cells per unit mass present per unit time, even though the actual specific growth rates in the individual reactors of the chains may differ. Consequently, the MCRT is still a parameter of fundamental value in the description of recycle systems, and may be used as the independent variable in the graphs.

Figure 13.9 presents curves for the substrate concentration, the cell concentration, and the viability in the fourth tank. The most striking difference between Fig. 13.9 and Fig. 13.2 is that the substrate concentration leaving the chain is much lower than the concentration in a single CSTR. It was seen in Chapters 4 and 12 that the extent of reaction for a first-order type reaction will be greater in a chain than in a single CSTR, when neither contains recycle. Even with recycle, because the substrate is soluble, its residence time in the system is equal to the space time of the chain and consequently the extent of reaction that can be achieved will be influenced strongly by the system configuration, thereby giving more substrate removal in the chain. Thus for all practical purposes all biodegradable substrate will be removed by a four tank chain at relatively short mean cell residence times, provided that the substrate has parameters similar to those in Table 12.1. The second thing of interest about the effluent substrate concentration is that the total system space time influences it. Surprisingly, at a fixed MCRT, the substrate concentration increases as τ_T is increased, although the effect is small for the parameter values used in the computation. The viability and cell concentration curves for the simple chain are very similar to those for a single CSTR. For example, when the MCRT is 100 hours, the total cell concentration in the four tank chain with τ_T equal to 6.0 hours is almost exactly the same as it is in a single CSTR with a τ of 6.0 hours. This result may appear surprising, considering the effects observed in a simple chain without recycle as given in Fig. 12.10. Further reflection, however, will show that the observed result is indeed logical. The residence time of the cells within a recycle system is equal to the MCRT and is independent of the residence time of the fluid within the system. Because the cells are recycled around the chain many times within one cell residence time the cells approach a completely mixed condition even though the fluid passes

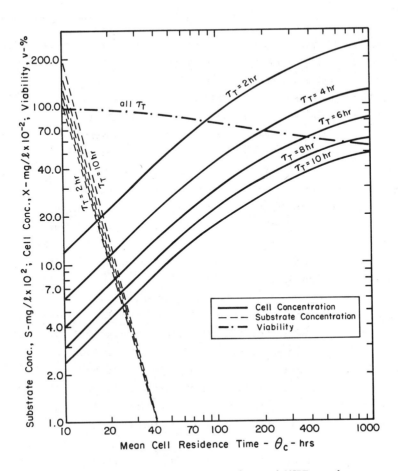

Figure 13.9. Effects of total space time and MCRT on the
              performance of four equivolume CSTR's in series
              with cell recycle and feed to tank No. 1;
              α = 0.25. All concentrations for tank No. 4.
              Parameters are listed in Table 12.1.

through a cascade. As long as the MCRT is larger than τ_T the mixing effect of the
cells within the recycle will overpower the cascade effect of the fluid in the
chain in determining the extent of reaction of the cells by death and decay. Hence
the cell concentration and the viability will be very similar to those in a single
CSTR with the same MCRT. An increase in the extent of reaction will only be seen
when τ_T is very large so that the difference between τ_T and the MCRT is small.

It will be recalled from Fig. 13.2 that the recycle ratio, α, had no influence
on S, X, and v in a single CSTR. Examination of Fig. 13.10 reveals that α also has
no significant effect upon X and v in the four-tank chain. It does however influence S, and the higher the recycle ratio the greater the substrate concentration
in the effluent. However, for the parameter values used to generate the curves the
effect is small and would be of little practical significance. Similarly, α also

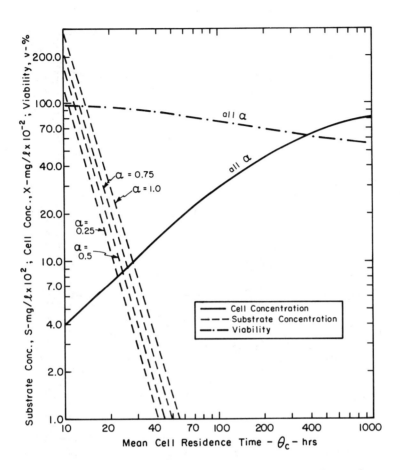

Figure 13.10. Effects of recycle ratio and MCRT on the performance
of four equivolume CSTR's in series with feed and
cell recycle to tank No. 1. τ_T = 6.0 hrs. All
concentrations for tank No. 4. Parameters are
listed in Table 12.1.

has no effect upon the total oxygen requirement and the excess microorganism production rate, as shown in Fig. 13.11. Neither does τ_T. Thus both RO_T and P_x are
functions of only the MCRT. Furthermore, comparison of the curves with the ones in
Fig. 13.3 shows that they are almost exactly the same, suggesting that the system
configuration has no effect. This is because at cell residence times greater than
60 hours almost all soluble substrate has been removed in both system configurations,
so that RO_T and P_x depend primarily upon decay. However, the extent of reaction
achieved by decay is independent of the system configuration as discussed previously,
so that RO_T and P_x are also. When the MCRT is small, however, both RO_T and P_x are
greater in the chain than in the single CSTR but this is due to the greater degree
of substrate removal in the chain and not to differences in decay.

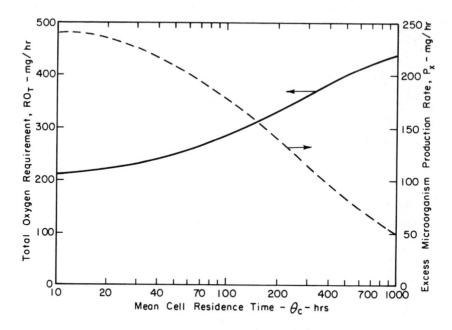

Figure 13.11.   Effect of MCRT on the total oxygen requirement and excess
                microorganism production rate in four equivolume CSTR's in
                series with feed and cell recycle to tank No. 1.
                $0.25 \leq \alpha \leq 1.00$; 2.0 liters $\leq V_T \leq$ 10.0 liters;
                F = 1.0 liter/hr. Parameters are listed in Table 12.1.

So far we have considered only the concentrations in the last reactor, and the
cell production and oxygen requirement for the entire chain, but have not investigated individual reactors. Therefore examine Fig. 13.12, where the substrate concentrations, cell concentrations, and oxygen requirements in each reactor are shown
for three different MCRT's. In light of the preceding discussion it is not
surprising to see that the cell concentration is approximately the same in all
reactors. There is a very slight decline due to decay as the cells progress down
the chain, but because τ_T is small with respect to the MCRT it is not significant.
A significant decline would occur only if τ_T and θ_c were both large and of approximately the same magnitude. There is, however, a significant decrease in S as the
fluid moves down the chain. All substrate is added to the first tank where it is
mixed with cells which have been returned from the settler. The presence of substrate in the tank allows the cells to attain a fairly high specific removal rate
(Eq. 10.45), which, when combined with the high cell concentration gives a rapid
total reaction rate (Eq. 10.41). Thus there is a considerable reduction of substrate in the first tank. Much less substrate enters the second tank so that the
specific rate is reduced, but the high cell concentration allows the maintenance of
a significant total rate so that more of the substrate is removed. In the final

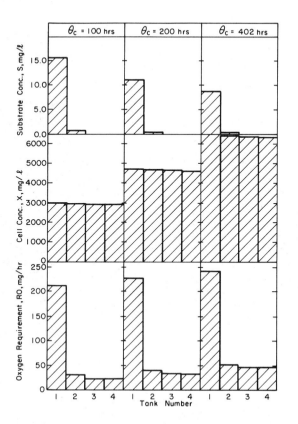

Figure 13.12.   Effect of MCRT on the substrate concentration,
                 cell concentration and oxygen requirement in
                 each tank of a four-tank chain. V_T = 6.0 liters,
                 F = 1.0 liter/hr, α = 0.50. All feed and cell
                 recycle to tank No. 1. Parameters are listed
                 in Table 12.1.

tanks the concentration approaches zero, making the specific removal rate quite low.
However the presence of a high cell concentration keeps the total rate high enough
to scavenge any remaining substrate from the system. If a plot were made of the
specific growth rate of the cells in each tank, it would follow the substrate plot
(Eq. 10.42), i.e., the cells would be growing fairly rapidly in tank 1, more slowly
in tank 2, and very little in tanks 3 and 4, thereby illustrating why the MCRT is
not representative of the actual specific growth rates in the individual reactors.
The oxygen requirements in the tanks reflect the metabolic activity in them. Because
the majority of the substrate removal occurs in tank 1, that is where the most
oxygen is needed. Some substrate removal occurs in tank 2 so its oxygen requirement is slightly higher than the requirements in tanks 3 and 4, which are due
solely to endogenous metabolism or decay.

Comparison of the diagrams in Fig. 13.12 allows recognition of the effects of MCRT. Increasing the MCRT increases the concentration of cells in the system, which in turn increases the total substrate removal rate (Eq. 10.41). Consequently the concentration of substrate in each tank drops. The increased substrate removal in the first tank is reflected by a rise in the oxygen requirement. Furthermore, the increase in MCRT allows more time for decay within the system thereby causing an increase in the oxygen requirements in all of the reactors including the first.

Figure 13.13 shows that the recycle ratio, α, has no significant effects upon S, X, and RO in each of the tanks. Consequently recycle cannot be used as a tool to alter significantly the substrate removal or oxygen uptake patterns in a simple chain. Furthermore, because α has no influence upon the solids concentrations in the tanks at a fixed MCRT, it will have no effect upon the volume of cells which must be wasted from tank 4 in order to maintain a desired MCRT. This is shown by Fig. 13.14 which is very similar to Fig. 13.4.

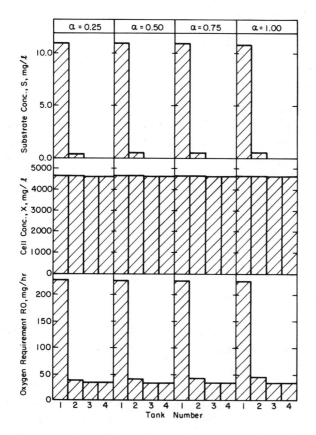

Figure 13.13.   Effect of recycle ratio on the substrate concentration, cell concentration and oxygen requirement in each tank of a four-tank chain. V_T = 6.0 liters, F = 1.0 liter/hr, θ_c = 200 hrs. All feed and cell recycle to tank No. 1. Parameters are listed in Table 12.1.

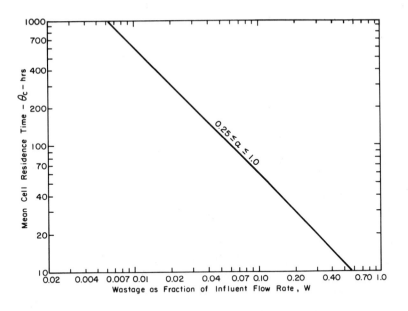

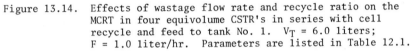

Figure 13.14.   Effects of wastage flow rate and recycle ratio on the
MCRT in four equivolume CSTR's in series with cell
recycle and feed to tank No. 1. V_T = 6.0 liters;
F = 1.0 liter/hr. Parameters are listed in Table 12.1.

The question arises as to the generality of these observations about the
performance of a simple chain of CSTR's with cell recycle. One way to test their
generality is to repeat the simulations using different sets of kinetic parameters.
Consequently this was done using the parameters listed in Table 12.3, which were
given as possible values for high and low temperatures. The effects of the changes
in parameter values upon the overall performance of the system are similar to those
noted for a single CSTR, both in this chapter and in Chapter 12, and thus the graphs
will not be presented. It is instructive, however, to examine the concentrations
of S and X, as well as RO, in the individual tanks, so these are shown in Fig. 13.15.
Inspection of the figure shows that the preceding generalizations (i.e., that most
of the substrate removal and oxygen uptake occurs in the first tank, and that there
is little change in the cell concentration along the chain) are indeed valid over a
wide range of parameter values. Figure 13.15 also illustrates a point made originally
in Section 12.1.4, i.e., that one must think in terms of all of the parameters and
not just single ones. For example, because Y_g is higher in the high temperature
case, the oxygen requirement in the first tank is less than in the first tank of the
low temperature case. The cell concentration in the high temperature case is lower,
however, but this is due to more decay, as reflected by the higher oxygen requirements in tanks 2, 3, and 4. A note of caution should be added here. It should not
be assumed that the profiles presented in Fig. 13.15 are typical of all low and high

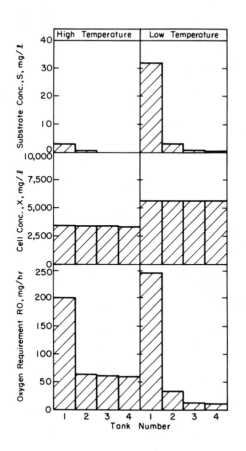

Figure 13.15.   Effect of temperature on the substrate
                concentration, cell concentration, and
                oxygen requirement in each tank of a
                four-tank chain. All feed and cell
                recycle to tank No. 1. V_T = 6.0 liters,
                F = 1.0 liter/hr, α = 0.5, θ_c = 200 hrs.
                Parameters are listed in Table 12.3.

temperature conditions. Rather, they represent possible profiles and are given
primarily to show the generality of the system response. In addition, however, they
serve well to illustrate that all situations must be investigated by a designer.
Microbial reactors with cell recycle are of sufficient complexity to make reliance
upon intuition dangerous.

One important characteristic of a simple chain with recycle is that most of
the oxygen must be supplied to the first reactor. Furthermore the relative quantities of oxygen which must be supplied to the individual tanks may change as the
environmental conditions or waste characteristics change. Consequently any system
employing this configuration must be built to ensure operational flexibility. The

main benefit associated with a chain, on the other hand, is its ability to remove
most of the biodegradable substrate, even under adverse conditions such as low
temperatures.

13.2.3 Chain with Recycle and Feed Distributed Evenly Among the Tanks

One way to reduce the oxygen requirement in the first reactor and increase it in the
others would be to distribute the feed directly to some of the others. A logical
technique would be to distribute it evenly among the four tanks. The question then
arises as to what to do with the recycle, and a logical choice would be to distribute
it evenly as well. When these changes are made and the simulations repeated the
performance of the chain becomes identical with that of a single CSTR with cell
recycle. The specific growth rate in each tank is the same and is related to MCRT
in the same way that it was in Eq. 13.13. Because performance curves for a single
CSTR have already been presented in Figs. 13.2 through 13.4 they will not be repeated.
Furthermore, there are no changes in concentration along the chain, and the oxygen
requirement in each tank is the same. Thus, while use of this flow and recycle
distribution has served to even out the oxygen requirement, it has also reduced the
quality of the effluent to that of a single CSTR, thereby removing the main benefit
associated with a chain. The significance of the fact that this case behaves exactly
like a single CSTR is that it is possible to operate a chain so that it behaves in
either manner, thereby providing considerable flexibility.

13.2.4 Chain with Recycle to Tank 1 and Feed Distributed Evenly Among Tanks

To simulate the performance of a chain with feed evenly distributed to all tanks all
values of ϕ should be 0.25. Because all recycle is to tank 1, ρ_1 is 1.0 while the
others are zero. These changes were made in the system equations and they were
solved again using the parameters in Table 12.1. Because all cells enter the first
tank, the specific growth rates of the cells in the various tanks will not be the
same so that the MCRT does not represent μ as it did in a single tank. Nevertheless,
the MCRT is a fundamental variable that can be used to describe the system and therefore, Fig. 13.16 presents performance curves for the system as a function of the MCRT.
Consider first the substrate concentration curve and compare it to the ones in
Figs. 13.2 and 13.9. The first thing that will be noticed is that the concentration
is higher than it was in a simple chain. This is caused by the addition of substrate
to the last tank. Furthermore, the substrate concentration in Fig. 13.16 is slightly
higher than in Fig. 13.2, showing that the performance of the configuration is not
quite as good as a single CSTR. The difference is not large, however, and would not
be very significant for the parameters listed in Table 12.1, although it could be
for others. The reason for this performance will be explained later. Finally, it
will be seen that τ_T has a slight effect upon substrate removal which is opposite to
the effect noted with a simple chain although its practical significance is small.
There is little difference in viability between this case and the other two, although

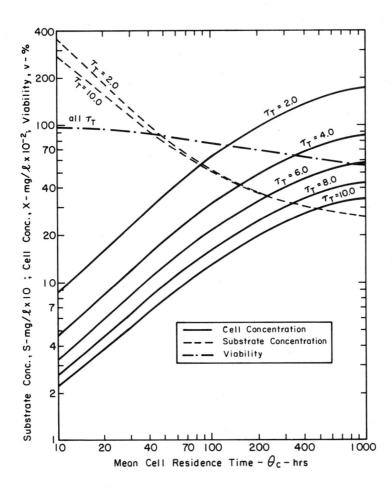

Figure 13.16.   Effects of total space time and MCRT on the
                 performance of four equivolume CSTR's in series
                 with cell recycle to tank No. 1 and feed evenly
                 distributed to all tanks. $\alpha = 0.5$. All
                 concentrations for tank No. 4. Parameters
                 are listed in Table 12.1.

there is between the cell concentrations. For all τ_T values the concentration of
cells in the last reactor is less than in either a single CSTR or simple chain with
the same MCRT. The reason for this will be seen when we examine the concentration
profiles within the chain.

Figure 13.17 shows the effect of the recycle ratio upon the performance of this
system and it should be compared with Fig. 13.10 (it will be recalled that α did not
effect a single CSTR). Examination of Fig. 13.17 shows that α influences both the
cell and substrate concentrations in the fourth tank, to a degree greater than
observed in a simple chain. An increase in the recycle ratio will cause an increase

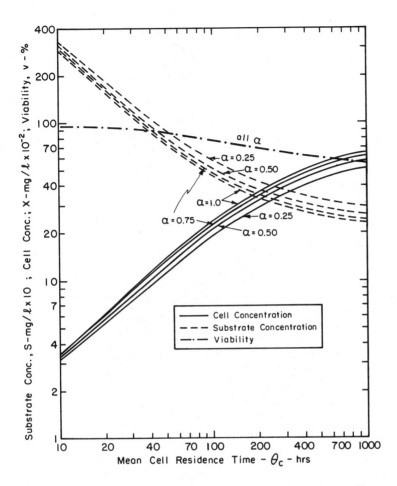

Figure 13.17.   Effects of recycle ratio and MCRT on the performance
                of four equivolume CSTR's in series with cell recycle
                to tank No. 1 and feed evenly distributed to all tanks.
                V_T = 6.0 liters, F = 1.0 liter/hr. All concentrations
                for tank No. 4. Parameters are listed in Table 12.1.

in the cell concentration in the fourth tank, with a concomitant decrease in the
substrate concentration. Although the effects upon S for the parameters used here
are small, they would be different for other parameter values, and thus this represents the first case in which α has a significant bearing upon system performance.
This increases the operational variables which may be altered to control the system,
but at the same time makes the control problem more difficult.

Figure 13.18 shows the effects of MCRT upon the oxygen requirements and the
excess microorganism production rate. Neither of these factors is influenced by
α or τ_T, and the curves are, in fact, almost identical with those for a single CSTR
shown in Fig. 13.3. For MCRT's less than 60 hours they differ from those of the

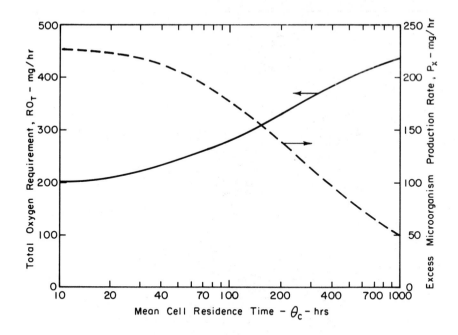

Figure 13.18. Effect of MCRT on the total oxygen requirement and excess
microorganism production rate in four equivolume CSTR's in
series with cell recycle to tank No. 1 and feed evenly
distributed among tanks. $0.25 < \alpha < 1.00$;
2.0 liters $\leq V_T \leq 10.0$ liters; $\overline{F} = \overline{1}.0$ liter/hr.
Parameters are listed in Table 12.1.

simple chain only because less substrate is being removed. This similarity between
the P_x curves for all three reactor types is quite important and is the key to
understanding some of the characteristics of this reactor configuration.

Consider the definitions of MCRT as given by Eq. 13.89 and of P_x as given by
Eq. 13.109. For MCRT's greater than 60 hours, P_x depends only on the MCRT and is
independent of the reactor configuration. Since the numerator of Eq. 13.89 is the
mass of microorganisms contained within the system and P_x is the same for all
reactor configurations when the MCRT is the same, they must all contain the same
mass of cells. This fact can be used to explain why the concentration of microorganisms in the fourth reactor of this chain is less than the concentration in a
single CSTR or in a simple chain. The concentration of cells in the recycle flow
is greater than the concentration in tank 4 because the recycle flow is less than
the flow entering the settler. When tank 1 receives all of the recycle but only
one-fourth of the inflow, the cell concentration in it will be considerably higher
than the concentration in tank 4. As the flow leaves tank 1 and enters tank 2 the
cell concentration will be diluted by the feed to that tank. Dilution continues as
the flow progresses down the chain, causing a decrease in cell concentration as shown

in Fig. 13.19. Because the masses of cells in all of the reactor systems are the
same when the MCRT's are equal, and because the total tank volumes are the same, the
average concentration of cells in the systems must be the same. However, if this
system has a concentration gradient down the chain, the concentration in the last
tank must be less than the concentration in a comparable CSTR in order for the
average concentrations to be equal. Consequently, it can be seen that the lower
cell concentration observed in Fig. 13.17 is a result of the fact that the first
tank receives primarily recycle flow with relatively little inflow.

Figure 13.19 also shows that the concentration profile for substrate is the
opposite of that for the cells, although it is not large. The increase in substrate
concentration is the result of the decrease in the cell concentration and the fact
that new flow and substrate are introduced into each tank. That this reactor configuration is able to accomplish what was desired, i.e., a smoothing of the oxygen
requirement profile is also shown in Fig. 13.19. There it can be seen that the

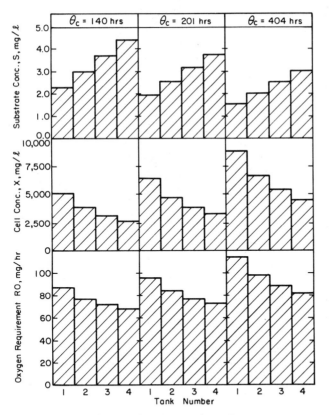

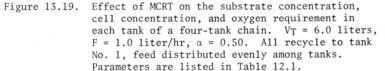

Figure 13.19. Effect of MCRT on the substrate concentration,
cell concentration, and oxygen requirement in
each tank of a four-tank chain. V_T = 6.0 liters,
F = 1.0 liter/hr, α = 0.50. All recycle to tank
No. 1, feed distributed evenly among tanks.
Parameters are listed in Table 12.1.

oxygen requirement follows the same pattern as the cell concentration, and thus
declines slightly toward the latter stages. Comparison of the diagrams in Fig. 13.19
allows an evaluation of the effects of MCRT, which are similar to those in the
simple chain, i.e., the increase in cell concentration allows removal of more substrate. Because most substrate has been removed even at an MCRT of 140 hours, the
increase in the oxygen requirement brought on by the increase in MCRT is due primarily to increased decay. This is particularly evident when you examine the changes
in RO within individual tanks for then it can be seen that the reactors with the
highest cell concentrations (i.e., those first in the chain) show the greatest increase in the oxygen requirement as the MCRT is increased.

Figure 13.20 shows that the recycle ratio has a much greater effect upon the
profiles in this system than it did in a simple chain. This, too, can be explained
by considering the mass of cells in the system. Since the MCRT is constant for the

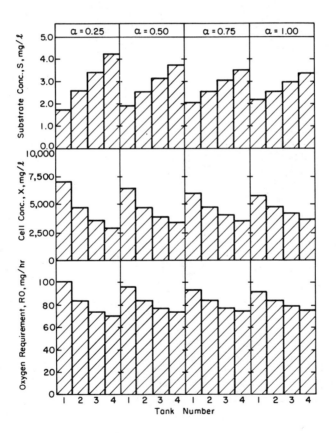

Figure 13.20. Effect of recycle ratio on the substrate
concentration, cell concentration and oxygen
requirement in each tank of a four-tank chain.
V_T = 6.0 liters, F = 1.0 liter/hr.,
θ_c = 201 hours. All cell recycle to tank
No. 1, feed distributed evenly among tanks.
Parameters are listed in Table 12.1.

four cases pictured, the masses of cells in each will be the same. As the recycle
ratio is increased the difference between X and X_r will decrease, thereby making the
difference in solids concentration between reactors 1 and 4 smaller and reducing the
concentration gradient. Also, since the total mass of cells is constant, if the
concentration in reactor 1 goes down the concentration in reactor 4 must go up, which
explains why an increase in α causes the increase in X shown in Fig. 13.17.
Conversely, if the concentration of cells in a tank goes up, the concentration of
substrate must come down, which explains why S in Fig. 13.17 decreases as α is
increased. Finally, if increasing α reduces the concentration profile through the
chain it will also reduce the differences in the oxygen requirements among the
reactors in the chain. Thus the degree to which this system configuration is able
to reduce the differences in oxygen requirements among the tanks is influenced by
α; the larger α, the more homogeneous the system will be. Since α influences the
concentration of cells in the last reactor we would expect it to affect the volume
of sludge which must be wasted from that reactor to maintain a desired MCRT.
Figure 13.21 shows that it does. The smaller α is, the smaller will be the cell
concentration in the last tank and thus the larger F_w must be in order to waste the
same amount of cells. Thus for this flow scheme the recycle ratio becomes another
control parameter which can be manipulated to achieve a desired MCRT or effluent
quality.

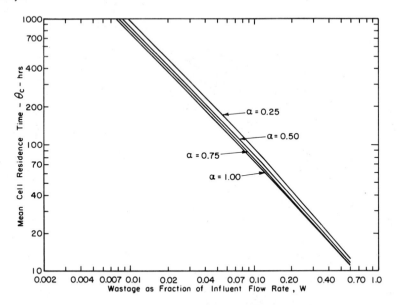

Figure 13.21. Effects of wastage flow rate and recycle ratio on
              the MCRT in four equivolume CSTR's in series with
              cell recycle to tank No. 1 and feed equally distri-
              buted among all tanks. V_T = 6.0 liters,
              F = 1.0 liter/hr. Parameters are listed in Table 12.1.

13.2.5 Chain with Recycle to Tank 1 and Feed to Tank 3

Another flow pattern which is of interest is to return all recycle to the first tank
but to delay adding the feed until the third reactor. Because no feed is added to
the last tank this configuration has the potential of achieving lower substrate concentrations than the previous one. Furthermore, because cells are added to a tank
which has no feed, ample opportunity for cellular decay should be provided. This
pattern may be simulated by setting ρ_1 and ϕ_3 equal to 1.0. Because some tanks
receive substrate and others don't, the specific growth rates in the reactors will
differ and thus the MCRT will not represent μ as it did for a single CSTR. Nevertheless, the MCRT is still a variable of fundamental importance since it represents the
overall net growth rate of the organisms in the system. Figure 13.22 shows the

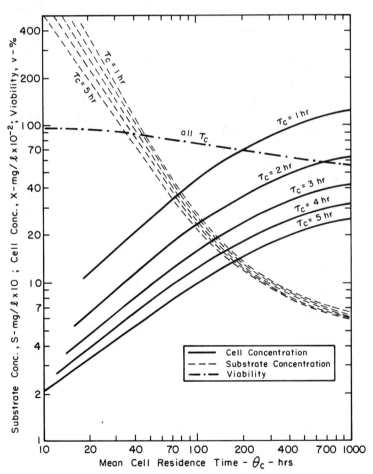

Figure 13.22. Effects of MCRT and the combined space time in
               tanks 3 and 4 on the performance of four equi-
               volume CSTR's in series with cell recycle to
               tank No. 1 and feed to tank No. 3. α = 0.5.
               All concentrations for tank No. 4. Parameters
               are listed in Table 12.1.

effect of MCRT on the system performance as evidenced by the concentrations in the
fourth reactor. Comparison of the substrate concentration curves with the ones in
Figs. 13.22, 13.9, and 13.16 shows that this system is able to remove more substrate
than a simple CSTR or a system with feed distributed evenly to all tanks, but not as
much as would be removed in a simple chain. Furthermore, the amount of substrate
removed is more sensitive to the space time than it is in any of the other systems.
In Fig. 13.22 the symbol τ_c is used to represent the space time of the reactors which
receive feed $((V_3 + V_4)/F)$. Because all reactors have the same volume but only the
last two receive feed, τ_c is one-half of τ_T for a system with the same total tank
volume. The effect of τ_c on S is similar to that in the previous configuration,
i.e., a decrease in τ_c will cause an increase in the effluent substrate concentration.
Comparison of the cell concentration curves shows that the concentrations in the
fourth tank for this system are less than the concentrations in any of the other
systems. The reason for this is similar to the reason for the low concentrations in
the previous configuration. In this case the concentration is even lower, however,
because the cells recycled to the first two tanks are not diluted by feed and thus
are at a higher concentration. Since the total mass of cells is fixed for a given
MCRT an increase of the cell concentration in the first tanks will require a decrease
in the concentrations in the last reactors. As has been the case in all other flow
patterns examined, the viability of the culture depends only on the MCRT and is unaffected by τ_c or the flow pattern.

Figure 13.23 demonstrates that the recycle ratio, α, has a much larger effect
upon the performance of this system than it has had upon any of the others. Both the
cell and substrate concentrations in the fourth reactor are affected, although the
viability is not. As α is increased the cell concentration in the fourth tank increases and the substrate concentration decreases for the same reasons that they did
in Section 13.2.4. As in the previous configuration, therefore, the recycle ratio
represents an operational variable that may be manipulated to alter performance independently from the MCRT. The recycle ratio will not affect the excess cell production
rate or the oxygen requirement however, as shown in Fig. 13.24. Comparison of
Fig. 13.24 with Figs. 13.18, 13.11 and 13.3 shows that the curves for RO and P_x are all
about the same for MCRT's greater than 60 hours. For cell residence times less than
60 hours the curves are similar to those of a simple chain shown in Fig. 13.11. Consequently, for a given MCRT the mass of cells contained within this system will be
the same as in the others considered previously.

The concentrations of substrate and cells in each of the reactors are shown in
Fig. 13.25 along with the oxygen requirement. Because the cell recycle enters the
first tank undiluted, the cell concentration there will be high, and there will be
little or no substrate remaining, even if some were present in the recycle stream.
The oxygen requirement in the first two reactors will reflect endogenous decay, but
there will be little drop in the cell concentration between tanks 1 and 2 because the

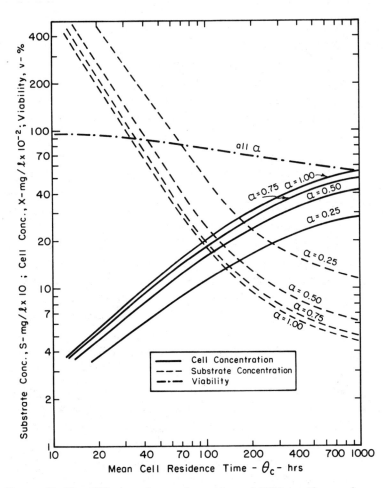

Figure 13.23. Effects of recycle ratio and MCRT on the performance
of four equivolume CSTR's in series with cell recycle
to tank No. 1 and feed to tank No. 3. V_T = 6.0 liters,
F = 1.0 liter/hr. All concentrations for tank No. 4.
Parameters are listed in Table 12.1.

space time in the tanks is small in comparison to the MCRT of the system. Substrate
enters the third tank, diluting the cells but causing a large increase in the oxygen
requirement due to substrate removal. This requirement is of a magnitude similar to
that of a simple chain and thus this configuration has one of the disadvantages of a
simple chain. Not all substrate will be removed in the third reactor due to the
short residence time, however much of that remaining will be removed in the fourth
tank. The oxygen requirement in the fourth reactor will thus be due to both removal
of the last amounts of substrate and to decay. The effects of MCRT upon the profiles
are also shown in Fig. 13.25. A greater mass of organisms must be present at higher
MCRT's, and thus the cell concentrations all increase as the MCRT is increased. The

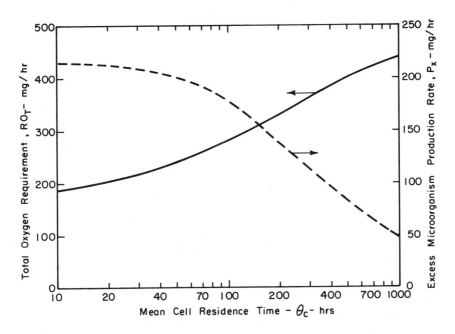

Figure 13.24. Effect of MCRT on total oxygen requirement and excess
microorganism production rate in four equivolume CSTR's
in series with cell recycle to tank No. 1 and feed to
tank No. 3. $0.25 < \alpha < 1.00$; 2.0 liters $< V_T < 10.0$ liters;
F = 1.0 liter/hr. Parameters are listed in Table 12.1.

presence of a greater mass of organisms in the reactors causes the oxygen requirement
due to decay to increase as the MCRT is increased. It also causes the rate of substrate removal in tank 3 to increase so that less is washed to tank 4. The entrance
of less substrate into tank 4 causes its oxygen requirement for synthesis to decrease;
however the presence of more cells causes the oxygen requirement for decay to increase,
with the net effect of keeping the oxygen requirement almost constant.

The effects of the recycle ratio upon the profiles are shown in Fig. 13.26,
where it can be seen that they are greater than for any of the other reactor configurations considered. An increase in α decreases the concentration of cells in the
recycle flow, and therefore decreases the concentration in the first tank. Furthermore, since all systems with the same MCRT have the same mass of cells present, a
decrease in the concentration in the first two tanks must be accompanied by an increase in the concentration in the last two tanks. These changes in concentration
are also reflected in the oxygen requirement profiles; as α is increased the oxygen
requirement in the first two tanks is decreased while the requirement in the last
two tanks is increased. The substrate removal pattern is also influenced by α. As
α is increased the cell concentration in the reactors receiving substrate increases,
which increases the substrate removal rates, thereby reducing the amount of substrate
leaving the system. Consequently, even for a fixed MCRT, manipulation of the recycle

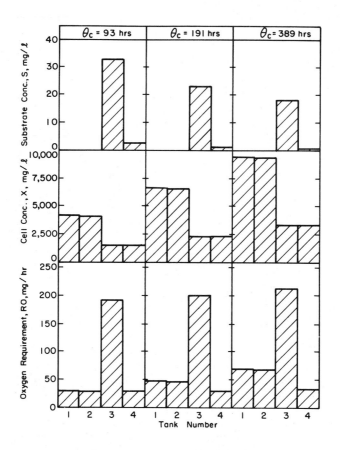

Figure 13.25.   Effect of MCRT on substrate concentration, cell
                 concentration, and oxygen requirement in each
                 tank of a four-tank chain. V_T = 6.0 liters,
                 F = 1.0 liter/hr, α = 0.50. All cell recycle
                 to tank No. 1 and all feed to tank No. 3.
                 Parameters are listed in Table 12.1.

ratio allows slight alterations in the system performance. Furthermore, because a
change in α will change the concentration of cells in the fourth reactor, the amount
of flow which must be wasted from that reactor to maintain a certain MCRT is also
sensitive to changes in α, as shown in Fig. 13.27. Thus consideration must be given
to the recycle ratio when decisions concerning cell wastage are made, even though the
wastage is performed directly from the fourth reactor. Comparison of all the performance curves presented in the preceding sections shows that each system configuration
has its own unique characteristics. We have considered only five cases out of an
infinite number, and, thus the question arises as to whether any particular case can
be considered to be the best.

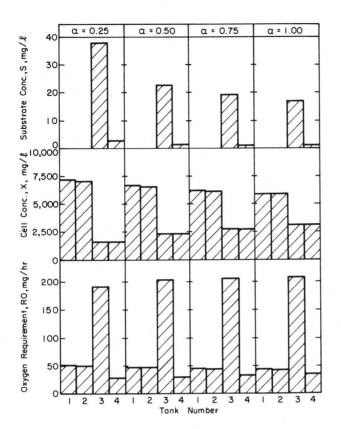

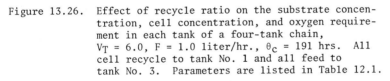

Figure 13.26.   Effect of recycle ratio on the substrate concen-
                 tration, cell concentration, and oxygen require-
                 ment in each tank of a four-tank chain,
                 V_T = 6.0, F = 1.0 liter/hr., θ_c = 191 hrs.   All
                 cell recycle to tank No. 1 and all feed to
                 tank No. 3.   Parameters are listed in Table 12.1.

13.2.6 Determination of Optimal Input, Recycle, and Volume Distributions

The determination of the best reactor configuration is not an easy problem and requires the use of sophisticated optimization techniques. Consequently relatively few
studies have been conducted, and few generalizations can be made. One conclusion
that is quite clear, however, is that the optimum distribution of tank volume, influent flow, and recycle flow depends upon the ease with which the waste can be degraded (as reflected by the parameter K_s), and the percentage removal of soluble substrate required [16,17,18]. Erickson and coworkers sought to find mathematically the
volume and feed distributions that would yield the minimum total volume for a fixed
degree of treatment. A multitank system always gave a smaller volume than a single
CSTR, with the degree of reduction increasing as K_s became larger [17]. Furthermore,

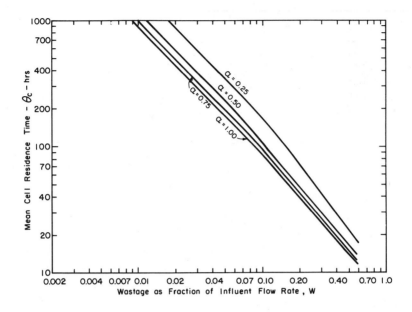

Figure 13.27.   Effects of wastage flow rate and recycle ratio
on the MCRT in four equivolume CSTR's in series
with cell recycle to tank No. 1 and feed to
tank No. 3. V_T = 6.0 liters; F = 1.0 liters/hr.
Parameters are listed in Table 12.1.

for a four-tank system with recycle to the first tank, the optimum design resulted
in progressively smaller tanks downstream, with all inflow to the first two
tanks [16]. Finally, the benefits to be derived from distribution of the feed among
the tanks increased as K_s became smaller.

One problem associated with the determination of the best combination is the
selection of the proper objective function. One choice might be to select the combination giving the minimum tank volume for a desired effluent concentration as
Erickson et al. did. Alternatively, one might be interested in determining the
system with the minimum total cost (i.e., the cost of the reactors, settlers, recycle
pumps, and blowers) required to achieve a desired effluent since the choice of τ_T
also influences the required settler size [14]. As will be discussed in Chapter 16,
CSTR's with recycle often must be designed to achieve a particular MCRT so that it
might be necessary to define the objective functions in terms of the volumes or costs
required to achieve a desired MCRT rather than S. That the objective function will
influence the optimum combination can be seen by returning to the performance figures
given in this chapter. For the parameter values used it is apparent that a simple
chain will result in the smallest MCRT (and thereby system size) to achieve a desired
value of S. However, if one wanted the least cost system to achieve a desired θ_c
even distribution of feed among all tanks might be the best because it would result
in a low cell concentration entering the settler, which would reduce the size of

settler required. Thus although no satisfactory answer can currently be given to the
question of the best reactor configuration to be used, the curves presented herein
serve well to illustrate how the configuration influences system performance, and why
generalizations are difficult. The determination of the best configuration depends
upon many factors, and is a systems analysis problem which must be solved for each
unique situation.

13.3 VERIFICATION OF MODELS

The basic concepts incorporated into the models presented in this chapter are identical with those of Chapter 12, and consequently the comments made in Section 12.3
about the validity of the simple CSTR models are applicable here as well. The primary extension of the theory presented in this chapter was due to the recycle of
organisms; however, unlike the simple CSTR, for which extensive experimental verification was available, little research has been done with pure cultures employing cell
recycle. Whereas the literature confirming the applicability of the models to the
prediction of the performance of a pure cultures in simple CSTR's was too extensive
to list, only one study has reported on the application of the recycle model to pure
culture reactors [19]. In it continuous filtration was used to recycle the yeast
Saccharomyces cerevisiae to a CSTR, and the results were found to agree qualitatively
with the concepts developed herein.

When one turns to the literature in the field of wastewater treatment much more
information is available confirming the validity of the trends predicted by the
models. Of primary importance is the work of Lawrence and McCarty [6] mentioned in
Section 12.3 because they were among the first to recognize that the behaviors of
microbial cultures in CSTR's with and without cell recycle were the same and could
be expressed in a unified manner by use of the MCRT. Recently, Stensel and Shell [20]
applied the models to the design of an industrial waste facility and discussed the
advantages inherent in the use of MCRT. Garrett [1], and Walker [21] have successfully used the concept of MCRT to operate treatment plants, while others have applied
models similar to those presented herein to improve the performance of domestic [22]
and industrial [23] wastewater treatment facilities. Furthermore, although some of
their equations are written differently, the models of McKinney [24] and Eckenfelder [25] are based on similar concepts and have found wide usage in the design of
wastewater treatment systems employing CSTR's with cell recycle. Consequently, it
may be concluded that the models are sufficiently accurate to allow their use for
design and control of biochemical operations as discussed in Chapter 16.

Although less information is available concerning the validity of the models
when applied to CSTR's in series with cell recycle, that which is available suggests
good qualitative agreement. For example, a study with 4, 8, and 16 tanks in
series [26] found substrate and oxygen requirement profiles similar to those in
Fig. 13.13 while another with 8 tanks in series [27] presented a MLSS profile like

it as well. Furthermore, the latter study [27] showed that almost all soluble substrate was removed in the first two reactors, a finding which is in agreement with
Fig. 13.12.

13.4 TECHNIQUES FOR ASSESSING KINETIC PARAMETERS

As was discussed in Section 12.4 the main use of the mathematical models is in design.
Consequently, the kinetic parameters must be evaluated under conditions that simulate
as closely as possible the conditions that will be imposed upon the microbial culture
in the final installation. This not only applies to physical parameters such as pH
and temperature, but also the type of reactor used, which Busch [28] has discussed in
detail. It is important that the tests be performed in a continuous-flow, completely
mixed reactor. Flow should be provided at a constant rate so that a steady state is
achieved, and sampling should be continued over a long enough period at each steady
state to allow statistical evaluation of the mean values of the variables (cell concentration, substrate concentration, etc.). The length of the sampling period depends
upon the MCRT of the reactor because it determines how rapidly changes in the cell
concentration will occur. As a rule of thumb, sample collection should continue for
at least two MCRT's although four would be preferable. This makes the time required
for assessment of the kinetic parameters much longer than the time needed for a
simple CSTR. At this point you might ask why you can't just use a simple CSTR to
assess the parameters since the same ones are used for both systems. The reason that
you can't is that the microbial community developed in a simple CSTR will not be the
same as that in a recycle reactor because the microorganisms in the simple CSTR are
not subjected to the selective pressure imposed by the settling chamber. If the
populations are not the same, the parameter values will not be the same and the data
will be useless. In fact, the settling regime used to recycle the organisms in the
lab study must be similar to that anticipated in the field (i.e., similar settling
rates should be employed) or the laboratory population may not be indicative of the
one which will develop in the full-scale facility. Thus, studies employing recycle
reactors are more complicated than those with simple CSTR's and the reader should
consult more detailed reports on the subject for further information on
methodology [28,29].

Whereas with the simple CSTR the independent variable was τ, when recycle is
employed the independent variable is θ_c. Consequently, most studies are performed
with a constant space time and the MCRT is varied by changing the sludge wastage
rate, F_w. Because the final settler is not perfect, when the MCRT is calculated, the
amount of suspended solids lost in the effluent must be included:

$$\theta_c = \frac{VX}{F_w X_w + (F - F_w)X_e} \tag{13.110}$$

If it is not, the MCRT will be in error, thereby leading to difficulties in the estimation of the parameters. With the exception of the use of the MCRT as the independent variable, the parameters are determined in the same manner that they were in Section 12.4. Consequently, only a brief review of the techniques will be given here, with emphasis upon the ones that are different. As in Chapter 12, two situations will be considered.

13.4.1 Soluble Substrate Situation

If the waste contains only soluble substrate the following data should be collected during the experiments:

C_o - Soluble COD concentration of the influent waste.

C - Soluble COD concentration in the reactor.

X - Total cell concentration in the reactor.

X_w - Total cell concentration in the waste sludge.

X_e - Total cell concentration in the final effluent.

v - Viability of the microbial solids.

V - Reactor volume under aeration.

F - Influent flow rate.

F_w - Waste sludge flow rate.

T - Total COD concentration in the reactor.

In order to determine S and S_o expressed as $T_b OD$, the nondegradable COD, C_i, must be estimated as shown in Fig. 12.18. Because a large range of viabilities will be encountered when recycle is employed, it is best to use Eq. 12.65 to calculate q if viability data are available. If viability data are not available, it will be necessary to use Eq. 12.66 although it will result in an overestimate of C_i. S_o and S can be calculated with Eqs. 12.67 and 12.68, respectively. The values of Y_g and b may be found from the linearized form of Eq. 13.38:

$$\left(\frac{S_o - S}{X\tau}\right) = \frac{1}{Y_g} \left(\frac{1}{\theta_c}\right) + \frac{b}{Y_g} \tag{13.111}$$

When a plot is made of $(S_o - S)/X\tau$ as a function of $(1/\theta_c)$, the result is a straight line with a slope of $1/Y_g$ and an ordinate intercept of b/Y_g, as shown in Fig. 13.28. Compare that figure with Fig. 12.19 to see the effect that recycle has. The reciprocal of θ_c is the independent variable and a least squares line of best fit may be used to estimate the slope and the intercept. The specific bacterial death rate, γ, may be determined by putting Eq. 13.40 in double reciprocal form:

$$\frac{1}{v} = 1 + \gamma \left(\frac{\theta_c}{1 + b\theta_c}\right) \tag{13.112}$$

This is analogous to Eq. 12.70, with the only difference being the substitution of θ_c for τ. Consequently, γ may be determined by plotting $1/v$ versus $[\theta_c/(1 + b\theta_c)]$ in a manner similar to Fig. 12.20. After b and γ have been determined μ_m and K_s may

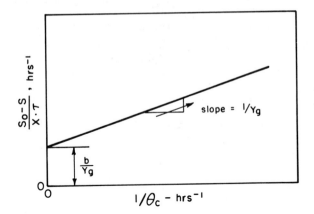

Figure 13.28.   Plot for the determination of b and Y_g in a CSTR with cell recycle.

be estimated from a linearized form of Eq. 13.27. The only difference between Eq. 13.27 and Eq. 12.6 is the substitution of θ_c for τ as the independent variable. Consequently, either technique discussed in Section 12.4.1 may be used, although the limitations of both should be kept in mind. The plots were shown in Figs. 12.21 and 12.22. The oxygen demand coefficient of the cell material, β, may be determined from the values of T, C, and X:

$$\beta = \frac{T - C}{X} \tag{13.113}$$

Finally, when all values of S are small it may be necessary to use the first-order approximation of the Monod equation. Thus the mean reaction rate coefficient, k_e, must be determined. This may be done by plotting q versus S as described in Eq. 12.73 and shown in Fig. 12.18b. Alternatively, k_e may be determined from the relationship between S and θ_c,

$$S = \frac{1}{k_e Y_g} (1/\theta_c + \gamma + b) \tag{13.61}$$

by plotting S as a function of $(1/\theta_c + \gamma + b)$ as was shown in Fig. 12.23. The latter technique is often preferable because it allows determination of C_i and k_e at the same time since $S = C - C_i$ (note: C_i is not needed for the estimation of b, Y_g, and γ). Furthermore, if no viability data is available, γ may be estimated from the literature and C_i determined from a plot of C versus $(1/\theta_c + \gamma + b)$ with greater accuracy than is possible with Eq. 12.66.

EXAMPLE 13.4.1-1

A culture of microorganisms was grown in a CSTR with cell recycle using a complex soluble medium containing a COD of 500 mg/liter, and the data in Table E13.2 were collected.

TABLE E13.2
Data Collected from a CSTR with Cell Recycle

F_w liter/hr	C mg/liter	X mg/liter	X_e mg/liter	Viability percent	T mg/liter
0.125	68.7	1255	16.7	86.6	1640
0.0625	50.8	2286	19.3	78.6	2910
0.0417	45.4	3074	22.0	73.5	3890
0.03125	43.1	3603	29.1	70.4	4550
0.025	41.3	4176	27.5	67.3	5260
0.0208	40.6	4467	34.1	65.8	5620

The reactor volume was 6.0 liters and the influent flow rate was 1.0 liters/hr. The excess microorganisms were wasted directly from the reactor at rate F_w. Using the data determine μ_m, K_s, Y_g, b, and γ for the culture.

(a)   First find C_i, the nondegradable COD so that the substrate concentration may be expressed as T_bOD. This requires the computation of q by Eq. 12.65:

$$q = \frac{F(C_o - C)}{VX_v} \tag{12.65}$$

The values are then plotted as a function of the soluble COD(C) as shown in Fig. E13.1. A linear extrapolation of the lower four data points shows C_i to be 25 mg/liter. This must be subtracted from all of the COD values to express the concentration as T_bOD. Thus S_o is 475 mg/liter and S is as given in Table E13.3.

TABLE E13.3
Biodegradable Substrate Concentration in a CSTR with Cell Recycle

F_w liter/hr	S mg/liter T_bOD
0.125	43.7
0.0625	25.8
0.0417	20.4
0.03125	18.1
0.025	16.3
0.0208	15.6

(b)   Next find Y_g and b from a plot of Eq. 13.111. This requires knowledge of θ_c which must be calculated from Eq. 13.110. The calculated data for θ_c are given in Table E13.4 and the plot is shown in Fig. E13.2 where it can be seen that Y_g is 0.45 mg cells/mg T_bOD and b is 0.003 hr^{-1}.

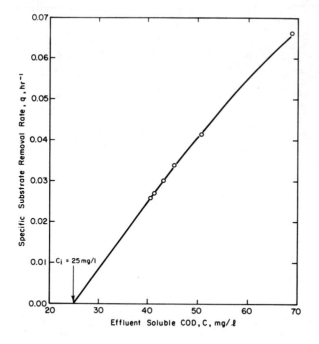

Figure E13.1.   Determination of C_i.

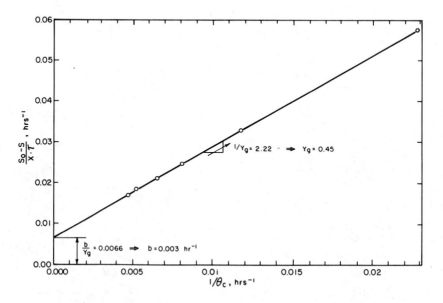

Figure E13.2.   Determination of Y_g and b.

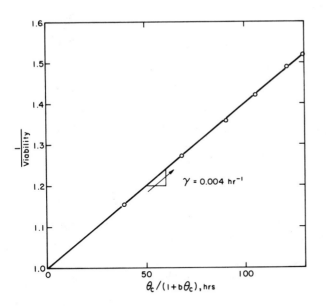

Figure E13.3.   Determination of γ.

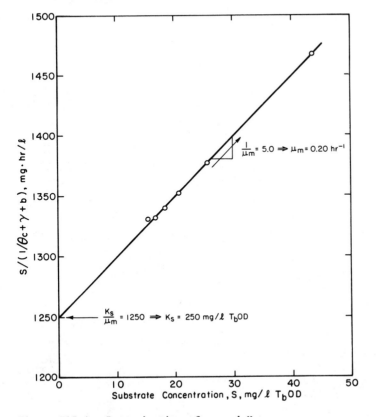

Figure E13.4.   Determination of μ_m and K_s.

TABLE E13.4
MCRT's in a CSTR with Cell Recycle

F_w liter/hr	θ_c hr
0.125	43.9
0.0625	85.2
0.0417	123.6
0.03125	153.6
0.025	190.9
0.0208	212.2

(c) The value of γ must be determined from a plot of Eq. 13.112 as given in Fig. E13.3. There it can be seen that γ is 0.004 hr^{-1}.

(d) Now that γ and b are known, μ_m and K_s may be determined from a Hanes plot as described in Chapter 12. The plot is presented in Fig. E13.4 where it can be seen that μ_m is 0.20 hr^{-1} and K_s is 250 mg/liter T_bOD.

(e) Determine the value of β, the oxygen demand coefficient of the cells in the reactor. β must be calculated with Eq. 13.113. The calculated values are shown in Table E13.5, where it can be seen that β is independent of the MCRT.

TABLE E13.5
Computation of β for a CSTR with Cell Recycle

θ_c hrs	T mg/liter COD	C mg/liter COD	X mg/liter	β
43.9	1640	69	1255	1.25
85.2	2910	51	2286	1.25
123.6	3890	45	3074	1.25
153.6	4550	43	3603	1.25
190.9	5260	41	4176	1.25
212.2	5620	41	4467	1.25

13.4.2 General Situation--Soluble Substrate plus Inert and Biodegradable Solids
As was pointed out in Section 12.4.2, most wastewaters contain suspended solids so that some technique must be available for taking their presence into account. This is particularly important with recycle reactors because of the build-up of inert solids within the system as indicated by Eq. 13.66. Thus the technique described in Sections 12.1.3 and 13.1.3 may be used to correct the soluble substrate equations. This requires assessment of the character of the solids in the influent, as described in Section 12.4.2. The same technique is required with a recycle reactor, with only a few exceptions. Consequently, the exceptions will be emphasized herein.

   During operation of a CSTR with recycle the following data should be collected while operating the system as described in the preceding section.

C_o - Soluble COD concentration of the influent waste.

C - Soluble COD concentration in the reactor.

T_o - Total COD concentration of the influent waste.

T - Total COD concentration in the reactor.

Z_o - Total suspended solids concentration in the influent waste.

M - MLSS concentration in the reactor.

M_w - Suspended solids concentration in the waste sludge.

M_e - Suspended solids concentration in the final effluent.

v - Viability of the microbial solids.

V - Reactor volume under aeration.

F - Influent flow rate.

F_w - Waste sludge flow rate.

In addition, long term batch experiments should be performed to determine the following:

f - Fraction of influent solids which is inert.

f' - Fraction of the particulate COD in the influent which is inert.

Comparison of this list with the one in Section 13.4.1 shows that the only additional data required are those related to the influent; T_o, Z_o, f, and f'. The techniques for estimating f and f' were described in detail in Section 12.4.2 and thus need not be repeated here. However, it should be recalled that both are assumed to be constant and thus should be estimated several times to get an average value.

The soluble, nonbiodegradable COD, C_i, may be determined through the use of q or q' as described earlier, using only the soluble COD's, C_o and C. Alternatively, Eq. 13.61 may be used as discussed in the preceding section. The latter technique will be the most accurate if no viability data is available. The MLSS concentration, M, must have the concentration of inert solids subtracted from it in order to determine X', the effective cell concentration, for use in calculating q or q'. Remembering that the concentration of inert solids in the reactor is given by Eq. 13.66, X' may be calculated by:

$$X' = M - Z_{io}(\theta_c/\tau) \qquad (13.114)$$

Since Z_{io} is a constant fraction of the solids in the influent, Z_o, the equation may be rewritten:

$$X' = M - fZ_o(\theta_c/\tau) \qquad (13.115)$$

Use of either Eq. 13.114 or 13.115 requires knowledge of the MCRT. This should be calculated by taking into account the suspended solids lost in the effluent, M_e. It is generally assumed that all suspended solids are a homogeneous mixture of cells and other solids so that knowledge of X and X_e is not required. Instead the MCRT may be calculated from:

$$\theta_c = \frac{VM}{F_w M_w + (F - F_w)M_e}$$

(13.116)

After C_i is known, the effective parameters Y'_g and b' may be determined from a plot of Eq. 13.111'

$$\frac{S'_o - S}{X'\tau} = \frac{1}{Y'_g} \left(\frac{1}{\theta_c}\right) + \frac{b'}{Y'_g}$$

(13.111')

This can be done exactly as described for a soluble substrate after the effective in-fluent substrate concentration has been determined with Eq. 12.80. The value of γ, μ_m, and K_s can all be determined exactly as described in preceding sections. The mean reaction rate coefficient, k_e, should only be determined through use of Eq. 13.61 as shown in Fig. 12.23, however, due to difficulties in determining the viable cell concentration for use in Eq. 12.75.

The value of β', the oxygen demand coefficient of the effective cell concentra-tion, X', may be found by COD measurements on the reactor during the continuous ex-periments. The effective cell concentration was given by Eq. 13.115. The total COD in the reactor, T, is made up of three components: the COD due to the effective cells, the COD due to soluble material, C, and the COD due to inert solids, P_i. Because recycle is employed in the system, the inert solids will build up in accordance with Eq. 13.66. The COD due to inert solids will thus build up in the same way so that:

$$P_i = P_{io}(\theta_c/\tau)$$

(13.117)

or

$$P_i = f'P_o(\theta_c/\tau)$$

(13.118)

However, P_o was given by Eq. 12.78 so that:

$$\beta' = \frac{T - C - f'(T_o - C_o)(\theta_c/\tau)}{M - fZ_o(\theta_c/\tau)}$$

(13.119)

Comparison of this with Eq. 12.82 shows the effect that recycle has upon the system. A value of β' should be calculated for each mean cell residence time used because in this situation β' is likely to be a function of θ_c.

13.5 KEY POINTS

1. Like a simple CSTR, if the specific growth rate (μ) of the microorganisms in a CSTR with cell recycle can be controlled, so can the concentration of growth limiting substrate (S) in the effluent. In a CSTR with cell recycle, μ is con-trolled by the mean cell residence time (MCRT) and is independent of the space time (τ); thus so is S. This fact is true regardless of the point from which cells are wasted.

2. Operationally, control of the MCRT is easier to achieve by wastage of cells directly from the reactor than by wastage from the settler underflow.

3. The concentration of cells in a CSTR with cell recycle depends upon the space time, the MCRT, and the amount of substrate removed. For a fixed MCRT the product of the cell concentration and the space time is constant.

4. The observed yield, the viability and the oxygen requirement in a CSTR with cell recycle depend only upon the MCRT and are independent of τ.

5. The recycle ratio required around a CSTR will depend upon the degree of concentration desired in the settler.

6. The excess microorganism production rate in a CSTR with cell recycle depends only upon the MCRT.

7. The MCRT is inversely proportional to the process loading factor for a CSTR with cell recycle.

8. The first-order approximation of the Monod equation may be used when the substrate concentration is much less than K_s. The only equation which this affects, however, is the one for the substrate concentration in the reactor.

9. When the influent to a CSTR with cell recycle contains inert solids, the fraction of inert solids in the mixed liquor suspended solids will increase as the MCRT is increased.

10. The effect of microbial cells in the influent to a CSTR is to reduce the specific growth rate. This requires alteration of the equations describing the process.

11. The performance of a CSTR with recycle receiving soluble substrate is depicted by Figs. 13.2 and 13.3. The curves for S, v, and RO are identical to those for a CSTR, except that the MCRT rather than τ determines the performance.

12. The product of the cell concentration and the space time $(X\tau)$ is fixed by the MCRT in a CSTR with recycle. Thus a small reactor requires a high concentration of cells and a large reactor requires a low concentration.

13. When the MCRT is small the excess microorganism production rate will be low because substrate removal is incomplete. When the MCRT is large it will be low because of the importance of cell decay. It will reach a peak at moderate values of the MCRT.

14. The recycle ratio, α, has no effect upon the performance of a CSTR with cell recycle although it will determine the volume of cells which must be wasted from the settler underflow to maintain a desired MCRT.

15. Wastage of cells directly from the reactor makes the wastage rate independent of the recycle ratio. Thus it is easier to control the MCRT by wastage directly from the reactor than by wastage from the settler underflow.

16. The performance of a chain of CSTR's with recycle can only be determined by simultaneous solution of the mass balance equations for the various components in each reactor.

17. A chain of CSTR's with all inflow and recycle to the first reactor is able to remove more soluble substrate than a single CSTR operated with the same MCRT.

18. Most of the soluble substrate removal and oxygen utilization occurs in the first tank in a series of CSTR's with all inflow and recycle to the first reactor, although environmental factors will influence the exact distribution. The cell concentration is approximately the same in all reactors, however.

19. Compared to a CSTR with recycle or to a chain with all feed and recycle to the first tank, a chain of CSTR's with cell recycle to the first tank and the feed distributed evenly to all tanks will have a lower cell concentration in the fourth tank provided all systems have the same MCRT and τ_T. A similar statement can be made about a chain with cell recycle to the first tank and all feed to the third tank.

20. When all cell recycle is to the first tank of a four-tank chain of CSTR's and the feed is either distributed evenly among the tanks or introduced only into the third tank, the recycle ratio is an important variable affecting system performance.

21. The total oxygen requirement and the excess microorganism production rate depend primarily upon the MCRT of the reactor system and are relatively independent of the manner in which the feed and recycle are distributed among the reactors in a four-tank chain.

22. Even distribution of the feed to a chain of CSTR's causes a more even distribution of the oxygen requirement than when the feed is all introduced into one reactor of a chain.

23. No generalization can be made about the best volume, feed, and recycle distribution for a system of CSTR's with cell recycle. This is a complex problem which must be solved for each particular situation.

24. The parameters μ_m, K_s, Y_g, b and γ may all be determined from linear plots of the data collected while running a CSTR with recycle at different mean cell residence times. The plots for determining μ_m, K_s, and γ are similar to those given in Chapter 12, whereas the plot for Y_g and b is different.

25. When the influent to a CSTR with cell recycle contains solids, the fraction of those solids which is inert (f) and the fraction of the COD contributed by those solids which is inert (f') must be determined before the kinetic parameters can be estimated.

26. When the influent to a CSTR with recycle contains biodegradable solids, μ_m and K_s are still determined from soluble T_bOD data but Y_g, b and β all become pseudo-parameters which incorporate the effects of the influent biodegradable solids.

13.6 STUDY QUESTIONS

The study questions are arranged in the same order as the text.

Section	Questions
13.1	1-14
13.2	15-20
13.3	none
13.4	21-23

1. List and discuss the significance of the three special assumptions made in the development of the mathematical models for a CSTR with cell recycle.

2. Explain the significance of cell recycle to the control of the specific growth rate in a CSTR.

3. Explain how and why the MCRT is an important parameter for determining the performance of a CSTR with cell recycle.

4. Explain why it is easier to control the MCRT in a CSTR with cell recycle which employs cell wastage directly from the reactor than in one which wastes cells from the cell recycle flow.

5. Why is the concentration of soluble substrate in a CSTR with cell recycle controlled by the MCRT of the system?

6. Define the "process loading factor" and explain how it is related to the MCRT.

7. A feed containing a soluble substrate with a T_bOD of 1000 mg/liter is flowing at a rate of 100 liters/hr into a 1000 liter aerobic CSTR with cell recycle which contains a mixed population of microorganisms. The parameters characterizing the culture are given in Table SQ13.1.

TABLE SQ13.1
Kinetic Parameters for a Microbial Culture

Parameter	Value
μ_m	0.40 hr^{-1}
K_s	120 mg/liter T_bOD
b	0.004 hr^{-1}
γ	0.003 hr^{-1}
Y_g	0.45 mg cells/mg T_bOD
β	1.20 mg O_2/mg cells

(a) If θ_c = 100 hours, determine the following:

(1) The concentration of soluble T_bOD in the effluent.

(2) The total cell concentration in the reactor.

(3) The viability of the culture.

(4) The mg/hr of oxygen which must be supplied to the reactor.

(5) The excess microorganism production rate.

(6) The wastage flow rate if cells are wasted directly from the reactor.

(7) The wastage flow rate if cells are wasted from the settler underflow and α = 0.75.

(8) The process loading factor.

(b) Would it be possible to achieve an effluent soluble T_bOD of 1.5 mg/liter with this substrate?

(c) What MCRT would be required to reduce the soluble substrate concentration to 20 mg/liter? What concentration of viable cells would be present in the reactor?

(d) If 150 mg/liter of inert solids are added to the influent, what fraction of the MLSS will be inert for the condition of part a? What fraction of the MLSS will be viable cells?

8. Explain why MCRT is preferable to the process loading factor as a means of controlling a CSTR with recycle when the influent contains inert solids.

9. Rework Study Question 7a, but include 110 mg/liter of cells with a viability of 80% in the influent. Assume β_o = 1.20 mg O_2/mg cells.

10. A CSTR with cell recycle is being used to degrade microbial cells. A stream
     containing cells at a concentration of 800 mg/liter is flowing at a rate of
     2 liters/hr into a CSTR with a volume of 48 liters which is being operated with
     a MCRT of 240 hours. The oxygen demand coefficient for the cells, β_o, is
     1.40 mg T_bOD/mg cells, and the specific decay rate, b, is 0.007 hr^{-1}.

     (a) What is the cell concentration in the reactor?

     (b) What is the oxygen consumption rate in mg/hr if β of the cells in the
          reactor is 1.20?

11. Describe and explain the effects of τ and MCRT upon the performance of a CSTR
     with cell recycle.

12. Explain why the recycle ratio affects the rate of cell wastage from a CSTR with
     cell wastage from the settler underflow but does not affect it when cells are
     wasted directly from the reactor.

13. When a CSTR with cell recycle receives inert solids in the influent, their
     contribution to the MLSS increases as the MCRT increases. Explain why.

14. Write the mass balance equations for viable cells, dead cells, and soluble
     substrate in each of two equivolume CSTR's in series and in the settler; 75%
     of the influent flow and recycle go to the first tank and 25% to the second.

15. Explain why a chain of CSTR's with all inflow and recycle to the first reactor
     is able to remove more soluble substrate than a single CSTR with the same MCRT,
     yet still has approximately the same excess microorganism production rate.

16. Describe the effect that an increase in MCRT would have upon the oxygen require-
     ment in each tank of a four tank chain of CSTR's in which all inflow and recycle
     were directed to the first reactor. Explain the effect, also.

17. Explain why the concentration of cells in the last reactor of a chain with cell
     recycle to the first tank and inflow either evenly distributed among the tanks
     or put all into the third tank is less than the concentration in a single CSTR
     with recycle.

18. Why does a change in the recycle ratio change the relative magnitudes of S, X,
     and RO in each tank of a chain of CSTR's with cell recycle to tank 1 and with
     the feed either evenly distributed to all tanks or all put in tank 3?

19. Why does the recycle ratio affect the rate at which cells must be wasted from
     the last tank of a four reactor chain (with all recycle to tank 1 and the inflow
     either all to tank 3 or evenly distributed) in order to maintain a fixed MCRT?

20. Describe how to determine the concentration of soluble nonbiodegradable COD in
     a CSTR with recycle.

21. Describe how to use data from a CSTR with recycle to determine μ_m, K_s, Y_g, b,
     γ, and β for both the soluble substrate and the general situations.

22. A treatability study was performed on a wastewater using a CSTR with cell re-
     cycle. The wastewater was totally soluble and had a T_bOD of 450 mg/liter. The
     studies were run in a lab-scale reactor which had a volume of 6.0 liters. The
     flow rate was maintained at a constant value of 1.0 liter/hr and the MCRT was
     maintained at the desired values by wasting excess cells directly from the
     reactor.

     (a) Using the data in Table SQ13.2 determine the kinetic parameters μ_m, K_s,
          Y_g, γ, and b. Use all three techniques to determine μ_m and K_s and com-
          pare their effectiveness.

(b) Using the data in Table SQ13.2 determine the kinetic parameters k_e, Y_g, γ, and b.

(c) Which set of parameters gives the best estimate in this case? Why?

TABLE SQ13.2
Data from a CSTR with Cell Recycle

MCRT hours	S mg/liter T_bOD	X mg/liter	Viability % of X
48	147	788	82.8
96	92	1458	75.4
144	76	1877	71.4
192	69	2163	68.8
288	61	2543	65.7
384	58	2772	63.9

23. A treatability study was performed on a wastewater by using a CSTR with cell recycle. The wastewater had a total COD of 740 mg/liter, a soluble COD of 500 mg/liter, and a suspended solids concentration of 200 mg/liter. Long term batch experiments indicated that 40% of the suspended solids were nonbiodegradable, as was 40% of the COD due to the suspended solids. The studies were run in a lab-scale reactor which had a volume of 8 liters. The flow rate was maintained at a constant value of 1 liter/hr, and the MCRT was maintained at the desired values by wasting excess cells directly from the reactor at rate F_w.

(a) Using the data in Table SQ13.3, determine the kinetic parameters μ_m, K_s, Y'_g, γ, and b'. Use all three techniques to determine μ_m and K_s. Compare their effectiveness.

(b) Using the data in Table SQ13.3, determine the kinetic parameters k_e, Y'_g, γ, and b'.

(c) Which set of parameters gives the best estimate in this case? Why?

TABLE SQ13.3
Data from a CSTR with Cell Recycle

F_w liters/hr	C mg/liter	M mg/liter	M_e mg/liter	Viability % of X
0.168	115.9	2000	15	90
0.0845	109.6	3550	17	83
0.0488	107.2	5280	25	78
0.0387	106.4	6210	25	76
0.0306	105.9	7130	31	74
0.0241	105.5	8060	40	72

REFERENCES AND FURTHER READING

1. M. T. Garrett, Jr., "Hydraulic control of activated sludge growth rate," *Sewage and Industrial Wastes,* 30, 253-261, 1958.

2. D. E. F. Harrison, "Physiological effects of dissolved oxygen tension and redox potential on growing populations of microorganisms," *Journal of Applied Chemistry and Biotechnology,* 22, 417-440, 1972.

3. D. G. Deaner and S. Martinson, "Definition and calculation of mean cell resi-
    dence time," *Journal of the Water Pollution Control Federation,* 46, 2422-2424,
    1974.

4. D. Herbert, "A theoretical analysis of continuous culture systems," *Continuous
    Culture of Microorganisms,* Society of Chemical Industry, London, Monograph #12,
    21-53, 1960.

5. A. F. Gaudy Jr. and E. T. Gaudy, "Biological concepts for design and operation
    of the activated sludge process," *Environmental Protection Agency Water Pollu-
    tion Research Series,* Report #17090 FQJ 09/71, Sept. 1971.

6. A. W. Lawrence and P. L. McCarty, "Unified basis for biological treatment design
    and operation," *Journal of the Sanitary Engineering Division, ASCE,* 96,
    757-778, 1970.

7. D. Jenkins and W. E. Garrison, "Control of activated sludge by mean cell
    residence time," *Journal of the Water Pollution Control Federation,* 40,
    1905-1919, 1968.

8. C. E. Adams, *et al.* "A kinetic model for design of completely-mixed activated
    sludge treating variable-strength industrial wastewaters," *Water Research,* 9,
    37-42, 1975.

9. R. I. Dick, "Role of activated sludge final settling tanks," *Journal of the
    Sanitary Engineering Division, ASCE,* 96, 423-436, 1970.

10. R. I. Dick and A. R. Javaheri, "Discussion of 'Unified basis for biological
    treatment design and operation' by A. W. Lawrence and P. L. McCarty," *Journal
    of the Sanitary Engineering Division, ASCE,* 97, 234-238, 1971.

11. A. W. Lawrence and T. R. Milnes, "Biokinetic approach to least cost design of
    activated sludge systems," paper presented at the 162nd National Meeting of the
    American Chemical Society, Washington, D.C., Sept. 1971.

12. A. W. Lawrence, "Modeling and simulation of slurry biological reactors," in
    Mathematical Modeling of Water Pollution Control Processes, Edited by T. M.
    Keinath and M. Wanilista, Ann Arbor Science, Ann Arbor, Mich, 1975.

13. M. D. Mynhier and C. P. L. Grady Jr., "Design graphs for activated sludge
    process," *Journal of the Environmental Engineering Division, ASCE,* 101, 829-846,
    1975.

14. C. P. L. Grady Jr., "Simplified optimization of activated sludge process,"
    Journal of the Environmental Engineering Division, ASCE, 103, 413-429, 1977.

15. M. J. Stewart, "Activated sludge process variations. The complete spectrum,"
    Water and Sewage Works, Reference No., 111, 241-262, 1964.

16. L. E. Erickson, *et al.* "Modeling and optimization of step aeration waste treat-
    ment systems," *Journal of the Water Pollution Control Federation,* 40, 717-732,
    1968.

17. L. E. Erickson and L. T. Fan, "Optimization of the hydraulic regime of activated
    sludge systems," *Journal of the Water Pollution Control Federation,* 40, 345-362,
    1968.

18. L. E. Erickson, *et al.* "Modeling and optimization of biological waste treatment
    systems," *Chemical Engineering Progress Symposium Series,* 64, No. 90, 97-110,
    1968.

19. S. J. Pirt and W. M. Kurowski, "An extension of the theory of the chemostat with feedback of organisms. Its experimental realization with a yeast culture," *Journal of General Microbiology, 63*, 357-366, 1970.

20. H. D. Stensel and G. L. Shell, "Two methods of biological treatment design," *Journal of the Water Pollution Control Federation, 46*, 271-283, 1974.

21. L. F. Walker, "Hydraulically controlling solids retention time in the activated sludge process," *Journal of the Water Pollution Control Federation, 43*, 30-39, 1971.

22. T. W. Keyes and T. Asano, "Application of kinetic models to the control of activated sludge process," *Journal of the Water Pollution Control Federation, 47*, 2574-2585, 1975.

23. H. J. Campbell, Jr. and R. F. Rocheleau, "Waste treatment at a complex plastics manufacturing plant," *Journal of the Water Pollution Control Federation, 48*, 256-273, 1976.

24. R. E. McKinney and R. J. Ooten, "Concepts of complete mixing activated sludge," *Transactions of the 19th Annual Conference on Sanitary Engineering,* University of Kansas, 32-59, 1969.

25. B. L. Goodman and A. J. Englande, Jr., "A unified model of the activated sludge process," *Journal of the Water Pollution Control Federation, 46*, 312-332, 1974.

26. J. Chudoba, *et al.* "Control of activated sludge filamentous bulking - I - Effect of the hydraulic regime or degree of mixing in an aeration tank," *Water Research, 7*, 1163-1182, 1973.

27. W. F. Milbury, *et al.* "Compartmentalization of aeration tanks," *Journal of the Sanitary Engineering Division, ASCE, 91*, No. SA3, 45-61, 1965.

28. A. W. Busch, *Aerobic Biological Treatment of Waste Waters - Principles and Practice,* Oligodynamics Press, Houston, Texas, 317-328, 1971.

29. W. W. Eckenfelder, Jr., *et al.* "Scale-up of biological wastewater treatment reactors," *Advances in Biochemical Engineering, 2*, 145-180, 1972.

Further Reading

Andrews, J. F., "Kinetic models of biological waste treatment," *Biotechnology and Bioengineering Symposium No. 2*, 5-33, 1971.

Christensen, D. R. and McCarty, P.L., "Multi-process biological treatment model," *Journal of the Water Pollution Control Federation, 47*, 2652-2664, 1975.

Gaudy, A. F. Jr. and Gaudy, E. T., "Mixed microbial populations," *Advances in Biochemical Engineering, 2*, 97-143, 1972.

Gaudy, A. F. Jr. and Kincannon, D. F., "Comparing design models for activated sludge," *Water and Sewage Works, 123*, #2, 66-70, 1977.

Gaudy, A. F. Jr. and Kincannon, D. F., "Functional design of activated sludge process," *Water and Sewage Works, 123*, #9, 76-81, 1977.

Hultman, B., "Modeling microbial growth in wastewater treatment," *Journal of the Water Pollution Control Federation, 47*, 843-850, 1975.

Malek, I. and Fencl, Z. (editors), *Theoretical and Methodological Basis of Continuous Culture of Microorganisms,* Academic Press, N.Y. 1966. Chapter 3 has a discussion of reactors with recycle.

Metcalf and Eddy, Inc., *Wastewater Engineering: Treatment/Disposal/Reuse,* 2nd Edition, McGraw-Hill Book Co., New York, N.Y. 1979. See Chapter 9.

Powell, E. O. and Lowe, J. R., "Theory of multi-stage continuous culture," in *Continuous Cultivation of Microorganisms,* Proceedings of the *2nd* Symposium held in Prague, June 1962. Edited by I. Malek, *et al.* Academic Press, London, 45-57, 1964.

Rich, L. G. *Unit Processes of Sanitary Engineering,* John Wiley & Sons, Inc., New York, N.Y., 1963. See Chapter 2.

Tyteca, D., *et al.* "Mathematical modeling and economic optimization of wastewater treatment plants," *CRC Critical Reviews in Environmental Control,* 8, 1-89, 1977.

CHAPTER 14

FIXED-FILM REACTORS

In Chapter 1 reference was made to fixed-film reactors such as packed towers and rotating-disc reactors, in which the microorganisms grow in a thin film on solid media while removing organic matter from the liquid flowing past them. In these hetero-geneous reactor systems the organic matter in the liquid must be transported into the biofilm before it can be utilized by the microorganisms residing there. This makes the response of these reactors fundamentally different from the response of a homo-geneous reactor like a CSTR. In homogeneous reactors the microbial cells are uni-formly dispersed and therefore the reaction may be assumed to take place in one phase with negligible mass transfer resistance (i.e., the substrate concentration to which the microbes respond is assumed to be the same as the concentration in the liquid). In fixed-film reactors, on the other hand, the requirement for movement of the organic matter from the liquid phase into the biofilm causes the substrate concentration surrounding the microorganisms to be less than the concentration in the liquid phase. Since the rates of microbial reactions are determined by the concentration of sub-strate surrounding the microbes, it is necessary to combine physical mass transport with microbial reactions when modeling fixed-film reactors.

In order to understand the influence of mass transfer resistance we will begin with the case in which substrate is consumed only at the surface of the biofilm so that the only resistance to mass transfer occurs in the liquid film adjacent to the biofilm. Next, we will consider the case in which substrate is transported and con-sumed throughout the thickness of a biofilm without any resistance to mass transfer in the adjacent liquid. Then, we will consider the general situation in which the two simpler cases are combined. Finally, we will apply the models to packed towers and rotating-disc reactors.

14.1 INFLUENCE OF EXTERNAL MASS TRANSFER RESISTANCE ON REACTIONS OCCURRING AT A SOLID SURFACE

Consider a flat plate covered with such a thin film (monolayer) of microorganisms that the substrate is consumed entirely at the surface. If this plate is placed into a substrate solution, the concentration of the substrate at the surface of the biofilm will be smaller than the concentration in the bulk of the fluid because of the sub-strate consumption. In order for this consumption to continue, substrate must be transported from the bulk fluid to the liquid-solid interface by diffusive and con-vective mass transfer processes. Consequently, the substrate concentration varies throughout the liquid phase, the variation being greatest near the interface and least far away from it, as shown in Fig. 14.1a. In this situation, the observed re-action rate depends on the rate of mass transport outside of the biofilm as well as on the true, intrinsic rate of the reaction at the biofilm so that only when the mass transfer resistance is negligible is the observed reaction rate equal to the intrinsic rate of the reaction. Thus the external mass transfer resistance obscures the true reaction rate and any attempt to model the situation without incorporating the effects of mass transfer would be futile.

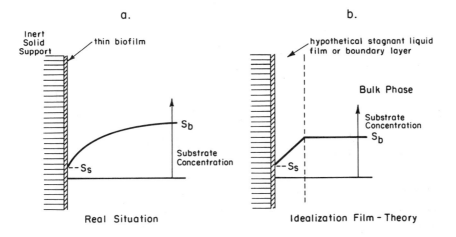

Figure 14.1.   Actual and idealized substrate concentration profiles
                for reaction at a surface with external mass transfer
                resistance.

14.1.1 Mass Transfer Between the Bulk Phase and the Liquid-Solid Interface

As shown in Fig. 14.1b we may idealize the substrate concentration profile by hypothesizing a stagnant liquid film, through which the substrate must be transported to
reach the biofilm. The variation in substrate concentration is restricted to the
stagnant liquid film so that the substrate concentration throughout the remaining
fluid (i.e., the bulk phase) is constant. The rate of mass transfer of substrate
from the bulk phase across a unit area of the stagnant liquid film to the biofilm is
called the flux, N_s. Under the idealized conditions of Fig. 14.1b it is proportional
to the change in substrate concentration across the stagnant layer, $S_b - S_s$:

$$N_s = k_L(S_b - S_s) \tag{14.1}$$

The proportionality constant, k_L, which is called the mass transfer coefficient and
which has units of length per time, incorporates all of the effects of the diffusive
and convective mass transfer processes into one parameter. Thus its value depends
upon the properties of the fluid (such as its viscosity, μ, and its density, ρ), the
diffusivity of the substrate in the fluid, D, and the nature of the turbulence, which
can be represented in part by the bulk fluid velocity, u. Because of the importance
of mass transfer to many engineering problems considerable effort has gone into the
development of equations which allow the prediction of k_L under a variety of conditions. For example, Mixon and Carberry [1] have developed correlations for fluid
flow past a flat plate:

    1. Constant velocity in axial and perpendicular direction

$$k_L = 1.13u S_c^{-\frac{1}{2}} R_e^{-\frac{1}{2}} \tag{14.2}$$

2. Velocity is given by laminar boundary layer theory

$$K_L = 0.817 u S_c^{-2/3} R_e^{-1/2}$$

(14.3)

where S_c is the Schmidt number ($\mu/\rho D$) and R_e is the Reynolds number ($u\rho d/\mu$). Note that these relationships predict that the mass transfer coefficient will increase with the square root of the velocity. In reality, however, it is generally necessary to determine experimentally how the mass transfer coefficient depends upon the linear velocity. Other authors have reported correlations for the dependence of k_L upon the mixing intensity in an agitated vessel and upon the speed of rotation of a flat disc in a quiescent fluid.

For more complicated situations, such as packed beds containing films of micro-organisms growing on discrete media, the flow passages are so complex that it has been necessary to develop semiempirical correlations between the mass transfer coefficients and the fluid velocity. These situations will be discussed briefly later. For detailed treatments consult the work of Satterfield [2].

14.1.2 Reaction on a Surface in the Presence of External Mass Transfer Resistance

GENERAL TREATMENT OF EFFECT. When a steady state exists at the surface of the biofilm there is no accumulation of substrate and hence the rate of substrate supply through the stagnant liquid layer by mass transfer must equal the rate of substrate consumption by reaction at the surface of the biofilm. Assuming that Monod kinetics apply at the surface, the rate of consumption of substrate is

$$-r_s'' = \frac{q_m'' S_s}{(K_s + S_s)}$$

(14.4)

where $-r_s''$ is the mass of substrate consumed per unit time per unit surface area of biofilm. Little is known about cell yield, death and decay within fixed films, con-sequently the specific substrate removal rate is used in Eq. 14.4 rather than the specific growth rate. Furthermore, the rate is expressed in terms of a unit area of film rather than a unit mass of microorganisms because the cells are attached to the solid surface and it is difficult to determine their concentration except in relation to that surface area. Furthermore, the active mass per unit area tends to be rela-tively constant for any given situation. Thus, q_m'' represents a maximum specific sub-strate removal rate which incorporates the cell mass per unit area.

The problem with Eq. 14.4 is that S_s, the concentration of substrate at the surface of the biofilm, is difficult to measure. Thus, it would be advantageous to express the rate in terms of the bulk concentration, S_b, which is easy to measure. We can do this by noting that at steady state $N_s = -r_s''$ so that

$$k_L(S_b - S_s) = q_m'' S_s/(K_s + S_s)$$

(14.5)

Then we can solve Eq. 14.5 for S_s in terms of S_b:

$$S_s = \{S_b - K_s - q_m''/k_L + [(S_b - K_s - q_m''/k_L)^2 + 4K_s S_b]^{0.5}\}/2 \qquad (14.6)$$

and substitute the result into Eq. 14.4

$$-r_s'' = \frac{q_m''\{S_b - K_s - q_m''/k_L + [(S_b - K_s - q_m''/k_L)^2 + 4K_s S_b]^{0.5}\}}{K_s + S_b - q_m''/k_L + [(S_b - K_s - q_m''/k_L)^2 + 4K_s S_b]^{0.5}} \qquad (14.7)$$

It is clear from Eq. 14.7 that the presence of external mass transfer resistance prevents the rate expression from exhibiting a Monod type dependence on the bulk concentration, S_b, even when the true kinetics are assumed to do so. It should also be noted that in the absence of mass transfer resistance (i.e., k_L approaching infinity) Eq. 14.6 reduces to the case where the substrate concentration at the interface equals the bulk substrate concentration ($S_s = S_b$) and Eq. 14.7 reduces to Eq. 14.4.

When the intrinsic reaction rate is first order, the imposition of mass transfer resistance does not change the functional form of the rate equation; it only changes the rate constant. This is true because the mass transfer phenomenon as given by Eq. 14.1 is also first order and thus the two events are additive. For example when $K_s \gg S_s$, Eq. 14.4 reduces to the first-order rate expression,

$$-r_s'' = (q_m''/K_s)S_s \qquad (14.8)$$

and Eq. 14.7 reduces to

$$-r_s'' = [(q_m''/K_s)/(1 + q_m''/k_L K_s)]S_b = k_o S_b \qquad (14.9)$$

thereby demonstrating this principle. The first-order rate constant, k_o, contains both the reaction rate constant and the mass transfer coefficient. It can be written as

$$\frac{1}{k_o} = \frac{1}{k_L} + \frac{1}{q_m''/K_s} \qquad (14.10)$$

Since Eq. 14.9 can be looked upon as a driving force, S_b, divided by a resistance, $1/k_o$, Eq. 14.10 suggests that the overall resistance is equal to the sum of two resistances in series; the external mass transfer resistance, $1/k_L$, and the resistance due to reaction, $1/(q_m''/K_s)$.

For zero-order kinetics, such as when the saturation constant is very small relative to the substrate concentration, the reaction rate is independent of the substrate concentration. Therefore, the observed rate is not influenced by the mass transfer effect.

EFFECTIVENESS FACTOR CONCEPT. Although Eq. 14.7 demonstrates that the existance of an external mass transfer resistance changes the form of the equation for the substrate removal rate, the exact effect of that resistance is difficult to visualize. In chemical engineering this problem has been reduced through use of the effectiveness factor. The effectiveness factor, η, is defined as the ratio of the actual, observed reaction rate to the theoretical rate that would occur in the absence of mass transfer resistance, i.e., the rate evaluated when S_s is equal to S_b. In this way the actual reaction rate can be obtained by multiplying the theoretical rate at the bulk substrate concentration by η. In other words, the effectiveness factor may be regarded as a correction factor, which, when multiplied by the rate without mass transfer resistance, gives the actual rate in the presence of mass transfer resistance. For Monod kinetics at the biofilm, the substrate removal rate would be given by

$$-r_s'' = [q_m'' S_b / (K_s + S_b)] \eta_e \qquad (14.11)$$

where the subscript e on η_e reflects the fact that the effectiveness factor is correcting for external mass transfer resistance. Comparison of Eq. 14.11 with Eq. 14.7 reveals that η_e is a function of the kinetic parameters q_m'' and K_s, as well as the mass transfer coefficient, k_L. Furthermore, it can be shown that for this situation η_e is also a function of the bulk substrate concentration, S_b. Once the exact functional form of η_e has been determined, the actual rate expression becomes a function of only the measurable bulk substrate concentration, S_b, and can be used in conjunction with a substrate mass balance on any type of reactor.

The evaluation of the effectiveness factor can be simplified through the use of dimensionless variables. Therefore, we will use the following:

$$\overline{S} = S_s / S_b \qquad (14.12)$$

$$\overline{K} = K_s / S_b \qquad (14.13)$$

$$D_a = q_m'' / k_L S_b \qquad (14.14)$$

The variable, D_a, has such important physical significance that it has been given a special name, the Damköhler number. Careful examination of it reveals that it is the ratio of the maximum possible substrate removal rate at the interface (q_m'') to the maximum possible substrate transfer rate across the stagnant liquid layer ($k_L S_b$). When D_a is greater than unity the maximum substrate removal rate exceeds the maximum rate of substrate transport and the situation is said to be transport limited. When D_a is less than unity, the opposite is true and the situation is said to be reaction limited.

The equation for η_e for Monod kinetics can be derived by substituting Eqs. 14.11 through 14.14 into Eq. 14.7 and rearranging:

$$\eta_e = \frac{(\bar{K} + 1)\{1 - \bar{K} - D_a + [(1 - \bar{K} - D_a)^2 + 4\bar{K}]^{0.5}\}}{(1 + \bar{K} - D_a) + [(1 - \bar{K} - D_a)^2 + 4\bar{K}]^{0.5}} \tag{14.15}$$

The effectiveness factor is plotted against D_a in Fig. 14.2 using $\bar{K}$ as a parameter. When D_a is very small, so that the maximum mass transfer rate is greater than the maximum substrate removal rate, the effect of mass transfer resistance is negligible and the effectiveness factor is approximately equal to unity. This means that the rate of substrate removal can be expressed by the Monod equation with substrate concentration equal to that in the bulk of the fluid. In other words, the substrate concentration at the biofilm is approximately equal to the concentration in the bulk of the fluid. As D_a exceeds unity and the maximum mass transfer rate becomes slower with respect to the maximum substrate removal rate the effectiveness factor decreases rapidly, thus demonstrating that mass transfer becomes quite important in determining the performance of the system. Under these circumstances the rate of substrate removal must be determined with Eq. 14.11 because the substrate concentration at the biofilm is less than the concentration in the bulk of the fluid. Figure 14.2 also

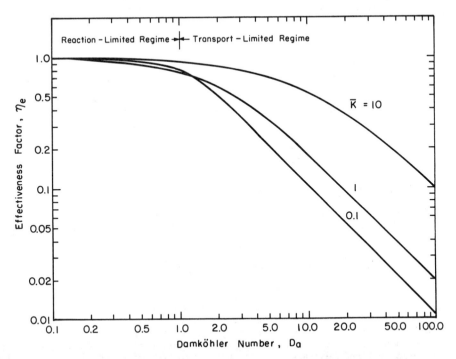

Figure 14.2. Effect of Damköhler number of the effectiveness factor for external mass transfer resistance as predicted by Eq. 14.15.

shows that η_e increases as $\overline{K}$ increases (i.e., as K_s increases with respect to the bulk substrate concentration) which is to be expected since an increase in K_s acts to reduce the rate of substrate removal by the biofilm.

14.1.3 Implications of External Mass Transfer Resistance for Reactor Performance

CSTR. Assume that steady-state conditions prevail over a reasonable period of time in a CSTR containing a solid surface covered with a thin biofilm. Then, in order to maintain a constant cell concentration on the solid surface the cells generated by consumption of substrate must be detached from the biofilm, dispersed throughout the bulk liquid phase, and washed out of the reactor by the effluent. Because cells are located in the bulk of the liquid as well as in the biofilm, they are consuming substrate from both locations. Thus the steady-state mass balance equation for substrate is

$$FS_o - FS_b + r''_s A_s + r_s V = 0 \tag{14.16}$$

where $-r''_s$ is the substrate consumption rate per unit area of biofilm, $-r_s$ is the usual substrate consumption rate per unit volume in a CSTR, F is the flow rate of influent and effluent, S_o is the influent substrate concentration, S_b is the effluent or bulk substrate concentration, A_s is the surface area of the biofilm, and V is the volume of liquid in the reactor. Substitution of Eq. 14.11 for $-r''_s$ and Eq. 10.39 for $-r_s$ yields

$$\frac{1}{\tau}(S_o - S_b) - \frac{q''_m S_b}{K_s + S_b} \eta_e (\frac{A_s}{V}) - \mu X_b/Y = 0 \tag{14.17}$$

where X_b is the concentration of cells in the bulk fluid, which is given by

$$X_b = X_o + Y(S_o - S_b)* \tag{14.18}$$

Therefore, Eq. 14.17 may be written as

$$\frac{1}{\tau} = \frac{\mu_m S_b}{K_s + S_b} [1 + X_o/Y(S_o - S_b)] + \frac{q''_m S_b}{(K_s + S_b)(S_o - S_b)} \eta_e (A_s/V) \tag{14.19}$$

where η_e is given by Eq. 14.15. For given values of τ, A_s/V, S_o, X_o, K_s, q''_m, μ_m, Y, and k_L, Eqs. 14.15 and 14.19 represent two equations in two unknowns, S_b and η_e. Therefore, they must be solved simultaneously for S_b. One method is to use an

---

*
  As mentioned in Section 14.1.2, little is known about death, decay, and yield in reactors containing fixed films, thus it is difficult to incorporate those concepts into the mass balance equations. The most straightforward approach is to base the rates upon a total cell concentration, X, knowing that the active, viable cell concentration will be less and to base the computation of the cell concentration upon some observed yield, Y, knowing that it will be less than the true growth yield. Thus, throughout the remainder of this chapter this approach will be used.

iterative procedure. First, guess a value for S_b and use it to calculate the corresponding value of η_e from Eq. 14.15. Then substitute the value for η_e into Eq. 14.19
and solve for S_b. If the value of S_b thus determined is not the same as the value
originally assumed, a new value of S_b must be assumed and the process repeated until
the calculated value agrees with the assumed value.

In the reaction limited regime where the Damköhler number, D_a, approaches zero,
the effectiveness factor, η_e, approaches unity so that no iteration is required to
solve Eq. 14.19 for S_b. At the other extreme, where the Damköhler number approaches
infinity it can be shown that

$$\eta_e = (1 + \overline{K})/D_a \tag{14.20}$$

and

$$-r''_s = k_L S_b \tag{14.21}$$

Consequently, Eq. 14.19 reduces to

$$\frac{1}{\tau} = \frac{\mu_m S_b}{K_s + S_b}[1 + X_o/Y(S_o - S_b)] + \frac{k_L S_b}{S_o - S_b}(A_s/V) \tag{14.22}$$

which can also be solved for S_b without iteration. Only when the maximum mass
transfer rate is significant in comparison to the maximum possible substrate removal
rate need one resort to an iterative approach for the computation of S_b.

Another case in which an iterative solution to Eq. 14.19 is not needed is when
the bulk substrate concentration is fixed at a desired value and the reactor space
time (τ) or surface area per unit volume (A_s/V) required to achieve that concentration is being calculated. Under that circumstance the effectiveness factor may
be determined directly from Eq. 14.15 or Fig. 14.2 for use in Eq. 14.19.

In order to demonstrate the effect of external mass transfer resistance upon the
performance of a CSTR containing a thin biofilm Fig. 14.3a was prepared by using
Eq. 14.19 with the parameter values listed in Table 14.1. Three curves

TABLE 14.1
Values of Kinetic Parameters and Variables Used to Plot Figure 14.3

Symbols	Value	Symbols	Value
μ_m	0.216 hr^{-1}	k_L	20 cm/hr in Part b, as noted in Part a
K_s	30 mg/liter	V	1000 cm^3
Y	0.5	A_s	100 cm^2 in Part a, as noted in Part b
q''_m	0.696 mg/(cm$^2 \cdot$hr)	S_o	200 mg/liter
		X_o	0

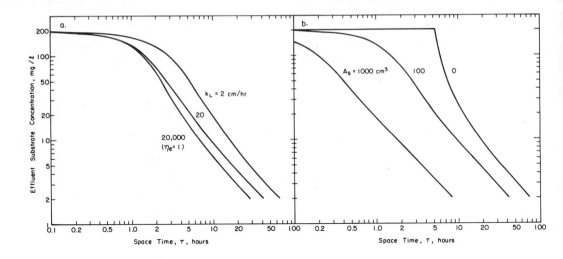

Figure 14.3.   Effects of external mass transfer and biofilm area on the
removal of substrate by a CSTR: (a) Effects of external
mass transfer; (b) Effects of biofilm area.

are presented, one with no external mass transfer resistance (so that η_e = 1 or
k_L = 20,000 cm/hr), one with an external mass transfer coefficient of 20 cm/hr, and
one with a coefficient of 2 cm/hr. Because of relationships like Eqs. 14.2 and 14.3,
each of these curves is representative of the effects that might be caused by changes
in the velocity of the fluid past the biofilm. Examination of the figure shows that
the effect of a decrease in k_L is to reduce the activity of the microbial film,
thereby making the effluent substrate concentration higher than it would be in a
reactor with less mass transfer resistance. Figure 14.3b shows the effect of the
surface area of microbial film available in a reactor with an external mass transfer
coefficient of 20 cm/hr. There it can be seen that reactors with more biofilm can
remove more substrate, but that the effects diminish as the space time is made larger
because of the effects of the bulk substrate concentration upon the effectiveness
factor. Another important point to be seen in Fig. 14.3b is that the presence of a
biofilm prevents washout of the reactor. Consequently, CSTR's with biofilm are
capable of removing substrate at space times well below those which would cause washout in a simple CSTR. Furthermore, the larger the area of biofilm, the shorter the
space time may be.

   PFR. Consider the PFR shown in Fig. 14.4 and place a solid surface containing
a thin biofilm into it. The total surface area of the biofilm is still denoted by
A_s but the area is distributed uniformly along the reactor length, L, so that the
area available per unit length is a_s (= A_s/L). Just as with the CSTR, cells generated
by the consumption of substrate must be continuously detached from the biofilm if
steady state is to be attained. Consequently, the mass balance equation for substrate

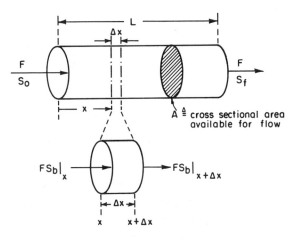

Figure 14.4. Schematic diagram of a PFR containing
a thin, flat biofilm.

must contain terms for substrate consumption by cells on the surface and in the bulk
of the fluid. Assuming perfect plug flow, a steady-state substrate balance around
an infinitesimal section gives

$$FS_b\Big|_x - FS_b\Big|_{x + \Delta x} + r''_s a_s \Delta x + r_s A \Delta x = 0 \qquad (14.23)$$

where A is the cross sectional area of the reactor available for fluid flow. The
usual limiting process yields

$$- F \frac{dS_b}{dx} + r''_s a_s + r_s A = 0 \qquad S_b = S_o \quad at \quad x = 0 \qquad (14.24)$$

Substitution of Eq. 14.11 for $-r''_s$, Eq. 10.39 for $-r_s$, and Eq. 14.18 for the cell
concentration in the bulk of the liquid yields:

$$\frac{dS_b}{dx} + \left[\frac{\mu_m AL}{FL} (\frac{X_o}{Y} + S_o - S_b) + (\frac{q''_m a_s L}{FL}) \eta_e \right] \frac{S_b}{K_s + S_b} = 0 \qquad (14.25)$$

$$S_b = S_o \quad at \quad x = 0$$

It should be noted that AL is the total liquid volume in the reactor, V, and that
$a_s L$ is the total surface area of the biofilm, A_s. Before Eq. 14.25 can be solved,
Eq. 14.15 must be substituted for the effectiveness factor. However, η_e is a complex
function of S_b, which varies along the reactor length. Thus Eqs. 14.25 and 14.15
must be combined and the integration performed numerically. It is only possible to
obtain an analytical solution for Eq. 14.25 when η_e is constant; thus one might be
tempted to assume a constant value. This could lead to serious error, however,

because the substrate concentration may reach low values within the reactor thereby
causing significant variation in n_e.

In order to illustrate the effects of external mass transfer resistance on the
performance of a PFR containing a thin biofilm, Fig. 14.5 was generated by numerically
integrating Eq. 14.25 plus Eq. 14.15 using the parameter values given in Table 14.2.

TABLE 14.2
Base Values of Kinetic Parameters and Variables Used to
Plot Figures 14.5 Through 14.7

Symbols	Value	Symbols	Value
μ_m	0.216 hr^{-1}	k_L	20 cm/hr (in Figs. 14.6 and 14.7)
K_s	30 mg/liter	A_s	2970 cm^2
Y	0.5	V	3000 cm^3
q''_m	0.69588 mg/(cm^2·hr)	X_o	10 mg/liter
		S_o	200 mg/liter (in Figs. 14.5 and 14.7)
		F	3600 cm^3/hr (in Figs. 14.5 and 14.6)

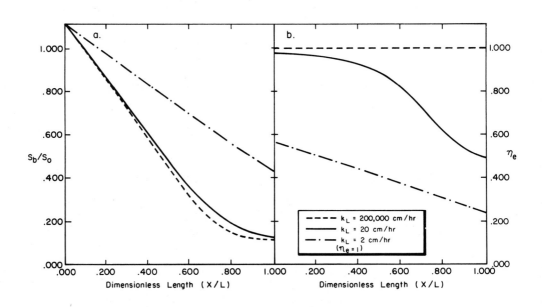

Figure 14.5. Effect of external mass transfer on the substrate concentration
and effectiveness factor profiles in a PFR containing a thin
biofilm: (a) Effects on substrate concentration profiles;
(b) Effects on effectiveness factor profiles.

There it can be seen that the main effect of the mass transfer resistance is manifested toward the effluent end of the reactor where the substrate concentration is
low, thereby making the effectiveness factor low. Furthermore, it is apparent from
the figure that PFR's with significant mass transfer resistance are not capable of
removing as much substrate as those without. Two other significant points arise
from Fig. 14.5. One is that there is a significant change in the effectiveness
factor along the reactor length, thereby demonstrating that the assumption of a constant value would be incorrect. The other is that as the substrate concentration
approaches zero the effectiveness factor approaches a nonzero limit. The reason for
this is that the reaction rate becomes first order as the bulk substrate concentration approaches zero. Under that condition, the definition of the effectiveness
factor gives

$$\eta_e = k_o S_b / (q_m''/K_s) S_b \tag{14.26}$$

Substitution of Eq. 14.10 for k_o yields

$$\eta_e = 1/(1 + q_m''/k_L K_s) \tag{14.27}$$

Thus the limiting value of η_e depends only upon the values of the parameters describing the rates of substrate removal and mass transfer.

The effects of two other variables may be seen in Figs. 14.6 and 14.7, which
were also generated from Eqs. 14.15 and 14.25 using the parameters in Table 14.2.

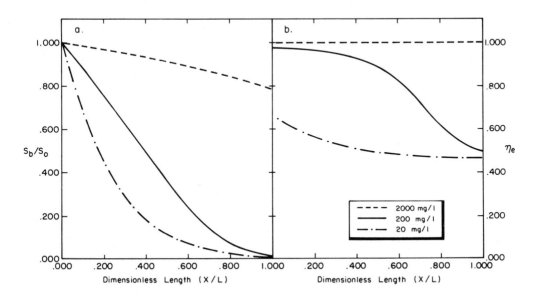

Figure 14.6. Effect of influent substrate concentration on the substrate
              concentration and effectiveness factor profiles in a PFR con-
              taining a thin biofilm: (a) Effects on substrate concentration
              profiles; (b) Effects on effectiveness factor profiles.

Figure 14.6 illustrates the effect of influent substrate concentration. When it is high, the decrease in substrate concentration along the reactor length is almost linear. This is caused by two factors: (1) the organisms are operating at a rate approaching their maximum throughout the entire reactor and (2) the effectiveness factor is almost unity so that no mass transfer limitations occur. At lower influent concentrations the rate of substrate removal declines along the reactor length because of the effects of substrate concentration upon both the rate of substrate removal and the effectiveness factor. These cause the substrate concentration to decrease in an exponential fashion. As expected the effectiveness factor is highest at the inlet where the substrate concentration is highest and lowest at the outlet. Nevertheless, when the influent substrate concentration is very low, the effectiveness factor at the inlet may be less than unity. For example, when S_o is only 20 mg/liter, the effectiveness factor at the inlet is only 0.65. It should be noted that even though an increase in influent concentration causes a decrease in the fraction of the influent removed, the mass removed is increased. This results from the fact that both the substrate removal rate and the rate of mass transport increase as the bulk substrate concentration is increased.

Figure 14.7 shows the effect of the influent flow rate, F, on the removal of substrate along the length of the PFR. An increase in flow rate causes a decrease in the percentage removal of substrate although not in the mass. As far as substrate

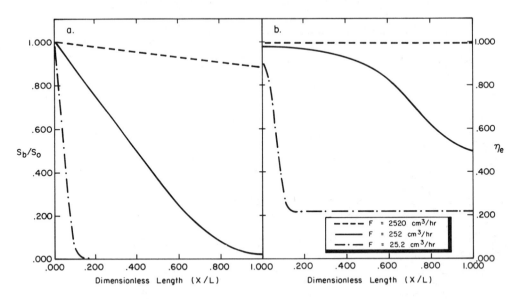

Figure 14.7. Effect of flow rate on the substrate concentration and effectiveness factor profiles in a PFR containing a thin biofilm: (a) Effects on substrate concentration profiles; (b) Effects on effectiveness factor profiles.

removal by the suspended organisms is concerned, the organisms must grow faster to stay in the reactor at the higher flow rate, and thus the concentration of substrate surrounding them must be higher. As far as removal by the fixed organisms is concerned, a higher flow rate brings a higher mass of substrate per unit time into contact with them. Since the mass of organisms in the biofilm is constant, the only way the rate of substrate removal can be increased is for the specific rate to increase, which requires an increase in the concentration of substrate available. The increase in flow rate also acts to increase the external mass transfer coefficient, thereby increasing the effectiveness factor associated with a given reactor substrate concentration. This, in turn, increases the specific substrate removal rate for a given substrate concentration, which helps to alleviate some of the increase in substrate concentration caused by the increased flow.

14.2 INFLUENCE OF INTERNAL MASS TRANSFER RESISTANCE ON REACTIONS OCCURRING WITHIN A FLAT FIXED FILM

We now want to turn our attention to the situation in which substrate must be transported to microorganisms living within a fixed film. In order to focus on what is happening within the biofilm we will assume for the time being that there is no external mass transfer resistance limiting the transport of substrate from the bulk liquid to the liquid-solid interface, i.e., we will assume that the concentration of substrate at the interface is S_b.

14.2.1 Mass Transfer Within a Flat Biofilm

Consider a flat solid support covered with a film of microorganisms as shown in Fig. 14.8. Even though the cells are held together in a complex geometric arrangement it may be assumed that they are uniformly distributed throughout the film. Because the biofilm has a finite thickness, however, substrate can only reach the organisms within it by the process of diffusion, which is characterized by Fick's law

$$N_s = - D \, dS/dx \qquad (14.28)$$

in which D is a diffusion coefficient with units of area per time and dS/dx is a concentration gradient. Fick's law was developed for free diffusion in an aqueous solution, but the biofilm is more complex than that due to the gelatinous matrix holding the cells together. Because of it, the substrate must diffuse through a complex network of tortuous passages before it can be utilized by the cells. This situation is similar to a porous catalyst having interconnected channels through which the reactant must diffuse to reach the active surface where it reacts. Consequently many authors have taken advantage of this similarity and have described diffusion through biofilms by the techniques developed to describe diffusion through catalysts. We will do the same.

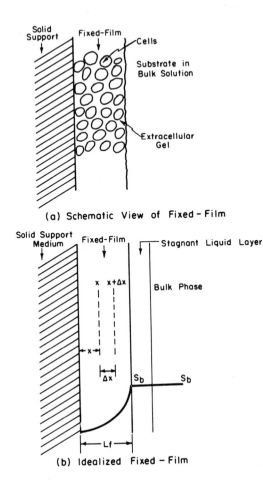

(a) Schematic View of Fixed-Film

(b) Idealized Fixed-Film

Figure 14.8.   Substrate concentration profiles for
reaction within a thick biofilm.

The usual way to describe mass transport by diffusion through catalysts is with Fick's law in which the free diffusion coefficient, D, has been replaced with an effective diffusion coefficient, D_e

$$N_s = - D_e \, dS/dx \qquad\qquad (14.29)$$

The effective diffusion coefficient is usually less than the free diffusion coefficient due to the retardent effect of movement through tortuous passages, and can be related to the free diffusion coefficient by two parameters, Λ and Ξ, which are the void volume fraction of the catalyst and the tortuosity factor, respectively:

$$D_e = D(\Lambda/\Xi) \qquad\qquad (14.30)$$

Although a discussion of the techniques for estimating D, Λ, and Ξ is beyond the scope of this text, a detailed treatment including experimentally determined corre-lations is available elsewhere [3]. Suffice it to say here that Eq. 14.29 can be used to describe mass transport within a biofilm.

14.2.2   Diffusion and Reaction Within a Flat Biofilm

Just as with the thin film, we will assume here that the system reaches a steady state when the biofilm is exposed for a reasonable time interval to a solution whose substrate concentration is maintained at S_b. This implies that the film will attain a constant thickness, L_f, and that there will be no accumulation of substrate within the film. Consequently, a mass balance on substrate around a differential element in the film, as shown in Fig. 14.8, yields

$$ - D_e A_s \left. \frac{dS}{dx} \right|_x + D_e A_s \left. \frac{dS}{dx} \right|_{x + \Delta x} + r''_s a'_s A_s \Delta x = 0 \tag{14.31} $$

where A_s is the total surface area normal to the direction of diffusion, x is the dis-tance into the film from the solid support, and a'_s is the surface area of micro-organisms per unit volume of microbial mass. If D_e is constant, dividing both sides by A_s and Δx and taking the limit as Δx approaches zero, yield

$$ D_e \frac{d^2 S}{dx^2} + r''_s a'_s = 0 \tag{14.32} $$

If r''_s is given by Eq. 14.4, this becomes

$$ D_e \frac{d^2 S}{dx^2} - a'_s q''_m S / (K_s + S) = 0 \tag{14.33} $$

which must be solved with two boundary conditions, one at the liquid-biofilm inter-face ($x = L_f$) and the other at the biofilm-support interface ($x = 0$). At the liquid-biofilm interface the substrate concentration is maintained at S_b. At the biofilm-support interface there is no transfer of substrate. Thus, the appropriate boundary conditions are

$$ S = S_b \text{ at } x = L_f \tag{14.34} $$

$$ \frac{dS}{dx} = 0 \text{ at } x = 0 \tag{14.35} $$

Due to the nonlinear nature of Monod kinetics, no analytical solution can be ob-tained for Eqs. 14.33 through 14.35. Instead, numerical techniques must be employed. Before presenting the numerical results, however, it would be instructive to consider the limiting case of first-order kinetics (Eq. 14.8) because an analytical solution can be obtained which will allow the illustration of certain concepts.

EFFECTIVENESS FACTOR CONCEPT. For first-order kinetics the substrate mass balance equation is

$$D_e \frac{d^2S}{dx^2} - a_s'(q_m''/K_s)S = 0 \tag{14.36}$$

which must be solved with Eqs. 14.34 and 14.35. In terms of dimensionless variables

$$\overline{S} = S/S_b \tag{14.37}$$

$$z = x/L_f \tag{14.38}$$

Eqs. 14.34 through 14.36 may be written as

$$\frac{d^2\overline{S}}{dz^2} - \phi^2\overline{S} = 0 \tag{14.39}$$

$$\overline{S} = 1 \text{ at } z = 1 \tag{14.40}$$

$$\frac{d\overline{S}}{dz} = 0 \text{ at } z = 0 \tag{14.41}$$

The parameter ϕ is called the Thiele modulus and is defined as

$$\phi = [a_s'(q_m''/K_s)L_f^2/D_e]^{0.5} \tag{14.42}$$

The square of the Thiele modulus may be rewritten to show the physical significance of the parameter.

$$\phi^2 = a_s'(q_m''/K_s)L_f^2/D_e = a_s'L_f(q_m''/K_s)/(D_e/L_f)$$

$$= a_s'(A_sL_f)(q_m''/K_s)S_b/[(A_s/L_f)D_eS_b] \tag{14.43}$$

The numerator represents a maximum total first-order reaction rate, while the denominator represents a maximum diffusion rate. Thus the Thiele modulus for reaction within a biofilm is analogous to the Damköhler number for reaction at the surface of a biofilm. Like the Damköhler number, a large value of the Thiele modulus represents a situation which is diffusion limited whereas a small value represents a reaction limited situation.

The solution to Eqs. 14.39 through 14.41 is:

$$\overline{S} = \cosh \phi z/\cosh \phi \tag{14.44}$$

Figure 14.9 shows profiles of substrate concentration versus depth within the biofilm for several values of the Thiele modulus. As would be anticipated, the substrate concentration drops as the observer moves away from the liquid-biofilm interface ($z = 1$). Furthermore, the higher the Thiele modulus, the greater the drop. This follows from the fact that a high Thiele modulus represents a situation in which the

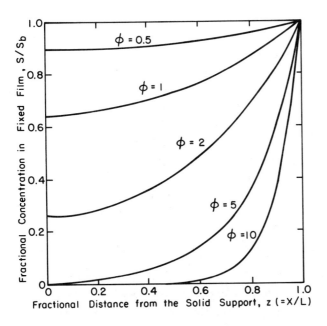

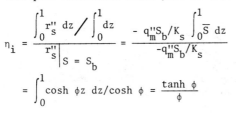

Figure 14.9. Substrate concentration profile within a flat
biofilm undergoing a first-order reaction.

microorganisms are utilizing substrate rapidly in relation to the rate at which it
is diffusing to them. When the substrate concentration within the film is less than
the concentration at the liquid-biofilm interface, the organisms within the film are
not utilizing the substrate as rapidly as those at the interface. This lowering of
the reaction rate by the need for substrate transport can be quantified through
application of the effectiveness factor concept.

In Section 14.1.2 the effectiveness factor was defined as the observed rate
divided by the rate that would exist if there were no mass transport limitation.
In this situation, the actual substrate removal rate decreases as we look at locations
deeper within the biofilm. Thus the observed rate will be some average value which
can be obtained by integrating over the entire film depth. Since there is no external mass transport limitation, the substrate concentration at the liquid-biofilm
interface is the bulk liquid concentration, S_b, and thus the reaction rate without a
mass transport limitation would just be that occurring at S_b. Consequently:

$$\eta_i = \frac{\int_0^1 r''_s \, dz \Big/ \int_0^1 dz}{r''_s \Big|_{S = S_b}} = \frac{- q''_m S_b / K_s \int_0^1 \overline{S} \, dz}{-q''_m S_b / K_s}$$

$$= \int_0^1 \cosh \phi z \, dz / \cosh \phi = \frac{\tanh \phi}{\phi} \qquad (14.45)$$

The subscript i on η refers to the fact that the effectiveness factor is correcting for internal mass transfer effects.

The effectiveness factor as given by Eq. 14.45 is shown in Fig. 14.10. For small values of the Thiele modulus (< 0.5), $\eta_i \cong 1$ indicating that the diffusion resistance is negligible so that the substrate concentration does not drop appreciably within the biofilm. Physically, a small value for ϕ implies a thin biofilm, slow reaction, or fast diffusion, all of which tend to reduce the effect of the mass transfer resistance. For large values of ϕ, (> 5), tanh ϕ approaches unity so that $\eta_i = 1/\phi$, indicating that the substrate concentration drops rapidly within the film. In this region diffusion strongly influences the overall reaction rate.

The actual average reaction rate can be written in terms of the substrate concentration at the liquid-biofilm interface:

$$-r''_s = (q''_m/K_s)S_b\eta_i = (q''_m/K_s)[(\tanh \phi)/\phi]S_b \tag{14.46}$$

Thus, just as with external mass transfer resistance, the actual rate is still first order with respect to S_b and the internal diffusion acts only to modify the rate constant. It should be emphasized, however, that this happens only for first-order reactions. For any other kinetics, the effectiveness factor is a function of S_b and therefore diffusional resistance changes the functional form of the rate expression as well as the parameter value.

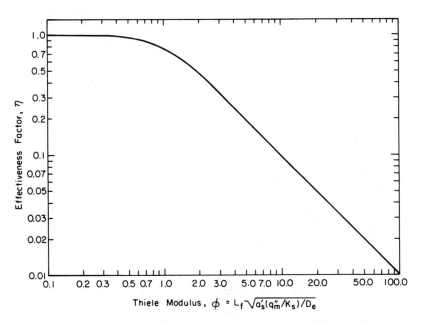

Figure 14.10.   Internal effectiveness factor for a first-order reaction within a flat fixed film.

EFFECTIVENESS FACTOR FOR MONOD KINETICS. Let us now return to the original nonlinear problem, Eq. 14.33, and apply the concepts of the Thiele modulus and the effectiveness factor to its solution. Using the dimensionless variables defined by
Eqs. 14.37, 14.38 and 14.42 and defining a new one,

$$\kappa = S_b/K_s \tag{14.47}$$

Eqs. 14.33 through 14.35 may be rewritten:

$$\frac{d^2\overline{S}}{dz^2} - \frac{\phi^2\overline{S}}{1 + \kappa\overline{S}} = 0 \tag{14.48}$$

$$\overline{S} = 1 \text{ at } z = 1 \tag{14.49}$$

$$\frac{d\overline{S}}{dz} = 0 \text{ at } z = 0 \tag{14.50}$$

The parameter κ indicates the degree to which Monod kinetics deviate from first-order
kinetics, i.e., a small value of κ indicates practically first-order behavior while
a large value indicates nearly zero-order behavior. By recognizing that the actual
reaction rate must equal the transport of substrate across the liquid-solid interface,
the internal effectiveness factor can be expressed as

$$\eta_i = \frac{-D_e \left.\frac{d\overline{S}}{dz}\right|_{z=1}}{-a_s'q_m''L_f^2/(K_s + S_b)} = (\frac{1 + \kappa}{\phi^2}) \left.\frac{d\overline{S}}{dz}\right|_{z=1} \tag{14.51}$$

which shows that the parameters affecting it are ϕ and κ. However, since κ is a
function of S_b, η_i is also a function of S_b. Nevertheless, the most convenient way
of representing the effectiveness factor is in terms of the parameters ϕ and κ, as
was done by Pitcher [4] who used a generalized modulus (suggested by Bischoff [5])

$$\phi_p = [\phi\kappa/(1 + \kappa)][2\kappa - 2 \ln(1 + \kappa)]^{-0.5} \tag{14.52}$$

as the abscissa in his graphs (See Fig. 14.11).

Comparison of Fig. 14.11 with Fig. 14.10 shows that the main effects which can
be attributed to the nonlinearity of Monod kinetics occur at intermediate values of
the generalized modulus. At low values of ϕ_p, the effectiveness factor approaches 1.0
(just as it did with first-order kinetics), indicating that the diffusional resistance
is negligible with respect to the maximal reaction rate. At high values of ϕ_p (> 3),
$\eta_i = 1/\phi_p$ (like first-order kinetics), indicating that diffusion strongly influences
the reaction rate. In the intermediate region, when κ is very small the substrate
removal rate reduces to first order so that the lower curve in Fig. 14.11 is very
similar to Fig. 14.10. When κ is very large, on the other hand, the substrate removal
rate approaches zero order and the effectiveness factor is inversely proportional

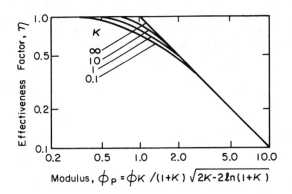

Figure 14.11.   Internal effectiveness factor
                 for Monod kinetics within a
                 flat biofilm.
                 (Adapted from Pitcher [4]).

to ϕ_p for any value of ϕ_p greater than 1.0.   Intermediate values of κ give η_i values
intermediate between those extremes.

Although graphical representations like Fig. 14.11 are convenient for some applications, there are other occasions when it would be more convenient to be able to determine the effectiveness factor analytically.   Consequently Atkinson and Davies [6] developed both complex and simplified functional relationships which agree quite well with numerical results.   In the interest of brevity, their equations will not be presented here, although they can be found elsewhere [6,7].

In summary, it has been shown that the effectiveness factor concept can be applied to Monod kinetics within a biofilm of finite thickness so that the actual substrate removal rate can be expressed by:

$$-r''_s = [q''_m S_b / (K_s + S_b)] \eta_i \qquad (14.53)$$

Therefore, let us use this expression to investigate the performance of a biofilm with internal mass transfer resistance when placed into a CSTR and a PFR.

14.2.3   Implications of Internal Mass Transfer Resistance for Reactor Performance

CSTR.   In a manner similar to that employed in Section 14.1.3 assume that steady-state conditions prevail in a CSTR containing a solid surface of area A_s covered with a biofilm of thickness L_f, so that the cells generated by consumption of the substrate are detached from the biofilm, dispersed throughout the bulk liquid phase, and washed out of the reactor by the effluent.   Under these circumstances the mass balance on substrate within the CSTR will be the same as Eq. 14.16.   When substitution is made into it, however, the substrate utilization rate per unit area of biofilm will be given by Eq. 14.53 rather than Eq. 14.11, thereby reflecting the stated condition

that there is mass transfer resistance within the biofilm but none external to it.
Since the only difference between Eq. 14.11 and Eq. 14.53 is the effectiveness factor
employed, substitution of the appropriate terms into the mass balance equation
(Eq. 14.16) leads to an equation identical to Eq. 14.19 with the exception that η_i
is used in place of η_e:

$$\frac{1}{\tau} = \frac{\mu_m S_b}{K_s + S_b} [1 + X_o/Y(S_o - S_b)] + \frac{q_m'' S_b}{(K_s + S_b)(S_o - S_b)} \eta_i (A_s/V) \qquad (14.54)$$

Equation 14.54 may be used to investigate the effects of various parameters upon
the performance of a CSTR containing a biofilm with internal mass transfer resistance.
Like η_e, however, η_i must be evaluated from Fig. 14.11 (or an appropriate equation)
after estimation of the Thiele modulus, ϕ, and the generalized modulus, ϕ_p.

In order to demonstrate the effect of internal mass transfer resistance upon
the performance of a CSTR containing a biofilm of thickness L_f, Fig. 14.12a was prepared by using Eq. 14.54 with the effectiveness factor equations and the parameter
values listed in Table 14.3. Three curves are presented, one with no internal mass
transfer resistance (so that $\eta_i = 1$), one with an effective diffusivity of
2.484×10^{-2} cm^2/hr, and one with an effective diffusivity of 2.484×10^{-3} cm^2/hr.
Examination of the figure and comparison to Fig. 14.3 show that the general effects
of internal mass transfer resistance are similar to those of external mass transfer
resistance, i.e., a reduction in the amount of substrate that can be removed by the
biofilm. One difference that should be noted, however, is that whereas the external
mass transfer resistance is subject to change by engineering factors such as the
velocity of flow past the biofilm, the internal mass transfer resistance is not.
Instead, it is a function of the physical and chemical properties of the wastewater
and the microorganisms in the system. Consequently, the curves in Fig. 14.12a represent the types of results that might be expected with different types of substrate

TABLE 14.3
Values of Kinetic Parameters and Variables Used to Plot Figure 14.12

Symbol	Value	Symbol	Value
μ_m	0.216 hr^{-1}	D_e	0.02484 cm^2/hr in Part b; as noted in Part a
K_s	30 mg/liter	L_f	0.05 cm
Y	0.5	A_s	100 cm^2 in Part a; as noted in Part b
q_m''	0.696 $mg/(cm^2 \cdot hr)$	a_s'	20 cm^2/cm^3
		S_o	200 mg/liter
		X_o	0

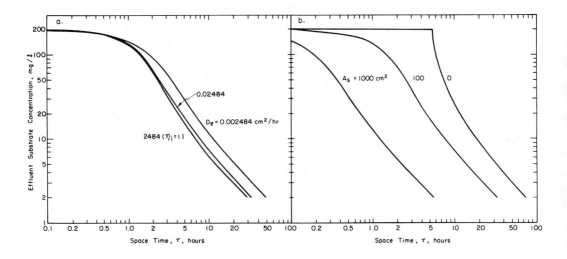

Figure 14.12.   Effects of internal mass transfer and biofilm area on the
                removal of substrate by a CSTR: (a) Effects of internal
                mass transfer; (b) Effects of biofilm area.

rather than with different liquid flow rates. Figure 14.12b shows the effects of
the surface area of biofilm available within the reactor. Again, the similarity to
Fig. 14.3 is apparent.

PFR. As with the CSTR, the development of the equations describing the performance of a PFR containing a biofilm with thickness L_f parallels the development in
Section 14.1.3 for a PFR with a thin biofilm. The schematic of the reactor was given
in Fig. 14.4 and the nomenclature is the same. The only difference is that there is
now internal mass transfer resistance but no external resistance. The mass balance
equation for substrate is exactly the same as Eq. 14.23, thereby yielding Eq. 14.24
when the limit is taken on Δx. When substitution is made into Eq. 14.24, however,
Eq. 14.53 must be used for the reaction rate in the biofilm rather than Eq. 14.11.
This leads to a differential equation like Eq. 14.25, except that the internal effectiveness factor, η_i, is used in place of the external one:

$$\frac{dS_b}{dx} + \left[\frac{\mu_m AL}{FL} \left(\frac{X_o}{Y} + S_o - S_b \right) + \left(\frac{q''_m a_s L}{FL} \right) \eta_i \right] \frac{S_b}{K_s + S_b} = 0 \qquad (14.55)$$

$$S_b = S_o \quad \text{at} \quad x = 0$$

Equation 14.55 may be used to investigate the effects of various parameters upon
the performance of a PFR containing a biofilm with internal mass transfer resistance
but no external resistance. We saw from Fig. 14.11, however, that η_i was a complex
function of the generalized modulus, ϕ_p, which was a function of the bulk substrate
concentration, S_b. Furthermore, we also saw that the functional relationship between

η_i and ϕ_p depended upon S_b, through the parameter κ. Since S_b varies along the reactor length, η_i will also. This means that the appropriate functional relationships for η_i must be utilized with Eq. 14.55 in order to solve for the change in substrate concentration with reactor length. Furthermore, since Eq. 14.55 and the equations describing η_i are all nonlinear, numerical techniques must be employed to solve the initial value problem.

In order to illustrate the effects of internal mass transfer resistance on the performance of a PFR containing a biofilm with thickness L_f, Fig. 14.13 was generated by numerically integrating Eq. 14.55 plus the appropriate equations for η_i and the parameter values given in Table 14.4. Comparing the substrate concentration curves for the two values of D_e with the curve for $\eta_i = 1$, shows that the effect of the mass transfer resistance is to reduce the rate of substrate removal, thereby reducing the amount of substrate that can be removed within the reactor, just as was the case with the external resistance shown in Fig. 14.5. Unlike that case, however, in which the main effects were exhibited toward the end of the reactor when the substrate concentration was low, the effects of the internal resistance are manifested all along the reactor length. This can be seen more clearly by comparing the variation in the internal effectiveness factor shown in Fig. 14.13b with the variation in the external one shown in Fig. 14.5b. The external effectiveness factor is very sensitive to S_b so that it is high initially and declines along the reactor length. The internal effectiveness factor, on the other hand, is low along the entire reactor and relatively insensitive to changes in S_b (for the parameter values utilized herein). This suggests that if internal mass transfer resistance were the only form present, that the assumption of a constant effectiveness factor would not lead to appreciable error.

TABLE 14.4
Base Values of Kinetic Parameters and Variables
Used to Plot Figures 14.13 Through 14.15

Symbol	Value	Symbol	Value
μ_m	0.216 hr^{-1}	a_s'	$5 \text{ cm}^2/\text{cm}^3$
K_s	30 mg/liter	F	$3600 \text{ cm}^3/\text{hr}$
Y	0.5	X_o	10 mg/liter
q_m''	$0.696 \text{ mg}/(\text{cm}^2 \cdot \text{hr})$	S_o	200 mg/liter
D_e	$0.02484 \text{ cm}^2/\text{hr}$		
L	180 cm	η_i	$1 - \dfrac{\tanh \phi}{\phi} \left[\dfrac{\phi_p}{\tanh \phi_p} - 1\right]$ for $\phi_p \leq 1$
A	16.66 cm^2		
L_f	0.2 cm		$\dfrac{1}{\phi_p} - \dfrac{\tanh \phi}{\phi} \left[\dfrac{1}{\tanh \phi_p} - 1\right]$ for $\phi_p > 1$
a_s	16.5 cm		
		ϕ_p	$[\phi\kappa/(1 + \kappa)][2\kappa - 2\ln(1 + \kappa)]^{-0.5}$
		ϕ	$[a_s'(q_m''/K_s)L_f^2/D_e]^{0.5}$

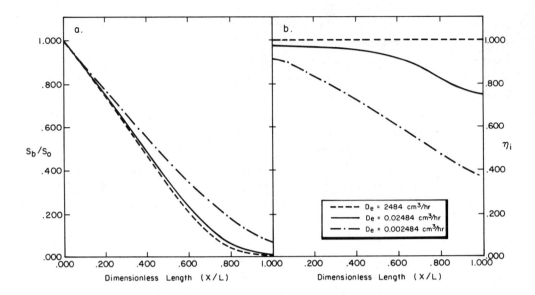

Figure 14.13. Effect of internal mass transfer on the substrate concentration
              and effectiveness factor profiles in a PFR containing a thick
              biofilm; (a) Effects on substrate concentration profiles;
              (b) Effects on effectiveness factor profiles.

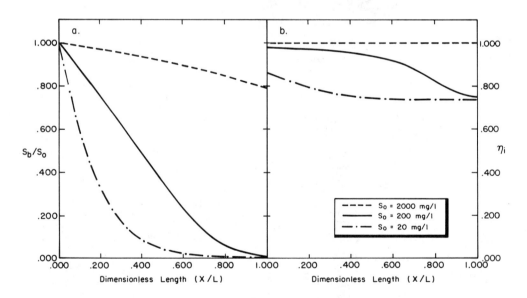

Figure 14.14. Effect of influent substrate concentration on the substrate
              concentration and effectiveness factor profiles in a PFR
              containing a thick biofilm: (a) Effects on substrate concen-
              tration profiles; (b) Effects on effectiveness factor profiles.

   Figure 14.14 shows the effect of the influent substrate concentration on the
performance of a PFR with only internal mass transfer resistance. These curves were
generated in the same manner as those in Fig. 14.13. As was the situation with only
external mass transfer resistance, high values of S_o cause almost linear removal of
substrate along the length of the reactor whereas lower values cause an exponential
decline in S_b. Furthermore, the percentage removal of substrate decreases as the
influent is increased although the mass removed increases. The reason for this was
discussed earlier. Again, the main difference between the effects of the two types
of mass transfer resistance may be seen in the behavior of the effectiveness factors.
   Figure 14.15 shows the effects of the influent flow rate. Again, as the flow
rate is increased the fractional removal of substrate decreases. Comparison of this
figure with Fig. 14.7 shows that the effects are more severe when the mass transfer
resistance is internal. This is because the flow rate does not directly affect the
rate of internal mass transfer as it does the rate of external mass transfer. Thus
the small changes in η_i shown in Fig. 14.15 are the result of the changes in the
concentration of substrate within the reactor.

14.3 INFLUENCE OF COMBINED EXTERNAL AND INTERNAL MASS TRANSFER
RESISTANCES ON REACTIONS OCCURRING WITHIN A FLAT FIXED FILM

Let us now combine the two previous situations and consider the general case in which
the substrate must be transported from the bulk liquid phase through a stagnant liquid
layer to the surface of a biofilm of thickness L_f within which both diffusion and

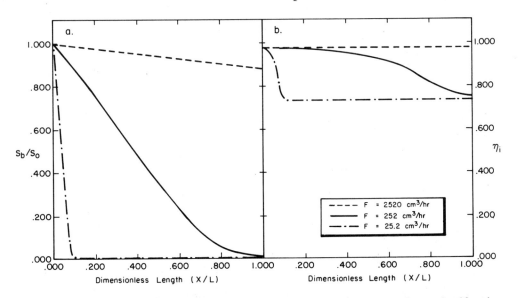

Figure 14.15. Effect of flow rate on the substrate concentration and effective-
           ness factor profiles in a PFR containing a thick biofilm: (a)
           Effects on substrate concentration profiles; (b) Effects on
           effectiveness factor profiles.

reaction occur, as shown in Fig. 14.16. Again, we will consider a simple first-order reaction first to illustrate the concept of an overall effectiveness factor.

14.3.1 Concept of Overall Effectiveness Factor

As discussed in Section 14.1.2, when a steady state exists at the surface of a biofilm there will be no accumulation of substrate and hence the rate of substrate supply through the stagnant liquid layer must equal the flux into the fixed film, which in turn must equal the overall rate of reaction in the film. The flux through the stagnant liquid layer was given by Eq. 14.1

$$N_s = k_L(S_b - S_s) \qquad\qquad (14.1)$$

Furthermore, as discussed in Section 14.2.2 when both diffusion and reaction occur within a biofilm of finite thickness, the average reaction rate for the biofilm can be written in terms of the substrate concentration at the surface of the film and the internal effectiveness factor for the film, η_i. Thus, for a first-order reaction in a biofilm which has substrate at concentration S_s at its surface, Eq. 14.46 may be modified and used for the reaction rate

$$-r_s'' = (q_m''/K_s)S_s\eta_i \qquad\qquad (14.56)$$

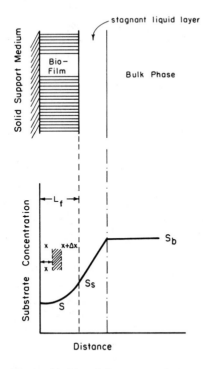

Figure 14.16.   Substrate concentration profiles for
                reaction within a thick biofilm with
                external mass transfer resistance.

The difficulty with Eq. 14.56 in this form is that S_s is hard to measure. It is easy to measure the bulk substrate concentration, S_b, however, and thus it would be advantageous to express Eq. 14.56 in terms of S_b. Using the technique employed in Section 14.1.2 this can be done by setting Eq. 14.56 equal to Eq. 14.1, solving for S_s in terms of S_b, and substituting the result into Eq. 14.56. Performing these steps leads to:

$$-r_s'' = k_o' S_b \qquad (14.57)$$

where

$$\frac{1}{k_o'} = \frac{1}{k_L} + \frac{1}{(q_m''/K_s)\eta_i} \qquad (14.58)$$

Equation 14.57 reveals that just as in the cases of each of the mass transfer resistances alone, the imposition of both resistances upon a first-order reaction does not change the functional form of the rate equation. It only changes the rate constant. Furthermore, examination of Eq. 14.58 reveals that the overall resistance is equal to the sum of two resistances in series; the external mass transfer resistance, $1/k_L$, and a combined resistance of internal mass transfer and reaction, $1/(q_m''/K_s)\eta_i$. It should also be noted that Eq. 14.58 reduces to Eq. 14.10 when there is no internal resistance to mass transfer, or when the film is extremely thin so that $\eta_i = 1$.

Drawing upon the effectiveness factor concept developed for each of the two resistances alone, we may define an overall effectiveness factor, η_o, so that Eq. 14.57 may be rewritten in terms of the intrinsic reaction rate:

$$-r_s'' = (q_m''/K_s)S_b\eta_o \qquad (14.59)$$

The relationship of the overall effectiveness factor to the other terms may be obtained by combining Eqs. 14.57 and 14.59:

$$1/\eta_o = (q_m''/K_s)/k_L + 1/\eta_i \qquad (14.60)$$

Equation 14.60 reveals that for a first-order reaction the inverse of the overall effectiveness factor is equal to the sum of two terms; the inverse of the internal effectiveness factor $1/\eta_i$, and a term representing the external effectiveness factor, $(q_m''/K_s)/k_L$. Consequently, η_o may be estimated from knowledge about each of the individual resistances. Unfortunately, for nonlinear kinetics like the Monod equation, this is not true and the overall effectiveness factor must be developed separately. Nevertheless it should be recognized that even in that more complex situation, the overall effectiveness factor incorporates the effects of each of the individual mass transfer resistances.

14.3.2 Overall Effectiveness Factor for Monod Kinetics

In order to develop the relationship for the overall effectiveness factor with Monod kinetics we must return to the approach used in Section 14.2.2 and write a mass

balance on substrate for a differential element within the biofilm. When this is
done the basic equation is the same

$$D_e \frac{d^2S}{dx^2} - a_s' q_m'' S/(K_s + S) = 0 \qquad (14.33)$$

as is the boundary condition at x = 0

$$\frac{dS}{dx} = 0 \quad \text{at} \quad x = 0 \qquad (14.35)$$

The boundary condition at $x = L_f$ is different, however, because the substrate concentration at the liquid-biofilm interface is not known. Consequently, it must be
expressed in terms of the transport of substrate across the stagnant liquid layer.

$$N_s = D_e \frac{dS}{dx} = k_L (S_b - S_s) \quad \text{at} \quad x = L_f \qquad (14.61)$$

Thus the development of an equation for the overall effectiveness factor requires the
solution of Eq. 14.33 with Eqs. 14.35 and 14.61 as boundary conditions.

Fink et al. [8] solved the equations as a special limiting case of more general
boundary conditions, using the following dimensionless quantities:

$$\psi = k_L L_f/D_e \qquad (14.62)$$

$$\phi_f = \phi [1/(1 + \kappa)]^{\frac{1}{2}} \qquad (14.63)$$

where ψ is a Sherwood number and ϕ_f is a modified Thiele modulus. The expressions
for ϕ and κ are given by Eqs. 14.42 and 14.47, respectively. The results of
Fink et al. [8] showing the overall effectiveness factor as a function of these
groups are given in Fig. 14.17. These values of the overall effectiveness factor
may be used to calculate the actual rate of substrate removal under conditions where
both internal and external mass transfer resistances exist. Thus:

$$-r_s'' = [q_m'' S_b/(K_s + S_b)]\eta_o \qquad (14.64)$$

A logical question to arise when viewing Fig. 14.17 concerns the relative con-
tributions of the internal and external resistances to the overall effectiveness
factor. One way to answer that question is to set $\psi = \infty$, which is equivalent to removing the external mass transfer resistance. Thus, the two curves for $\psi = \infty$ represent a situation in which the only resistance to mass transfer is internal. Comparison of those curves to ones with smaller values of ψ demonstrates that the existence of external mass transfer resistance reduces the effectiveness factor dramati-
cally. For example, when ψ lies between 0.01 and 1, a ten-fold reduction in ψ results in an approximately ten-fold reduction in the effectiveness factor.

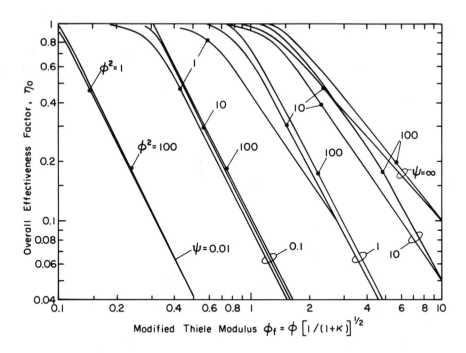

Figure 14.17. Overall effectiveness factor for Monod kinetics within a flat
biofilm with external mass transfer resistance.
(Adapted from Fink et al. [8]).

Inspection of Fig. 14.17 reveals that for all practical purposes plots of the
log of the overall effectiveness factor as a function of the log of the modified
Thiele modulus form straight lines. Therefore, the overall effectiveness factors
may be approximately correlated to the modified Thiele modulus by equations of the
form

$$\eta_o = a\phi_f^{\,b} \tag{14.65}$$

where a and b are constants for fixed values of ψ and ϕ^2. Such correlations will
make it easier to use the effectiveness factor approach in numerical solutions, although care must be exercised to ensure that the correlations are applied only to the
linear portions of the curves. Thus, it would be advisable to always state the
limits on the modified Thiele modulus within which the equations are applicable.

14.3.3 Implication of Combined External and Internal Mass
        Transfer Resistances for Reactor Performance

CSTR. If we were to again place a flat biofilm into a CSTR as was done in
Sections 14.1.3 and 14.2.3, and if we were to invoke the same assumptions regarding
steady state, etc., we would arrive at an equation similar to Eqs. 14.19 and 14.54.

The only difference would be that the overall effectiveness factor, n_o, would be used rather than the external, n_e, or internal, n_i, factors. Thus,

$$\frac{1}{\tau} = \frac{\mu_m S_b}{K_s + S_b}[1 + X_o/Y(S_o - S_b)] + \frac{q_m'' S_b}{(K_s + S_b)(S_o - S_b)} n_o(A_s/V) \tag{14.66}$$

Just as with each of the resistances alone, this equation can be utilized to investigate the effects of various parameters upon the performance of a CSTR containing a biofilm with both external and internal mass transfer resistance. The approach would be similar to that applied with the individual resistances alone because n_o is a function of S_b. In this case, however, n_o must be evaluated from Fig. 14.17 or an appropriate empirical equation after estimation of the Thiele modulus, ϕ, and the modified Thiele modulus, ϕ_f.

By making use of the fact that the curves for $\psi = \infty$ in Fig. 14.17 represent the case of internal resistance alone, it can be seen that the response of a system containing both resistances will be qualitatively similar to that shown in Fig. 14.13. The major difference, however, will be to make the overall effectiveness factor smaller than the internal effectiveness factor, thereby reducing the amount of substrate removed at any particular reactor space time. Because this difference is fairly easily visualized additional graphs will not be presented and the reader is instead urged to review Section 14.2.3.

PFR. Following a similar argument for the PFR, the equation describing its performance will be similar to Eqs. 14.25 and 14.55; the only difference will be the use of the overall effectiveness factor, n_o:

$$\frac{dS_b}{dx} + \left[\frac{\mu_m AL}{FL}\left(\frac{X_o}{Y} + S_o - S_b\right) + \left(\frac{q_m'' a_s L}{FL}\right)n_o\right]\frac{S_b}{K_s + S_b} = 0 \tag{14.67}$$

$$S_b = S_o \quad \text{at} \quad x = 0$$

Likewise, the performance will be qualitatively similar to a PFR containing a biofilm with only internal mass transfer resistance, only less substrate will be removed. Thus, no performance graphs will be presented, although the reader is urged to review Section 14.2.3.

14.4 PACKED TOWER

Two types of fixed-film reactors were shown in Fig. 1.1, a packed tower and a rotating-disc reactor. In a packed tower the microorganisms grow as a film on an immobile support media over which the liquid flows in thin sheets. Substrate must be transported from the bulk of the liquid into the interior of the biofilm so that it is likely that both external and internal resistances to mass transfer will be encountered. This implies that both resistances must be considered when models are written.

14.4.1 Model for Substrate Removal in a Packed Tower

With one minor exception, the approach used in Section 14.3 may be used to model a
packed tower. The exception arises from the fact that the media in a packed tower
may be of any shape. Consequently, the surface of the biofilm is seldom flat. This
means that the correlations for the external mass transfer coefficient given in
Section 14.1.1 are not applicable and that proper ones must be utilized. At first
glance it would also appear that the entire approach developed in Section 14.2 for
modeling the effects of mass transfer into a flat biofilm would not be valid for a
packed tower. Careful investigation of the situation, however, reveals that this is
not the case. The thickness of the active biofilm is around 0.1 mm while the equivalent radius of the support media ranges from approximately 31 mm to 38 mm [9]. Thus,
the biofilm may be looked upon as an annulus whose thickness is negligibly small in
comparison to its total radius so that it may be approximated as a flat plate with
the same thickness. In other words, even though the surface of the biofilm is not
actually flat, it may be looked upon as a flat plate, thereby allowing the mass
balance equation for substrate within the biofilm to be written in planar geometry
as in Eq. 14.32. Even if this flat plate approximation could not be made, however,
the overall effectiveness factor could still be used because it has been found that
the numerical values of the effectiveness factors for curved shapes like cylinders
and spheres are very close to those of flat plates if the proper value of the characteristic length, L_f, is used [10]. The proper value for arbitrary shapes is

$$L_f = \frac{\text{volume of biofilm}}{\text{external surface area for reactant penetration}} \qquad (14.68)$$

Thus, even if the surface cannot be considered to behave exactly like a flat plate,
proper choice of L_f will minimize the error, thereby allowing us to account for the
effects of both internal and external mass transfer by using the overall effectiveness factor approach, with values of η_o as given by Fig. 14.17.

The complex nature of the flow patterns in a packed tower causes another problem
because it prevents the flow regime from being plug flow in the strict sense. Unfortunately, it is not currently possible to predict precisely the extent of dispersion
in the axial and radial directions and thus an attempt to rigorously model the flow
regime would be premature. Instead it would be preferable to model the packed tower
with a relatively simple model which retains the major macroscopic characteristics
and has the ability to predict the performance of an ideal tower. A plug-flow model
will meet those criteria and thus we will use it.

EXTERNAL MASS TRANSFER. The support media in a packed tower might be rocks,
special plastic rings, or formed plastic sheets. Consequently, the flow patterns
within a tower are extremely complex, reflecting interactions among fluid elements
flowing over different support elements, variations in the cross sectional area available for flow, and irregularities caused by channeling and short circuiting. In fact,

the flow patterns are so complex that it has been necessary to develop empirical correlations for the mass transfer coefficients within such towers [2,11].

As discussed in Section 14.1.1 one common form of correlation relates the mass transfer coefficient to the Schmidt number ($S_c = \mu/\rho D$) and the Reynolds number ($R_e = u\rho d/\mu$). Recognizing that the product of the velocity (u) and the density (ρ) of a fluid is the mass velocity (G), the Reynolds number is often written as Gd/μ, in which d is the equivalent sphere diameter of the support media and μ is the fluid viscosity. Using this concept, Wilson and Geankoplis [12] reported the following correlations for mass transfer to liquids in packed beds:

$$k_L = (0.25 \; G/\epsilon\rho) \; S_c^{-2/3} \; R_e^{-0.31} \qquad\qquad 55 < R_e < 1500 \qquad\qquad (14.69)$$

$$k_L = (1.09 \; G/\epsilon\rho) \; S_c^{-2/3} \; R_e^{-2/3} \qquad\qquad 0.0016 < R_e < 55 \qquad\qquad (14.70)$$

These equations are restricted to $0.35 < \epsilon < 0.75$, where ϵ is the void space between media elements as a fraction of the total bed volume. Because the Reynolds number is directly proportional to G, Eq. 14.69 predicts that the mass transfer coefficient increases with $G^{0.69}$ while Eq. 14.70 predicts that it increases with $G^{0.33}$, when G is the superficial mass velocity based upon the entire bed cross sectional area normal to the direction of flow. Unfortunately, many of the plastic media used in biological packed towers have void fractions as high as 0.95 and thus correlations are needed for predicting k_L in such systems. Dimensional analysis, as well as the above equations, suggests the use of a correlation of the form:

$$k_L = a'G^{b'} \qquad\qquad\qquad\qquad\qquad\qquad\qquad\qquad (14.71)$$

in which a' depends upon the properties of the fluid (e.g., ρ,μ,D) and of the media (e.g., ϵ,d) while b' depends upon the flow range employed. To date (1980) little data has been made available on typical values of a' and b' for media used in biological packed towers, and thus they will have to be determined experimentally.

GENERAL MODEL. The schematic diagram of a packed tower with recirculation is given in Fig. 14.18. The effluent from the tower is sent to a clarifier where the microbial solids are removed from it prior to discharge. A portion of the overflow, which is essentially free of cells, may be recirculated to the inlet of the tower where it is mixed with the untreated feed stream to form the influent to the tower.

Following the procedure used in the previous sections we will assume that steady-state conditions prevail in the tower, i.e., that the mass of organisms remains constant. Consequently the cells generated in the biofilm by the consumption of substrate must be detached from it and dispersed throughout the bulk fluid phase, which carries them on down the tower to the clarifier. Therefore, the possibility exists that substrate will be consumed by the cells dispersed in the liquid phase as well as by those in the biofilm.

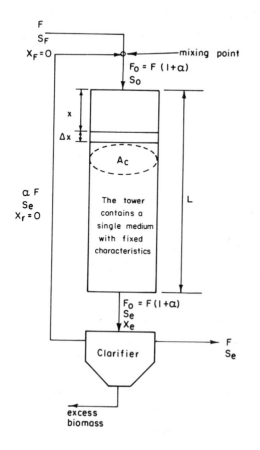

Figure 14.18.   Schematic diagram of a
packed tower with recirculation.

Under the assumption of plug flow the steady-state substrate balance over an
infinitesimal volume element of the packed tower as shown in Fig. 14.18 will be similar to that for a plug-flow reactor developed earlier. The only differences will be
that the total flow through the tower is $F(1 + \alpha)$ and that the cross sectional area
for fluid flow, A, is equal to hA_c where A_c is the empty tower cross sectional area
and h is the fraction of that area occupied by the liquid film. Consequently, the
resulting differential equation will be similar in form to Eq. 14.67:

$$\frac{dS_b}{dx} + \left\{ \frac{\mu_m h A_c L}{F(1 + \alpha)L} \left(\frac{X_o}{Y} + S_o - S_b \right) + \left(\frac{q''_m a_s L}{F(1 + \alpha)L} \right) \eta_o \right\} \frac{S_b}{K_s + S_b} = 0 \qquad (14.72)$$

$$S_b = S_o \quad \text{at} \quad x = 0$$

Because of the circulation around the tower, the influent concentrations S_o and X_o
must be related to the feed concentrations S_F and X_F by writing steady-state

substrate and cell balances around the recirculation mixing point. These balances
yield

$$X_o = (X_F + \alpha X_r)/(1 + \alpha) \qquad\qquad (14.73)$$

and

$$S_o = (S_F + \alpha S_e)/(1 + \alpha) \qquad\qquad (14.74)$$

These equations show that the feed concentrations are reduced by the recirculation
flow.

The performance of an ideal packed tower may be predicted by numerically integrating Eq. 14.72 using the overall effectiveness factor given by Fig. 14.17 or an
empirical correlation such as Eq. 14.65. When a large number of simulations are to
be performed it is more convenient to use the empirical correlation because it can be
substituted into Eq. 14.72 prior to integration. A technique must be used, however,
which will allow choice of the proper values of a and b, depending upon the values
of ψ, ϕ^2, and ϕ_f.

14.4.2 Performance of an Ideal Packed Tower

Predictions of ideal tower performance were obtained by numerically integrating
Eq. 14.72 with a fourth-order Runga-Kutta method using the parameter values given in
Table 14.5. Equations, such as Eq. 14.65 were used to allow estimation of η_o as a
function of ψ, ϕ^2, and ϕ_f. For the situations in which the curves in Fig. 14.17 were
nonlinear, the curves were divided into sections within which the error in the use of
Eq. 14.65 was minimal. Furthermore, because the value of ψ is affected by the feed
flow rate, F, the recirculation ratio, α, and the superficial cross sectional area of
the tower, A_c, interpolation formulas were developed to allow estimation of η_o for
any value of ψ between 0.1 and 10.0. When there was no recirculation ($\alpha = 0$)

TABLE 14.5
Base Values of Kinetic Parameters and Variables
Used to Plot Figures 14.19 Through 14.28

Symbol	Value	Symbol	Value
q''_m	0.4175 mg/(cm^2·hr)	a'_s	10 cm^2/cm^3
K_s	30 mg/liter	A	hA_c = 70 cm^2
μ_m	0.1296 hr^{-1}	L	200 cm
Y	0.5	A_c	200 cm^2
D_e	0.13918 cm^2/hr	F	20 liters/hr
ε	0.35	S_F	300 mg/liter
L_f	0.1 cm	X_F	0
a_s	200 cm	k_L	1.3918 $[F(1 + \alpha)/(A_c)(100)]^{0.69}$

Eq. 14.72 was integrated directly starting with the known value of the influent substrate concentration, $S_o = S_F$. When there was recirculation, however, the influent substrate concentration was not known because the effluent substrate concentration was unknown (see Eq. 14.74). Therefore, an iterative technique was used. First, the effluent substrate concentration, S_e, was guessed and used in Eq. 14.74 to calculate the influent substrate concentration, S_o, needed to begin the integration of Eq. 14.72. At the end of the integration, the effluent substrate concentration was checked against the guessed value. If they did not agree, the process was repeated using as the next trial value the effluent substrate concentration just calculated. This numerical scheme was found to converge very rapidly, giving a better than 0.1% agreement within a few iterations.

Typical profiles of substrate concentration and effectiveness factor as functions of tower depth are shown in Figs. 14.19 and 14.20 for towers without and with recirculation, respectively. In both towers the substrate concentration drops exponentially

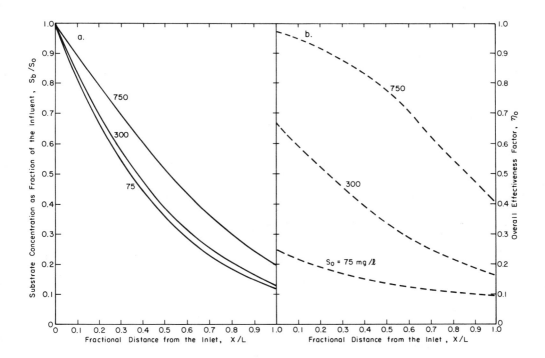

Figure 14.19. Effect of influent substrate concentration on the substrate concentration and effectiveness factor profiles in a packed tower with no recirculation: (a) Effects on substrate concentration profiles; (b) Effects on effectiveness factor profiles.

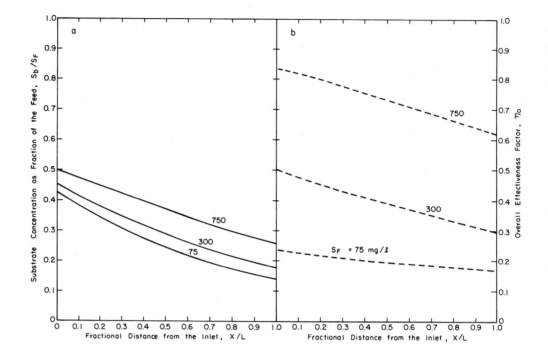

Figure 14.20.   Effect of influent substrate concentration on the substrate
concentration and effectiveness factor profiles in a packed
tower with recirculation: (a) Effects on substrate concentration profiles; (b) Effects on effectiveness factor profiles.

with tower depth. This decrease in removal rate with depth is typical of plug-flow
reactors (See Figs. 14.12 and 14.15) and is due to the dependency of the rate upon
the substrate concentration through the Monod equation and to the declining effectiveness factor. When no recirculation is employed the effectiveness factor also decreases exponentially with tower depth. When the feed substrate concentration is
high, on the other hand, it acts to keep the effectiveness factor large over a greater
portion of the tower depth. Only after the substrate concentration has dropped considerably does the effectiveness factor start to decline in an exponential fashion.
When recirculation is employed, the influent substrate concentration is reduced considerably due to dilution by the recirculation stream. Consequently, the effectiveness factor at the tower inlet is also reduced. These factors reduce the average
reaction rate thus causing the effluent substrate concentration to be higher than it
was in the tower without recirculation. It should be noted that it was assumed that
the settler was perfect so that $X_r = 0$. If there had been cells in the recirculation
flow, it is possible that the reaction term for substrate removal by suspended organisms would have been large enough to make the effluent substrate concentration lower
than it was without recirculation [13]. A comparison of Figs. 14.19 and 14.20 reveals

that the variation of the effectiveness factor from the top to the bottom of the tower is much less when recirculation is practiced. This is because the variation in substrate concentration is smaller. Furthermore, Fig. 14.20 shows that the drop in the effectiveness factor in the tower with recirculation is almost linear with depth.

Figure 14.21 shows the effects of the feed substrate concentration, S_F, on the removal of substrate. As the feed substrate concentration is increased the fractional substrate removal decreases although the mass removal rate, $F(S_F - S_e)$, increases. At first the mass removal rate increases almost linearly with the feed concentration because both the intrinsic reaction rate and the effectiveness factor increase with increases in substrate concentration. Eventually, however, η_o approaches unity and the substrate removal rate approaches its maximal value so that the mass removal rate approaches an upper limit. An important point to note is that the variation in the effectiveness factor over the tower length declines as the feed substrate concentration is increased. This indicates that the substrate concentration is so high that the mass transfer resistance becomes negligible and the biofilm becomes reaction limited.

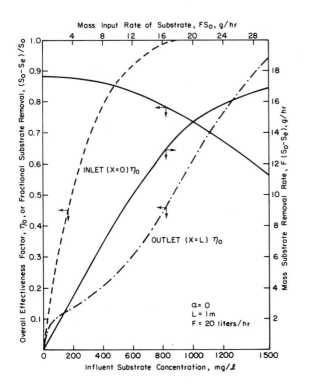

Figure 14.21.   Effect of influent substrate concen-
                tration on the performance of a
                packed tower with no recirculation.

The effects of feed flow rate are shown in Fig. 14.22. As it is increased the fractional substrate removal decreases whereas the mass substrate removal rate increases. This behavior is typical of plug-flow reactors containing biofilms and was discussed in Section 14.1.3. For the parameter values used herein, changes in flow rate have a greater effect on system performance than do comparable changes in feed substrate concentration. This can be seen most easily by comparing the substrate removal curves using the upper abscissa scale. Other parameter values can cause the opposite effect or can make the effects of changes in flow rate and concentration the same. Because the external mass transfer coefficient depends upon the velocity of flow over the media, the effectiveness factor will increase as the flow rate is increased. Because the influent substrate concentration is constant in Fig. 14.22, the curve for η_o at the tower inlet reflects these effects. The curve for the outlet, however, reflects changes in substrate concentration as well as in flow rate. Nevertheless, the increase in the overall effectiveness factor with its associated increase in the reaction rate at any point in the tower is not large enough to compensate for the decrease in the residence time so that the total substrate removal within the tower is decreased. Consequently, faster flow rates generally require greater tower depths to achieve a fixed effluent concentration, as shown in Fig. 14.23.

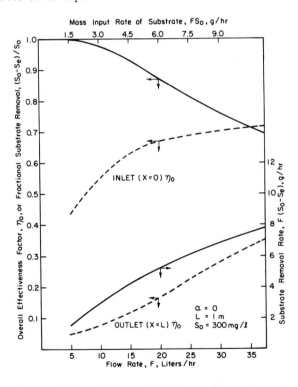

Figure 14.22. Effect of influent flow rate on the
               performance of a packed tower with
               no recirculation.

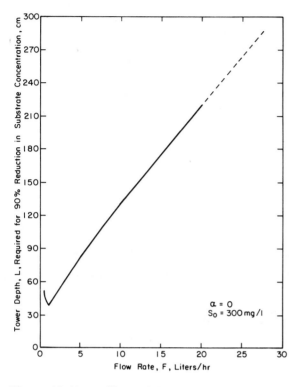

Figure 14.23.   Effect of influent flow rate on the
                depth of packed tower required to
                achieve a 90% reduction in substrate
                concentration.

However, when the flow rate is very low, increases in the flow rate will result in
decreases in the required tower depth. Under those circumstances the mass transfer
coefficient is very small, so that the external mass transfer resistance is very
large. Apparently, for the mass transfer correlation employed, the effectiveness
factor increases faster than the mass application rate of substrate for the low
values of flow rate, thereby decreasing the required tower height. Once that region
has been passed, however, further increases in the flow rate still cause the effectiveness factor to increase, but not as fast as the increase in the mass application
rate of substrate, so that a taller tower is required. It should be realized that
other mass transfer correlations and other parameter values can cause the initial
decrease in tower height to be manifested at either higher or lower flow rates.

The effects of recirculation from a perfect settler ($X_r = 0$) are shown in
Figs. 14.24 through 14.26. Figure 14.24 shows that higher recirculation ratios result in flatter substrate concentration profiles, which would be expected because a
PFR behaves like a CSTR as the recirculation ratio approaches infinity. As a result
of the reduction in the substrate concentration profiles the effectiveness factor

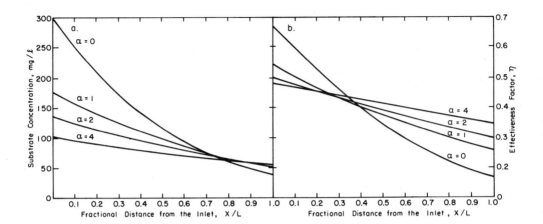

Figure 14.24.   Effect of recirculation on the substrate concentration and
effectiveness factor profiles in a packed tower: (a) Effects
on substrate concentration profiles; (b) Effects on effectiveness factor profiles.

profiles also become flatter. Note that the reduction in the effectiveness factors
at the inlet of the tower is not as pronounced as the reduction in the influent substrate concentration. This is because an increase in the recirculation ratio increases the flow rate through the tower, which in turn increases the mass transfer
coefficient, thereby acting to increase the overall effectiveness factor. Note also
that the effectiveness factor at the outlet tends to increase as the recirculation
ratio is increased. In this particular instance recirculation has little effect upon
the effluent substrate concentration. Consequently, the major factor influencing
the overall effectiveness factor is the mass transfer coefficient, which is increased
by the increase in recirculation, thereby increasing the effectiveness factor. The
effects of recirculation on the percent substrate removal from a tower of fixed depth
receiving a fixed flow rate are shown in Fig. 14.25. Introduction of recirculation
dilutes the influent substrate concentration thereby reducing the reaction rate. At
low recirculation ratios this decrease in reaction rate offsets the effect of recirculation upon the rate of mass transfer causing the percent substrate removal to decrease. However, as the recirculation ratio is increased, the effect of dilution
eventually becomes minimal so that it is offset by the reduction in the external mass
transfer resistance and the percent substrate removal increases slightly. Again it
should be recognized that other choices of parameter values or of mass transfer correlations may not show the subsequent increase in substrate removal. Figure 14.26
shows the effect of the recirculation ratio on the tower depth required to remove 90%
of the substrate in a tower of fixed cross sectional area. At flow rates greater than
1.0 liter/hr, the increase in the overall effectiveness factor caused by the increased
flow does not exceed the decrease in the overall effectiveness factor caused by

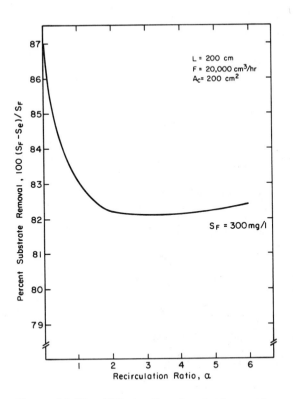

Figure 14.25. Effect of recirculation ratio on the
performance of a packed tower.

dilution of the substrate so that the required depth increases as the recirculation
ratio is increased. For flow rates between 2.5 to 7.5 liters/hr the required depth
increases initially, reaches a maximum, and finally decreases slightly as the recirculation ratio is increased indicating that eventually the reduction in the external mass transfer resistance overcomes the effect due to dilution of the substrate.
At extremely low flow rates, for example 0.5 liter/hr, recirculation can reduce the
tower depth required. Under those circumstances the external mass transfer resistance is so high that its reduction by recirculation increases the overall effectiveness factor more than the associated dilution of the substrate would tend to reduce
it, thereby giving a smaller tower. What is suggested by Fig. 14.26, is that when
the tower is limited by external mass transfer, recirculation can reduce the tower
depth necessary to remove a given fraction of the substrate introduced, whereas when
the external mass transfer resistance does not control, recirculation will merely act
to dilute the substrate, thereby reducing the reaction rate and increasing the required tower height. Using a plug-flow model in which the saturation constant in
the Monod equation was empirically correlated to vary with the flow rate,
Kornegay [14] showed qualitatively similar results. It should be realized that with

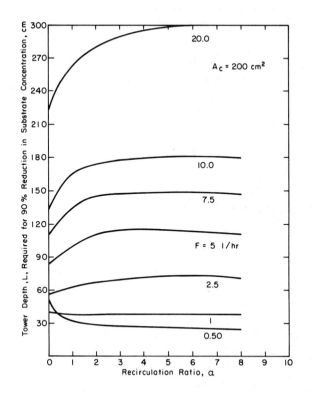

Figure 14.26. Effect of recirculation ratio on the
depth of packed tower required to
achieve a 90% reduction in substrate
concentration.

other mass transfer correlations and other parameter values, the flow rate at which
recirculation reduces the tower depth may be either higher or lower than that found
here.

The effects of tower cross sectional area are illustrated in Fig. 14.27. An increase in the cross sectional area, A_c, results in an increase in the surface area
of the biofilm per unit length of the tower, a_s, and a decrease in the fluid velocity.
Thus, it is apparent from Eq. 14.72 that such an increase should decrease the depth
of tower required to remove a given fraction of substrate. However, the decrease in
the depth is not in proportion to the increase in the area so that the net result is
an increase in tower volume. This happens because the decrease in fluid velocity increases the external mass transfer resistance thereby decreasing the overall effectiveness factor. The results in Fig. 14.27 suggest that the total tower volume will
be minimized by choosing a tall, thin tower rather than a short, fat one. Because a
similar conclusion has been reached with other models [14], as well as experimentally [15,16], it appears to be general and thus would be expected to be true for other

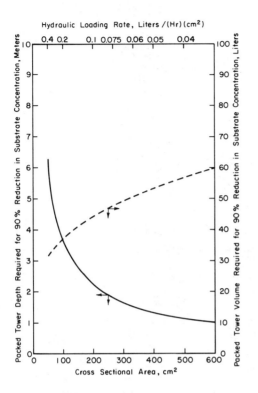

Figure 14.27.   Effect of the cross sectional area
                of a packed tower upon the depth and
                volume of media required to achieve
                a 90% reduction in substrate
                concentration.

parameter values as well. Increasing the cross sectional area for a fixed flow rate
is equivalent to decreasing the hydraulic loading on a tower. To illustrate this
point the hydraulic loading is given along the upper abscissa of Fig. 14.27.

The effects of changes in the surface area of biofilm per unit tower length, a_s,
are shown in Fig. 14.28. In performing the calculations for this figure it was assumed that the tower dimension remained unchanged and that only the type of support
media was changed. An increase in a_s results in an increase in the amount of biomass available for substrate removal; therefore more substrate is removed.

14.4.3 Parameter Estimation

Before the model, Eq. 14.72, can be used to predict the performance of a particular
tower numerical values must be available for all of the parameters in it. Unfortunately, research in this area is in its infancy and many questions remain to be
answered. For example, consider the two mass transfer parameters, D_e and k_L. Even
though much information is available concerning the diffusivity of organic compounds

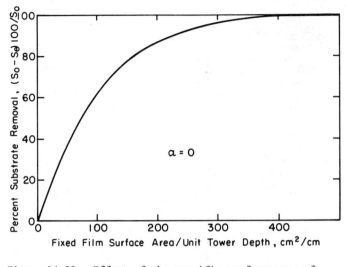

Figure 14.28. Effect of the specific surface area of
fixed film on the performance of a
packed tower without recirculation.

in water, little work has been done related to diffusion through biofilms. Furthermore, that which has been done suggests that effective diffusivity values depend
upon the growth conditions of the microorganisms in the biofilm [17]. Many correlations are available for mass transfer coefficients in absorption and heat exchanger
towers. Nevertheless, it is not clear whether those correlations can be used directly
for packed microbial towers; e.g., does the presence of the biofilm alter the surface
characteristics of the packing sufficiently to change the relationship between the
flow rate and the external mass transfer coefficient? Similar questions can be
raised about many of the other parameters which must be evaluated.

     Even though a number of techniques can be suggested whereby the parameters in
Eq. 14.72 can be evaluated, none have been verified. Consequently, it was decided
that their inclusion would be premature. Hopefully future research will provide the
needed verification and thus the reader is urged to follow the literature in this
area. In the meantime, use must be made of simpler empirical models which do not
attempt to account for all of the physical and biochemical phenomena occurring in
packed towers.

14.4.4 Other Packed Tower Models
It is apparent by now that a packed tower is a very involved biochemical operation
which is difficult to model theoretically. Consequently, empirical and semiempirical
approaches have been used to model it and a number of design equations have been
developed. The NCR [15] and Galler-Gotaas [18] equations are strictly empirical because they are based on regression analysis techniques. Consequently, they are valid
only in the limited region over which data were available and can only be used for

interpolation within that region. They cannot be used for extrapolation. Semi-empirical models are also primarily interpolative, but can be successfully employed for extrapolation if the extrapolations do not extend them beyond the range over which their simplifying assumptions are valid. In order to recognize the assumptions within the semiempirical models of Velz [19], Eckenfelder [20], Kornegay [14], and Schroeder [21] let us compare them with the theoretical model (Eq. 14.72).

VELZ EQUATION. Velz' [19] equation states that

$$\frac{S_e}{S_o} = \exp[-KL] \tag{14.75}$$

where K is a constant. Comparison of Eq. 14.75 with Eq. 14.72 shows that the Velz equation is limited to the situation in which the removal of substrate by suspended organisms is negligible, the substrate consumption rate is first order ($S_b \ll K_s$), and the overall effectiveness factor is constant throughout the tower. Under these circumstances Eq. 14.72 can be integrated to yield

$$S_e/S_o = \exp\{-[(q_m''/K_s)a_s\eta_o/F(1 + \alpha)]L\} \tag{14.76}$$

from which we see that $K = (q_m''/K_s)a_s\eta_o/F(1 + \alpha)$. Thus the Velz equation would be limited to one flow rate.

ECKENFELDER EQUATION. Eckenfelder [20] has explicitly accounted for the effects of flow rate

$$S_e/S_o = \exp\{-K'a_v^{1 + m}[A_c/F(1 + \alpha)]^n L\} = \exp\{-K''[A_c/F(1 + \alpha)]^n L\} \tag{14.77}$$

where m and n are parameters which are dependent upon the specific support media, a_v is the surface area per unit volume of media ($= a_s/A_c$) and K' is a pseudorate constant. This equation is a modification of the Velz equation, in which the effluent concentration is allowed to depend explicitly on F, A_c and a_v, therefore, we may look upon it as a limiting case of Eq. 14.72 in which the removal of substrate by suspended organisms is negligible, the substrate consumption rate is first order ($K_s \gg S_b$), and the overall effectiveness factor remains constant throughout the tower but varies with the superficial velocity, $F(1 + \alpha)/A_c$. The last point becomes clear if we rearrange Eqs. 14.76 and 14.77 and compare them. Equation 14.76 becomes:

$$S_e/S_o = \exp\{-(q_m''/K_s) a_v[A_c/F(1 + \alpha)]\eta_o L\} \tag{14.78}$$

whereas Eq. 14.77 becomes

$$S_e/S_o = \exp\{-K'a_v^{1 + m}[A_c/F(1 + \alpha)][A_c/F(1 + \alpha)]^{n - 1}L\} \tag{14.79}$$

Thus, if η_o is proportional to $[A_c/F(1 + \alpha)]^{n - 1}$ they are equivalent for a fixed media. According to Liptak [9] most media have an n value of 0.7 to 0.8. Consequently, a 28-fold increase in $F(1 + \alpha)/A_c$ would result in a 1.95- to 2.72-fold increase in $[A_c/F(1 + \alpha)]^{n - 1}$. Likewise, for the values of ψ, ϕ^2, and ϕ_f likely to be found in a packed tower, a 28-fold increase in $F(1 + \alpha)/A_c$ will increase the overall effectiveness factor by a factor of 2.5. Thus, it appears that the changes in $[A_c/F(1 + \alpha)]^{n-1}$ in the Eckenfelder model are similar to the changes in the overall effectiveness factor caused by flow.

KORNEGAY EQUATION. The differential equation used by Kornegay [14] is

$$\frac{dS_b}{dx} = - K_1 [A_c/F(1 + \alpha)] \frac{S_b}{K_g + S_b} \qquad (14.80)$$

$$S_b = S_o \quad \text{at} \quad x = 0$$

where K_1 depends upon the substrate and type of media and K_g is a pseudosaturation constant which decreases asymptotically to K_s as the superficial velocity $(F(1 + \alpha)/A_c)$ becomes very large. Equation 14.80 may be rewritten as

$$\frac{dS_b}{dx} = - K_1 [A_c/F(1 + \alpha)] \frac{S_b}{K_s + S_b} (\frac{K_s + S_b}{K_g + S_b}) \qquad (14.81)$$

where K_s is the intrinsic saturation constant which remains invariant for a specific organism and a specific substrate. If there is no substrate removal by suspended organisms, Eq. 14.72 reduces to:

$$\frac{dS_b}{dx} = - K_2 [A_c/F(1 + \alpha)] \frac{S_b}{K_s + S_b} \eta_o \qquad S_b = S_o \text{ at } x = 0 \qquad (14.82)$$

in which K_2 depends upon the substrate and type of media. Comparison of Eq. 14.81 with Eq. 14.82 reveals that they are equivalent for a given media if

$$\eta_o = (\frac{K_s + S_b}{K_g + S_b}) \qquad (14.83)$$

Let us examine the requirement on η_o in more detail. Inspection of Eq. 14.83 reveals that the overall effectiveness factor: (1) approaches unity as the substrate concentration becomes very large; (2) approaches unity when the superficial velocity becomes very large so that K_g approaches K_s; (3) decreases as the substrate concentration decreases, approaching an asymptotic value of K_s/K_g as the substrate concentration approaches zero; and (4) decreases as K_g is increased (i.e., as the superficial velocity is decreased). These asymptotic and general trends are consistent with those that can be predicted from Fig. 14.17. Thus, it appears that the Kornegay equation may be interpreted as a limited case of Eq. 14.72 in which substrate removal by suspended organisms is negligible and the overall effectiveness factor is given by Eq. 14.83.

SCHROEDER EQUATION. Schroeder's [21] equation states that

$$\frac{dS_b}{dx} = - \frac{q''_m a_s}{F(1 + \alpha)} \frac{S_b}{K_s + S_b} \eta_i \qquad (14.84)$$

where η_i is the internal effectiveness factor. Comparison of Eq. 14.84 with
Eq. 14.72 reveals that the Schroeder equation is a limiting case in which both the
removal of substrate by suspended organisms and the external mass transfer resistance
are assumed to be negligible. The assumption of negligible external mass transfer
resistance may be valid only when the superficial velocity is very high. Furthermore, Schroeder assumed that the internal effectiveness factor is directly proportional to the substrate concentration. Figures 14.19, 14.20 and 14.24 suggest,
however, that this assumption is a poor one.

14.5 ROTATING-DISC REACTOR

The other common type of fixed-film reactor is the rotating-disc reactor (RDR) in
which closely spaced discs are mounted on a common horizontal shaft placed very near
to or touching the water surface. The shaft is rotated at constant speed thereby
allowing any point on a disc to be alternately submerged and exposed to the air.
When water containing organic matter and nutrients flows through the reactor, microorganisms consume the substrate and grow attached to the disc. The rotating action
imparts a shear force to the biofilm, keeping its thickness relatively constant by
removing the cells generated by assimilation of the substrate. The turbulence
generated by the rotation provides for oxygen transfer to the bulk of the liquid and
also keeps the sloughed organisms in suspension so that they are carried out in the
effluent. The most common arrangement of the discs is parallel to the direction of
fluid flow as shown in Fig. 14.29. Under those conditions the turbulence is sufficient to make the substrate concentration the same throughout the tank, i.e., for all
practical purposes the tank can be considered to be completely mixed. In Chapter 4
it was seen that for many types of reactions CSTR's in series give better performance
than single CSTR's of equal total volume. Because of this and because of the modular
character of RDR's, most applications utilize a series of reactors. Consequently,
the performance of an RDR system can be modeled as a series of CSTR's containing
biofilms.

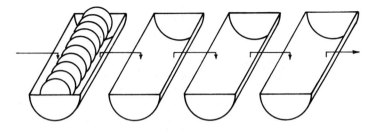

Figure 14.29. Schematic diagram of a rotating-disc reactor system.

14.5.1 Model for Substrate Removal in a Single RDR

Because of the complex nature of an RDR, certain simplifying assumptions must be made in order to model it. The first is that steady-state conditions prevail so that organisms are sheared from the surface of the biofilm at a rate equal to their growth. The second assumption is that the turbulence in the reactor fluid is sufficient to keep the detached organisms in suspension so that they can be washed out with the effluent. The third is that both the fixed and detached organisms contribute to substrate removal. The fourth is that oxygen and other nutrients are present in excess so that the organic substrate is the growth limiting nutrient. The fifth is that the thickness of the liquid film is uniform over the aerated sector of the disc. The final assumption is that the substrate concentration in the liquid film on the aerated sector depends only on the circumferential angle θ and not upon the radial position. In other words we treat the liquid film on the aerated sector as a PFR on top of the biofilm.

Just as with a packed tower, the development of the model for this reactor must incorporate the effects of both external and internal mass transfer resistances through the overall effectiveness factor, η_o. Although the surface of a disc is generally not flat, the diameter of the undulations is large in comparison to the thickness of the active biofilm so that the effectiveness factor for a flat biofilm developed in Section 14.3 should be applicable. Because of the rotation of the disc, however, new correlations are required for the external mass transfer coefficient.

EXTERNAL MASS TRANSFER. As shown in Fig. 14.30, each disc may be divided into two sectors: submerged and aerated. Because the biomass is attached to the disc, it moves through the bulk fluid in the submerged sector, thereby making the external

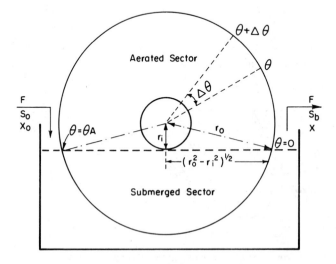

Figure 14.30.   Schematic diagram of a single disc
in a rotating-disc reactor.

mass transfer coefficient, k_L, dependent upon the rotational speed, ω. As a point on
the surface of the disc leaves the submerged sector and enters the aerated sector a
thin film of liquid adheres to it and is carried along with it. Although this film
may be assumed to have no motion relative to the biofilm on the disc, its thickness
is a function of the rotational speed of the disc and consequently the mass transfer
coefficient for the substrate within it, k_{LA}, also depends upon the speed.

Mass transfer from a fluid in laminar flow to the surface of a rotating disc was
analyzed by von Karman and given by Levich [22]:

$$k_L = 1.55 \ D^{2/3} (\mu/\rho)^{-1/6} (\omega/r)^{1/2} \tag{14.85}$$

where k_L is the mass transfer coefficient in cm/sec, r is the radius of the disc in
cm, D is the diffusion coefficient in cm^2/sec, ρ is the fluid density in g/cm^3, μ is
the fluid viscosity in g/(cm·sec), and ω is the rotational speed of the rotating
disc in revolutions/sec. Equation 14.85 indicates that the mass transfer coefficient
will increase with the square root of the rotational speed. In practice, however,
both the proportionality constant and the power on the rotational speed may be different due to deviations from the assumptions made in deriving the equation. Nevertheless, it is possible to obtain experimental correlations of the form

$$k_L = e \ \omega^f \tag{14.86}$$

where e and f are constants with e depending upon the physical properties of the
fluid and the radius of disc.

In the aerated sector, the entrained liquid forms a stagnant layer on top of
the biofilm. Thus it will be assumed that the mass transfer coefficient is equal
to the diffusivity divided by the thickness of the stagnant liquid film. The thickness of a film entrained on a flat plate withdrawn vertically from a quiescent
liquid has been analyzed by Landau and Levich [23], and found to be

$$\delta_L = a \ v^{2/3} \tag{14.87}$$

where δ_L is the film thickness in cm, v is the withdrawal velocity in cm/sec, and a
is a parameter which is dependent upon the fluid properties. Since the withdrawal
velocity of a point on a rotating disc depends upon its radial position, some average
value should be used, such as

$$v = 2\pi\omega(r_o + r_i)/2 \tag{14.88}$$

where r_o and r_i are the outer and inner radii of the submerged sector as shown in
Fig. 14.30. This suggests that the average film thickness is given by

$$\delta_L = a_1 \omega^{2/3} \tag{14.89}$$

where a_1 is a constant which depends upon both the fluid properties and the size of the disc. Hartmann [24] has reported a stagnant liquid film thickness of 40 μm for a smooth rotating disc and Eq. 14.89 predicts a thickness of 60 μm under similar conditions. Grieves [25] on the other hand, reported that the thickness of the stagnant liquid film on top of a rotating biofilm ranged from 50 to 200 μm and was not reproducible. Thus, there is a need to add an arbitrary amount to the thickness predicted by Eq. 14.89 to account for the retention of fluid by the biofilm. Hence, a more appropriate form might be

$$\delta_L = a_2 + a_3\omega^{a_4} \tag{14.90}$$

where a_2, a_3 and a_4 must be determined experimentally. Once an expression has been obtained for the stagnant film thickness, the mass transfer coefficient for the aerated sector may be estimated by dividing the diffusivity of the solute in water by the liquid film thickness:

$$k_{LA} = D/(a_2 + a_3\omega^{a_4}) \tag{14.91}$$

EFFECTIVENESS FACTORS. Once the external mass transfer coefficients have been estimated for each of the sectors the overall effectiveness factors may be determined from Fig. 14.17. They will not be the same. Consequently we will use η_{oa} and η_{os} to denote the overall effectiveness factors for the aerated and submerged sectors, respectively. The substrate consumption rates per unit area of biofilm in each of the sectors are given by Eq. 14.64 with the appropriate values for the overall effectiveness factor.

GENERAL MODEL. First let us consider the case of a single reactor as shown in Fig. 14.30. The assumptions are the same as those employed in Section 14.3. Taking the liquid volume in the trough as the control volume, V, we see that the substrate is brought in by two streams; the influent flow, F, and the liquid film on the aerated sector of each disc, F_L. Two streams also comprise the output; the effluent flow, F, and the liquid film entrained by the rotating discs, F_L. Substrate is removed by the biofilm within the submerged sector as well as by the suspended organisms. Thus, the steady-state mass balance equation on substrate in the trough is

$$FS_o + F_LS_{LR} - FS_b - F_LS_b - \frac{q_m''S_bA_s}{K_s + S_b}\eta_{os} - \frac{\mu_m S_b[X_o + Y(S_o - S_b)]V}{Y(K_s + S_b)} = 0 \tag{14.92}$$

which yields

$$\frac{1}{\tau} = \frac{\mu_m S_b}{K_s + S_b}[1 + X_o/Y(S_o - S_b)] + \frac{q_m''S_b}{(K_s + S_b)(S_o - S_b)}(A_s/V)\eta_{os} - \frac{(S_{LR} - S_b)}{(S_o - S_b)}(\frac{F_L}{V})$$

$$\tag{14.93}$$

In these equations S_b is the substrate concentration in the trough and in the entrained liquid film at the point of entrainment, S_{LR} is the substrate concentration in the entrained liquid film at the point of return and, A_s is the total surface area in the submerged sector. The other terms retain their usual meanings. The volumetric flow rate of the entrained liquid film in cm^3/sec is given by

$$F_L = 2N\pi(r_o^2 - r_i^2)\delta_L\omega \tag{14.94}$$

where N is the number of discs in the reactor and therefore takes on integer values only. If we ignore the edges of the discs the total surface area in the submerged sector is (see Fig. 14.30)

$$A_s = 2N[\pi r_o^2(2\pi - \theta_A)/2\pi - 2r_i(r_o^2 - r_i^2)^{1/2}/2]$$

$$= N(2\pi r_o^2)\{\cos^{-1}(r_i/r_o)/\pi - (r_i/r_o)[1 - (r_i/r_o)^2]^{1/2}/\pi\} \tag{14.95}$$

where N is the number of discs, $2\pi r_o^2$ is the total surface area of one disc, and the expression in the braces, {}, represents the fractional submergence of the disc, f_s. Equation 14.95 may now be substituted into either Eq. 14.92 or 14.93 thereby making the performance of the RDR a function of the number, size, and degree of submergence of the discs.

Equations 14.92 and 14.93 both contain a term for the concentration of substrate in the liquid film being returned to the trough by the rotation of the disc. Thus, before either of them can be solved information must be available about that concentration. As the disc rotates through the aerated sector the microorganisms in the biofilm remove substrate from the entrained liquid film thereby making S_{LR} less than S_b. Furthermore, since the entrained film acts like a PFR, S_{LR} will be dependent upon S_b. This dependency can be shown by writing a steady-state mass balance equation for substrate in a differential element of the entrained film as shown in Fig. 14.30. When performing the balance the assumption is made that the substrate concentration at any point in the aerated sector of the disc, S_L, is a function of only θ and is independent of the radial position of the point. Ignoring the removal of substrate by organisms suspended in the liquid film, the mass balance is

$$(F_L/2N)S_L\Big|_\theta - (F_L/2N)S_L\Big|_{\theta + \Delta\theta} - \frac{q_m''S_L}{K_s + S_L}\eta_{oa}\pi(r_o^2 - r_i^2)\Delta\theta/2\pi = 0 \tag{14.96}$$

where η_{oa} is the overall effectiveness factor for the biofilm in the aerated sector. The usual limiting process leads to

$$\frac{dS_L}{d\theta} = -[q_m''(r_o^2 - r_i^2)N/F_L][S_L/(K_s + S_L)]\eta_{oa} \tag{14.97}$$

Substitution of Eq. 14.94 for F_L gives

$$\frac{dS_L}{d\theta} = - (q''_m/2\pi\delta_L\omega) [S_L/(K_s + S_L)]\eta_{oa} \tag{14.98}$$

Because of the assumption that the substrate concentration in the liquid film is a function of θ alone the boundary condition may be approximated by

$$S_L \cong S_b \quad \text{at} \quad \theta = 0 \tag{14.99}$$

and the concentration in the returning liquid film may be approximated by

$$S_{LR} \cong S_L \quad \text{at} \quad \theta = \theta_A \tag{14.100}$$

Thus, Eqs. 14.93 and 14.98 are the mass balance equations which describe the removal of substrate by the rotating-disc reactor and they must be solved simultaneously. For a given set of reactor conditions the unknowns are S_{LR}, S_b and η_{os} in Eq. 14.93 and S_b and η_{oa} in Eq. 14.98. As we have seen in previous sections, however, η_{oa} and η_{os} are functions of S_b as well as the physical and operational characteristics of the reactor. The only difference is that η_{oa} depends on k_{LA} while η_{os} depends on k_L. Thus, the value of η_{oa} is expected to decrease with increasing angular velocity while that of η_{os} is expected to increase.

Because η_{oa} is a function of S_b, ψ, and ϕ^2, Eq. 14.98 cannot be integrated analytically. Therefore, it must be integrated numerically using the approach in Section 14.4.3. Equation 14.93 is a nonlinear algebraic equation which is similar to Eq. 14.66 and therefore a numerical technique is applicable to it also. It must be recognized, however, that these two equations are coupled by the boundary conditions, Eqs. 14.99 and 14.100.

If the kinetic and mass transfer parameters are known, Eqs. 14.93 and 14.98 may be used to calculate the number of discs and the reactor liquid volume required to reduce the substrate concentration to some desired value, S_b. Using the known S_b value, η_{oa} is estimated from Fig. 14.17 and the result used in Eq. 14.98 to obtain the value of S_L at the point of return. This value is then substituted into Eq. 14.93 to obtain either the required disc surface area or the reactor liquid volume. There will be many combinations of disc area and reactor volume which satisfy the equations, thus one is usually used as a parameter and the other as the dependent variable. Once the area is known, the number of discs can be calculated with Eq. 14.95.

If one wishes to calculate the substrate concentration remaining in a reactor with known characteristics, an iterative approach must be used. Because η_{os} and η_{oa} are functions of S_b it is convenient to iterate on S_b by assuming a value. After using the assumed value to estimate η_{oa}, Eq. 14.98 may be integrated to obtain S_{LR}. The assumed value is also used to estimate η_{os}, which is then used along with S_{LR} to calculate a new value of S_b with Eq. 14.93. The iteration ends when the calculated value of S_b is the same as the assumed one.

SIMPLIFIED MODEL. Inspection of Eqs. 14.98 through 14.100 shows that the substrate concentration in the liquid film in the aerated sector, S_L, has a maximum value equal to the substrate concentration in the liquid in the trough, S_b, and a minimum value of S_{LR}. Thus, when the substrate concentration in the liquid in the trough is small in comparison to the saturation constant, i.e., $S_b < K_s$, (which is the usual case encountered in practice) the substrate concentration in the liquid film on the aerated sector would be much smaller than the saturation constant, i.e., $S_L << K_s$. Therefore, the Monod kinetics in Eq. 14.98 may be assumed to reduce to first-order kinetics without significant error. It will be recalled that for first-order reactions both the internal and overall effectiveness factors are independent of the substrate concentration. Thus it is not necessary to use Fig. 14.17 to estimate η_{oa}. Rather, it may be determined by a combination of Eqs. 14.45 and 14.60:

$$\eta_{oa} = \frac{1}{q_m''/K_s k_{LA} + \phi/\tanh \phi} \tag{14.101}$$

where ϕ is given by Eq. 14.42. Once η_{oa} is known it may be substituted into the first-order form of Eq. 14.98

$$\frac{dS_L}{d\theta} = -(q_m'' \eta_{oa}/2\pi\delta_L \omega K_s) S_L \qquad S_L = S_b \quad \text{at } \theta = 0 \tag{14.102}$$

which may be integrated analytically to obtain

$$S_L = S_b \exp[-(q_m'' \eta_{oa}/2\pi\delta_L \omega K_s)\theta] \tag{14.103}$$

Therefore, the substrate concentration in the returning liquid film is

$$S_{LR} = S_b \exp\{-q_m'' \eta_{oa}[\cos^{-1}(-r_i/r_o)]/\pi\delta_L \omega K_s\} \tag{14.104}$$

Substitution of Eq. 14.104 into Eq. 14.93 yields

$$\frac{1}{\tau} = \frac{\mu_m S_b}{K_s + S_b}[1 + X_o/Y(S_o - S_b)] + \frac{q_m'' S_b}{(K_s + S_b)(S_o - S_b)}(A_s/V)\eta_{os}$$

$$+ \frac{S_b\{1 - \exp[-q_m'' \eta_{oa} \cos^{-1}(-r_i/r_o)/\pi\delta_L \omega K_s]\}(F_L/V)}{S_o - S_b} \tag{14.105}$$

Finally, the most generalized equation for RDR performance may be obtained by substituting Eq. 14.94 for F_L, the amount of liquid entrained by the rotating discs, and Eq. 14.95 for A_s, the total disc surface area in the submerged sector. The result is:

$$\frac{1}{\tau} = \frac{\mu_m S_b}{K_s + S_b}[1 + X_o/Y(S_o - S_b)]$$

$$+ \frac{q_m'' S_b N(2\pi r_o^2)\{\cos^{-1}(r_i/r_o)/\pi - (r_i/r_o)[1 - (r_i/r_o)^2]^{1/2}/\pi\}\eta_{os}}{(K_s + S_b)(S_o - S_b)V}$$

$$+ \frac{S_b N[2\pi(r_o^2 - r_i^2)]\delta_L\omega\{1 - \exp[- q_m'' \eta_{oa}\cos^{-1}(- r_i/r_o)/\pi\delta_L\omega K_s]\}}{(S_o - S_b)V} \qquad (14.106)$$

For a given set of reactor conditions the only two unknowns in Eq. 14.106 are S_b and η_{os} (since η_{oa} is given by Eq. 14.101). As we have seen in the previous sections, η_{os} is a function of S_b and the physical characteristics of the reactor as shown in Fig. 14.17. Thus we can use the same iterative approach that was used before. The only difference here is that the mass transfer coefficient, k_L is related to the size and rotational speed of the disc through an expression such as Eq. 14.86.

14.5.2 Performance of a Single RDR

Predictions of the performance of a single, ideal rotating-disc reactor were obtained by numerically solving Eq. 14.106 using the parameter values given in Table 14.6. Equations such as Eq. 14.65 were used to allow estimation of η_{os} as a function of ψ, ϕ^2, and ϕ_f, and the particular forms employed are given in Table 14.6. In addition, the overall effectiveness factor for the aerated sector, η_{oa}, was estimated by Eq. 14.101 using the expression for k_{LA} given in Table 14.6.

The effects of flow rate upon the removal of substrate in a RDR with a fixed number of discs rotating at a fixed speed are shown in Fig. 14.31. As would be anticipated from Fig. 14.3 and 14.12 in which the effects of τ upon the performance of a CSTR were presented, increases in flow rate (which cause a decrease in τ) cause an increase in the effluent substrate concentration. The reasons for this have been discussed previously, and thus will not be repeated here. As in a packed tower, the hydraulic loading (based upon the total wetted disc surface area) is often used to characterize the performance of a RDR. Consequently, that parameter is given on the upper abscissa.

The effects of influent substrate concentration are also shown in Fig. 14.31. Examination of the curves shows that the effluent substrate concentration will increase as the influent concentration is increased. This was also observed in packed towers and the reasons have already been discussed. It is interesting to note, however, that the percentage substrate removal is relatively independent of the influent substrate concentration at low flow rates. This is due to the fact that at low flow

TABLE 14.6
Values of Kinetic Parameters and Variables
Used for Figures 14.31 Through 14.39

Symbol	Value
a'_s	$100 \text{ cm}^2/\text{cm}^3$
D_e	$0.02 \text{ cm}^2/\text{hr} = 5.556 \times 10^{-6} \text{ cm}^2/\text{sec}$
F	$3000 \text{ liters/hr} = 8.333 \times 10^2 \text{cm}^3/\text{sec}$
k_L	$0.1863 \ (\omega/r_o)^{1/2} \text{ cm/sec}$
k_{LA}	$D_e/\delta_L \text{ cm/sec}$
L_f	0.01 cm
N	44
q''_m	$1 \text{ mg}/(\text{cm}^2 \cdot \text{hr}) = 2.778 \times 10^{-7} \text{ g}/(\text{cm}^2 \cdot \text{sec})$
r_o	150 cm
r_i	30 cm
S_o	$100 \text{ mg/liter} = 1 \times 10^{-4} \text{ g/cm}^3$
V	$3000 \text{ liters} = 3 \times 10^6 \text{ cm}^3$
X_o	0
Y	0.5
δ_L	$2.144 \times 10^{-3} \ [\omega(r_o + r_i)/2]^{2/3} \text{cm}$
μ_m	$0.2 \text{ hr}^{-1} = 5.556 \times 10^{-5} \text{ sec}^{-1}$
K_s	$50 \text{ mg/liter } T_bOD$
ω	$2 \text{ rpm} = 2/60 \text{ revolutions/sec}$

$$\eta_{os}$$

at $\psi=1$ $0.66\phi_f^{-1.864}$

$\psi=5$ $0.1987\phi_f^{-1.869} + 0.832\phi_f^{-1.3681}; \ 1.65 < \phi_f < 10$

$0.1987\phi_f^{-1.869} + 0.7045\phi_f^{-1.0362}; \ 1.25 < \phi_f < 1.65$

$\psi=10$ $1.1904\phi_f^{-1.3681}$

rates the effluent substrate concentrations are low relative to the saturation constant (K_s = 50 mg/liter) so that the Monod kinetics reduce to first-order kinetics.
Therefore, the percentage substrate removal becomes independent of the influent substrate concentration as was seen in Chapter 4. At higher flow rates the effluent
substrate concentrations are significant in comparison to the saturation constant so
that Monod kinetics prevail. Consequently, the percentage substrate removal decreases as the influent substrate concentration is increased, just as it does in
other fixed-film reactors.

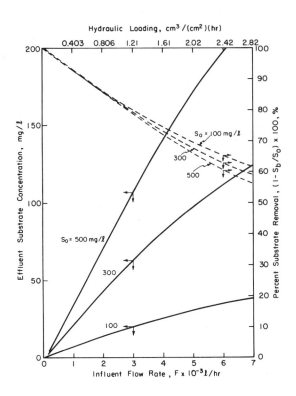

Figure 14.31.   Effects of influent flow rate and
influent substrate concentration
on the performance of a single
rotating-disc reactor.

Figure 14.32 shows the effects of rotational speed upon the performance of an RDR. As the rotational speed is increased the percentage substrate removal increases up to an upper limit characterized by the other system parameters. Several factors interact to cause this response. First, as shown in Eq. 14.85 mass transfer in the bulk liquid increases as the rotational speed of the discs is increased. Consequently the overall effectiveness factor in the submerged sector is increased, which indicates that the substrate is consumed more rapidly by the submerged biofilm as the speed is increased. The events occurring in the aerated sector are more compli-cated, however. Examination of Eq. 14.87 reveals that an increase in rotational speed will cause the thickness of liquid film on the aerated sector to increase, which will, in turn, make the mass transfer coefficient in the liquid film, k_{LA}, de-crease (see Eq. 14.91) thereby decreasing the overall effectiveness factor in the aerated sector. This coupled with the increase in the rotational speed and the thickness of the liquid film, will cause an increase in the concentration of sub-strate returning to the bulk fluid from the face of the discs. The increase in the thickness of the liquid film, on the other hand, causes the volume of fluid carried

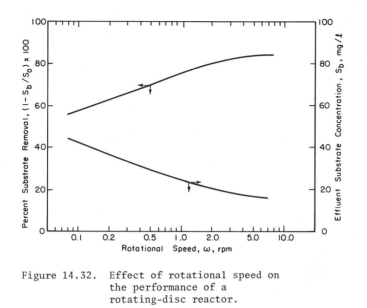

Figure 14.32. Effect of rotational speed on
the performance of a
rotating-disc reactor.

with the discs into the aerated region to increase as the rotational speed is increased. As a consequence, even though the reduction in substrate concentration
across the aerated region decreases as ω is increased, the mass of substrate removed per unit time by the aerated sector increases slightly. Hence the rate of
substrate removal by both the submerged and the aerated sectors increases as the
rotational speed is increased with the result that the effluent substrate concentration will decrease as ω is increased. Figure 14.32 shows that the effect is more
significant when the rotational speed is small, e.g., up to 2 rpm. Beyond that, the
improvement is marginal. Apparently, at a low value of ω the mass transfer resistance in the submerged sector is quite important so that an increase in it makes a
significant improvement in reactor performance. Beyond the value of 2 rpm, however,
the mass transfer resistance is not as important and hence the performance does not
improve significantly with increasing values of ω. It should be noted that very high
rotational speed could cause excessive shear of the biomass from the discs, thereby
disrupting performance. This possible effect is not reflected in the models.

Another effect that is related to rotational speed is shown in Fig. 14.33 where
the influence of disc size upon performance is shown. In this case the influent
flow rate was increased as the disc size was increased so as to maintain a constant
hydraulic loading. Furthermore, the rotational speed was decreased in proportion to
disc size in order to maintain a constant peripheral velocity. Consequently the loss
of performance associated with an increase in disc size is primarily due to the effects of rotational speed upon the rate of mass transfer and the movement of liquid
into the aerated sector as discussed in the preceding paragraph.

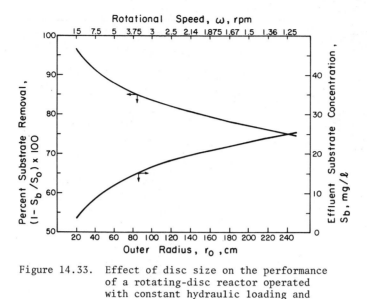

Figure 14.33.   Effect of disc size on the performance
                of a rotating-disc reactor operated
                with constant hydraulic loading and
                constant peripheral velocity.

The effect of the number of discs on the performance of a RDR is shown in
Fig. 14.34. An increase in the number of discs causes a corresponding increase in
both the submerged biofilm area and the volume of fluid carried with the discs per
unit time into the aerated region. Consequently, the substrate consumption rates in
both the aerated and submerged sectors will increase with an increase in N. Thus,
for a given influent substrate concentration an increase in N will result in an

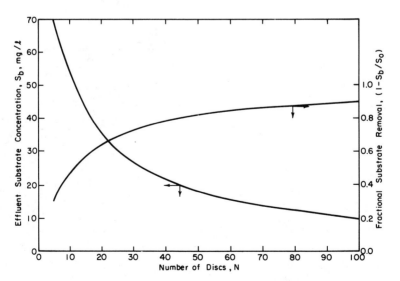

Figure 14.34.   Effect of the number of discs on the
                performance of a rotating-disc reactor.

increase in the substrate removal rate and a decrease in the effluent substrate concentration.

Figure 14.35 shows the effect of the fractional submergence of the discs, f_s, upon the percent substrate removal in an RDR with a fixed number of discs rotating at a constant speed. By reducing the inner radius, r_i, the discs may be submerged to any fraction up to 0.5. An increase in the fractional submergence allows more organisms to grow on a disc of fixed size and consequently causes the substrate removal rate in the submerged sector to increase. It also increases the volume of fluid carried with the rotating discs into the aerated sector, thereby causing the substrate removal rate in that sector to increase as well. Consequently the percent substrate removal will increase as the degree of submergence of the discs is increased. Although not reflected in the model, submergences in excess of 0.5 will decrease the rate of oxygen transfer in the system, thereby hurting reactor performance.

14.5.3 Model for Substrate Removal in a Chain of RDR's

In most situations a number of rotating-disc reactors will be used in series in order to economically achieve a desired effluent substrate concentration. Such a system may be modeled as a number of CSTR's in series and mass balance equations may be written in turn for each RDR in the system just as was done in Chapter 4. Therefore, for a system of M RDR's in series there are M equations like Eq. 14.106. If all necessary parameter values are given and it is desired to know the effluent

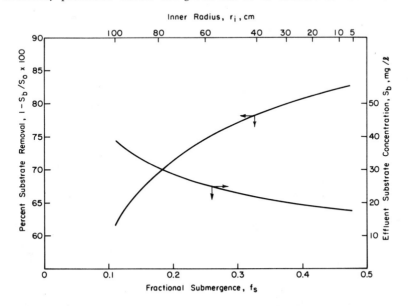

Figure 14.35. Effect of the degree of submergence of the discs on the performance of a rotating-disc reactor.

concentration from the Mth RDR, we would solve the equations sequentially, starting
with the first reactor and working our way down the chain as was done in Chapter 4.
If we wish to determine the total disc area needed to achieve a desired effluent
concentration from a chain of M reactors, we need to work backwards up the chain,
ultimately forming M equations which must be solved simultaneously.

14.5.4 Performance of a Chain of RDR's

Using the approach outlined in the preceding section, it is possible to investigate
the performance of a number of rotating-disc reactors in series. Since the effects
of the individual parameters will be the same for a chain as they were for a single
reactor, we will limit this discussion to the effects of the influent characteristics
on the substrate removal profile through a six-reactor chain. For convenience we will
again assume that a first-order reaction occurs in the aerated sector of the discs so
that the simplified model may be used, although it should be recognized that this assumption may be weak for the first one or two reactors in the chain. The parameter
values used are given in Table 14.6 and the influent conditions are given on the
graphs.

As was done for tanks in series in Chapter 12 and 13, bar graphs are used to
show the profiles in substrate concentration along the reactor chain. Figure 14.36

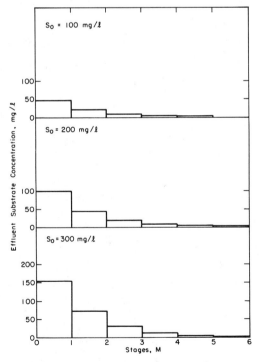

Figure 14.36.   Effect of influent substrate
concentration on the substrate
concentration profile through a
series of six rotating-disc reactors.

shows the effect of influent substrate concentration upon that profile. There it
can be seen that the majority of the substrate is removed in the first two reactors
and that the others contribute relatively little. Furthermore, as would be expected,
at higher influent concentrations more removal occurs in the latter stages. This is,
of course, a direct consequence of the behavior that was observed with a single stage.
Even though this is a stagewise operation, it would be instructive to make a semilog
plot of substrate concentration as a function of stage number, as shown in Fig. 14.37.
Examination of this figure reveals that the substrate concentration decreases in an
almost exponential manner as the liquid moves from stage to stage. In fact, this is
to be expected from the performance of a PFR as observed in the earlier sections of
this chapter. This particular behavior makes it an easy matter to predict the effect
of additional stages upon the performance of a chain. Furthermore, it suggests that
if the performance of a single stage were known, a reasonable prediction of the performance of a chain could also be made. Another interesting observation from
Fig. 14.37 is that the lines for different influent substrate concentrations are almost parallel. This suggests that data from one substrate concentration can be
extrapolated with reasonable accuracy to another, provided that care is taken to
ensure that the range over which the extrapolation is made is reasonable.

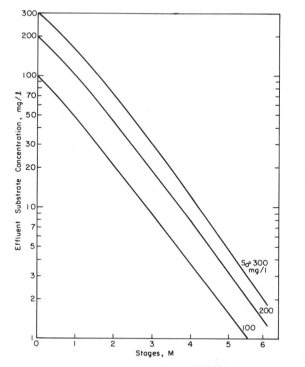

Figure 14.37.   Effect of influent substrate concentration
                on the substrate concentration profile through
                a series of six rotating-disc reactors. The
                data are plotted on a semilog scale as if
                removal were continuous rather than stagewise.

The effects of hydraulic loading upon the substrate profile within a chain of RDR's is shown in Fig. 14.38 where the hydraulic loading decreases as the number of discs, N, is increased. As would be anticipated from the discussion in prior sections, an increase in hydraulic loading shifts the removal of substrate to reactors further down the chain. Figure 14.39 shows the same data plotted as a semilog function of the reactor number. Examination of that figure reveals that the linearity of the plot is best at lower hydraulic loading rates. This suggests that the extrapolation procedure in the preceding paragraph will work best at lower hydraulic loadings. It is also apparent from the figure that the slope of the semilog plot is a function of the hydraulic loading. Unfortunately, examination of Fig. 14.39 does not reveal any correspondingly simple way of extrapolating from one hydraulic loading to another.

14.5.5 Parameter Estimation

Before the model, Eq. 14.106, can be used to predict quantitatively the performance of an RDR, numerical values must be available for all of the parameters which appear in it. Even though a number of techniques can be suggested by which those parameters can be evaluated, most of them have not been verified experimentally. Consequently, it was concluded that their inclusion would be premature.

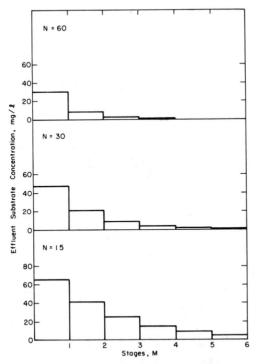

Figure 14.38. Effect of hydraulic loading (F/N) on the substrate concentration profile through a series of six rotating-disc reactors.

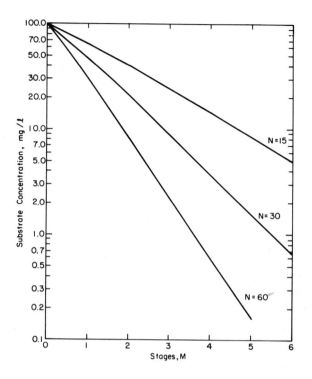

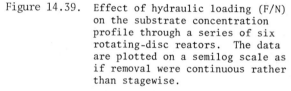

Figure 14.39.   Effect of hydraulic loading (F/N)
on the substrate concentration
profile through a series of six
rotating-disc reators. The data
are plotted on a semilog scale as
if removal were continuous rather
than stagewise.

14.5.6 Other Rotating-Disc Reactor Models

Because widespread use of the RDR is relatively recent, and because of its complexity, few models have been proposed to describe its performance. Rather, it is a common practice to use a graphical interpolation technique based on pilot plant data for design [26]. Nevertheless, quantitative models have been proposed by Kornegay [14] and Hansford *et al.* [27] among others. These are based on a number of simplifying assumptions and thus we shall briefly compare them with Eq. 14.106 to illustrate those assumptions.

KORNEGAY MODEL. The model proposed by Kornegay [14] is

$$F(S_o - S_b) - 2N\pi(r_o^2 - r_i^2)q_m'' S_b / (K_g + S_b) = 0 \qquad (14.107)$$

This model ignores the two distinct sectors, aerated and submerged, and assumes that the entire biofilm is exposed to the substrate concentration in the bulk liquid. It also assumes that the rate of substrate removal by suspended cells is negligible.

Although K_g is a function of the rotational speed of the discs, no other explicit accounting of mass transfer effects is made. It will be recalled from Section 14.4.5 that the use of K_g to account for mass transfer effects is equivalent to making the overall effectiveness factor dependent upon the substrate concentration and K_g. This concept will apply to the submerged sector of the RDR so that

$$\eta_{os} = (K_s + S_b)/(K_g + S_b) = 1 - \frac{(K_g/K_s - 1)}{(K_g/K_s + S_b/K_s)} \tag{14.108}$$

Thus, Kornegay's model may be viewed as a special case of Eq. 14.106, in which $\delta_L = 0$, η_{os} is given by Eq. 14.108 and the area of the submerged sector is equal to the area of disc covered by biofilm, $A_s = 2N\pi(r_o^2 - r_i^2)$.

MODEL OF HANSFORD et al. The steady-state model proposed by Hansford et al. [27] has many similarities to the one used herein. It recognizes that there are two sectors, submerged and aerated, and that a rotating disc carries a liquid film from the trough into the aerated sector as it turns. Their model was developed by writing four substrate mass balance equations: for the liquid film and the biofilm on the aerated sector, the fixed biofilm in the submerged sector and the liquid in the trough. The following assumptions were made: (1) the liquid in the trough is completely mixed so that the substrate concentration is uniform; (2) the liquid film and biofilm on the aerated sector are assumed to be completely mixed in both the axial and radial directions so that the substrate concentrations are independent of radial position and depth and therefore are functions of only the angular position θ; (3) the substrate concentration in the biofilm on the submerged sector is uniform (independent of the radial, axial or angular position); (4) the thickness of the liquid film on the aerated sector is assumed to be constant and independent of the rotational speed; and (5) the suspended cells in the trough are assumed to consume no substrate. The major difference between this model and the one employed herein is that it assumes that the substrate concentration in the biofilm is independent of depth. This is equivalent to saying that there is no mass transfer resistance within the biofilm or that the thickness of the biofilm is so small that the reaction takes place only on its surface (see Section 14.2). In either case the internal effectiveness factor is unity. Furthermore, this model assumes that the reaction takes place only in the biofilm rather than in both the biofilm and the bulk liquid. Finally, the thickness of liquid film in the aerated sector is assumed constant, whereas the model herein allows variations due to rotational speed. In spite of these differences, the trends predicted by the model of Hansford et al. [27] are qualitatively similar to those in this chapter as well as trends observed in the field.

MODEL OF FAMULARO et al. The RDR model proposed by Famularo et al. [28] is the most complete one published to date because it incorporates the transport of both oxygen and substrate up to and into the biomass. The basic components of the model (i.e., the mass transfer effects, reaction rates, etc.) are very similar to the ones in the model developed in this chapter, although the effectiveness factor approach is not employed. Consequently they must use a finite-difference procedure to solve the nonlinear coupled differential equations. A minor difference is that they modeled the aerated sector as four mixed regions in series whereas a plug-flow model was used herein. The differences in the results caused by this are likely to be slight, and indeed the trends predicted by the model are in agreement with the trends presented in this chapter. The major benefit of the model of Famularo *et al.* is that the inclusion of oxygen transfer effects allows prediction of the loading conditions under which oxygen limitations are likely to occur.

Compared to the modeling of slurry reactors, the modeling of fixed-film reactors is in its infancy. Consequently large advances are likely to be made in the near future and thus the interested reader is encouraged to consult the current literature in this area.

14.6 KEY POINTS

1. In a fixed-film reactor the substrate concentration at which the reaction takes place is less than the measured bulk phase concentration due to the resistances encountered in the transport of substrate from the bulk phase into the fixed film (biofilm). Therefore, it is necessary to combine physical mass transport with microbial reactions when modeling fixed-film reactors.

2. For mass transfer between the bulk liquid phase and the liquid-biofilm interface the substrate concentration is allowed to vary only within the hypothetical stagnant liquid film adjacent to the liquid-biofilm interface, through which the substrate must be transported to reach the biofilm, and assumed constant throughout the remaining bulk phase.

3. The rate of mass transfer of substrate from the bulk phase across a unit area of the stagnant liquid film to the biofilm is proportional to the drop in the substrate concentration across the stagnant film. The proportionality constant, k_L, is called the mass transfer coefficient and it has the units of length per time.

4. Due to the mass transfer resistance, the actual substrate concentration at the surface of the biofilm is less than the bulk substrate concentration. This concentration can be determined by equating the flux across the stagnant film to the reaction rate per unit area of the biofilm.

5. For first-order reactions the imposition of mass transfer resistance does not change the functional form of the rate equation. It is still first order with respect to the bulk phase substrate concentration although the rate constant is different. The apparent rate constant, k_o, contains both the reaction rate constant and mass transfer coefficient. For any other kinetics the imposition of mass transfer resistance changes both the functional form and the constants of the rate equation. This effect is best handled by the concept of effectiveness factor.

6. The effectiveness factor, η, is defined as the ratio of the actual, observed rate to the theoretical rate that would occur in the absence of mass transfer resistance. Thus, the effectiveness factor is a correction factor, which, when multiplied by the homogeneous rate without mass transfer resistance gives the actual rate in the presence of mass transfer resistance.

7. For Monod kinetics the effectiveness factor in the presence of external mass transfer resistance (through the stagnant liquid film) is a function of $\overline{K}$, a dimensionless saturation constant and D_a, the Damköhler number, both of which depend on the bulk phase substrate concentration. The effectiveness factor decreases with increases in D_a and decreases in $\overline{K}$.

8. The Damköhler number is the ratio of the maximum possible substrate removal rate (reaction rate) at the interface to the maximum possible substrate transfer rate (mass transfer rate) across the stagnant liquid layer.

9. The design equation for a CSTR with biofilm in the presence of external mass transfer resistance contains two reaction terms, one for the removal of substrate by suspended microorganisms and the other for the removal due to the biofilm.

10. A decrease in k_L reduces the activity of the biofilm and makes the effluent substrate concentration higher than it would be in a reactor with less external mass transfer resistance.

11. Reactors with more biofilm remove more substrate.

12. The presence of biofilm prevents washout of a CSTR. As a result CSTR's with biofilm are capable of removing substrate at space times well below those which cause washout in a CSTR without biofilm.

13. The design equation for a PFR with biofilm is a differential equation which contains two reaction terms, one homogeneous and the other heterogeneous. In general, it must be integrated numerically.

14. The external effectiveness factor varies significantly along the length of a PFR, being the highest at the inlet where the substrate concentration is highest and decreasing to a nonzero limit at the outlet.

15. An increase in flow rate through a PFR causes a decrease in the percentage removal of substrate but not in the mass. The increase in flow rate increases the external mass transfer coefficient thus increasing the external effectiveness factor.

16. The transport of substrate within a biofilm may be described by Fick's law with the free diffusion coefficient, D, replaced by an effective diffusion coefficient, D_e.

17. For first-order reactions an analytical expression is obtained for the effectiveness factor accounting for internal mass transfer effects, $n_i = \tanh \phi/\phi$. It is independent of the substrate concentration. For small values of $\phi (< 0.5)$, $n_i \cong 1$ and for large values of $\phi (> 5)$, $n_i = 1/\phi$.

18. For Monod kinetics the internal effectiveness factor depends on the Thiele modulus ϕ and the parameter κ. Therefore, it depends on the substrate concentration. It is best represented numerically as a function of the generalized modulus ϕ_p. At low values of ϕ_p it approaches 1.0, while at high values of $\phi_p (> 5)$, $n_i = 1/\phi_p$.

19. Internal mass transfer resistance reduces the amount of substrate that can be removed by a biofilm. Its effects on the performance of a CSTR are very similar to those of the external mass transfer resistance except that the flow rate does not affect the internal mass transfer.

20. The effects of internal mass transfer resistance on the performance of a PFR are very similar to those of the external mass transfer resistance.

21. The overall effectiveness factor accounts for the presence of both external and internal mass transfer resistances. For first-order kinetics, the inverse of the overall effectiveness factor, $1/\eta_o$, is equal to the sum of two terms; the inverse of the internal effectiveness factor, $1/\eta_i$, and a term representing the external effectiveness factor, $(q_m''/K_s)k_L$. Thus, it is independent of the substrate concentration.

22. For Monod kinetics the overall effectiveness factors must be determined numerically and are presented graphically in terms of a Sherwood number, ψ, and a modified Thiele modulus ϕ_f. Therefore, they depend on the substrate concentration. In general, the overall effectiveness factor decreases rapidly with the modified Thiele modulus, increases slightly with ϕ^2, and increases rapidly with ψ.

23. A packed tower may be modeled by assuming plug flow and allowing the substrate to be consumed by both the dispersed and the attached microorganisms. For the substrate consumption by the biofilm the overall effectiveness factor must be used to account for external and internal mass transfer resistances.

24. The substrate concentration drops exponentially with depth in packed towers either with and without recirculation.

25. Recirculation reduces the inlet substrate concentration considerably thereby reducing the effectiveness factor at the inlet and making the drop in the effectiveness factor almost linear throughout the tower depth.

26. When there is no recirculation and when the feed substrate concentration is high, the effectiveness factor remains large over a portion of the tower depth and then drops exponentially thereafter with the tower depth.

27. As the feed substrate concentration to a packed tower is increased the fractional substrate removal decreases, although the mass removal rate increases almost linearly until it approaches its maximal value. The effectiveness factor also increases and its variation over the tower depth declines.

28. As the feed flow rate to a packed tower is increased the fractional substrate removal decreases whereas the mass substrate removal rate increases. The effectiveness factor increases with the flow rate but not fast enough so that faster flow rates generally require greater tower depths to achieve a fixed effluent concentration.

29. Higher recirculation ratios result in flatter substrate and effectiveness factor profiles in a packed tower. The percent substrate removal decreases at low recirculation ratios whereas it increases slightly as the recirculation ratio is increased.

30. The required depth to achieve a fixed effluent concentration in a tower of fixed cross sectional area generally increases as the recirculation ratio is increased. However, at extremely low flow rates where the external mass transfer resistance is limiting, recirculation may reduce the required tower depth.

31. An increase in the cross sectional area results in a decrease in the depth of tower required to remove a given fraction of substrate, but not in proportion to the increase in the area. The total tower volume may be minimized by choosing a tall, thin tower rather than a short, fat one.

32. An increase in the surface area of biofilm per unit tower length results in an increase in the amount of biomass and therefore a corresponding increase in the amount of substrate removed.

33. Other models proposed by Velz, Eckenfelder, Kornegay, and Schroeder are limited versions of the model proposed herein.

34. The performance of a rotating-disc reactor (RDR) system may be modeled as a series of CSTR's containing biofilm.

35. In a rotating-disc reactor there are two external mass transfer coefficients that need to be evaluated, one in the submerged sector, k_L, and the other in the aerated sector k_{LA}. The former increases with the rotational speed of the rotating disc in accordance with a power law, while the latter decreases in accordance with an inverse power law.

36. A general model for substrate removal in a single RDR is obtained by writing a steady-state mass balance on the substrate around the liquid volume in the trough (CSTR) and a steady-state mass balance on the substrate around an infinitesimal liquid volume on the aerated sector of the rotating disc (PFR).

37. A simplified model is obtained by assuming that the substrate concentration in the liquid film on the aerated sector is much smaller than the saturation constant, which allows analytical integration of the PFR equation.

38. Increases in flow rate cause increases in the effluent substrate concentration from an RDR.

39. The effluent substrate concentration increases as the influent concentration to an RDR is increased. At low flow rates the percentage substrate removal is relatively independent of the influent substrate concentration, while at high flow rates it decreases as the influent substrate concentration is increased.

40. As the rotational speed of an RDR is increased the percentage substrate removal increases up to an upper limit characterized by the system parameters.

41. The effluent substrate concentration from an RDR increases as the disc size is increased provided that the reactor is operated at constant hydraulic loading and constant peripheral velocity.

42. An increase in the number of discs in an RDR results in an increase in the substrate removal rate and a decrease in the effluent substrate concentration.

43. An increase in the fractional submergence results in an increase in the substrate removal rate in an RDR.

44. Rotating-disc reactors are usually run in series in order to economically achieve a desired effluent substrate concentration.

45. A model for substrate removal in a chain of RDR's is obtained by writing mass balance equations for each RDR in the system.

46. In a chain of RDR's in series the majority of substrate is removed in the first few reactors and the others contribute relatively little. At higher influent concentrations more removal occurs in the latter stages. The substrate concentration decreases in an almost exponential manner as the liquid moves from stage to stage.

47. An increase in hydraulic loading shifts the removal of substrate to reactors further down the chain.

14.7 STUDY QUESTIONS

The study questions are arranged in the same order as the text.

Section	Questions
14.1	1-13
14.2	14-22
14.3	23-27
14.4	28-32
14.5	33-38

1. Define the mass transfer coefficient, k_L.

2. How do you determine the actual substrate concentration at the surface of the biofilm?

3. Show that when the intrinsic reaction rate is first order the imposition of the external mass transfer resistance does not change the functional form of the rate equation for the reaction which takes place on the surface of the biofilm.

4. Show that Eq. 14.7 reduces to Eq. 14.9.

5. Define the effectiveness factor. What is its utility?

6. What is the physical significance of the Damköhler number?

7. Describe briefly the dependence of the external effectiveness factor on the Damköhler number and $\overline{K}$.

8. An extremely thin film of microorganisms capable of growing with $\mu_m = 0.2 \text{ hr}^{-1}$ was deposited on a flat plate and a kinetic study was done in the absence of mass transfer resistance using a solution containing a high concentration of substrate moving past the plate at extremely high flow rates. The kinetic parameters were found to be: $q_m'' = 7.2 \text{ mg}/(\text{cm}^2 \cdot \text{hr})$ and $K_S = 30 \text{ mg/liter}$. A study of the effect of flow rate, F, upon the mass transfer coefficient k_L (where k_L is in cm/hr and F is in cm^3/hr) revealed that k_L could be correlated with F by the equation $k_L = 1.095(F)^{0.5}$. The thin biofilm (total area = 10^4cm^2) was then placed into a CSTR with a volume of 1 liter and sterile feed ($X_0 = 0$) with a substrate concentration of 200 mg/liter was allowed to flow through.

   (a) Calculate the flow rate necessary to reduce the substrate concentration to 100 mg/liter. Compare this value with that which would be required should there be no mass transfer resistance.

   (b) If the flow rate is set to be one half of that determined in part (a), what would be the effluent substrate concentration?

   (c) If the microbial surface area is doubled, what would be the flow rate that would be needed to reduce the effluent substrate concentration to 100 mg/liter?

9. Describe the effects of k_L and the area of biofilm on the substrate removal rate.

10. Why does the presence of biofilm prevent washout of a CSTR?

11. In a PFR containing a biofilm the external effectiveness factor approaches a nonzero limit as the substrate concentration approaches zero. What is its value?

12. Describe the effects of flow rate and influent substrate concentration on the external effectiveness factor and the substrate profile in a PFR containing a biofilm.

13. Verify the results in Figure 14.7 using the parameter values given in Table 14.2.

14. Describe the transport of substrate within a biofilm. What is the effective diffusivity?

15. Verify that Eq. 14.44 is the solution to Eqs. 14.39 through 14.41.

16. Describe the substrate profile in a biofilm in the presence of internal mass transfer resistance alone.

17. What is the analytical expression for the internal effectiveness factor for a first-order reaction? What is the value when the Thiele modulus is 10?

18. Define the Thiele modulus for a first-order reaction in a flat biofilm. What is its physical significance?

19. What are the two parameters which characterize the internal effectiveness factor for Monod kinetics in a flat biofilm? What are their physical significances? What is the value of the internal effectiveness factor when the generalized Thiele modulus, ϕ_p is 10?

20. A fixed film of microorganism, whose kinetic and mass transfer characteristics are given in Table SQ14.1, was deposited on a flat plate and placed into a CSTR. A sterile feed ($X_0 = 0$), with a substrate concentration of 200 mg/liter, was then allowed to flow through at 1 liter/hr. The external mass transfer resistance may be ignored.

    (a) Determine the surface area of fixed film required to reduce the effluent substrate concentration to 100 mg/liter when the liquid volume in the reactor is one liter.

    (b) Determine the substrate concentration that would be in the effluent if the surface area of the fixed film and the reactor liquid volume were both doubled.

TABLE SQ14.1
Kinetic Parameters for a Biofilm

Parameter	Value	Parameter	Value
q_m''	0.7 mg/(cm^2)(hr)	μ_m	0.2 hr^{-1}
K_s	20 mg/liter = 0.02 mg/cm^3	L_f	0.1 cm
a_s	10 cm^{-1}	D_e	2.484 x 10^{-2} cm^2/hr
η_i	given in Table 14.4		

21. Reduce Eq. 14.55 to a situation in which the homogeneous reaction and the internal mass transfer resistance are both negligible. Integrate the resulting mass balance equation and rearrange it so that the required parameter values $q_m''a_s$ and K_s may be determined experimentally. Describe an experimental method to determine these parameters.

22. Describe the effects of the internal mass transfer resistance, the influent substrate concentration, and the influent flow rate on the substrate concentration and the internal effectiveness factor profiles in a PFR containing biofilms in the absence of external mass transfer resistance.

23. What is the role of the overall effectiveness factor?

24. Derive Eq. 14.60 from Eqs. 14.1, 14.56 and 14.59.

25. Is the overall effectiveness factor for a first-order reaction independent of the bulk substrate concentration?

26. What are the three parameters that characterize the overall effectiveness factor for Monod kinetics in a flat biofilm? Describe briefly the effects of these parameters on the value of the overall effectiveness factor.

27. In Study Question 20 the external mass transfer resistance was assumed to be negligible. Now consider the situation of the same biofilm in a CSTR with the presence of an external resistance to mass transfer. A study of the effect of flow rate, F, on the mass transfer coefficient k_L (where k_L is in cm/hr and F is in liters/hr) revealed that k_L could be correlated with F by the equation $k_L = 2.484 \ F^{0.5}$. For a reactor with parameters as given in Table SQ12.1 determine the surface area of biofilm necessary to reduce the effluent substrate concentration to 100 mg/liter when the liquid volume in the reactor is one liter and compare that area to the one required in the absence of external mass transfer resistance.

28. Describe briefly the effects of the recirculation ratio, the feed substrate concentration, and the feed flow rate on the performance of a packed tower.

29. Will the depth required to achieve a fixed effluent concentration in a packed tower of fixed cross sectional area always increase as the recirculation ratio is increased? Why?

30. What are the effects of the cross sectional area and the surface area of biofilm per unit of tower length on the performance of the packed tower?

31. Verify the results given in Fig. 14.21 by integrating Eq. 14.72 using the nominal parameter values given in Table 14.5.

32. How is the model proposed by Kornegay related to Eq. 14.72?

33. How is the mass transfer coefficient in the liquid in the aerated sector of an RDR related to the rotational speed?

34. Why is there a difference between η_{oa} and η_{os} for an RDR?

35. The nominal physical and chemical characteristics of an RDR are given in Table 14.6.

    (a) Determine the total surface area and the number of discs needed to reduce the substrate concentration from 100 mg/liter to 20 mg/liter.

    (b) Double the number of discs and determine the effluent substrate concentration.

    (c) Repeat Part (a) by increasing the value of ω to 8 rpm.

36. What is the difference between the general and the simplified models of an RDR?

37. Describe the effects of the flow rate, the influent substrate concentration, and the rotational speed on the performance of a single RDR.

38. The nominal physical and chemical characteristics of two identical RDR's in series are given in Table 14.6. The liquid volume in each RDR is one half of the total, however, i.e., $V_1 = V_2 = 1.5 \times 10^6 \ cm^3$.

(a) Determine the number of discs in each RDR needed to reduce the substrate concentration from 100 mg/liter to 20 mg/liter.

(b) Determine the effluent substrate concentration that would result if the number of discs in each RDR were only one half of the value determined in Part (a).

REFERENCES AND FURTHER READING

1. F. O. Mixon and J. J. Carberry, "Diffusion within a developing boundary layer - A mathematical solution for arbitrary velocity distribution," *Chemical Engineering Science,* 13, 30-33, 1970.

2. C. N. Satterfield, *Mass Transfer in Heterogeneous Catalysis,* MIT Press, Cambridge, Massachusetts, 79-128, 1970.

3. C. R. Wilke and P. Chang, "Correlation of diffusion coefficients in dilute solutions," *American Institute of Chemical Engineers Journal,* 1, 264-270, 1955.

4. W. H. Pitcher, Jr., "Engineering of immobilized enzyme systems," *Catalysis Reviews - Science and Engineering,* 12, 37-69, 1975.

5. K. B. Bischoff, "Effectiveness factors for general reaction rate forms," *American Institute of Chemical Engineering Journal,* 11, 351-355, 1965.

6. B. Atkinson and I. J. Davies, "The overall rate of substrate uptake (reaction) by microbial films, Part I. A biological rate equation," *Transaction of Institution of Chemical Engineers,* 52, 248-259, 1974.

7. B. Atkinson, *Biochemical Reactors,* Pion Limited, London, 80-88, 1974.

8. D. J. Fink, *et al.* "Effectiveness factor calculations for immobilized enzyme catalysts," *Biotechnology and Bioengineering,* 15, 879-888, 1973.

9. B. G. Liptak, *Environmental Engineer's Handbook,* Volume I, Chilton Book Co., Randor, PA, 1974.

10. R. Aris, "On shape factors for irregular particles - I - The steady state problem. Diffusion and reaction," *Chemical Engineering Science,* 6, 262-268, 1957.

11. T. K. Sherwood, *et al.* *Mass Transfer,* McGraw-Hill Book Co., New York, New York, Chapter 6, 1975.

12. E. J. Wilson and C. J. Geankoplis, "Liquid mass transfer at very low Reynolds numbers in packed beds," *Industrial Engineering Chemistry Fundamentals,* 5, 9-14, 1966.

13. G. J. Kehrberger and A. W. Busch, "The effects of recirculation on the performance of trickling filter models," *Proceedings of the 24th Industrial Waste Conference,* Purdue University Engineering Extension Series No. 138, 37-52, 1969.

14. B. H. Kornegay, "Modelling and simulation of fixed film biological reactors for carbonaceous waste treatment," in *Mathematical Modelling for Water Pollution Control Processes,* edited by T. M. Keinath and M. Wanielista, Ann Arbor Science, Ann Arbor, Mich., 271-315, 1975.

15. NRC Subcommittee Report, "Sewage treatment at military installations," *Sewage Works Journal,* 18, 897, 1946.

16. J. H. Sorrels and P. S. A. Zeller, "Two-stage trickling filter performance,"
     Sewage and Industrial Waste, 28, 943-954, 1956.

17. D. M. Pipes, *et al.* "Discussion of substrate removal mechanism of trickling
     filters," *Journal of the Environmental Engineering Division ASCE,* 100, 225-226,
     1974.

18. W. S. Galler and H. B. Gotaas, "Analysis of biological filter," *Journal of the
     Sanitary Engineering Division, ASCE,* 90, SA6, 59-79, 1964.

19. C. J. Velz, "A basic law for the performance of biological filters," *Sewage
     Works Journal,* 20, 607-617, 1948.

20. W. W. Eckenfelder, Jr., "Trickling filter design and performance," *Journal of
     the Sanitary Engineering Division, ASCE,* 87, SA6, 33-40, 1961.

21. E. D. Schroeder, *Water and Wastewater Treatment,* McGraw-Hill, Inc., New York,
     294, 1977.

22. V. G. Levich, *Physicochemical Hydrodynamics,* Prentice Hall, Englewood Cliffs,
     New Jersey, 69, 1962.

23. L. D. Landau and V. G. Levich, "Dragging of a liquid by a moving plate," *Acta
     Physicochimica, U.R.S.S.,* 17, 42-54, 1942.

24. H. Hartmann, "Undersuchung uber die biologische Reinigung von Abwasser mit
     Hilfe von tauchtop Korperanlagen," *Stuttgarter Berichte zur
     Siedlungswasserwirtschaft Kommissionsverlag,* Band 9, R., Oldenbourg, Munich,
     1960.

25. C. G. Grieves, "Dynamic and steady state models for the rotating biological
     disc reactor," Ph.D. Thesis, Clemson University, 1972.

26. R. L. Antonie, *Fixed Biological Surfaces-Wastewater Treatment, The Rotating
     Biological Contactor,* CRC Press, Cleveland, Ohio, 47-53, 1976.

27. G. S. Hansford, *et al.* "A steady state model for the rotating biological disc
     reactor," *Water Research,* 12, 855-868, 1978.

28. J. Famularo *et al.* "Application of mass transfer to rotating biological
     contactors," *Journal of the Water Pollution Control Federation,* 50, 653-671,
     1978.

Further Reading

Wu, Y. C. (Editor), *Advanced Wastewater Treatment by Rotating Biological Contactor
     Processes,* Ann Arbor Science, Ann Arbor, Mich. 1980.

PART V

APPLICATIONS: NAMED BIOCHEMICAL OPERATIONS

Part IV presented the theoretical equations describing microbial growth in three
types of reactors; CSTR's, CSTR's with recycle, and fixed-film reactors. In this
part we will see how those same equations can be used to describe the performance
of the named biochemical operations listed in Chapter 1. The intent of Part V is
four-fold. First it will provide a physical description of each named biochemical
operation and give a brief discussion of the circumstances in which it may be used.
Next, it will describe the performance of each operation, in terms of both the
theory of Part IV and the departures from that theory. Third, it will consider the
process design for each operation, i.e., the determination of reactor volume, mixing
power requirements, oxygen requirements, recycle requirements, etc. This text will
discuss neither detail design nor system design, i.e., how the biochemical operation
interacts with other unit operations in the system. That material is covered in
texts on wastewater treatment system design. It is important, however, that the
reader be familiar with the practical constraints which must be considered when
the theoretical equations are applied to design. Consequently, they too will be
presented. Finally, it will look at the operation of each of the biochemical
operations in order to see how they can be controlled once they are built.

A common characteristic of all of the named biochemical operations is that they
utilize complex, heterogeneous microbial communities. Consequently, all of them are
subject to numerous microbial interactions which can shift the balances among the
various populations in the community. Such interactions are not reflected in the
models of Part IV, and this constitutes a potential weakness to their application in
the design of the named operations. Unfortunately, our level of understanding of
microbial interactions is not yet sufficiently advanced to allow us to model them
simply and accurately. This subject, however, is one upon which much research is now
being performed so that we can expect considerable progress in the future. In the
meantime, experience has taught us that the models of Part IV can be used for design
with excellent accuracy and reliability as long as the user recognizes their limitations and abides by them. One should not interpret those equations as being
capable of providing an absolute design. Rather they should be used as indicators
of reasonably expected performance. The key to the successful design of biochemical
operations is flexibility, and the equations herein should be used by the designer
to provide a facility with sufficient operational flexibility to allow it to accomodate the circumstances not accounted for by the models.

CHAPTER 15

COMPLETELY MIXED AERATED LAGOON

A wastewater treatment lagoon is a relatively shallow basin with a large surface area, generally several acres, which operates on a flow-through basis, i.e., there is no recycle of material within the system. There are six types of lagoons, with classification into a specific type being based upon such factors as the complexity of the microbial community employed, the degree of mixing, and the reaction environment. The types are facultative oxidation lagoon, aerated facultative lagoon, high-rate aerobic lagoon, tertiary lagoon, anaerobic lagoon, and completely mixed aerated lagoon (CMAL) [1]. The first four all employ mixtures of heterotrophic and autotrophic organisms living in a complex ecosystem and consequently the design basis for them is highly empirical. They will not be discussed herein. The fifth type is anaerobic and consequently will be discussed in Chapter 21. The last type, the CMAL,

is the simplest, from both a physical and an ecological point of view, and is the
subject of this chapter.

15.1 DESCRIPTION OF CMAL SYSTEMS

15.1.1 Physical Characteristics

Completely mixed aerated lagoons usually have a depth between 5 and 15 feet and a
surface area of several acres. They are generally earthen basins although it is
common for them to be lined with an impervious material such as asphalt or plastic.
The banks are covered with grass except near the water line where they are protected
from wave erosion by concrete aprons or rip-rap. The surface configuration of CMAL's
is highly varied and is often chosen to make maximum utilization of the available
land area. Ideally a basin should be square or round for those shapes lend themselves well to the attainment of completely mixed conditions. Figure 15.1 illustrates four different types of surface configuration. Shapes such as those in Fig.
15.1 a, b and c have been analyzed for their mixing characteristics using inert

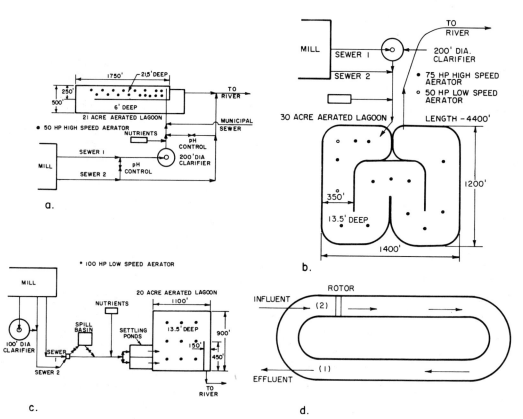

Figure 15.1. Typical layouts for aerated lagoon treatment systems.
(a, b and c are adapted from Murphy and Wilson [2]).

tracers [2]. As might be anticipated, the approximately square lagoon came the
closest to being completely mixed. The mixing patterns of the other two could both
be approximated as three unequal CSTR's in series. It is also common for actual
lagoons in series to be employed, particularly for industrial wastewater treatment.
The configuration shown in Fig. 15.1d is referred to as an oxidation ditch because
the cross-sectional profile of the channel is very similar to that of a large drainage ditch. If the time of flow around the basin is small in comparison to the
residence time of fluid within it, then the mixing characteristics will approach
complete mixing.

The mixing energy supplied to a CMAL must be large enough to keep all solids
in uniform suspension, provide an adequate supply of oxygen throughout the basin,
and give a tracer washout curve that approximates complete mixing (provided that
the reactor configuration will allow this to happen). With the exception of the
oxidation ditch, CMAL's are generally mixed and aerated by mechanical surface
aerators of either the fixed or floating type. If the basin is shallow, adequate
mixing can be maintained with only surface aerators, whereas a submerged turbine
must be provided if it is deep. Occasionally deep mixing is accomplished by the use
of draft-tubes attached to surface aerators. In all cases, however, it is common
practice to place concrete pads beneath the mixers to prevent scour of the basin
bottom. Oxidation ditches are almost universally mixed and aerated by cage rotor
aerators, often referred to as Kessener brushes.

The microbial population in a CMAL is predominately heterotrophic and metabolism is carried out by aerobic respiration as long as the oxygen supply is sufficient. If the space time provided is sufficiently long, i.e., in excess of 1 day,
then a relatively complete food chain is likely to be present, with bacteria
utilizing the organic matter and higher life forms feeding on them. Occasionally
algae will be found in the system, although their growth should be discouraged as
it is contrary to the objective of the operation. As a general rule, the ecology
of the microbial community is quite complex with many interactions among the
species, as discussed in Section 7.4. These interactions are not considered in the
mathematical models used for design, but if proper care is taken during the treatability studies required to characterize the overall kinetics of the system,
reasonable designs will result. A photograph of a typical CMAL installation is
shown in Fig. 15.2.

15.1.2 Applications

CMAL's are usually designed to meet one of two objectives: conversion of soluble
organic matter into insoluble microbial cells, or stabilization of the organic
material [3]. The first objective finds two applications. One is the pretreatment
of an industrial waste prior to discharge to a municipal sewer. By converting the
soluble organic matter to microbial cells, two things are accomplished. First, the
total organic load discharged from the industry to the municipal sewer will be

Figure 15.2. Typical CMAL installation. *(Courtesy of Clow-Yeomans Corporation).*

reduced by about 40%, i.e., by the amount of energy for synthesis required by the cells. Sometimes such a reduction in load is sufficient to meet the requirements of the city. Second, such pretreatment converts the organic matter to a common form, i.e., microbial cells. Prior to pretreatment, the organic matter might have been some exotic chemical; after pretreatment it is something that can be handled easily by the municipal plant. This makes the character of the wastewater entering the municipal plant more uniform, thereby alleviating the problems associated with a diversity of organic compounds from many sources.

A second application of the first objective is as the total treatment system for a small community. In that case it would be necessary to remove the cells prior to discharge. This can often be accomplished in settling basins, provided that the space time in the CMAL is sufficient to produce microbial cells that will settle. Some problems have been associated with this application, although work is continuing to correct them [4].

The second objective for which a CMAL may be designed is stabilization of the organic matter. As was seen in Chapter 12, stabilization occurs when sufficient time is provided to allow a large fraction of the microbial cells to undergo decay. Since very long space times would be required to achieve high degrees of stabilization it is not usually economic to rely upon a CMAL as the sole means of treatment. It is often desirable, however, to achieve a moderately high degree of stabilization prior to sedimentation of the solids in order to reduce the quantity of sludge which must be disposed of. For example, it is common practice in the pulp and paper industry to follow a CMAL by one of the other types of lagoons. In that case the degree of stabilization achieved in the CMAL must be chosen by balancing the cost of the CMAL against the cost and performance of the second lagoon. As another example, a CMAL being used for pretreatment of an industrial waste may be required to achieve a reduction in organic load in excess of that associated with only soluble substrate removal. The design engineer may decide to achieve it through stabilization of the biomass rather than through its removal by sedimentation.

15.1.3 Merits and Defects

One of the main reasons that CMAL's are used is their ease of operation, which makes them particularly attractive for installations with small operating budgets [5]. The waste simply flows in on one side and out on the other without any recycle of biological solids or effluent. Furthermore, because the space times employed are normally in excess of one day, CMAL's are relatively resistant to upsets by organic and pH shock loads [6]. High strength wastes can also be handled easily because the reactor volume is usually so large that oxygen transfer is not a problem. Finally, because lagoons act like cooling ponds, wastes with relatively high temperatures can be treated without the requirement for additional cooling.

Surprisingly, some of the merits attributed to CMAL's can also be defects under certain conditions. For example, as discussed in Chapters 9 and 10, the values of the kinetic parameters vary with temperature, with the result that in the winter, when the cooling properties of the lagoon reduce the waste temperature, the degree of treatment attained will also be reduced. Coupled with this, and contributing to the problem, is the fact that no operational control can be exerted over the values of the kinetic parameters. If anything happens to change the values of the parameters there is nothing that can be done to compensate, and therefore the degree of treatment will change. This is a severe limitation which should be considered during design. Along these same lines, since τ influences reactor performance the process is subject to disturbance by hydraulic shock loads, because they cause the space time to fluctuate. In spite of these defects there are many instances in which the merits inherent in CMAL's make them the reactor of choice so that they have found an accepted place among the named biochemical operations.

15.2 PERFORMANCE OF CMAL SYSTEMS

Because a CMAL is a completely mixed basin operating on a continuous flow, single pass basis, it is essentially a CSTR and therefore has a theoretical performance similar to that presented in Sections 12.1.4 and 12.2.2. There are, however, complicating factors which can cause significant deviations from that theoretical performance.

15.2.1 Theoretical Performance

SINGLE CMAL. Theoretically, a single CMAL will perform in a manner similar to
that depicted by the curves in Figs. 12.2 and 12.4. Thus for wastewaters containing
organic compounds which are relatively easy to degrade there will be little biodegradable organic matter left in solution at reactor space times in excess of 24
hours. Consequently long space times have little effect upon the efficiency of
soluble substrate removal, it being relatively complete for all of them. That this
occurs in practice is shown in Fig. 15.3 where the performance curves for a 3.7
million gallon (14,000 m^3) CMAL treating domestic wastewater are presented [7].
Even though BOD$_5$ was used to measure lagoon performance the similarity between the
soluble substrate removal curves in Figs. 15.3 and 12.4 is apparent. Because
essentially all of the soluble organic matter is removed, most of the organic matter
in the effluent will be associated with the suspended solids [8,9]. Consequently,
the efficiency of total BOD removal in Fig. 15.3 is analogous to the percent
stabilization curve shown in Fig. 12.4, and again the similarity between the curves
is apparent. It can also be seen that long space times are required to achieve high
degrees of stabilization. This would necessitate a large land area (since the
basins are relatively shallow) and consequently CMAL's are only used for waste
stabilization where land is inexpensive. The third curve in Fig. 15.3 demonstrates
one of the points made in Section 15.1.2, i.e., that high effective degrees of
stabilization can be achieved by removing the microbial cells from the final effluent.

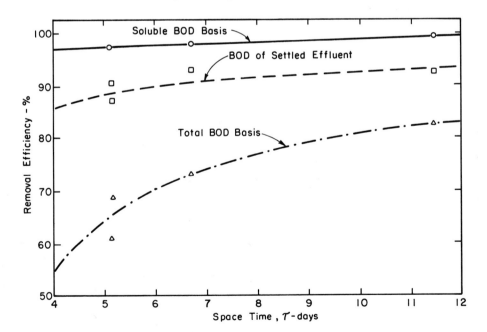

Figure 15.3. Performance of CMAL on domestic sewage.
(Based on data from Balasha and Sperber [7]).

Since the percent soluble substrate removal is essentially constant, the percent removal in the settled effluent will be primarily a reflection of the efficiency of suspended solids removal.

Because a single CMAL behaves like a single CSTR the other factors discussed in Section 12.1.4 will also apply here, and thus there is no need to repeat all of them. There are three additional significant points that should be emphasized, however. The first is that the amount of oxygen which must be supplied to the reactor will increase as τ is increased, in a manner similar to that shown in Fig. 12.3. It will be recalled that this was due to the increased importance of cellular decay as τ is increased. The second point is that there will be a minimum attainable soluble substrate concentration due to the necessity for driving the growth reaction to balance the loss of viable cells by death and decay. Consequently, the percent soluble substrate removal will always reach a limit less than 100% as shown in both Figs. 12.4 and 15.3. Finally, there will be a minimum space time required before growth and substrate removal can occur, as shown in Fig. 12.2. CMAL's should be sized sufficiently large to keep τ well above the minimum, particularly if the reactor will be subjected to hydraulic shock loads. High flow rates which force τ to approach τ_{min} can lead to process failure, or at the very least, to a serious loss of efficiency as shown in Figs. 12.2 and 12.4.

CMAL's IN SERIES. In general CMAL's in series behave as CSTR's in series, which were described in Section 12.2. It is generally best to put all feed into the first reactor, consequently CMAL's in series are normally built as simple chains. Referring to Figs. 12.10a and 12.12 it can be seen that there is little difference in soluble substrate concentration between a single CSTR and a chain, as long as the substrate is easily degradable and τ_T is in excess of 24 hours. This has been found to be true in practice. Benefits for soluble substrate removal are only achieved when the waste is difficult to degrade. Examination of Fig. 12.10d shows that some benefit can be attained with respect to waste stabilization, although the theoretical curves may overestimate it. The reason why will be discussed in Section 15.2.2.

Figure 15.4 shows the oxygen requirement in each reactor of a three reactor chain as a function of τ. It can be seen that most of the oxygen will be required in the first reactor. This, too, has been observed in practice, and is an important point, because if the oxygen supply system is not designed to take this distribution into account, performance will suffer. Examination of Figs. 12.12, 12.13, and 12.16 shows that CMAL's in series can be designed with most of the system volume in the first lagoon without significantly affecting soluble substrate removal or waste stabilization, while reducing the mass of oxygen required per unit volume per unit time. This has little effect on the total amount of oxygen which must be supplied to the reactors, however, thus the total aerator HP would be about the same. What it does accomplish is to distribute the required power among the reactors in a more economical fashion. This will be discussed more in Section 15.3.3.

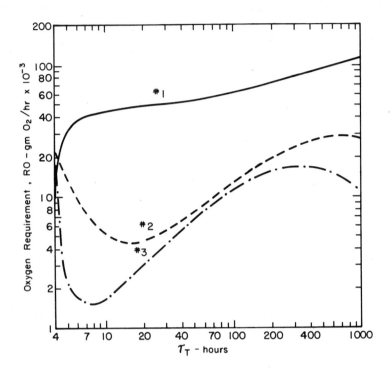

Figure 15.4. Effects of total space time on the oxygen
              requirement in each lagoon of a 0.33:0.33:0.33
              series. Theoretical curves calculated with
              Eq. 12.48 using typical parameters.

The similarity of CMAL performance with the theoretical CSTR performance
presented in Chapter 12 means that the CSTR equations can be used to model and
design CMAL's. However, there are some significant inadequacies of the theory
which should be recognized before those equations are applied.

15.2.2 Inadequacies of Theory

A significant deviation from theory was discussed in Section 12.1.1 where it was
pointed out that even though Eq. 12.6 shows no dependence of the effluent soluble
substrate concentration upon the influent concentration, one does, in fact,
exist [10,11]. The effect is essentially linear, so that at a fixed τ the percent
soluble substrate removal will stay constant as the influent substrate concentration,
S_o, is changed. In practice, this means that the kinetic constants describing
degradation of a wastewater should be determined using samples containing the same
concentration of organic matter as the actual wastewater.

The graphs and equations in Chapter 12 imply that high degrees of stabilization
can be achieved by the use of lagoons with long space times. Because of an
assumption made in the derivation of the equations, care should be exercised in

extrapolating that conclusion to practice, however. For the sake of simplicity, it was assumed in Eq. 10.54 that the decay rates of live and dead cells are equal. In reality, however, we should expect them to be different, and for the dead cells to decay at a lower rate. As τ is increased dead cells make up a larger percentage of the biomass (see the viability curve in Fig. 12.2). Consequently, if they decay more slowly, the effect would be to cause a decrease in the apparent decay rate of the total biomass, b, as τ is increased [12]. Such a decline in b with increases in τ has been observed in practice [13]. The effect of such a decline is to make the actual percent stabilization less than would be anticipated from the equation. This error could be eliminated by using separate values for the decay rates of live and dead cells, b_v and b_d, however, as discussed in Section 10.5.2 the current evidence is not sufficient to allow this to be done. Hopefully, the future will provide the needed data. In the meantime, the use of the equations in practice should be limited to space times less than 400 hours. An alternative procedure would be to employ equations similar to those used in Chapter 17 to express the loss of solids during aerobic digestion.

Another reason that the degree of stabilization may be overestimated is that algae may grow in the lagoon [4]. Because lagoons are relatively shallow, they will have a large surface area whenever τ is made large to achieve stabilization. If the mixing energy is not sufficient to keep all of the biosolids in suspension, thereby keeping the fluid turbid and reducing light penetration, lagoons will provide a perfect environment for algal growth. Because algae are autotrophic and fix carbon dioxide to form cell mass, they reverse the stabilization process and negate the results desired from the employment of a long space time. Thus even though algal growth is not included in the equations of Chapter 12, the possibility of its occurrence should be recognized when consideration is being given to the use of a CMAL as a stabilization device.

It was stated in Section 15.1.2 that CMAL's could be used as the complete means of wastewater treatment at moderate space times if the cells were separated from the fluid prior to discharge. The usual means of cell separation is gravity settling, and its use places a restriction on CMAL design. When bacteria are growing fairly rapidly (i.e., when τ is small) they tend to grow as individual, discrete organisms. Because of their small size, the cells will settle extremely slowly, and thus such growth is characterized as dispersed. As the specific growth rate is decreased, however, (i.e., as τ is increased) bacteria have a tendency to flocculate and form clumps which settle more rapidly. Although the possible reasons for this will be discussed in Chapter 16, it should be recognized at this point that space times in excess of five days are often required to obtain efficient removal of suspended solids from CMAL effluents [4]. The exact effect of τ on settling would have to be determined during treatability studies.

All of the inadequacies of theory discussed above apply both to single CMAL's and CMAL's in series. In addition, other factors may affect the predictive ability of the equations as applied to CMAL's in series. The first concerns the decay rates of live and dead cells, as discussed earlier. Examination of Fig. 12.10c shows that the viability of the culture in the third reactor of a three reactor chain will be less than it is in a single CSTR with τ equal to the total space time of the chain. If the specific decay rate of dead cells is less than that of live cells, then the overall specific decay rate constant, b, in the third reactor of the chain will be less than the value in the single reactor. This has the effect of making the total cell concentration in the third reactor greater than predicted by the model. In other words, the cell concentration in the third reactor would not be as far below the value in a single reactor as is indicated by the curves in Fig. 12.10b. Thus the benefits upon waste stabilization to be obtained by the use of reactors in series are not as great as indicated by the models. This deviation from theory has been observed in practice.

The benefits to be gained with respect to soluble substrate removal from the use of CMAL's in series are also not as great as predicted by the equations in Chapter 12 due to the fact that the kinetic parameters change as the waste moves down the chain [14,15,16]. This is caused by the heterogeneous nature of the organic matter in wastewaters. The organic compounds which are most easily degraded will be removed most rapidly, causing the character of the wastewater to change as it moves from reactor to reactor, thereby bringing about a change in the kinetic parameters.

While these inadequacies of the theory should be recognized and accounted for, they do not detract from the usefulness of the equations in design. When treatability studies are performed to evaluate the kinetic parameters, reactors should be run at space times bracketing the conditions anticipated in the final installation. Because curve fits are performed to estimate the kinetic parameters, the performance equations become tools for interpolating between experimental conditions. For this they are well suited. The danger of error arises when the equations are used to extrapolate over a range of conditions outside of that covered in the treatability studies.

15.3 DESIGN OF CMAL SYSTEMS

The process design of a CMAL includes three decisions: the volume of reactor required, the minimum power input to achieve adequate mixing, and the amount of oxygen which must be supplied. There are a multitude of other decisions which must be made in order to accomplish a complete system design (e.g., whether to use a single CMAL or CMAL's in series; the layout of the lagoon; type of construction; type of aeration system) but these are all dependent upon the specific conditions and requirements peculiar to a given installation. They will not be covered here

because the intent of this text is to provide general information about biochemical
operations which will aid the reader in incorporating them into large-scale systems.

15.3.1 Process Design Equations

As seen in Section 15.2 a CMAL behaves very much like a simple CSTR. Consequently,
its performance can be modeled with the equations presented in Chapter 12. Process
design requires that those equations be utilized to determine the physical characteristics of a system which will produce a desired degree of performance. This
means that the values of the appropriate kinetic parameters must be available to the
designer. These must be determined during lab- or pilot-scale studies using model
reactors. The design equations are thus just a means of scaling up the system from
the small reactors to the full-scale one, under the assumption that the performance
of the two will be similar. Thus it is imperative that the conditions imposed upon
the small-scale system mirror as closely as possible the conditions anticipated for
the full-scale one. The only variable which influences the performance of a CMAL
is the space time, τ. Consequently, the model reactors must be run at a number of
different space times to allow evaluation of the kinetic parameters as described
in Section 12.4. Many applications of lagoons to the pretreatment of industrial
wastes involve soluble substrates. Most other wastewaters, however, including all
domestic wastewaters, contain a mixture of soluble and solid organics. Thus the
appropriate set of equations should be used. It will be recalled from Chapter 12
that the only difference between the two was in nomenclature, with the more general
model simplifying to the one for soluble substrates in the absence of solid substrates. In order to minimize the number of equations which must be presented, only
the more general model will be given here.

SINGLE CMAL. The equations of Chapter 12 can be used directly to determine the
reactor volume and oxygen requirement once the kinetic parameters are known.

Design for Conversion of Soluble Organic Matter. If the CMAL is being designed
to achieve a soluble substrate concentration, S, the reactor volume can be calculated by substituting the definition of τ into Eq. 12.16 and rearranging:

$$V = \frac{F(K_s + S)}{S(\mu_m - \gamma - b') - K_s(\gamma + b')} \tag{15.1}$$

(It will be recalled from Section 12.1.3 that the use of a prime on a kinetic
parameter indicates that it is an effective parameter based on a mixture of cells
and undegraded solid substrate.) The flow rate, F, must be chosen to ensure proper
system performance but is often taken as the average daily flow. If γ cannot be
conveniently determined during the treatability studies, then the assumption of a
value of 0.004 hr^{-1} appears reasonable from the limited data in Table 10.5. Alternatively, it could be assumed equal to zero with relatively little error. If the
first-order approximation is applicable, then the volume can be calculated from the

rearranged form of Eq. 12.26:

$$V = \frac{F}{SY'_g k_e - \gamma - b'} \tag{15.2}$$

If settling of the biosolids from the CMAL effluent is planned, then the reactor volume must be sufficient to give a space time in excess of 5 days [4]. If the volume as calculated from Eq. 15.1 or 15.2 does not do this, then the volume associated with a τ of 5 days must be used, thereby lowering the soluble substrate concentration beneath the design value.

Even though the reactor volume is usually computed on the basis of the average daily flow, the rate of oxygen uptake by the microbial culture will respond rapidly to changes in both the rate of flow and the concentration of substrate in the influent so that provision must be made for supplying oxygen to meet peak demands [17]. The large space times act to minimize fluctuations in the cell and soluble substrate concentrations in a CMAL by two mechanisms: (1) the frequency of the influent variations is small with respect to τ, and (2) the cells possess a potential for utilizing substrate in excess of the steady-state rate, provided that the influent variations are of short duration [18]. Because the cell concentration is relatively constant around the average value, the utilization of oxygen for maintenance energy generation is also relatively constant. Thus the fluctuations in oxygen utilization are caused by changes in the rate associated with the generation of energy for synthesis. This rate will go both above and below the average rate although for design purposes the primary interest is in the maximum rate. The average rate of oxygen uptake for generation of energy for synthesis plus the oxygen uptake rate associated with maintenance energy requirements was given by Eq. 12.23':

$$RO_{SS} = \frac{F(S'_o - S)(1 + b'\tau - \beta'Y'_g)}{1 + b'\tau} \tag{12.23'}$$

F is the average daily flow rate and S'_o is the average daily influent substrate concentration as defined by Eq. 12.80. During the peak loading period approximately 40% of the substrate in excess of the average amount will be oxidized to provide the energy needed for synthesis. Thus, in addition to the steady-state oxygen requirement given by Eq. 12.23', an amount of oxygen equivalent to 40% of the increase in demand above the average steady-state loading should be provided to meet the transient-state demand:

$$RO_{TS} = 0.40[(FS'_o)_{peak} - (FS'_o)_{avg}] \tag{15.3}$$

$(FS'_o)_{avg}$ is the product of the average daily influent flow and concentration, and $(FS'_o)_{peak}$ refers to the product of influent flow and concentration resulting in the maximum mass loading of substrate to the CMAL. The latter could be caused by a change in flow, a change in concentration, or both. The total oxygen requirement is the sum of RO_{SS} and RO_{TS}, and it is what determines the size of the aeration system.

Design for Stabilization of Organic Matter. The degree of stabilization achieved by biological oxidation is equal to the fraction of the average influent oxygen demand $[(FS_o')_{avg}]$ that is satisfied. Since the oxygen demand satisfied is the steady-state oxygen requirement as given by Eq. 12.23', the fraction of the waste stabilized, ζ, is:

$$\zeta = \frac{RO_{SS}}{(FS_o')_{avg}} \tag{15.4}$$

A simple equation allowing direct computation of the reactor volume required to achieve a desired degree of stabilization, ζ, can be obtained by recognizing that S will be much smaller than S_o' for degrees of stabilization in excess of 0.40. Thus if S is dropped from Eq. 12.23', and the equation is then substituted in Eq. 15.4, the result is:

$$V = \frac{F[\beta'Y_g' - (1 - \zeta)]}{b'(1 - \zeta)} \tag{15.5}$$

If the assumption is not valid due to a low value of S_o' or a high value of S, then a more complicated quadratic equation will have to be used. For most practical cases, however, Eq. 15.5 will suffice. The steady-state oxygen requirement for a desired degree of stabilization is given by rearrangement of Eq. 15.4:

$$RO_{SS} = \zeta(FS_o')_{avg} \tag{15.6}$$

The total oxygen requirement is given by the sum of the steady-state and transient-state oxygen requirements as described earlier.

If the suspended solids in the CMAL effluent are to be settled prior to discharge, it would be advantageous to know the mass rate of production of dry solids so that plans could be made for their disposal. The production rate of solids, P_M, was defined in Chapter 13 as the product of the flow rate and solids concentration of the waste sludge stream. With a CMAL, all solids leaving in the effluent are wasted from the system, consequently the flow rate of the waste sludge stream is equal to the total flow rate, F. The concentration of solids in the effluent stream is the same as in the reactor, and is the sum of the inert solids, Z_i, and biosolids, X', as given by Eq. 12.28. Consequently the production rate of solids is given by:

$$P_M = F[Z_{io} + \frac{Y_g'(S_o' - S)}{1 + b'\tau}] \tag{15.7}$$

CMAL's IN SERIES. The design of a chain of CMAL's requires that another
decision be made in addition to the choice of system volume and oxygen requirement,
i.e., the system configuration. Because CMAL's behave like CSTR's in series we may
turn to the discussions in Sections 12.2 and 15.2 for guidance. As was seen there
the most advantageous performance is achieved when the system is constructed as a
simple chain with all flow entering the first reactor. Consequently this is how
CMAL's in series are normally designed. It was also seen that the critical space
time causing washout in the system depends upon the first reactor in the chain.
Because it is desirable to keep the space time of the first reactor considerably
greater than the critical value to reduce the possibility of system failure, and
because most of the oxygen must be supplied to the first reactor, it is advantageous
to design the chain as two or three unequal tanks in series with the majority of the
volume in the first reactor. For any given waste and treatment objective there will
be one volume distribution which costs less than all of the others and it may be
determined through application of systems analysis techniques. In the absence of
system optimization, it appears that a volume distribution of 0.6:0.2:0.2 would
often be reasonable.

Design for Conversion of Soluble Organic Matter. Because of the sequential
nature of the calculations with reactors in series it is not possible to calculate
directly the total system volume required to achieve a desired effluent soluble
T_bOD for a given system configuration. The most straightforward approach is to use
Eq. 12.43 to calculate the effluent soluble T_bOD from the last reactor as a function
of the total system space time and then choose the space time required from that
information. As was pointed out in Section 15.2.2 this equation may overestimate the
effects obtained from reactors in series if the parameter values used were all
obtained from studies on a single CSTR. Thus judgement must be employed in the
application of the equations. Once the sizes of the reactors have been chosen,
Eq. 12.48 may be used to calculate the steady-state oxygen requirement for each
reactor. Provisions should also be made for handling the oxygen requirement for
shock loads, although no well established guidelines are yet available for distributing the requirement among the reactors. If the main objective of the system is
the removal of soluble organic matter, the space times of the reactors are likely
to be short, with the result that some of the organic matter from the shock load
will pass through the first reactor untreated. Consequently, it would appear
prudent to distribute the transient-state oxygen requirement among the basins in
proportion to their volumes.

Design for Stabilization of Organic Matter. It is also difficult to calculate
directly the total system volume needed to achieve a desired degree of stabilization.
The total steady-state oxygen requirement for the system is a reflection of the
degree of stabilization achieved as shown in Eq. 15.4. Consequently the easiest
way to determine the total system volume is to use Eq. 12.48 to calculate the

steady-state oxygen requirement for each reactor at each of several space times, and then sum them to get the total steady-state oxygen requirement as a function of the total system space time. Examination of the data will then allow selection of the system space time required to achieve the desired degree of stabilization, ζ. Because large space times are generally required to achieve values of ζ in excess of 0.40, it is unlikely that much soluble organic matter will pass through the first reactor, even during shock loads. Consequently, the entire transient-state oxygen requirement, as calculated from Eq. 15.3, should be supplied to the first reactor.

15.3.2 Determination of Parameter Values

The values of the kinetic parameters should be determined through laboratory or pilot plant treatability studies using heterogeneous microbial cultures growing on the particular wastewater for which the lagoon is being designed. All studies should be run using small scale CSTR's and ample oxygen and other nutrients should be supplied to ensure that organic carbon is rate limiting. The temperature of the reactors should be held constant and if possible two or more temperatures should be studied so that the temperature effects can be quantified. Reactors should be run at several different space times and data should be collected to allow evaluation of the kinetic parameters as described in Section 12.4. Care should be exercised, however, to ensure that the space times completely bracket the values anticipated in the full-scale design. If this is done then the errors associated with the use of the design equations should be small.

15.3.3 Required Constraints

The equations of Section 15.3.1 cannot be applied blindly, rather judgment must be used to ensure that the process design is reasonable and consistent with good engineering practice. Consequently constraints have been established which aid the designer. The most important constraint on CMAL design has to do with the minimum allowable space time. Equation 12.9 gave the space time at which washout would occur in a single reactor or in the first reactor of a chain:

$$\tau_{min} = \frac{K_s + S_o}{S_o(\mu_m - \gamma - b) - K_s(\gamma + b)} \tag{12.9}$$

The larger τ is with respect to τ_{min} the more stable the reactor will be. Consequently it has been recommended that no reactor be built with a space time less than 10 times τ_{min} [19]. Seldom will single CMAL's be designed with space times near that limit because longer ones are usually required to accomplish the treatment objective. It may be necessary to apply the constraint to the first reactor in a chain, however, and this offsets part of the theoretical advantage of series reactors.

If the cells are to be removed from the effluent of a CMAL (or chain of CMAL's) by gravity sedimentation, then the total space time will have to be in excess of 5 days, as was discussed in Section 15.2.2.

15.3.4 Factors Influencing the Design

The equations in Section 15.3.1 do not completely define the process design of a CMAL, because factors not expressed in those equations must be considered. The two most important ones are the mixing requirements and the effects of temperature.

MIXING. It is important that completely mixed conditions be maintained in a CMAL. The first and most important reason is that all of the performance equations are based upon the assumption of complete mixing. In addition, if mixing is not adequate, solids will settle to the bottom and undergo anaerobic degradation thereby impairing system performance. Finally, inadequate mixing will tend to cause algae to grow in the lagoon [4].

Most aerated lagoons are mixed and aerated by mechanical surface aerators. The rate of oxygen transfer by those devices depends upon many factors so that manufacturers rate them according to the mass of oxygen transferred per unit time per unit power input under standard conditions [20]. The standard transfer rate must be converted to an actual transfer rate by consideration of the conditions existing in the installation, as described in other texts [20]. Once the oxygen requirement and the standard transfer rate have been established, however, the designer may calculate the total power input required to supply the oxygen. He then must compare that figure with the power needed to give adequate mixing and provide the larger of the two.

There is not yet complete agreement in the literature concerning the amount of power required to achieve complete mixing. For example, 6-10 HP/10^6 gal (0.0012 - 0.002 kw/m^3) will provide a uniform distribution of oxygen throughout a basin 8-10 ft deep [21,22] whereas 30-50 HP/10^6 gal (0.006 - 0.01 kw/m^3) are required before a soluble tracer curve will approach the one typical of complete mixing (see Section 6.1.2) [8,14]. These figures do not give the complete picture, however, because a critical factor is the achievement of uniform suspension of the microbial solids. The power level to keep solids in suspension depends upon their density and concentration [22], but for the concentrations normally found in aerated lagoons an input of 50-60 HP/10^6 gal (0.01 - 0.012 kw/m^3) appears to be adequate [21], although higher values have been recommended [23].

The power required for aeration depends upon both the influent substrate concentration and the reactor volume (see Eq. 12.23) whereas the power required for mixing depends only on the reactor volume. If the power requirements for mixing and aeration are both plotted as functions of reactor volume on the same graph they will cross with the crossover point depending upon the influent substrate concentration. For example, when S_o is 200 mg/liter, crossover occurs at a reactor volume giving a space time of around 1 day [8,24]. For reactor volumes larger than the crossover volume the power requirement for mixing will be larger than the one for oxygen transfer. This concept has not been recognized by all designers. When Beychock [21] conducted a study of aerated lagoon practice he found that most lagoons did not have sufficient power for mixing.

Consideration must also be given to the number and placement of the aerators in a CMAL. An important factor in the achievement of complete mixing is the turnover time in the basin, i.e., the length of time that it takes for the mixers to pump a volume equal to the basin volume. The smaller the turnover time, the closer the basin will approach complete mixing. The pumping rate per unit power input to a mixer is inversely proportional to the power supplied to it and approaches a maximum of around 50 gal/(min·HP) (4.2 x 10^{-3} m^3/(s·kw)) [15]. Thus it is normally better to use several small units than one large one. The effective distance of an aerator depends upon the amount of power supplied to it, with a reported maximum of 2 ft/HP (0.8 m/kw) [25]. The spacing between adjacent aerators, however, should not exceed 250 ft (76 m) [16]. Although guidelines are not yet well established, it is now recognized that a minimum spacing should also be observed to avoid adverse inter-actions which can reduce the oxygen transfer capabilities and depth of mixing. One study reported adverse interactions when 100 HP (75kw) units were placed closer together than 100 ft (30m) [26]. It is possible, however, that the use of draft tubes to improve in-depth mixing could have reduced that distance [27]. Regardless of spacing, however, when mechanical surface aerators are used in basins with depths greater than 10 feet (3.05m) draft tubes or deep secondary impellers should be used to ensure adequate bottom mixing [19].

An important factor in the design of series lagoon systems is that the majority of the oxygen must be supplied to the first basin. This requirement, when coupled with the need for adequate mixing in all of the lagoons, can have an effect upon the total power needed in the system. For example, if three equal sized lagoons are used, the first one may have a power input per unit volume for aeration which is over twice as large as a single CMAL achieving comparable treatment, whereas the aeration power input to the latter two lagoons may be less than half that of a single lagoon. It is thus likely that the power requirement to the latter two will be controlled by mixing even when the power to a single lagoon is controlled by aeration. The net effect could be to make the total power requirement to the series system larger than the single CMAL even though the total volume might be smaller. Conse-quently, when a series of lagoons is being designed, consideration must be given to the power requirements for aeration and mixing in each individual lagoon and the aerator/mixers must be chosen to meet the larger of the two. It is not sufficient to calculate the total power requirement for aeration and then divide it evenly among the basins, nor is it sufficient to assume that because mixing is adequate in the first basin it will be in the others as well. From this discussion another benefit to be derived from a volume distribution of 0.6:0.2:0.2 becomes apparent; it is similar to the split in total oxygen requirement and thus is likely to give an economic power design.

EFFECTS OF TEMPERATURE. As seen in Sections 9.3.3 and 10.6.1 temperature affects all of the parameters describing CMAL performance. Both μ_m and b increase with increasing temperature within the range of 10-35°C, however the effects of

temperature upon K_s and Y_g depend upon the type of system so that generalizations
are difficult. It does appear, however, that K_s increases as the temperature is
increased for heterotrophic growth such as will occur in CMAL's. As yet, there is
insufficient evidence to determine what happens to γ. The possibility of temperature
effects on diffusion was also discussed in Section 10.6.1 and was cited as a possible
cause of discrepancy between lab and field data. Because of the relatively low
concentration of biological solids encountered in CMAL's, diffusional limitations
are less likely than they are in other biochemical operations. Nevertheless, when
studies on the effects of temperature are performed during lab- or pilot-scale tests
care should be taken to ensure that the mixing energy levels are similar to those
anticipated in the full-scale installation so that the observed temperature effects
are characteristic.

As was seen in Fig. 12.8 the removal of soluble substrate in a CSTR is relatively insensitive to temperature at long space times but is sensitive at shorter
ones. This is consistent with observations from operational lagoons [9,28].
Stabilization is more sensitive to temperature over the entire range of space times,
however, as seen from the biological solids curve in Fig. 12.8.

A lagoon which has been properly sized can give adequate performance at low
temperatures [15], but this means that the lagoon must be sized for the worst
condition (i.e., lowest temperature). This can be accomplished in two ways. The
first would be to run all treatability studies at two temperatures: those anticipated in winter and summer operation. The parameters obtained from winter operation
would be used to size the lagoon whereas those from summer operation would be used
to determine the oxygen requirements. The second way would be to use reports from
the engineering literature to predict the effects of temperature. Most engineering
reports use the modified van't Hoff-Arrhenius equation to express the effects of
temperature upon the parameters describing aerated lagoon performance:

$$k_1 = k_2 \; \theta^{(T_1 - T_2)} \qquad\qquad\qquad (10.57)$$

Malina *et al.* [8] have summarized values of θ which describe soluble substrate
removal and have concluded that 1.08 is generally accepted, although the range was
from 1.035 to 1.1. That number could thus be applied to μ_m with a reasonable degree
of certainty. Even though K_s appears to increase as the temperature is increased
for heterotrophic cultures, the data from continuous growth systems is too limited
to allow generalization. Thus in the absence of actual data, K_s should be considered to remain constant because the resulting winter design will be more conservative than the one that would result from assuming that K_s decreases as the temperature drops. It was stated in Section 10.6.1 that the magnitude of the effects of
temperature upon b is similar to that upon μ_m for rapidly growing cultures, but that
for slowly growing cultures such as in a CMAL it is less. There is also difficulty
in fitting the temperature effects of slowly growing cultures to an Arrhenius

equation. Nevertheless, in the absence of data it should be possible to apply
Eq. 10.57 with a value of θ at the lower end of the range, for example, around 1.04.
The trends in Y_g are not yet clear, nor are they for γ. Thus, in the absence of
data it would be best to consider them to be constant. It should be clear to the
reader that there is still a good deal of confusion in the literature concerning the
effects of temperature and therefore that the above recommendations are tentative at
best. It cannot be emphasized too much, however, that the effects of temperature
should be determined during treatability studies if at all possible.

One reason that consideration of the effects of temperature upon the performance of a CMAL is so important is that CMAL's are excellent heat dissipating devices. Consequently, in cold weather their temperatures are likely to drop severely.
The engineer must, therefore, be able to predict the lagoon operating temperature
before he can establish the correct kinetic parameters. Even though higher degrees
of agitation could cause greater heat losses, most equations which have been used to
predict heat losses do not incorporate it [3,8,29,30]. Instead, they rely more upon
the surface area of the lagoon available for transfer. Malina *et al.* [8]
recommended

$$\frac{T_L}{T_i} = \frac{62.4(F/A_L) + 145(T_A - 2)/T_i}{62.4(F/A_L) + 145} \tag{15.8}$$

where.

   T_L = weekly average lagoon temp., °F

   T_i = weekly average influent temp., °F

   T_A = weekly average air temp., °F

   F = wastewater flow rate, ft^3/day

   A_L = surface area of lagoon, ft^2

   145 = average heat transfer coefficient, BTU/($ft^2 \cdot$day$\cdot$°F)

They found that the equation is relatively insensitive to changes in the heat
transfer coefficient. This indicates that predictions using the equation should be
adequate for a number of geographical locations and that agitation has little effect,
thereby justifying the exclusion of power input from the calculation.

15.3.5 Example Problems
The easiest way to demonstrate the process design procedure is through examples.

EXAMPLE 15.3.5-1
Consider treatment of the waste characterized in Example 12.4.2-1. Laboratory
treatability studies using a CSTR operated at several space times have revealed
the average characteristics listed in Table E15.1.

TABLE E15.1
Wastewater Characteristics at 68°F (20°C)

Symbol	Value	Symbol	Value
T_o	640 mg/liter as COD	μ_m	0.40 hr^{-1}
C_o	520 mg/liter as COD	K_s	120 mg/liter as T_bOD
Z_o	120 mg/liter	Y'_g	0.5 mg "cells"/mg T_bOD removed
C_i	80 mg/liter	b'	0.0035 hr^{-1}
f	0.40	γ	not determined
f'	0.30	β	1.25 mg T_bOD/mg "cells"

The wastewater has a temperature of 77°F (25°C) and the lowest weekly average
air temperature in the area is expected to be 23°F (-5°C). No studies on the effects
of temperature were performed, but the temperature characteristic, θ, is assumed to
be 1.08 and to be applicable to μ_m. The value of θ for b' is assumed to be 1.04 and
the other parameters are assumed to be independent of temperature. The bacterial
specific death rate, $\dot{\gamma}$, was not determined and is assumed to be 0.004 hr^{-1}. The
waste flows at a constant rate of 634,000 gal/day (100,000 liters/hr = 100 m^3/hr),
but receives quantitative shock loads which can raise the soluble COD to 800 mg/liter
and the total COD to 920 mg/liter for periods of no more than two hours. Size an
aerated lagoon to reduce the soluble T_bOD to 10 mg/liter, and determine the total
aerator/mixer HP required if the mechanical surface aerators to be used can deliver
2.5 lbs O_2/(HP·hr) (1.52 kg O_2/(kw·hr)) under field conditions.
Solution:
(a) The reactor volume must be determined for cold weather conditions because that
will give the lowest value of μ_m. Since the lagoon temperature depends on the
reactor volume it cannot be determined until the volume is known. Thus an iterative
approach must be used. To begin, estimate the temperature to be 68°F (20°C). The
reactor volume may be found by use of Eq. 15.1 with S set to 10 mg/liter T_bOD.

$$V = \frac{(100,000)(120 + 10)}{(10)(0.40 - 0.004 - 0.0035) - (120)(0.004 + 0.0035)}$$

$$V = 4,300,000 \text{ liters} = 4300 \text{ m}^3 = 1.14 \times 10^6 \text{ gal} = 1.52 \times 10^5 \text{ ft}^3$$

If the lagoon is made with an average depth of 10 ft, the surface area will be
$1.52 \times 10^4 ft^2$ (1910 m^2). Estimate the cold weather lagoon temperature using
Eq. 15.8:

$$F = 634,000 \text{ gal/day} = 84,800 \text{ ft}^3/\text{day}$$

$$\frac{T_L}{77} = \frac{62.4(84,800/15,200) + 145(23 - 2)/77}{62.4(84,800/15,200) + 145}$$

$$T_L = (0.786)(77) = 60.5°F \ (15.8°C)$$

The original estimate of the temperature was too high, therefore assume a winter lagoon temperature of 54°F (12°C) and repeat the process. Adjust μ_m and b' using Eq. 10.21:

$$\mu_{m12} = (0.40)(1.08)^{(12 - 20)} = 0.22 \text{ hr}^{-1}$$

$$b'_{12} = (0.0035)(1.04)^{(12 - 20)} = 0.0026 \text{ hr}^{-1}$$

Using these parameters, determine the required volume again using Eq. 15.1. The answer is:

$$V = 9,690,000 \text{ liters} = 9,690 \text{ m}^3 = 2.56 \times 10^6 \text{ gal} = 3.42 \times 10^5 \text{ ft}^3$$

If the depth of the lagoon is kept at 10 ft, the area will be $3.42 \times 10^4 \text{ ft}^2$, which gives a winter temperature of 50°F (10°C). This is close to the assumed temperature of 54°F (within the accuracy of Eq. 15.8). Thus the CMAL must have a volume of $3.42 \times 10^5 \text{ ft}^3$ (9690 m^3), which corresponds to a space time of 97 hrs (4.0 days).

(b) The aerator power should be calculated for summer operation because that is when the greatest oxygen requirement will occur. The steady-state oxygen requirement may be calculated with Eq. 12.23 by using S'_o, b', and β' since the waste contains insoluble biodegradable organic matter. The effective influent substrate concentration, S'_o, may be calculated with Eq. 12.80:

$$S'_o = 640 - 80 - 0.30 (640 - 520)$$

$$S'_o = 524 \text{ mg/liter } T_bOD$$

The oxygen requirement depends on S, the soluble T_bOD in the effluent. This will not be 10 mg/liter, but will be lower because little cooling of the waste will occur in the summer. If the summer lagoon temperature is 77°F (25°C), then the kinetic parameters can be found from Eq. 10.21:

$$\mu_{m25} = (0.40)(1.08)^{(25 - 20)} = 0.59 \text{ hr}^{-1}$$

$$b'_{25} = (0.0035)(1.04)^{(25 - 20)} = 0.0043 \text{ hr}^{-1}$$

The summer time effluent soluble substrate concentration, S, may be estimated from Eq. 12.6:

$$S = \frac{(120)[(1/97) + 0.004 + 0.0043]}{(0.59) - [(1/97) + 0.004 + 0.0043]}$$

$$S = 4 \text{ mg/liter } T_bOD$$

The steady-state oxygen requirement can be calculated with Eq. 12.23:

$$RO_{SS} = \frac{(100,000)(524 - 4)[1 + (0.0043)(97) - (1.25)(0.50)]}{1 + (0.0043)(97)}$$

$$RO_{SS} = 2.91 \times 10^7 \text{ mg } O_2/\text{hr} = 29.1 \text{ kg } O_2/\text{hr} = 64.2 \text{ lb } O_2/\text{hr}$$

Allowance must also be made for the transient-state oxygen requirement since the soluble COD, C_o, reaches 800 mg/liter. The nonbiodegradable soluble COD, C_i, also probably increases during this period, and it would be reasonable to assume that it remains a constant fraction of C_o. Thus,

$$C_{iT} = (800)(80/520)$$

$$C_{iT} = 123 \text{ mg/liter}$$

The effective influent substrate concentration during the transient can be calculated with Eq. 12.80. Since the total COD increases to 920 mg/liter during the transient, the effective influent concentration is:

$$S'_{oT} = 920 - 123 - (0.30)(920 - 800)$$

$$S'_{oT} = 761 \text{ mg/liter } T_b OD$$

The transient-state oxygen requirement can be calculated with Eq. 15.3:

$$RO_{TS} = 0.40 \left[(100,000)(761) - (100,000)(524) \right]$$

$$RO_{TS} = 0.95 \times 10^7 \text{ mg } O_2/\text{hr} = 9.5 \text{ kg } O_2/\text{hr} = 20.9 \text{ lb } O_2/\text{hr}$$

The total oxygen requirement will be the total of the two, or 85.1 lb/hr (38.6 kg/hr) of oxygen.

The total power for aeration is found by dividing the lb/hr of oxygen by the transfer efficiency of 2.5 lb O_2/HP·hr, giving an answer of 34.1 HP (25.4 kw). To provide adequate mixing at least 50 HP/10^6 gal (0.01 kw/m^3) must be provided. Since the basin volume is 2.56 x 10^6 gal (9690 m^3), the mixing power must be 128 HP (97 kw). The mixing power is greater, thus approximately 130 HP must be supplied which will provide all of the oxygen that the microorganisms will need while maintaining a high DO level in the basin.

EXAMPLE 15.3.5-2

The waste in Example 15.3.5-1 is to be treated in a CMAL and the cells are to be removed by sedimentation prior to discharge. Determine the effluent soluble $T_b OD$ and the mass of sludge produced per day during summer operation.

Solution:

(a) The space time of the reactor must be at least 5 days to obtain efficient settling. Thus, design the system to give a τ of 5 days (120 hrs). The effluent soluble T_bOD will be given by Eq. 12.6 using the kinetic parameters for 77°F (25°C) determined in the preceding example.

$$S = \frac{(120)\,[(1/120) + 0.004 + 0.0043]}{(0.59) - [(1/120) + 0.004 + 0.0043]}$$

$S = 3.5$ mg/liter T_bOD

(b) The mass of dry sludge solids produced per day is given by Eq. 15.7. The concentration of inert solids in the influent is

$$Z_{io} = (0.4)(120) = 48 \text{ mg/liter}$$

$$P_M = (100,000)\,[48 + \frac{(0.50)(524 - 4)}{1 + (0.0043)(120)}]$$

$$P_M = 2.20 \times 10^7 \text{ mg/hr} = 22 \text{ kg/hr} = 527 \text{ kg/day} = 1160 \text{ lb/day}$$

115 kg/day (256 lb/day) of this is due to the inert solids in the influent.

EXAMPLE 15.3.5-3

Determine the size of CMAL required to bring about 50% stabilization during the winter of the waste in Example 15.3.5-1. What percent stabilization will occur in the summer?

Solution:

(a) Assume the winter temperature of the lagoon to be 41°F (5°C), and adjust b' using Eq. 10.57:

$$b_5' = (0.0035)(1.04)^{(5 - 20)} = 0.0019 \text{ hr}^{-1}$$

Use Eq. 15.5 to determine the required volume.

$$V = \frac{(100,000)\,[(1.25)(0.5) - (1 - 0.50)]}{(0.0019)(1 - 0.50)}$$

$V = 13,158,000$ liters $= 13,158$ m^3 $= 3.48 \times 10^6$ gal $= 4.65 \times 10^5$ ft^3

This volume corresponds to a space time of 132 hrs (5.5 days).

Check the temperature to see if the assumed value of b' is correct. Assuming a depth of 10 ft the surface area will be 4.65×10^4 ft^2 (4320 m^2). Using Eq. 15.8 the lagoon temperature is estimated to be 46°F (7.6°C). This is not close enough. Assume the temperature to be 46.4°F (8°C), adjust b' again and recalculate the volume.

$$b_8' = (0.0035)(1.04)^{(8 - 20)} = 0.0022 \text{ hr}^{-1}$$

Applying Eq. 15.5 again leads to:

$$V = 11,364,000 \text{ liters} = 11,364 \text{ m}^3 = 3.00 \times 10^6 \text{ gal} = 4.01 \times 10^5 \text{ ft}^3$$

This corresponds to a τ of 114 hrs (4.7 days) and gives a temperature close to 8°C.
(b) The degree of stabilization occurring in the summer can be found by rearranging
Eq. 15.5:

$$\zeta = \frac{1 + b'\tau - \beta'Y'_g}{1 + b'\tau} \tag{15.9}$$

and using b' for 25°C, which was shown in Example 15.3.5-1 to be 0.0043 hr^{-1}.
Substitution in Eq. 15.9 gives

$$\zeta = \frac{1 + (0.0043)(114) - (1.25)(0.5)}{1 + (0.0043)(114)} = 0.58$$

Thus there will be a slight improvement in performance during the summer.

15.4 OPERATION OF CMAL SYSTEMS

There is little operational control that can be exerted over aerated lagoons. Most
operational problems which have been reported deal with short circuiting, solids
deposition and freezing of the aerators. The first two are actually design shortcomings and can only be overcome by modification of the facility. Ice buildup on
surface aerators during the winter is a common problem in northern areas [16,29,31].
One study reported no problems with high speed, floating surface aerators because
constant contact with the water prevented serious depositions, although where low
speed, pontoon type aerators were used, ice formation on the aerator platform and
superstructure sometimes caused the units to tilt or sink [29]. These problems were
solved with the addition of electrical trace heaters to the units. Thus ice formation should be considered in the selection of the aerators for a CMAL.

15.5 EXAMPLES OF CMAL's IN PRACTICE

The CMAL was developed originally for waste treatment in the pulp and paper industry,
and it wasn't until 1960 that it was applied to domestic waste treatment [15]. Most
applications have been to industrial wastes. McKinney et al. [15] reviewed industrial applications of aerated lagoons and reported on the following industries:
pulp and paper, 5 installations; textile, 2; refinery, 2; chemical, 2; roofing, 1;
food processing, 6; and packinghouse or animal wastes, 5. In addition, Beychock [21]
conducted a survey of CMAL application for the A.I.Ch.E. and reported the following:
pulp and paper, 8 installations; textile, 2; refinery, 4; chemical, 4; wood
treating, 1; and distillery, 1. While there is some overlap between these reviews,

the large number of applications and the diverse nature of the wastewaters attest
to the acceptance of this means of treatment by environmental engineers.

Table 15.1 summarizes the most important characteristics of some of the CMAL's
which have been reported upon in the literature. Although a value is given for the
influent organic concentration (expressed as BOD_5) none is given for the effluent
because of a lack of consistency in the literature. Some authors reported soluble
BOD_5, reflecting the conversion of soluble substrate, while others reported a BOD_5
on settled effluent, reflecting the combined effects of the lagoon and a subsequent
settling basin. In many cases the conditions of the effluent sample were not stated,
making it impossible to determine the category under which it would fall. Reviewing
the information in the table, it will be seen that a large range of space times has
been employed in practice: from 0.67 to 17.3 days. This reflects several things:
the different objectives employed by the various designers (i.e. conversion or
stabilization); the ease of degradation of the wastewater; and the temperature of
treatment. Many designers have chosen a τ of around 2 days because almost all soluble
substrate will have been converted to cells within that time. Relatively few of the
lagoons were designed with mixing energy levels over 30 $HP/10^6$ gal (0.006 kw/m^3),
the lower limit to achieve complete mixing, consequently occurrences of solids
deposition have been reported [5,33,34]. A few of the lagoons did not have power
levels above 6 $HP/10^6$ gal (0.0012 kw/m^3), the level considered necessary to keep
oxygen distributed evenly throughout the lagoon, thus much of their volume could be
considered wasted since they have marginal oxygen levels. Three of the four series
designs reported in Table 15.1 utilized equal sized reactors. All of them supplied
the majority of the oxygen to the first basin: #12, 49%; #20, 85%; #21, 63%; and
#22, 75%. This is consistent with the concept of rapid removal of soluble organics
in the first tank.

In general, where performance data were available, they followed the concepts put
forth in this chapter, i.e., rapid removal of soluble organics, with the slower
degradation of the resulting biomass. In many instances the soluble substrate concentration was inversely proportional to the space time implying that the linear approximation given by Eq. 12.24 would be sufficiently accurate for design. Good agreement
was generally found between pilot plant data and the actual plant performance, except
when mixing energy levels were insufficient. If well mixed lab- or pilot-scale
reactors are run to obtain design data, it is imperative that the full scale installation have sufficient mixing to make the assumption of complete mixing valid. The
equations of this chapter could not be expected to predict performance in systems
that are not completely mixed. The ability of the lagoons to handle shock loads
was mentioned by some authors [5,32]. The longer the space time the more stable the
system.

Table 15.1
Examples of Industrial Uses of CMAL's

Example No.	Type of Plant	τ days	$\frac{HP}{10^6 \text{ gal}}$	Depth ft.	Area acres	S' mg/l BOD$_5$	Temp. °F	Comments	Ref. No.
1	Pulp/Paper Mill	8	4.7	10-12	21	200	64-86		32
2	Pulp/Paper Mill	7	3.5	10-12	23	130	61-84		32
3	Pulp/Paper Mill	11	15	8	8	470	55-80		33
4	Pulp/Paper Mill	7.5	11	14	4.7	-	-	Plug-flow	5
5	Pulp/Paper Mill	3.6	24	5	3.7	530	76		34
6	Pulp/Paper Mill	4	8.8	5	11	-	60-79		25
7	Pulp/Paper Mill	2.3	11.2	10	1.6	250	83		21
8	Pulp/Paper Mill	4	33	12	1.4	225	79		21
9	Pulp/Paper Mill	2.6	4	8	50	130	90		21
10	Pulp/Paper Mill	3.8	-	8	0.02	400	65		21
11	Pulp/Paper Mill	1.8	16	13	1.2	325	75		21
12	Pulp/Paper Mill	2.3	17.3	12	4.2	300	78	Basin #1 of 3 in series	21
12	Pulp/Paper Mill	2.4	10.6	12	4.3	-	74	Basin #2 of 3 in series	21
12	Pulp/Paper Mill	2.3	7.3	12	4.2	-	72	Basin #3 of 3 in series	21
13	Textile Plant	2.0	20	12	1.3	210	-		35
14	Textile Plant	0.67	51	10	64	200	85		21
15	Chemical Plant	2.75	460	17	0.9	2100	75	O$_2$-limited	6
16	Chemical Plant	7.0	24	9	0.7	710	47		21
17	Chemical Plant	2.1	40	5-6	1.7	600	80		21
18	Chemical Plant	17.3	50	15	6.1	2500	70		21
19	Oil Refinery	3.5	8.4	5	1.1	100	85		21
20	Oil Refinery	1.2	33.5	10	2.4	160	94	Basin 1 of 2 in series	21
20	Oil Refinery	1.2	5.8	10	2.4	-	90	Basin 2 of 2 in series	21
21	Oil Refinery	1.0	79	14	0.8	270	90	Basin 1 of 2 in series	21
21	Oil Refinery	1.0	45	14	0.8	-	90	Basin 2 of 2 in series	21
22	Fiberglass Products	4.5	46	5	0.4	933	55	Basin 1 of 2 in series	21
22	Fiberglass Products	1.4	50	5	0.12	-	55	Basin 2 of 2 in series	21

The need for adequate nutrients was covered in Chapters 9 and 10. Nevertheless it should be emphasized at this point that unless adequate nutrients are provided, the system will not be carbon limited as assumed in the development of the kinetic theory and thus the effluent $T_b OD$ will not respond to the space time in the manner expected. Haynes [33] demonstrated this in a pulp and paper lagoon when he showed that the efficiency of substrate removal suffered when nitrogen addition was stopped. Because the net yield depends upon the space time of the reactor, it would be anticipated that less nutrients would be required at long space times [36], and this has been observed to be true with pulp and paper wastes [28]. Finally, oxygen is a nutrient and insufficient oxygen will reduce performance. This was demonstrated clearly by Amberg *et al.* [25] who operated two lagoons in parallel. One was deficient in oxygen and gave significantly poorer results. Similar results have been presented by others [6,37].

In summary it can be stated that properly designed completely mixed aerated lagoons offer a reasonable alternative for the treatment of readily biodegradable wastewaters. The most important design parameter is the space time but care must be exercised to ensure that the mixing energy input, oxygen supply, and nutrient addition are all sufficient to meet the assumptions made in the development of the design theory. The decision to use a CMAL must be made on the basis of economics after comparing the total system cost with that of alternative systems. Where land costs are cheap, aerated lagoons are often the alternative of choice, and in situations where little skilled operation will be available their simplicity may override other considerations. Their simplicity, however, is their greatest weakness since no operational control can be exerted to alter process performance.

15.6 KEY POINTS

1. Completely mixed aeration lagoons (CMAL) are large shallow (5 - 15 ft) single pass reactors, generally mixed and aerated by mechanical surface aerators, in which mixed communities of heterotrophic microorganisms are maintained in an aerobic condition.

2. CMAL's are generally used for two major purposes:

    (a) To convert soluble organic matter into insoluble microbial cells, either as a means of pretreatment of an industrial waste prior to discharge to a municipal sewer or prior to sedimentation as the major treatment of the wastewater from a small community.

    (b) To stabilize organic matter thereby reducing the mass of microbial solids remaining when a CMAL is used as the major means of wastewater treatment.

3. Although CMAL's are the simplest of the biochemical operations they are not amenable to operational control.

4. Theoretically, a single CMAL behaves like a single CSTR as depicted in Figs. 12.2 and 12.4, and a chain of CMAL's behaves like a chain of CSTR's as depicted in Figs. 12.10 and 12.12.

5. The actual performance of a CMAL differs from theory in several ways:

   (a) The concentration of soluble organic matter in the effluent is not
       independent of the concentration in the influent.

   (b) The degree of stabilization achieved at long space times will not be as
       great as predicted by theory.

   (c) CMAL's in series do not have as large an advantage over a single CMAL as
       theory indicates.

6. The volume required to achieve a desired degree of substrate removal in a CMAL
   may be calculated from the equation for soluble substrate removal in a CSTR.
   The amount of oxygen supplied must be sufficient to meet peak demands in
   addition to the steady-state requirement. The volume required to achieve a
   desired degree of stabilization may be calculated from an equation which is
   based on the fact that the degree of stabilization is equal to the amount of
   influent oxygen demand satisfied within the lagoon.

7. CMAL's in series should be designed as two or three unequal tanks in series
   with the majority of the volume to the first reactor. All flow should enter
   the first reactor.

8. The kinetic parameters for the design of a CMAL may be obtained from treatability studies using small scale CSTR's. Care should be exercised to see that
   the space times used completely bracket the values anticipated in the full-
   scale design.

9. Both mixing and aeration must be considered when sizing the mechanical surface
   aerators for a CMAL. The power level to keep solids in suspension depends
   upon their density and concentration, but for the concentrations normally found
   in aerated lagoons an input of 50 - 60 HP/10^6 gal (0.01 - 0.012 kw/m^3) appears
   to be adequate. The power level to transfer oxygen will depend upon the design
   of the particular unit. The larger of the two requirements must be provided.

10. Temperature effects must be considered when calculating the required volume for
    a CMAL. Y'_g, K_s, and γ should be considered to be independent of temperature
    but μ_m and b' increase as the temperature is increased. The effects of temper-
    ture upon μ_m and b' may be described by the modified van't Hoff-Arrhenius
    equation if the temperature range is small. The temperature characteristic,
    θ, for μ_m appears to be larger than the value for b', although both should be
    determined experimentally.

11. There is little operational control that can be exerted over aerated lagoons.

12. Many of the CMAL designs reported in the literature did not have sufficient
    mixing energy to keep all solids in suspension. However, in those cases where
    mixing was adequate the performance of the lagoons followed the concepts of
    this chapter.

15.7 STUDY QUESTIONS

The study questions are arranged in the same order as the text:

Section	Questions
15.1	1-3
15.2	4-5
15.3	6-11

1. Describe the essential physical characteristics of a CMAL.

2. Why is a CMAL better suited to the conversion of soluble organic matter into microbial cells than it is to the stabilization of organic matter?

3. Why are CMAL's not subject to operational control?

4. Explain the difference in the three performance curves of Fig. 15.3.

5. Why is the degree of stabilization achieved in a CMAL with a long space time not as great as predicted by CSTR theory?

6. Why must the aeration system for a CMAL be capable of supplying oxygen at a rate in excess of the steady-state requirement?

7. Explain the basis for Eq. 15.5 for calculating the CMAL volume required to achieve a desired degree of waste stabilization.

8. Discuss the factors which must be considered when sizing mechanical surface aerators for a CMAL.

9. Explain why it is important to consider the effects of temperature when designing a CMAL.

10. A lab treatability study of a wastewater revealed the characteristics listed in Table SQ15.1.

    TABLE SQ15.1
    Values of Kinetic Parameters and Waste Characteristics
    Determined by a Treatability Study

Symbol	Value	Symbol	Value
T_o	800 mg/liter as COD	μ_m	$0.30 \ hr^{-1}$
C_o	600 mg/liter as COD	K_s	300 mg/liter as T_bOD
Z_o	150 mg/liter	Y_g'	0.50 mg "cells"/mg T_bOD removed
C_i	50 mg/liter as COD	b'	$0.0045 \ hr^{-1}$
f	0.40	γ	not determined
f'	0.40	β	1.35 mg T_bOD/mg "cells"

    The study was run at 20°C. The wastewater has a temperature of 77°F (25°C) and the lowest weekly average air temperature in the area is expected to be 32°F (0°C). The waste flows at a constant rate of 1,270,000 gal/day (200 m³/hr) and has a relatively constant concentration. Size a CMAL 10 ft (3m) deep capable of achieving at least 50% stabilization of the waste in both summer and winter, and determine the total aerator/mixer HP required if the mechanical surface aerators to be used can deliver 2.0 lbs O_2/(HP·hr) (1.22 kg O_2/(kw·hr)) under field conditions. What is the soluble T_bOD concentration in the effluent during the summer?

11. Given the wastewater and the CMAL in Study Question 10, what flow rate could be treated to achieve an effluent soluble COD of 70 mg/liter from the lagoon during the winter?

REFERENCES AND FURTHER READING

1. G. Barson, "Lagoon performance and the state of lagoon technology," *Environmental Protection Agency Environmental Protection Technology Series*, Report No. EPA-R2-73-144, June, 1973.

2. K. L. Murphy and A. W. Wilson, "Characterization of mixing in aerated lagoons," *Journal of the Environmental Engineering Division, ASCE*, 100, 1105-1117, 1974.

3. J. L. Mancini, and E. L. Barnhart, "Industrial waste treatment in aerated lagoons," in *Advances in Water Quality Improvement*, edited by E. F. Gloyna and W. W. Eckenfelder, Jr., Univ. Texas Press, Austin, 313-324, 1968.

4. S. C. White and L. G. Rich, "How to design aerated lagoon systems to meet 1977 effluent standards - experimental studies," *Water and Sewage Works*, 123, #3, 85-87, 1976.

5. W. M. Laing, "New secondary aerated stabilization basins at the Moraine Division," *Proceedings of the 23rd Industrial Waste Conference*, Purdue University Engineering Extension Series, No. 132, 484-492, 1968.

6. F. D. Bess and R. A. Conway, "Aerated stabilization of synthetic organic chemical wastes," *Journal of the Water Pollution Control Federation*, 38, 939-956, 1966.

7. E. Balasha and H. Sperber, "Treatment of domestic wastes in an aerated lagoon and polishing pond," *Water Research*, 9, 43-49, 1975.

8. J. F. Malina, *et al.* "Design guides for biological wastewater treatment processes," *Environmental Protection Agency Water Pollution Control Research Series*, Report No. 11010 ESQ 08/71, August, 1971.

9. C. N. Sawyer, "New concepts in aerated lagoon design and operation," in *Advances in Water Quality Improvement*, edited by E. F. Gloyna and W. W. Eckenfelder, Jr., Univ. Texas Press, Austin, 325-335, 1968.

10. C. P. L. Grady Jr., *et al.* "Effects of growth rate and influent substrate concentration on effluent quality from chemostats containing bacteria in pure and mixed cultures," *Biotechnology and Bioengineering*, 14, 391-410, 1972.

11. C. P. L. Grady Jr. and D. R. Williams, "Effects of influent substrate concentration on the kinetics of natural microbial populations in continuous culture," *Water Research*, 9, 171-180, 1975.

12. C. P. L. Grady Jr. and R. E. Roper, Jr., "A model for the bio-oxidation process which incorporates the viability concept," *Water Research*, 8, 471-483, 1974.

13. B. L. Goodman and A. J. Englande, "A unified model of the activated sludge process," *Journal of the Water Pollution Control Federation*, 46, 312-332, 1974.

14. H. R. Flockseder, "Performance of the aerated lagoon process," thesis presented to University of Texas, at Austin, Texas, in 1970, in partial fulfillment of the requirements for the degree of Master of Science in Environmental Health Engineering.

15. R. E. McKinney, *et al.* "Waste treatment lagoons - State of the art," *Environmental Protection Agency Water Pollution Control Research Series*, Report No. 17090 EHX, 07/71, July 1971.

16. D. J. Thimsen, "Biological treatment in aerated lagoons - Theory and practice," paper presented the 12th Annual Waste Engineering Conference, Univ. of Minn., Dec. 1965.

17. W. J. O'Brien and C. E. Burkhead, "Oxygen consumption in continuous biological culture," *Environmental Protection Agency Water Pollution Control Research Series,* Report No. 17050 DJS 05/71, May 1971.

18. J. C. McLellan and A. W. Busch, "Hydraulic and process aspects of reactor design - II - Response to variations," *Proceedings of the 24th Industrial Waste Conference,* Purdue University Engineering Extension Series, No. 135, 493-506, 1969.

19. A. W. Lawrence and P. L. McCarty, "Unified basis for biological treatment design and operation," *Journal of the Sanitary Engineering Division, ASCE,* 96, 757-778, 1970.

20. H. W. Parker, *Wastewater System Engineering,* Prentice-Hall, Inc., Englewood Cliffs, N.J., 272-299, 1975.

21. M. R. Beychock, "Performance of surface-aerated basins," *Chemical Engineering Progress Symposium Series,* 67, No. 107, 322-339, 1971.

22. E. L. Barnhart, "The treatment of chemical wastes in aerated lagoons," *Chemical Engineering Progress Symposium Series,* 64, No. 90, 111-114, 1968.

23. W. Von der Emde, "Entwurf Beluftungssysteme," in: *Wiener Mitteilungen,* Vol. 4, edited by W. Von der Emde, Abwasserreingungsanlagen - Entwurf, Bau, Betrieb, Vienna, 1969.

24. H. Benjes, Jr., "Theory of aerated lagoons," *Proceedings 2nd International Symposium for Waste Treatment Lagoons,* University of Kansas, Lawrence, Kansas, edited by R. E. McKiney, 210-217, 1970.

25. H. R. Amberg, *et al.* "Aerated lagoon treatment of sulfite pulping effluents," *Environmental Protection Agency Water Pollution Control Research Series,* Report No. 12040 ELW 12/70, Dec. 1970.

26. K. S. Price, *et al.* "Surface aerator interactions," *Journal of Environmental Engineering Division, ASCE,* 99, 283-300, 1973.

27. G. F. Williams, "Discussion to 'Surface aerator interactions'," *Journal of Environmental Engineering Division, ASCE,* 100, 768-770, 1974.

28. I. Gellman, and H. F. Berger, "Waste stabilization pond practices in the pulp and paper industry," in *Advances in Water Quality Improvement,* edited by E. F. Gloyna and W. W. Eckenfelder, Jr., Univ. Texas Press, Austin, 492-496, 1968.

29. P. L. Timpany, *et al.* "Cold weather operation in aerated lagoons treating pulp and paper mill wastes," *Proceedings of the 26th Industrial Waste Conference,* Purdue University Engineering Extension Series, No. 140, 776-790, 1971.

30. E. H. Bartsch and C. W. Randall, "Aerated lagoons - A report on the state of the art," *Journal of the Water Pollution Control Federation,* 43, 699-708, 1971.

31. A. R. Pick, *et al.* "Evaluation of aerated lagoons as a sewage treatment facility in the Canadian prairie provinces," *Proceedings of the International Symposium on Water Pollution Control in Cold Climates, Environmental Protection Agency Water Pollution Control Research Series,* Report No. 16100 EXH 11/71, 191-212, Nov. 1971.

32. K. R. Devones, *et al.* "Experience with low rate biological treatment processes," *Proceedings of the 23rd Industrial Waste Conference,* Purdue University Engineering Extension Series, No. 132, 10-17, 1968.

33. F. D. Haynes, "Three years operation of aerated stabilization basins for paperboard mill effluent," *Proceedings of the 23rd Industrial Waste Conference,* Purdue University Engineering Extension Series, No. 132, 361-373, 1968.

34. H. R. Amberg, "Aerated stabilization of board mill white water," *Proceedings of the 20th Industrial Waste Conference,* Purdue University Engineering Extension Series, No. 118, 525-537, 1965.

35. S. W. Williams, Jr. and G. A. Hutto, Jr. "Treatment of textile mill wastes in aerated lagoons," *Proceedings of the 16th Industrial Waste Conference,* Purdue University Engineering Extension Series, No. 109, 518-529, 1961.

36. J. H. Sherrard, and E. D. Schroeder, "Importance of cell growth rate and stoichiometry to the removal of phosphorous from the activated sludge process," *Water Research,* 6, 1051-1057, 1972.

37. W. C. Boyle and L. B. Polkowski, "Treatment of cheese processing wastewater in aerated lagoons," *Proceedings of the Third National Symposium on Food Processing Wastes, Environmental Protection Agency Environmental Protection Technology Series,* Report No. EPA-R2-72-018, 323-370, Nov. 1972.

Further Reading

Benefield, L. D. and Randall, C. W., *Biological Process Design for Wastewater Treatment,* Prentice-Hall, Inc., Englewood Cliffs, N.J., 1980, See Chapter 6.

Busch, A. W., *Aerobic Biological Treatment of Wastewaters,* Oligodynamics Press, Houston, Texas, 1971.

McKinney, R. E. (editor), *Proceedings of the 2nd International Symposium for Waste Treatment Lagoons,* University of Kansas, Lawrence, Kansas, June 1970.

Metcalf & Eddy, Inc., *Wastewater Engineering: Treatment/Disposal/Resuse,* 2nd Edition, McGraw-Hill Book Co., New York, N.Y., 1979. See Chapter 10 for Aerated Lagoons.

Parker, H. W., *Wastewater Systems Engineering,* Prentice-Hall, Inc., Englewood Cliffs, N.J., 1975. See Chapters 6 and 9.

Rich, L. G. and White, S. C., "How to design aerated lagoon systems to meet 1977 effluent standards," *Water and Sewage Works,* 123, #3, 85; #4, 82; #5, 88; #6, 90, 1976.

CHAPTER 16

ACTIVATED SLUDGE

A brief description of the biochemical operation known as activated sludge was given in Section 1.2.2, where it was seen that the name is applied to a number of different reactor configurations. The factors which are common to them all are: (1) they utilize a flocculent slurry of microorganisms to remove soluble organic matter from wastewater; (2) they employ sedimentation to remove the organisms prior to discharge, thereby producing an effluent low in microbial solids; (3) they recycle a concentrated slurry of microorganisms from the settler underflow to the reactor; and

(4) they perform in a manner which is primarily dependent upon the mean cell residence time (MCRT) of the organisms within the system. Thus when reduced to its essential components, the activated sludge process consists of a reactor (often called the aeration basin) and a settler, with provisions for the recycle of organisms from the settler underflow to the reactor and for wastage of a portion of them to maintain a desired MCRT. It is this minimal combination which is the subject of this chapter.

16.1 DESCRIPTION OF ACTIVATED SLUDGE PROCESS

Since its inception by Ardern and Lockett [1] in 1914 the activated sludge process has grown in popularity until today it is the biochemical operation most widely used in wastewater treatment. During that time it has undergone much experimentation and modification so that today a number of variations are in use. The most popular ones will be emphasized in this chapter.

16.1.1 Physical Characteristics

GENERAL CHARACTERISTICS. Activated sludge reactors are generally open tanks around 15 feet deep constructed of concrete with vertical side walls, although steel and earthen basins are also sometimes used. Oxygen for the metabolism of the organisms is provided by any of a number of means. The oldest method is by diffused air, using either coarse or fine bubble diffusers. Mechanical surface aerators are also used, in both fixed and floating positions, as are units which combine mechanical turbines with diffused air. This traditional use of air as the oxygen source is why activated sludge reactors are usually called aeration basins. More recently, pure oxygen has been supplied in enclosed vessels, thereby greatly increasing the maximum rate at which oxygen can be transferred to the biomass. Mixing is generally accomplished by the same device that is transferring the oxygen, so that dual design criteria must be applied to the oxygen transfer system.

Two streams enter the reactor. One is untreated wastewater and the other contains the concentrated slurry of microorganisms which is being recycled from the final settler. The relative concentrations of soluble and particulate organic matter in the wastewater stream depend upon the character of the wastewater being treated and the type of treatment that it has received prior to reaching the reactor. The concentration of organisms in the recycle stream depends upon the concentration entering the settler and the rate of the recycle flow in proportion to the raw waste flow rate. The concentration of solids entering the settler, however, depends upon the MCRT of the system, the reactor space time, and the reactor configuration. Upon entering the reactor the microorganisms remove the soluble substrate by assimilation, thereby providing carbon and energy for their growth. Particulate organic matter is physically entrapped in the flocculent biomass where it is attacked by exocellular enzymes and solubilized to make it available for assimilation by the microorganisms. This mixture of microorganisms and undegraded particulate organic matter is called the mixed liquor suspended solids (MLSS) or activated sludge. Upon passing to the settler the MLSS move to the bottom for recycle to the reactor while the clarified effluent exits from the top. A portion of the MLSS must be wasted from the system to balance the growth of microorganisms and thereby maintain a constant MCRT. This wastage may be removed either from the settler underflow or directly from the reactor, but in either case the waste organisms must be disposed of. Two techniques which are used to prepare the organisms for disposal are aerobic and anaerobic digestion, discussed in Chapters 17 and 21, respectively.

The microbial characteristics of the MLSS were discussed in Section 7.4.2. To review briefly, the organisms primarily responsible for the degradation of organic matter are aerobic and facultative heterotrophic bacteria. Because of the selective pressures exerted by sludge settling the culture is flocculent with most organisms growing in large clumps or flocs. Higher organisms, such as protozoa, abound in the sludge, feeding upon the bacteria. Fungi are sometimes present, although they tend to be a nuisance organism because their filamentous morphology prevents the formation of dense floc, thereby reducing the settling velocity. The interactions among the species present in the population are not reflected in the mathematical models used for design. It is for this reason that the laboratory studies to determine the various kinetic parameters must simulate as closely as possible the environmental conditions anticipated in the final facility. Only by doing this can the engineer be confident that the population in the pilot culture is representative of the one that will develop in the full-scale facility. If the proper growth conditions exist in the reactor, autotrophic nitrifying bacteria will also be present. Since they do not use organic matter for carbon and energy they do not affect the primary role of the activated sludge process; however, they are responsible for the oxidation of ammonia and thus are important in determining the effluent quality. The emphasis in this chapter will be upon the heterotrophic organisms; however, if conditions exist which allow nitrification then the amount of oxygen supplied must be sufficient to meet the needs of both types of organisms. The method for determining the amount of oxygen needed by the nitrifiers will be discussed in Chapter 20.

PROCESS VARIATIONS.

Conventional Activated Sludge (CAS). Originally the activated sludge process employed long rectangular aeration basins with the influent and recycle introduced at one end and effluent removed at the other as shown in Fig. 16.1a. Consequently this is usually referred to as conventional activated sludge (CAS), although the reactor has also been called a flow-through aeration basin. Another version, shown in Fig. 16.1b, makes better use of space in the plant lay-out. The usual practice with CAS was to use air diffusers spaced evenly along one wall of each basin with

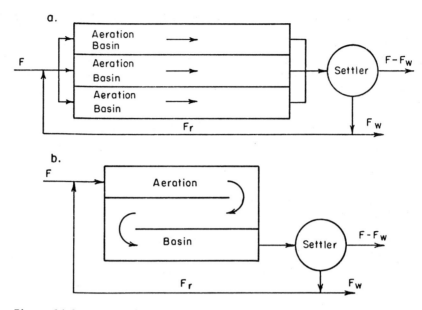

Figure 16.1. Two techniques for achieving flow-through
aeration as used in CAS and TAAS.

the same amount of air supplied to each diffuser. The rising bubbles imparted a
spiral flow to the liquid thereby minimizing longitudinal mixing. Although it was
felt that this would provide plug flow, tracer studies on full-scale installations
have shown the mixing pattern to be equivalent to three to five CSTR's in series [2].
Usually CAS is designed with a MCRT between 3 and 15 days, which with typical
domestic sewage gives a reactor space time (V/F) between 4 and 8 hours if the MLSS
concentration is around 2000 mg/liter. The recycle is generally between 10 and 30%
of the influent flow and sludge wastage is from either the settler underflow or the
aeration basin. Because of the mixing pattern in the reactor, the reaction environment changes from point to point with the result that more oxygen is needed near the
influent end of the basin where the substrate concentration is high than near the
effluent end where it is low. Early designs did not take this changing oxygen
requirement into account with the result that portions of the basin were oxygen
limited, thereby hurting process efficiency.

    Tapered Aeration Activated Sludge (TAAS). The first major variation in
activated sludge design came as a result of the recognition of the implications of
the variable nature of the oxygen uptake rate in the conventional process. Consequently, in 1936 Kessler *et al.* [3] proposed that the rate of oxygen supply be
matched with the oxygen requirement at different points along the reactor. Most
oxygen must be supplied at the influent end where substrate removal is rapid while
only small amounts are needed toward the effluent end where endogenous metabolism
and decay are the major reactions. Because the rate of oxygen supply diminished

gradually along the length of the basin, this modification was named tapered aeration. It is now the standard procedure whenever tanks of the flow-through type are used.

Step Aeration Activated Sludge (SAAS). The next major change in the activated sludge process came in 1942 when Gould [4] proposed the step aeration process as a means whereby more efficient utilization could be made of the aeration capacity in plants with conventional aeration basins. As shown in Fig. 16.2 the influent waste flow is split into several portions which are then fed into the aeration basin at different points, thereby giving a more even distribution of oxygen demand. The flow diagram shown in Fig. 16.2a is as Gould originally proposed it, with the initial portion of the aeration basin receiving only concentrated sludge from the final settler. The purpose of this was to regenerate the sludge prior to mixture with the incoming waste. Currently, most flow diagrams which are depicted as SAAS no longer have sludge reaeration, but instead mix the sludge and influent immediately as shown in Fig. 16.2b. SAAS was proposed as a way of achieving high degrees of treatment with short space times although it utilizes MCRT's within the same range as CAS. It also provides considerable operational flexibility and consequently is still in wide use today.

Contact Stabilization Activated Sludge (CSAS). The idea of reaeration of the sludge prior to mixing with the raw waste was carried one step further in 1951 with the introduction of the contact stabilization process [5]. As shown in Fig. 16.3 the underflow from the settler is pumped to a reactor (stabilization tank) prior to being mixed with the raw waste in the contact tank. The space time in the aerated contact tank is relatively small (0.5 - 2 hrs) so that the process relies upon the ability of the microorganisms to remove substrate rapidly to achieve a good quality effluent. After the short contact period the sludge is removed from the effluent,

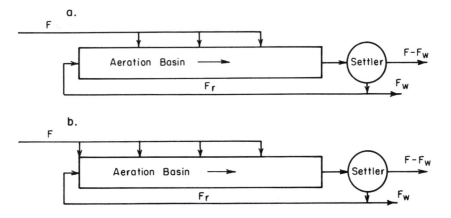
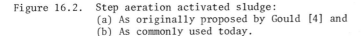

Figure 16.2. Step aeration activated sludge:
(a) As originally proposed by Gould [4] and
(b) As commonly used today.

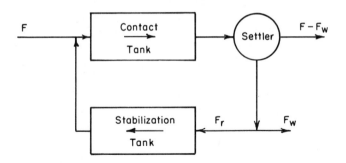

Figure 16.3.   Contact stabilization activated sludge (CSAS).

which is then discharged. The sludge, however, is returned to the stabilization
tank where it is aerated long enough to keep the system MCRT within the range of
3-15 days. This generally requires a reactor with a space time (based on F_r) of
less than 6 hours. Because the flow rate of the recycle stream is only 40 to 70
percent of the influent flow, the reactor volume required for reaeration is smaller
than a conventional system with equal space time. Consequently, even though contact
stabilization plants use two reactors, the total reactor volume is usually less than
a conventional plant with equal MCRT. As a result, it is often possible to increase
the capacity of a conventional plant without having to add any reactor volume,
simply by converting to the contact stabilization mode.

    The mechanism by which CSAS works has been the subject of much debate.
Originally it was thought that the rapid removal of substrate during the contact
period was due to adsorption of the organic matter onto the sludge particles,
leading to its also being called the biosorption process [5]. While this, together
with physical entrapment in the floc, no doubt contributes substantially to the
removal of colloidal and particulate substrate, it has little effect upon soluble
substrate [6,7]. One mechanism by which soluble substrates can be removed rapidly,
however, is through their incorporation into the cells as storage products, particularly glycogen and poly-β-hydroxybutyrate, and laboratory studies have shown that
this can occur [8,9]. Furthermore, when organisms are grown on a feed-starve cycle,
as they are in CSAS, replication of new cells can occur after the removal of soluble
substrate [10]. Consequently, CSAS has been described as a process in which rapid
substrate removal occurs in the contact tank due to storage, either between the
floc particles (when the substrate is colloidal or particulate) or internally (when
the substrate is soluble). Separation of the sludge from the wastewater stream,
therefore, removes the substrate from it even though the organic matter has not been
metabolized and converted into new cell material. Then when the sludge is reaerated
in the stabilization tank the storage products are metabolized and cell replication
occurs, preparing the organisms for more substrate storage in the contact tank.
Thus, this mechanism could be called the storage/metabolism mechanism.

While there is good evidence that the storage/metabolism mechanism is valid
for some substrates, particularly the simple carbohydrates and fatty acids with
which most research studies have been performed, the general applicability of the
mechanism is doubtful [11,12]. For example, Miller [13] was able to find no
evidence for the storage of metabolites during the contact phase while McCarty [14]
showed that rapid substrate removal could occur by growth alone and postulated that
the stabilization period was required to maintain the MCRT in a region conducive to
flocculation. Subsequently, Orhon and Jenkins [15] proposed a growth/decay model
to explain how CSAS works and postulated that the normal kinetic relationships for
substrate removal and cell growth could be used to explain the high rates of substrate removal observed in the contact tank. Furthermore, the phenomena of cell
death and decay could be used to explain the decline in the concentration of MLSS
that occurs in the stabilization tank. They were able to verify their basic concept
in lab- and pilot-scale reactors treating domestic sewage [15], and a subsequent
study, also utilizing domestic sewage, confirmed and extended it [16]. Thus, while
there is little doubt that the removal of some substrates can occur by storage with
subsequent metabolism in the stabilization tank, it appears that the growth/decay
mechanism is more general and can adequately describe the manner in which treatment
occurs during contact stabilization.

Completely Mixed Activated Sludge (CMAS). Conventional activated sludge and
its various modifications were able to meet most wastewater treatment needs until
the 1950's when the growth of the chemical process industry resulted in the generation of high strength wastes containing organic compounds not found in domestic
sewage. When CAS was used to treat some of those wastes the concentration gradient
inherent in it caused toxicity problems which made the achievement of a stable
process difficult. Thus in the late 1950's McKinney [17], in studies performed
concurrently with the ones by Garrett and Sawyer [18], put forth the idea of
microbial population dynamics as a key factor in the performance of the activated
sludge process. He reasoned that smooth operation could only be obtained if the
microbial population was maintained in a relatively constant condition, and that
this condition could be achieved only if the incoming wastes were mixed with the
entire contents of the reactor as they are in a completely mixed reactor. This
increase in process stability led to the rapid acceptance of CMAS by design
engineers so that by 1968 over 5000 plants were in operation. Completely mixed
conditions are easily obtained by the proper choice of aeration tank shape, method
of introducing the influent, and aeration equipment. Although this subject will be
considered in more detail in Section 16.3.4, Fig. 16.4 shows two common reactor
configurations. Figure 16.4a shows that a plant with a reactor configuration like
CAS can be converted to CMAS by distributing the influent and recycle flows evenly
along its length. Figure 16.4b is more indicative of the reactor configuration used
in new plants. Mechanical aerators are commonly employed because they give excellent mixing. The MCRT's, space times, and MLSS concentrations employed in CMAS

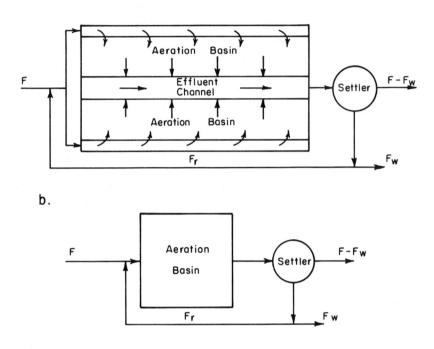

Figure 16.4. Two aeration basin configurations for completely mixed
             activated sludge (CMAS): (a) Cross-flow, commonly used
             to convert CAS to CMAS; (b) Square, sometimes used in
             new construction.

reactors are similar to those in CAS systems. The recycle flows are usually higher
though, being in the range of 35 to 100% of the influent rate. This is necessary
because the sludges produced in CMAS reactors often do not compact as well as
sludges from CAS reactors.

  Extended Aeration Activated Sludge (EAAS). The extended aeration version of
the activated sludge process was conceived to overcome two objections to the other
variations: (1) the large amount of waste sludge which must be disposed of and (2)
the need for close operational control of the process. The original idea was to
use a 24 hour space time and to waste no sludge from the reactor because under those
conditions the sludge production from normal domestic sewage would reach an equilibrium with that lost in the effluent when the MLSS concentration was between 5000
and 7000 mg/liter, thereby giving a MCRT between 20 and 30 days. Often however, the
concentration of suspended solids in the final effluent was excessive so that
periodic wasting of sludge was required. Consequently it is now common practice to
waste a small quantity of sludge periodically to maintain the MCRT between 20 and
30 days, and thereby prevent excess solids discharges in the effluent. The waste
sludge is usually well oxidized and can be disposed of directly without further

biochemical treatment. Most EAAS systems employ completely mixed reactors but the recycle ratios are generally higher (0.75-1.5) because of the higher MLSS concentrations. Many package plants employ this concept and thus it finds considerable use as the wastewater treatment system in small communities.

  Pure Oxygen Activated Sludge (POAS). Wastewater treatment engineers have long been interested in the use of pure oxygen as a way of transferring large amounts of oxygen to activated sludge reactors. When air is used as the carrier gas the relatively low partial pressure of oxygen, combined with a limitation on the mixing energy that the activated sludge floc particles can tolerate, places a severe restriction on the maximum rate at which oxygen can be transferred to the liquid. This, in turn, limits the substrate removal rate, $-r_s$, thereby restricting designs to a relatively narrow range. By using pure oxygen, the oxygen transfer rate can be increased by a factor of up to 5 for any given mixing energy input, thus enlarging the feasible design region. The impediment to the application of pure oxygen, however, was attainment of its economic use. If pure oxygen is just bubbled into the bottom of an open tank the amount escaping to the atmosphere is so great that the process is uneconomic. During the late 1960's however, Albertson *et al.* [19] thought of using a stage-wise operation as shown in Fig. 16.5. Three or four CSTR's in series are used, with influent and recycle to the first stage. Highspeed submerged turbines are used for mixing and gas transfer. Pure oxygen is introduced into the first stage, which is a closed vessel so that its off-gas, which is a mixture of oxygen and carbon dioxide, can be passed on to stage 2. This scheme is continued down the entire chain until almost all of the oxygen is used, thereby making very efficient

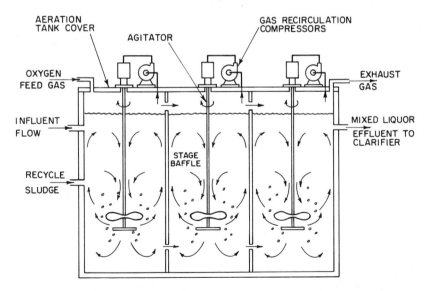

Figure 16.5.   Pure oxygen activated sludge.   (POAS).
              (Adapted from Albertson et al. [19]).

use of the oxygen generating equipment. It is also often necessary to scrub part of the CO_2 from the gas to prevent large drops in the pH of the liquid as it flows from stage to stage.

Pure oxygen activated sludge reactors commonly are covered tanks, usually of concrete, that are built with common-wall construction. Although submerged turbines were first used as the oxygen transfer devices, subsequent designs have used mechanical surface aerators as well. Because high oxygen transfer rates are possible, POAS systems use MLSS concentrations in the range of 4000-8000 mg/liter, with MCRT's in the range of 3-15 days. This results in reactor space times on the order of 2 hours for many wastes. The sludges compact well, however, so that even with the high MLSS concentration recycle flows are still between 25 and 50% of the influent flow.

16.1.2 Applications

With the exception of extended aeration activated sludge, the primary objective of the activated sludge process is the removal of soluble organic matter. Much of the particulate organic matter in wastewaters can be removed prior to the biochemical operation, and optimization studies have shown that more economic systems will result when this is done [20,21]. There will always be some particulate matter entering the process, however, and it will become entrapped in the flocculent sludge where it will undergo aerobic decomposition. As stated in Section 1.2.2 activated sludge should not be used when the concentration of soluble T_bOD is less than 50 mg/liter because physical-chemical means will be more economic. At the other extreme, if the concentration of soluble T_bOD is greater than 4000 mg/liter it is likely to be more economic to use an anaerobic biochemical operation. If an aerobic one is preferred, however, activated sludge can be used to treat wastes with soluble T_bOD's up to 15,000 mg/liter. Beyond that it will be difficult to perform cell recycle and a CMAL will have to be used instead.

Activated sludge can attain an effluent of high quality because the cells are removed from the wastewater prior to discharge and therefore it is often used as the primary means of wastewater treatment for both municipalities and industries. It does require a reasonably high degree of operator skill, however, and thus is not usually used in small communities where skilled operators are not available. With the exception of EAAS, high degrees of waste stabilization are not achieved in the activated sludge process and thus provision must be made within the rest of the wastewater treatment system for the treatment and disposal of the excess sludge produced. This limits the use of activated sludge to installations which have a complete wastewater treatment system, unless provisions can be made for treatment and disposal of the sludge elsewhere. Extended aeration activated sludge seeks to attain a high degree of stabilization by providing long MCRT's for decay of the resulting sludge. Economics usually limit its application to small facilities with low strength wastes, however.

16.1.3 Merits and Defects

The chief merit associated with activated sludge is that it produces an effluent of high quality at reasonable cost. Compared to a CMAL it is capable of achieving equal reductions in soluble substrate in reactors of much smaller volume, while producing an effluent relatively free of suspended solids. Furthermore, it is controllable, because through adjustment of the amount of sludge wasted the operator is able to regulate the MCRT to obtain the desired effluent quality. Activated sludge reactors are relatively resistant to shock loads and can achieve acceptable effluents in spite of dynamic inputs, although they do perform better under more stable conditions. Because their volumes are smaller than those of CMAL's, activated sludge reactors are less subject to atmospheric cooling and thus have a more uniform temperature. If the temperature does change, however, the MCRT may be adjusted to compensate for it.

The main defect associated with the activated sludge process is a direct result of its controllability, i.e., it is a complicated process which requires relatively sophisticated operation in order to achieve the desired results. It is, however, relatively sluggish in response to transient changes, which limits the applicability of some automatic control techniques. Capital and operating costs, though reasonable, are among the highest for biochemical operations, so that the decision to install an activated sludge system requires a considerable commitment on the part of an industry or municipality. The degree of this commitment has often not been recognized, with the result that most problems which are associated with the process are the direct result of inadequate operation.

In addition to the general merits and defects associated with activated sludge, each of the variations has special ones. Conventional activated sludge had such a poor distribution of oxygen that when the total amount of oxygen provided was equal to the total amount needed by the system, portions of the aeration basin were oxygen limited, causing process performance to suffer. On the other hand, when the oxygen transfer rate was set equal to the maximum utilization rate in the system, excess oxygen was supplied to parts of the reactor, making operation uneconomic. These defects were so great that plants are no longer built using equal distribution of oxygen throughout a flow-through basin. Instead, tapered aeration is used. TAAS systems have a substrate concentration gradient through them which is responsible for both a merit and a defect. The merit is that the gradient provides an ecological environment which is favorable to the growth of floc forming bacteria and unfavorable to filamentous forms [22,23]. As a consequence, sludges develop which settle and compact well. This merit is shared by pure oxygen activated sludge which uses tanks in series with feed to the first tank. In addition, the high dissolved oxygen levels in the POAS systems are apparently beneficial to the growth of floc forming bacteria because sludges from such systems tend to have excellent settling characteristics [24]. The problem associated with the substrate concentration gradient in a TAAS reactor is that it makes the system susceptible to upset by toxic materials. If a toxic component enters the reactor, the organisms are exposed to it at high concentration whereas in a more highly mixed basin it would be distributed throughout the basin, thereby reducing its concentration and lessening the effect. The problem is not as critical in POAS because each tank is completely mixed so that the highest concentration encountered is that of the effluent from the first reactor. The problem is also less severe in SAAS because the wastewater is introduced into the long reactor at several points, thereby reducing the maximum concentration that will be attained. The main benefit associated with SAAS, however, is that for a given MCRT the concentration of cells entering the final settler will be less than for a TAAS, POAS, or CMAS system of equal reactor volume. Thus if the settler in one of those systems was overloaded in thickening the problem could be alleviated by switching to the SAAS configuration. By the same token, if TAAS and SAAS systems were being designed with the same MCRT and the same MLSS concentration entering the settler, the SAAS reactor could be smaller, thereby reducing the capital costs, although the concentration of soluble substrate in the effluent may not be as low. Contact stabilization activated sludge carries this benefit even further, because recycled cells are aerated without being diluted at all. Consequently for a given MCRT the required reactor volume in a CSAS system will be less than the volume of any other activated sludge system. Again, this savings is usually accomplished at the price of more soluble substrate in the effluent, although the magnitude of the increase depends upon the characteristics of the wastewater being treated. One of the most popular activated sludge systems for industrial waste treatment is the completely mixed one because of its resistance to toxic shock loads. Because the entire reactor contents are at the concentration of the effluent, CMAS is also well suited for the treatment of wastes which are toxic or inhibitory at high concentration. Because the organisms are never in contact with the concentrated waste, no toxic effects can be exerted. The homogeneity of the tank contents also makes it amenable to automatic control of dissolved oxygen, pH, and nutrient addition and to modular installation, all factors which make it attractive for industrial uses. About the only disadvantage associated with CMAS is its tendency to produce a sludge which settles more slowly than that from reactors which have a substrate concentration gradient through them, a defect which it shares with SAAS and CSAS. A modification to the process has been suggested which would provide the concentration gradient when needed while still maintaining many of the advantages of CMAS [25]. Finally EAAS provides the added benefit of reduced quantities of excess sludge, although this is done at the expense of a greater oxygen requirement and an effluent which contains a higher concentration of suspended solids.

16.2 ACTIVATED SLUDGE PERFORMANCE

Examination of Figs. 16.1 through 16.5 reveals that all activated sludge systems
employ one or more reactors with cell recycle. Although CAS and TAAS are physically
built as long aeration chambers, their mixing patterns can be represented as 3 to
5 CSTR's in series [2], so that one method of predicting their performance would be
as a simple chain of CSTR's with cell recycle around the entire chain. SAAS also
uses a long aeration chamber but feed is introduced at a number of different points.
Thus, by extension of the above idea, this reactor could be modeled as several
CSTR's in series with feed distributed to each but with all recycle to the first.
Although early CSAS systems were designed with long aeration basins it is now common
practice to employ completely mixed reactors for both the contact and stabilization
tanks, and therefore CSAS could be modeled as two CSTR's in series with all recycle
to the first and all influent to the second. Both CMAS and EAAS utilize single
CSTR's with cell recycle, the only difference between the two being the range of
MCRT's employed. Consequently, both can be modeled easily. Finally POAS employs
several CSTR's in series with all feed and recycle to the first tank, and thus its
performance can also be predicted from a tanks in series model. From this examination of all the various reactor types it can be seen that the performance of all
of them can be estimated by models employing either CSTR's or CSTR's in series with
recycle of organisms. Accordingly, we can obtain a good understanding of how these
activated sludge systems perform by returning to the curves presented in Chapter 13.

As was noted in Chapter 13, the mean cell residence time is the parameter of
primary importance in determining the performance of reactors with cell recycle.
That is why the MCRT's of the various alternative activated sludge processes were
stressed in Section 16.1. Examination of the figures in Chapter 13 shows that the
theoretical performance curves were plotted with MCRT's ranging from 10 to 1000
hours. In Section 16.1, however, the range of MCRT's listed for all activated
sludge variations except EAAS was usually from 3-15 days (72-360 hours), considerably smaller than the range employed in Chapter 13. Thus before we investigate the
theoretical performances of the various activated sludge modifications let us see
why this constraint is applied to the MCRT.

16.2.1 Constraints on MCRT

As currently operated, activated sludge is dependent upon the development of a
flocculent mass of microorganisms which can be separated from the wastewater by
gravity settling. Although its mechanism is poorly understood, natural biological
flocculation occurs under specific growth conditions and the maintenance of those
conditions is imperative if the process is to function properly. Bacteria are
generally colloidal sized particles when growing as individual cells so that if
flocculation does not occur, their settling velocity will be so small that sedimentation will be impossible. Under that condition the concentration of bacteria in

the final effluent will be large, thereby causing the system to behave as a simple
CSTR, with all of its inherent disadvantages.

Several theories have been proposed to explain the mechanism of bioflocculation
and two appear to be particularly applicable to activated sludge [26,27]. One
theory states that protozoa play an active role in the flocculation of bacteria,
perhaps through the excretion of some substance into the medium [28,29]. The other
theory is based upon the excretion by the bacteria themselves of polymeric substances which form bridges between them causing agglomeration [30,31]. Even before
these theories were proposed, however, many workers had observed that flocculation
of activated sludge occurred only at MCRT's above a threshold value. There are at
least two possible reasons for this. One is that protozoa have lower values of μ_m
than bacteria so that relatively long MCRT's are required to establish them in the
activated sludge community [29]. The other involves the relative rates of biomass
and biopolymer production by a microbial culture [30]. According to this theory,
the polymeric substances are always excreted, but under conditions of prolific
growth new surfaces are generated faster than the polymer and thus little bridging
occurs.

The effects of the threshold MCRT on the settling characteristics of activated
sludge can be manifested in two ways: in the amount of dispersed growth in the
culture (the clarification capacity) and in the zone settling rate of the concentrated slurry (the thickening capacity). The first way will influence directly the
quality of the effluent which can be achieved because it will determine the concentration of suspended solids in it. The second will determine the effectiveness of
the final clarifier in thickening the sludge for recycle, thereby influencing the
economy of the system. Bisogni and Lawrence [32] measured the percent of dispersed
cells in cultures grown in CSTR's with cell recycle at various MCRT's and the
results are shown in Fig. 16.6 where it can be seen that the threshold MCRT is

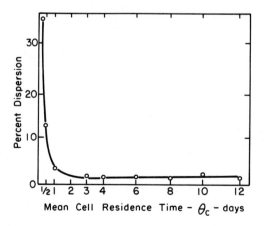

Figure 16.6.   Effect of MCRT on the amount of dispersed
               growth in activated sludge effluent. (From
               J. J. Bisogni and A. W. Lawrence, *Water
               Research*, 5, 753-763, 1971 [32].
               Reproduced by permission of Pergamon Press Ltd.).

around 2 days. Observations of the suspended solids in the effluent from a full
scale treatment plant support this finding [33]. Two types of results have been
presented to characterize the effects of MCRT on the thickening characteristics of
activated sludge. Ford and Eckenfelder [34] investigated the effects of MCRT upon
the sludge volume index (SVI), a parameter which gives some indication of the degree
of compaction that can be obtained in the final settler. Their results indicate a
deterioration in compactability at MCRT's below 3 days. Similar results have been
reported by Magara *et al*. [35]. Bisogni and Lawrence [32] sought to obtain information which would be useful in the application of thickening theory so they
measured the effect of MCRT on the zone settling velocity of activated sludge at a
concentration of 2000 mg/liter. Their results are presented in Fig. 16.7 and show
clearly that an increase in the MCRT causes an increase in the settling velocity.
At MCRT's below 2 days the large quantity of dispersed growth prevented an accurate
determination of the settling velocity. Thus on the basis of both of these factors
it appears that the MCRT should be at least 3 days (72 hours) to avoid undue
problems in either clarification or thickening of the activated sludge.

The upper limit of approximately 15 days imposed upon the MCRT is also a result
of sludge settling considerations. Once that limit is passed, there is sometimes a
deterioration in the settling characteristics of the culture as a result of very
small floc particles, called pin floc [36]. These should not be confused with
dispersed growth, which is the result of insufficient polymer production, because it
is possible that pin floc is caused by the presence of excess biopolymer which
restabilizes the colloid [37]. This condition has also been attributed to excess
decay of the cells at long MCRT's [34], although further studies are needed to
elucidate the mechanism. In addition to pin floc, long MCRT's may also act to
decrease the compactibility of the sludge [34]. In spite of these potential

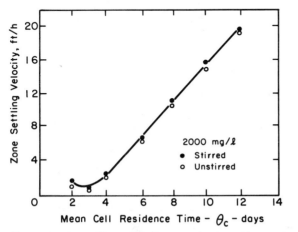

Figure 16.7.   Effect of MCRT on the settling velocity
of activated sludge. (From J. J. Bisogni
and A. W. Lawrence, *Water Research*, 5,
753-763, 1971 [32]. Reproduced by
permission of Pergamon Press, Ltd.).

problems, MCRT's in excess of 15 days are often required, particularly, in industrial wastewater treatment. In those situations the designer should give special
attention during treatability studies to the settling characteristics of the sludge
so that special provisions can be made in the design of the settler if the need
exists.

16.2.2 Theoretical Performance
Although it is one of the newest variations of the activated sludge process, CMAS
is the easiest to understand, thus let us start with it.

   COMPLETELY MIXED ACTIVATED SLUDGE (CMAS). Since CMAS is just a CSTR with cell
recycle its theoretical performance can be depicted by the curves in Figs. 13.2,
13.3, and 13.4. Those curves have no constraints upon the MCRT, however, so bounds
must be placed at 72 and 360 hours (3 and 15 days) in order to get a true picture
of CMAS performance. Accordingly, Fig. 13.2 has been reproduced as Fig. 16.8, upon
which the MCRT bounds are superimposed. Examination of that figure shows that the
soluble substrate concentration is relatively independent of MCRT within the practical limits. In fact, for a waste with parameters like those used to generate the

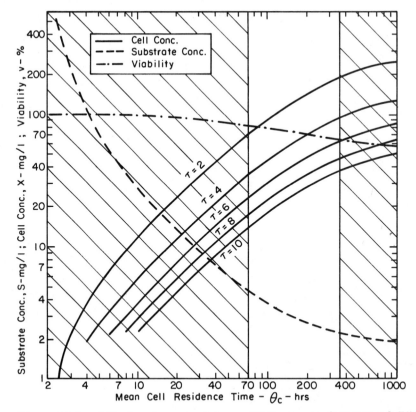

Figure 16.8. Effects of constraints on MCRT upon performance of CMAS.
              Compare with Fig. 13.2.

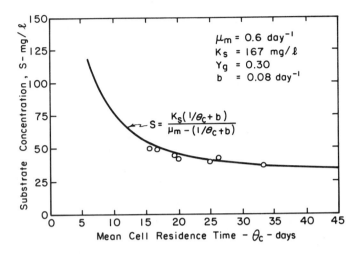

Figure 16.9.   Effect of MCRT on effluent COD from a full-scale
CMAS reactor.   *(Adapted from Campbell and Rocheleau [38])*.

curves in Fig. 16.8 it is unlikely that any effect of MCRT upon S could be observed
due to the normal variability associated with BOD or COD determinations. That this
occurs in practice is shown in Fig. 16.9 where the soluble COD in the effluent from
a 1.7 MGD (0.074 m^3/sec) CMAS system treating a plastics manufacturing waste is
plotted as a function of the MCRT. Even though the values plotted are COD, and not
T_bOD, it can be clearly seen that there was little change in S over the broad range
of MCRT's investigated. Wastes which are more difficult to degrade will show more
variation in S but in many cases the MCRT used in a design will be governed by
sludge settling considerations rather than by soluble substrate removal. Once the
MCRT is made large enough to obtain sludge that settles well the soluble substrate
concentration is likely to be considerably below the effluent discharge standard at
which the design is aimed. If a waste is so difficult to degrade that the effluent
soluble T_bOD does not meet the effluent standard at a MCRT of 15 days then consideration should be given to the use of tanks in series or special provisions should be
made for clarifying the final effluent.

The MLSS curves in Fig. 16.8 emphasize the point that for a given wastewater
the MLSS concentration will depend upon both the space time and the MCRT. The
generality of the effect of MCRT at a fixed τ is indicated by Fig. 16.10 which contains data from a lab-scale reactor treating a soluble, complex substrate [39].
Although τ may be chosen by the designer, consideration must be given to the MLSS
concentration that will result. The factors which must be considered when making
that decision will be discussed in Section 16.3.3. Not all of the cells in the
mixed liquor are viable, as indicated in Fig. 16.8. The validity of this trend has
been confirmed in both pilot- [40] and lab- [41] scale studies. Consequently, one

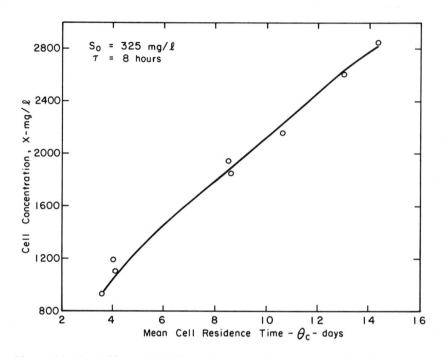

Figure 16.10.   Effect of MCRT on microorganism concentration in
                 aeration basin. *(Adapted from Stall and Sherrard [39]).*

result of increasing the MCRT is a reduction in the percentage of cells within the
process that actually contribute to substrate removal.

     Since for many wastewaters the MCRT has little effect upon the soluble substrate
concentration in the effluent, the decision regarding a design value is often based
upon the effects that the MCRT has upon the excess sludge production and the oxygen
requirement. These were presented in Fig. 13.3, which is reproduced in Fig. 16.11
with the constraints applied to the MCRT. For that example, changes in MCRT over
the practical range can cause as much as a 47% reduction in the amount of sludge
to be disposed of while increasing the oxygen requirement by as much as 42%.
Excess sludge production is proportional to the observed yield (compare Eqs. 13.39
and 13.51) and thus we would expect similar changes in that parameter as the MCRT is
changed. Such changes have been observed [39,42,43]. The oxygen requirement must
increase as the MCRT is increased because more of the carbon which had been stored
in the biomass is being oxidized to CO_2, and this is a common observation in practice.
This inverse relationship between sludge production and oxygen requirement suggests
a method of choosing the MCRT when S is relatively constant over the practical range,
i.e., to minimize the total cost of sludge disposal and oxygen supply.

     Figures 13.2 and 13.3 showed that the recycle ratio, α, had no effect upon the
performance of a CSTR with recycle. Thus during design of a CMAS system α may be
selected on the basis of economics without regard to any effects upon process

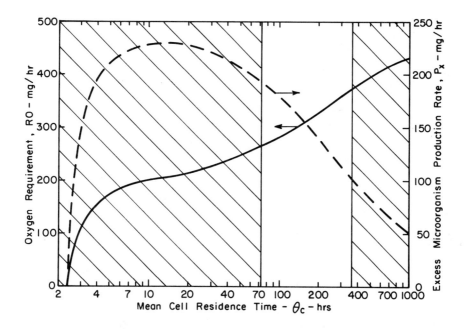

Figure 16.11. Effects of constraints on MCRT upon performance of CMAS. Compare with Fig. 13.3.

performance. The only thing that α will influence directly is the volume of sludge which must be wasted from the settler underflow to maintain a constant MCRT. If sludge is wasted from the reactor no consideration need be given to α in determining the wastage rate required, as shown in Fig. 13.4.

Because CMAS behaves like a single CSTR with cell recycle the other factors discussed in Section 13.1.4 apply here as well, and thus there is no need to repeat all of them. Two points should be emphasized however. The first is that the minimum MCRT will seldom be of practical significance since for most wastewaters it will be well below 3 days, the minimum practical value. Nevertheless the ratio of θ_c to θ_{cmin} can be used as an indication of the stability of the process. The second is that inert solids will build up in the system as shown in Fig. 13.7 and thus their presence must be accounted for during design.

EXTENDED AERATION ACTIVATED SLUDGE (EAAS). EAAS also uses a single completely mixed reactor with cell recycle and thus its theoretical performance can also be depicted by Figs. 13.2 and 13.3. In this case, however, the normal range for the MCRT is 480 to 720 hours (20 to 30 days). Within that range it can be seen that the soluble substrate concentration will approach S_{min}, the lowest concentration that can be achieved in a CSTR. For easily degraded wastewaters this value will be very small so that essentially all soluble substrate is removed. The reason for

utilizing EAAS, however, is to reduce the amount of excess sludge produced, and
Fig. 13.3 shows that this is accomplished at long MCRT's, thereby requiring a large
amount of oxygen. The long MCRT reduces the viability of the culture as well,
however. The general trends depicted by the graphs have been substantiated in
practice although there are significant quantitative deviations with respect to the
reductions in excess sludge production to be expected. These will be discussed in
Section 16.2.3. It should be noted, however, that the MCRT range employed in EAAS
is outside of the range associated with good thickening and clarification. Thus
difficulties are likely to be encountered in sludge removal and recycle [44].
Consequently, special consideration should be given to those factors during treatability studies for design.

TAPERED AERATION AND PURE OXYGEN ACTIVATED SLUDGE (TAAS & POAS). Both TAAS
and POAS can be considered to be CSTR's in series with all influent and recycle to
the first tank so that their theoretical performance can be depicted by Figs. 13.9
through 13.15. Examination of those figures shows that in comparison to CMAS the
only significant effect caused by the series reactor configuration is a reduction
in the concentration of soluble substrate in the effluent. In reality, however,
when the MCRT is constrained within the practical limits of 72 to 360 hours,
significant differences in soluble substrate concentration are difficult to detect
[45]. Figure 13.9 also indicated that τ exerts an effect on effluent quality in a
tanks in series system but again it is so small that it would be difficult to
observe in practice so that substrate removal in TAAS can be considered to be
independent of τ as long as the MCRT is held within the normal range [45]. Figure
13.12 showed that more oxygen is needed in the first reactor of a chain or at the
influent end of a long aeration basin. The generality of this requirement has been
accepted for many years, and is, in fact, the basis of TAAS [3]. The major difference between CMAS and TAAS is not depicted in the performance curves at all because
it is in the settling characteristics of the sludge, which have been shown to be
better for the tanks in series configuration [22,23]. This has been attributed to
the development of a different microbial population in the staged system, caused by
the substrate concentration gradient through the tanks. This can be of significant
practical importance, particularly if the wastewater is of the type which favors the
growth of filamentous organisms.

The curves in Figs. 13.9 through 13.15 also give a good indication of the
qualitative nature of POAS performance. It will be recalled from Eq. 13.38 that the
product of X and τ is fixed by the MCRT for a particular wastewater undergoing treatment in CMAS, and thus that one of them may be chosen freely, thereby fixing the
value of the other. It was also seen in Chapter 13 that $X\tau$ for tanks in series is
the same as for a single CSTR with recycle at the same MCRT and that the changes in
X from tank to tank in a chain are insignificant. Furthermore, the discussion in
the preceding paragraph showed that the theoretical influence of τ upon system

performance is of little practical significance. Consequently, the implication of
Eq. 13.38, that either X or τ may be chosen freely, also applies to POAS. As we will
see in Section 16.3.3 one factor which must be considered when choosing X or τ is
whether the rate of oxygen transfer possible within the resulting reactor volume is
sufficient to supply the needed oxygen. An advantage of using pure oxygen is that
for a given mixing energy the rate of oxygen transfer can be as much as five times
greater than it would be with air, implying that the reactor volume (and τ) may be
much smaller, thereby reducing the capital cost for tankage. Whether it will allow
the design of a less expensive system depends, of course, on many factors. The
significant point, however, is that the only certain advantages of POAS over an air
system with the same reactor configuration are its ability to transfer oxygen
efficiently to smaller liquid volumes and its capability for maintaining higher
concentrations of dissolved oxygen in the liquid. Nevertheless a number of other
benefits have been claimed for POAS and it would be beneficial to consider them
briefly.

In a review article, Boon [46] cited seven sources as attributing the following
benefits to POAS: (1) increased rate of treatment, (2) increased density of sludge
and rate of sludge settlement, and (3) reduced rate of sludge production. The first
is easy to understand. The rate of substrate removal is the product of the specific
rate and the cell concentration (Eq. 10.41). By being able to use a larger value
of X, POAS is able to increase $-r_s$. The better sludge settlement could be due to
the tanks in series configuration and to the fact that the lower mixing energy
levels in the pure oxygen system produce a denser sludge with better settling
characteristics. The third possible benefit is more difficult to explain, and yet
is potentially quite significant. In Chapter 13 comparison of Figs. 13.3 and 13.11
showed that for a given MCRT, the amount of sludge produced by cell growth in the
tanks in series configuration is essentially the same as that in a single CSTR.
Thus the decreased sludge production cannot be attributed to the reactor configuration. The only way that the pure oxygen system could produce less sludge would be
if the kinetic parameters for it were different. One study [47], run under carefully
controlled laboratory conditions showed no differences in the kinetic parameters
when the dissolved oxygen concentrations were kept the same. Nevertheless several
studies in systems oxygenated with air and pure oxygen showed that the oxygen system
produced less sludge when the systems were grown at the same MCRT and thus the
benefit appears to be real [46,48,49]. The reason lies in the ability of POAS to
maintain a higher concentration of dissolved oxygen in the medium [50]. Because the
microbial population in activated sludge is flocculent, materials must reach the
cells in the inner regions of the floc particles by diffusion. Consequently, high
concentrations of dissolved oxygen could allow deeper and more rapid penetration of
oxygen into the floc, thereby removing an oxygen limitation on the inner cells [50,51].
This could have the effect of increasing the effective specific decay rate, b,

thereby allowing a greater amount of decay at a given MCRT and causing a reduced
excess microorganism production rate, P_x. When air and pure oxygen systems are
operated at the same dissolved oxygen level, no differences are observed [50].
Consequently, it has been recommended that all activated sludge systems be operated
with a minimum dissolved oxygen level of 2 mg/liter [50]. This points out the
necessity for performing treatability studies in the environment anticipated in the
final facility. Because factors such as dissolved oxygen concentration and mixing
energy can affect the apparent kinetic characteristics of the sludge, careful
consideration should be given to them. If this is not done, the resulting parameters
will not adequately reflect the true nature of the system.

In summary, the curves in Figs. 13.9 through 13.15 reflect qualitatively the
anticipated response from a TAAS or POAS system. The use of a tanks in series
configuration will allow more removal of soluble substrate than in a CMAS system
with equal MCRT, but unless the waste is very difficult to degrade the improvement
will not be significant. Other than that, the tanks in series configuration has
little effect upon performance although it does often produce a sludge which settles
better. The use of pure oxygen will allow more cell decay to occur than in a
comparable system utilizing air because the concentration of dissolved oxygen can be
kept higher. This will happen no matter what the reactor configuration, however.

STEP AERATION ACTIVATED SLUDGE (SAAS). SAAS was defined in Section 16.1.1 as
a long narrow basin with feed distributed to several points within it. Since the
hydraulic regime of a long narrow basin is equivalent to four CSTR's in series,
SAAS can be modeled as four CSTR's in series with feed distributed to each. Because
equal feed distributions are commonly employed, the theoretical curves in Figs.
13.16 through 13.21 can be used to gain an understanding of the way in which SAAS
performs. Again, only MCRT's between 72 and 360 hours are used in practice so
consideration of the graphs should be limited to that region. From the standpoint
of substrate removal, SAAS will perform very much like CMAS as long as no more than
one-fourth of the feed is added to the last reactor, and any differences between the
two would be difficult to detect with current methods. Figure 13.16 indicates that
τ will affect the soluble substrate removal, whereas it has no effect in CMAS.
Within the normal range of MCRT's, however, the effect will be small and of little
practical significance. The recycle ratio will have a much larger effect upon the
system, however, influencing both the cell and substrate concentrations in the last
reactor. This suggests that α is a more important parameter in the design of SAAS
than in the design of CMAS. The reason for this can be seen in Fig. 13.20, and was
discussed in Section 13.2.4. In general, the larger α is, the more the performance
of SAAS will approach that of CMAS, i.e., the more uniform the various concentrations
will be from tank to tank.

Of particular significance to the design of a SAAS system is the fact that the
excess sludge production rate and oxygen requirement (Fig. 13.18) are the same as
those for CMAS (Fig. 16.11) over the practical range of MCRT. This means that the

mass of sludge contained within a SAAS system will be the same as that in a CMAS system treating the same waste at the same MCRT. Thus if the two systems are designed to give the same MLSS concentration entering the settler, the total reactor volume of the SAAS system will be smaller, which could reduce the capital costs of the facility. Conversely, if the two systems are designed with the same volume reactors, SAAS will have a lower concentration of MLSS entering the final settler, which could reduce the size of settler required, also lowering the capital cost.

Compared to CAS or TAAS, SAAS will have a more uniform oxygen requirement from tank to tank, (compare Fig. 13.19 with Fig. 13.12). Substrate removal performance will not be quite as good, but for most wastewaters the effect will be small within the practical range of MCRT's. For wastewaters which are more difficult to degrade, a better effluent can be achieved by reducing the quantity of feed added to tank 4 in relation to the others. In this way many of the benefits associated with CMAS can be retained, while effluent quality can be improved. This operational and design flexibility gives SAAS a great potential for future development which has been essentially unexploited to date, probably because rather sophisticated optimization techniques are required to take full advantage of it. Perhaps as more wastewater treatment engineers employ optimization this variation will find more application in practice.

CONTACT STABILIZATION ACTIVATED SLUDGE (CSAS). The mechanisms responsible for substrate removal in CSAS were discussed in Section 16.1.1 where it was concluded that the growth/decay mechanism can adequately describe the performance for most wastewaters. This suggests that the tanks in series model of Section 13.2.5, in which all feed is to reactor 3 and all recycle to reactor 1, can describe qualitatively the performance of CSAS, since decay will be the predominant reaction in tanks 1 and 2 and growth will be the major one in tanks 3 and 4. Therefore examination of Figs. 13.22 through 13.27 will give a good indication of how CSAS responds to various operational parameters. Again, because of limitations on cell separation, the MCRT must be constrained to the region between 72 and 360 hours.

The most significant point to note when examining Figs. 13.22 through 13.27 is the important role that the recycle ratio, α, plays in determining system performance. This importance has been confirmed in lab and pilot plant studies [15,52] and is the key to the operational flexibility inherent in this process. The space time in the contact tank is also important; in fact, even more so than is indicated in Fig. 13.22. The theoretical substrate concentration curves are lower than they would be in a typical CSAS system because most plants employ only a single CSTR for the contact tank. Thus in general, CSAS will produce an effluent higher in soluble substrate than CMAS with the same MCRT [16]. Like SAAS, Fig. 13.24 shows that over the practical range of MCRT the excess microorganism production rate and oxygen requirement are essentially the same as in CMAS and this has been confirmed experimentally [16]. Furthermore, experimental observations also confirmed the inverse relationship between sludge production and oxygen requirement [16]. Since the mass of sludge contained within a CSAS system is the same as that in a CMAS system treating the same waste at the same MCRT, the total tank volume needed to give the same MLSS concentration entering the final settler will be smaller in a CSAS. This is because of the high concentration in the stabilization tank, as seen in Fig. 13.25. Consequently, it is usually more economic to design a CSAS system than a CMAS system, particularly when the waste is easy to degrade so that the MCRT is controlled by sludge settling requirements rather than by substrate removal.

In general, the trends presented in Figs. 13.22 through 13.27 are supported by observations in the literature. For example, the oxygen requirement in the contact reactor will be due primarily to substrate removal while that in the stabilization reactor will be due to decay [16]. If the reactors are of equal volume the amount of oxygen required in the contact tank will be higher, as indicated in Figs. 13.25 and 13.26. However, if the size of the stabilization tank is made much larger than the contact tank, then the difference between the two requirements will be less. It should be realized that the curves presented in Section 13.2.5 were for reactors of equal volume. Changes in the relative volumes of the reactors will have effects upon performance not depicted by the changes in τ_c in the figures, because they will change the relative times available for substrate removal and decay [15,16,52]. The recycle ratio can also be used to change those times and that is why it has the large effect noted in the diagrams. In summary, it can be concluded that CSAS performance is influenced by three process variables: the mean cell residence time, the recycle ratio, and the relative masses of sludge in the two reactors [16]. This is in contrast to CMAS which is influenced by only the MCRT. Thus, like SAAS, CSAS design has the potential for optimization although few such studies have yet been performed.

16.2.3 Inadequacies of Theory

When the equations of Chapter 12 were applied to the prediction of CMAL performance in Chapter 15 several differences between theory and reality were noted. Because the same kinetic expressions were used to develop the theoretical equations in Chapter 13, many of the inadequacies listed in Section 15.2.2 apply here as well and thus the reader should review that section. A few, however, do not. In this section we will list only those inadequacies which are of particular importance, or which do not apply to CSTR's with cell recycle.

A major difference noted in Section 15.2.2 was the approximately linear dependence of effluent substrate concentration upon the influent concentration. Such a dependence has also been observed in reactors with cell recycle [53,54], but should pose no difficulties as long as laboratory studies for the determination of parameter values are made using waste of the same concentration as that anticipated in the full-scale installation. It was also noted in Section 15.2.2 that the assumption of equal decay rates for live and dead cells could lead to an overestimate of the amount of sludge lost to decay at very long MCRT's. Since most activated sludge systems

operate at MCRT's less than 360 hours, any error due to this will be minimal. Such
errors may become quite important, however, during application of the equations to
EAAS because it is normally operated at very long MCRT's [55]. No completely satisfactory technique is available to overcome this problem. The most pragmatic
approach may be to apply techniques similar to those presented in Chapter 17 for
aerobic digestion, because EAAS is simply a process in which soluble substrate
removal and aerobic digestion are carried out simultaneously. It was also stated
in Section 15.2.2 that algal growth could cause the equations to overpredict the
amount of solids destruction occurring in a CMAL. This is not a problem in activated
sludge systems, however, because the small reactor surface area and highly turbid
liquid prevent significant light penetration for algal growth.

Two problems were listed in Section 15.2.2 concerning the application of the
CSTR equations to CMAL's in series. These dealt with substrate removal and cell
decay. The comments about soluble substrate removal are equally applicable here
and help explain the lack of observed difference between TAAS and CMAS systems [45].
The comment that the decay rate will decline as cells move down a chain is not true
in a system with recycle as long as the MCRT is large with respect to τ_T because the
recycle keeps the cell mass relatively homogeneous and prevents any large differences
in viability among the tanks. There will be some difference in viability between
the contact tank and the stabilization tank in CSAS [15] but it will not be large
enough to cause any problems with the models.

The most important possible weakness of the model when applied to CSAS concerns
the removal of soluble substrate. As discussed in Section 16.1.1 two mechanisms
have been proposed to explain how soluble substrate is removed in CSAS. The
growth/decay mechanism can be considered to be more general in nature than one based
on storage/metabolism. If the latter mechanism is operative for a particular wastewater, but the model is based on the growth/decay mechanism, the result is likely to
be an underestimation of system performance, thereby yielding a conservative design.
In the opposite situation, however, the result would probably be an overestimation
of performance, leading to an inadequate design. Thus in the absence of concrete
evidence concerning the mechanism in a particular wastewater, the use of the growth/
decay model would be preferable.

Another possible deviation in the application of models to CSAS occurs in the
prediction of the oxygen requirement in each tank. Gujer and Jenkins [16] found
that the oxygen equivalence (β) of microbial cells in the contact tank was different
from that of the cells in the stabilization tank. Thus the values should be measured
during treatability studies if possible.

As stated in Section 15.2.2, while these inadequacies of theory should be
recognized and accounted for, they do not detract from the usefulness of the
equations in design. When treatability studies are performed to evaluate the
kinetic parameters, reactors should be run at MCRT's bracketing the conditions

anticipated in the final installation. The space times employed are immaterial. Because curve fits are performed to estimate the kinetic parameters, the performance equations become tools for interpolating between experimental conditions. For this they are well suited. The danger of error arises when the equations are used to extrapolate over a range of conditions outside of that covered in the treatability studies.

16.2.4 Modification of Models

No modifications need be made to the model in order to apply it to the design of CMAS. Unless the value of the decay rate constant, b, is measured over the range of MCRT's anticipated in practice it may be necessary to modify the model in a manner similar to that discussed in Chapter 17 before it is applied to EAAS. Because that modification will be discussed in Section 17.2.3 it will not be considered further here. The equations from which the performance curves for TAAS, POAS, SAAS, and CSAS were obtained had to be solved simultaneously using a computer. Because explicit solutions are not possible, the easiest way to design a TAAS, POAS, or SAAS system is to use a computer code to simulate the performance of a number of alternative systems and then choose the one which does the best job of meeting the performance objectives. By making some simplifying assumptions it is possible to reduce the complexity of the equations, making it possible to solve them with a calculator. However, the large number of equations involved makes this a very tedious procedure and thus the use of a code is recommended. As pointed out in Section 16.1.1 CSAS usually employs only two tanks, one for contact and one for stabilization as shown in Fig. 16.3. This reduces the number of equations and makes it possible to perform a direct design after making some simplifying assumptions.

Important to direct solution of the CSAS equations is the fact that the viability of the culture depends only upon the MCRT and is relatively independent of α and τ. Furthermore, the viability was almost exactly the same as in CMAS when the MCRT was the same, as was seen in Section 13.2.5 when the viability curves in Figs. 13.22 and 13.23 were compared with the curve in Fig. 13.2. Thus the assumption will be made that the viability of the cells in the CSAS contact tank can be approximated by the viability in a CMAS reactor of equal MCRT. A second assumption that facilitates a direct solution is that no substrate removal occurs in the stabilization tank [16]. While this is not actually true, the assumption will introduce little error as long as the system is being designed to give a low effluent soluble substrate concentration because then the amount of substrate entering the stabilization basin will be quite small. Using those assumptions and the nomenclature in Fig. 16.12, let us now develop explicit design equations for CSAS. The technique used is similar to that of Orhon and Jenkins [15], modified to include cell viability. Benefield and Randall [56] have also presented equations for CSAS using a somewhat similar procedure. The differences between the two approaches have been detailed elsewhere [57].

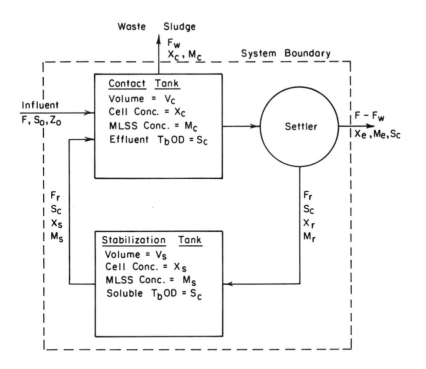

Figure 16.12. Schematic diagram of contact stabilization activated
sludge showing nomenclature.

The concentration of soluble substrate in the fluid leaving the contact tank
is controlled by the specific growth rate of the organisms in it while all particulate substrate is removed through entrapment in the biological floc. The settling
characteristics of the organisms in the system, however, are determined by their
overall net growth rate, which is the reciprocal of the MCRT. Therefore, the
contact tank may be designed to give the desired effluent soluble substrate concentration while the stabilization tank may be sized to produce the MCRT required to
make the sludge settle. Unlike CMAS in which S and the MCRT are intimately related,
CSAS allows that relationship to be broken so that an effluent of any quality may
be produced, even though the MCRT is maintained at a value which will produce a
sludge that settles well. This allows CSAS to achieve partial treatment of a wastewater, something that cannot be accomplished with most other activated sludge systems.

The effluent soluble substrate concentration and the specific growth rate of the
organisms in the contact tank are related by the Monod equation:

$$\mu_c = \frac{\mu_m S_c}{K_s + S_c} \tag{10.42}$$

where the subscript c refers to contact tank. Thus once S_c is selected, μ_c is fixed.

By assuming that no substrate is removed in the stabilization tank, so that the concentration of substrate in the recycle stream is S_c, a mass balance on soluble substrate in the contact tank leads to

$$X_{vc} V_c = \frac{FY_g (S_o - S_c)}{\mu_c} \qquad (16.1)$$

Defining the viability of the cells in the contact tank as:

$$v_c = \frac{X_{vc}}{X_c} \qquad (16.2)$$

the equation becomes:

$$X_c V_c = \frac{FY_g (S_o - S_c)}{v_c \mu_c} \qquad (16.3)$$

Thus once v_c and S_c are known, either X_c or V_c may be selected at will, thereby fixing the other one. Usually X_c is chosen to give good settler performance. The viability may be estimated by the equation for CMAS after the MCRT has been selected:

$$v_c = \frac{1 + b\theta_c}{1 + b\theta_c + \gamma\theta_c} \qquad (13.40)$$

Now we must derive the equations which will allow us to determine the size of the stabilization basin required to achieve a desired MCRT. The derivation will be made by assuming a perfect settler (i.e., one in which X_e is zero), although in practice some solids will be lost in the final effluent. This unintentional solids loss may be compensated for by reducing the rate of sludge wastage and therefore does not affect any equations except the ones for F_w. Solids wastage will be assumed to be from the contact tank, although where appropriate, equations will also be given for wastage from the settler underflow. The MCRT will be defined by:

$$\theta_c = \frac{V_c X_c + V_s X_s}{F_w X_c} \qquad (16.4)$$

Furthermore Γ and λ will be defined as the fraction of cells in the contact and stabilization tanks, respectively:

$$\Gamma \equiv \frac{V_c X_c}{V_c X_c + V_s X_s} \qquad (16.5)$$

$$\lambda \equiv \frac{V_s X_s}{V_c X_c + V_s X_s} \qquad (16.6)$$

By definition,

$$\Gamma + \lambda = 1.0 \tag{16.7}$$

Perform a mass balance on cells within the dotted boundary of Fig. 16.12. The following reactions are assumed to occur: viable cells are formed by growth and lost by death and decay in the contact tank; viable cells are lost by death and decay in the stabilization tank; dead cells are formed by death of viable cells and lost by decay in both the contact and stabilization tanks. It is assumed that no reactions occur in the settler. The mass balance can be rearranged and simplified by invoking the definition of the total cell concentration (Eq. 12.10), giving:

$$F_w X_c = \mu_c X_{vc} V_c - b X_c V_c - b X_s V_s \tag{16.8}$$

Substituting the definitions of Γ and v_c yields:

$$\frac{F_w X_c}{V_c X_c + V_s X_s} = \Gamma \mu_c v_c - b \tag{16.9}$$

or

$$1/\theta_c = \Gamma \mu_c v_c - b \tag{16.10}$$

But λ is equal to $1 - \Gamma$ so that Eq. 16.10 can be rearranged to give:

$$\lambda = \frac{\mu_c v_c - 1/\theta_c - b}{\mu_c v_c} \tag{16.11}$$

Thus, once μ_c and v_c are known, and the desired MCRT is chosen, then λ, the fraction of cells in the stabilization tank can be calculated. The definition of λ may be rearranged to give V_s, the stabilization tank volume:

$$V_s = V_c [\lambda/(1 - \lambda)](X_c/X_s) \tag{16.12}$$

However, before it can be used, the ratio X_c/X_s must be known. That ratio can be found by combining a mass balance on cells in the settler with a mass balance on cells in the stabilization tank. After using the definition of the MCRT to remove F_w, and after substitution of the definition of λ, the following expression is obtained:

$$\frac{X_c}{X_s} = \frac{F_r}{F + F_r - [V_c/(1 - \lambda)][(1/\theta_c) + b\lambda]} \tag{16.13}$$

Substitution of Eq. 16.13 into 16.12 gives:

$$V_s = \frac{F_r V_c [\lambda/(1 - \lambda)]}{F + F_r - [V_c/(1 - \lambda)][(1/\theta_c) + b\lambda]} \tag{16.14}$$

Thus the recycle flow F_r, must be selected before V_s can be calculated. Once V_s has been determined, however, X_s may be calculated from the definition of λ:

$$X_s = [\lambda/(1 - \lambda)](V_c/V_s)X_c \qquad\qquad (16.15)$$

So far we have equations which allow determination of V_c or X_c, V_s, and X_s for a desired S_c, θ_c, and F_r. All that is left is an equation to allow calculation of F_w, the wastage rate required to maintain the desired MCRT. For a perfect settler this can be done by combining Eqs. 16.4, 16.5 and 16.7 :

$$F_w = \frac{V_c}{\theta_c(1 - \lambda)} \qquad\qquad (16.16)$$

This is for wastage from the contact tank. If sludge is wasted from the settler underflow, the wastage rate may be determined by performing a mass balance on cells in the settler and then simplifying:

$$F'_w = \frac{V_c F_r}{\theta_c(1 - \lambda)(F + F_r) - V_c} \qquad\qquad (16.17)$$

The next task is to calculate the oxygen requirement for each tank by performing a balance on oxygen demanding material entering and leaving. Letting β_c be the oxygen equivalence of cells in the contact tank and β_s the value for cells in the stabilization tank, the equation for RO in the contact tank is:

$$RO_c = F(S_o - S_c) - F\beta_c X_c + F_r(\beta_s X_s - \beta_c X_c) \qquad\qquad (16.18)$$

Similarly, the equation for the oxygen uptake in the stabilization basin can be shown to be

$$RO_s = \beta_c X_c(F - F_w) - F_r(\beta_s X_s - \beta_c X_c) \qquad\qquad (16.19)$$

when wastage is from the contact tank. The total oxygen uptake in the system is just the sum of these two equations. The resulting expression is similar to that for CMAS.

If inert solids are present in the influent they will build up in the system just as they did in CMAS. A mass balance on inert solids within the system boundary reveals that the concentration in the contact tank, Z_{ic}, is:

$$Z_{ic} = Z_{io}(F/F_w) \qquad\qquad (16.20)$$

Substitution of Eq. 16.16 for F_w reveals that

$$Z_{ic}V_c = FZ_{io}[\theta_c(1 - \lambda)] \qquad\qquad (16.21)$$

It should be noted that θ_c and λ are defined as the mean *cell* residence time and the fraction of *cells* within the system that are housed in the stabilization tank. Because inert solids do not undergo decay in the stabilization tank they will make up a different percentage of the solids there than they do in the contact tank.

Consequently, if θ_c and λ were calculated on the basis on MLSS rather than cells, the values would be slightly different. As long as cells are wasted from the contact tank or settler underflow, however, the wastage rates required to maintain a desired MCRT can still be calculated from Eqs. 16.16 and 16.17 because no knowledge of the cell concentrations is required. The concentration of inert solids in the stabilization tank is the same as the concentration in the settler underflow. A mass balance on the settler, with substitution of Eq. 16.20 for Z_{ic} gives

$$Z_{ir} = Z_{is} = (\frac{F - F_w}{F_r} + 1)(\frac{F}{F_w})Z_{io} \qquad (16.22)$$

These equations provide a complete description of the CSAS system and will give performance curves similar to those shown in Section 13.2.5 provided that S_c is fairly small with respect to S_o so that the assumption of insignificant growth in the stabilization tank is valid. If particulate substrate is present it will be removed in the contact tank by entrapment in the floc so that S_c is still the concentration of soluble substrate alone. In order for the predictions of the "cell" concentrations to be accurate, the effective substrate concentration, S_o', should be employed, as should the "effective" parameters, as described in Sections 12.1.3 and 13.1.3. Special care should be used in the determination of β_c' and β_s', however, because they are more likely to be different in this case.

16.3 DESIGN

The process design for activated sludge requires many decisions, the first of which must be the type to be employed. Economics will guide that choice, but many other factors relating to the characteristics of the wastewater and the objectives of the entire treatment system must be considered as well. Consequently, that is a system design decision, and as such is beyond the scope of this book. Once it has been made, however, then the techniques considered herein may be employed to perform the actual process design, which involves selecting the MCRT, the reactor volume with its associated MLSS concentration, the recycle ratio, and the wastage flow rate. In addition the mixing and oxygen requirements must be determined so that a suitable aeration system may be designed.

The theoretical equations in Chapter 13 were based on the assumption of steady-state conditions even though the influents to most activated sludge systems are subject to both diurnal and random variations. Considerable research is currently in progress on transient-state modeling of activated sludge systems, but the techniques are not yet sufficiently developed to allow an integrated approach to design and control. Furthermore, research on automatic control of activated sludge systems has shown their response to be sluggish, suggesting that the most effective way the designer can deal with transient conditions is to minimize the magnitudes of the perturbations before they reach the process. Actually, this is one of the major

axioms of automatic control and therefore many designers are incorporating equalization prior to the activated sludge process in wastewater treatment systems. As a
result, the steady-state equations can be used along with daily average flows and
concentrations when performing an activated sludge process design. Nevertheless,
since complete equalization is not realistic, some allowances must be made for
transients and these will be discussed where appropriate.

16.3.1 Process Design Equations

Definitive equations are available for the design of completely mixed and contact
stabilization activated sludge using the techniques of Sections 13.1 and 16.2.3,
respectively. The tanks in series configurations associated with TAAS, POAS, and
SAAS prevent the development of such equations for them, however, because even when
simplifying assumptions are made which allow direct computation of the various
concentrations, the large number of equations involved makes the procedure tedious.
Consequently the easiest way to design those activated sludge variations is to
simulate the performance for various reactor sizes, MCRT's, etc. using the computer,
and then to select the design from the resulting output. For this reason, design
equations will only be given for CMAS and CSAS.

   COMPLETELY MIXED ACTIVATED SLUDGE (CMAS). The equations of Section 13.1 can be
used directly for the design of CMAS using the average daily values for the influent
flow rate, F, and substrate concentration, S_o. Because many wastewaters contain
particulate substrate in addition to soluble substrate the equations will be written
in terms of effective parameters to remind the reader that such material will be
removed. The concentration of particulate substrate in the influent will depend
upon the effectiveness of upstream unit operations, such as sedimentation, and that
must be considered when determining S'_o.

   The first task of the designer is to calculate the MCRT required to give a
desired effluent soluble T_bOD concentration, S. The value of S chosen for design
purposes should take into account the normal statistical variation expected in the
final effluent and should be selected to allow the process to achieve the required
effluent quality with a desired probability [20]. After S has been selected, the
MCRT may be calculated with a rearranged form of Eq. 13.27:

$$\theta_c = \frac{K_s + S}{S(\mu_m - \gamma - b') - K_s(\gamma + b')} \qquad (16.23)$$

Alternatively, if the first-order approximation of the Monod equation is used, then
the MCRT may be calculated from a rearranged form of Eq. 13.61:

$$\theta_c = \frac{1}{Y'k_g S_e - \gamma - b'} \qquad (16.24)$$

As was discussed in Section 16.2.1, to achieve proper settling of the activated
sludge, the MCRT must be between 3 and 15 days. If the value determined with
Eq. 16.23 or Eq. 16.24 is less than 3 days, then it must be increased to fall

within the desired range. In that case the choice for the MCRT is often made by
considering the effects it will have upon excess sludge production. Consideration
might also be given to the fact that systems with longer MCRT's have a better response to shock loads [58,59]. If the MCRT chosen for design is greater than the
one determined by Eq. 16.23 or 16.24, the actual effluent soluble T_bOD must be
calculated using Eq. 13.27 or Eq. 13.61

$$S = \frac{K_s(1/\theta_c + \gamma + b')}{\mu_m - (1/\theta_c + \gamma + b')} \tag{13.27}$$

$$S = \frac{1/\theta_c + \gamma + b'}{Y'_g k_e} \tag{13.61}$$

If the MCRT calculated with Eq. 16.23 or 16.24 exceeds 15 days then consideration
should be given to the use of an activated sludge modification which employs
reactors in series. If that is not desirable, then the final settler should be
designed to accomodate the anticipated pin floc formation.

After the MCRT has been selected, the next task is to calculate the $X\tau$ product
using Eq. 13.38:

$$X\tau = \frac{Y'_g(S'_o - S)}{1/\theta_c + b'} \tag{13.38}$$

If the wastewater contains inert solids, then Eq. 13.67 should be used instead to
calculate the $M\tau$ product:

$$M\tau = \theta_c \left[\frac{Y'_g(S'_o - S)}{1 + b'\theta_c} + Z_{io} \right] \tag{13.67}$$

Once $X\tau$ or $M\tau$ has been computed, alternative designs can be considered by selecting
various MLSS concentrations and determining the associated space times. The reactor
volumes can be calculated from the definition of τ:

$$V = \tau F \tag{4.38}$$

Any MLSS concentration within reason may be chosen (see Section 16.3.3). Generally,
the MLSS concentration will also affect the size of settler needed in the installation so that it is possible to arrive at a least cost combination of aeration basin
volume and final settler area [60,61].

The recycle ratio required will depend upon the degree of concentration of
solids desired in the settler (X_r/X), and can be calculated from Eq. 13.43:

$$\alpha = \frac{1 - (\tau/\theta_c)}{(X_r/X) - 1} \tag{13.43}$$

If inert solids are present use

$$\alpha = \frac{1 - (\tau/\theta_c)}{(M_r/M) - 1} \tag{16.25}$$

Both equations are applicable regardless of the point from which sludge is wasted. The recycle ratio can affect the size of final settler required and thus can be included as a parameter in the decisions leading to a least cost design [21]. The usual approach is to provide recycle pumps which are capable of providing any flow between one-half and twice the design recycle value.

The waste sludge flow rate, F_w, required to maintain a desired MCRT depends upon the point from which sludge is wasted from the system because that determines the waste sludge concentration. If sludge is wasted from the settler underflow, then

$$F_w = F\left[\frac{\tau \alpha}{\theta_c(1 + \alpha) - \tau}\right] \tag{13.14}$$

If, on the other hand, it is wasted from the reactor, then

$$F_w = V/\theta_c \tag{13.23}$$

The values calculated from Eqs. 13.14 and 13.23 are nominal values based upon the assumption that no solids are lost in the final effluent. In reality, some solids will always escape from the system so the actual wastage rate will be less than the nominal values [62]. If solids are lost in the effluent at concentration M_e, then the wastage rate must be determined from a rearranged form of Eq. 13.116:

$$F_w = \frac{(VM/\theta_c) - M_e F}{M_w - M_e} \tag{16.26}$$

For wastage from the settler underflow M_w is equal to M_r whereas for wastage from the reactor it is equal to M. The effect of solids loss in the effluent is quite important and cannot be ignored during operation of the activated sludge process [62]. Simplified techniques for determining the actual waste sludge flow rate will be discussed in Section 16.4.2. Regardless of the actual rate of sludge wastage that will be used during operation, provision should be made during design for wasting sludge at whatever rate may be required to maintain the MCRT at any value within the normally accepted range, thereby providing operational flexibility for the plant.

The excess sludge production rate is independent of the point of sludge wastage since it is based on the dry weight of solids wasted. If the wastewater contains no inert solids, then P_x may be calculated with Eq. 13.51:

$$P_x = \frac{FY_g'(S_o' - S)}{1 + b'\theta_c} \tag{13.51}$$

If inert solids are present, they must be included:

$$P_M = F[\frac{Y'_g(S'_o - S)}{1 + b'\theta_c} + Z_{io}] \qquad (13.68)$$

Both of these equations assume that perfect clarification occurs in the final settler, because in that way the calculated excess sludge production rates will be the maximum possible for a given MCRT. Because of the cost of sludge disposal, serious consideration should be given to P_x or P_M when selecting a design value for the MCRT.

All of the above factors should be calculated using the average daily flow for F and the average daily influent substrate concentration for S'_o. The determination of the oxygen requirement, however, should be made using peak values, just as it was for a CMAL in Section 15.3.1 [63]. Thus the oxygen requirement will be the sum of the steady-state and transient-state requirements. The steady-state carbonaceous oxygen requirement can be determined from Eq. 13.47:

$$RO_{SS} = \frac{F(S'_o - S)(1 + b'\theta_c - \beta'Y'_g)}{1 + b'\theta_c} \qquad (13.47)$$

If nitrification occurs in the reactor oxygen must be provided for it as well. This will be discussed in Chapter 20. The transient-state oxygen requirement is given by Eq. 15.3:

$$RO_{TS} = 0.40 \ [(FS'_o)_{peak} - (FS'_o)_{avg}] \qquad (15.3)$$

where $(FS'_o)_{peak}$ refers to the product of flow and concentration resulting in the maximum mass loading of substrate to the reactor. The aeration system must be capable of supplying a quantity of oxygen equal to the sum of RO_{SS} and RO_{TS} while maintaining a minimum dissolved oxygen concentration of 2.0 mg/liter in the reactor. The design of aeration systems is beyond the scope of this book, but is discussed elsewhere [64]. Suffice it to say that the aeration system must also provide sufficient mixing. This will be discussed in Section 16.3.4.

CONTACT STABILIZATION ACTIVATED SLUDGE (CSAS). CSAS may be designed by the equations developed in Section 16.2.3 using the daily average influent flow and substrate concentration. The first task is to calculate the specific growth rate in the contact tank required to give the desired effluent soluble T_bOD, S_c:

$$\mu_c = \frac{\mu_m S_c}{K_s + S_c} \qquad (10.42)$$

Next the MCRT for the process is selected to give the desired sludge settling characteristics and the resulting viability of the sludge is calculated:

$$v_c = \frac{1 + b'\theta_c}{1 + b'\theta_c + \gamma\theta_c} \qquad (13.40)$$

Once the viability is known, the fraction of cells in the stabilization tank, λ, may be calculated using Eq. 16.11:

$$\lambda = \frac{\mu_c v_c - 1/\theta_c - b'}{\mu_c v_c} \tag{16.11}$$

If the wastewater contains no inert solids, the $X_c V_c$ product may be calculated using Eq. 16.3

$$X_c V_c = \frac{FY'_g (S'_o - S_c)}{\mu_c v_c} \tag{16.3}$$

The concentration of solids in the contact tank, X_c, may be selected to give an economic design and the contact tank volume, V_c, may be calculated. If the wastewater contains inert solids the $M_c V_c$ product may be calculated by recognizing that M_c is the sum of X_c and Z_{ic} and therefore combining Eqs. 16.3 and 16.21:

$$M_c V_c = F \left[\frac{Y'_g (S'_o - S_c)}{\mu_c v_c} + Z_{io} \theta_c (1 - \lambda) \right] \tag{16.27}$$

M_c may likewise be selected for economy of design and V_c calculated. A realistic sludge recycle rate, F_r, may be selected to ensure proper operation of the settler and the required stabilization tank volume calculated with Eq. 16.14:

$$V_s = \frac{F_r V_c [\lambda/(1 - \lambda)]}{F + F_r - [V_c/(1 - \lambda)][(1/\theta_c) + b'\lambda]} \tag{16.14}$$

Although it is not required for the design, the concentration of solids in the stabilization tank may be calculated next. If the influent contains no inert solids, the concentration will be given by Eq. 16.15. If inert solids are present, then their concentration, as given by Eq. 16.22, must be added to the cell concentration, X_s, to get the MLSS concentration, M_s. As shown in Fig. 16.12, solids may be wasted directly from the contact tank to maintain the desired MCRT. The rate should be:

$$F_w = \frac{V_c}{\theta_c (1 - \lambda)} \tag{16.16}$$

If sludge is wasted from the settler underflow, the rate must be:

$$F'_w = \frac{V_c F_r}{\theta_c (1 - \lambda)(F + F_r) - V_c} \tag{16.17}$$

The flow rates given by Eqs. 16.16 and 16.17 are nominal values that would be required if no solids were lost in the final effluent. Just as with CMAS, these are the values upon which the design should be based even though lower ones will be used in operation to compensate for solids lost in the effluent. The excess cell production rate may be calculated with Eq. 16.28 which comes from the definition of P_x and substitution of the appropriate equations for $F_w X_w$

$$P_x = \frac{FY'_g (S'_o - S_c)}{1 + b'\theta_c} \qquad (16.28)$$

If inert solids are present, P_M may be found by adding the mass inflow rate of inert solids, FZ_{io}, to P_x.

Just as with CMAS, all of the above factors should be calculated using the average daily flow, F, and the average daily influent substrate concentration, S'_o. When determining the oxygen requirements, however, consideration must again be given to the peak values. Transient conditions will have their largest impact upon the contact tank, so that is where the transient-state oxygen requirement should be supplied. Thus the oxygen supply rate to the contact tank should be equal to the sum of the steady-state requirement as given by Eq. 16.18

$$RO_{SS} = F(S'_o - S_c) - F\beta'_c X_c + F_r(\beta'_s X_s - \beta'_c X_c) \qquad (16.18)$$

and the transient-state requirement as given by Eq. 15.3. The oxygen requirement for the stabilization basin will be given by Eq. 16.19:

$$RO_{SS} = \beta'_c X_c (F - F_w) - F_r(\beta'_s X_s - \beta'_c X_c) \qquad (16.19)$$

Just as with CMAS the oxygen transfer system must be capable of supplying the needed amount of oxygen while maintaining a minimum dissolved oxygen concentration of 2.0 mg/liter in the reactors and providing sufficient mixing.

16.3.2 Determination of Parameter Values

The values of the kinetic parameters characteristic of a particular wastewater may be determined during laboratory treatability studies using bench-scale CMAS reactors operated with MCRT's of from 3 to 15 days. The techniques discussed in Section 13.4 should be employed and tests should also be conducted to determine the effects of the MCRT upon the settling characteristics of the sludge and upon β, its oxygen demand coefficient. If the full-scale facility is to be a CMAS plant, then the kinetic parameters may be used directly in the equations for the system design. If, on the other hand, one of the other activated sludge variations is to be employed, then an additional set of experiments should be performed. Recall that in Section 13.4 the importance of the ecological characteristics of the sludge was discussed. Because of those factors, it is possible that the population which develops in one of the other variations will be different from that in CMAS, thereby causing the kinetic parameters to be slightly different. Thus after completion of the CMAS studies the parameters should be used with the proper equations to design a system of the desired configuration, i.e., CSAS, SAAS, etc, and then a bench-scale model of that system should be set up in the lab using the same MCRT, space times, recycle

flows, etc. as planned for the final design. Operation of this system at the design
point will allow verification of the model predictions and indicate any changes which
need to be made in the final design. If additional experiments are required, they
can be selected using the model, thereby minimizing the amount of lab work required.
If a POAS system is to be designed, then the CMAS studies should be run with dissolved
oxygen concentrations similar to those anticipated in the final facility. As discussed in Section 16.2.2 failure to do this is likely to cause the kinetic parameters
to be incorrect. In the equations for CSAS two values of β are needed, β_c and β_s.
If only CMAS experiments are run in the lab, β_c should be estimated by β at the
shortest MCRT employed whereas β_s should be estimated as β at the design MCRT.

One point should be emphasized about activated sludge design and the use of the
equations presented herein: the variable nature of the microbial population makes
it impossible to arrive at "the best" design. Rather, the designer should strive to
arrive at a logical and economic design, consistent with the probable kinetic characteristics of the wastewater, which has the flexibility to permit operation under a
variety of different conditions. The equations should be used to aid the engineer
in the practice of the art of engineering. They cannot replace the use of sound
engineering judgement; rather they should augment it.

16.3.3 Required Constraints

Several choices were left to the designer in the application of the equations of
Section 16.3.1. Although considerable freedom exists in the making of those choices,
there are practical constraints which must be adhered to. One, discussed in Section
16.2.1, is upon the MCRT. Another is upon the choice of the MLSS concentration,
either in the aeration basin of CMAS or the contact tank of CSAS. Equations 13.38
and 16.3 suggested that the concentrations could be chosen freely because the reactor
volumes would then be fixed to complement the choice. Upper and lower limits must
be placed on the MLSS concentration, however, because of the requirements for mixing,
oxygen transfer, and liquid-solid separation.

First consider mixing and oxygen transfer. In most activated sludge systems
both needs are met by the aeration system. The oxygen requirement is fixed for a
given MCRT, thereby fixing the power needed to transfer it to the liquid. Mixing, on
the other hand, requires a certain amount of power per unit volume, so that the
larger the reactor, the greater the power required to mix it. At high MLSS concentrations, when the reactor volume is small, the power needed for oxygen transfer is
likely to be larger than that for mixing. As the MLSS concentration is reduced, however, the power required for mixing will increase due to the need for a larger reactor,
until at some concentration the two are equal. At MLSS concentrations below that
value more power is required to mix the system than to supply the oxygen, a situation
which may be uneconomic. For most wastewaters, when the MLSS concentration is 1500
mg/liter the power for mixing is much larger than that for oxygen transfer so that
further reductions in MLSS concentration are difficult to justify economically.

Looking at the other extreme, as the MLSS concentration is made larger, so that the
reactor may be smaller, the power per unit volume applied for oxygen transfer increases, thereby increasing the turbulence in the vessel. A point will be reached
at which fluid shear will damage the biological floc, thereby making sedimentation
of the sludge difficult. When air is used as the source of oxygen this limitation
on turbulence is generally reached at a MLSS concentration of around 6000 mg/liter.
If pure oxygen is used instead of air, oxygen transfer rates two to four times as
large can be achieved with the same degree of turbulence, so that the MLSS could
conceivably be two to four times as large if mixing and oxygen transfer were the
only consideration [19]. Increases of that magnitude are not generally possible,
however, because of the other limitation: liquid-solid separation.

The upper limit on the MLSS concentration is generally imposed by the requirement for sludge thickening in the settler. At low MLSS concentrations the size of
the settler is generally determined by the requirement for producing a clear effluent,
i.e., the clarification capacity, so that the size of the settler is relatively independent of the MLSS concentration entering it. As the concentration is increased,
however, a point will be reached at which the ability of the settler to thicken the
sludge from the influent concentration, M, to the underflow concentration, M_r, begins
to govern its size [65]. From that point on, the size of the settler will increase
as the MLSS concentration is increased so that even though the size of the aeration
basin and its associated cost will decrease, the cost of the settler will increase.
This suggests that there is a least cost combination of the two units [60,61]. The
optimum MLSS concentration will depend upon the characteristics of the system and
thus no generalizations can be made about it [20]. Nevertheless it can be stated
that seldom will economic designs result when the MLSS concentration exceeds 5000
mg/liter for systems with oxygen transfer by air [66]. POAS systems result in sludges
that settle better than those from air systems so that the upper bound will be higher,
although its value is not yet well established.

In summary, considerations of mixing, oxygen transfer, and liquid-solid separation suggest that the MLSS concentration in CMAS, TAAS, the contact tank of CSAS,
and the last tank of SAAS should not be less than 1500 mg/liter nor greater than
5000 mg/liter. Violation of the lower limit will generally result in less economic
designs but may sometimes be necessary in the early years of operation of a new
facility. Violation of the upper limit may result in both an uneconomic design and
one that does not work well. The upper limit is higher for POAS.

Another factor which should be considered in the decision for a least cost CMAS
design is the recycle ratio, α, because it affects the size of the final settler
without influencing the size of the aeration chamber [21]. No generalizations can
be made about the best ratio because that will depend upon the sludge settling characteristics. As a rough guide, however, the design value for α should generally be
between 0.2 and 1.0, and the recycle pumps should be capable of delivering flows

between one-half and twice the design value. In CSAS design the size of the stabilization tank depends directly upon α, but the normal range employed is the same as in
CMAS.

16.3.4 Factors Influencing the Design

Just as with the design of a CMAL, the process design equations do not fully define
the process design for an activated sludge system. Other factors must be considered;
among them are the attainment of the proper mixing conditions, the effects of temperature, and the providing of proper nutrients.

   MIXING. CMAS, CSAS, and POAS all utilize completely mixed reactors, either
singly or in series, whereas TAAS and SAAS often use flow-through aeration basins.
As discussed in Section 16.1.3 there are some disadvantages associated with the use
of a flow-through basin, particularly when the wastewater is high in concentration,
but many of them can be overcome by using three or four CSTR's in series, as was done
in the models. Thus, since all activated sludge systems can be designed with completely mixed reactors, let us look at the attainment of completely mixed conditions
in full-scale facilities.

   The classical definition of complete mixing requires that the influent be instantaneously dispersed throughout the vessel contents and that the concentrations
at all points in the tank be the same. Unfortunately, instantaneous mixing is not
practical in vessels of the size used for activated sludge aeration. Therefore a
practical definition of a completely mixed tank meets only the second requirement,
i.e., the concentrations of MLSS, substrate, and dissolved oxygen, as well as the
oxygen uptake rate, are fairly uniform throughout. Furthermore, the turnover time
of the vessel should be no more than 1/5 of the space time [67]. These requirements
can usually be met with a power input of around 0.5 HP/1000 ft^3 (13.8 kW/1000 m^3),
which would require an air flow rate of approximately 20 scfm/1000 ft^3
(20 m^3/(min·1000 m^3)) of reactor volume for a diffused air system. These values
should be sufficient for any MLSS concentration normally encountered in air CMAS or
TAAS but may not be sufficient for the stabilization tank of CSAS or the first tank
in SAAS. In those cases, the technique outlined in Section 17.3.3 for aerobic digestion should be employed. Higher values may also be needed in POAS if high MLSS
concentrations are used, in which case the power input may also be determined using
the technique described for aerobic digestion. As an upper limit, the power input
should not exceed about 2.25 HP/1000 ft^3 (60 kW/1000 m^3) or 90 scfm/1000 ft^3 in order
to avoid floc shear which will be detrimental to the settling characteristics of the
sludge. As discussed previously, when the MLSS concentration and reactor volume are
chosen, the power input for aeration should be checked against the power input required to achieve complete mixing. In systems which use reactors in series, the
aeration and mixing power should be computed for each vessel and the larger of the
two requirements should be met in each tank. One of the advantages of CMAS is that
the reactors can usually be designed so that the requirement for aeration controls,

thereby minimizing power inputs. This may not always be possible in each tank of
the other modifications.

Aeration system design is a problem in physical unit operation design and as such
is beyond the scope of this book [64]. An estimate of the power requirements for
aeration can be made easily, however, and this will be sufficient for checking the
mixing conditions. This should be done using the steady-state oxygen requirement because it will be supplied at all times whereas the transient-state requirement will
only be supplied as needed. If mechanical surface aerators are used, the power required can be calculated by assuming an average mass of oxygen transferred per HP·hr
as was done in Example 15.3.5-1. If bubble aeration is used, the scfm of air required may be estimated by Eq. 16.29 which assumes an average oxygen transfer efficiency of 10%:

$$Q = (RO \text{ in } lbs/min)(575) \tag{16.29}$$

where Q is in cfm. Just as was discussed in Section 15.3.4, consideration must be
given to the placement of the aerators in the aeration basins. Particular attention
should be paid to the manufacturer's recommendations concerning spacing because
aerator interactions are likely to be severe in activated sludge systems due to small
tank volumes. Busch [67] has suggested that performance criteria, such as uniform
suspended solids or dissolved oxygen concentrations within the zone of influence of
the aerator, be used to ensure that the devices perform in the manner needed.

There are currently no well-established guidelines concerning the most suitable
reactor configuration for attaining complete mixing. Generally the reactors are of
three types: square, round, or rectangular, as shown in Fig. 16.13. Square and round
basins tend to be used more with mechanical aerators whereas rectangular ones are used
with either bubble or mechanical aerators. In the square vessel, influent is brought
in on one side and effluent is discharged on the other, with the number of inlets
and outlets depending upon the vessel size. Baffles should be provided to guard
against short circuiting and the direction of rotation of mechanical mixers should
be selected to prevent direct short circuiting between the inlet and outlet as shown
in Fig. 16.13a. The influent usually enters the center of circular basins and the
effluent is discharged at the periphery, as shown in Fig. 16.13b. In small basins
with a single central mixer, introduction of the influent into the highly turbulent
region beneath the turbine will help to ensure its uniform distribution throughout
the basin. Basin depths are generally 12-15 ft but no guidelines exist for the other
dimensions of square or round tanks. Rectangular basins are usually of the same depth
and have a width (from inlet to outlet) of 25 to 50 ft. There is no theoretical limit
on the tank length because flow is across the tank, but four inlets (two for the wastewater and two for return sludge) are needed for each 50 ft of length [45]. These inlet ports should be located symmetrically on one side of the tank and the effluent
should flow over a continuous weir on the other side. If aeration is by diffused
air, a minimum of 20 diffusers should be used per 100 ft^2 of floor area to ensure

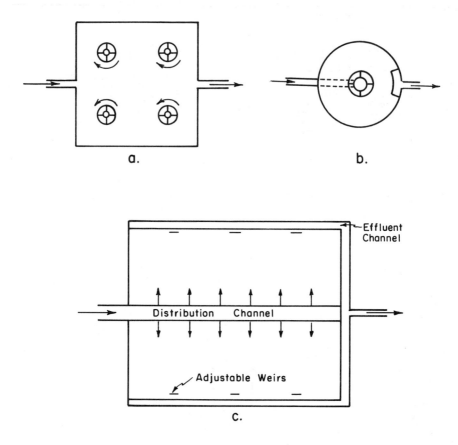

Figure 16.13.   Reactor configurations for CMAS: (a) Square;
(b) Circular; and (c) Rectangular.

that air bubbles cover the entire tank [45].   In addition, it has been found that
the placement of more diffusers on the effluent side yields a mixing pattern which
conforms better to complete mixing.   They should not be placed directly under the
effluent channel, however, because the air-water mixture tends to cause uneven flow.

In Section 13.2.3 it was seen that several completely mixed tanks in series with
an equal distribution of influent and recycle among them give an overall performance
equivalent to a single completely mixed reactor with recycle.   This suggests that the
most advantageous reactor configuration for CMAS would be three or four tanks in
series because by redistribution of the influent or recycle the operator could convert the system from CMAS to SAAS, TAAS, or CSAS.   This provides a tremendous amount
of operational flexibility for relatively little cost, thereby greatly improving the
chances that an effluent of excellent quality can be attained under a large variety
of influent conditions.

TEMPERATURE. Temperature effects were discussed in Sections 9.3.3, 10.6.1, and 15.3.4. The comments of the last section are particularly pertinent to activated sludge, but in order to save space they will not be repeated. Because of the flocculent nature of the biomass, diffusional effects appear to be important in determining the influence of temperature upon activated sludge performance. Consequently, it is important that temperature studies be performed with mixing energy levels similar to those used in the field. The removal of soluble substrate by activated sludge has been found to be relatively insensitive to changes in temperature [68], but this is probably due more to the long MCRT's employed than to any insensitivity of the kinetic parameters themselves. Figure 13.15 showed that there would be no detectable differences in substrate removal in a four tank chain even though there had been drastic changes in the parameters. The biggest effect of temperature will be upon the amount of excess sludge produced and the amount of oxygen required.

In light of the uncertainty surrounding the effects of temperature, the safest approach would be to run the laboratory studies at the anticipated minimum and maximum temperatures and then use the observed kinetic parameter values in design. If that is impossible, then values must be selected from the literature for the temperature characteristics, just as in Section 15.3.4. In this case, however, θ for μ_m will lie between 1.00 and 1.04 [68], suggesting that a value of 1.02 could be used as a rough estimate. The value of θ for b, however, should still be taken to be around 1.04, just as it was for a CMAL. The grossly approximate nature of these estimates should be recognized, however.

One thing that reduces the danger of using rough estimates of the temperature effects is that temperature variations in activated sludge reactors are not nearly so great as they are in CMAL's [69]. For example, one treatment plant located in the midwest experienced a temperature differential from winter to summer of only 11°C, reaching a high of 25.4°C and a low of 14.1°C [70]. For a range no larger than that, performance estimates based on the average temperature would be adequate because operational changes in the MCRT could be made to compensate for any small losses in actual performance. The main reasons for the smaller temperature variations in activated sludge are that the basins are smaller than a CMAL and have much shorter space times [69]. In addition, diffused air systems can actually add heat to the reactors through the compressed air, thereby maintaining a higher temperature in the winter. For example, one plant experienced temperature changes from influent to effluent of the aeration basin of only 3 to 4°F, even though the ambient air temperature reached a low of -15°F [69]. This suggests that much of the winter to summer temperature difference observed in an activated sludge plant is due to changes in the temperature of the influent wastewater rather than to heat loss from the reactor. Until a definitive study of heat losses in reactors aerated by diffused air is performed, about all a designer can do is to estimate the change in temperature by Eq. 15.8, taking into account the changes in influent temperature as well. Greater heat losses occur in

aeration basins with mechanical surface aerators because not only is no heat added,
but heat is lost by transfer to the atmosphere from the water spray as well. Equation
15.8 included no terms for heat loss by spray because the surface areas of lagoons
are so large that the majority of the heat loss is across the pond surface. This is
not true for activated sludge aeration basins, however, and consideration must be
given to the effects of spray. A technique has been presented which allows this to
be done [71].

NUTRIENTS. As discussed in Section 8.2 inorganic nutrients are required in the
proper amounts if organic carbon is to control the rates of substrate removal in
activated sludge systems. The two most likely to be deficient are nitrogen and
phosphorus and in Section 9.4.3 we saw how the equation for microbial growth can be
used to estimate the quantities needed once the observed cell yield is known.
Equation 13.39 allows prediction of Y as a function of the MCRT:

$$Y = \frac{Y_g}{1 + b\theta_c} \qquad (13.39)$$

Thus once the MCRT is chosen for a design, the predicted value of Y may be used to
write the equation for microbial growth (as shown in Example 9.4.2-1), which can then
be employed to determine the nutrient requirements. A similar technique has been
utilized to show how the COD:N:P ratio changes as the MCRT is changed [72] and to
demonstrate how much N and P will be removed from a wastewater containing excess
nutrients (such as domestic sewage) during treatment by activated sludge [73].

Another technique allows estimation of the nitrogen requirements without writing
the stoichiometric equation. Microbial cells are approximately 10.5% nitrogen on a
dry weight basis (assuming an empirical formula of $C_5H_7O_2N$). The units of yield are
mass of cells formed per unit mass of COD removed so that multiplication of the yield
by 0.105 gives the mass of nitrogen which will be incorporated into cell material per
unit mass of COD removed. Applying this concept to Eq. 13.39 gives an equation for
determining the nitrogen requirement as a function of θ_c:

$$NR = \frac{0.105 \; Y_g}{1 + b\theta_c} \qquad (16.30)$$

where NR is the nitrogen requirement in mg N per mg COD removed. The phosphrous requirement will be approximately one-fifth of the nitrogen requirement. Nutrients
should be added in slight excess of the theoretical amounts to ensure that carbon is
rate limiting.

Although most studies on nutrient requirements have dealt with the need for
nitrogen and phosphorus many other inorganic nutrients are also necessary, as discussed in Section 8.2. Failure to supply them in proper amounts can result in operational difficulties, particularly with respect to sludge settling [74,75]. Thus
consideration should be given to the presence of trace elements during design.

16.3.5 Example Problems

The easiest way to demonstrate the process design procedure is through examples.

EXAMPLE 16.3.5-1

Determine the size of a diffused air CMAS system capable of reducing the soluble T_bOD of the waste described in Example 15.3.5-1 to no more than 10 mg/liter. Assume a constant temperature of 20°C, and that the treatability studies were done in lab-scale CMAS reactors.

The kinetic characteristics of the waste are described in Table E15.1. The average biodegradable substrate concentration, S_o' is 524 mg/liter T_bOD and the peak value is 761 mg/liter T_bOD. The flow rate is 634,000 gal/day (10^5 liters/hr = 100 m^3/hr = 3532 ft^3/hr).

Solution:

(a) The first task is to estimate the MCRT required to reduce S to 10 mg/liter. This is determined with Eq. 16.23:

$$\theta_c = \frac{(120 + 10)}{10(0.40 - 0.004 - 0.0035) - 120(0.004 + 0.0035)}$$

$$\theta_c = 43 \text{ hrs} = 1.8 \text{ days}$$

This is not long enough to provide a sludge that will settle well. Therefore, based on sludge settling characteristics and other factors, a MCRT of 7 days (168 hrs) is chosen. The resulting substrate concentration can be estimated from Eq. 13.27:

$$S = \frac{120(1/168 + 0.004 + 0.0035)}{0.40 - (1/168 + 0.004 + 0.0035)}$$

$$S = 4.2 \text{ mg/liter}$$

(b) We must now determine the $M\tau$ product using Eq. 13.67. The concentration of inert solids in the influent, Z_{io}, is 48 mg/liter (Example 15.3.5-2).

$$M\tau = 168[\frac{0.5(524 - 4)}{1 + (0.0035)(168)} + 48]$$

$$M\tau = 35,570 \text{ mg/(liter·hr)}$$

We may now choose either M or τ and calculate the other. After considering the size of the final settler, we decide to use a MLSS concentration of 3557 mg/liter, making τ equal to 10 hours. The reactor volume may be calculated from the definition of τ:

$$V = (10 \text{ hrs})(3532 \text{ } ft^3/hr)$$

$$V = 35,320 \text{ } ft^3 = 1000 \text{ } m^3$$

(c) It is estimated that a settler can be designed to give an underflow solids concentration, M_r of 12,000 mg/liter. What recycle ratio would be required to give
that concentration? Determine α from Eq. 16.25

$$\alpha = \frac{1 - (10/168)}{(12,000/3557) - 1}$$

$\alpha = 0.40$

Thus the recycle flow rate must be 253,600 gal/day (40 m^3/hr). To provide
flexibility in the plant, the recycle pumps should have the capability of
pumping at any rate up to twice this value.

(d) At what rate would sludge have to be wasted from the reactor to maintain the
desired MCRT if the final settler were perfect? This can be determined from
Eq. 13.22:

$$F_w = \frac{35,320 \text{ ft}^3}{168 \text{ hrs}}$$

$F_w = 210 \text{ ft}^3/\text{hr} = 26.2 \text{ gal/min} = 5.95 \text{ m}^3/\text{hr}$

At what rate would sludge have to be wasted if it were from the settler underflow? This can be determined from Eq. 13.14:

$$F_w = 3532 \left[\frac{(10)(0.40)}{(168)(1 + 0.40) - 10} \right]$$

$F_w = 62.7 \text{ ft}^3/\text{hr} = 7.86 \text{ gal/min} = 1.77 \text{ m}^3/\text{hr}$

(e) The excess sludge production rate may be determined from Eq. 13.68:

$$P_M = 10^5 \left[\frac{0.5(524 - 4)}{1 + (0.0035)(168)} + 48 \right]$$

$P_M = 21.2 \times 10^6$ mg/hr of dry solids = 21.2 kg/hr = 46.7 lb/hr

(f) The steady-state oxygen requirement can be estimated with Eq. 13.47:

$$RO_{SS} = \frac{(10^5)(524 - 4)[1.0 + (0.0035)(168) - (1.25)(0.5)]}{1 + (0.0035)(168)}$$

$RO_{SS} = 3.15 \times 10^7$ mg O_2/hr = 31.5 kg O_2/hr = 69.5 lb O_2/hr

The air flow rate required to meet the steady-state oxygen requirement can be
estimated with Eq. 16.29:

$Q = (69.5/60)(575)$

$Q = 666$ cfm

The additional amount of oxygen which must be added during the peaks in loading
can be estimated with Eq. 15.3:

$RO_{TS} = 0.40[(10^5)(761) - (10^5)(524)]$

$RO_{TS} = 0.95 \times 10^7$ mg O_2/hr = 9.5 kg O_2/hr = 20.9 lb O_2/hr

Thus the total oxygen requirement is 90.4 lb/hr (41.0 kg/hr) during peak loading periods. This requires 866 cfm of air. In order to attain complete mixing in the reactor, air must be supplied at a rate of 20 cfm/1000 ft^3. Since the reactor volume is 35,320 ft^3, the air flow required to provide adequate mixing is 706 cfm. Thus the amount of air needed to mix the tank will be more than adequate for steady-state operation although additional air must be added during peak loads.

(g) The amount of nitrogen required to treat the waste can be estimated with Eq. 16.30:

$$NR = \frac{(0.105)(0.5)}{1 + (0.0035)(168)}$$

NR = 0.033 mg N/mg COD removed

The amount of COD removed per hour is $(10^5)(524 - 4) = 5.20 \times 10^7$ mg/hr, which means that 1.72 kg (3.79 lb) of N will be required per hour.

EXAMPLE 16.3.5-2

Compare the total reactor volume of a CSAS system treating the waste in Example 15.3.5-1 to a concentration of 10 mg/liter with the reactor volume for CMAS as determined in Example 16.3.5-1. Use the same MLSS concentration, M, and the same recycle flow rate, F_r.

Solution:

(a) The desired effluent substrate concentration is 10 mg/liter, which gives a contact tank specific growth rate of:

$$\mu_c = \frac{(0.40)(10)}{120 + 10} \tag{10.6}$$

$$\mu_c = 0.031 \text{ hr}^{-1}$$

The MCRT used in example 16.3.5-1 was 7 days (168 hrs). Using the same MCRT the viability may be calculated from Eq. 13.40:

$$v_c = \frac{1 + (0.0035)(168)}{1 + (0.0035)(168) + (0.004)(168)}$$

$$v_c = 0.70$$

Next the fraction of cells in the stabilization tank must be calculated using Eq. 16.11

$$\lambda = \frac{(0.031)(0.70) - (1/168) - (0.0035)}{(0.031)(0.70)}$$

$$\lambda = 0.564$$

The influent contains 48 mg/liter of inert solids so the M_cV_c product must be calculated with Eq. 16.27:

$$M_cV_c = 10^5[\frac{0.5(524 - 10)}{(0.031)(0.70)} + (48)(168)(1 - 0.564)]$$

$$M_cV_c = 1.54 \times 10^9 \text{ mg solids}$$

The MLSS concentration from Example 16.3.5-1 was 3557 mg/liter, thus the contact reactor volume is

$$V_c = \frac{1.54 \times 10^9}{3557}$$

$$V_c = 4.329 \times 10^5 \text{ liters} = 433 \text{ m}^3 = 15{,}300 \text{ ft}^3$$

(b)   The recycle rate is to be the same as in Example 16.3.5-1. The recycle ratio was 0.40, thus F_r is 4.0×10^4 liters/hr. The stabilization tank volume may now be calculated with Eq. 16.14

$$V_s = \frac{(4.0 \times 10^4)(4.329 \times 10^5)[0.564/(1 - 0.564)]}{10^5 + (4.0 \times 10^4) - [(4.329 \times 10^5)/(1 - 0.564)][(1/168) + (0.0035)(0.564)]}$$

$$V_s = 1.70 \times 10^5 \text{ liters} = 170 \text{ m}^3 = 6000 \text{ ft}^3$$

The total reactor volume is 21,300 ft^3 which is 60% of the volume of the CMAS reactor. Thus the CSAS system would have a smaller capital cost.

The space time of the contact tank in Example 16.3.5-2, V_c/F, is 4.3 hrs, which is larger than the values normally associated with contact stabilization. This was caused by two things: the MLSS concentration of 3557 mg/liter, which was chosen to be the same as in the CMAS, and the kinetic parameters of the waste, which represent a substrate which is moderately difficult to degrade. The economic advantages associated with CSAS are greater for easily degraded wastewaters. For example, if μ_m had been 0.60 hr^{-1} and K_s had been 50 mg/liter, the value of μ_c required to give an effluent of 10 mg/liter would be 0.1 hr^{-1}, which would increase the fraction of cells in the stabilization tank to 0.865 and decrease the M_cV_c product to 4.76×10^8 mg of solids. If M_c were kept at 3557 mg/liter, V_c would be reduced to 4730 ft^3 (134 m^3) thereby reducing τ_c to 1.34 hrs, a value more characteristic of contact stabilization. In order to maintain a MCRT of 7 days V_s would have to be increased to 9,250 ft^3 (262 m^3), but the total reactor volume will be reduced to only 40% of the CMAS volume, demonstrating that greater savings can be accomplished when the waste is easy to degrade.

16.4 ACTIVATED SLUDGE OPERATION

Information presented in this chapter and in Chapter 13 has established the importance of mean cell residence time to the proper operation of the activated sludge
process. No matter what process modification is employed, the MCRT is the primary
factor influencing the settling characteristics of the activated sludge. It is the
sole operational variable determining the concentration of soluble substrate in the
effluent from CMAS and is by far the most important one in TAAS and POAS. Furthermore, even though the recycle ratio, α, has an effect upon the concentration of
soluble substrate leaving the SAAS and CSAS processes, the MCRT is still an important
determinant of process performance. Thus, an important objective of activated sludge
operation is the maintenance of a relatively constant MCRT even though plant loading
conditions change. Two types of loading changes can occur: long-term and short-term.
Long-term changes are the result of growth within the industry or city being served,
or of seasonal adjustments in load, such as in a resort town. During such long-term
changes in plant loading, some action must be taken to ensure that the MCRT stays
relatively constant so that the average plant performance will remain satisfactory.
Short-term changes are superimposed upon the long-term ones and are caused by the
normal diurnal pattern in waste flow and strength and by shock loads. These short-
term changes are responsible for the dynamic nature of wastewater treatment plants
and action must be taken to minimize their impact without disrupting the maintenance
of the MCRT required to achieve the desired long-term plant performance. In this
section we will briefly investigate control procedures for each situation. Before
doing that, however, it would be advantageous to discuss visual observations because
they can provide an operator or engineer with valuable information concerning plant
performance.

16.4.1 Visual Observations

Observations of the physical characteristics of the sludge in the aeration basin and
final settler can provide helpful clues about the MCRT [36,76]. When a plant is
operating at a reasonable MCRT and producing a good final effluent there will be only
a modest accumulation of fresh, crisp, white foam on the surface of the aeration tank.
On the other hand, if the aeration tanks are covered by thick voluminous billows of
white sudsy foam, the operator can be relatively certain that the MCRT is too small
or that the plant is being subjected to large shock loads of organic matter. Very
often when microorganisms are growing fairly rapidly they will release surface active
metabolic products which will act like detergents, thereby causing excessive foaming.
At the other extreme, the presence of a dense, somewhat greasy, scummy layer of tan
to brown foam covering the entire aeration tank surface is a good indication that the
MCRT is too long. If this scum cannot be collapsed by water sprays there is a good
possibility that it is being caused by the growth of actinomycetes in the sludge [77].
Actinomycete scum is particularly troublesome because it tends to be carried over
into the final effluent where it causes excessively high suspended solids

concentrations. The only successful solution identified so far is to reduce the
MCRT below nine days. At times, large clumps of sludge may be seen rising, then
breaking up and spreading over the clarifier surface. This is called clumping and
is usually associated with denitrification in the final settler. When a long MCRT is
employed, much of the ammonia nitrogen in the wastewater may be converted to nitrate
through nitrification (see Section 8.4.1). If the residence time of the solids in the
final settler is excessive, or if the concentration of dissolved oxygen in the aeration basin effluent is low, the oxygen in the sludge blanket will be depleted and
the nitrate will be used as the terminal hydrogen acceptor by the bacteria, releasing
nitrogen gas (see Section 9.1.3), which becomes trapped in the sludge, making it rise.
This problem can often be eliminated by increasing either the rate of recycle flow or
the concentration of dissolved oxygen in the effluent, but if those don't work it will
be necessary to reduce the MCRT until nitrification no longer occurs. Other problems
associated with excessive MCRT's are ashing and pin floc. Ashing is the rise of
small dark brown to gray sludge particles in the final settler. Pin floc occurs when
the sludge becomes granular, rather than flocculant, resulting in the suspension of
small, compact floc particles ($\sim$1/32 of an inch) throughout a moderately turbid effluent. Both are usually the result of excessive cell decay although pin floc may
also be caused by high turbulence in the aeration tank. At the other extreme is
straggler floc, made of small, almost transparent, very light, fluffy, bouyant
sludge particles (1/8 - 1/4 inch) which rise to the clarifier surface through an
otherwise clear effluent. These signal the need for a slight increase in MCRT. On
the other hand, if the MCRT is much too small, dispersed growth will occur and the
effluent will be very turbid.

If a homogeneous sludge blanket extends throughout the entire final clarifier,
so that the activated sludge solids are escaping over the weirs, the plant is either
being operated with an insufficient return sludge rate or the sludge is bulking. The
two situations may be differentiated easily by performing an SVI test with sludge at
a concentration of around 2000 mg/liter. If the SVI is less than 150 or so then the
sludge is not bulking, in which case the problem may usually be alleviated by increasing the sludge recycle rate. If that doesn't alleviate the problem then the
settlers were probably not designed to handle the mass flow rate of solids entering
them and the only way to overcome the problem (short of building additional settlers)
is to reduce the concentration of MLSS entering them. This will require either a
reduction in the MCRT or an alteration of the process flow sheet. If the problem is
bulking, then its solution is more difficult. In general, there are two types of
bulking, filamentous and nonfilamentous, and the two may be differentiated easily by
microscopic examination. If the bulking is nonfilamentous, the chemical composition
of the waste should be checked to see if all nutrients are available in the proper
amount, and corrective action taken if need be. If nutrients are not the cause of
the problem, then the MCRT should be increased. This may be difficult to do if the
sludge won't settle properly, and thus the temporary addition of polymers or

coagulants may be necessary to improve the settling characteristics enough for an increase in MCRT to be achieved. If the plant is designed to be operated in the SAAS mode, then another alternative exists [78]. It will be recalled from Section 13.2.4 that a plant operated as step aeration has a lower concentration of MLSS entering the settler than does CMAS or TAAS with the same space time and MCRT. Thus conversion of the plant to SAAS will reduce the mass loading of solids to the settler, thereby alleviating the sludge settling problem as the MCRT is increased. The same technique may be employed to reduce the MLSS concentration entering an overloaded final settler. The alleviation of filamentous sludge bulking is usually more difficult because extensive experimentation may be required to determine the cause. Although there is no consensus concerning the factors responsible for the excessive growth of filamentous organisms in activated sludge, it appears likely that the protrusion of filaments from the bulk of the floc particles into the surrounding fluid gives them a kinetic advantage under conditions of low substrate, oxygen, or nutrient concentration [79]. Although the critical dissolved oxygen level to prevent bulking will depend upon many factors, including the MCRT and the reactor configuration, it appears that the concentration should never be allowed to fall below 2.0 mg/liter. If sufficient nutrients and oxygen are provided, and if the pH is near neutrality, yet filamentous bulking persists, it may be necessary to take additional action to shift the predominance away from filamentous microbial forms. Among the possibilities are: (1) shift the flow regime to the TAAS mode which establishes a substrate concentration gradient [22,23]; (2) add an oxidizing agent such as hydrogen peroxide which breaks up the protruding filaments [80]; or (3) add ferrous iron which adsorbs to the filaments and interferes with the transport of substrate [81]. It should be recognized, however, that the latter two will give only temporary alleviation of the problem and that the root cause must still be found and corrected.

16.4.2 Long-Term Control of MCRT

The major concern in the preservation of good effluent quality in the face of long-term changes in plant loading is the maintenance of a constant MCRT. Equation 13.116 defined the MCRT of CMAS in terms of the MLSS concentration:

$$\theta_c = \frac{VM}{F_w M_w + (F - F_w)M_e} \tag{13.116}$$

For the process variations using more than one reactor, the numerator must be the sum of $V_i M_i$ for each of the n tanks in the process:

$$\theta_c = \frac{\sum_{i=1}^{n} V_i M_i}{F_w M_w + (F - F_w)M_e} \tag{16.31}$$

As mentioned in Chapter 13, there has been some question about the definition of MCRT for control purposes; specifically some have felt that the mass of sludge in the final settler should be included in the numerator of Eqs. 13.116 and 16.31. Stall and Sherrard [82], however, have demonstrated that such an addition is unnecessary and have suggested that Eqs. 13.116 and 16.31 may be preferable because of the ease with which the variables in them may be determined. In Section 13.2.2 it was shown that the MLSS concentration does not change greatly down a simple chain so that Eq. 13.116 may also be used for TAAS and POAS if V is taken to mean the total reactor volume. The following discussion about MCRT control is based on Eq. 13.116 and is directly applicable to CMAS, TAAS, and POAS. A similar approach can be used to develop control strategies for SAAS and CSAS, although the equations will be slightly more complex in form and special techniques of sludge wasting may be required [83].

The denominator of Eq. 13.116 represents the rate of solids loss from an activated sludge system and can be divided into two components. The first term, $F_w M_w$, represents the intentional loss of solids from the system by the operator. The other term, $(F - F_w)M_e$, is the loss of solids in the final effluent, which is unintentional and not under operational control. The smaller the unintentional wastage the better the chance for control of the MCRT through manipulation of the intentional wastage. If the unintentional wastage were negligible, it would be possible to control the MCRT by simple hydraulic control strategies, (i.e., those based only upon the manipulation of flows without recourse to suspended solids measurements). Consequently, several such hydraulic control techniques have been developed using the assumption that the amount of unintentional wastage is negligible [84,85,86]. Unfortunately, the assumption is not generally true. In fact, if the unintentional term is ignored the actual MCRT of the system may be 40-45% smaller than that estimated from the intentional wastage alone [62]. This suggests that if proper control of the MCRT is to be achieved, consideration must be given to the accidental loss of solids from the system. An easy way to accomplish this is by application of an empirical correction factor to the hydraulic control formulas [62].

Hydraulic control formulas are those from which the solids concentrations have been removed so that the wastage rate required to maintain a desired MCRT can be calculated on the basis of hydraulic terms alone. Because they assume that the unintentional wastage is negligible, they are based on Eq. 16.32

$$\theta_{ch} = \frac{VM}{F_w M_w} \tag{16.32}$$

where θ_{ch} stands for the hydraulic estimate of the MCRT. If wastage is from the settler underflow the concentration of MLSS in the wastage flow will be M_r, the concentration in the recycle flow. If Eq. 13.41 is used to remove the ratio M/M_r from Eq. 16.32, the result is:

$$\theta_{ch} = \frac{V}{F_w} \left[\frac{\alpha + (F_w/F)}{1 + \alpha} \right] \tag{16.33}$$

Furthermore, since F_w/F is usually much smaller than α, Eq. 16.33 may be simplified to:

$$\theta_{ch} \cong \frac{V\alpha}{F_w(1 + \alpha)} \qquad (16.34)$$

Equation 16.34 is just a simplified version of Eq. 13.14 which allows easy determination of the wastage rate required to maintain a given MCRT when the unintentional wastage is negligible. It shows that for the maintenance of a fixed MCRT, adjustment of the wastage flow rate is required any time the recycle ratio, α, is changed. An even simpler hydraulic control formula may be obtained by using the alternative scheme whereby sludge is wasted directly from the aeration chamber, because then M_w is equal to M so that Eq. 16.32 becomes:

$$\theta_{ch} = \frac{V}{F_w} \qquad (16.35)$$

All that is required to maintain a desired MCRT in this case is the maintenance of a fixed wastage flow, regardless of the recycle flow. This has been called the Garrett flow scheme [84].

The application of an empirical factor to a hydraulic control formula to correct for the unintentional loss of solids is called tuned hydraulic control. A calibration coefficient, K, which is a function of θ_c, M_e, and S_o, is used to make a formula specific to a particular installation [87]:

$$\theta_c = K\theta_{ch} \qquad (16.36)$$

If historical operating data are available, K may be calculated by dividing the actual MCRT by its hydraulic approximation, and then plotted as a function of M_e/S_o to yield a calibration chart which can be used in operation. The sensitivity of K to MCRT is not great so that a single chart may be made for a range of MCRT's, e.g., one chart might apply to all MCRT's between 10 and 15 days. Figure 16.14 shows a calibration chart prepared from actual operating data from a municipal wastewater treatment plant [87]. As more data are compiled, the chart can be updated periodically so that it reflects the current operating conditions in the plant.

To use tuned hydraulic control of MCRT, the following steps would be followed.

(1) M_e and S_o must be measured. Average daily values are sufficient since this is a long-term control strategy.

(2) The ratio M_e/S_o is calculated and K is read from the calibration chart.

(3) θ_{ch} is calculated from K and the desired MCRT using Eq. 16.36.

(4) The required wastage rate is computed using either Eq. 16.34 or 16.35 depending upon the point of sludge wastage.

(5) The wastage rate is maintained constant until the next day when the procedure is repeated. If sludge wastage is from the settler underflow

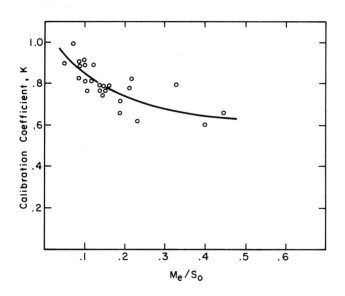

Figure 16.14. Calibration chart prepared from data
from a municipal activated sludge plant.
(Adapted from Roper and Grady [87]).

adjustments must be made to F_w whenever α is changed. These may be
determined with Eq. 16.34.

To obtain economic operation of a treatment plant the proper recycle ratio, α,
must be selected because the cost of operating the recycle pumps is directly proportional to α. Furthermore, if sludge wastage is from the settler underflow, α
should be as small as possible in order to maximize the waste sludge concentration
thereby reducing disposal costs. The minimum feasible recycle ratio for a given
settler will depend upon both the mass flow rate of solids to it and the thickening
characteristics of the sludge. If the recycle ratio is made smaller than the minimum, it will be impossible to remove solids from the bottom of the settler as rapidly
as they are entering the top, so that they will accumulate in the settler and eventually be lost in the effluent, causing process failure. A settler which is being
operated to maximize the concentration of solids in the underflow is poised on the
verge of failure and thus when an operator chooses a recycle ratio he must be certain
that it is sufficient to allow satisfactory operation of the settler over the entire
range of input conditions anticipated. In other words, he will have to sacrifice a
degree of sludge thickening in order to provide a reasonable margin of safety against
settler failure. The task of the operator can be made simpler if the design engineer
provides a chart like that shown in Fig. 16.15 [62,87]. The solid lines are the
MLSS concentrations to be expected for given wastewater flows and concentrations at
a fixed MCRT. They are generated from Eq. 13.67. The dashed lines represent the
maximum MLSS concentrations that the settler can handle at a given plant flow and

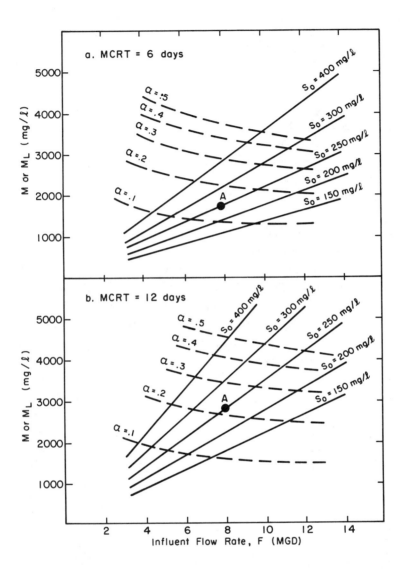

Figure 16.15. Operational diagrams for two MCRT's: (a) Six days and (b) Twelve days. *(Adapted from Roper and Grady [62]).*

recycle ratio. They must be generated from a sludge thickening analysis [62,65] and cannot be generated from Eq. 13.42 which just came from a mass balance on the settler without regard to whether the desired underflow concentration could actually be attained. The allowable operating region for a given recycle ratio is that area which lies below the dashed line corresponding to it. Conversely, the region of settler failure is the area above that line. Thus, in order for the system to operate properly, the operating point set by the influent flow and concentration must lie below the dashed line of the recycle ratio employed. Furthermore, the closer the operating point lies to the dashed line, the greater the underflow sludge concentration will be, but the closer the system will be to failure. This allows the operator to choose easily a recycle ratio that will give him a good balance between maximum thickening and safety.

Operational charts for two different MCRT's are shown in Figs. 16.15a and b, where it can be seen that the enhanced settling at longer MCRT's causes the dashed lines to be shifted upward. However, the solid lines are also shifted upward because of the greater mass of sludge carried in the system. If wastage is from the settler underflow, the MCRT may be increased either by reducing F_w or by increasing α (Eq. 16.34). If the former is done, the operating point might move above the dashed line, leading to system failure. Compare point A in Fig. 16.15a to A in 16.15b; an α of 0.2 is sufficient at an MCRT of 6 days but not at 12. This suggests that the safest way to increase the MCRT is by changing the recycle ratio, α. After the plant has reached the new MCRT, the new operational diagram can be used to reduce α, thereby minimizing pumping costs. Control, therefore, becomes a three step action: (1) use of the calibration chart to determine the value of the calibration coefficient; (2) use of a tuned hydraulic control formula to maintain or adjust the MCRT; and (3) use of the operational diagram to protect against system failure due to settler overload while minimizing pumping costs.

There are some common operational practices which are counterproductive and which, therefore, should not be used. One of these is the return of waste activated sludge to the primary settler. This practice allows organisms which have been removed from the activated sludge system to return to it through inefficiencies in the primary settler, thereby making determination of the MCRT extremely difficult. Another is to maintain a constant MLSS concentration in the reactor even though the influent conditions have changed. Equation 13.31 showed that the cell concentration in a reactor of volume V is directly proportional to the influent flow rate and concentration as long as the MCRT is held constant. Consequently it is impossible to maintain both the MCRT and the MLSS concentration constant when the influent conditions change. Finally, many plants are operated so that the waste sludge flow rate is a constant percentage of the recycle flow. Unfortunately, this can lead to gross variations in MCRT as the influent conditions change. The easiest and simplest procedure is just to employ tuned hydraulic control.

16.4.3 Short-Term (Transient) Operational Control

Superimposed upon the long-term changes in process loading will be diurnal fluctuations and shock loads. Although control of MCRT at a constant value has a beneficial effect upon the long-term performance of the process it does little to dampen short-term perturbations. This is not to imply that it has no effect, however, because activated sludge reactors with long MCRT's have a better response to short-term variations in loading than reactors with short MCRT's [58,59]. It should be emphasized, however, that maintenance of a given MCRT is a long-term control strategy and that attempts should not be made to vary the MCRT in response to short-term changes in the influent. The fact that the MCRT is on the order of 3 to 15 days whereas short-term changes occur in a matter of hours will doom such attempts to failure and may even disrupt the long-term stability of the process. Any control action to offset the effects of short-term changes must be made within the constraints imposed by the maintenance of a desired long-term MCRT.

The most effective way of reducing the effects of short-term perturbations in wastewater flow and strength is to dampen their magnitude through equalization. Total equalization is economically infeasible, however, so there are two alternative means by which the effects of the diminished short-term variations can be reduced: simple hydraulic techniques and computer aided control. Relatively little can be accomplished with simple hydraulic control because of the sluggish transient response of activated sludge and because of the limited amount of extra sludge that may be stored in the system. The only variable subject to hydraulic control is the recycle flow so the question arises as to what recycle policy should be followed during the day as the influent flow changes. Should α be held constant throughout the day so that the recycle flow rate changes in response to changes in the influent flow, or should the recycle flow rate be held constant at a value proportional to the average influent flow? Both Lech [88] and Busby [89] have studied this problem using dynamic simulation and have shown that a constant recycle ratio will always give performance equal to or better than that achieved with a constant recycle flow. Better control of the activated sludge process can be achieved by the use of computer aided control provided that it is coupled with changes in the contacting pattern (such as changing from CMAS to SAAS or shifting the distribution of influent and recycle within SAAS) or with the use of a reservoir of excess sludge which can be returned to the reactor as needed [88,89,90]. Although this topic is currently the subject of considerable research effort, a discussion of the many proposed techniques should be based upon an understanding of the transient-state response of activated sludge, a subject beyond the scope of this book. Thus we will not consider control further here. Suffice it to say that considerable improvement in activated sludge performance could be achieved if all plants at least practiced long-term control of MCRT. The engineer's responsibility does not stop with the design of the plant. He must ensure that it can be operated and that control techniques are available which are appropriate to the size and complexity of the installation.

16.5 EXAMPLES OF ACTIVATED SLUDGE IN PRACTICE

A list of the wastewaters upon which the activated sludge process has been used would include almost all biodegradable wastewaters because this biochemical operation is by far the most popular of all. Since its inception it has found wide usage for domestic wastewater treatment. In addition, its ability to produce an effluent of excellent quality, coupled with its resistance to upset by shock loads, has made it quite popular in the field of industrial wastewater treatment. For example, a review of wastewater treatment in the paper and allied products industry references 20 articles which appeared in 1975 alone dealing with the use of activated sludge for the treatment of those wastewaters [91]. A further indication of the use of this process is the fact that by 1968 over 5000 CMAS plants were in operation. By now the total number of activated sludge plants of all types far exceeds that value.

General guidelines have already been given concerning the selection of reactor type, mixing conditions, MLSS concentration, and the many other factors which must be considered when applying activated sludge to a specific wastewater. Many of the decisions concerning design are specific to a particular situation and thus the selection of a few examples as typical activated sludge installations could be misleading. Therefore, even though an extensive literature on activated sludge exists, no specific examples have been selected for presentation here. Instead, the reader is referred to the review articles which are listed under Further Reading.

Before a fundamental theory of activated sludge kinetics was developed, design was generally based upon empirical guidelines devised from experience, and many have been adopted as standards [92]. The problem with such standards is that they attempt to specify factors such as the reactor space time for a typical plant, even though a typical plant is difficult to define. For example, Ten States Standards [92] calls for a space time of 5 hours for SAAS. This may or may not be sufficient, depending upon the characteristics of the wastewater and upon the designer's choice of the MCRT and MLSS concentration. Similarly, they specify a minimum air flow rate in terms of the cubic feet of air which must be supplied per pound of BOD entering the aeration basin. Again, this may provide adequate mixing conditions, or it may not, depending upon the concentration of organic matter in the wastewater and the volume of the aeration basin. The point is that design standards specified in terms of typical conditions can be misleading just as it would be misleading to present a few examples of typical plants herein. An engineer should not think that designing a plant to meet standards such as Ten States Standards is sufficient to ensure adequate performance. Rather, he should trust his own engineering judgement based upon the fundamental principles derived from research over the past twenty years. His guidelines should be in the form of a checklist.

    (1)   Is the MCRT adequate to produce the desired soluble substrate concentration?

    (2)   Does the MCRT fall within the range that will give a sludge which settles well?

    (3)   Does the waste contain adequate nutrients?

    (4)   Has the MLSS concentration been chosen to allow proper performance of the final settler?

    (5)   Has the reactor volume been selected to provide the needed $M\tau$ product?

(6)   Is the design recycle ratio adequate for the desired return sludge
      concentration and final settler size?

(7)   Are the recycle pumps capable of pumping a variety of rates both above
      and below the design value?

(8)   Is the oxygen transfer system capable of providing oxygen at a rate
      sufficient to meet the peak needs of the culture?

(9)   Is the mixing energy supplied to the aeration basin sufficient to keep
      all sludge in suspension and in a well-mixed condition?

(10) Is the mixing energy below the limit that will cause floc shear?

(11) Is the sludge wastage rate sufficient to maintain the desired MCRT?

If the engineer can answer all of these questions positively, he can be reasonably
certain that the plant will perform in the way that it was intended to.

16.6   KEY POINTS

1. Regardless of the reactor configuration, all activated sludge systems share
    four common characteristics: (a) they employ aerobic flocculent slurries of
    microorganisms; (b) they separate the microorganisms from the effluent by
    sedimentation; (c) they recycle the microorganisms from the settler to the
    reactor; and (d) their performance is dependent on the MCRT.

2. There are seven major types of activated sludge: conventional (CAS), tapered
    aeration (TAAS), step aeration (SAAS), contact stabilization (CSAS), completely
    mixed (CMAS), extended aeration (EAAS), and pure oxygen (POAS).

3. With the exception of extended aeration activated sludge, all of the activated
    sludge variations normally have an MCRT of between 3 and 15 days.

4. Two mechanisms have been proposed to explain the way in which contact stabili-
    zation activated sludge works: storage/metabolism and growth/decay. The
    latter one appears to be the more general.

5. The primary objective of the activated sludge process is the removal of soluble
    organic matter, although colloidal and suspended organics will be entrapped in
    the floc and removed as well. Its most economic application is to wastewaters
    in which the soluble T_bOD lies between 50 and 4000 mg/liter.

6. The chief merit associated with the activated sludge process is that it produces
    an effluent of high quality at reasonable cost. Its main defect is that it is
    a complicated process which requires relatively sophisticated operation.

7. Because of the requirement for sedimentation of the sludge, the MCRT in the
    activated sludge process is generally held between 3 and 15 days, although
    longer values are often needed for the treatment of industrial wastewaters.
    When larger values are used, special consideration should be given to the
    settling characteristics of the sludge.

8. The theoretical performances of all of the activated sludge process variations
    can be obtained from models of either single CSTR's or CSTR's in series with
    cell recycle.

9. Because of the lower limit on the MCRT, there is usually little soluble substrate
    remaining in the effluent from the activated sludge process. Consequently, in
    practice, there is usually little difference in performance between variations
    which employ one CSTR and those which have several in series.

10. Compared with TAAS and CMAS, SAAS and CSAS can achieve the same MCRT with less total reactor volume, although the excess sludge production and oxygen requirement will be about the same.

11. Errors are likely to result if the models of Chapter 13 are applied directly to EAAS, thus the decay of solids in EAAS should be modeled as it is in aerobic digestion, as discussed in Chapter 17.

12. The equations for a CSTR with cell recycle developed in Chapter 13 may be applied without modification to the design of CMAS.

13. Because it will give a conservative design, the growth/decay model can be used to develop design equations for CSAS.

14. Explicit design equations for CSAS can be developed after assuming that: (a) the viability of the cells in the CSAS contact tank can be approximated by the viability in a CMAS reactor of equal MCRT and (b) no substrate removal occurs in the stabilization tank. The equations can then be used to size the contact tank to give the desired effluent soluble substrate concentration and to size the stabilization tank to produce the MCRT required to make the sludge settle.

15. The process design for activated sludge requires that the following items be specified: The MCRT, the reactor volume and the associated MLSS concentration, the recycle ratio, the wastage flow rate, and the mixing and oxygen requirements.

16. Kinetic parameters may be determined using lab-scale CMAS reactors and applied directly to the design of CMAS systems. If the parameters are applied to the design of one of the other types of activated sludge, however, the performance of the particular system configuration should be verified by running a prototype unit.

17. Considerations of mixing, oxygen transfer, and liquid-solid separation suggest that the MLSS concentration in CMAS, TAAS, the contact tank of CSAS, and the last tank of SAAS should not be less than 1500 mg/liter nor greater than 5000 mg/liter. The upper limit is higher for POAS.

18. In order to achieve proper mixing in a CMAS reactor without undue floc shear the power input should be between 0.5 and 2.25 HP/1000 ft^3 (13.8 - 60 kw/1000 m^3). If diffused air is used, the air flow rate should lie between 20 and 90 scfm/1000 ft^3.

19. Heat losses through activated sludge aeration basins are generally not large; consequently seasonal temperature changes reflect primarily the changes in the temperature of the influent wastewater. Effects of temperature upon performance may be compensated for by shifting the MCRT.

20. By making use of the fact that microbial cells are approximately 10.5% nitrogen, the nutrient requirements for an activated sludge reactor can be predicted from the equation for the yield.

21. The following steps should be followed when designing CMAS:

    (a) Estimate the MCRT required to give the desired effluent soluble substrate concentration.

    (b) Check the MCRT against the lower constraint of 3 days and adjust it if necessary.

    (c) Calculate the Mτ product using the MCRT.

    (d) Choose either M or τ and calculate the aeration tank volume.

(e)   Determine the required recycle ratio.

(f)   Determine the required sludge wastage rate.

(g)   Determine the excess microorganism production rate.

(h)   Calculate the oxygen requirement and estimate the air flow rate or power required to supply the needed oxygen.

(i)   Using the aeration tank volume, determine the air flow rate or power needed to ensure complete mixing.

(j)   Compare the requirements of (h) and (i) and provide the larger of the two.

22. The following steps should be followed when designing CSAS:

(a)   Estimate the contact tank specific growth rate required to give the desired effluent soluble substrate concentration.

(b)   Choose an MCRT to give a sludge that will settle well.

(c)   Estimate the viability of the sludge.

(d)   Calculate the fraction of cells in the stabilization tank.

(e)   Calculate the $M_c V_c$ product.

(f)   Choose either M_c or V_c.

(g)   Choose the recycle ratio.

(h)   Calculate V_s.

(i)   Determine the required sludge wastage rate.

(j)   Determine the excess microorganism production rate.

(k)   Calculate the oxygen requirement for each reactor and estimate the air flow rate or power required to supply the needed oxygen.

(l)   Determine the air flow rate or power required to provide complete mixing in each reactor.

(m)   Compare the requirements of (k) and (l) for each reactor and provide the larger of the two.

23. Visual observations can be a valuable aid to the control of an activated sludge plant.

24. The unintentional loss of suspended solids in the final effluent from the activated sludge process contributes materially to the total quantity of MLSS wasted daily and failure to account for it in the calculation of the MCRT can cause a substantial error.

25. An operational diagram such as in Fig. 16.15 allows a plant operator to visualize whether a particular recycle ratio can be safely used, thereby allowing him to minimize sludge pumping costs while maintaining a reasonable margin of safety against settler failure.

26. The maintenance of a given MCRT is a long-term control strategy and no attempt should be made to vary the MCRT in response to short-term changes in the influent.

27. Rather than rely upon "typical" design standards, an engineer designing an activated sludge plant should ensure that the eleven questions in Section 16.5 can all be answered positively.

16.7 STUDY QUESTIONS

The study questions are arranged in the same order as the text:

Section	Questions
16.1	1-5
16.2	6-11
16.3	12-17
16.4	18-21
16.5	None

1. Describe the general characteristics of the activated sludge process.

2. Prepare a table comparing the following factors for the seven major variations of the activated sludge process: MCRT, space time, MLSS concentration, recycle ratio, reactor configuration, and type of mathematical model.

3. Compare the two mechanisms by which contact stabilization activated sludge is thought to operate.

4. List the main merits and defects of the activated sludge process.

5. List the relative merits and defects of each of the variations of the activated sludge process.

6. Explain why it is necessary to hold the MCRT between 3 and 15 days in order to obtain an activated sludge which settles well.

7. Three benefits have been attributed to pure oxygen activated sludge: (a) increased rate of treatment; (b) increased density of sludge and rate of sludge settlement; and (c) reduced rate of sludge production. Discuss and evaluate these claims.

8. Why will all variations of the activated sludge process have approximately the same efficiency of soluble substrate removal?

9. Why is it preferable to use the growth/decay concept rather than the storage/metabolism concept when developing design equations for CSAS in the absence of concrete evidence concerning the actual mechanism in a particular wastewater?

10. Why can CSAS be designed to achieve partial treatment of a wastewater but CMAS cannot?

11. What reactions are assumed to occur in the contact and stabilization tanks during the derivation of the CSAS equations?

12. Why is it wise to run a prototype reactor to verify a design based upon kinetic parameters determined in lab-scale CMAS reactors, when the activated sludge variation being designed is something other than CMAS?

13. Why should the choice of MLSS concentration in a CMAS reactor normally lie between 1500 and 5000 mg/liter?

14. Why must both an upper and a lower limit be placed on the mixing energy supplied to an activated sludge reactor?

15. Why are heat losses from diffused air activated sludge systems less than from CMAL's?

16. An industrial wastewater has the characteristics listed in Table SQ16.1.

TABLE SQ16.1
Characteristics of an Industrial Wastewater

Characteristic	Value	Characteristic	Value
Flow	5×10^3 m^3/hr	Y_g	0.55 mg cells/mg T$_b$OD
Soluble T$_b$OD	400 mg/liter	b	0.007 hr^{-1}
Inert solids	60 mg/liter	γ	0.003 hr^{-1}
μ_m	0.30 hr^{-1}	β	1.25 mg O$_2$/mg cells
K_s	60 mg/liter as T$_b$OD		

Assuming that the flow and concentration are constant, and that the parameter values are for the most probable waste temperature, perform a process design for a CMAS system to produce a soluble T$_b$OD of no more than 10 mg/liter. Give a justification for each arbitrarily selected value.

17. Repeat Study Question 16 for a CSAS system to produce a soluble T$_b$OD of 10 mg/liter. Assume that β is the same for cells in the contact and stabilization tanks.

18. Define each of the following terms and tell what each indicates about the operation of an activated sludge plant:

    (a) Clumping

    (b) Ashing

    (c) Pin floc

    (d) Straggler floc

19. If the CMAS process in Example 16.3.5-1 has an effluent suspended solids concentration of 15 mg/liter, at what rate must sludge be wasted from the settler underflow to maintain the MCRT at 168 hrs? What will be the calibration coefficient? If the recycle ratio is changed to 0.5, what wastage rate will be required to maintain the MCRT at 168 hours if K remains constant?

20. Assume that the calibration chart in Fig. 16.14 describes the performance of an activated sludge system with a volume of 100 m^3 receiving a flow of 25 m^3/hr at a concentration of 300 mg/liter T$_b$OD. At what rate must sludge be wasted from the settler underflow to maintain an MCRT of 200 hours if the concentration of suspended solids in the final effluent is 25 mg/liter and the recycle ratio is 0.5?

21. Explain how the various lines in Fig. 16.15 are obtained and what they represent.

REFERENCES AND FURTHER READING

1. E. Ardern and W. T. Lockett, "Experiments on the oxidation of sewage without the aid of filters," *Journal of the Society of Chemical Industries,* 33, 523, 1914.

2. K. L. Murphy and B. I. Boyko, "Longitudinal mixing in spiral flow aeration tanks," *Journal of the Sanitary Engineering Division, ASCE,* 96, 211-221, 1970.

3. L. H. Kessler, *et al.* "Tapered aeration of activated sludges," *Municipal Sanitation,* 7, 268, 1936.

4. R. H. Gould, "Operating experiences in New York City," *Sewage Works Journal,* 14, 70, 1942.

5. A. H. Ulrich and M. W. Smith, "The biosorption process of sewage and waste treatment," *Sewage and Industrial Wastes,* 23, 1248, 1951.

6. C. Smallwood, Jr., "Adsorption and assimilation in activated sludge," *Journal of the Sanitary Engineering Division, ASCE,* 83, #SA4, 1-12, 1957.

7. P. H. Jones, "A mathematical model for contact stabilization modification of the activated sludge process," *Advances in Water Pollution Research,* 5, paper II-5, 1970.

8. P. Krishnan and A. F. Gaudy Jr., "Mechanisms and kinetics of substrate utilization at high biological solids concentrations," *Proceedings of the 21st Industrial Waste Conference,* Purdue University Engineering Extension Series No. 121, 495-510, 1966.

9. C. F. Walters, *et al.* "Microbial substrate storage in activated sludge," *Journal of the Sanitary Engineering Division, ASCE,* 94, 257-269, 1968.

10. R. E. Speece, *et al.* "Cell replication and biomass in the activated sludge process," *Water Research,* 7, 361-374, 1973.

11. J. Chudoba, "Residual organic matter in activated sludge process effluents - I - Degradation of saccharides, fatty acids, and amino acids under batch conditions," *Scientific Papers of the Institute of Chemical Technology, Prague, Technology of Water,* 12, 39-76, 1967.

12. H. A. Khararjian and J. H. Sherrard, "Batch aerobic treatment of a colloidal wastewater," *Journal of the Water Pollution Control Federation,* 49, 1985-1992, 1977.

13. W. J. Miller, "A study of the mechanism and kinetics of the removal of COD from domestic wastewaters by the activated sludge process," thesis presented to the University of California at Berkeley in 1970, in partial fulfillment of the requirements for the degree Doctor of Philosophy.

14. P. L. McCarty, "Discussion of 'The role of enzymes in the contact stabilization process'," *Advances in Water Pollution Research,* 2, 373-377, 1966.

15. D. M. Orhon and D. Jenkins, "The mechanism and design of the contact stabilization activated sludge process," *Advances in Water Pollution Research,* 6, 353-362, 1972.

16. W. Gujer and D. Jenkins, "The contact stabilization activated sludge process - Oxygen utilization, sludge production, and efficiency," *Water Research,* 9, 553-560, 1975.

17. R. E. McKinney, *et al.* "Design and operation of a complete mixing activated sludge waste treatment plant," *Sewage and Industrial Wastes,* 30, 287-295, 1958.

18. M. T. Garrett and C. N. Sawyer, "Kinetics of removal of soluble BOD by activated sludge," *Proceedings of the 7th Industrial Waste Conference,* Purdue University Engineering Extension Series No. 79, 51-77, 1952.

19. J. G. Albertson, *et al.* "Investigation of the use of high purity oxygen aeration in the conventional activated sludge process," *USFWQA Water Pollution Control Research Series,* Report No. 17050 DNW, May 1970.

20. A. R. Tarrer, *et al.* "Optimal activated sludge design under uncertainty," *Journal of the Environmental Engineering Division, ASCE,* 102, 657-673, 1976.

21. C. P. L. Grady Jr., "Simplified optimization of activated sludge process," *Journal of the Environmental Engineering Division, ASCE,* 103, 413-429, 1977.

22. J. Chudoba, *et al.* "Control of activated sludge filamentous bulking - I - Effect of the hydraulic regime or degree of mixing in an aeration tank," *Water Research,* 7, 1163-1182, 1973.

23. J. H. Rensink, "New approach to preventing bulking sludge," *Journal of the Water Pollution Control Federation,* 46, 1888-1894, 1974.

24. K. W. Young *et al.* "Sludge considerations of oxygen-activated sludge" in *Applications of Commercial Oxygen to Water and Wastewater Systems,* edited by R. E. Speece and J. F. Malina Jr., Center for Research in Water Resources, The University of Texas at Austin, 254-267, 1973.

25. J. Chudoba, *et al.* "Control of activated sludge filamentous bulking - II - Selection of microorganisms by means of a selector," *Water Research,* 7, 1389-1406, 1973.

26. W. C. Boyle *et al.* "Flocculation phenomena in biological systems," *Advances in Water Quality Improvement,* 1, 287-312, 1968.

27. R. H. Harris and R. Mitchell, "The role of polymers in microbial aggregation," *Annual Review of Microbiology,* 27, 27-50, 1973.

28. C. R. Curds, *et al.* "An experimental study of the role of ciliated protozoa in the activated sludge process," *Water Pollution Control,* 67, 312-329, 1968.

29. C. R. Curds, "The flocculation of suspended matter by *Paramecium caudatum,*" *Journal of General Microbiology,* 33, 357-363, 1963.

30. M. W. Tenney and W. Stumm, "Chemical flocculation of microorganisms in biological waste treatment," *Journal of the Water Pollution Control Federation,* 37, 1370-1388, 1965.

31. J. L. Pavoni, *et al.* "Bacterial exocellular polymers and biological flocculation," *Journal of the Water Pollution Control Federation,* 44, 414-431, 1972.

32. J. J. Bisogni and A. W. Lawrence, "Relationships between biological solids retention time and settling characteristics of activated sludge," *Water Research,* 5, 753-763, 1971.

33. J. F. Malina *et al.* "Design guides for biological wastewater treatment processes," *Environmental Pollution Agency Water Pollution Control Research Series,* Report No. 11010 ESQ, August 1971.

34. D. L. Ford and W. W. Eckenfelder, Jr., "Effect of process variables on sludge floc formation and settling characteristics," *Journal of the Water Pollution Control Federation,* 39, 1850-1859, 1967.

35. Y. Magara, *et al.* "Biochemical and physical properties of an activated sludge on settling characteristics," *Water Research,* 10, 71-77, 1976.

36. A. W. West, "Operational control procedures for the activated sludge process, Part I, Observations," *Environmental Protection Agency Report* No. EPA-330/9-74-001-a, April 1973.

37. R. H. Harris and R. Mitchell, "Inhibition of the flocculation of bacteria by biopolymers," *Water Research,* 9, 993-999, 1975.

38. H. J. Campbell Jr. and R. F. Rocheleau, "Waste treatment at a complex plastics manufacturing plant," *Journal of the Water Pollution Control Federation,* 48, 256-273, 1976.

39. T. R. Stall and J. H. Sherrard, "Effect of wastewater composition and cell residence time on phosphorus removal in activated sludge," *Journal of the Water Pollution Control Federation,* 48, 307-322, 1976.

40. C. L. Weddle and D. Jenkins, "The viability and activity of activated sludge," *Water Research,* 5, 621-640, 1971.

41. A. K. Upadhyaya and W. W. Eckenfelder Jr., "Biodegradable fraction as an activity parameter of activated sludge," *Water Research,* 9, 691-694, 1975.

42. J. H. Sherrard and E. D. Schroeder, "Cell yield and growth rate in activated sludge," *Journal of the Water Pollution Control Federation,* 45, 1889-1897, 1973.

43. G. W. Kumke, *et al.* "Performance of internally clarified activated sludge process treating combined petrochemical - municipal waste," *Proceedings of the 23rd Industrial Waste Conference,* Purdue University Engineering Extension Series No. 132, 567-582, 1968.

44. J. J. Westrick, *et al.* "A field study of the extended aeration process," *Proceedings of the 21st Industrial Waste Conference,* Purdue University, Engineering Extension Series No. 121, 424-439, 1966.

45. E. D. Toerber, *et al.* "Comparison of completely mixed and plug flow biological systems," *Journal of the Water Pollution Control Federation,* 46, 1995-2014, 1974.

46. A. G. Boon, "Technical review of the use of oxygen in the treatment of waste water," *Water Pollution Control,* 75, 206-213, 1976.

47. M. J. Humenick and J. E. Ball, "Kinetics of activated sludge oxygenation," *Journal of the Water Pollution Control Federation,* 46, 735-747, 1974.

48. N. Banks, "U.K. work on the use of oxygen in the treatment of waste water, associated with the CCMS advanced waste water treatment project," *Water Pollution Control,* 75, 214-220, 1976.

49. T. D. Chapman, *et al.* "Effect of high dissolved oxygen concentration in activated sludge systems," *Journal of the Water Pollution Control Federation,* 48, 2486-2510, 1976.

50. D. S. Parker and M. S. Merrill, "Oxygen and air activated sludge: Another view," *Journal of the Water Pollution Control Federation,* 48, 2511-2528, 1976.

51. J. W. Matson, *et al.* "Oxygen supply limitations in full scale biological treatment systems," *Proceedings of the 27th Industrial Waste Conference,* Purdue University Engineering Extension Series No. 141, 894-903, 1972.

52. W. Gujer and D. Jenkins, "A nitrification model for the contact stabilization activated sludge process," *Water Research,* 9, 561-566, 1975.

53. C. E. Adams, *et al.* "A kinetic model for design of completely-mixed activated sludge treating variable-strength industrial wastewaters," *Water Research,* 9, 37-42, 1975.

54. G. T. Daigger and C. P. L. Grady Jr., "Factors affecting effluent quality from fill-and-draw activated sludge reactors," *Journal of the Water Pollution Control Federation,* 49, 2390-2396, 1977.

55. B. L. Goodman and A. J. Englande, Jr., "A unified model of the activated sludge process," *Journal of the Water Pollution Control Federation,* 46, 312-332, 1974.

56. L. D. Benefield and C. W. Randall, "Design procedure for a contact stabilization activated sludge process," *Journal of the Water Pollution Control Federation,* 48, 147-152, 1976.

57. D. Orhon, "Discussion of 'Design procedure for a contact stabilization activated sludge process," *Journal of the Water Pollution Control Federation,* 49, 865-869, 1977.

58. C. P. L. Grady Jr., "A theoretical study of activated sludge transient response," *Proceedings of the 26th Industrial Waste Conference,* Purdue University Engineering Extension Series No. 140, 318-335, 1971.

59. J. H. Sherrard and A. W. Lawrence, "Response of activated sludge to step increase in loading," *Journal of the Water Pollution Control Federation,* 47, 1848-1856, 1975.

60. R. I. Dick and A. R. Javaheri, "Discussion of 'Unified basis for biological treatment design and operation'," *Journal of the Sanitary Engineering Division, ASCE,* 97, 234-238, 1971.

61. M. D. Mynhier and C. P. L. Grady Jr., "Design graphs for activated sludge process," *Journal of the Environmental Engineering Division, ASCE,* 101, 829-846, 1975.

62. R. E. Roper Jr. and C. P. L. Grady, Jr., "Activated sludge hydraulic control techniques evaluation by computer simulation," *Journal of the Water Pollution Control Federation,* 46, 2565-2578, 1974.

63. J. B. Duggan and J. L. Cleasby, "Effect of variable loading on oxygen uptake," *Journal of the Water Pollution Control Federation,* 48, 540-550, 1976.

64. H. W. Parker, *Wastewater Systems Engineering,* Prentice-Hall, Inc., Englewood Cliffs, N.J., 166-172; 272-299, 1975.

65. R. I. Dick, "Role of activated sludge final settling tanks," *Journal of the Sanitary Engineering Division, ASCE,* 96, 423-436, 1970.

66. W. W. Eckenfelder, Jr., "Comparative biological waste treatment design," *Journal of the Sanitary Engineering Division, ASCE,* 93, No. SA6, 157-170, 1967.

67. A. W. Busch, *Aerobic Biological Treatment of Waste Waters, Theory and Practice,* Oligodynamics Press, Houston, Texas, 317-328, 1971.

68. W. W. Eckenfelder, Jr. and A. J. Englande, "Temperature effects on biological waste treatment processes," *International Symposium on Water Pollution Control in Cold Climates, Water Pollution Control Research Series,* Report No. 16100 EXH, 180-190, 1971.

69. J. D. Boyle, "Biological treatment process in cold climates," *Water and Sewage Works Reference Number,* 123, R-28 - R-50, 1976.

70. D. E. Bloodgood, "The effect of temperature and organic loading upon activated sludge plant operation," *Sewage Works Journal,* 16, 913-924, 1944.

71. D. L. Ford, *et al*. "Temperature prediction in activated sludge basins using mechanical aerators," *Proceedings of the 27th Industrial Waste Conference*, Purdue University Engineering Extension Series No. 141, 587-598, 1972.

72. J. H. Sherrard and E. D. Schroeder, "Stoichiometry of industrial biological wastewater treatment," *Journal of the Water Pollution Control Federation*, 48, 742-747, 1976.

73. J. H. Sherrard and L. D. Benefield, "Elemental distribution diagrams for biological wastewater treatment," *Journal of the Water Pollution Control Federation*, 48, 562-569, 1976.

74. J. L. Carter and R. E. McKinney, "Effects of iron on activated sludge treatment," *Journal of the Environmental Engineering Division, ASCE*, 99, 135-152, 1973.

75. D. K. Wood and G. Tchobanoglous, "Trace elements in biological waste treatment," *Journal of the Water Pollution Control Federation*, 47, 1933-1945, 1975.

76. J. Nemke, "Visual observations can be process control aids," *Deeds and Data*, 1-8, Sept. 1975.

77. W. O. Pipes, "Actinomycete scum production in activated sludge process," *Journal of the Water Pollution Control Federation*, 50, 628-634, 1978.

78. J. F. Andrews and C. R. Lee, "Dynamics and control of a multi-stage biological process," *Fermentation Technology Today*, 4, 55-63, 1972.

79. M. Sezgin, *et al*. "A unified theory of filamentous activated sludge bulking," *Journal of the Water Pollution Control Federation*, 50, 362-381, 1978.

80. C. A. Cole, *et al*. "Hydrogen peroxide cures filamentous growth in activated sludge," *Journal of the Water Pollution Control Federation*, 45, 829-836, 1973.

81. J. T. Pfeffer and Y. Chang, "Use of iron salts for control of activated sludge bulking caused by *Sphaerotilus*," University of Illinois Water Resources Center Report #127, Urbana, Ill., 1977.

82. T. R. Stall and J. H. Sherrard, "Evaluation of control parameters for the activated sludge process," *Journal of the Water Pollution Control Federation*, 50, 450-457, 1978.

83. M. E. Burchett and G. Tchobanoglous, "Facilities for controlling the activated sludge process by mean cell residence time," *Journal of the Water Pollution Control Federation*, 46, 973-979, 1974.

84. M. T. Garrett Jr., "Hydraulic control of activated sludge growth rate," *Sewage and Industrial Wastes*, 30, 253-261, 1958.

85. D. Jenkins and W. E. Garrison, "Control of activated sludge by mean cell residence time," *Journal of the Water Pollution Control Federation*, 40, 1905-1919, 1968.

86. L. F. Walker, "Hydraulically controlling solids retention time in the activated sludge process," *Journal of the Water Pollution Control Federation*, 43, 30-39, 1971.

87. R. E. Roper Jr. and C. P. L. Grady Jr., "A simple effective technique for controlling solids retention time in activated sludge plants," *Journal of the Water Pollution Control Federation*, 50, 702-708, 1978.

88. R. F. Lech, *et al.* "Automatic control of the activated sludge process - II - Efficacy of control strategies," *Water Research,* 12, 91-99, 1978.

89. J. B. Busby and J. F. Andrews, "Dynamic modeling and control strategies for the activated sludge process," *Journal of the Water Pollution Control Federation,* 47, 1055-1080, 1975.

90. M. K. Stenstrom, "A dynamic model and computer compatible control strategies for wastewater treatment plants," thesis presented to Clemson University, at Clemson, South Carolina, in 1975, in partial fulfillment of the requirements for the degree of Doctor of Philosophy.

91. F. W. Gove and I. Gellman, "Paper and allied products," *Journal of the Water Pollution Control Federation,* 48, 1234-1263, 1976.

92. Great Lakes - Upper Mississippi River Board of State Sanitary Engineers, *Recommended Standards for Sewage Works,* Health Education Service, Albany, N.Y., 1971.

Further Reading

Andrews, J. F., "Dynamic models and control strategies for wastewater treatment processes," *Water Research,* 8, 261-289, 1974.

Atkinson, B. and Daoud, I. S., "Microbial flocs and flocculation in fermentation process engineering," *Advances in Biochemical Engineering,* 4, 41-124, 1976.

Benefield, L. D. and Randall, C. W., *Biological Process Design for Wastewater Treatment,* Prentice-Hall, Inc., Englewood Cliffs, N.J., 1980. See Chapter 4.

Eckenfelder, W. W. Jr., *et al.* "Scale-up of biological wastewater treatment reactors," *Advances in Biochemical Engineering,* 2, 145-180, 1972.

Eckenfelder, W. W. Jr. and O'Connor, D. J., *Biological Waste Treatment,* Pergamon Press, New York, N.Y., 1961.

Gaudy, A. F. Jr. and Kincannon, D. F., "Comparing design models for activated sludge," *Water and Sewage Works,* 123, #2, 66-70, 1977.

Gaudy, A. F. Jr. and Kincannon, D. F., "Functional design of activated sludge process," *Water and Sewage Works,* 123, #9, 76-81, 1977.

Lawrence, A. W. and McCarty, P. L., "Unified basis for biological treatment design and operation," *Journal of the Sanitary Engineering Division, ASCE,* 96, 757-778, 1970.

McWhirter, J. R. *The Use of High-Purity Oxygen in the Activated Sludge Process,* Vols I and II. CRC Press, Inc., West Palm Beach, Florida, 1978.

Metcalf and Eddy, Inc., *Wastewater Engineering: Treatment/Disposal/Reuse,* 2nd Edition, McGraw-Hill Book Co., New York, N.Y., 1979, See Chapter 10.

Rich, L. G., *Unit Processes of Sanitary Engineering,* John Wiley & Sons, Inc., New York, N.Y., 1963. See Chapter 2.

Sawyer, C. H., "Milestones in the development of the activated sludge process," *Journal of the Water Pollution Control Federation,* 37, 151-162, 1965.

Schroeder, E. D., *Water and Wastewater Treatment,* McGraw-Hill, Inc., New York, 1977.

Speece, R. E. and Malina, J. F. Jr., (editors), *Applications of Commercial Oxygen to Water and Wastewater Systems,* Center for Research in Water Resources, The University of Texas at Austin, 1973.

Stewart, M. J., "Activated Sludge Process Variations, The Complete Spectrum," *Water and Sewage Works, Reference Number,* 111, R241-R262, 1964.

Tyteca, D., *et al.* "Mathematical Modeling and Economic Optimization of Wastewater Treatment Plants," *CRC Critical Reviews in Environmental Control,* 8, 1-89, 1977.

CHAPTER 17

AEROBIC DIGESTION

The major by-product from the aerobic degradation of soluble organic matter is excess
microorganisms, commonly referred to as secondary sludge. In fact, so much sludge is
produced that the disposal of primary plus secondary sludge constitutes a major expenditure in wastewater treatment. Part of this expenditure arises from the need to
stabilize and dewater the sludge prior to disposal. Although anaerobic digestion has
been the traditional means of sludge stabilization, the desire for a more easily
operated process led to the development of aerobic digestion, which has been defined
as "...a process wherein thickened sludges separated from their associated liquors
undergo stabilization during prolonged aeration" [1].

17.1 DESCRIPTION OF AEROBIC DIGESTION

17.1.1 Physical Characteristics

Although primary sludges may have appreciable soluble organic matter associated with them the predominant organic matter in both primary and secondary sludges is insoluble. Consequently, the main objective of aerobic digestion is the destruction of insoluble organic matter in an aerobic environment--an activity which is normally performed in three types of reactors: a CSTR, a CSTR with recycle, and a batch reactor (See Table 1.2). Actually, batch reactors are seldom used in the field, their primary use being to provide design data. Furthermore, as was seen in Chapter 13, the performance of a CSTR with recycle is the same as that of a simple CSTR as long as the MCRT of the former is equal to the space time of the latter. Because of those facts, plus the fact that simple CSTR's are used more than recycle reactors for aerobic digestion [2], the discussion in this chapter will be limited to reactors without recycle. It should be clear to the reader by this time, however, that recycle could be incorporated by a simple extension of the concepts presented herein.

Because of the relatively long MCRT's associated with aerobic digestion, a complex microbial community will normally exist, ranging from bacteria to higher organisms such as nematodes. This complete food chain is not reflected explicitly in the mathematical model used for design. Nevertheless, as long as the parameter values used in the equations are determined on the actual waste, the model should accurately reflect the average performance to be expected from the system.

There are no special physical characteristics associated with aerobic digesters although many are concrete tanks around 15 feet deep, similar in design to activated sludge aeration chambers. Many, in fact, are converted activated sludge basins. Since the design is usually based on the assumption of complete mixing the only requirement with respect to reactor configuration is the fulfillment of that assumption. Mixing and oxygen transfer are accomplished by both bubble and mechanical aeration using air as the oxygen source although pure oxygen has been utilized when it is used in the remainder of the plant. With the exception of the pure oxygen systems the tanks are usually uncovered. In small plants it is common for the digesters to be operated on a fill-and-draw schedule so that the tanks are equipped with decant pipes. In large plants, on the other hand, operation is continuous so that a settling basin is usually provided to thicken the treated sludge and separate it from the bulk of the liquid prior to discharge.

17.1.2 Applications

Aerobic digestion is generally most applicable to secondary sludges such as those generated by the activated sludge and trickling filter processes. Because these sludges are predominantly biological solids the most important reaction will be microbial decay. Aerobic digestion has also been used on primary sludge although for best results the amount of organic material in the sludge should exceed 60% [3]. Generally, however, it has been found to be more economic to use anaerobic digestion

on primary sludge even when aerobic digestion is used for the secondary sludge [4].
This is because the large amount of nonmicrobial organic matter present will be
converted to biomass, thereby exerting a large oxygen requirement during aerobic
digestion and forming much more residual sludge than would be formed by anaerobic
decomposition.

17.1.3 Merits and Defects

Several authors [4,5,6] have enumerated the beneficial characteristics of aerobic
digestion.

(1) Aerobic digestion produces an odorless, humus-like, biologically stable
end product which can be disposed of easily.

(2) Aerobic digestion produces a supernatant which contains little soluble
biodegradable organic matter.

(3) The digested sludge usually has good dewatering characteristics, although
this should be qualified by noting that the dewatering characteristics depend upon
the digestion time.

(4) Aerobic digestion can economically handle low waste sludge concentrations,
thereby eliminating the need for sludge thickening devices associated with anaerobic
digestion of secondary sludges.

(5) When applied to secondary sludges, aerobic digestion can achieve organic
solids reductions similar to those of anaerobic systems (i.e., around 45-50%).

(6) In comparison to anaerobic digestion, more of the sludge's basic fertilizer
value is retained, making it beneficial for land application.

(7) Capital costs for aerobic digesters are less than for anaerobic digesters
because the reactors are simpler.

(8) Aerobic digesters experience fewer operational difficulties than anaerobic
ones, therefore less skillful labor can be used to operate the facility.

Aerobic digestion also suffers from some defects, however [4,5,6].

(1) Energy must be expended to supply oxygen to the system, thus operating
costs are higher than for anaerobic digestion.

(2) There is no recovery of energy in the form of methane.

(3) Aerobically digested sludges do not always clarify well, consequently,
the supernatant may be high in suspended solids.

(4) Aerobic digestion of primary sludge produces more residual sludge than
anaerobic digestion.

(5) The efficiency is dependent upon temperature and therefore it will vary
unless temperature control is practiced.

17.2   AEROBIC DIGESTER PERFORMANCE

17.2.1   Theoretical Performance

An aerobic digester receiving secondary sludge is essentially a CSTR receiving only
microbial cells and thus its theoretical performance is given by the equations in
Chapter 12. Because such an influent contains only an insignificant amount of
soluble, biodegradable organic matter the only reactions occurring will be those of
cellular death and decay. Because cell decay can be represented by a first-order
rate equation, the concentration of cells in the reactor will decline as τ is increased, although the incremental effects of increases in τ also decrease. This was
depicted in Fig. 12.7 where it was seen that the viability of the culture also
declines as τ is increased. The amount of oxygen required to bring about digestion
was also depicted in Fig. 12.7. Because the only need for oxygen is to supply the
endogenous respiration requirement, the shape of the oxygen utilization curve is the
reverse of the cell concentration curve, i.e., more oxygen will be required at longer
space times. In general, this first-order depiction of aerobic digestion is widely
accepted in the literature [1,2,3,4,5,7,8,9,10]. Furthermore, data presented by
Eckenfelder [11] indicate that when waste activated sludge is digested, β remains
relatively constant during the course of digestion, indicating that Eq. 12.41 can be
used directly to calculate the oxygen requirement as was done in Fig. 12.7.

Figure 12.14 showed that the first-order nature of cell decay will allow CSTR's
in series to achieve more solids degradation than a single tank of equal total space
time. The first tank in the chain will require slightly more oxygen as shown in
Fig. 12.15, however, the differences between tanks will not be as great as for a
system receiving soluble substrate. Finally, Fig. 12.16 demonstrated that tanks in
series should all have the same volume. It should be noted, however, that the above
statements will not be true for reactors with cell recycle in which the space time is
much less than the MCRT. The reasons for this were discussed in Section 13.2.2.

If an aerobic digester is receiving primary solids, insoluble substrate will be
present. As was pointed out in Section 12.1.3 the usual assumption is that solids
destruction occurs at a rate which is first order with respect to the concentration
of biodegradable solids. The effect of τ on the concentration of biodegradable
solids is thus similar to the cell concentration curve in Fig. 12.7, although the
curve for the oxygen requirement will be somewhat different [7,12]. Equation 12.41
can still be used to calculate the oxygen requirement, but β will probably depend
upon τ, thereby causing higher oxygen requirements at low values of τ. Such a change
in β has been suggested by data in the literature [13].

17.2.2   Inadequacies of Theory

The assumption was made in Eq. 10.54 that the specific decay rates of live and dead
cells, b_v and b_d, respectively, were equal. Although this assumption did not detract
materially from the ability of the model to predict CMAL and CMAS performance (as
long as θ_c was held within practical limits) it is primarily responsible for the

deviation of the model predictions from observations of aerobic digester performance.
Figure 17.1 contains a plot of the fraction of volatile (organic) suspended solids
(VSS) remaining as a function of τ for a CSTR receiving waste activated sludge [2].
Also included in the figure is the solids curve from Fig. 12.7. Comparison of the
two curves shows that although the observed cell concentration approaches a lower
limit, the theoretical concentration continues to decline. As cells undergo decay
a residual fraction is left which does not break down as easily as the original cell
material [1,4,14] and consequently a stable VSS concentration appears to develop in
the reactor. This material corresponds to a portion of the dead cell material in
the models of Chapter 12, and consequently b_d would have to be set lower than b_v in
order to predict the actual performance. The improvement in the model which can be
obtained by setting b_d less than b_v is indicated by the third curve in the figure.
Consequently, consideration must be given to the existence of the two types of
materials or the models will predict removals in excess of those actually obtainable.

Although the models could be employed by using two different decay rates, their
evaluation would require the measurement of viabilities during treatability studies,
a very tedious task. Instead, an alternative model has been developed which is much
easier to use. It is based on the assumption that the decay rate of the residual
material is so low that the material may be considered to be nonbiodegradable.
Although it has been shown that this material is indeed subject to biodegradation
[15,16,17], this pragmatic approach has found wide application to the modeling of

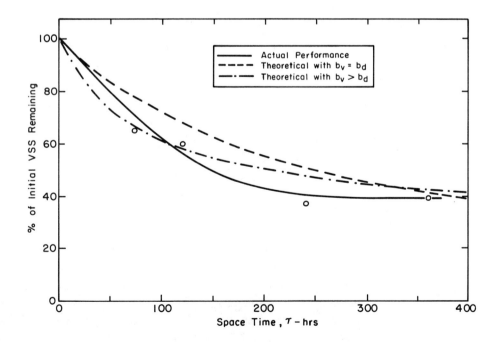

Figure 17.1. Comparison of actual response of a CSTR receiving waste activated
sludge to theoretical response using model from Chapter 12.
Based on data from Adams, et al. [2].

aerobic digestion because of its simplicity. Thus, the usual approach is to assume
that the rate of degradation of the biodegradable cell material, X_b, is first order
with respect to the concentration of X_b [1]:

$$-r_{dX_b} = bX_b \qquad (17.1)$$

where

$$X_b = X - X_n \qquad (17.2)$$

in which X_n is the nonbiodegradable cell material.

Equations 17.1 and 17.2 use the usual definition of X, i.e., the total concentration of microorganisms as measured by a suspended solids determination. When
modeling aerobic digestion, however, most workers have expressed X on a volatile
suspended solids basis on the assumption that only the volatile fraction of the
cells will be lost by decay while the fixed (inorganic) fraction will remain in
suspension. It appears, however, that this assumption is incorrect [12,18,19,20].
The inorganic fraction of bacterial cells is intimately associated with the organic
(volatile) fraction. Consequently, as cells undergo decay one would expect the
inorganic constituents to be released and resolubilized as the organic material was
degraded, thereby maintaining an essentially constant ratio between fixed and volatile suspended solids. This is exactly what Randall has found [12,18,19,20].
Furthermore, data by Reynolds, [9], Eckenfelder (activated sludge grown on pulp and
paper waste) [11], and Reece *et al.* [21] support this view point. Consequently, the
aerobic digestion of microbial solids can be satisfactorily modeled on a total
suspended solids basis, as has been done by Benefield and Randall [22].

17.2.3 Modification of Model

Most waste sludges contain solids other than bacteria, and the model must depict
what happens to all of them. Thus most models for aerobic digestion are written in
terms of the MLSS concentration, M. In this case it will be necessary to distinguish
between volatile and fixed solids, however, because the MLSS may contain some of each
which are totally inert to degradation. Furthermore, because the MLSS may contain
organic solids which have no inorganic solids associated with them (in contrast to
cells) the specific rate of resolubilization of fixed solids may be different from
the specific rate of decay of the volatile solids. Evidence for this can be found in
studies in which both the volatile and fixed suspended solids concentrations changed
during digestion, but by different percentages [11]. Consequently, different rate
expressions should be written for the degradable volatile and fixed mixed liquor
suspended solids, and the two rate constants should be evaluated during treatability
studies. Although the rate constant for the destruction of fixed solids may turn out
to be so small that it may be safely taken to be zero, this assumption should never
be made *a priori* because it could lead to serious errors in the prediction of total
solids losses during digestion.

In light of the above, two rate expressions may be written:

$$-r_{dM_{bg}} = b_g M_{bg} \tag{17.3}$$

$$-r_{dM_{bf}} = b_f M_{bf} \tag{17.4}$$

The symbol M refers to the MLSS concentration and the subscripts b,n,g and f refer
to biodegradable, nonbiodegradable, volatile (gaseous), and fixed, respectively.
(It should be recognized that the fixed solids are not truly biodegradable, but are
resolubilized as a result of biodegradation of the associated organic solids).
Consequently, the following identities hold:

$$M = M_g + M_f \tag{17.5}$$

$$M = M_b + M_n \tag{17.6}$$

$$M_b = M_{bg} + M_{bf} \tag{17.7}$$

$$M_n = M_{ng} + M_{nf} \tag{17.8}$$

$$M_g = M_{bg} + M_{ng} \tag{17.9}$$

$$M_f = M_{bf} + M_{nf} \tag{17.10}$$

It is important that the rate equations be written in terms of only the biodegradable
solids, otherwise there will be an apparent change in the rate constant during
digestion, making its evaluation difficult [1]. If the rate equations are written
properly, however, it is possible to predict the performance of an aerobic digester,
as shown in Fig. 17.2. The data points are the same as those plotted in Fig. 17.1
for a CSTR. The curve, however, comes from application of Eq. 17.3 to a CSTR using
the value of b_g obtained from a batch reactor.

The presence of nonbiodegradable solids in a waste sludge limits the degree of
solids reduction that can be achieved during aerobic digestion. Because the nonbiodegradable solids are related in part to the dead cells in activated sludge, and
because the fraction of dead cells is related to the MCRT of the activated sludge,
one would expect the MCRT of an activated sludge to influence the amount of nonbiodegradable solids present. Determination of M_b and M_n for an activated sludge
grown on soluble substrate showed that this was indeed the case [21]. Consequently,
when waste activated sludge is being stabilized by aerobic digestion, the MCRT in
the activated sludge process will determine the maximum degree of stabilization that
can be achieved during subsequent digestion. The greater the MCRT, the lower the
maximum possible degree of stabilization. The specific decay rate constant for
volatile solids, b_g, used in aerobic digestion is essentially the value for viable
cells alone. Thus b_g for the digestion of waste activated sludge should be relatively independent of the MCRT of the activated sludge. This has been confirmed

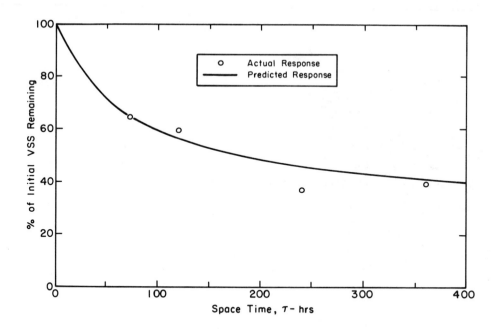

Figure 17.2.   Comparison of actual response of a CSTR receiving waste
activated sludge to theoretical response using modified
rate equation. *Based on data from Adams et al. [2].*

experimentally [21]. Consequently, the rate of digestion of the biodegradable
volatile solids should be relatively insensitive to the MCRT of the preceding
activated sludge process.

In summary, aerobic digestion can bring about a reduction in both volatile and
fixed suspended solids although a residual, nondegradable fraction of each will
remain. The magnitude of the nondegradable fraction depends upon the character of
the sludge as well as upon the operational characteristics of the upstream biological treatment process and it determines the maximum degree of stabilization that
can be achieved during aerobic digestion. The actual degree of stabilization
achieved, however, depends upon the values of the rate constants b_g and b_f as well
as upon the space time in the reactor. For many sludges the degree of stabilization
will approach the maximum at space times around 15 to 20 days [4,18,23,24] although
longer periods may be required [18]. Process design, therefore, becomes a matter of
choosing the space time required to achieve a desired degree of stabilization. If
reactors with solids recycle are employed for digestion, then the performance will
depend upon the MCRT [25].

17.3 DESIGN OF AN AEROBIC DIGESTER

Three items must be determined in arriving at the process design for an aerobic
digester: the reactor volume, the amount of oxygen required, and the minimum power
input to achieve adequate mixing. In this section we will look at the equations
required to determine those items and discuss other factors which must also be
considered.

17.3.1 Process Design Equations

Much of the early work on aerobic digestion used volumetric loading (mass of
VSS/unit volume·day) as the parameter for determining the required reactor volume
[23,24]. As the kinetics of the process became better understood, however, it
became possible to use a reactor engineering approach in which a rate equation is
combined with a mass balance on the reacting component in the reactor. There are
two reacting components in an aerobic digester, the biodegradable volatile (M_{bg}) and
fixed (M_{bf}) solids, and equations for their concentrations in a CSTR may be derived
in a manner similar to that used to derive Eq. 12.39:

$$M_{bg} = \frac{M_{bgo}}{1 + b_g \tau} \tag{17.11}$$

and

$$M_{bf} = \frac{M_{bfo}}{1 + b_f \tau} \tag{17.12}$$

where the subscript o indicates the concentration of each component in the influent.
These equations describe the effects of the space time upon the extent of reaction.

Most designers want to select the space time required to give a desired degree
of destruction of either the volatile suspended solids or all suspended solids.
Thus it is convenient to have equations which relate those parameters to τ. Letting
ξ stand for the fractional destruction of suspended solids and ξ_g for the fractional
destruction of volatile suspended solids, their defining equations are:

$$\xi \equiv \frac{M_o - M}{M_o} \tag{17.13}$$

$$\xi_g \equiv \frac{M_{go} - M_g}{M_{go}} \tag{17.14}$$

Substitution of Eqs. 17.9 and 17.11 into 17.14 and recognition of the fact that the
concentration of nonbiodegradable volatile solids in the reactor is the same as the
concentration in the influent lead to:

$$\xi_g = \frac{M_{bgo}}{M_{go}} \left(\frac{b_g \tau}{1 + b_g \tau} \right) \tag{17.15}$$

The maximum possible destruction of volatile solids is just M_{bgo}/M_{go} and would occur when τ is very large. The space time required to achieve a desired value of ξ_g can be obtained by rearranging Eq. 17.15:

$$\tau = \frac{\xi_g M_{go}}{b_g (M_{bgo} - \xi_g M_{go})} \tag{17.16}$$

Equations similar to 17.15 and 17.16 could be written for fixed solids by an appropriate change in notation should the need arise. Using a similar approach an equation relating the destruction of the total suspended solids to τ may be derived:

$$\xi = \frac{M_{bgo}[b_g \tau/(1 + b_g \tau)] + M_{bfo}[b_f \tau/(1 + b_f \tau)]}{M_o} \tag{17.17}$$

The maximum possible destruction of total solids is just $(M_{bgo} + M_{bfo})/M_o$. The space time required to achieve a desired value of ξ can be obtained by rearranging Eq. 17.17:

$$b_g b_f (\xi M_o - M_{bo})\tau^2 + [b_g (\xi M_o - M_{bgo}) + b_f (\xi M_o - M_{bfo})]\tau + \xi M_o = 0 \tag{17.18}$$

Once the space time is known the reactor volume may be calculated from the definition of τ.

It is sometimes desirable to use reactors in series. Equations 17.11 and 17.12 apply to each reactor in the chain with the output from one reactor serving as the input to the next. The easiest way to use the equations is to just apply them serially down the chain. As was shown in Fig. 12.14, it is seldom necessary to use more than two reactors.

Oxygen will be required for the destruction of volatile suspended solids and nitrification if it occurs. If microbial solids are being digested it will be necessary to provide 0.56 gms of oxygen per gram of volatile solids destroyed to meet the needs of the nitrifiers. If other volatile solids are being digested the amount of oxygen needed for nitrification will depend upon the nitrogen content of the solids. The determination of the oxygen requirement for nitrification under those circumstances will require a nitrogen balance and the technique for performing one will be discussed in Chapter 20. The oxygen required for the destruction of volatile solids may be determined by a mass balance on oxygen demanding material, leading to an equation similar to Eq. 12.41:

$$RO = F\{ [M_{ngo}(\beta_{go} - \beta_g)] + \frac{M_{bgo}[(\beta_{go} - \beta_g) + \beta_{go} b_g \tau]}{1 + b_g \tau} \} \tag{17.19}$$

The term $M_{ngo}(\beta_{go} - \beta_g)$ is required because β_{go} and β_g are the oxygen demand coefficients of all volatile suspended solids in the influent and in the reactor, respectively. If β_{go} and β_g could be measured on the basis of only the biodegradable volatile solids the term would drop out. If the influent solids are all waste cells from a biological reactor then β_{go} and β_g are likely to be the same and approximately equal to 1.41 gm oxygen/gm volatile solids, i.e., the theoretical oxygen demand of cell material [4]. If other solids are included in the sludge the values are likely to be different and thus it is good practice to determine them during treatability studies.

17.3.2 Determination of Parameter Values

Before the design equation can be used, the values of the rate constants, b_g and b_f, and the oxygen demand coefficients, β_{go} and β_g, must be determined. In addition, the fractions of nonbiodegradable volatile and fixed suspended solids in the sludge must be determined. Batch tests are required for the latter. The first-order nature of the rate equations makes it convenient to use batch tests to determine the rate constants. Consequently, it is common practice to use batch decay tests for the design of aerobic digesters. Continuous culture techniques are not required since analysis of data presented by Adams et al. [2] indicated that any potential increase in accuracy is not sufficient to warrant the extra labor required.

The procedure is quite simple. Well aerated and mixed batch reactors should be set up at constant temperature and filled with waste sludge at the concentration anticipated in the final facility. All water lost by evaporation should be replaced and samples should be analyzed daily to determine the following concentrations: suspended solids, M, volatile suspended solids, M_g, fixed suspended solids, M_f, soluble COD, C, and total COD, T. The procedure should be continued until little change is noted in M.

The COD data will be used to calculate the value of β_g as a function of digestion time:

$$\beta_g = (T - C)/M_g \tag{17.20}$$

If β_g changes appreciably during digestion the data must be used in conjunction with an exit age distribution curve for a CSTR (see Section 6.2.1) to determine how β_g would be affected by τ.

The reaction rate coefficients may be found by making use of the equation that results when a first-order reaction is carried out in a batch reactor. Modifying Eq. 4.2 to the nomenclature of this chapter we have:

$$M_{bg}/M_{bgo} = \exp(-b_g t) \tag{17.21}$$

where t is the time of aeration. Substituting Eq. 17.9 into 17.21 and noting that the nonbiodegradable volatile suspended solids concentration (M_{ng}) doesn't change, the equation becomes:

$$\frac{M_g - M_{ng}}{M_{go} - M_{ng}} = \exp(-b_g t) \tag{17.22}$$

This can be used to determine the rate constant by either of two techniques. The preferred method is to use the data describing M_g as a function of time to find M_{ng} and b_g by performing a nonlinear least squares curve fit using the computer, thereby achieving the best possible fit of the model to the data. The alternative procedure is to determine M_{ng} and b_g graphically. First an arithmetic plot must be made of M_g versus time and M_{ng} estimated by eye as the lower limit on M_g. The technique of Isaacs and Gaudy [26] can then be used to refine the estimate. Then a plot must be made of $\ln(M_g - M_{ng})$ as a function of time so that b_g may be determined from the slope. A similar procedure may be used to find b_f.

Although they are similar, the rate constant b_g is not the same as the b used in Chapters 15 and 16, but will probably be larger because b incorporates both b_v and b_d whereas b_g includes only b_v. Consequently, the value of b measured during continuous culture treatability studies for activated sludge design (as described in Section 13.4) will not be applicable to subsequent aerobic digestion. Separate studies must be performed using the waste sludge as the feed.

17.3.3 Factors Influencing the Design

MIXING. If adequate mixing is not provided, solids will settle in the reactor thereby reducing the effective reactor volume and allowing anaerobic conditions to develop in the sludge. Both of these effects will cause a reduction in process efficiency. Reynolds [5] has related the power which must be dissipated to mix an aerobic digester to the concentration of suspended solids in it:

$$P/V = 0.00475 \, \mu_c^{0.3} \, M^{0.298} \tag{17.23}$$

where

P/V = HP/1000 gals

μ_c = viscosity of water in centipoise

M = MLSS concentration in mg/liter

To express P/V in $kW/1000m^3$ change 0.00475 to 0.935. The equation is applicable to digesting and digested sludges (e.g., as in a CSTR) but about twice as much power is required for undigested sludge. If mixing and aeration are carried out by bubble diffusion, the required air flow rate may be calculated by [5]:

$$\frac{Q}{V} = 50.5\,[(P/V)/\log_{10}(\frac{H+34}{34})] \tag{17.24}$$

where

Q/V = gas flow rate in $cfm/1000ft^3$ or $m^3/min\cdot1000m^3$.

P/V = HP/1000 gal which must be dissipated in the fluid.

H = submergence depth of diffusers, in ft.

At 25°C a concentration of 10,000 mg/liter requires a power input of 0.07 HP/1000 gal (13.8 kW/1000m^3) and an air flow rate of 22 cfm/1000ft^3 (22m^3/min·1000m^3) both common values in practice [4]. Because of the low oxygen uptake rates associated with aerobic digestion the mixing requirement generally governs the power input to the reactor.

   TEMPERATURE. There is no doubt that temperature affects aerobic digestion or that the effect becomes more important as the space time is made shorter [4]. There is considerable doubt, however, about the best way to represent the effect. Several workers [2,7,10] use the equations from Chapter 10 (10.56 through 10.58) to quantify the effects of temperature upon b_g in the mesophilic range. When using Eq. 10.57 Adams *et al.* [2] stated that θ lay between 1.02 and 1.07, with 1.05 being a reasonable value for waste activated sludge. Eikum *et al.* [7], whose data are shown in Fig. 17.3, found θ to equal 1.114 for primary sludge. Other workers [13,20,27] have not been able to express the effects of temperature by a simple equation, although good correlations have been obtained between percent solids destruction and the product of temperature and τ [27]. Thus the safest approach would be to experimentally determine the effects of temperature during the treatability studies. If that is not possible, then one must assume that Eq. 10.57 is valid with θ equal to 1.05. Care should be exercised in the application of the equation to ensure that it is only used within the mesophilic range. Different microbial populations will develop in the mesophilic and thermophilic ranges, resulting in a discontinuity in the data as the temperature is shifted from one range to another. A different value of θ is likely to be found in the thermophilic range.

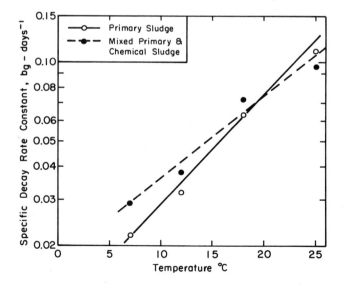

Figure 17.3. Effect of temperature on decay rate
             constant. *(Adapted from Eikum et al. [7].)*

Consideration of temperature effects during design is important because the relatively long space times required for complete digestion (e.g., 15-20 days) cause open, uninsulated, unheated digesters to become quite cold in the winter. In fact Adams *et al.* [2] have stated that such digesters will normally be at ambient temperature as long as it is above freezing, and at 1°C if the ambient temperature is below freezing. The rate constant at 1°C is generally so low that designers in northern climates should consider covering and insulating the digesters as a means of reducing the volume. Another way in which the heat losses from a digester may be reduced is through the use of pure oxygen, which reduces the quantity of gas passing through the digester and the associated heat loss by evaporation. In fact, by using pure oxygen in systems containing high solids concentrations, heat conservation is so good that digesters can operate in the thermophilic range using microbially generated waste heat as the only heat source [28,29]. This has the benefit of allowing smaller reactors to be used (there is no drastic drop in temperature during the winter) and providing a treated sludge which contains a low concentration of pathogenic bacteria. These benefits will not result, however, unless operation is such that a truly thermophilic population develops. Otherwise most solids reduction will just result from lysis of the mesophilic bacteria with the return of much organic matter to the soluble state [30].

pH. The rate constant, b_g, is a function of pH. In one study a pH of 6.5 gave the maximum value, followed in order by 8.0, 5.0, and 3.5 [19]. Changes in pH occur during digestion, primarily as a result of nitrification (see Chapter 20), and the magnitude of the change will depend upon the relative concentrations of organic nitrogen and alkalinity in the sludge. Consequently provision should be made during digester design for the monitoring and adjustment of pH if needed. Adjustment of pH to neutrality is quite important during the batch tests for the determination of b_g and b_f. Without pH adjustment it is not unusual for the pH to drop as low as 5.5 which can cause the rate of digestion to decrease thereby causing inaccurate estimates of the rate constants and the nondegradable fraction.

DEWATERING CHARACTERISTICS. Because one objective of aerobic digestion is the reduction of the mass of sludge to be disposed of, an important consideration during design is the effect that the digestion has on the dewatering characteristics of the sludge. Although there is agreement in the literature that the space time affects both the settleability and dewaterability of digested sludge, few other generalizations can be made [23,24,31,32]. One reason for this is the large effect exhibited by the sludge itself. Nevertheless it can be stated that the settleability and dewaterability will improve as the space time is increased until some optimum value is reached after which they will deteriorate. The exact nature of the curve must be determined during the treatability studies so that those effects can be considered when choosing a design value for τ.

17.3.4 Example Problems

The easiest way to demonstrate the process design procedure is through examples.

EXAMPLE 17.3.4-1

A batch aeration test was performed on waste activated sludge from a pulp and paper mill waste [11]. The temperature was 32°C. The data are given in Table E17.1.

Table E17.1
Data Collected during Batch Aeration of the Waste
Activated Sludge from a Pulp and Paper Mill Waste[a]

Time days	M mg/liter	M_g mg/liter	M_f mg/liter
0	7642	6434	1208
1	7416	6160	1256
2	5921	4940	981
3	6195	5320	875
5	5690	4790	900
6	4940	4140	800
8	4780	3840	940
9	4630	3890	740
14	4150	3550	600

[a] *Adapted from Eckenfelder [11].*

(a) Determine the concentrations of the volatile and fixed nondegradable suspended solids, M_{ng} and M_{nf}.

Plot the concentrations of volatile and fixed suspended solids as functions of time as shown in Fig. E17.1. Draw curves through the data and estimate the nondegradable concentrations as the final stable concentrations. From the figure the following values are obtained:

M_{ng} = 3550 mg/liter

M_{nf} = 710 mg/liter

(b) Estimate the values of the rate constants b_g and b_f.

To do this it is first necessary to subtract M_{ng} from M_g to get M_{bg} and to subtract M_{nf} from M_f to get M_{bf}. Then the natural logs of M_{bg} and M_{bf} must be found. These values are listed in Table E17.2. Now, $\ln M_{bg}$ and $\ln M_{bf}$ must both be plotted as functions of time so that b_g and b_f may be determined from the slopes. This is shown in Fig. E17.2. The values obtained are:

b_g = 0.011 hr^{-1}

b_f = 0.013 hr^{-1}

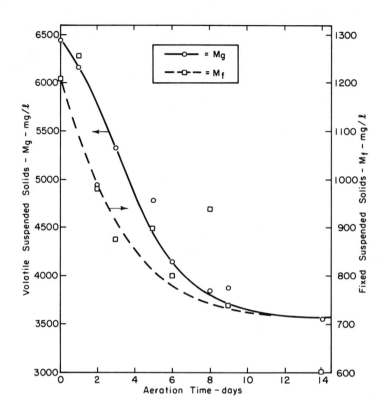

Figure E17.1.   Arithmetic plots of M_g and M_f as functions
of time to determine M_{ng} and M_{nf} in
Example 17.3.4-1.

TABLE E17.2
Calculated Data for the Determination of b_g and b_f

Time days	M_{bg} mg/liter	$\ln M_{bg}$	M_{bf} mg/liter	$\ln M_{bf}$
0	2884	7.97	498	6.21
1	2610	7.87	546	6.30
2	1390	7.24	271	5.60
3	1770	7.48	165	5.11
5	1240	7.12	190	5.25
6	590	6.38	90	4.50
8	290	5.67	230	5.44
9	340	5.83	30	3.40

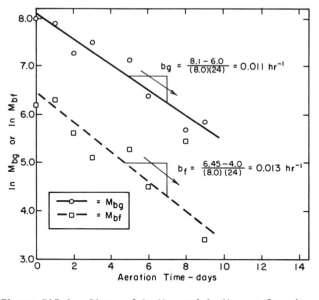

Figure E17.2. Plots of ln M_{bg} and ln M_{bf} as functions
of time to determine b_g and b_f in
Example 17.3.4-1.

EXAMPLE 17.3.4-2

The waste activated sludge characterized in Example 17.3.4-1 is to be treated by
aerobic digestion in a CSTR. The following things are known or assumed:

M_o = 8000 mg/liter

The percent volatile solids in the sludge is the same as in Example 17.3.4-1.

The percent degradable solids in the sludge is the same as in Example 17.3.4-1.

Summer temperature of sludge = 32°C.

Winter temperature of sludge = 5°C.

θ = 1.04

Prepare graphs showing the fractional destruction of solids, ξ, during summer
and winter.

Solution:

To prepare the graphs we must use Eq. 17.17, which requires knowledge of M_{bgo}
and M_{bfo}. Making use of the fact that the percent volatile and degradable solids
in the sludge are the same as in Example 17.3.4-1 we find:

$$M_{go} = (6434/7642)8000 = 6735 \text{ mg/liter}$$

$$M_{fo} = (1208/7642)8000 = 1265 \text{ mg/liter}$$

$$M_{bgo} = [(6434 - 3550)/6434]6735 = 3019 \text{ mg/liter}$$

$$M_{bfo} = [(1208 - 710)/1208]1265 = 522 \text{ mg/liter}$$

The summer values of b_g and b_f were found in Example 17.3.4-1. The winter values may be estimated by using Eq. 10.21:

$$b_{g5} = (0.011) \, 1.04^{(5 - 32)} = 0.0038 \text{ hr}^{-1}$$

$$b_{f5} = (0.013) \, 1.04^{(5 - 32)} = 0.0045 \text{ hr}^{-1}$$

The curves may now be prepared by using Eq. 17.17, which is shown below for each temperature condition:

Summer:
$$\xi = \frac{3019\left(\dfrac{0.011\tau}{1 + 0.011\tau}\right) + 522\left(\dfrac{0.013\tau}{1 + 0.013\tau}\right)}{8000}$$

Winter:
$$\xi = \frac{3019\left(\dfrac{0.0038\tau}{1 + 0.0038\tau}\right) + 522\left(\dfrac{0.0045\tau}{1 + 0.0045\tau}\right)}{8000}$$

The curves are shown in Fig. E17.3 where it can be seen that considerably longer space times are needed in winter to achieve the same destruction as in the summer. Furthermore, it should be noted that the fairly high fraction of nondegradable solids limits the maximum solids destruction to around 40%.

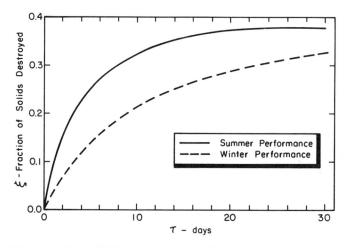

Figure E17.3.   Effect of τ on the fraction of solids destroyed in a CSTR aerobic digester, as calculated in Example 17.3.4-2.

EXAMPLE 17.3.4-3

Further data on the waste sludge described in Examples 17.3.4-1 and 2 showed that β of the untreated sludge was 1.41 mg O_2/mg VSS whereas β of the treated sludge was relatively constant at 1.30 mg O_2/mg VSS for τ greater than 4 days. A τ of 14 days was chosen for treatment of 1000 gal/day (3.785 m^3/d) of the sludge in order to achieve a 25% destruction of solids in the winter (see Fig. E17.3).

(a) Determine the mass of oxygen required per day during the summer (not counting nitrification).

To do this, use Eq. 17.19 with the following values:

β_{go} = 1.41 mg O_2/mg VSS

β_g = 1.30 mg O_2/mg VSS

M_{bgo} = 3019 mg/liter

M_{ngo} = 6735 - 3019 = 3716 mg/liter

F = 1000 gal/day = 3785 liters/day

τ = 14 days = 336 hrs

b_g = 0.011 hr^{-1}

Thus,

$$RO = 3785\{3716(1.41 - 1.30) + \frac{3019[(1.41 - 1.30) + (1.41)(0.011)(336)]}{1 + (0.011)(336)}\}$$

$RO = 1.45 \times 10^7$ mg/day = 14.5 kg/day of oxygen.

(b) Determine the HP required to mix the tank.

This must be done for the winter condition because that is when the MLSS concentration and the viscosity will be the highest. Use Eq. 17.23 with the following values:

M = (8000)(1 - 0.25) = 6000 mg/liter

μ_c = 1.5 centipoise

Thus,

$$P/V = (0.00475)(1.5)^{0.3}(6000)^{0.298}$$

P/V = 0.072 HP/1000 gals

At τ = 14 days with F = 1000 gal/day, V = 14,000 gal.

Thus

P = 1.00 HP

At least 1.00 HP must be dissipated in the basin for mixing, whether it is needed for oxygen transfer or not.

17.4   OPERATION OF AEROBIC DIGESTERS

The manner in which an aerobic digester is operated depends upon the size of the
wastewater treatment facility.   If the facility is large, then sludge is likely to
be wasted continuously from the biological treatment process so that the digester
can operate in a continuous flow mode.   A separate settler will be used for the
digester and the treated sludge must be dewatered.   Generally the dewatering
facility will only be operated periodically, so that something must be done with the
treated sludge when it is shut down.   If the digester is a CSTR with cell recycle,
the sludge may simply be recycled when the dewatering facility is not operating.
The amount of sludge which should be released from the digester each day will depend
upon the digester volume and the desired MCRT, and may be determined in a manner
analogous to that discussed in Section 16.4 for activated sludge.   If the digester
is a CSTR without recycle then an aerated holding tank must be provided for the
sludge to keep it aerobic from the time it settles until it is dewatered.   Sludge
should not be allowed to accumulate in the settler because even short periods of
anaerobiosis will cause the dewatering characteristics to deteriorate [32].

If the wastewater treatment facility is fairly small, then sludge is likely to
be wasted from the biological process only once a day.   In that case the aerobic
digester may be operated on a fill-and-draw basis, thereby removing the need for a
separate settling basin.   The mixing system should have sufficient power to bring
about complete resuspension of the solids following settling, however.   The steps
generally followed are [33]:

(1)   Shut off the air supply.

(2)   Allow the sludge to settle.

(3)   Withdraw the desired volume of supernatant.

(4)   Withdraw the desired volume of settled sludge for dewatering.

(5)   Turn the air back on.

(6)   Waste sludge from the biological process to the digester.

The time required for each step will depend upon the nature of the facility and the
characteristics of the sludge.   Because the fraction of the total volume added each
day is small, these reactors will behave much like a CSTR so that CSTR performance
curves may be used to determine the required volume.   Any error in estimated
performance will be on the conservative side.

17.5   EXAMPLES OF AEROBIC DIGESTION IN PRACTICE

Although the literature on aerobic digestion is extensive, most of it deals with
laboratory and pilot plant studies performed to determine the efficacy of this biochemical operation for a specific waste sludge.   Papers reporting its performance
after construction of the full-scale facility, however, are scarce.   Nevertheless,
on the basis of the reported studies it is apparent that aerobic digestion can bring
about the stabilization of primary sludge, trickling filter humus, and waste

activated sludge as well as various mixtures of the three. The rate constant for
the degradation of the sludge will depend upon the type of material in it, the pH,
and the temperature. Furthermore, because the character of a wastewater determines
the nature of the microbial population in the biological treatment process, it also
influences the rate constant for the aerobic digestion of the resultant sludge.
Consequently, it is impossible to generalize about the value of the rate constant.
Indeed, it must be emphasized that treatability studies are required to determine
its value. In fact, recognizing that the rate constant is liable to change with
time, Adams et al. [2] recommend that enough values of it be determined to allow the
design to be statistically based. In light of the ease with which batch tests can
be run, this is an excellent suggestion.

One paper has dealt extensively with the performance of full-scale plants
treating waste activated sludge [33]. The study took place in Canada and covered
seven digesters, all completely mixed tanks operated on a fill-and-draw sequence.
The space times were long, ranging from 14 to 360 days. The air supply rates ranged
from 8.4 to 30 cfm/1000ft^3 (m^3/(min·1000 m^3)) but because the solids concentrations
were in excess of 20,000 mg/liter these did not provide adequate mixing, particularly
at the lower levels. The problem was particularly acute because of the fill-and-
draw operation with the result that solids deposition occurred, thereby reducing the
effective volume. Consequently, the authors recommended that the aeration devices
in such digesters be capable of maintaining a solids concentration of 60,000 mg/liter
in suspension. According to the equations given earlier, this would require around
40 cfm/1000 ft^3 at 25°C. One digester received sludge from an extended aeration
plant, and as anticipated the degree of volatile solids destruction was not high in
the digester. Three of the digesters employed a single tank and none obtained a
volatile solids destruction that exceeded 25%, reflecting the cold temperatures at
which they were operating. Four of the digesters employed two tanks in series and
these were able to achieve volatile solids reductions as high as 50%, confirming the
theoretical effects in Fig. 12.16. Thus it appears that two tanks in series can be
beneficial, especially in cold climates. In addition to the effects on process
efficiency the low temperatures encountered during winter operation caused freezing
which damaged an aeration system. Consequently, plants subject to extremely cold
weather should be designed to minimize heat losses. Most of the organic matter in
the effluents from the digesters was due to solids which didn't settle. These
concentrations were high enough to prevent direct discharge of the effluents and to
require their return to the treatment plants. This points out the necessity for
determining the effects of τ on effluent quality during treatability studies.

In general, aerobic digestion offers a reasonable alternative for the stabilization of biological sludges. Like all biochemical operations, however, successful
design requires well planned and performed treatability studies coupled with the
application of good common sense and sound engineering judgement.

17.6 KEY POINTS

1. The objective of aerobic digestion is the destruction of insoluble organic matter in an aerobic environment.

2. Aerobic digestion is generally most applicable to secondary sludges such as those generated by activated sludge and trickling filters.

3. An aerobic digester receiving only secondary sludge essentially behaves like a CSTR receiving only microbial cells as depicted in Fig. 12.7. However, because a portion of those cells will only be degradable at a very slow rate there will be a residual fraction remaining after digestion which is usually considered to be nonbiodegradable.

4. Aerobic digestion may be modeled accurately by assuming that the rate of solids destruction is first order with respect to the concentration of biodegradable solids.

5. Fixed (inorganic) solids will be solubilized during aerobic digestion and thus their losses from suspension must be accounted for during modeling of aerobic digester performance.

6. Three items must be determined in arriving at the process design for an aerobic digester: the reactor volume, the amount of oxygen required, and the minimum power input to achieve adequate mixing.

7. Batch tests are convenient for the determination of the rate constants b_g and b_f as well as for estimation of the fractions of nonbiodegradable volatile and fixed suspended solids in a sludge.

8. The power which must be dissipated to keep the solids in suspension in an aerobic digester depends upon the concentration of the solids.

9. The rate constants for aerobic digestion are functions of temperature and decrease as the temperature decreases. This has a severe effect upon winter performance in northern climates.

10. The settleability and dewaterability of aerobically digested sludge will improve as τ is increased until some optimum τ is reached, after which it will deteriorate.

11. Aerobic digesters may be fed either continuously or on a fill-and-draw basis, with small facilities employing the latter.

17.7 STUDY QUESTIONS

The study questions are arranged in the same order as the text.

Section	Questions
17.1	1
17.2	2-5
17.3	6-10
17.4	11

1. List the merits and defects of aerobic digestion.

2. Explain how and why the models of Chapter 12 do not accurately predict the performance of an aerobic digester.

3. Define each of the following symbols and describe what will happen to those constituents during aerobic digestion: M_{bg}, M_{nf}, M_{ng}, M_{bf}.

4. Explain the pragmatic modeling approach used to describe aerobic digester performance.

5. Why is the maximum possible fractional destruction of volatile suspended solids equal to M_{bgo}/M_{go}?

6. Describe the batch technique for determining the rate constants and other parameters for aerobic digestion of a sludge.

7. Why is it important to maintain a constant temperature and pH during batch tests for determination of the rate constants for aerobic digestion?

8. A batch aeration test was performed on waste activated sludge from a pilot plant treating a soluble waste. The temperature was 20°C. The data are given in Table SQ17.1.

TABLE SQ17.1
Data from a Batch Aeration Test

Time days	M mg/liter	% Volatile in M
0.00	3080	87.0
0.34	3000	86.6
0.84	2890	85.2
1.9	2630	85.7
2.9	2320	85.9
3.9	2140	87.3
4.9	1926	87.5
5.9	1710	88.0
6.9	1690	87.9
7.9	1590	88.5
8.9	1520	87.7
9.9	1490	88.2
10.9	1320	88.0
11.9	1260	87.3
12.9	1280	87.8
13.9	1140	88.6
14.9	980	86.6
15.9	1060	87.7
16.9	1030	87.3

(a) Determine the concentrations of the volatile and fixed nondegradable suspended solids, M_{ng} and M_{nf}.

(b) Estimate the values of the rate constants b_g and b_f.

9. The sludge in Study Question 8 is to be thickened to a concentration of 10,000 mg/liter and then treated by aerobic digestion. The winter temperature in an open reactor is estimated to be 5°C, and the temperature factor θ is assumed to be 1.04. The flow rate of the thickened sludge is 10,000 liters/day.

(a) What space time will be required in an open CSTR to get a 33% destruction of suspended solids in the winter?

(b) By covering and insulating the reactor the temperature can be kept at 25°C, even in the winter. What space time would be required for 33% destruction of suspended solids in that case?

(c) If β_{go} and β_g are both 1.35 mg O_2/mg VSS, how much oxygen will be required in (a) and (b) above?

(d) How much power is required to mix the tanks in (a) and (b) above?

10. List the steps in fill-and-draw operation of an aerobic digester.

REFERENCES AND FURTHER READING

1. E. L. Barnhart, "Application of aerobic digestion to industrial waste treatment," *Proceedings of the 16th Industrial Waste Conference,* Purdue University Engineering Extension Series No. 109, 612-618, 1961.

2. C. E. Adams, *et al.* "Modification to aerobic digester design," *Water Research,* 8, 213-218, 1974.

3. R. M. Stein, *et al.* "A practical model of aerobic sludge stabilization," *Proceedings of the Third Annual Environmental Engineering and Science Conference,* University of Louisville, Kentucky, 781-795, 1973.

4. R. C. Loehr, "Aerobic digestion: Factors affecting design," *Water and Sewage Works, Reference Number,* 112, R-169 - R-180, 1965.

5. T. D. Reynolds, "Aerobic digestion of thickened waste activated sludge," *Proceedings of the 28th Industrial Waste Conference,* Purdue University Engineering Extension Series No. 142, 12-37, 1973.

6. R. S. Burd, "A study of sludge handling and disposal," Publication WP-20-4, U.S. Department of the Interior, Federal Water Pollution Control Administration, May 1968.

7. A. S. Eikum, *et al.* "Aerobic stabilization of primary and mixed primary-chemical (alum) sludge," *Water Research,* 8, 927-935, 1974.

8. S. D. Bokil and J. K. Bewtra, "Influence of mechanical blending on aerobic digestion of waste activated sludge," *Advances in Water Pollution Research,* Proceedings of the 6th International Conference, Jerusalem, 421-432, 1972.

9. T. D. Reynolds, "Aerobic digestion of waste activated sludge," *Water and Sewage Works,* 114, No. 2, 37-42, 1967.

10. P. Benedek, *et al.* "Kinetics of aerobic sludge stabilization," *Water Research,* 6, 91-97, 1972.

11. W. W. Eckenfelder, Jr., "Studies on the oxidation kinetics of biological sludges," *Sewage and Industrial Wastes,* 28, 983-990, 1956.

12. C. W. Randall, *et al.* "Aerobic digestion of trickling filter humus," *Proceedings of the Fourth Annual Environmental Engineering and Science Conference,* University of Louisville, Kentucky, 557-577, 1974.

13. P. L. Bishop and L. R. LePage, "Aerobic digestion of wastewater sludge under low temperature conditions," *Proceedings of the Fourth Annual Environmental Engineering and Science Conference,* University of Louisville, Kentucky, 539-556, 1974.

14. W. J. Jewell and P. L. McCarty, "Aerobic decomposition of algae," *Environmental Science and Technology,* 5, 1023-1031, 1971.

15. D. R. Washington, *et al.* "Long-term adaptation of activated sludge organisms to accumulated sludge mass," *Proceedings of the 19th Industrial Waste Conference,* Purdue University Engineering Extension Series No. 117, 655-666, 1964.

16. A. F. Gaudy Jr. *et al.* "Studies on the operational stability of the extended aeration process," *Journal of the Water Pollution Control Federation,* 42, 165-179, 1970.

17. A. W. Obayashi and A. F. Gaudy, Jr., "Aerobic digestion of extracellular microbial polysaccharides," *Journal of the Water Pollution Control Federation,* 45, 1584-1594, 1973.

18. C. W. Randall, *et al.* "Biological and chemical changes in activated sludge during aerobic digestion," *Proceedings of the 18th Southern Water Resources and Pollution Control Conference,* North Carolina State University, 155-174, 1969.

19. C. W. Randall, *et al.* "Effects of detention time and pH on the performance of continuous flow aerobic digestion," Paper presented at the 9th Air, Land and Water Pollution Control Conference, Mississippi State University, Dec. 1974.

20. C. W. Randall, *et al.* "Temperature effects on aerobic digestion kinetics," *Journal of the Environmental Engineering Division, ASCE,* 101, 795-811, 1975.

21. C. S. Reece, *et al.* "Aerobic digestion of waste activated sludge," *Journal of the Environmental Engineering Division, ASCE,* 105, 261-272, 1979.

22. L. D. Benefield and C. W. Randall, "Design relationships for aerobic digestion," *Journal of the Water Pollution Control Federation,* 50, 518-523, 1978.

23. D. E. Drier, "Aerobic digestion of solids," *Proceedings of the 18th Industrial Waste Conference,* Purdue University Engineering Extension Series No. 115, 123-140, 1963.

24. N. Jaworski, *et al.* "Aerobic sludge digestion," *International Journal of Air and Water Pollution,* 4, 106-114, 1961.

25. G. W. Lawton and J. D. Norman, "Aerobic sludge digestion studies," *Journal of the Water Pollution Control Federation,* 36, 495-504, 1964.

26. W. P. Issacs and A. F. Gaudy Jr., "A method for determining constants of first-order reactions from experimental data," *Biotechnology and Bioengineering,* 10, 69-82, 1968.

27. D. A. Koers and D. S. Mavinic, "Aerobic digestion of waste activated sludge at low temperatures," *Journal of the Water Pollution Control Federation,* 49, 460-468, 1977.

28. J. E. Smith, Jr., *et al.* "Biological oxidation and disinfection of sludge," *Water Research,* 9, 17-24, 1975.

29. L. C. Matsch and R. F. Drnevich, "Autothermal aerobic digestion," *Journal of the Water Pollution Control Federation,* 49, 296-310, 1977.

30. C. W. Randall, *et al.* "The aerobic digestion of activated sludge at elevated temperatures," *Proceedings of the Fifth Annual Environmental Engineering and Science Conference,* University of Louisville, Kentucky, 451-472, 1975.

31. Q. B. Graves, *et al.* "Aerobic sludge digestion at a trickling filter waste treatment plant," *Proceedings of the 27th Industrial Waste Conference,* Purdue University Engineering Extension Series No. 141, 501-512, 1972.

32.   C. W. Randall, *et al.* "Optimal procedures for the processing of waste activated
      sludge," Bulletin 61, Virginia Water Resources Research Center, Virginia Poly-
      technic Institute and State University, Blacksburg, Va., 1973.

33.   N. R. Ahlberg and B. I. Boyko, "Evaluation and design of aerobic digesters,"
      Journal of the Water Pollution Control Federation, <u>44</u>, 634-643, 1972.

Further Reading

Adams, C. E., "Partial volatile solids destruction: Aerobic digestion,"
      Environmental Engineers Handbook, Vol. I, Edited by B. G. Liptak, Section
      I.5.16, Chilton Book Company, 858-872, 1972.

Benefield, L. D. and Randall, C. W., *Biological Process Design for Wastewater
      Treatment,* Prentice-Hall, Inc., Englewood Cliffs, N.J., 1980. See Chapter 8.

Clark, J. W., *et al. Water Supply and Pollution Control,* 2nd Edition, International
      Textbook Co., Scranton, Pa., 1971. See Section 11-23 for a general description
      of aerobic digestion.

Eckenfelder, W. W. and Ford, D. L., *Water Pollution Control,* The Pemberton Press,
      Austin, Texas, 1970.

Metcalf & Eddy, Inc., *Wastewater Engineering: Treatment/Disposal/Reuse,* 2nd
      Edition, McGraw-Hill Book Co., New York, N.Y., 1979. See Chapter 11.

Rich, L. G., "Rational design of aerobic digestion systems," *Water and Sewage Works,*
      <u>124</u>, No. 4, 94-95, 1977.

CHAPTER 18

TRICKLING FILTER

A brief description of the biochemical operation known as a trickling filter was given in Section 1.2.4 where it was seen to be a fixed-film reactor in the shape of a packed tower. In such reactors, which were discussed in Chapter 14, bacteria and other microorganisms grow as a thin film on solid media while removing soluble organic matter from the liquid flowing over them. Thus filtration, in the classical sense of the word, does not occur. In fact, because of the detachment of microorganisms from the film surface, the concentration of suspended solids in the effluent from the reactor is in excess of the concentration in the influent. This occurs because, like activated sludge, a trickling filter converts soluble organic matter into an insoluble form which can be removed by sedimentation. Although the total population served by trickling filters is not as large as that served by

activated sludge, the number of installations is quite large, making them second
only to activated sludge as a means of removing soluble organic matter from
wastewaters [1].

18.1 DESCRIPTION

18.1.1 Physical Characteristics

Figures 18.1 and 18.2 show typical trickling filters. In spite of the apparent
differences between these two reactors, both possess the same major components: a
filter media, an enclosure, a distribution system, and an underdrain system.

MAJOR COMPONENTS. The filter media provides the surface upon which the microorganisms grow. Pearson [2] and Chipperfield [3] have listed the following characteristics of an ideal media: (1) it provides a large surface area for microbial
film growth; (2) it allows the liquid to flow evenly in a thin sheet over the
microbial film; (3) it has sufficient void space for the free flow of air; (4) it
has sufficient void space to allow the organic solids sloughed from the microbial
film to be carried away; (5) it is biologically inert (i.e., it neither undergoes
biodegradation nor inhibits the growth of microorganisms); (6) it is chemically
stable; and (7) it is mechanically stable. Although a large number of materials
have been tried, the most successful have been plastic and crushed granite or limestone. The recommended size of stone media is between two and four inches because
that provides the best compromise between surface area and void space. For example,
2 inch rock will provide around 30 ft^2 of surface area per cubic foot of reactor
volume (98 m^2/m^3) and will have around 50% voids [1]. Smaller sizes have a larger
specific surface area (the surface area per unit volume of reactor), but their void
space is so small that the microbial film will interfere with the flow of liquid
and air. In order to obtain a media with both a larger specific surface area and a
larger void volume, manufacturers turned to plastic. Figures 18.3 and 18.4 show
two common types. Those shown in Fig. 18.3 are dumped in place and are available
with specific surface areas between 30 ft^2/ft^3 (98 m^2/m^3) and 104 ft^2/ft^3
(340 m^2/m^3) with void ratios of 93-95%. Modular media (Fig. 18.4) are stacked in
place and have specific surface areas from 25 to 60 ft^2/ft^3 (81 to 195 m^2/m^3),
depending upon the manufacturer. Although the void space is also around 93-95%, the
flow path is straighter and more open than in dumped packing. Details about
specific plastic media can be obtained from the manufacturers.

If stone media is used the enclosure must be of reinforced concrete or some
other material capable of holding it in place vertically. The enclosure must also
be strong enough to be filled with water since it is sometimes necessary to flood a
rock filter. The enclosure for a filter filled with dumped plastic media must also
support it vertically, although its low bulk density (less than 6 lb/ft^3 (96 kg/m^3))

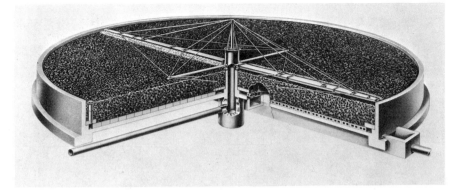

Figure 18.1. Typical rock media trickling filter.
(Courtesy of Dorr-Oliver, Inc.).

Figure 18.2. Typical plastic media trickling filter.
(Courtesy of BSP Division of Envirotech Corp.).

Figure 18.3.   Typical dumped plastic media.
          (Courtesy of Norton Pollution Control Products).

requires less structural support than rock media. Modular plastic media is self
supporting. Consequently the enclosures serve primarily to minimize splashing and
heat loss, and to improve the esthetics of the reactor. Thus many types of light
weight materials are used.

The function of the distributor is to provide a uniform hydraulic load (flow
per unit of cross-sectional area) over the entire reactor. Two common types are
used, rotary and fixed. Rotary distributors with either one or two arms are used
on rock filters and most are driven by the reaction of the water flowing out of them,
which requires a pressure head of around 2 feet (0.61 m) measured from the center
of the arms. A common rotational speed is one revolution in 10 minutes because it
provides intermittent dosage to any point in the filter thereby allowing air to
flow through when the liquid isn't. If the liquid flow rate is insufficient to keep
the rotor turning at the proper speed then a dosing siphon is used. Wastewater is
distributed across plastic filters with either four arm rotary distributors or with
fixed nozzles. Because of the larger void space in plastic media, the liquid and
air can move simultaneously, thereby allowing continuous application of the

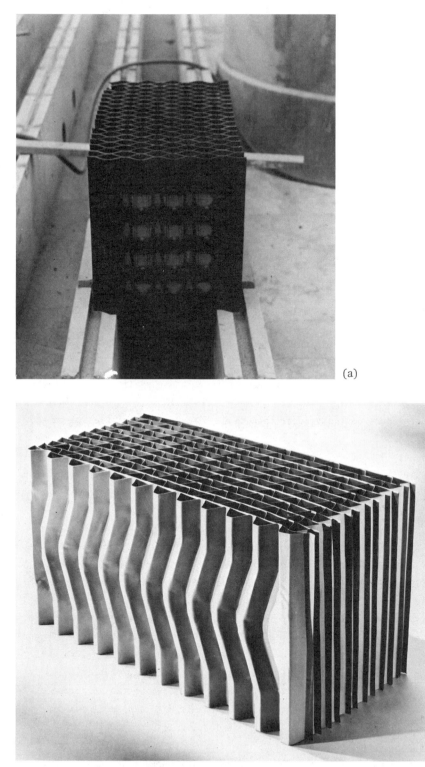

Figure 18.4. Typical modular plastic media.

   (a) Surfpac [C] *(Courtesy of BSP Division of Envirotech Corp.);*

   (b) Vinyl Core [C] *(Courtesy of B. F. Goodrich).*

wastewater. Many types of distribution systems are available and one job of the engineer is to ensure that the distributor and media are compatible.

The underdrain system must do three things: (1) support the media; (2) provide an uninterrupted path of flow for the liquid to the floor drain; and (3) allow access of air to the bottom of the filter so that a continuous supply of oxygen may be brought to the microorganisms in it. Prior to the advent of plastic media almost all underdrain systems were made of vitrified clay because of the weights which they had to support. Plastic media are not as heavy, however, so that the underdrain systems supporting many of them are simple metal gratings. In either type of installation the floor slopes toward the collection channel at a 1 to 2 percent gradient and the underdrains may be open at both ends to allow inspection and provide ventilation. Detailed specifications for underdrain design can be obtained from the Trickling Filter Floor Institute [4].

SYSTEM LAYOUT. Examination of Figs. 18.1 and 18.2 reveals an important difference in the configurations of rock and plastic media filters: rock media filters tend to be shallow with a large cross-sectional area whereas plastic media filters tend to be tall with a small cross-sectional area. Rock media has a high density and a low porosity, both of which require the filters to be shallow. For example, within the U.S. it is standard practice to make rock media filters 5 to 7 feet (1.5-2.1m) deep because depths greater than that cause uneconomic stresses on the underdrain system and have excessive resistance to air flow. Modular plastic media, on the other hand, is self supporting to depths of 20 ft (6.1 m) and offers little resistance to air flow; thus plastic media filters tend to be tall. In fact plastic media filters greater than 20 ft tall can be built by providing intermediate structural supports for the media.

There is, of course, no requirement that all of the media be placed into a single filter. The designer has the option of using filters in series or in parallel and the decision will be an economic one dictated by the type of media employed and the local conditions of the design. Thus when deep filters are required it has been common practice to use filters in series, especially with rock media. It is sometimes necessary to employ filters in parallel when large flows are being treated. For example, the maximum allowable diameter for a filter with a rotary distributor is 200 ft [5]. Thus if the total flow were more than could be handled with a filter of that size it would be necessary to use two or more filters in parallel.

As was seen in Chapter 14, an important factor influencing the performance of a packed tower is the hydraulic application rate per unit of cross-sectional area. In trickling filter design this is called the hydraulic loading and is usually expressed as $gal/(day \cdot ft^2)$, $ft^3/(day \cdot ft^2)$ or $m^3/(day \cdot m^2)$. Although the designer has considerable latitude in choosing the hydraulic loading, upper and lower constraints do exist. Generally, the lower limit is dictated by the quantity of liquid required to wet all of the media because media which is not wet is not being used effectively.

Malina *et al*. [6] have suggested that the minimum hydraulic loading for rock media
(expressed as $ft^3/(day \cdot ft^2)$) should be 1.33 times the specific surface area
(expressed as ft^2/ft^3). Because plastic media is more open and has larger void
spaces, it requires higher hydraulic loadings. One manufacturer of dumped media
recommends a minimum of 94 $ft^3/(day \cdot ft^2)$ (29 $m^3/(day \cdot m^2)$) [7] whereas a manufacturer
of modular media suggests 150 $ft^3/(day \cdot ft^2)$ (45 $m^3/(day \cdot m^2)$) [8]. The upper limit
in rock media filters is governed by the ability of the liquid to move through the
tortuous voids in thin films. A commonly accepted value for coarse rock media is
135 $ft^3/(day \cdot ft^2)$ (45 $m^3/(day \cdot m^2)$) [1]. The upper limit on modular plastic media
is governed both by thin film flow and by scouring of the microbial film. Although
this limit is not well established, flows as high as 1150 $ft^3/(day \cdot ft^2)$
(350 $m^3/(day \cdot m^2)$) have been used with good results. One manufacturer [8] recommends
that hydraulic loadings in excess of 770 $ft^3/(day \cdot ft^2)$ (234 $m^3/(day \cdot m^2)$) not be used.

Another factor which is important to the performance of a trickling filter is
the mass application rate of organic matter per unit volume of reactor. This is
called the organic loading and is expressed as the pounds of biodegradable
COD/(day·1000 ft^3) or lbs T_bOD/(day·1000 ft^3) (kg T_bOD/(day·1000 m^3)). Because the
organic loading is related to the rate at which the microorganisms must utilize
substrate, many investigators have correlated it with system performance. It will
be seen later, however, that other factors can also influence performance so that
the specification of an organic loading is not a sufficient criterion for design.
Nevertheless, knowledge of the organic loading can be quite helpful. For example,
high organic loadings must be accompanied by high hydraulic loadings in order to
continually wash the microorganisms from the media. If this is not done (particularly with rock media) excessive film thicknesses will clog the pores of the filter,
causing failure of the system.

Once a filter has been built, the only way that the hydraulic and organic
loadings may be adjusted independently is through the use of recirculation. Otherwise a change in the hydraulic loading will result in a proportional change in the
organic loading, and vice versa. Recirculation returns a portion of the treated
effluent for mixture with the influent, thereby breaking that proportionality. If a
high strength wastewater is being treated, the organic loading required to achieve
the desired effluent may result in a hydraulic loading which is below the minimum
recommended by the manufacturer or which is too low to wash away the growing biofilm.
By employing recirculation of treated effluent, however, the hydraulic loading may
be increased to an appropriate value while maintaining a constant organic loading.
Figure 18.5 shows some common recirculation patterns. Care must be exercised in
choosing a pattern because of its effect upon the system. For example, if clarified
effluent is recirculated, the size of the final settler must be sufficient to handle
the influent flow plus the recirculated flow. If unsettled effluent is recirculated,

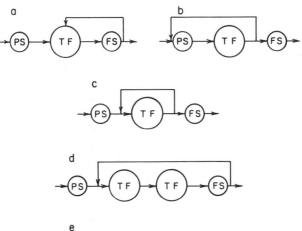

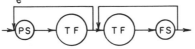

Figure 18.5. Recirculation patterns in trickling filters:
              (a) Single filter, clarified effluent;
              (b) Single filter, clarified effluent;
              (c) Single filter, unclarified effluent;
              (d) Filters in series, clarified effluent
                  from entire process chain; and
              (e) Filters in series, clarified effluent
                  from individual filters.

on the other hand, the filter media must have sufficient void space to prevent the
build up of suspended solids from clogging the filter. Regardless of the pattern
employed, however, the ratio of recirculated flow to influent flow is generally kept
less than 4.

    The flow diagrams in Fig. 18.5 all include both a primary and a final settler
in the process train. Because a trickling filter works by converting soluble
organic matter to insoluble cell material, a final settler is always needed to
remove the cells. The necessity for a primary settler depends upon the type of media
employed. If it is rock, a primary settler should be used to remove suspended solids
which might clog the filter. If it is modular plastic, the possibility of clogging
is small and thus there is little need for the primary settler. Little degradation
of the influent solids will occur in the filter, however, and thus they must be
removed in the final settler.

18.1.2 Applications
Like activated sludge, the primary objective of a trickling filter is the removal
of soluble organic matter through its conversion to microbial cells. Consequently,
trickling filters are widely used for the treatment of municipal and industrial

wastewaters. Whereas activated sludge is generally used when the concentration of organic matter is between 50 and 4000 mg/liter as T_bOD, trickling filters can often be applied economically to wastewaters with concentrations below that range. Care must be exercised in their application to wastewaters at the upper end of the range, however, because limitations on oxygen transfer may cause odor and performance problems. Nevertheless, because trickling filters can achieve partial removal of organic matter they are often used to pretreat high strength wastewaters prior to discharge to municipal sewers or prior to treatment by activated sludge. Furthermore, because plastic media filters can be constructed as tall towers, they are particularly useful where land area is limited. Finally, because they are simple to operate, trickling filters are used extensively by small communities which cannot afford the highly skilled operators required for activated sludge. Like activated sludge, however, they must be part of a complete treatment system because provision must be made for disposal of the biosolids produced.

18.1.3 Merits and Defects

The primary merits associated with trickling filters stem from their simplicity and ease of operation, thus making them ideal for remote sites or for small communities [9]. Furthermore, because large masses of organisms must be present for them to achieve effluents of high quality, they possess substantial reserve capacities which make them very robust and tolerant to changes in the influent. In addition, the dense nature of the microbial films which slough from the media produces sludges of relatively constant character which can be readily removed by sedimentation. This can be particularly beneficial for the treatment of wastewaters which produce bulking sludges in the activated sludge process. One merit which has long been claimed for trickling filters is an ability to survive shock loads of toxic wastes [1]. It has been noted, however, that if this is true, it is not because the microorganisms are any more hardy [10,11,12]. Rather it is because of the relatively short retention time of the wastewater in the reactor [10] or because only organisms on the surface may be killed. Thus, as the dead organisms are removed by sloughing, a layer will be exposed which has not been subjected to the toxic material [11]. If the shock load of toxic material is of long duration, however, or of a type which will be adsorbed onto the biofilm, then a trickling filter can be severely affected, as was shown in recent studies [13,14]. Consequently the purported resistance of trickling filters to toxic shock loads should not be considered to be a general characteristic, although it will probably be true in some cases.

Like completely mixed aerated lagoons the factors responsible for the chief merits associated with trickling filters are also the factors responsible for their principal defects, i.e., their simplicity and ease of operation. Because the microorganisms grow attached to a fixed surface there is nothing that can be done to alter their mass in a reactor in response to environmental changes and therefore, there is no effective way of controlling the effluent quality. Consequently, if increases occur in the concentration or flow rate of the wastewater being treated, the quality of the effluent will deteriorate. Likewise, if the temperature drops, the rate of substrate removal will also drop, and the effluent quality will suffer. In a similar vein, large reactors are required to ensure the achievement of effluents containing low concentrations of organic matter over a broad range of influent or environmental conditions. Thus the designer of a trickling filter is faced with the choice of a variable effluent quality on the one hand versus an overdesign on the other. In addition to the above problems, the changing seasons cause others. For example, in the summer, rock media filters may serve as a breeding ground for *Psychoda* flies, thereby creating a nuisance condition in the immediate area. In the winter, icing can be a problem in northern climates unless the filters are enclosed.

Like all of the other named biochemical operations, the merits associated with trickling filters outweigh the defects for certain applications, and thus they find extensive use under those circumstances.

18.2 TRICKLING FILTER PERFORMANCE

18.2.1 Theoretical and Observed Performance

A trickling filter is basically a packed tower; consequently it should be possible to characterize its performance by the mathematical model and theoretical curves presented in Section 14.4. Such a reactor is extremely complicated, with many events occurring simultaneously. Among those expressed in the model are substrate transport through the liquid film, substrate transport through the biofilm, and substrate consumption. Among those not expressed are oxygen transport and the intermittent sloughing of microorganisms from the biofilm. The model, therefore, is a first generation one which attempts to consider the major events occurring in a trickling filter. It would be instructive to review the theoretical findings from Section 14.4 and to compare them with observations from field- and pilot-scale reactors because that will give us a better understanding of how trickling filters can be expected to perform.

Consider first the general performance as depicted in Fig. 14.25, which shows an exponential decrease in substrate concentration as the wastewater moves deeper into the filter. This particular characteristic, illustrated in Fig. 18.6 with data collected by Cook and Kincannon [15], has long been recognized and has been incorporated into many mathematical models [16-19]. The reduced ability of each successive unit of filter depth to remove substrate is a result of the decrease in substrate concentration because that concentration controls the rates of both physical transport and biological reaction.

Since the concentration of substrate at any point in a filter influences the rate at which the substrate is removed at that point, it would be logical to expect the concentration of substrate entering a trickling filter to influence its performance. Figure 14.25 showed the theoretical effect of influent concentration upon

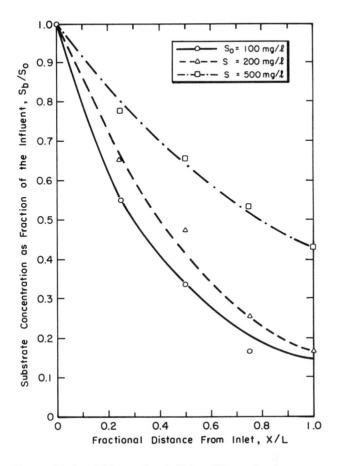

Figure 18.6.   Effect of trickling filter depth
on the removal of substrate.
(Data from Cook and Kincannon [15]).

the removal of substrate in a packed tower. Although the mass of substrate removed
increases as the influent concentration is increased, the fractional removal (based
upon the influent) decreases because the rate at which bacteria can remove substrate
reaches a maximum value as its concentration is increased. The qualitative
similarity between the curves in Figs. 18.6 and 14.25 attests to the validity of
that theoretical effect. From the above discussion we would expect the percent
substrate removal by a filter of fixed depth to decline as the influent concentration
was increased, even though the total amount removed per unit time increased. This
was shown in Fig. 14.27 along with the fact that the total amount removed approaches
a maximum as the total amount applied is increased. Figure 18.7 contains experimental curves drawn from the data of Cook and Kincannon [15] as presented in the
thesis of Cook [20]. The similarity between the two figures is apparent. These

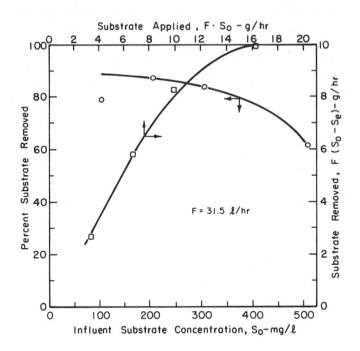

Figure 18.7. Effect of influent substrate concentration
               on substrate removal in a trickling filter.
               (Data from Cook and Kincannon [15,20]).

effects of substrate concentration have not always been recognized although some
empirical models have incorporated a term for them [19].

The other factor which can increase the mass of substrate applied to a filter
per unit time is the flow rate, or hydraulic loading. Since the mass of organisms
in the filter is fixed, the only way that the rate of substrate removal can be
increased is for the substrate concentration to increase. Thus, as the flow rate
is increased, the percent of the applied substrate which can be removed must
decrease. This was shown in Fig. 14.28. Figure 18.8 shows a similar effect for an
experimental system. It will be recalled from Chapter 14 that one effect of an
increase in flow rate is a decrease in the external resistance to mass transfer.
This acts to increase the rate of substrate removal associated with any given
concentration of substrate in the bulk of the fluid. When the applied flow rate
is low, an increase in flow rate will cause a relatively large reduction in the
resistance to mass transfer; consequently, the rate of substrate removal can
increase with little increase in substrate concentration. As further increases are
made in the flow rate, however, additional reductions in the mass transfer resistance
become negligible so that no benefits accrue to offset the deleterious effect of the
decreased time of contact. Consequently, the decrease in performance becomes greater
at higher flow rates. This, too, is shown in Figs. 14.28 and 18.8. This variable

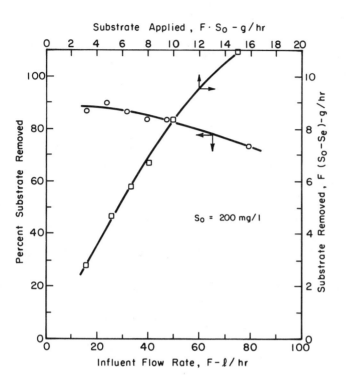

Figure 18.8.   Effect of influent flow rate on substrate
removal in a trickling filter.
(Data from Cook and Kincannon [15,20]).

effect of flow rate has been incorporated into some empirical models for trickling
filters [17,19].

As defined earlier, the organic loading is the mass of substrate applied per
unit time divided by the volume of the filter. Examination of Figs. 18.7 and 18.8
reveals that the curves of substrate removed versus substrate applied are superimposable over the application range of 3-9 g/hr. Since the same reactor was used
to obtain both curves, plots of substrate removed versus organic loading would be
the same over that range, regardless of whether the loading was altered by changing
the flow or by changing the concentration. Consequently, some observers have concluded that organic loading is the primary parameter controlling the performance of
trickling filters [15,21,22,23], and some groups have incorporated that concept into
their design procedures [7,24]. As long as the valid range for such a relationship
is not exceeded, it provides a useful means for predicting the performance of a
trickling filter. Care must be exercised in its use, however, because changes in
influent flow and concentration influence the reactor in fundamentally different
ways and thus will not have the same effect over the entire span of possible values.

That this is true may be seen by comparing the theoretical curves in Figs. 14.27 and 14.28 and the experimental curves in Figs. 18.7 and 18.8, as well as by examining other data collected over a greater range of values [25].

Another factor which can influence the performance of a trickling filter is the specific surface area of the media. Theoretically, substrate removal will increase as the specific surface area is increased because the availability of more surface means that more microorganisms can be contained within each unit volume of filter. This was shown in Fig. 14.34. A similar plot, made from the data of Fleming and Cook [26] is shown in Fig. 18.9. An important factor is the physical configuration of the media because it will determine the flow patterns that are established with their resultant effects on wetted surface area, mass transfer, etc. The media with a specific surface area of 37 ft^2/ft^3 (121 m^2/m^3) in Fig. 18.9 had a configuration that differed the most from the other media utilized and this no doubt contributed to its low performance. The physical configuration of the media will also influence the holdup of microbial slime within the reactor, thereby affecting performance. Factors such as these must be considered when the media is being selected.

Of all of the factors which influence the performance of a trickling filter, perhaps the least understood is recirculation. One possible reason for this is that several schemes can be employed (see Fig. 18.5), each with a different effect. Usually the recirculated liquid passes through one or more settling basins before entering the filter. Consequently the concentration of microorganisms in it will be low, thereby making the removal of substrate by the biofilm predominate. Under those conditions, the model predicts that the application of recirculation will

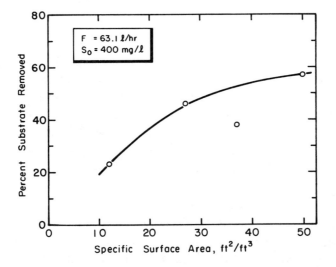

Figure 18.9. Effect of a specific surface area on substrate removal in a trickling filter. (Data from Fleming and Cook [26]).

generally result in a slight deterioration in performance (see Fig. 14.31). The
reason for this can be seen by considering what recirculation does. First, it
dilutes the influent, which reduces the concentration of substrate throughout the
reactor thereby retarding the rates of both mass transport and substrate removal.
Second, it increases the turbulence in the liquid film, which decreases the external
resistance to mass transfer. If the decrease in mass transfer resistance is less
important than the reduction in reaction rate caused by dilution, then recirculation
will cause the filter performance to deteriorate. Conversely, if the decrease in
mass transfer resistance is more important, performance will be improved. Finally,
if the effects on mass transfer and reaction rate are of the same magnitude,
recirculation will have little effect. The available experimental evidence indicates that recirculation of clarified effluent has little or no effect upon
trickling filter performance [27,28,29]. Relatively small ranges of recirculation
flow were considered in those studies, however, and additional experiments are
needed to delineate fully the effects of recirculation.

In contrast, recirculation of unclarified effluent around plastic media filters
is likely to improve their performance by increasing the importance of substrate
removal by organisms in suspension [30]. Evidence to this effect has come from
studies in which sludge from the bottom of the secondary settler was returned to
the influent of the primary settler, thereby increasing the concentration of
acclimated organisms entering the filter [27], and in which effluent from an
aeration unit was used for pseudorecirculation [28]. Care should be exercised in
utilizing this fact, however, because an increase in substrate removal by suspended
organisms might interfere with oxygen transfer to the fixed film, thereby limiting
the effectiveness of the reactor.

A limitation on oxygen transfer to the fixed film is an important factor which
is not considered in the model. Other models have incorporated simultaneous
diffusion of oxygen and substrate into the biofilm and have shown that oxygen can
become limiting when the substrate concentration is high [31,32]. Figure 18.10
shows calculated oxygen and substrate profiles in a biofilm under three possible
situations. In general, it appears that influent substrate concentrations in excess
of 400 mg/liter (as $T_b OD$) could cause an oxygen limitation [31,32]. Consequently,
if odors and substandard performance are to be prevented, the concentration of
substrate entering the filter should be kept below that value. One way to do that
is with recirculation.

Another important factor, which is not explicitly stated in mathematical
models, is the dosing schedule, i.e., the period over which liquid is applied to the
filter in combination with the period without flow. Cook and Crame [33] investigated its effects on the performance of a plastic media trickling filter and their
results are shown in Fig. 18.11. In this case, the program consisting of five

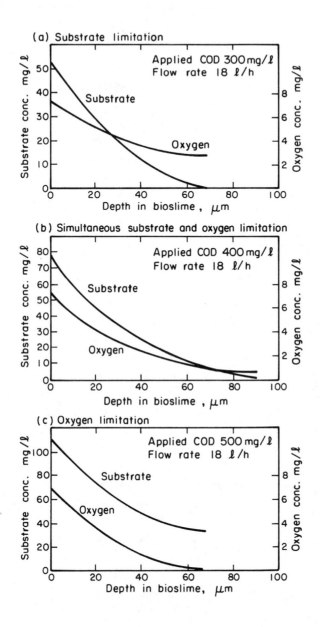

Figure 18.10. Theoretical oxygen and substrate concen-
               tration profiles in a microbial film.
               (Reprinted by permission from H. P. Harris
               and G. S. Hansford, "A study of substrate
               removal in a microbial film reactor,"
               Water Research, 10, 935-943, 1976,
               Pergamon Press Ltd.).

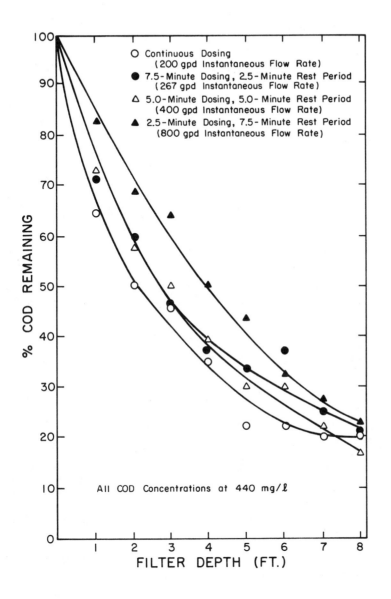

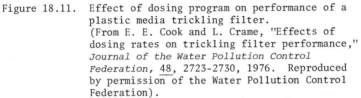

Figure 18.11.   Effect of dosing program on performance of a
                plastic media trickling filter.
                (From E. E. Cook and L. Crame, "Effects of
                dosing rates on trickling filter performance,"
                *Journal of the Water Pollution Control
                Federation,* 48, 2723-2730, 1976. Reproduced
                by permission of the Water Pollution Control
                Federation).

minutes of application followed by five minutes of rest proved to be the best.
Application periods in excess of five minutes had insufficient velocities to wash
the excess microbial growth from the upper reaches of the filter, leading to con-
striction of flow and ponding. With the five minute application period, however,
film thickness was controlled without harming the efficiency of substrate removal
across the entire reactor. Several important points arise from this work. First
the application rate must be high enough to cause sufficient sloughing of the biofilm to prevent clogging and ponding. If it is not high enough, it must be
increased, either by changing the dosing schedule or by using recirculation.
Furthermore, when thick biofilms occur, the kinetics of substrate removal are
likely to be altered, probably because of oxygen and substrate transport limitations.
Finally, when pilot plant studies are performed to generate design data, the dosing
schedule should be the same as that anticipated for the full-scale facility. If it
is not, the kinetic characteristics observed may not accurately reflect the performance of the full-scale facility.

One consequence of allowing a thick biological film to accumulate on the media
is a reduction in the wetted surface area available for substrate removal [34].
This can have an important effect upon the performance of the system because as
shown in Figs. 14.34 and 18.9 performance will decrease as the surface area available
for substrate removal is decreased. Atkinson and Ali [34] proposed an equation of
the form:

$$\frac{a_w}{a_v} = 1 - \exp\left[-\frac{TF}{a_v}\right] \qquad (18.1)$$

to relate the actual wetted surface area per unit volume, a_w, to the total surface
area of media per unit volume, a_v. Performance would then depend upon a_w rather
than a_v. The value of the parameter T must be determined for the particular media
in question.

From the preceding it is apparent that the proper performance of a trickling
filter is dependent upon control of the microbial film thickness. Furthermore,
controlled sloughing of the biofilm appears to be important because uncontrolled
sloughing can cause variations in effluent quality [35]. Unfortunately, our
knowledge of sloughing is insufficient to allow prediction of the effect of
hydraulic loading or dosing schedule upon it. Thus all that can be done is to
observe the effects during pilot plant studies.

18.2.2 An Empirical Model of Trickling Filter Performance
Rigorous theoretical models seek to express mathematically the major physical,
chemical, and biochemical events occurring in a system, thereby allowing us to
learn about its performance under operational conditions outside of our experience.
Empirical models, on the other hand, merely express our experience mathematically
and allow us to extend it to new conditions within our experiential space. Because

empirical models are interpolative rather than extrapolative, they are usually
simpler and easier to work with. Furthermore, as long as they are used as they
were intended, empirical models are well suited to the mathematical expression of
pilot plant data for use in system design. The theoretical modeling of packed
towers is in its infancy and much work remains to be done on the events which should
be included, the methods of expressing them, and the evaluation of the parameters
describing them. These factors, coupled with the complicated solution techniques
required, suggest that it would be premature to use theoretical models for the
design of trickling filters. Consequently, we will rely upon an empirical model
which is similar in form to some of the widely used models presented in
Section 14.4.4 [17,19].

Consider a trickling filter with length, L, and cross-sectional area, A_c,
receiving feed at flow rate, F, and concentration, S_F, as shown in Fig. 18.12. The

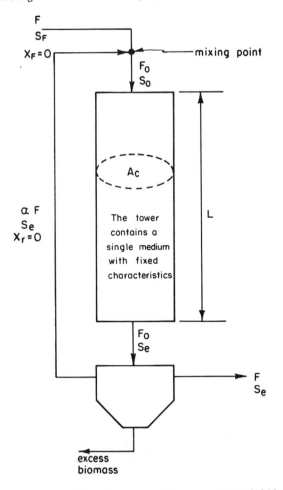

Figure 18.12. Schematic diagram of a trickling filter.

concentration of microorganisms in the feed is assumed to be negligible (i.e., $X_F = 0$). Furthermore, only clarified effluent is recirculated (i.e., $X_r = 0$). In other words, no term will be included in the empirical model for substrate removal by suspended organisms. In addition, the model will be restricted to one type of media with fixed shape, specific surface area, and porosity receiving wastewater on a fixed dosing schedule. Thus the pilot plant upon which data is collected for evaluation of the parameters in the model must contain the filter media and dosing schedule which will be used in the full-scale filter.

The first performance characteristic which should be incorporated into the empirical model is the exponential decrease in substrate concentration as the flow moves downward through the filter. Thus:

$$S/S_o = \exp(-Kx) \tag{18.2}$$

where S_o is the concentration of substrate actually applied to the filter, S is the concentration at any distance x down the filter, and K is an apparent rate constant. Such an equation has been used in many models [16-19]. Other factors which influence the removal of substrate and which should be stated explicitly in the model are the flow rate and concentration of the influent to the filter, F_o and S_o. Examination of Figs. 18.6, 18.7, and 18.8, reveals that increases in either variable retard the fractional decrease in substrate concentration; in other words the apparent rate constant, K, must be a function of those variables. Further consideration of the effect of flow rate suggests that the real variable is the hydraulic loading, F_o/A_c, since it is what influences turbulence and mass transfer. The question now arises concerning the kind of function which should be used to represent the effects of influent substrate concentration and hydraulic loading. Examination of other empirical models suggests a simple exponential form [17,19]:

$$K = K'S_o^m(F_o/A_c)^n \tag{18.3}$$

in which K' is another apparent rate constant whose value depends upon the units of S_o, F_o, and A_c. Substituting Eq. 18.3 into Eq. 18.2 we obtain the empirical model:

$$S/S_o = \exp[-K'S_o^m(F_o/A_c)^n x] \tag{18.4}$$

The final factor which must be incorporated explicitly into the model is the effect of recirculation. Experimental evidence indicates that recirculation of clarified effluent does not affect the rate constant [29]. Furthermore, theory suggests that the main effects would be to dilute the feed and increase the turbulence in the filter. If we limit our use of Eq. 18.4 to the overall performance of the filter by restricting x to the total length, L, then S becomes S_e and we may incorporate the effects of recirculation simply by employing mass balances about the mixing point:

$$F_o = (1 + \alpha)F \tag{18.5}$$

$$S_o = (S_F + \alpha S_e)/(1 + \alpha) \tag{14.74}$$

Substitution into Eq. 18.4 (along with the restriction that x = L and S = S_e)
yields:

$$\frac{S_e(1 + \alpha)}{S_F + \alpha\ S_e} = \exp\ \{-K'[(S_F + \alpha S_e)/(1 + \alpha)]^m[(1 + \alpha)F/A_c]^n L\} \qquad (18.6)$$

Equation 18.6 may now be used to choose the filter length, L, cross-sectional area,
A_c, and recirculation ratio, α, required to treat a wastewater with flow, F, and
concentration, S_F, to effluent concentration, S_e.

The ability of the empirical model to represent the performance of a trickling
filter can be assessed by using the data of Cook [20], who investigated the effects
of flow rate and influent substrate concentration on a prototype filter. Figure
18.13a shows his data when plotted simply as a function of filter depth. Figure
18.13b shows the same data when plotted according to Eq. 18.4. Although the model
isn't perfect, the reduction in scatter is apparent. Thus, the empirical model is
seen to be useful for reducing data from a broad range of conditions into a useable
form.

18.3 DESIGN

The process design for a trickling filter includes three decisions: the filter
depth, the filter surface area, and the recirculation ratio. Many combinations of
these variables can result in an effluent of the desired quality and process design
equations are used by the engineer to delineate a number of candidate systems to
which he may apply engineering judgement and economics to arrive at a final design.

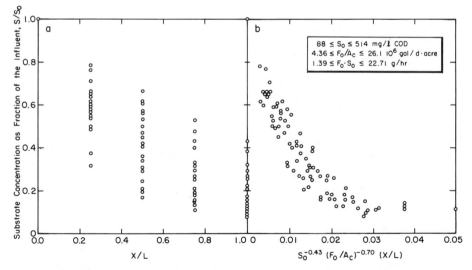

Figure 18.13. Plots of the data of Cook [20]:
      (a) Raw data showing the changes in concentration with
          depth for a broad range of influent conditions;
      (b) The same data plotted in accordance with Eq. 18.6.
          (10^6 gal/(d·acre) = 23.0 gal/(d·ft^2)
          = 3.07 ft/day = 0.936 m/day.)

In addition, there are a number of nonquantifiable factors which influence the
design, such as the type of media, dosing schedule, type of underdrains, and
provisions for ventilation. All of these must enter into the decision-making
process as well. Thus, just as with all of the other biochemical operations in this
text, the process design equations provide the base line from which the engineer
works.

18.3.1 Process Design Equations

The design equations for trickling filters are usually arranged to allow calculation
of the depth required to reduce the substrate concentration in a specific wastewater
to a desired value. Because the depth will depend upon the hydraulic loading and
recirculation ratio, the results are often expressed in graphical form with hydraulic
loading as the independent variable and recycle ratio as a parameter. When the
designer undertakes the economic analysis of the facility he will compute the filter
volume associated with each hydraulic loading and recirculation ratio because the
volume is a major factor in determining the capital costs. Consequently, similar
plots are also often made of filter volume. Operating costs will depend primarily
upon the depth and the recirculation ratio because the former influences the pumping
head and the latter the flow rate. A decision concerning the depth, area, and
recirculation for the final design can then be made by minimizing the total cost of
the facility.

An equation of the type required may be obtained by rearranging Eq. 18.6:

$$L = \frac{\ln[(S_F + \alpha S_e)/S_e(1 + \alpha)]}{K'[(S_F + \alpha S_e)/(1 + \alpha)]^m[(1 + \alpha)F/A_c]^n} \tag{18.7}$$

If no recirculation is used the equation reduces to

$$L = \frac{\ln(S_F/S_e)}{K'S_F^m(F/A_c)^n} \tag{18.8}$$

18.3.2 Determination of Parameter Values

Three parameters must be determined before Eq. 18.7 or 18.8 can be applied to
design: K', m, and n. The only realistic way of doing this is with data from a
pilot plant study using the media and dosing schedule that have been selected for
the final facility. Usually, a single tower is employed, with height equal to the
maximum feasible for the media. Recirculation is not normally practiced, however,
although it may be during verification studies. Data are collected on the change
in substrate concentration with depth for a number of influent flows and concentrations. Different influent concentrations can be obtained by dilution of the
wastewater with tap water but the maximum should not exceed 400 mg/liter T_bOD to
avoid oxygen limited conditions. The minimum flow rate employed should give a
hydraulic loading equal to the minimum recommended by the media manufacturer whereas

the maximum should not exceed the maximum recommended loading. The number of combinations of flow and concentration investigated will depend upon the technique employed to estimate the parameters. The minimum number of experiments will be five: one for each of the combinations of the extreme values and one in the center. This will supply sufficient information to allow someone well-versed in statistics to simultaneously evaluate the three parameters. The maximum number will be nine: a complete matrix covering three flows and three concentrations. This will supply sufficient data to allow the parameters to be evaluated sequentially by stepwise application of linear least-squares techniques. Sequential estimation of the parameters is less desirable than simultaneous evaluation because in spite of the additional experiments, the parameters will not fit the data as well. If the concentration of the wastewater is less than 400 mg/liter and recirculation will not be used, then the only variable which need be studied is flow rate, which greatly simplifies the task.

If the parameters are evaluated sequentially, the first step is to estimate K for each experiment. Linearization of Eq. 18.2 yields

$$\ln(S/S_o) = -Kx \tag{18.9}$$

Thus a plot of $\ln(S/S_o)$ as a function of depth should result in a straight line with slope -K. After the values of K have been obtained, m and n may be evaluated by taking advantage of the linearized form of Eq. 18.3:

$$\ln(K) = \ln(K') + m \ln(S_o) + n \ln(F_o/A_c) \tag{18.10}$$

Thus a plot of $\ln(K)$ versus $\ln(S_o)$ for a constant flow rate should yield a straight line with slope m. If several flow rates are investigated, m should be evaluated for each and the average taken. The value of n may be estimated by plotting $\ln(K)$ versus $\ln(F_o/A_c)$ for a constant influent concentration. Again, if several influent concentrations are employed, n should be evaluated for each and averaged. Finally, K' can be estimated by using the linearized form of Eq. 18.4:

$$\ln(S/S_o) = - K'S_o^m(F_o/A_c)^n \tag{18.11}$$

Thus a plot of $\ln(S/S_o)$ versus $S_o^m(F_o/A_c)^n$ will allow K' to be evaluated from the slope of the resulting line. Although these techniques have been used successfully to evaluate the parameters from pilot-scale trickling filters [19], it is apparent that they are subject to considerable error. Consequently, the use of statistical techniques is highly recommended.

No matter which technique is employed, the substrate concentration should be expressed as T_bOD. Since the easiest way to measure the concentration of soluble organic matter is as COD, it will be necessary to estimate the nonbiodegradable COD. This may be done in a manner similar to that employed to estimate the nonbiodegradable solids remaining after aerobic digestion (See Figure E17-1). The T_bOD concentration is then merely the initial COD minus the residual value.

After the parameters have been evaluated, the design equations can be used to size filters capable of producing the desired effluent and economic analysis can be applied to narrow the number of alternatives. Finally, after a design has been selected, the pilot plant should be operated in the selected configuration to confirm its ability to produce the desired effluent quality. The utility of the design equations, therefore, is to allow the designer to investigate a large design space using simulation. The uncertainty associated with the answer can then be reduced through a second application of the pilot plant.

18.3.3 Factors Influencing the Design

Unless it is feasible to conduct a large pilot plant study, the media will normally be chosen before the study. Because of its relatively low specific surface area and low porosity, the success of rock media is generally best when it is applied to wastes of low concentration, such as domestic sewage. Consequently, most modern trickling filters treating industrial wastewaters utilize plastic media, of which there are several. Once the choice has been made, the media manufacturer should be consulted concerning the minimum and maximum hydraulic loadings, the type of distribution system, and dosing schedule to be employed during the pilot study.

After the parameters have been evaluated, the designer is ready to consider alternative systems. One of the major decisions which must be made at that point is whether to use recirculation. As discussed in Section 18.2.1 the effect of recirculation of clarified effluent depends very much upon the system in question. This point may be further amplified by noting that modeling studies dealing with total treatment costs have concluded that: (1) costs are greater with recirculation [36]; (2) costs are less with recirculation [37]; (3) the effect of recirculation on costs is dependent upon the degree of treatment required [38]; (4) the effect is dependent upon the performance model employed [39]; or (5) the effect is dependent upon the parameter values in a single model [40]. Thus, unless other factors influence it, the decision concerning recirculation can only be made after a complete economic analysis of the particular situation.

There are three situations in which clarified effluent would be recirculated, regardless of economics. The first is when the influent substrate concentration exceeds 400 mg/liter T_bOD because concentrations in excess of that are likely to result in oxygen limitations which could either impair performance or cause odors [31,32]. Consequently, sufficient recirculation must be provided to keep the influent concentration below that value. The second situation is when the minimum flow rate to the filter is insufficient to keep the hydraulic loading above the minimum recommended value or to provide enough scour to keep the media pores open. In this case, one approach would be to design the filter with optimum recirculation under maximum flow conditions and then to increase the recirculation as needed to maintain proper operational conditions during periods of low flow. The third situation is when the waste is inhibitory at high concentration because then a decrease

in substrate concentration would increase the substrate removal rate, thereby
improving performance.

Another task facing the designer is the selection of the underdrain system.
Like the distribution system, the underdrain system must be compatible with the
media employed. Consequently the designer should consider the recommendations of
the media manufacturer when choosing the underdrains. If rock media is employed,
particular attention should be paid to the specifications prepared by the Trickling
Filter Floor Institute [4].

An important aspect of the underdrain system is its influence on the flow of
air through the filter. A trickling filter acts much like a cooling tower; consequently there will be both heat and mass transfer between the liquid and the air.
Both of these transfers affect the density of the air within the filter so that a
convection current will be established, with the direction of flow dependent upon
the temperatures of the air and the wastewater [41]. It will be downward when the
ambient air temperature is greater than the wastewater temperature and upward when
the reverse is true. Under some conditions, stagnation can occur. Since proper
operation of the filter is dependent upon adequate oxygen transfer, care must be
taken to minimize the occurrence of stagnation. Schroeder [42] has presented
techniques for calculating air flow rates through filters and if such calculations
indicate that natural ventilation will be inadequate, forced ventilation must be
provided. Generally, however, natural ventilation will be adequate provided the
following precautions are taken [5,8]: (1) Underdrains and collecting channels are
designed to flow no more than half full; (2) Ventilating manholes with open grating
covers are installed at both ends of the central collection channel; (3) Large
diameter filters are provided with branch collecting channels with ventilating manholes or vent stacks; (4) If filter blocks are used, the open area of the slots in
top is not less than 15 percent of the cross-sectional area of the filter; (5) For
rock media, the open area of ventilating manholes or vent stacks equals or exceeds
$1 \, ft^2$ per $250 \, ft^2$ of filter cross-sectional area; and (6) For plastic media, the
ventilation area is sufficient to allow about $90 \, ft^3/min$ of air per 100 lbs/day of
T_bOD applied ($5.6 \, m^3/min$ per 100 kg/day).

Because the air flow through the filter acts to cool the wastewater, and
because the efficiency of substrate removal will decline as the temperature drops,
consideration must be given to the effects of temperature during design. One way
to correlate performance with temperature is to make the rate constant, K', a
function of temperature using Eq. 10.57:

$$K_1' = K_2' \theta^{T_1 - T_2} \qquad\qquad (10.57)$$

The ideal procedure would be to determine θ during the pilot plant studies. If
that is impossible, a value of 1.035 has been recommended (when T_2 is 20°C) [43],
and could be used. Estimates can be made of the anticipated change in water

temperature by using cooling tower correlations. If they indicate that severe losses in performance are likely during the winter, then consideration should be given to covering the filters and controlling the air flow rate provided that such plans are consistent with adequate oxygen transfer.

Finally, it must be remembered that the trickling filter is only one unit operation in the process so that proper consideration can be given to its impact upon the other operations. For example, if recirculation of clarified effluent is to be practiced, the surface area of the final settler must be determined on the basis of the total flow entering it. Generally, the concentration and character of the sloughed microorganisms entering the final settler are such that it will be limited by clarification rather than thickening. Nevertheless, knowledge is needed of the amount of excess microorganisms produced so that the sludge disposal facility can be designed. The easiest way to predict the excess microorganism production rate is to multiply the change in substrate concentration across the filter by the flow rate and the observed yield, which can be estimated during the pilot plant study. One problem in establishing the yield is the variable rate of sloughing, which causes considerable fluctuation in yield values calculated over short periods of time. Thus data for determination of the yield should be collected over the duration of the pilot plant study. Because the microorganisms have a reasonably long residence time on the media, the observed yields will be less than the Y_g values in Chapter 9. For example, one study [31] reported an average observed yield of 0.31 gm cells/gm COD removed when glucose was the substrate.

18.3.4 Example Problems

The easiest way to demonstrate the process design procedure is through examples.

EXAMPLE 18.3.4-1

A wastewater with a T_bOD concentration of 340 mg/liter and an average flow rate of 84,800 ft^3/day (100 m^3/hr) is to be treated on a plastic media trickling filter. The minimum flow rate anticipated is 40,000 ft^3/day (47.3 m^3/hr) and the maximum is 107,000 ft^3/day (126 m^3/hr). The media to be employed has a minimum allowable hydraulic loading of 120 ft^3/(day·ft^2) (36.6 m^3/(day·m^2)) and a maximum allowable hydraulic loading of 1000 ft^3/(day·ft^2) (304 m^3/(day·m^2)). Evaluation of the data from a pilot plant study run with that media yielded the following parameter values:

> $K' = 100$ (when S_o is in mg/liter and F/A_c is in ft^3/(day·ft^2))
>
> $m = -0.40$
>
> $n = -0.65$

Investigate the sizes of full-scale trickling filters that are capable of producing an effluent with a soluble T_bOD equal to or less than 20 mg/liter under any anticipated flow condition.

Solution:

The design strategy will be to size the filter to produce the desired effluent under maximum flow conditions without recirculation, and then to check the performance of the system under average and minimum flow conditions to be sure that it will be adequate. The first task is to investigate the effect of hydraulic loading on the size of filter required using Eq. 18.8:

$$L = \frac{\ln(340/20)}{(100)(340)^{-0.40}(F/A_c)^{-0.65}} = \frac{0.292}{(F/A_c)^{-0.65}}$$

Figure E18.1 shows the area, depth, and volume of filter required for hydraulic loading rates between the minimum (120 ft^3/(day·ft^2)) and the maximum (1000 ft^3/(day·ft^2)). For this media the maximum unsupported depth is 20 ft. Thus for required depths greater than 20 ft, either intermediate structural supports or two filters in series must be utilized. These measures are not generally economic if the required effluent can be produced with a single filter of reasonable size because the additional capital costs involved more than offset the savings associated with the reduction in filter volume. Consequently a horizontal constraint line has been drawn at L = 20 ft, which produces the vertical constraint line at a hydraulic loading of 680 ft^3/(day·ft^2). A vertical line has also been drawn at the minimum acceptable hydraulic loading and the feasible region lies between the two.

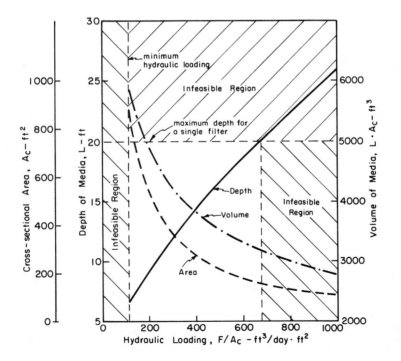

Figure E18.1. Effect of hydraulic loading on the size of trickling filter required under maximum flow conditions in Example 18.3.4-1.

Any of the paired values of area and depth corresponding to hydraulic loadings within the feasible region will produce the desired effluent. The designer must investigate the total cost associated with all feasible filters in order to decide which one to use. Examination of the figure reveals that filters with high hydraulic loadings require less total volume, which reduces their capital costs. They are also taller, however, which increases the pumping head, thereby increasing their operating costs.

The curves in Fig. E18.1 were computed for the maximum flow condition (107,000 ft^3/day). Using the upper and lower limits on hydraulic loading shown in the figure, the filter could have any cross-sectional area between 157 and 892 ft^2. However, in investigating the economics of filters between those limits, the designer must also consider the low flow condition (40,000 ft^3/day). If a filter of maximum area were chosen, the hydraulic loading under low flow conditions would be only 45 ft^3/(day·ft^2), which is beneath the minimum allowable specified by the manufacturer. Thus a filter of that size would require recirculation under low flow conditions. However, since recirculation is not needed to dilute the influent, it probably would not be economic to provide it just to maintain the hydraulic loading under minimum flow conditions. Instead, it would be preferable to reduce the maximum possible cross-sectional area to meet the constraint of the minimum allowable hydraulic loading under minimum flow conditions. Thus the maximum possible cross-sectional area would be 40,000/120 = 333 ft^2 and the designer would be limited to filters with areas between 157 and 333 ft^2. This means that the feasible region in Fig. E18.1 must be reduced to lie between hydraulic loadings (under maximum flow conditions) of 320 and 680 ft^3/(day·ft^2). (It should be noted that if the difference between the minimum and maximum flow rates is great enough, it may be impossible to meet the constraints of both the minimum and maximum hydraulic loadings without the use of recirculation.)

It will be recalled that the design criterion required that the effluent sub-strate concentration never exceed 20 mg/liter, regardless of the flow conditions. If the filter is designed to produce the desired effluent under maximum flow without recirculation, this criterion will be met for all lower flows. This can be seen by examining Eq. 18.4. As the flow rate is reduced, all terms in the exponent except $(F_o/A_c)^n$ remain constant. Because n is negative, reducing the flow rate causes an increase in the absolute value of the exponential term, which causes S/S_o to decrease.

EXAMPLE 18.3.4-2

A plastic media trickling filter is to be used to treat a wastewater with a T_bOD of 700 mg/liter which is flowing at a rate of 250,000 ft^3/day (7080 m^3/day). The hydraulic loading on the media must be at least 80 ft^3/(day·ft^2) (24.4 m^3/(day·m^2)) but must not exceed 800 ft^3/(day·ft^2) (244 m^3/(day·m^2)). Evaluation of the data from a pilot-plant study yielded the following parameter values:

$K' = 75$ (when S_o is in mg/liter and F/A_c is in $ft^3/(day \cdot ft^2)$)

$m = -0.45$

$n = -0.55$

Determine the characteristics of a full-scale trickling filter capable of reducing the soluble T_bOD to at least 15 mg/liter.

Solution:

Because the substrate concentration in the wastewater exceeds 400 mg/liter, recirculation must be used to reduce it to at least that value before application to the filter. A recirculation ratio of 1 will reduce the applied T_bOD concentration to 358 mg/liter whereas a ratio of 2 will reduce it to 243 mg/liter (Eq. 14.74). Thus we will investigate the use of each of these ratios. Since recirculation is to be used, we must employ Eq. 18.7 in the design. For a recirculation ratio of 1, we use:

$$L = \frac{\ln[(700 + 15)/15(1 + 1)]}{75[(700 + 15)/(1 + 1)]^{-0.45}[(1 + 1)F/A_c]^{-0.55}}$$

$$L = \frac{0.596}{[2F/A_c]^{-0.55}}$$

For a recirculation ratio of 2 we use

$$L = \frac{\ln[(700 + 30)/15(1 + 2)]}{75[(700 + 30)/(1 + 2)]^{-0.45}[(1 + 2)F/A_c]^{-0.55}}$$

$$L = \frac{0.440}{(3F/A_c)^{-0.55}}$$

Figure E18.2 shows the filter depths and media volumes required to produce the desired effluent for hydraulic loadings within the acceptable limits. Note that the hydraulic loadings in the figure are based upon the total flow applied. Consequently, if a recirculation ratio of 1.0 were used, the minimum allowable cross-sectional area would be $(250,000)(2)/800 = 625$ ft^2 whereas the maximum would be 6250 ft^2. If a recirculation ratio of 2.0 were used, the minimum area would be $(250,000)(3)/800 = 937.5$ ft^2 and the maximum 9375 ft^2. Thus, although a filter with a recirculation ratio of 2 will be shorter than one with a ratio of 1, it will have a slightly greater total media volume. Since the differences between the media volumes required for the two systems are small, the configuration of choice will probably be dictated by operating costs. Thus the final design can only be completed after an economic analysis of the alternatives.

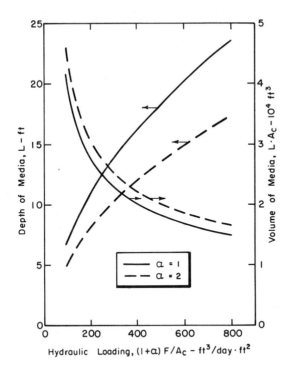

Figure E18.2.   Effects of hydraulic loading and
                recirculation ratio on the size
                of trickling filter required in
                Example 18.3.4-2.

18.4   TRICKLING FILTER OPERATION

Recirculation is the only variable that can be adjusted to meet changes in the
incoming wastewater flow rate and concentration. Because there is a minimum
hydraulic loading required to keep all of the media wet and to prevent clogging
due to excessive biofilm thicknesses, it may be necessary to vary the amount of
recirculation throughout the day to maintain the hydraulic loading above that value.
Unfortunately, it is not clear what effect this will have upon the efficiency of
substrate removal. Unlike the activated sludge process, there is no strategy for
maintaining a constant effluent quality during changes in the applied organic
loading and thus the only operation that can be utilized is that required to prevent
or overcome problems.

    One problem which results from insufficient hydraulic loadings to rock media
filters is ponding of the wastewater on the surface. Since the active film thickness is less than 100 μm [44], no additional substrate removal potential is obtained
by allowing thick films to accumulate. Thus either the hydraulic loading or the

dosing schedule should be adjusted to provide the fluid shear needed to keep the films thin, thereby preventing ponding. If those changes don't alleviate the problem, it may be necessary to replace the media with one with a larger void volume or to add additional filters, thereby reducing the organic loading.

Another problem associated with rock media filters is the growth of *Psychoda* flies, which lay their eggs during warm weather in the biofilm under the stones near the surface. Although they don't normally harm the operation of the filter they are a nuisance and a potential health hazard. Sometimes the problem can be alleviated by increasing the hydraulic loading to reduce the film thickness but this technique may not be effective near the enclosure walls. An alternative is to flood the filter periodically, thereby killing the larvae. If neither technique works, periodic spraying with insecticide may be necessary.

Odors from trickling filters generally result from anaerobic conditions in the microbial film. If the anaerobic conditions are the result of excessive film thicknesses the hydraulic loading or dosing schedule should be adjusted as discussed earlier. If the conditions are the result of excessive influent substrate concentrations, recirculation should be used. Finally, if the conditions are caused by inadequate air flow, forced draft ventilation should be installed. Because odors are a symptom of a serious problem, every effort should be made to diagnose and correct it.

In northern climates, total enclosure of a trickling filter may be desirable even if odors are not present because it will reduce the cooling of the wastewater, thereby giving better cold weather performance [1]. If it is not possible to cover the filters then the amount of heat lost may be reduced by decreasing the amount of recirculation, provided this does not cause other problems.

18.5 EXAMPLES OF TRICKLING FILTERS IN PRACTICE

Rock media trickling filters were developed for the treatment of domestic wastewater and literally thousands of them have been used for that purpose [1]. Originally they were designed with very low organic and hydraulic loadings, thereby producing effluents which were highly nitrified and very low in organic content. Problems were experienced with some of them, however, because their hydraulic loadings were too low to keep all of the media wet and to prevent clogging. As more experience was gained, the necessity for maintaining adequate hydraulic loads was recognized and recirculation was adopted as a way of doing that. Furthermore, when nitrification was not desired, higher loadings were applied until finally organic loadings in excess of 25 lb/(1000 ft^3·day) (0.4 kg/(m^3·day)) became the rule. Good performance was generally obtained with domestic wastewaters.

As the need for treatment of industrial wastes grew, rock media filters were applied to them as well. However problems often developed because of high influent concentrations and low void ratios in the media. These problems became particularly

Table 18.1
Examples of Trickling Filters

Example No.	Type of Waste	Type of Media	Depth Ft.	Organic Load lb BOD/day·1000 ft³	Hydraulic Load ft³/day·ft²	Recirculation Ratio	Influent Conc. mg/l BOD₅	Effluent Conc. mg/l BOD₅	Comments	Reference
1	Sewage + Textile	Rock	6	137	27	None	493	195	Pilot Plant	45
2	Sewage + Textile	Rock	6	143	36	None	381	134	Pilot Plant	45
3	Sewage + Textile	Rock	6	19.5	39	2:1	146	62	2nd Stage Pilot Plant	45
4	Sewage + Textile	Rock	6	10.3	33	4:1	154	38	2nd Stage Pilot Plant	45
5	Chipboard	Rock	5	15.4	2.7	None	459	66	Pilot Plant	46
6	Chipboard	Rock	5	15.2	5.3	1:1	454	29	Pilot Plant	46
7	Brewery	Plastic	8.5	226	115	3.3:1	1150	320	1st of three	9
8	Brewery	Plastic	6	78	115	3.9:1	320	150	2nd of three	9
9	Brewery	Plastic	8	67	115	0.95:1	150	---	3rd of three	9
10	Distillery	Plastic	18	140	115	1.7:1	1000	312	1st of three	9
11	Distillery	Plastic	18	44	115	1.7:1	312	67	2nd of three	9
12	Distillery	Plastic	18	9.6	115	1.7:1	67	< 20	3rd of three	9
13	Potato Processing	Plastic	20	170	115	0.5:1	500	260		9
14	Petrochemical	Plastic	19.8	100	366	3:1	300	37	Average values	47
15	Pulp and Paper	Plastic	18.4	196	144	None	400	96	Pilot Plant	48
16	Pulp and Paper	Plastic	18.4	392	288	None	400	168	Pilot Plant	48
17	Synthetic Fiber	Plastic	15.5	255*	116	6.6:1	4150*	994*	1st of three	9
18	Synthetic Fiber	Plastic	15.5	61*	116	6.6:1	994*	341*	2nd of three	9
19	Synthetic Fiber	Plastic	15.8	41*	116	2.8:1	341*	182*	3rd of three	9
20	Textile	Plastic	22	8.1	193	8:1	134	14	Pilot Plant	49
21	Textile	Plastic	22	28	193	2:1	154	29	Pilot Plant	49
22	Textile	Plastic	22	68	385	1:1	124	34	Pilot Plant	49

* COD Basis.

acute when rock media filters were used for partial treatment of highly concentrated wastewaters. This led to the development of plastic media, which allowed the application of trickling filters to a wider range of problems. Thus today they find use for both pre- and complete treatment of municipal and industrial wastewaters.

So many examples of trickling filter applications are in the literature that no attempt will be made to provide a complete summary herein. Nevertheless it would be instructive to review a few cases. First consider lines 5 and 6 in Table 18.1. The facility had very low organic and hydraulic loadings so that an increase in the hydraulic loading through the addition of recirculation improved performance. This was probably due to more complete utilization of the media since the original hydraulic loading was less than the minimum needed to keep all of it wet. Examination of the other examples shows that low effluent concentrations will generally only result at low organic loadings. From our knowledge of theory, however, we also know that low organic loadings do not guarantee an effluent low in organic matter, and this is illustrated by several of the examples. Trickling filters can be quite effective in reducing the concentration of high strength wastewaters as is illustrated by lines 7-9, 10-12, and 17-19. Such reductions will generally require deep filters, however, as in those cases. An interesting point made by those examples is that the character of a wastewater may change as it undergoes treatment, thereby preventing the last stage of a multi-stage filter from achieving as large an efficiency of substrate removal as an earlier stage. For example, the third-stage filter treating the synthetic fiber wastewater removes only 47% of the COD entering it whereas the second-stage filter removes 65%, in spite of having a higher influent concentration.

Perusal of Table 18.1 leads to the conclusion that no generalities can be made about trickling filter performance. This follows, of course, from the complex nature of fixed-film reactors, as discussed in Chapter 14. Consequently the application of handbook values or rules of thumb could lead to considerable error. This means that the only safe approach to design is through the application of pilot plant data. As long as that is done, there is a good chance that a trickling filter can be designed to meet the intended objective. Then its cost can be compared against other biochemical operations to determine which is best suited to the job.

18.6 KEY POINTS

1. Trickling filters contain four major components: filter media, an enclosure, a distribution system, and an underdrain system.

2. The hydraulic loading is the hydraulic application rate per unit of cross-sectional area. Its lower limit is dictated by the quantity of liquid required to wet all of the media while the upper limit is governed either by thin film flow or the scouring of the microbial film.

3. Recirculation of treated effluent to the inlet of a trickling filter allows independent control of the hydraulic and organic loadings.

4. Trickling filters remove soluble organic matter by its conversion to microbial cells and thus find wide application to the treatment of municipal and industrial wastewaters.

5. Trickling filters are simple, easy to operate, and tolerant to changes in the influent, however there is no effective way of controlling the quality of the effluent from them.

6. An increase in the concentration or flow rate of the influent to a trickling filter will cause an increase in the mass of organic matter removed per unit time (up to some maximum value) but will cause a decrease in the percent substrate removal.

7. If odors and substandard performance caused by oxygen transfer limitations are to be avoided the concentration of substrate entering a trickling filter should be kept below 400 mg/liter T_bOD. One way to do this is with recirculation.

8. The performance of a trickling filter can be adequately represented by an empirical model in which the fractional removal of substrate is an exponential function of the filter depth. The rate constant in the model is a function of the influent substrate concentration and the hydraulic loading. The effects of recirculation are incorporated through the latter two terms.

9. The process design for a trickling filter includes three decisions: the filter depth, the filter surface area, and the recirculation ratio. Many combinations of these variables can result in an effluent of the desired quality.

10. Three parameters must be evaluated before the design equations can be used. Simultaneous evaluation using a computer will require fewer experiments and will provide better estimates than sequential evaluation.

11. There are three situations in which recirculation of clarified effluent would be employed regardless of cost: (1) when the influent substrate concentration is high enough to result in oxygen limitations; (2) when the minimum flow rate to the filter is insufficient to wet all of the media or provide adequate scour; and (3) when the waste is inhibitory at high concentration.

12. Many factors, such as media type, influence the design of a trickling filter even though they are not explicitly included in the design equations. Thus careful consideration must be given to them during design.

13. There is currently no strategy for maintaining a constant effluent quality during changes in the organic loading applied to a trickling filter and thus the only operation that can be utilized is that required to prevent or overcome problems, such as ponding, *Psychoda* flies, odors, etc.

14. Because no generalities can be made about trickling filter performance the application of rules of thumb during design can lead to considerable error.

18.7 STUDY QUESTIONS

The study questions are arranged in the same order as the text:

Section	Questions
18.1	1-5
18.2	6-10
18.3	7-15
18.4	16

1. Describe the physical characteristics and functions of the major components of a trickling filter.

2. Explain what factors determine when trickling filters should be used: (a) in series; (b) in parallel.

3. Define hydraulic loading and organic loading; discuss the importance of each to trickling filter performance; and explain how the two may be made independent of each other.

4. Explain why primary and final settlers are used with trickling filters.

5. List the chief merits associated with trickling filters and explain why they are also responsible for their principal defects.

6. Describe the effects of influent substrate concentration and flow rate upon the performance of a trickling filter and explain why they are as they are.

7. Explain the physical effects caused by the application of recirculation of clarified effluent to the top of a trickling filter and how those effects can interact to influence the performance of the reactor.

8. Define "dosing schedule" and explain how it can influence the performance of a trickling filter.

9. State the factors which should be included in an empirical model of trickling filter performance and explain why they are important.

10. Explain how the effects of recirculation are incorporated into the empirical model.

11. Describe the experiments that should be run to obtain the data needed to evaluate the parameters K', m and n.

12. List and discuss the factors which influence the design of a trickling filter and which must be considered even though they are not explicitly stated in the empirical model.

13. A wastewater with a T_bOD concentration of 250 mg/liter and an average flow of 50,900 ft^3/day (60 m^3/hr) is to be treated on a plastic media trickling filter. The minimum flow rate anticipated is 20,000 ft^3/day (23.6 m^3/hr) and the maximum is 80,000 ft^3/day (94.3 m^3/hr). The media to be employed has a minimum allowable hydraulic loading of 100 ft^3/(day·ft^2) (30.5 m^3/(day·m^2)) and a maximum allowable hydraulic loading of 900 ft^3/(day·ft^2) (273.6 m^3/(day·m^2)). Evaluation of the data from a pilot-plant study run with that media yielded the following parameter values: K' = 80 (when S_o is in mg/liter and F_o/A_c is in ft^3/(day·ft^2)); m = -0.35; n = -0.55.

   (a) Prepare a graph showing the heights and volumes of trickling filters capable of producing an effluent with a soluble T_bOD equal to or less than 15 mg/liter under any anticipated flow condition.

   (b) Choose a design value for the filter size and explain why it was chosen. Tell whether recirculation will be employed, and why. Tell whether filters in series or filters in parallel will be employed, and why.

14. Repeat Study Question 13, but change the influent substrate concentration to 900 mg/liter T_bOD.

15. Repeat Study Question 14, but change the effluent concentration to 150 mg/liter T_bOD so that the filter will be employed only as a roughing filter.

16. Explain why it is difficult to devise a strategy for maintaining constant effluent quality out of a trickling filter during changes in loading.

REFERENCES AND FURTHER READING

1. Dow Chemical Co., Functional Products and Systems Department, "A literature search and critical analysis of biological trickling filter studies," Vols I & II, *Environmental Protection Agency Water Pollution Control Research Series,* Report No. 17050 DDY 12/71, Dec. 1971.

2. C. R. Pearson, "Use of synthetic media in the biological treatment of industrial wastes," *Journal and Proceedings of the Institute of Sewage Purification,* 519-524, 1965.

3. P. N. J. Chipperfield, "Performance of plastic filter media in industrial and domestic waste treatment," *Journal of the Water Pollution Control Federation,* 39, 1860-1874, 1967.

4. Trickling Filter Floor Institute, "Handbook of trickling filter design," *Public Works Magazine,* 1960.

5. Metcalf & Eddy, Inc., *Wastewater Engineering: Treatment/Disposal/Reuse,* 2nd Edition, McGraw-Hill Book Co., New York, N.Y., 534-548, 1979.

6. J. F. Malina *et al.* "Design guides for biological wastewater treatment processes," *Environmental Protection Agency Water Pollution Control Research Series,* Report No. 11010 ESQ 08/71, Aug. 1971.

7. "Actifil media for packed biological reactors," Bulletin ACT-30, Norton Chemical Process Products Division, Pollution Control Products, P.O. Box 350, Akron, Ohio 44309.

8. "Surfpac Design Manual," BSP Division, Envirotech Corp., One Davis Drive, Belmont, Calif., 94002, July 1978.

9. P. N. J. Chipperfield, *et al.* "Multiple-stage, plastic-media treatment plants," *Journal of the Water Pollution Control Federation,* 44, 1955-1967, 1972.

10. E. E. Cook and L. P. Herning, "Shock load attenuation, trickling filter," *Journal of the Environmental Engineering Division, ASCE,* 104, 461-469, 1978.

11. A. W. Busch, "Biochemical oxidation of process waste water," *Chemical Engineering,* 72, No. 5, 71-76, 1965.

12. W. W. Eckenfelder Jr. and D. J. O'Connor, Civil Engineering Department. Manhattan College, New York, 1965.

13. R. E. McKinney, "Biological treatment systems for refinery wastes," *Journal of the Water Pollution Control Federation,* 39, 346-359, 1967.

14. G. G. Cillie, *et al.* "The effect of toxic loads on effluent purification systems," *Water SA,* 3, 83-89, 1977.

15. E. E. Cook and D. F. Kincannon, "Organic concentration and hydraulic loading versus organic loading in evaluation of trickling filter performance," *Proceedings of the 25th Industrial Waste Conference,* Purdue University Engineering Extension Series No. 137, 230-238, 1970.

16. C. S. Velz, "A basic law for the performance of biological filters," *Sewage Works Journal, 20*, 607-617, 1948.

17. W. E. Howland, "Flow over porous media as in a trickling filter," *Proceedings of the 12th Industrial Waste Conference,* Purdue University Engineering Extension Series No. 94, 435-465, 1957.

18. K. L. Schulze, "Load and efficiency of trickling filters," *Journal of the Water Pollution Control Federation, 32*, 245-261, 1960.

19. S. Balakrishnan, *et al.* "Organics removal by a selected trickling filter media," *Water and Wastes Engineering, 6*, No. 1, A22-A25, 1969.

20. E. E. Cook, "Kinetics and mechanisms of fixed bed reactors," Thesis presented to Oklahoma State University, at Stillwater, Oklahoma in partial fulfillment of the requirements for the degree of Doctor of Philosophy, May 1970.

21. "Sewage treatment at military installations," Report of the Sub-committee on Sewage Treatment in Military Installations of the Committee on Sanitary Engineering, National Research Council, *Sewage Works Journal, 18*, 787-1028, 1946.

22. W. T. Ingram, "An investigation of the mechanisms of controlled filtration," *Sewage and Industrial Wastes, 31*, 1147-1158, 1959.

23. T. L. Bentley and D. F. Kincannon, "Application and comparison of activated sludge design and operational parameters to biological towers," *Water and Sewage Works, Reference Number,* R10-R13, 1976.

24. Great Lakes-Upper Mississippi River Board of State Sanitary Engineers, "Recommended standards for sewage works," Health Education Service, P.O. Box 7283, Albany, N.Y., 1971.

25. B. E. Jank and W. R. Drynan, "Substrate removal mechanism of trickling filters," *Journal of the Environmental Engineering Division, ASCE, 99*, 187-204, 1973.

26. M. L. Fleming and E. E. Cook, "The effect of the specific surface area provided by a synthetic medium on the performance of a trickling filter," *Proceedings of the 27th Industrial Waste Conference,* Purdue University Engineering Extension Series No. 141, 513-521, 1972.

27. W. A. Moore, *et al.* "Efficiency study of a recirculating sewage filter at Centralia, Mo.," *Sewage and Industrial Wastes, 22*, 184-189, 1950.

28. C. Lumb and P. K. Eastwood, "The recirculation principle in filtration of settled sewage - Some notes and comments on its application," *Journal and Proceedings of the Institute of Sewage Purification,* 380-398, 1958.

29. J. E. Germain, "Economical treatment of domestic waste by plastic-medium trickling filters," *Journal of the Water Pollution Control Federation, 38*, 192-203, 1966.

30. G. J. Kehrberger and A. W. Busch, "The effects of recirculation on the performance of trickling filter models," *Proceedings of the 24th Industrial Waste Conference,* Purdue University Engineering Extension Series No. 135, 37-52, 1969.

31. N. P. Harris and G. S. Hansford, "A study of substrate removal in a microbial film reactor," *Water Research, 10*, 935-943, 1976.

32. E. D. Schroeder and G. Tchobanoglous, "Mass transfer limitations on trickling filter design," *Journal of the Water Pollution Control Federation,* 48, 771-775, 1976.

33. E. E. Cook and L. Crame, "Effects of dosing rates on trickling filter performance," *Journal of the Water Pollution Control Federation,* 48, 2723-2730, 1976.

34. B. Atkinson and M. E. A. R. Ali, "The effectiveness of biomass hold-up and packing surface in trickling filters," *Water Research,* 12, 147-156, 1978.

35. J. A. Howell and B. Atkinson, "Sloughing of microbial film in trickling filters," *Water Research,* 10, 307-315, 1976.

36. J. F. Roesler and R. Smith, "A mathematical model for a trickling filter," *Proceedings of the 24th Industrial Waste Conference,* Purdue University Engineering Extension Series No. 135, 550-575, 1969.

37. W. C. Pisano and J. J. Harrington, "Discussion of 'Design optimization for biological filter models'," *Journal of the Environmental Engineering Division, ASCE,* 101, 167-170, 1975.

38. H. B. Gotaas and W. S. Galler, "Design optimization for biological filter models," *Journal of the Environmental Engineering Division, ASCE,* 99, 831-850, 1973.

39. C. R. Lee and T. Takamatsa, "Cost of trickling filter recirculation," *Water and Sewage Works,* 122; No. 1, 57-59; No. 2, 64-66, 1975.

40. J. A. Mueller, "Discussion of 'Oxygen theory in biological treament plant design'," *Journal of the Environmental Engineering Division, ASCE,* 99, 383-385, 1973.

41. E. L. Piret, *et al.* "Aerodynamics of trickling filters," *Industrial and Engineering Chemistry,* 31, 706-712, 1939.

42. E. D. Schroeder, *Water and Wastewater Treatment,* McGraw-Hill, Inc., New York, N.Y., 298-301, 1977.

43. S. Balakrishnan and W. W. Eckenfelder, Jr., "Discussion of 'Recent approaches for trickling filter design'," *Journal of the Sanitary Engineering Division, ASCE,* 95, No. SA1, 185-187, 1969.

44. B. H. Kornegay and J. F. Andrews, "Characteristics and kinetics of fixed-film biological reactors," Final Report, Federal Water Pollution Control Administration Research Grant No. WP 01181, Clemson University, Clemson, S.C. 1969.

45. F. M. Gibson and J. H. Wiedeman, "Treatment of mixed sewage and textile finishing wastes on trickling filters and activated sludge," *Proceedings of the 17th Industrial Waste Conference,* Purdue University Engineering Extension Series No. 112, 165-174, 1962.

46. R. T. Richey, "Pilot plant studies on treatment of chipboard waste by low-rate trickling filter," *Proceedings of the 17th Industrial Waste Conference,* Purdue University Engineering Extension Series No. 112, 748-757, 1962.

47. H. W. Harlow, *et al.* "A petro-chemical waste treatment system," *Proceedings of the 16th Industrial Waste Conference,* Purdue University Engineering Extension Series No. 109, 156-161, 1961.

48. W. W. Eckenfelder, Jr. and E. L. Barnhart, "Treatment of pulp and paper mill
     waste by high rate filtration using plastic filter media," *Proceedings of the
     17th Industrial Waste Conference,* Purdue University Engineering Extension
     Series No. 112, 105-115, 1962.

49. D. W. Snyder, "Dow Surfpac pilot study on textile waste," *Proceedings of the
     18th Industrial Waste Conference,* Purdue University Engineering Extension
     Series No. 115, 476-482, 1963.

Further Reading

Benefield, L. D. and Randall, C. W., *Biological Process Design for Wastewater
     Treatment,* Prentice-Hall, Inc., Englewood Cliffs, N.J., 1980. See Chapter 7.

Busch, A. W., *Aerobic Biological Treatment of Wastewaters,* Oligodynamics Press,
     Houston, Texas, 1971.

CHAPTER 19

ROTATING BIOLOGICAL CONTACTOR

Rotating biological contactors (RBC) are another form of fixed-film reactor and thus, in a general sense, their performance and utility are similar to that of trickling filters. Their physical characteristics, however, give them several unique attributes so that they have become accepted as a separate named biochemical operation. Therefore we will consider them in detail herein.

19.1 DESCRIPTION OF RBC

19.1.1 Physical Characteristics

In its simplest form, an RBC consists of parallel circular discs attached perpendicularly to a horizontal shaft which passes through their centers. The entire assembly is placed into a tank with the shaft slightly above the surface of the liquid so that the discs are approximately half immersed. Microorganisms grow on the surface of the discs and rotation of the shaft brings all of them into contact with the liquid allowing them to remove the organic matter from it. Although microbial growth results from this substrate utilization, the rotation of the discs through the liquid provides a constant shear force which causes continual sloughing of the culture, thereby maintaining a more or less constant film thickness. The rotation of the discs also mixes the liquid which keeps the stripped biomass in suspension and allows it to be carried from the reactor by the effluent. Aeration of the culture is accomplished by two mechanisms. As a point on a disc rises above the liquid surface a thin film of liquid remains attached to it and oxygen is transferred to that film as it passes through the air. Reimmersion of that point returns this highly aerobic liquid to the reactor, thereby increasing its dissolved oxygen content. In addition, a certain amount of air is entrained by the bulk of the liquid due to the turbulence caused by the rotation of the discs. The mixing within the reactor disperses this air, thereby maintaining a relatively uniform dissolved oxygen concentration.

The material and construction of the media used in the discs depend upon the manufacturer, of which there are several. One constructs its discs of alternating layers of flat and corrugated sheets of high-density polyethylene as shown in Fig. 19.1 [1]. Figure 19.2 is a photograph of a typical microbial culture as it has developed on that media. Although the exact composition of the microbial population on a disc depends upon the type of wastewater being treated and the relative position of the disc in the reactor, in general, the population tends to consist of more filamentous and fewer slime-forming organisms than that found on a trickling filter. Consequently, the sloughed biosolids tend to be relatively dense with good settling characteristics, although like waste activated sludge they are not particularly amenable to vacuum filtration [2].

As would be anticipated from the theory in Chapter 14 and the discussion of trickling filters in Chapter 18 the performance of an RBC depends upon the total surface area of media available for microbial growth. Thus RBC's are constructed with a large number of discs on the same shaft. As a general rule, full-scale RBC's have a disc diameter of 12 ft (3.6m) and maximum shaft length of 25 ft (7.62 m) [1]. Because of the nature of the kinetics of substrate removal most RBC systems are designed as reactors (stages) in series with each stage consisting of one shaft of discs mounted perpendicular to the direction of flow as shown in Fig. 19.3. The total disc surface area per stage and the number of stages required are factors which must be determined during the design process. When systems require more area per stage than can be provided by one 25 ft (7.62 m) shaft, modular construction is used so that the total system consists of stages in parallel as well as in series, as shown in Fig. 19.4. Each shaft is driven by a separate electric motor-gear reducer or by an air drive system. Although the usual rotational speed for 12 ft (3.6 m)

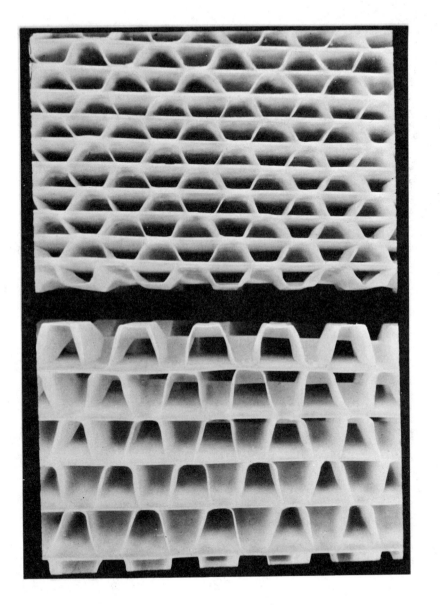

Figure 19.1.   Cross-section of RBC media showing alternating flat and corrugated sheets. Left-side, regular media; Right-side, high density media. *(Courtesy of Autotrol Corp.).*

Figure 19.2. Typical microbial culture on an RBC disc.
             (Courtesy of Autotrol Corp.).

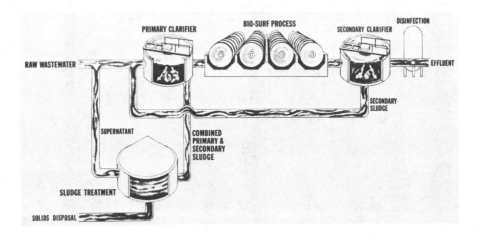

Figure 19.3. Typical flow diagram for an RBC wastewater treatment
             system. *(Courtesy of Autotrol Corp.).*

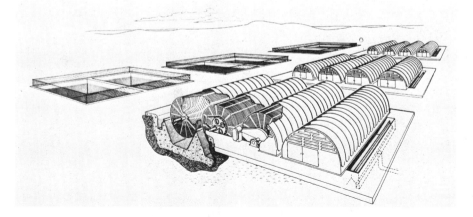

Figure 19.4.   RBC wastewater treatment system employing reactors
in parallel and in series.   *(Courtesy of Autotrol Corp.).*

diameter discs is 1.6 rpm (which gives a peripherial speed of 1 ft/sec (0.3 m/sec))
variable speed drives can be used for flexibility. The tankage housing the discs
is generally shallow (5 ft (1.5 m)) and may be of either concrete or steel. Light
weight plastic enclosures are used to protect the discs from adverse weather conditions and to reduce the loss of heat from the reactors as shown in Fig. 19.4.

19.1.2 Applications

The major objective of an RBC is the removal of soluble organic matter by its conversion to insoluble microbial cells which can be removed by sedimentation, thereby
providing an effluent of high quality. Consequently RBC's can be used as the major
means of treatment for both municipal and industrial wastewaters. Like trickling
filters, RBC's can also be used to achieve partial removal of soluble organic matter,
and thus they find application for the pretreatment of industrial wastewaters as
well. Although many RBC's have been used for low flow applications they are not
limited to that situation and plants have been built in the U.S. with capacities
exceeding 30 MGD [3]. Because the process does not employ a large mass of flocculent microorganisms within which to entrap colloidal particles, it is most applicable to wastewaters containing a high percentage of the organic matter in the
soluble state. It can be used with domestic sewage, however, because a certain
amount of bioflocculation of colloidal particles will occur, just as it does in a
trickling filter. Nevertheless, primary sedimentation should generally be used
ahead of an RBC to reduce the amount of bioflocculation required and special consideration should be given to the settling characteristics of the RBC effluent to
ensure that it will be acceptable. As with trickling filters, low temperatures
have an adverse effect upon RBC performance but the use of enclosures can essentially
eliminate any heat loss so that the temperature which controls process efficiency

becomes the temperature of the influent [4]. Consequently, RBC's may even be used
in northern climates.

19.1.3 Merits and Defects

Many of the merits associated with an RBC are similar to those of a trickling filter.
There are however, key differences which give the RBC important benefits [5]. For
example, the biomass is passed through the wastewater rather than passing the wastewater over the biomass. This ensures adequate wetting of all of the organisms regardless of the influent flow rate, thereby alleviating the need for recirculation.
In addition, the intensity of contact between the biomass and the wastewater can be
altered by adjusting the rotational speed of the discs, thus allowing direct control
of the film thickness. Adjustment of the rotational speed can also alter the rate
of oxygen transfer although supplemental aeration has been found to be beneficial in
some cases. The recycle of solids is not required in RBC systems thereby resulting
in several more benefits [1]. First, inert solids don't build up in the system and
thus the final settlers do not have to be designed to accomodate a high mass loading
of solids. Second, the underflow solids can be concentrated more than they can be
when they have to be recycled. Finally, in contrast to activated sludge in which
entrapped colloidal suspended matter is at least partially oxidized, little oxidation
of such material occurs in the RBC. This means that little of the oxygen demand of
the suspended organic matter must be met in the reactor although it does increase the
mass of sludge to be disposed of.

Perhaps one of the major merits associated with an RBC is its low power requirement. Because of the buoyancy of the discs, the bearing friction on the support
shafts is quite low. This means that the major power requirement for rotation is to
overcome the drag friction of the discs in the liquid. Consequently, the energy required to operate an RBC is just a fraction of that required to operate an activated
sludge reactor [3]. Because little degradation of entrapped colloidal organic matter
occurs, sludge production appears to be about the same as that from activated sludge
with a short MCRT, i.e., about 0.4 to 0.5 lbs of biosolids per lb of BOD_5 removed [3].
This means that sludge disposal costs may be greater than that for an activated sludge
process with a long MCRT. Conclusions about the relative power requirements for
alternative wastewater treatment systems must be based upon the total power required
for the entire waste treatment system, including sludge disposal, and not just upon
the liquid treatment system. Thus, while it is likely that RBC's will be more energy
efficient than activated sludge, each situation must be evaluated separately.

Like a trickling filter, the claim is often made that RBC's have a better shock
load or transient-state response than slurry reactor systems [6]. This is true in the
sense that toxic and other deleterious shock loads often pass through the reactor
rapidly enough to avoid killing all of the biomass so that recovery is fairly rapid.
As far as simple quantitative changes in flow and concentration are concerned, however, transient-state experiments have indicated that RBC's are sensitive to them

and provide little reserve capacity with which to minimize fluctuations in effluent quality [7]. Thus, if stringent effluent quality criteria must be met by an RBC receiving a variable influent, either equalization or extra treatment capacity will be required to reduce the fluctuations in effluent quality and ensure the desired level of treatment [7,8].

The major defects associated with RBC's are this lack of reserve capacity and the absence of operational controls. Like trickling filters and aerated lagoons, once an RBC is built there is little that can be done to adjust its performance to accomodate changes in the character of the influent or in the effluent criteria. Although this is a major defect, it is offset somewhat by the modular nature of RBC systems which makes the addition of new reactors relatively easy. Another defect associated with RBC's is that the rate of oxygen transfer from rotation of the discs places a limitation on how heavily they may be loaded. For example, with high strength wastewaters it may be impossible to supply sufficient oxygen from rotation alone to keep the reactors fully aerobic, thereby causing odor problems or the growth of nuisance organisms [9]. Consequently, one manufacturer has developed an air-drive system which also provides supplemental oxygen transfer. Another alternative would be to house the discs in a high oxygen environment and this has recently been the subject of investigation [10].

19.2 RBC PERFORMANCE

19.2.1 Theoretical and Observed Performance

A rotating biological contactor is basically a rotating disc reactor, and therefore, it should be possible to characterize its performance by the mathematical model presented in Section 14.5. As discussed in that section, many events occur simultaneously in such a reactor and consequently the model attempts to consider only the major ones while ignoring others. Among those considered are the transport of substrate to the biofilm from the bulk liquid in the submerged sector and from the entrained stagnant liquid layer in the aerated sector, and the subsequent transport and consumption of substrate within the biofilm in both sectors. On the other hand, the transport of oxygen with its potential effect on the substrate consumption rate, imperfect mixing in the tank, and the intermittent nature of biofilm release (rather than continuous as assumed) are among the events not considered in the model. Therefore, it would be instructive to review the theoretical predictions made in Section 14.5 and to compare them with experimental observations made with both pilot- and full-scale reactors.

Consider first the effect of influent flow rate as shown in Fig. 14.31. There it was seen that the effluent substrate concentration increases almost linearly with increases in flow rate while the percent substrate removal decreases almost linearly. This decrease is almost independent of the influent substrate concentration at low flow rates, but at high flow rates the percent substrate removal improves at lower influent concentrations. This is because at low flow rates the effluent concentrations are low so that substrate consumption follows first-order kinetics whereas at high flow rates the kinetics approach zero-order. These linear characteristics have been widely observed in practice and typical data are shown in Fig. 19.5 [11]. It should be noted that the abscissa in Fig. 19.5 is the hydraulic loading, which in RBC practice is defined as the influent flow divided by the total wetted disc surface area. This is slightly different from the definition used in Chapter 14, where only the wetted area in the submerged sector was included. The terms are used in an analogous manner, however, and differ only in magnitude. Since the fractional substrate removal is a linear function of the hydraulic loading (i.e., $(S_o - S)/S_o = 1 - a(F/A_s'))$, the effluent concentration should be a linear function of both the hydraulic loading and the influent substrate concentration (i.e., $S = aS_o(F/A_s'))$. Therefore, plots of effluent substrate concentration against hydraulic loading should form a family of straight lines with slope proportional to the influent concentration. This response was also predicted theoretically in Fig. 14.31.

Another factor which influences the performance of an RBC is the rotational speed of the discs because it affects the rate of aeration, the intensity of contact between the wastewater and the biofilm, and the intensity of mixing in each stage.

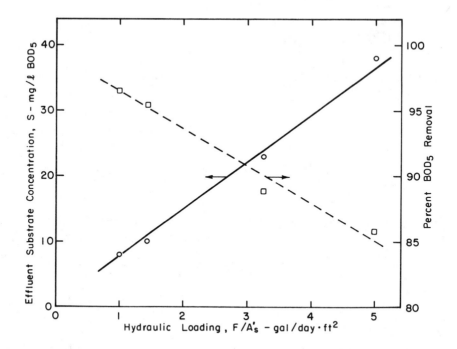

Figure 19.5.   Effect of hydraulic loading on the performance of an RBC. *(Adapted from Antonie [11])*.

Increases in these factors improve the rate of substrate transport between the bulk
liquid phase and the biofilm thereby increasing the substrate consumption rate and
the percent substrate removal. The theoretical prediction of the effects of rotational speed was shown in Fig. 14.32, where it was seen that the effects are more
pronounced at lower rotational speeds because the mass transfer resistance is more
important there. Similar effects have been observed experimentally, as shown in
Fig. 19.6. As predicted, increases in performance become marginal at higher speeds.
Hartman [13] reported that the percent substrate removal increases with the rotational
speed raised to the 0.1 power.

   The total wetted surface area of biofilm is another factor which is important
to the performance of an RBC. Both the number of discs and their fractional submergence determine that area and these effects were shown in Figs. 14.34 and 14.35,
respectively. Since the hydraulic loading is based upon the total wetted surface
area, the effects of the number of discs and the degree of submergence should be
analogous to those of hydraulic loading. Indeed increases in both factors improve
the percent substrate removal as they decrease the hydraulic loading. Thus we may
accept Fig. 19.5 as being consistent with the theoretical predictions of the effects
of the number of discs and the fractional submergence.

   In practice, RBC units are usually operated in series. The theoretical predictions of the performance of RDR units in series were shown in Figs. 14.37 and 14.39.
There it was seen that the effluent substrate concentrations from subsequent stages

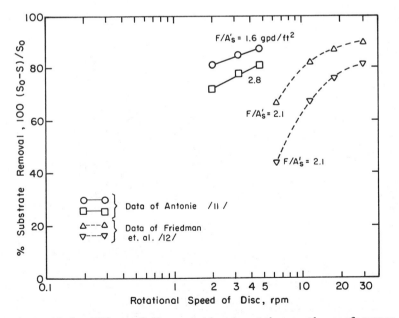

Figure 19.6.   Effect of disc rotational speed upon the performance of
              an RBC. (Data from Antonie [11] and Friedman et al. [12]).

decrease almost exponentially, so that when they are plotted against the stage number on semilog paper, an almost straight line results. This implies that the fractional removal from stage to stage is almost constant (i.e., $S_1/S_o = S_2/S_1 = S_3/S_2$ etc.). This is characteristic of first-order reactions. When the experimental data of Torpey *et al.* [14] are plotted against the stage number on semilog paper as shown in Fig. 19.7, they form a straight line, thereby confirming the theoretical prediction. On the other hand, when the reaction is not first order, such as when the effluent substrate concentration is high relative to the saturation constant or when different cultures are present in each stage, such plots should not form straight lines. This has been observed experimentally [15,16]. The effect of influent flow rate (or hydraulic loading) on the substrate profile in a series of RDR's was shown in Fig. 14.39 where it was seen that the profile becomes steeper with decreasing flow rate. This phenomenon has been also observed experimentally [17].

Among those factors not considered by the theoretical model is the rate of oxygen transport. Instead, it was assumed that it is much faster than the substrate consumption rate. However, as discussed in Section 18.2.1 the rate of oxygen transfer can become the limiting factor when the substrate concentration is high. Thus

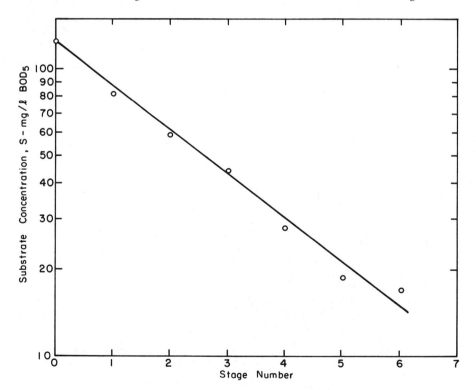

Figure 19.7. Effect of staging on substrate removal in an RBC system.
(Data from Torpey et al. [14]).

special attention should be given to the first stage where the substrate concentration and the substrate removal rate are highest to ensure that oxygen is not limiting. This can only be done experimentally as suggested by McAliley [18].

19.2.2 A Simplified Model of RBC Performance

As discussed in Chapters 14 and 18 the major aim of rigorous mathematical models is to express mathematically the major biochemical, chemical and physical phenomena that occur in a system so that its performance can be accurately predicted over a range of conditions broader than our experience. Empirical models, on the other hand, require no understanding of those phenomena and are therefore only capable of predicting the performance of a system within the experimental region upon which they are developed. However they are simpler to develop and easier to use. As was stated in Section 14.5.1 the theoretical modeling of RBC's is in its infancy and much more research is needed to verify the models and to establish reliable and efficacious techniques for evaluating the mass transfer, kinetic, and reactor parameters required to characterize them. Consequently just as was done with trickling filters, we will rely upon a semiempirical model which may be used to predict the performance of an RBC operating under conditions which lie within the experimental range covered in treatability studies. We will also apply the graphical procedure which was developed in Section 4.2.2 so that a reactor design may be performed with graphical information rather than with an explicit mathematical model.

It will be recalled from Section 14.5 that $-r_s''$, the rate of substrate consumption per unit surface area of wetted biofilm, is a function of the concentration of substrate in the bulk liquid and of the physical and operational characteristics of the reactor which control the mass transfer phenomena (such as disc size, rotational speed, etc.). However, if a pilot plant study is run using a reactor with characteristics similar to those which will be used in the full-scale facility then $-r_s''$ may be considered to be a function of the effluent substrate concentration only. Furthermore, since the quantity of attached biomass far exceeds that in suspension and since the space time of the fluid in the vessel housing the discs is small, substrate consumption by the suspended growth may be ignored [11,19]. This means that if we consider the vessel to be completely mixed we can write a very simple mass balance on substrate

$$FS_o - FS + r_s'' A_s' = 0 \tag{19.1}$$

in which F is the volumetric flow rate, S_o is the influent substrate concentration, S is the effluent substrate concentration, A_s' is the total wetted surface area (including that in both the aerated and submerged sectors), and $-r_s''$ is the experimentally determined reaction rate.

Because Eq. 19.1 is analogous to the usual CSTR equation it may be rearranged as

$$-r_s'' = F(S_o - S)/A_s' \tag{19.2}$$

to allow generation of the rate data using the methods described in Section 5.1.   In
fact, further rearrangement reveals that

$$F/A_s' = -r_s''/(S_o - S) \qquad\qquad (19.3)$$

which states that the hydraulic loading plays the same role for an RBC that $1/\tau$ does
for a CSTR. Thus the graphical and analytical procedures developed in Sections 4.1.2
and 4.2.2 for the determination of τ for a CSTR are directly applicable to the determination of the hydraulic loading for an RBC.

As discussed in Chapter 14 variations in the overall effectiveness factor often
make it difficult to develop a straightforward mathematical correlation between the
observed reaction rate and the bulk substrate concentration. Consequently the graphical procedure will often be the method of choice. On other occasions, however, the
effectiveness factor may be relatively constant, making it possible to correlate
$-r_s''$ with S using a form of the Monod equation:

$$-r_s'' = q_m'' S/(K_s + S) \qquad\qquad (14.6)$$

Under those circumstances Eq. 14.6 may be combined with Eq. 19.3 to obtain an analytical equation for design purposes:

$$F/A_s' = q_m'' S/[(K_s + S)(S_o - S)] \qquad\qquad (19.4)$$

Equation 19.4 requires knowledge of q_m'' and K_s which must be determined by some
linearization of the Monod equation, such as those discussed in Section 10.1.

The ability of the simplified model to represent the performance of an RBC can
be assessed from the works of Davies and Vose [8], McAliley [18], and
Clark et al. [19]. Davies and Vose investigated the use of RBC units in refinery
wastewater treatment. Figure 19.8 is a plot of substrate consumption rate per unit
surface area of wetted biofilm, $-r_s''$, versus the effluent substrate concentration.
Obviously the model is not perfect as indicated by the scatter in data. Nevertheless, the scatter is not bad in terms of predicting the effluent substrate concentration as shown in Fig. 19.9. The predicted values were obtained from the Monod
model of the rate, Eq. 14.6, with the values of q_m'' and K_s obtained from a Lineweaver-
Burk plot. Also shown in Fig. 19.8 is a curve representing Eq. 14.6 with the experimentally determined values of q_m'' and K_s. The prediction in Fig. 19.9 and the reasonable collapse of the experimental data around the model in Fig. 19.8 indicate that
the simplified model with its implied limitations is suitable for design purpose.
McAliley also showed that the simplified model is adequate for design purposes when
treating unbleached kraft mill wastes. Finally, it should be noted that the simplified model was also proposed by Kornegay [20].

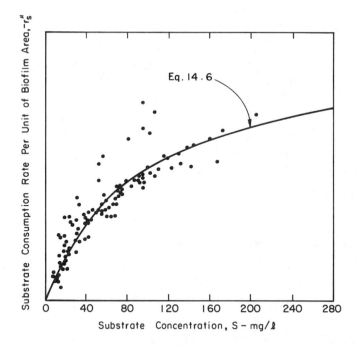

Figure 19.8. Effect of substrate concentration on the rate of
substrate consumption per unit of wetted biofilm
area. *(Adapted from Davies and Vose [8])*.

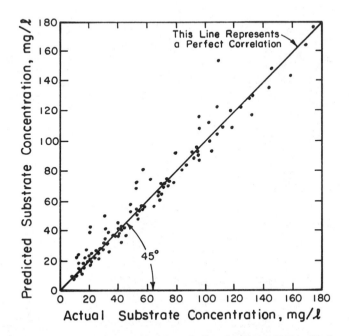

Figure 19.9. Demonstration of the ability of Eq. 19.4 to
predict the performance of an RBC.
(Adapted from Davies and Vose [8]).

19.3 DESIGN

As shown in Section 4.2.2, CSTR's in series are more efficient than a single large CSTR for a monotonically increasing rate function; consequently it follows that a system of small RBC's in series will perform better than one large RBC. In addition, the use of a number of stages in series allows the development of specific microbial cultures which are adapted to the wastewater characteristics in each of the successive stages so that the treatment efficiency can be further improved. Thus, RBC systems are usually designed and operated as four or more stages in series [11]. From Eq. 19.3 it is evident that the most important design variable for an RBC is the hydraulic loading required to obtain an effluent of the desired quality because once it is known the required wetted surface area can be readily calculated. Thus the process design for an RBC system usually involves the determination of the wetted surface area, tank volumes, and number of stages in series required to reduce the substrate concentration to a desired value.

19.3.1 Process Design Equations and Procedure

Because the simplified model for an RBC, Eq. 19.3, is analogous to a CSTR model the analytical and graphical procedures developed in Sections 4.1.2 and 4.2.2 are directly applicable here. Therefore, the reader is encouraged to review the material in those sections.

   ANALYTICAL PROCEDURE - DESIGN EQUATIONS. In order to apply the analytical procedure we will assume that a treatability study has revealed that $-r''_s$ can be represented adequately by the Monod equation, Eq. 14.6, and that the parameters q''_m and K_s have been determined. In this situation the design equation is Eq. 19.4, from which the total wetted surface area required to reduce the substrate concentration from S_o to S in one stage may be calculated as

$$A'_s = F(K_s + S)(S_o - S)/q''_m S \tag{19.5}$$

For n stages in series the surface area in each stage is

$$A'_{si} = F(K_s + S_i)(S_{i-1} - S_i)/q''_m S_i \qquad i = 1,2,\ldots,n \tag{19.6}$$

The usual practice is to have at least four stages, each with the same wetted surface area. This follows from the fact that the effluent substrate concentration from each stage is often low in comparison to the saturation constant, K_s, so that the RBC process can be considered to exhibit first-order kinetics [11]. It will be recalled from Section 4.2.2 that for a first-order reaction the relative size of each reactor is immaterial. Thus the use of equal sized reactors is not only convenient, but consistent with theory. If each reactor has the same wetted surface area, A'_s, Eq. 19.6 may be written as

$$A'_s = F(K_s + S_i)(S_{i-1} - S_i)/q''_m S_i \qquad i = 1,2,\ldots,n \tag{19.7}$$

To determine the total surface area required to reduce the substrate concentration to a desired level it is necessary to solve Eq. 19.7 iteratively. First a value of the area in each stage, A_s', is assumed and Eq. 19.7 is solved sequentially from the first to the last stage. The resulting effluent substrate concentration is then compared to the desired value. If it is greater (less), a larger (smaller) value of A_s' is assumed and the process repeated until the desired effluent concentration is obtained.

In some cases specific microbial cultures develop in each stage thereby causing the parameters q_m'' and K_s to vary from stage to stage. This makes it necessary to modify Eq. 19.7:

$$A_s' = F(K_{si} + S_i)(S_{i-1} - S_i)/q_{mi}'' \, S_i \qquad i = 1,2,\ldots,n \qquad (19.8)$$

where K_{si} and q_{mi}'' represent K_s and q_m'' in the ith stage. As before, this equation can be used sequentially to determine the required surface area provided that information is available about how the upstream treatment affects the kinetic parameters.

GRAPHICAL PROCEDURE. As was seen in Sections 4.1.2 and 4.2.2 it is not essential to have an analytical expression for the rate in order to choose τ for a CSTR. Likewise an analytical rate expression is not necessary for the determination of A_s' for an RBC. All that is needed is a plot of the substrate consumption rate per unit of wetted surface area versus the effluent substrate concentration. Such a plot is given in Fig. 19.10 and the graphical procedure is indicated. For a single reactor

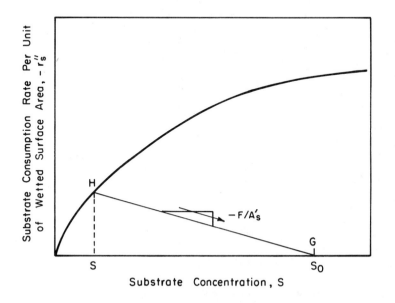

Figure 19.10. Graphical procedure for determining the hydraulic loading for a single RBC.

a straight line is drawn from the influent substrate concentration, S_o (point G) to the point on the rate curve corresponding to the effluent substrate concentration, S (point H). The slope of that line is the required hydraulic loading from which the needed area, A'_s, can easily be calculated.

If reactors in series are to be used an iterative procedure must be employed. This is demonstrated in Fig. 19.11. To determine the required surface area, a value of the hydraulic loading, F/A'_s, is guessed and a straight line is drawn through S_o (point G) with a slope of $-F/A'_s$ and its intersection with the rate curve (point H) is located. The value of the abscissa corresponding to H (point I) is located. This is the effluent concentration from stage 1 and the influent concentration to stage 2. The same procedure is repeated for the second stage utilizing a line of the same slope since all reactors are to have the same area. This procedure is repeated until the last stage is reached and the resulting effluent substrate concentration is compared with the desired value. If the resulting concentration is greater (less) than the desired, then a smaller (larger) value of hydraulic loading, F/A'_s, is picked, and the entire procedure is repeated until the resulting concentration agrees with the desired value.

When the microbial cultures in each stage are different, the graphical procedure outlined above must be modified by plotting a family of substrate consumption rate curves, with one for each stage as shown in Fig. 19.12. The graphical procedure is similar to that in Fig. 19.11 except that we must use the intersections of the

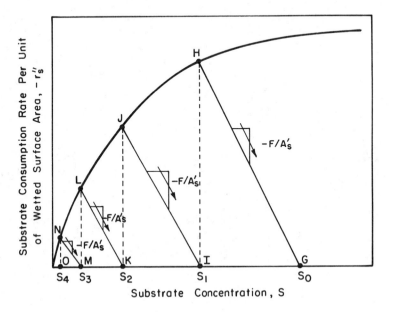

Figure 19.11.   Graphical procedure for determining the
                hydraulic loading for an RBC system.

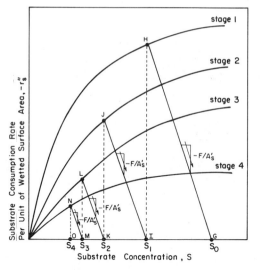

Figure 19.12. Graphical procedure for determining the hydraulic loading
               for an RBC system which has a specific microbial culture
               in each stage.

straight lines of slope $-F/A'_s$ with the appropriate rate curve for the stage under
consideration. First, the influent concentration, S_o is located on the abscissa
(point G). Then a straight line of slope $-F/A'_s$ is drawn through G and its intersection with the rate curve for stage 1 (point H), is located. The value of the
abscissa corresponding to H is located and identified as I. This is the effluent
concentration from stage 1 and the influent to stage 2. The same procedure is repeated starting from point I except that the rate curve for stage 2 is used to
locate J. The value of the abscissa corresponding to J is labeled as K, and this
is the effluent substrate concentration from stage 2. By the same procedure, points
L, M, N, and O are located. This type of graphical technique has been demonstrated
by Clark *et al.* [19].

19.3.2 Determination of Substrate Consumption Rate and Parameter Values

Before the procedures given in Section 19.3.1 can be used, it is necessary to evaluate the substrate consumption rates at various effluent substrate concentrations so
that the necessary graphical correlations or analytical parameter values can be obtained. This is best done with a pilot plant containing discs with diameter and
rotational speed equal to the values to be used in the actual facility. The reason
for this lies in the problems associated with scale-up of the mass transfer effects.
Several investigators have suggested that the results obtained from small-scale
pilot plants can be applied directly to full-scale systems provided that the peripheral velocity is held constant. Figure 14.33 showed, however, that the performance of a large-scale system will not be as good as that of a small-scale one when
this is done. Unfortunately, it is not yet clear exactly what criterion to apply to
scale-up, and furthermore, it is not even certain that a single criterion can be used

for all wastewaters because of the different relative effects associated with the
kinetic and mass transfer parameters. Thus, until more is learned about scale-up
in these systems, pilot plant studies should be done with full diameter discs.

A series of steady state runs should be made by varying the influent flow rate,
F, and/or the influent substrate concentration, S_o, (by diluting) and the corresponding effluent concentrations recorded. Then the substrate consumption rates, $-r_s''$,
may be calculated at the various effluent concentrations using Eq. 19.2. If the
graphical procedure is to be employed, a plot of $-r_s''$ versus S should be made. When
a series of RBC units is used in the pilot study, this should be done for each stage
and the rate versus effluent concentration plotted for each one separately. This
procedure for obtaining the rate as a function of the effluent concentration was discussed fully in Section 5.1 and no further elaboration is needed here. To use the
analytical procedure, the rates must be correlated with the effluent concentrations
through Eq. 14.6 and the parameter values, q_m'' and K_s, must be determined. A number
of ways of linearizing the Monod equation were considered in detail in Sections 10.1
and 12.4 and any one of them may be used to linearize Eq. 14.6. When specific microbial cultures are suspected of being present in each stage of a multi-stage system,
the rate data from each stage should be correlated separately as practiced by
Clark *et al.* [19]. It should be recalled from Chapter 14 that in the presence of
mass transfer limitations, the observed substrate consumption rate, i.e., the intrinsic rate multiplied by the effectiveness factor, may not be of the Monod form even
when the intrinsic rate is. Consequently, it may be in error to try and fit the data
to Eq. 14.6. Since the graphical approach makes no assumptions about the form of
the rate expression it may be more accurate in many situations.

Sometimes, because of various factors, a designer may have to perform a treatability study on a small-scale pilot plant. In that case it will be necessary to
correct for the scale-up effect. As discussed above, no general correction factor
exists for all wastewaters. Nevertheless, based upon studies with domestic wastewater, Wilson *et al.* [21] have recommended that the surface area requirements obtained from the procedures in Section 19.3.1 be increased by 25% when designing a
12 ft (3.6 m) diameter system from data collected on a 1.6 ft (0.5 m) diameter pilot
plant and by 10% when designing from data from a 6.6 ft (2.0 m) pilot plant.

Regardless of the approach used to arrive at the final design, the pilot plant
should be set up and operated in the selected configuration to confirm its ability
to produce the expected performance.

19.3.3 Factors Influencing the Design

Some of the factors which influence the design of RBC systems are temperature, oxygen
transfer, media rotational speed, and tank volume. Of these temperature and oxygen
transfer have the most significant effects and thus they must be considered in the
design of any system. An empirical method of correcting for the effects of temperature has been developed based on extensive pilot plant data [22]. This technique

employs a temperature correction factor by which the hydraulic loading must be multiplied in order to maintain the same performance at the reduced temperature. The base
temperature at which the correction factor is unity is 55°F since it was found that
the temperature of the wastewater had no effect above that value. When the wastewater temperature falls below 55°F however the substrate removal efficiency decreases and the hydraulic loading must be decreased to compensate for the effect.
Figure 19.13 gives the temperature correction factors which must be applied to the
hydraulic loading in order to maintain the same efficiency. For example, when the
wastewater temperature is 45°F the temperature correction factor is approximately
0.76. Therefore, the hydraulic loading calculated for temperatures of 55°F or above
must be reduced by this factor, resulting in a 32% increase in the wetted surface
area needed to achieve the desired performance.

        Although neither the model developed in Section 14.5.1 nor the design techni-
ques presented in Section 19.3.1 included the effects of oxygen transfer, they must
be considered during design of an RBC just as with a trickling filter. If an RBC
becomes oxygen limited, performance will suffer and, in extreme cases, nuisance
organisms may proliferate thereby preventing the growth of organisms capable of removing the substrate. One way to prevent oxygen limited situations is to keep the
substrate removal rate per unit of wetted surface area below the value at which
oxygen limitation is likely to occur. Although the onset of oxygen limitation can
be influenced by many factors, one manufacturer [22] recommends an upper limit on
$-r''_s$ of 4.0 lb/(day·1000 ft^2)(0.02 kg/(day·m^2)) for their mechanical drive RBC's.
Values in excess of that can be tolerated by using supplemental aeration.

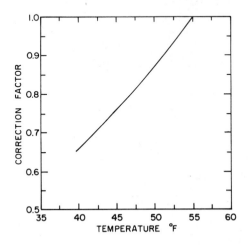

Figure 19.13. Empirical temperature correction factors by
               which the hydraulic loading on an RBC must
               be multiplied to maintain the same performance.
               (Courtesy of Autotrol Corp.).

As discussed in Section 14.5.1 the rotational speed of the media is an important factor because it influences the mass transfer characteristics of the reactor. Figure 14.32 showed the effect of rotational speed on the theoretical performance of a rotating-disc reactor and Figure 19.6 showed the effects which have been observed with RBC's. In both cases significant improvements are apparent up to a limiting speed and thereafter the improvements are marginal. For RBC systems treating domestic wastewater it has been found that the limiting speed occurs when the peripheral velocity of the disc is approximately 2.0 ft/sec (0.61 m/sec) [11]. Since the power requirements increase exponentially with increases in disc rotational speed, there will definitely be an optimal rotational speed. This value cannot be predicted, however, and must be determined experimentally once the facility has been built.

Another factor which will have some effect upon the performance of an RBC system is the tank volume because it determines the residence time of the wastewater in the reactor. When studies were done to determine the effects of tank volume it was found that no significant increase in performance was observed when the volume was made larger than 0.12 gallon/ft^2 (0.005 m^3/m^2) of wetted surface area [11]. Consequently for the treatment of domestic wastewaters with influent concentrations up to 300 mg/liter BOD$_5$ the tank volume in gallons (m^3) may be obtained by multiplying the ft^2 (m^2) of wetted surface area by 0.12 (0.005).

The design techniques have been based on the assumption that pilot plant data can be obtained for the wastewater to be treated. Although this is the preferred technique it is not always possible to carry out extensive pilot plant testing. In such a situation a rough design calculation may be carried out using an empirical procedure developed by the Autotrol Corporation for the design of RBC's treating municipal wastewater [22]. Figure 19.14 shows the correlation. The necessary

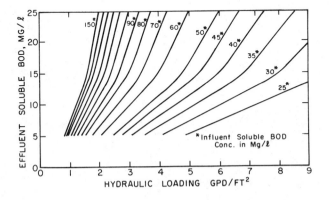

Figure 19.14. Design correlation for soluble BOD$_5$ removal in municipal wastewater treatment. *(Courtesy of Autotrol Corp.)*.

hydraulic loading can be determined from the figure by using the required effluent soluble BOD_5 and the known influent soluble BOD_5 level. Detailed procedures are also available to help the designer choose an economic system configuration once the total wetted surface area of media has been calculated from the hydraulic loading [22].

19.3.4 Example Problems

Examples are given to illustrate the design procedure for RBC systems.

EXAMPLE 19.3.4-1

A wastewater containing 100 mg/liter soluble T_bOD is to be treated by an RBC system. The flow rate is 500,000 gal/day (1.89×10^6 liters/day = 1.89×10^3 m^3/day). Evaluation of the data from a pilot plant study with the wastewater at 60°F yielded $q_m'' = 50,000$ mg/($m^2 \cdot$ day) and $K_s = 100$ mg/liter.

(a) How much surface area of wetted biofilm will be required to reduce the T_bOD to 20 mg/liter in a single-stage RBC at a temperature of 60°F?

Solution:

The wetted surface area for a single-stage RBC may be determined from Eq. 19.5:

$$A_s' = F(K_s + S)(S_o - S)/q_m''S$$

$$= (1.89 \times 10^6)(100 + 20)(100 - 20)/(50,000)(20)$$

$$= 18,100 \ m^2 \ (195,000 \ ft^2)$$

The substrate removal rate per unit of wetted area should now be checked to ensure that the system will not be oxygen limited. This may be done with Eq. 19.2:

$$-r_s'' = F(S_o - S)/A_s'$$

$$= (1.89 \times 10^6)(100-20)/18,100$$

$$= 8.35 \times 10^3 \ mg/(day \cdot m^2)$$

$$= 0.00835 \ kg/(day \cdot m^2)$$

This is well below the value of 0.02 kg/($day \cdot m^2$) given earlier and thus no problems are likely to exist with oxygen transfer. Thus a single-stage RBC with a total wetted surface area of 18,100 m^2 (195,000 ft^2) will be satisfactory. The number of discs required to obtain this area may be calculated from the disc diameter and the degree of submergence. If the required number of discs exceeds the maximum number normally placed on a single shaft by the manufacturer, then reactors in parallel must be used.

(b) How much surface area would be required for each stage of a four-stage RBC system if stages of equal size are used?

Solution:

The surface area requirement for each stage must be determined by an iteration technique using Eq. 19.7 sequentially. Equation 19.7 is rearranged and solved for S:

$$S_i = \{[(q_m''A_s'/F + K_s - S_{i-1})^2 + 4K_s S_{i-1}]^{\frac{1}{2}} - (q_m''A_s'/F + K_s - S_{i-1})\}/2 \qquad (19.9)$$

Begin by assuming a value for A_s' for each stage. Since a four-stage system should be more efficient than the single-stage system, assume a value of 2250 m^2 (24,200 ft^2) for each stage, which will give a total area of approximately one-half of the area determined in (a). Equation 19.9 must be solved sequentially to see if S_4 is equal to 20 mg/liter. If it is not, then a new area must be assumed and the process repeated. Equation 19.9 for stage 1 is:

$$S_1 = \{[(((50,000)(2250)/1.89 \times 10^6) + 100 - 100)^2 + (4)(100)(100)]^{\frac{1}{2}}$$

$$- [((50,000)(2250)/1.89 \times 10^6) + 100 - 100]\}/2$$

$$= 74.6 \text{ mg/liter}$$

This value is now used in Eq. 19.9 for stage 2, giving:

$$S_2 = 53.8 \text{ mg/liter}$$

Repeating the process twice gives:

$$S_3 = 37.5 \text{ mg/liter}$$

$$S_4 = 25.4 \text{ mg/liter}$$

The final effluent concentration is greater than the desired value of 20 mg/liter and thus a larger area must be used. Try a value of 2500 m^2 (26,900 ft^2). Repeating the process yields:

$$S_1 = 72.3 \text{ mg/liter, } S_2 = 50.2 \text{ mg/liter, } S_3 = 33.6 \text{ mg/liter, and } S_4 = 21.7 \text{ mg/liter}$$

The effluent is still too high. Additional trials lead to an area of 2650 m^2 (28,500 ft^2) per stage, which yields:

$$S_1 = 70.9 \text{ mg/liter, } S_2 = 48.1 \text{ mg/liter, } S_3 = 31.4 \text{ mg/liter, and } S_4 = 20 \text{ mg/liter}$$

Each stage must now be checked to be sure that it will not be oxygen limited. This will be done with Eq. 19.2

$$-r_{s1}'' = (1.89 \times 10^6)(100 - 70.9)/2650$$

$$= 0.02 \times 10^6 \text{ mg/(day·m}^2)$$

$$= 0.020 \text{ kg/(day·m}^2)$$

$$-r_{s2}'' = (1.89 \times 10^6)(70.9 - 48.1)/2650$$

$$= 0.016 \text{ kg/(day·m}^2)$$

$$-r''_{s3} = (1.89 \times 10^6)(48.1 - 31.4)/2650$$

$$= 0.012 \text{ kg}/(\text{day} \cdot \text{m}^2)$$

$$-r''_{s4} = (1.89 \times 10^6)(31.4 - 20)/2650$$

$$= 0.008 \text{ kg}/(\text{day} \cdot \text{m}^2)$$

All stages have rates equal to or less than the recommended value and thus all should perform satisfactorily. Therefore the required area for each stage is 2650 m^2 (28,500 ft^2), for a total area of 10,600 m^2 (114,000 ft^2), which is only 59% of the total area requirement for the single-stage RBC.

(c) Repeat (a) at a wastewater temperature of 45°F.

Solution:

The temperature correction factor at 45°F in Fig. 19.13 is 0.76. Therefore the surface area must be increased to:

$$A'_s = 18,100/0.76 = 23,800 \text{ m}^2 \text{ (256,000 ft}^2)$$

EXAMPLE 19.3.4-2

A wastewater with a T_bOD content of 200 mg/liter and a flow rate of 200,000 gal/day (7.57 $\times$ 10^5 liter/day = 7.57 $\times$ 10^2 m^3/day) is to be treated by an RBC system. A pilot plant treating the wastewater yielded the T_bOD removal rate versus concentration data given in Fig. E19.1.

(a) Determine the surface area needed to reduce the T_bOD to 20 mg/liter in a single-stage RBC.

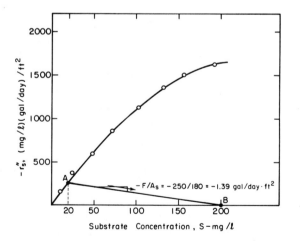

Figure E19.1. Graphical solution to part (a) of Example 19.3.4-2.

Solution:

· The desired effluent T_bOD is 20 mg/liter. Hence we locate the substrate removal rate at 20 mg/liter (point A) and the abscissa value of 200 mg/liter (the influent T_bOD (point B)). Now a straight line is drawn between points A and B. The slope of this line is equal to $-F/A'_s$ and is found to be -1.39 gal/(day·ft^2). Therefore, the required wetted surface area is

$$A'_s = 200,000/1.39 = 144,000 \text{ ft}^2 \ (13,380 \text{ m}^2)$$

This gives a substrate removal rate of 0.01 kg/(day·m^2) and thus the reactor should not be oxygen limited.

(b) Determine the surface area for each stage if a four-stage equisized RBC system is used.

Solution:

Figure E19.2 contains the same rate data as Fig. E19.1. Starting from point B we assume a value for the slope and draw a straight line to locate its intersection with the rate curve (point C). A straight line is drawn through point D with the same slope and its intersection with the rate curve is located and labeled as E. This procedure is repeated for lines through points F and H. If the assumed slope is the right value the line passing through point H should intersect the rate curve at point A. When this is done the slope is determined to be -15.63 gal/(day·ft^2). Therefore,

$$A'_{si} = \frac{200,000}{15.63} = 12,800 \text{ ft}^2 \ (1189 \text{ m}^2)$$

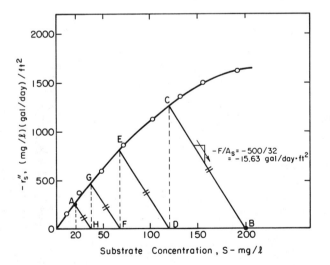

Figure E19.2. Graphical solution to part (b)
of Example 19.3.4-2.

The total area is therefore 51,200 ft^2 (4756 m^2), or 35.5% of a single-stage RBC.

When the substrate removal rate per unit of wetted area is calculated for each stage the values are found to be: $-r''_{s1}$ = 0.052, $-r''_{s2}$ = 0.032, $-r''_{s3}$ = 0.020, and $-r''_{s4}$ = 0.011 kg/(day·m^2). The rates in the first two stages are in excess of the recommended maximum value. Consequently they are likely to be oxygen limited. Thus, although a considerable reduction in media area could theoretically be achieved by using four equally sized stages, such a system probably would not work properly because of problems with oxygen transfer.

One alternative open to the designer would be to use a staged system with a greater amount of surface area per stage near the influent end. The value of $-r''_s$ in stage 1 would be set equal to the upper limit. When that limit is converted into the units of Fig. E19.2, its value is 480 (mg/liter)(gal/day)/ft^2. Consequently, the maximum practical removal rate in stage 1 would be at approximately point G in the figure. This suggests that a two-stage system would be satisfactory. The surface area required in the first stage would be determined by the slope of a line drawn from point B to point G. The surface area for the second stage would be determined by the slope of line AH and would be the same as that calculated earlier in this example.

19.4 RBC OPERATION

There is nothing that can be altered on an RBC system except the speed of rotation and degree of submergence of the discs. Thus once the proper settings have been established no further attention is needed. This can be both a merit and a defect. From the positive standpoint this means that RBC's are ideal reactors for small communities and other installations which cannot afford fulltime operators. From the negative standpoint, it means that little can be done to alter the effluent quality should the need arise. One area that is under investigation is alteration of the flow pattern as a means of process control, i.e., switching from parallel to series flow, and vice versa. Although no specific recommendations can yet be made in this regard, the reader should watch the technical literature for new developments.

Maintenance of RBC's is equally simple [4]. On a weekly basis, maintenance is limited to greasing the bearings and checking the level of lubricant in the chain guards. On a quarterly or semiannual basis, maintenance includes changing the lubricant in the gear reducer and inspecting the chains and sprockets for wear and slack.

19.5 EXAMPLES OF RBC's IN PRACTICE

RBC's originated in Europe and over 1000 installations were built there between 1960 and 1974 [4]. Their introduction into the United States did not occur until much later; nevertheless, more than 50 plants were either in operation or under

construction by 1974. Table 19.1 contains six examples from United States experience. Three are actual operating data whereas the other three are the estimates used
for design. Examples 1-4 represent situations in which the RBC was the major means
of wastewater treatment whereas examples 5 and 6 represent pretreatment prior to
discharge to a sewer. The organic loadings were all low in examples 1-4, but there
was considerable variation in hydraulic loading even though it is considered to be a
primary parameter during design [1]. All of the facilities used several stages in
series. In addition to these industrial waste applications, RBC's have been used
to treat a number of municipal wastewaters. As a result of the experience gained
with them Autotrol Corp. [22] has developed graphs like Fig. 19.14 which allow prediction of the performance of RBC's on typical domestic sewage. Such graphs allow
consideration to be given to the effects of hydraulic loading and influent concentration. In general, the correlation between predicted performance and actual performance in practice has been satisfactory [1].

In contrast to the paucity of data available relative to full-scale application,
a number of pilot plant studies have been reported [2,25,26,27]. One of the more complete investigated the application of RBC's to pulp and paper mill wastes and a summary of the results is presented in Table 19.2 [2]. For any particular wastewater
the hydraulic loading was found to be the most influential parameter, with a decrease
in performance at higher loadings, as would be anticipated from Section 19.2. In
addition, the data show a definite trend of decreasing performance with increasing
influent concentration, with a very strong effect exhibited by concentrations in excess of 500 mg/liter. This suggests that oxygen limitation could be an important
factor with high strength wastewater, as discussed earlier. RBC pilot plants have
also been run on a number of other wastewaters and Antonie and Hynek [25] have summarized the type of performance that can be obtained as shown in Table 19.3. Although no information was given about the hydraulic or organic loads used, the data
serve to illustrate the wide range of wastewaters which is amenable to treatment by
RBC's.

In summary, there is an emerging body of knowledge which indicates that RBC's
may be used to remove soluble organic matter from many types of wastewaters. Provided that sufficient pilot plant studies are conducted to ensure the development
of a satisfactory design, the RBC should prove to be a reliable and energy efficient
means of wastewater treatment.

19.6 KEY POINTS

1. The RBC consists of parallel circular discs attached perpendicularly to a
    horizontal shaft which passes through their centers. The entire assembly is
    placed into a tank so that the discs are approximately half immersed. Micro-
    organisms grow on the surfaces of the discs and rotation of the shaft brings
    them into contact with the liquid allowing them to consume the organic matter
    in the liquid. Most RBC systems are designed as reactors (stages) in series
    with each stage consisting of one shaft of discs mounted perpendicular to the
    direction of liquid flow.

TABLE 19.1
Examples of Full-Scale RBC's in Practice

Example No.	Type of Wastewater	Flow Rate 10^6 gal/day	Media Area ft^2	Number of Stages	Hydraulic Loading gal/(day·ft^2)	Influent Conc. mg/liter[a]	Organic Loading lb/(day·ft^2)[a]	Effluent Conc. mg/liter[a]	% Removal	Reference
1	Refinery	4.32	1,800,000	3 or 4	2.4	178	0.0036	34	81	8
2	Refinery	1.03	152,000	4	6.8	47	0.0026	32	32	8
3	Cheese Processing	0.003	13,800	4	0.22	1062	0.002	42	96	23
4	Winery[b]	0.44	630,000	6	0.70	600	0.0035	24	96	24
5	Dairy[b]	0.24	--	6	--	2270	--	227	90	25
6	Bakery[b]	0.05	--	4	--	2000	--	< 300	> 85	25

[a] BOD_5

[b] Estimates used for design

TABLE 19.2
Examples of RBC Pilot Plant Studies on Various Pulp and Paper
Wastewaters.[a] Values are Percent BOD_5 Removal.

Waste Type	Hydraulic Loading Rate - gpd/sq ft					
(Avg. Influent Conc. mg/liter)	0-0.9	1.0-1.9	2.0-2.9	3.0-3.9	4.0-4.9	5.0-5.9
Hard Board (2294)	62					
Comb. Board (A)(1513)	58	53				
Bl. Sulfite (1321)	75[b]					
Comb. Board (A)(778)	72	68	32			
Magnefite (543)		58				
Un. Bl. Kraft (257)				84	72	
Ins. Board (240)			88	83	77	
Bl. Kraft (A)(235)	85	81	61			
Bl. Kraft (B)(176)	93	89		80[b]		74
Coated Paper (198)		92	79	52		
Unc. Paper (138)			95			90
Comb. Board (B)(101)				88	76	

[a]From W.J. Gillespie, et al. "A Pilot Scale Evaluation of Rotating Biological
Surface Treatment of Pulp and Paper Mill Wastes," Proceedings of the 29th Industrial Waste Conference, Purdue University Engineering Extension Series No. 145,
1026-1037, 1974. Reprinted by permission of Purdue Research Foundation.

[b]Average of more than one set of observations in this range.

TABLE 19.3
Examples of RBC Pilot Plant Studies on Various Wastewaters[a]

Wastewater	Concentration		Removal
	Influent mg/liter BOD_5	Effluent mg/liter BOD_5	Percent
Chicken and Tuna Processing	4600	180	96
	3090	250	92
	1965	300	85
Fishery	480	140	71
	190	40	79
Meat Processing	860	130	85
Vegetable Canning	1200	48	96
Potato Processing	1800	180	90
Citrus	1500	225	85
Tomato	720	72	90
Yeast Manufacturing	1200	24	98
Animal Glue	1200	84	93
Chicken Waste	1000	30	97
Plastic Manufacturing	40	3	93
Chemical Processing	2900	170	94
Textile	150	20	87
Tannery + Domestic	300	21	93
Textile + Domestic	230	25	89
Packinghouse + Domestic	300	9	97

[a]Adapted from Antonie and Hynek [23].

2. The major objective of an RBC is the removal of soluble organic matter. Therefore RBC's can be used to treat both municipal and industrial wastewaters. They
can also be used to achieve partial removal of soluble organic matter, thus
finding applications in pretreatment of industrial wastewaters.

3. Many of the merits associated with an RBC are similar to those of a trickling
filter. In addition, due to adequate wetting of all of the organisms regardless of the influent flow rate there is no need for recirculation. The energy
and maintenance requirements are minimal. However, a lack of reserve capacity
and the absence of operational controls are the major shortcomings associated
with the RBC.

4. A rotating biological contactor is basically a rotating disc reactor (RDR), and
therefore its performance may be characterized by the mathematical model presented in Section 14.5 for an RDR. In general there is good agreement between
the theoretical predictions and the experimental observations made with both
pilot- and full-scale reactors.

5. A simplified model of RBC performance is analogous to the usual CSTR equation.
Therefore, the graphical and analytical procedures developed for a CSTR are
directly applicable to the design of an RBC.

6. In the simplified model of an RBC the hydraulic loading plays the same role
that $1/\tau$ does for a CSTR.

7. Using a pilot-scale reactor with characteristics similar to those which will be
used in the full-scale facility the rates of substrate consumption per unit of
wetted area can be generated as a function of the substrate concentration.

8. The process design for an RBC system involves determination of the number of
stages in series and wetted area per stage required to reduce the substrate
concentration to a desired value.

9. Once a treatability study has revealed that $-r''_s$ can be represented adequately
by the Monod equation and the parameters q''_m and K_s have been determined, the
wetted surface area in each stage can be determined iteratively from the mass
balance equation for each stage. If different microbial cultures develop in
each stage, the same procedure may be applied provided that q''_m and K_s for each
stage are known.

10. The graphical procedures developed for the determination of $1/\tau$ for a CSTR and
CSTR's in series apply directly to the determination of the hydraulic loadings
for an RBC system. This requires a graphical representation of $-r''_s$ as a function
of the substrate concentration. When the microbial culture in each stage is
different, a family of substrate consumption rate curves, one for each stage, is
needed.

11. Because the graphical approach to design makes no assumptions about the form of
the rate expression, it may be preferable to the analytical method.

12. Other factors which influence the design of RBC systems are temperature, oxygen
transfer, disc rotational speed, and tank volume.

19.7 STUDY QUESTIONS

The study questions are arranged in the same order as the text.

Section	Questions
19.1	1-2
19.2	3-4
19.3	5-11

1. What are some of the specific merits and defects associated with an RBC? What are the key differences between those of the RBC and those of a trickling filter?

2. Is the claim that RBC's have a better transient response than slurry reactor systems correct?

3. Elaborate on the possibility that the rate of oxygen transfer may be limiting in an RBC system.

4. Discuss the effects of influent flow rate, rotational speed of the discs, and total wetted surface area of biofilm on the performance of pilot- and full-scale RBC's.

5. A wastewater containing 300 mg/liter of T_bOD is to be treated by an RBC system. The flow rate is 1.0×10^6 gal/day. The kinetic parameters evaluated at 60°F are $q''_m = 3,000$ mg/(ft$^2\cdot$day) and $K_s = 150$ mg/liter. Determine the surface area of wetted biofilm and the tank volume necessary to reduce the T_bOD to 30 mg/liter in a single stage RBC at a temperature of 60°F.

6. How much wetted surface area would be required per stage if a four-stage equi-sized RBC system were used in Study Question 5? What would be the tank volume for each stage?

7. If in Study Question 5 the surface area of biofilm were 1,000,000 ft^2, what would be the effluent T_bOD?

8. If in Study Question 5 a four-stage equisized RBC system with a surface area of 125,000 ft^2 per stage were used, what would be the effluent T_bOD? Would any stages be oxygen limited?

9. A pilot plant RBC treating a wastewater at 60°F yielded the T_bOD removal rate curve given in Fig. E19.1.

    (a) Determine the surface area needed to reduce the T_bOD from 150 mg/liter to 10 mg/liter in a single stage RBC at 60°F.

    (b) Estimate the surface area requirement at 50°F.

10. In Study Question 9 a four-stage equisized RBC system is to be used. Each stage contains 15,000 ft^2 of biofilm. What would be the effluent T_bOD from each stage at 60°F? Would any stage be oxygen limited?

11. A pilot plant study using a single-stage RBC yielded the data in Table SQ19.1. Size a three-stage RBC system to reduce the substrate concentration from 200 mg/liter to 20 mg/liter T_bOD without having oxygen limitation in any stage.

TABLE SQ19.1
RBC Pilot Plant Data

S mg/liter	$-r''_s$ kg/(day·m^2)
11	0.0032
25	0.0072
48	0.0124
73	0.0180
104	0.0236
133	0.0284
156	0.0312
192	0.0336

REFERENCES AND FURTHER READING

1. R. L. Antonie, "Rotating biological contactor for secondary wastewater treatment," presented at Culp/Wesner/Culp WWT Seminar, South Lake Tahoe, Stateline, Nevada, Oct. 27-28, 1976.

2. W. J. Gillespie, *et al.* "A pilot scale evaluation of rotating biological surface treatment of pulp and paper mill wastes," *Proceedings of the 29th Industrial Waste Conference,* Purdue University Engineering Extension Series No. 145, 1026-1037, 1974.

3. R. L. Antonie, "Discussion of 'Evaluation of a rotating disk wastewater treatment plant'," *Journal of the Water Pollution Control Federation,* 46, 2792-2795, 1974.

4. R. L. Antonie, *et al.* "Evaluation of a rotating disk wastewater treatment plant," *Journal of the Water Pollution Control Federation,* 46, 498-511, 1974.

5. R. L. Antonie, "The Bio-Disc process: New technology for the treatment of biodegradable industrial wastewater," *Chemical Engineering Progress Synposium Series,* 67, No. 107, 585-588, 1971.

6. R. L. Antonie, "Response of the Bio-Disc process to fluctuating wastewater flow," *Proceedings of the 25th Industrial Waste Conference,* Purdue University Engineering Extension Series No. 137, 427-435, 1970.

7. M. P. Filion, *et al.* "Performance of a rotating biological contactor under transient loading conditions," *Journal of the Water Pollution Control Federation,* 51, 1925-1933, 1979.

8. B. T. Davies and R. W. Vose, "Custom designs cut effluent treating costs - Case histories at Chevron USA, Inc.," *Proceedings of the 32nd Purdue Industrial Waste Conference, May, 1977,* Ann Arbor Science Publishers, Inc., Ann Arbor, Mich., 1035-1060, 1978.

9. D. B. Johnson and W. P. Krill, "RBC pilot plant treatment of pretreated meat slaughtering/processing waste," *Proceedings of the 31st Purdue Industrial Waste Conference, May, 1976,* Ann Arbor Science Publishers, Inc., Ann Arbor, Mich., 733-742, 1977.

10. H. H. J. Bintanja, *et al.* "The use of oxygen in a rotating disc process," *Water Research,* 10, 561-565, 1976.

11. R. L. Antonie, *Fixed Biological Surfaces - Wastewater Treatment, The Rotating Biological Contactor,* CRC Press, Inc., Cleveland, Ohio, 1976.

12. A. A. Friedman, *et al.* "Effect of disk rotational speed on biological contactor efficiency," *Journal of the Water Pollution Control Federation,* 51, 2678-2690, 1979.

13. H. Hartmann, "Undersuchung uber die biologische Reinigung von Abwasser mit Hilfe von tauchtop Korperanlangen," *Stuttgarter Berichte zur Siedlungswasserwirschaft Kommissionsverlag,* Band 9, R. Oldenbourg, Munich, 1960.

14. W. N. Torpey, *et al.* "Rotating discs with biological growths prepare wastewater for disposal or reuse," *Journal of the Water Pollution Control Federation,* 43, 2181-2188, 1971.

15. J. A. Chittenden and W. J. Wells, "Rotating biological contactors following anaerobic lagoons," *Journal of the Water Pollution Control Federation,* 43, 746-754, 1971.

16. E. L. Stover and D. F. Kincannon, "Evaluating rotating biological contactor performance," *Water and Sewage Works,* 123, No. 3, 88-91, 1976.

17. R. L. Antonie and F. M. Welch, "Preliminary results of a novel biological process for treating diary wastes," *Proceedings of the 24th Industrial Waste Conference,* Purdue University Engineering Extension Series No. 135, 115-126, 1969.

18. J. E. McAliley, "A pilot plant study of a rotating biological surface for secondary treatment of unbleached kraft mill waste," *Technical Association of the Pulp and Paper Industry,* 57, No. 9, 106-111, 1974.

19. J. H. Clark, *et al.* "Performance of a rotating biological contactor under varying wastewater flow," *Journal of the Water Pollution Control Federation,* 50, 896-911, 1978.

20. B. H. Kornegay, "Modeling and simulation of fixed film biological reactors for carbonaceous wastewater treatment" in *Mathematical Modeling for Water Pollution Control Processes,* T. M. Keinath and M. Wanielista ed., Ann Arbor Science Publishers, Inc., Ann Arbor, Mich. 271-315, 1975.

21. R. W. Wilson, *et al.* "Scaleup in rotating biological contactor design," *Journal of the Water Pollution Control Federation,* 52, 610-621, 1980.

22. "Wastewater treatment systems - Design manual," Autotrol Corporation, Bio-Systems Division, 1701 West Civic Drive, Milwaukee, Wisconsin, 53209, 1979.

23. C. W. Birks and R. J. Hynek, "Treatment of cheese processing wastes by Bio-Disc process," *Proceedings of the 26th Industrial Waste Conference,* Purdue University Engineering Extension Series No. 140, 89-105, 1971.

24. S. A. LaBella, *et al.* "Treatment of winery wastes by aerated lagoon, activated sludge, and rotating biological contactor," *Proceedings of the 27th Industrial Waste Conference,* Purdue University Engineering Extension Series No. 141, 803-816, 1972.

25. R. L. Antonie and R. J. Hynek, "Operating experience with Bio-Surf process treatment of food-processing wastes," *Proceedings of the 28th Industrial Waste Conference,* Purdue University Engineering Extension Series No. 142, 849-860, 1973.

26. R. W. Corneille and J. C. O'Shaughnessy, "Treatment of apple wastes using rotating biological contactors," *Proceedings of the 30th Purdue Industrial Waste Conference, May 1975,* Ann Arbor Science Publishers, Inc., Ann Arbor, Michigan, 675-688, 1977.

27. J. W. Hudson, *et al.* "Rotating biological contactor treatment of shellfish processing wastewaters," *Proceedings of the 31st Purdue Industrial Waste Conference, May 1976,* Ann Arbor Science Publishers, Inc., Ann Arbor, Michigan, 193-205, 1977.

Further Reading

Benefield, L. D. and Randall, C. W., *Biological Process Design for Wastewater Treatment,* Prentice-Hall, Inc., Englewood Cliffs, N.J., 1980, See Chapter 7.

Schroeder, E. D., *Water and Wastewater Treatment,* McGraw-Hill Book Co., New York, N.Y., 1977. See Chapter 9.

Wu, Y. C. (Editor), *Advanced Wastewater Treatment by Rotating Biological Contactor Processes,* Ann Arbor Science, Ann Arbor, Mich., 1980. Proceedings of a conference on RBC's.

CHAPTER 20

NITRIFICATION

Within the United States the major effort of wastewater treatment engineers has been directed toward the removal of organic matter in order to reduce the carbonaceous oxygen demand entering surface waters. It has become apparent, however, that this is not sufficient to maintain water quality, so recently attention has been focused on the reduction of nitrogenous oxygen demand by conversion of ammonia nitrogen to the nitrate form through the operation called nitrification. Actually, as was pointed out in Section 1.2.6, nitrification is a type of biochemical operation, and as such it can be performed in any one of a number of reactor types (see Table 1.2). In this chapter we will consider the requirements for performing nitrification in CSTR's with cell recycle, in packed towers, and in rotating-disc reactors.

20.1 DESCRIPTION

20.1.1 Microbiological and Biochemical Characteristics

As defined in Section 7.1.3 nitrification is the conversion of ammonia nitrogen
(NH_4^+-N)* to nitrate nitrogen (NO_3^--N)* and it may be performed by either heterotrophic
or autotrophic bacteria [1]. In spite of the fact that over one hundred species
have been cited as forming nitrite (NO_2^-) from ammonia [2], it is doubtful that significant quantities of nitrate are generated heterotrophically in natural systems [3].
Consequently, most research into nitrification in wastewater treatment systems has
concentrated on the autotrophic organisms, as will this chapter. The major nitrifying bacteria are thought to be of the genera *Nitrosomonas* and *Nitrobacter* (See
Section 8.4.1). *Nitrosomonas* oxidizes NH_4^+-N to NO_2^--N through a relatively complex
pathway which is not yet completely understood, although the hypothesized route was
given in Eq. 8.5. *Nitrobacter* oxidizes NO_2^--N to NO_3^--N in a single step as given by
Eq. 8.6. Because both of these organisms obtain their energy for growth from inorganic oxidations they are classified as chemoautotrophs. This does not mean that
they cannot incorporate exogenous organic compounds while obtaining their energy
from inorganic oxidations because they can [4]. The amount of the uptake will be
small and will vary with the growth conditions, however, so that most equations depicting the stoichiometry of nitrification ignore it and utilize carbon dioxide as
the sole source of carbon.

   Using reasonable values for cell yields and the techniques described in
Section 9.4.2, it is possible to write theoretical stoichiometric equations for the
growth of *Nitrosomonas* and *Nitrobacter* [5]. For *Nitrosomonas* the equation is:

$$55 \ NH_4^+ + 76 \ O_2 + 109 \ HCO_3^- \rightarrow C_5H_7O_2N + 54 \ NO_2^- + 57 \ H_2O + 104 \ H_2CO_3 \qquad (20.1)$$

The equation for *Nitrobacter* is:

$$400 \ NO_2^- + NH_4^+ + 4 \ H_2CO_3 + HCO_3^- + 195 \ O_2 \rightarrow C_5H_7O_2N + 3 \ H_2O + 400 \ NO_3^- \qquad (20.2)$$

These equations are important for two reasons. First, they tell us that a large
amount of alkalinity (HCO_3^-) will be utilized during the oxidation of NH_4^+-N:
8.64 mg HCO_3^-/mg NH_4^+-N oxidized. A small part of this will be incorporated into the
cell material, but the majority will be used to neutralize the hydrogen ions released
during the oxidation. If the water contains insufficient alkalinity, nitrification
will be retarded because of the unavailability of a needed reactant and the deleterious effect of the resulting low pH (see Section 20.2.1). Secondly, they tell us that
approximately 3.22 mg O_2 will be required for each mg of NH_4^+-N oxidized to NO_2^--N, and
1.11 mg O_2 will be needed for each mg of NO_2^--N oxidized to NO_3^--N for a total of

---

*The use of NH_4^+-N or NO_3^--N is not intended to imply the ionic state of the nitrogen
 species. They simply mean ammonia and nitrate nitrogen.

4.33 mg O_2 per mg of NH_4^+-N oxidized all the way to NO_3^--N. In previous sections of this book, the oxygen requirement for a biochemical operation has been calculated from an oxygen demand balance rather than from the stoichiometric equation because of the large variations possible in the yield of heterotrophic organisms. The yield of autotrophs is very small, however, so that variations in its value have little effect upon the oxygen requirement. In fact, complete oxidation of NH_4^+-N to NO_3^--N without the formation of any cells requires only 4.57 mg O_2/mg NH_4^+-N. Consequently a value between 4.33 and 4.57 can be used in engineering calculations with little error, so that the oxygen requirement for a reactor can be estimated simply by knowing how much NH_4^+-N is reacted. This simplified procedure will be used herein.

20.1.2 Physical Characteristics

Nitrification may be performed in CSTR's with cell recycle, packed towers, and rotating-disc reactors, consequently the physical characteristics of the reactors are the same as those discussed in Chapters 16 (activated sludge), 18 (trickling filter) and 19 (rotating biological contactor). In some cases nitrification occurs concurrently with the removal of organic matter in combined carbon oxidation-nitrification reactors. In other cases the oxidation of organic matter occurs in one reactor followed by nitrification in a subsequent one. This is called separate-stage nitrification. There are no exact rules concerning the applicability of one type of system over the other and the engineer must make the choice of which to use by considering the unique circumstances of each case. We will consider some of the factors involved in this choice in Section 20.3.1.

20.1.3 Applications

Nitrification systems are designed to meet one objective, to convert NH_4^+-N to NO_3^--N. Currently their primary application is to domestic wastewaters. These contain approximately 30-40 mg/liter of nitrogen (TKN), which is converted to the ammonia form during normal degradation of the carbonaceous substrate. In that form the concentration is sufficiently high to be toxic to fish or to cause severe oxygen depletion in the receiving stream as the natural aquatic nitrifiers oxidize the nitrogen to the nitrate state. Generally, however, the resulting concentration of NO_3^--N in the receiving stream is not high enough to cause health problems so that if it could be formed during wastewater treatment its discharge would be permissible. Another application is to wastewaters containing so much nitrogen that even the discharge of NO_3^--N would cause problems. In this case nitrification is coupled with denitrification to give complete nitrogen removal (see Chapter 22). Special problems are often encountered with these wastewaters and they will be considered briefly in Section 20.2.1. Alternative nonbiological nitrogen removal systems are also available [6] but to date biological systems are the most popular.

20.2 PERFORMANCE

In this section we will compare the performance of nitrification systems to the
theoretical models of Chapters 13 and 14. Before doing that, however, it would be
helpful to review the kinetics of nitrification because an understanding of some of
their peculiar features will make it easier to understand why the reactors behave
as they do.

20.2.1 Kinetics of Nitrification

KINETICS OF GROWTH AND SUBSTRATE UTILIZATION. The kinetics of growth and substrate removal for autotrophic bacteria can be expressed with the same equations
used earlier for heterotrophic bacteria. For example, the specific growth rates of
Nitrosomonas and *Nitrobacter* with NH_4^+-N and NO_2^--N as the respective growth-limiting
substrates can be expressed by the Monod equation;

$$\mu = \frac{\mu_m S}{K_s + S} \qquad\qquad (10.42)$$

where S refers to the concentration of the particular form of nitrogen being utilized. Comparison of the values of μ_m for nitrifiers given in Table 10.4 with the
typical values for heterotrophs given in Table 10.1 shows the former to be at least
an order of magnitude smaller than the latter. Thus it is apparent that nitrification will only occur in reactors in which the specific growth rate is quite low.
Table 10.4 also contains values of K_s for each of the genera and they are seen to
be quite low. Other authors [3,6,7] have reported values as low as 0.25 mg/liter
NH_4^+-N and 0.07 mg/liter NO_2^--N. The importance of these low K_s values can be seen
by examining Eq. 10.42. First, when K_s is small with respect to S, the specific
growth rate becomes essentially independent of the substrate concentration over a
broad range; in other words it becomes zero order with respect to substrate. Such
zero-order kinetics have been observed by a number of workers [8,9,10], and have a
profound effect upon the performance of some of the reactor types. Second, when the
actual magnitude of K_s is very small, the concentration of S in a CSTR being operated
to give a low specific growth rate will also be quite low, and thus high degrees of
nitrification can be achieved. This, too, has been observed.

The other parameters of importance in characterizing microbial reactors are the
true growth yield, Y_g, the specific decay rate, b, and the specific death rate, γ.
Poduska and Andrews [7] and Stankewich [11] have summarized observed yields (Eqs.
12.17 and 12.18) from the literature and have reported them to vary from 0.03 to
0.13 mg cells/mg NH_4^+-N oxidized for *Nitrosomonas* and from 0.02 to 0.07 mg cells/mg
NO_2^--N oxidized for *Nitrobacter*. Consequently Y_g must lie within the same ranges.
These are considerably less than the values for heterotrophs listed in Table 9.3.
Few values of b have been reported in the literature but Poduska and Andrews [7]
have estimated it to be 0.005 hr^{-1} for both organisms. No values of γ have been reported. Thus for simplicity's sake it will be assumed to be zero. The key point

from this summary is that the kinetics of nitrification can be expressed by the same equations used to describe heterotrophic growth, but that because of the autotrophic nature of nitrification, the parameter values are considerably different.

FACTORS AFFECTING KINETICS. A number of factors affect the kinetics of nitrifying bacteria, just as several factors affected the heterotrophs discussed earlier. In a few cases investigators have attempted to quantify these effects, thereby suggesting equations which might be used. Some of these equations have been adopted herein to illustrate the general trends which might be anticipated. It should be emphasized, however, that the specific effects on a given system can only be determined experimentally.

In the other chapters of this book we have assumed that only one nutrient was rate limiting, thereby allowing Eq. 10.42 to be used. One of the key nutrients which was assumed to be supplied in excess was dissolved oxygen (DO) and this approach presented no problems in heterotrophic systems because the value of K_s for oxygen is generally small in comparison to the concentration of DO in the reactor, thereby making the rates independent of DO. For nitrification systems, on the other hand, K_s for DO is relatively high, having been estimated to be 2.0 mg/liter [12]. Other authors [3] have estimated values as low as 0.5 mg/liter, but the factors affecting the magnitude of K_s for DO are not yet well established. Consequently for illustrative purposes, Parker *et al.* [6] assumed K_s for DO to be 1.3 mg/liter. This relatively high value is quite important in light of the low K_s values for NH_4^+-N, because it means that unless special precautions are taken, the concentration of DO, rather than the NH_4^+-N or NO_2^--N concentrations, will control the rate of nitrification.

The effects of pH on the specific growth rate of bacteria have been mentioned in other sections of this book but because they are generally small over a rather broad band on either side of neutrality and because the degradation of organic matter produces CO_2 which tends to buffer the pH near neutrality, little kinetic consideration has been given to pH. When nitrification is occurring, however, two events require us to consider the effects of pH. The first was seen in Eq. 20.1, i.e., nitrification causes a destruction of alkalinity and thus the potential exists for a drastic drop in pH. The second is that nitrifying bacteria are very sensitive to pH, as can be seen in Fig. 20.1 [8]. These results are from enrichment cultures. The effects of pH on nitrifiers in mixed cultures, such as activated sludge, are not quite as severe. A wide range of pH optima have been reported, [11], but most workers agree that as the pH moves to the acid range the rate of ammonia oxidation declines [6]. Furthermore, if a culture is acclimated to a low pH the effect is less severe than if the pH is suddenly shifted. Downing *et al.* [13] proposed an equation to depict the effect of pH on μ_m for *Nitrosomonas* when the pH is below 7.2:

$$\mu_m = \hat{\mu}_m [1 - 0.833(7.2 - pH)] \tag{20.3}$$

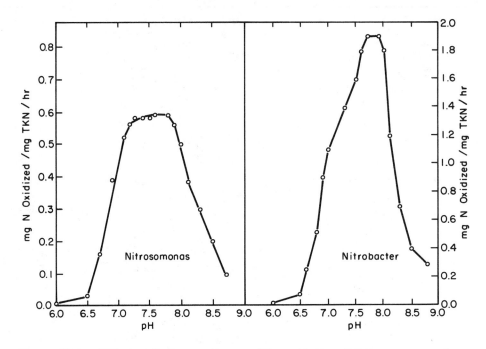

Figure 20.1.   Effects of pH upon the specific activity of *Nitrosomonas* and
                Nitrobacter enrichment cultures.
                (Adapted from Srinath et al. [8]).

where $\hat{\mu}_m$ is the maximum μ_m, which is assumed to be constant between pH 7.2 and 8.0.
Because nitrification tends to depress the pH, values above 8.0 were not considered.
Parker *et al.* [6] adopted Eq. 20.3 in their illustrative model. Figure 20.1b shows
that pH also affects *Nitrobacter*. An equation of similar form could be used for it,
although generally only the effect on *Nitrosomonas* is modeled since it must provide
the substrate for *Nitrobacter* and its rates are usually limiting.

Temperature has a strong effect upon the growth rate of nitrifying bacteria
just as it has upon heterotrophs. Several workers have shown that the effect of
temperature on μ_m fits an Arrhenius-type equation over the physiological range [14,15].
Wong-Chong and Loehr [14] found that deactivation of *Nitrobacter* occurred at lower
temperatures than did deactivation of *Nitrosomonas* and that the temperature dependency
of both genera was a function of pH. Stankewich [11] has summarized the results of
several authors with the general equation:

$$\mu_{mT} = \mu_{m15} \ \exp[K(T - 15)] \tag{20.4}$$

where μ_{mT} is the maximum specific growth rate at any temperature T (°C) and μ_{m15} is
the rate at 15°C. For *Nitrosomonas* the reported values for K lay between 0.095 and
0.12 whereas for *Nitrobacter* they were between 0.056 and 0.069. Knowles *et al.* [16]

reported that K_s also varied with temperature so that

$$K_{sT} = K_{s15} \exp[0.118(T - 15)] \tag{20.5}$$

for *Nitrosomonas* and

$$K_{sT} = K_{s15} \exp[0.146(T - 15)] \tag{20.6}$$

for *Nitrobacter*. The reported values of K_{s15} were 0.405 and 0.625 for *Nitrosomonas* and *Nitrobacter*, respectively. These equations can be used to approximate the effects of temperature, but as a general rule the actual effects should be determined experimentally during treatability studies.

Parker *et al.* [6] have expressed the combined effect of all of these quantifiable factors by putting them into one equation:

$$\mu = \mu_m \left(\frac{S}{K_s + S}\right)_N \left(\frac{S}{K_s + S}\right)_{DO} [1 - 0.833(7.2 - pH)] \tag{20.7}$$

The effect of temperature can be incorporated by using the appropriate form of Eq. 20.4 for μ_m and either Eq. 20.5 or 20.6 for K_s. Since little is known about it, K_s for dissolved oxygen will be assumed to be independent of temperature.

There are other factors which affect nitrification, but which are not yet fully quantifiable. For example, *Nitrosomonas* and *Nitrobacter* are subject to both substrate and product inhibition, i.e., if the concentration of either their substrates or their products is too high, there will be a decrease in the rates of activity of the microorganisms [3,6,7]. Studies with pure cultures have shown that free ammonia and undissociated nitrous acid are more inhibitory than NH_4^+ or NO_2^- [17,18,19], and studies with mixed cultures have suggested the same thing [20]. This means that pH has an influence on the degree of inhibition observed at any given substrate or product concentration. Furthermore, the interaction between pH and substrate or product concentration could help explain the various pH effects reported in the literature [11]. Equation 20.3 should be limited to situations where the substrate and product concentrations are relatively low since more work is needed before equations can be written depicting the combined effects of pH and substrate (or product) concentration.

Because of the autotrophic nature of nitrifying bacteria the concept developed that organic compounds display a general toxicity toward them. That this concept is fallacious has been demonstrated in pure [21] and mixed cultures [22,23]. Nitrification can proceed at rapid rates in the presence of organic matter, provided that other environmental factors such as pH, DO, etc., are adequate. In fact, under some circumstances the presence of organic matter can even enhance the rate of nitrification [23]. There are some organic compounds which are inhibitory, however, and therefore act to reduce the specific growth rate of the nitrifiers.

Tomlinson *et al.* [24] studied inhibition of nitrification in the activated sludge process and Table 20.1 summarizes their results. Examination of it shows that the most potent inhibitors (which were not also general inhibitors of bacterial growth) contained sulfur and thus acted as metal chelating compounds. Nitrifiers are particularly susceptible to inhibition by such compounds because many of their enzymes require metals for activation. Table 20.2 summarizes the results of another study [26]. Of the 52 compounds studied, 20 were found to be inhibitory to ammonia oxidation but only 3 inhibited nitrite oxidation. Thus it appears that *Nitrosomonas* is the weak link in the nitrification chain. Examination of the effects of dodecylamine, aniline, and ethylene diamine showed that the inhibition of ammonia oxidation did not conform to either classical competitive or noncompetitive kinetics. Instead it increased as the substrate concentration was increased [26].

There have also been suggestions in the literature that the presence of heterotrophic bacteria is deleterious to the activity of nitrifying bacteria but this has been shown to be false [23]. Any effect of heterotrophs is probably indirect, such as a reduction in the DO concentration or an alteration of the pH. Because of the sensitivity of the nitrifying bacteria to these factors, special care must be given to the design of facilities in which heterotrophs and autotrophs share the same space. As long as that is done the presence of the heterotrophs is not likely to cause problems.

TABLE 20.1
Organic Compounds Which Inhibit Activated Sludge Nitrification[a]

Compound	Conc.[b] mg/liter	Compound	Conc.[b] mg/liter
Acetone[c]	2000	Mercaptobenzothiazole	3.0
Allyl Alcohol	19.5	Methylamine hydrochloride	1550
Allyl chloride	180	Methyl isothiocyanate	0.8
Allyl isothiocyanate	1.9	Methyl thiuronium sulfate	6.5
Benzothiazole disulfide	38	Phenol[c]	5.6
Carbon disulfide[c]	35	Potassium thiocyanate	300
Chloroform	18	Skatol	7
o-Cresol	12.8	Sodium dimethyl	
Di-allyl ether	100	dithiocarbamate	13.6
Dicyandiamide	250	Sodium methyl	
Diguanide	50	dithiocarbamate	0.9
2,4-Dinitrophenol	460	Tetramethyl thiuram	
Dithio-oxamide	1.1	disulfide	30
Ethanol	2400	Thioacetamide	0.53
Guanidine carbonate	16.5	Thiosemicarbazide	0.18
Hydrazine	58	Thiourea	0.076
8-Hydroxyquinoline	72.5	Trimethylamine	118

[a] Data extracted from Tomlinson *et al.* [24].

[b] Concentration giving approximately 75% inhibition.

[c] In the list of industrially significant chemicals [25].

TABLE 20.2
Organic Compounds Which Inhibit Ammonia Oxidation
in an Enrichment Culture of Nitrifying Bacteria[a]

Compound	Conc.[b] mg/liter
Dodecylamine	<1
Aniline	<1
n-Methylaniline	<1
1-Naphthylamine	15
Ethylenediamine[c]	17
Napthylethylenediamine diHCl	23
2,2'-Bipyridine	23
p-Nitroaniline	31
p-Aminopropiophenone	43
Benzidine diHCl	45
p-Phenylazoaniline	72
Hexamethylene diamine[c]	85
p-Nitrobenzaldehyde	87
Triethylamine	127
Ninhydrin	>100
Benzocaine	>100
Dimethylgloxime	140
Benzylamine	>100
Tannic acid	>150
Monoethanolamine[c]	>200

[a] Data extracted from Hockenbury and Grady [26].

[b] Concentration giving approximately 50% inhibition.

[c] From the list of industrially significant chemicals [25].

20.2.2 Nitrification in CSTR's with Cell Recycle

Nitrification in CSTR's with cell recycle can be done in either of two ways: in
combination with carbon oxidation or in a separate system following carbon oxidation. It has been shown experimentally [9] that as far as the performance of the
nitrifying bacteria is concerned it makes no difference which technique is used,
provided that the proper pH and DO levels are maintained in the system and provided
that none of the organic compounds present are inhibitory. Consequently we will
first investigate performance without regard to the type of CSTR system. This distinction will be important during design, however, and consequently it will be considered in Section 20.3.2. We will, however, touch briefly upon the differences
when inhibitory materials are present.

Considering first a single reactor, the models in Chapter 13 predict that the
conversion of NH_4^+-N to NO_3^--N will depend upon the MCRT and this is exactly the case,
as shown in Fig. 20.2 [7]. The curve points out an interesting fact associated with
nitrification in CSTR's at steady state, i.e., they tend to give an all-or-none response. They will either give almost total nitrification or the nitrifying population will wash out, giving none. This is a result of the very low K_s values
associated with the nitrifying bacteria. Furthermore, the kinetic parameters are
such that the growth rate of *Nitrobacter* can easily exceed that of *Nitrosomonas*.

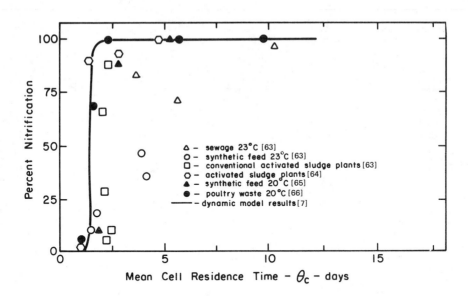

Figure 20.2.   Effect of MCRT on the steady-state nitrification performance
               of a CSTR with cell recycle. *(Adapted from Poduska and
               Andrews [7]).* The reference numbers refer to the sources
               of the data.

As a consequence of these factors, the choice of MCRT during design is made to ensure
that *Nitrosomonas* will not wash out, which in general also means that a high degree
of nitrification will occur.

There are some occasions in which this all-or-none phenomenon will not occur.
One is when the reactor is subjected to dynamic loads, such as the normal diurnal
flow pattern on a domestic wastewater treatment plant. Simulation results have shown
the effect of MCRT upon dynamic performance to be like that shown in Fig. 20.3, a
result which agrees with observations in practice. As shown in the figure complete
nitrification can still be achieved, although a slightly longer MCRT will be required.
Another occasion is when the wastewater contains a very high concentration of NH_4^+-N
[20,27]. In this situation the concentrations of NO_2^--N and NO_3^--N formed are likely
to be high enough to inhibit the bacteria so that complete nitrification cannot occur,
even at very long MCRT's. Thus a multistage system must be used in which partial
nitrification is allowed to occur, and then denitrification is used to destroy the
end products, thereby allowing nitrification to be completed in a final nitrification
step [20,27]. As pointed out earlier, inhibition effects cannot yet be reliably
modeled. Consequently, pilot studies should always be run on wastes containing high
concentrations of NH_4^+-N (several hundred mg/liter). A final circumstance in which a
curve with a different shape may be obtained is illustrated in Fig. 20.4 [28]. The
data were collected on an industrial waste containing organic compounds which were
resistant to biodegradation and inhibitory to the nitrifying bacteria. Because of

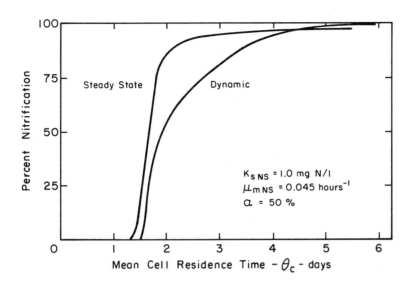

Figure 20.3.   Comparison of steady state and dynamic performance of
                nitrification in a CSTR with cell recycle.
                (Adapted from Poduska and Andrews [7]).

the inhibitors the MCRT's required to get a high level of nitrification were an order
of magnitude larger than the ones which would be computed from the typical kinetic
parameters in Table 10.4.

As pointed out in Chapter 16 the performance of several tanks in series with cell
recycle around the entire chain is similar to the performance of conventional or
tapered aeration activated sludge and is primarily a function of the MCRT. When used
with the proper kinetic parameters those models also depict the course of nitrification in such reactors. If the pH is held constant and if adequate dissolved oxygen
is present, NH_4^+-N oxidation will occur at a constant specific rate down the length of
the chain because of the low value of K_s. This generally means that the NH_4^+-N concentration will approach zero at some intermediate point in the chain. As a consequence
essentially complete nitrification will occur at MCRT's sufficient to maintain the
nitrifying population.

The other activated sludge modification modeled in Chapter 13 and discussed in
Chapter 16 was contact stabilization activated sludge (CSAS). One of the features of
CSAS discussed in those chapters was that it could achieve partial substrate removal
at normal design MCRT's by proper modification of the contact time and recycle ratio.
Although that tank configuration is not well suited to nitrification, the question
has often arisen concerning the possibility of achieving nitrification in conjunction
with the removal of organic compounds in such a system. Gujer and Jenkins [29]
studied that question and determined that the answer was yes. As would be expected,

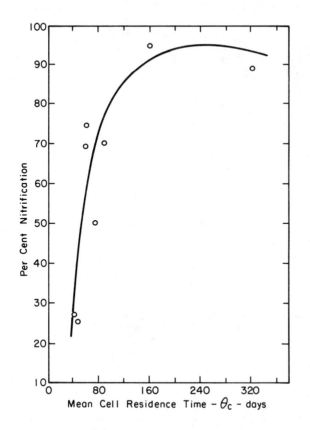

Figure 20.4. Effect of MCRT on the performance of
              nitrification in a reactor receiving
              a wastewater containing inhibitory
              compounds. *(Adapted from Bridle et al. [28]).*

nitrification was not complete, but they achieved between 30 and 80% conversion,
depending upon the specific growth rate of the nitrifying microorganisms in the contact tank. Furthermore, they developed the model after which the one in Chapter 16
was patterned and showed that it did an excellent job of predicting nitrification
efficiency.

     In summary, the models presented in Chapter 13 for CSTR's with cell recycle
have been shown to satisfactorily predict the degree of nitrification obtained in
reactor systems similar to those used for completely mixed activated sludge, conventional or tapered aeration activated sludge, and contact stabilization activated
sludge. One would therefore conclude that they will also predict the performance in
reactor systems like step aeration activated sludge as well. Furthermore, they will
model the performance of either separate-stage or combined carbon oxidation-nitrification systems. All of this discussion has been based upon the absence of inhibitory

organic compounds which are also biodegradable. If such compounds are present in the wastewater, then the performance of separate-stage and combined systems are likely to be different. The performance of a separate-stage system in which the inhibitory material has been degraded in the first stage will be similar to the performance dis- cussed above. The performance of a combined system may be poorer. If a single CSTR with recycle is used for the combined system, and the MCRT is high enough for essen- tially complete degradation of the inhibitory material, then nitrification will proceed as if the inhibitory material were not even in the influent, provided that the reactor operates at steady state. During dynamic conditions, however, nitrifi- cation is likely to be inhibited due to transient rises in the concentration of the inhibitory material. For tanks in series (CAS or TAAS) the concentration of inhibi- tory compound in the first tank (or at the head end of a long aeration basin) is like- ly to be high enough to inhibit nitrification completely. In that case nitrification will not proceed until the inhibitory material is degraded, and thus only part of the MCRT will be available for growth of the nitrifiers [30]. This means that a longer MCRT will be required for nitrification in a combined system than in the second stage of a separate-stage system. Because CSAS does not give complete nitrification even in the absence of inhibitors, its nitrification performance will be even worse when inhibitors are present and thus it is not suited to that situation.

20.2.3 Nitrification in Packed Towers

Unlike nitrification in CSTR's with cell recycle, there is no accepted kinetic theory for predicting the extent of nitrification likely in a packed tower. As was dis- cussed in Chapter 14, the mathematical models for packed towers are still in the developmental stage. Consequently, in this section experimental observations will be presented without trying to relate them directly to the mathematical models of that chapter. Then in a later section (20.3.3) the factors which must be considered during design from pilot plant data will be presented.

Unlike CSTR's in which all organisms grow together as a homogeneous mass, there are fundamental differences in packed towers between the ways in which nitrification will proceed in the presence and absence of organic matter. As discussed in Chapter 14, the specific growth rates of the organisms in a packed tower depend upon the concentration of substrate in the liquid passing over them and therefore the rates decline as the distance from the top of the tower increases. Furthermore, as the organisms grow, the film thickness will reach some maximal value determined by the fluid shear, at which point the loss of organisms from the film must balance their growth. Now, consider the situation in which a wastewater contains high concentra- tions of both biodegradable organic matter and NH_4^+-N. At the top of the tower the specific growth rates of both the heterotrophic and autotrophic bacteria will approach their maximal values. The value of μ_m for the autotrophs is approximately an order of magnitude lower than the value for the heterotrophs, however, so that as the two types of bacteria compete for space in the film the heterotrophs will displace the

autotrophs. Thus no nitrification will occur. As the liquid proceeds down the
tower, the specific growth rate of the heterotrophs will decline as the concentration of biodegradable organic matter falls. Eventually a point will be reached at
which the specific growth rate of the nitrifiers will be large enough in comparison
to that of the heterotrophs to allow them to compete for space in the film. Thus,
as the concentration of organic matter declines in proportion to the concentration
of NH_4^+-N, the autotrophs will make up a larger percentage of the biomass in the
film, thereby causing the rate of nitrification per unit area of film to increase.
This is illustrated in Fig. 20.5 in which the rates of nitrification in films grown
with different ratios of organic matter to nitrogen are compared with the rate obtained when the film was grown on a medium devoid of organic matter. The effect of
this competition for space is to prevent significant nitrification from occurring
until the majority of the organic matter has been removed, as has been observed in
practice [32]. When a wastewater containing both organic matter and NH_4^+-N is applied
to a packed tower, only a fraction of the tower height will be available for nitrification, and the magnitude of that fraction will depend upon both the absolute and
relative concentrations of the two substrates. On the other hand if the wastewater
contains little organic matter, nitrification will occur throughout the entire tower.
Because of these effects of wastewater composition, the approach used to express the
nitrification performance of a packed tower receiving a wastewater containing organic matter is different from the one used when the wastewater is free of organic
matter.

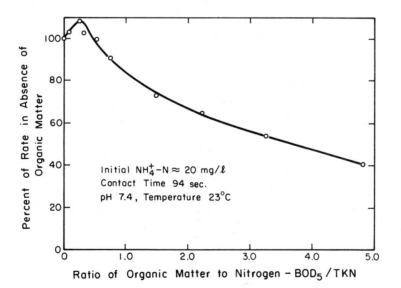

Figure 20.5.   Effect of the ratio of organic matter to nitrogen
               on the rate of nitrification observed in fixed
               films. *(Adapted from Huang [31]).*

Consider first the application of wastewaters containing both organic matter and NH_4^+-N. After reviewing the literature, Parker et al. [6] concluded that the primary parameter currently in use for characterizing nitrification performance is the organic loading [(mass of organic matter)/(unit volume·unit time)]. Figure 20.6 is a summary of results from eight different rock media trickling filters. Nitrification will only occur when the organic loading is low because only then will there be sufficient media for the nitrifying bacteria to become established. As seen in Chapter 14, the specific surface area of the media is an important factor in determining the amount of biomass present in a packed tower. Because plastic media have a higher specific surface area than rock media, one would expect nitrification to be achieved at higher loadings with it, and this is exactly what has been observed [6]. This dependency of performance upon specific surface area means that unique curves will result for each type of media as well as for each waste, thus requiring that designs be made from pilot plant data. In addition to specific surface area, there are other factors which affect performance. One of these is recirculation, which has been shown to improve nitrification [6,33]. As discussed in Chapters 14 and 18, recirculation reduces the concentration gradient through a packed tower. Because K_s for most organic wastes is relatively high, this reduces the specific growth rate of the heterotrophic bacteria at the top of the tower. On the other hand, because the K_s values for nitrification are so low, considerable dilution can occur without

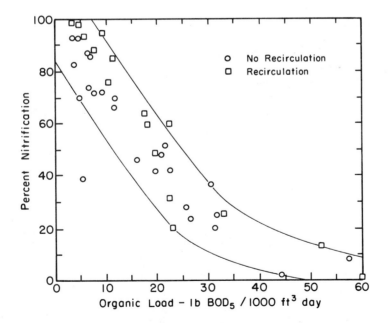

Figure 20.6. Effect of organic load on the nitrification efficiency of rock trickling filters performing combined carbon oxidation and nitrification. (Adapted from Parker et al. [6]. 1 lb BOD_5/1000 ft^3·day = 1.6 x 10^{-2} kg/m^3·day).

causing a reduction in the specific growth rate of the autotrophic bacteria. Thus
it is likely that recirculation exerts its effect by acting to reduce the specific
growth rate of the heterotrophs with respect to that of the autotrophs, thereby allowing the latter to compete more effectively for space in the microbial film.
Oxygen diffusion may limit the rate of nitrification in packed towers [34], and the
effect of recirculation may also be due to a reduction in that limitation via an increase in the DO to NH_4^+-N ratio. Regardless of its effect upon performance, however,
recirculation will probably be required in plastic media towers to keep all of the
media wet at the low organic loadings required for nitrification (see Section 18.3.3).
From Section 20.2.1 it is known that temperature will affect performance, although
little quantitative data are available for towers in which both carbon removal and
nitrification occur [6]. The data presented in Fig. 20.6 were for ∿20°C, and lower
loadings would be required for comparable performance at lower temperatures.

The most comprehensive study of nitrification in packed towers receiving wastewater from which the organic matter has been removed is that of Duddles et al. [35,36]
which was performed in a 21.5 feet (6.6 m) tall plastic media trickling filter pilot
plant. Figure 20.7 shows that the performance depended upon the hydraulic application rate, a finding which is consistent with the principles of Chapter 14.
Another study [37], performed on a similar tower at a different location, achieved
similar results. Duddles et al. [35,36] studied the effect of recirculation and
found it to be negligible on an average basis, although it did help reduce

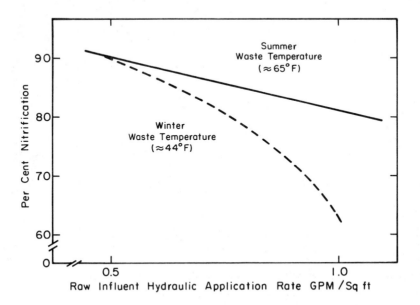

Figure 20.7. Effect of hydraulic loading on the nitrification
efficiency of a plastic media packed tower performing separate-stage nitrification. *(Adapted
from Duddles and Richardson [36])*.

fluctuations in performance during diurnal variations. As a consequence, a recirculation ratio of 1:1 has been considered adequate for maintaining the proper hydraulic
application during dry weather flow [6]. Although both towers achieved low effluent
NH_4^+-N concentrations at low application rates during the summer, performance deteriorated materially in the winter. Thus during cold weather it appears to be difficult
to achieve a high degree of nitrification in a packed tower with economical application rates.

20.2.4 Nitrification in Rotating-Disc Reactors

The discussion at the beginning of the preceding section concerning the growth of
heterotrophs and autotrophs in fixed-film reactors is equally applicable to rotating-
disc reactors. As a consequence we would expect to have sequential oxidation of the
organic matter and the NH_4^+-N, and this has been observed [38]. Figure 20.8 summarizes
the performance of several rotating-disc reactors at different locations. In this
graph the hydraulic loading is expressed as a function of the wetted surface area of
the discs since it determines the biomass available within the reactor. It will be
noted that for a given hydraulic loading, the degree of nitrification decreases as
the BOD_5 of the wastewater increases, which is in accordance with the discussion of

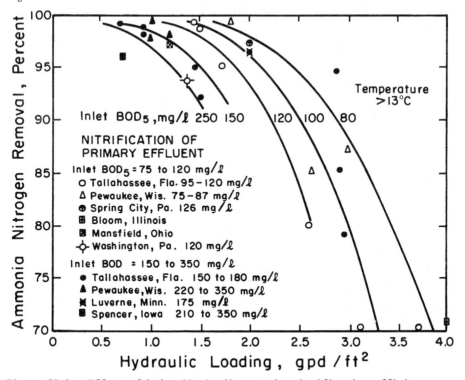

Figure 20.8.   Effect of hydraulic loading on the nitrification efficiency
               of rotating-disc reactors performing combined carbon oxidation
               and nitrification. *(Adapted from Antonie as given in
               Parker et al [6].* 1 gpd/ft^2 = 0.041 m^3/m$^2 \cdot$day.)

microbial growth in Section 20.2.3. Figure 20.8 was prepared for all cases in which
the temperature of the wastewater was greater than 13°C because temperature had little
effect above that value [6]. As discussed earlier, low temperatures retard nitrification rates so that correction factors must be applied during design. These will be
discussed in Section 20.3.4.

Antonie [39] has summarized data from the nitrification of wastewaters which
contain little or no organic matter (Fig. 20.9) [6]. This plot shows the NH_4^+-N
oxidation rate on a single stage of a rotating-disc reactor as a function of the
NH_4^+-N concentration in the stage. The ordinate should not be thought of as a rate
of removal per unit of biomass because stages operating at higher NH_4^+-N concentrations
contained larger quantities of biomass per unit area. This prevents the rate from
becoming independent of NH_4^+-N concentration (i.e., zero-order) in that plot. In
addition, other factors, such as oxygen transfer and substrate diffusion, influenced
the overall kinetics thereby also preventing zero-order kinetics from being observed.
Subsequent work has suggested that in most operating systems zero-order kinetics are
observed when the concentration of NH_4^+-N in a stage exceeds 5 mg/liter. It is

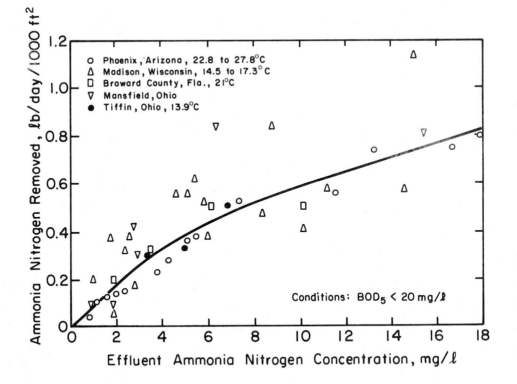

Figure 20.9. Effect of NH_4^+-N concentration on the rate of ammonia oxidation
in rotating-disc reactors performing separate-stage nitrification. *(Adapted from Antonie [39] as given in Parker et al. [6].*
1 lb/day·1000 ft^2 = 4.85 kg/day·1000 m^2.)

possible to relate nitrification performance in the absence of organic matter to the hydraulic loading so that it becomes a major parameter governing the performance, regardless of the type of wastewater being applied [39]. Such a relationship will be used for design in Section 20.3.4.

20.3 DESIGN

20.3.1 Selection of a Suitable System

The first decision facing the designer of a nitrification system is the type of reactor system to be used. Actually this involves two decisions, for not only must the designer decide upon the reactor configuration (i.e., CSTR, packed tower, or rotating-disc), he must also decide whether to use combined carbon oxidation and nitrification or separate-stage nitrification. No generalizations can be made concerning these decisions. Consequently, it will be necessary for the designer to consider all facets of the problem for his particular situation and to choose the most cost effective system. The information in Table 20.3 [6] can help the designer narrow the choices which must be given serious consideration. When deciding whether to use a combined or separate-stage system, the major tradeoff is between size and stability. Because the two reactions can be performed in a single reactor the size of the combined system is generally smaller (and therefore less expensive) than one in which carbon oxidation and nitrification occur in separate reactors. On the other hand, inhibitory materials may reach the nitrifiers in a combined system, thereby impairing performance. When carbon oxidation and nitrification are performed in separate stages, organic inhibitors will probably be degraded before they can reach the nitrifiers, and inorganic inhibitors (such as heavy metals) are likely to be adsorbed onto the biomass in the first-stage reactor, thereby reducing the concentration reaching the nitrifying bacteria. Thus a separate-stage system will be more stable when inhibitors are present. Stensel, *et al.* [40] have outlined a simple procedure for detecting the presence of inhibitory materials.

Table 20.3 also summarizes some performance characteristics of the individual reactor types. Chief among these is that packed towers normally cannot achieve low NH_4^+-N concentrations, regardless of system configuration. Rotating-disc reactors in separate-stage systems can, but often have difficulties meeting effluent guidelines during cold weather in combined systems. Finally, although CSTR's can achieve good nitrification in either type of system, their performance is strongly dependent upon the operation of the settler, as we will see in the next section.

20.3.2 Nitrification in CSTR's with Cell Recycle

SEPARATE-STAGE NITRIFICATION. A nitrification reactor may be designed as a separate-stage system whenever the wastewater being treated is low in organic matter, regardless of whether it has undergone previous treatment. In other words, separate-stage nitrification is characterized by a low T_bOD:TKN ratio, regardless of how the low value is achieved. Because of the many factors which can affect the kinetics

TABLE 20.3
Comparison of Nitrification Alternatives[a]

System Type	Advantages	Disadvantages
Combined carbon oxidation-nitrification		
CSTR's with cell recycle	Combined treatment of carbon and ammonia in a single system Very low effluent ammonia possible Inventory control of mixed liquor stable due to high T_bOD/TKN ratio	No protection against toxicants Only moderate stability of operation Stability linked to operation of secondary clarifier for biomass return Large reactors required in cold weather
Packed towers and rotating-disc reactors	Combined treatment of carbon and ammonia in a single system Stability not linked to secondary clarifier as organisms on media	No protection against toxicants Only moderate stability of operation Effluent ammonia normally 1-3 mg/liter (except RDR) Cold weather operation impractical in most cases
Separate-stage nitrification		
CSTR's with cell recycle	Good protection against most toxicants Stable operation Very low effluent ammonia possible	Sludge inventory requires careful control when low T_bOD/TKN ratio Stability of operation linked to operation of secondary clarifier for biomass return Greater number of unit processes required than for combined carbon oxidation-nitrification
Packed towers and rotating-disc reactors	Good protection against most toxicants Stable operation Less sensitive to low temperatures Stability not linked to secondary clarifier as organisms on media	Effluent ammonia normally 1-3 mg/liter Greater number of unit processes required than for combined carbon oxidation-nitrification

[a] *(Adapted from Parker et al. [6]).*

of nitrification it is risky to rely upon values of the kinetic parameters obtained
from the literature. It is best to perform laboratory or pilot plant treatability
studies to determine them for the particular wastewater being treated. Generally,
the growth of *Nitrosomonas* is rate limiting since it grows more slowly than
Nitrobacter. Consequently the values of μ_m and K_s are evaluated for *Nitrosomonas*
while Y_g and b are estimated for the entire population. Once the values have been
determined, the design may proceed exactly as described in Section 16.3 for activated sludge. The first thing the designer must do is to calculate the MCRT required to achieve the desired NH_4^+-N concentration. This may be done using Eq. 16.23.
Because K_s is small, the value may be close to θ_{cmin} as given by Eq. 13.28. Consequently many designers make the MCRT equal to θ_{cmin} times an appropriate safety
factor. Parker *et al.* [6] have recommended that the minimum safety factor should
equal the ratio of the peak mass flow rate of ammonia to the average mass flow rate
in order to prevent the discharge of excess NH_4^+-N during the peak. The more conservative the design, the larger the safety factor will be, but it usually is in excess
of 2.5. Since the objective of the safety factor approach to design is to provide
stable operation under all conditions, it follows that the values of the kinetic
parameters used to calculate θ_{cmin} should reflect the worst anticipated operational
conditions.

Once the design value of the MCRT has been chosen the design may continue as
outlined in Section 16.3. The next task is to calculate the $M\tau$ product using
Eq. 13.67. In this case S_o and S represent the concentrations of NH_4^+-N in the influent and effluent, respectively, and Y_g is the yield based upon the amount of NH_4^+-N
oxidized to NO_3^--N. If soluble organic matter is present in the influent, its contribution to the MLSS can be calculated by using Eq. 13.38 with the appropriate kinetic
parameters and organic substrate concentrations. The value of $X\tau$ calculated with it
may be added directly to $M\tau$ calculated from Eq. 13.67 to obtain a total $M\tau$ product.
Furthermore, if heterotrophic bacteria are present in the influent, their contribution to the MLSS may be calculated using the approach in either Chapter 16 or 17,
depending upon their concentration in relation to the soluble T_bOD present. If their
concentration is large then the approach of Chapter 17 should be used. After the
total $M\tau$ product has been calculated, either M or τ may be selected, thereby fixing
the other and allowing computation of the reactor volume. The recycle ratio, α,
may be calculated with Eq. 16.25 using an appropriate value for the concentration
of MLSS in the recycle flow. The wastage rate, F_w, may be calculated from either
Eq. 13.14 or 13.23, depending upon the point within the system from which sludge is
wasted. The amount of excess sludge to be disposed of (P_M) can be calculated from
either Eq. 13.48 or 13.49, depending upon the point of sludge wastage (M_r should be
substituted for X_r and M for X to get P_M). The oxygen requirement for oxidation of
the nitrogen can be calculated by multiplying the *peak* mass flow rate of nitrogen
by a factor of between 4.33 and 4.57 mg O_2 per mg N oxidized. If the safety factor is

small so that the design MCRT is near θ_{cmin}, then 4.33 should be used, but as the
safety factor is made larger, the factor should be made larger to account for the
subsequent cell decay. If soluble T_bOD or microbial cells are in the influent,
then their contribution to the total oxygen requirement can be calculated from
either Eq. 13.47 or Eq. 17.19, whichever is appropriate.

The constraints which must be imposed upon the design of an activated sludge
reactor, as discussed in Section 16.3.3, must also be imposed upon a nitrification
reactor.

The factors which influence the design of an activated sludge reactor, as discussed in Section 16.3.4, also influence the design of a nitrification reactor. In
addition, several other factors should be given special emphasis. One is the state
of the environment in which the nitrifiers will grow; specifically, the pH, temperature, and DO level. As discussed in Section 20.2.1, all influence the kinetics of
nitrification and thus must be considered when computing θ_{cmin}.

A second factor of importance is the loss of solids in the final effluent.
Equation 13.51 allows computation of P_x, the mass of nitrifying bacteria formed
during oxidation of NH_4^+-N, which is equal to the mass of cells which must be wasted
daily at steady state to maintain the desired MCRT. If the value of P_x is divided
by the influent flow rate, F, the result is the maximum concentration of nitrifying
bacteria that can be lost unintentionally in the final effluent while still maintaining the desired steady-state MCRT. Estimation of that concentration for nitrification of normal domestic sewage will illustrate the importance of this factor.
The value of Y_g for *Nitrosomonas* was seen earlier to lie between 0.03 and 0.13 mg
cells formed/mg NH_4^+-N oxidized whereas the value for *Nitrobacter* lies between 0.02
and 0.07 mg cells formed/mg NO_2^--N oxidized. Consequently, the overall Y_g for the
nitrifying population lies between 0.05 and 0.20 mg cells formed/mg NH_4^+-N oxidized
to NO_3^--N. Furthermore domestic wastewater contains approximately 25 mg/liter NH_4^+-N
after normal secondary treatment. If it were then treated in separate-stage nitrification the maximum concentration of nitrifying bacteria which could be lost in the
final effluent would therefore be approximately 5 mg/liter. Very few final settling
tanks can produce an effluent containing that few suspended solids. Consequently,
if nitrifying bacteria were the only source of suspended solids in the reactor, it
would be impossible to operate separate-stage nitrification. Complicating the issue
is the observation that nitrifying bacteria do not settle well [41]. In fact, it
has even been suggested that a high fraction of the MLSS must be heterotrophic to
maintain good bioflocculation [42]. Consequently, bacteria from some other source
must enter the reactor. One choice is to use all or part of the heterotrophic
bacteria which are wasted from the preceding activated sludge reactor. Alternatively,
the wastewater could be given only partial treatment prior to nitrification so that
heterotrophs would grow in the nitrification reactor, thereby contributing to the
sludge in it. If the fraction of nitrifiers in the suspended solids escaping from

a final settler is the same as the fraction in the MLSS, then the maximum allowable fraction of nitrifiers in the MLSS, ψ_{max}, is equal to the allowable concentration of nitrifiers in the effluent divided by the expected concentration of suspended solids in the effluent. Figure 20.10 shows the effect of the influent NH_4^+-N concentration on that fraction for a system in which the expected effluent suspended solids concentration is 20 mg/liter. For that situation, the concentration of NH_4^+-N entering the reactor must be in excess of 117 mg/liter before the MLSS can be composed entirely of nitrifiers. Different values would result from different conditions; nevertheless, this discussion has illustrated that great care must be used when designing a CSTR with recycle for separate-stage nitrification of wastewaters containing low concentrations of NH_4^+-N. Application of the techniques discussed herein in combination with those of Chapters 16 and 17 will allow the design of a suitable system for nitrification of such wastewaters, however.

Another problem arises when the concentration of NH_4^+-N in the influent is high; the fact that the nitrifying bacteria are subject to both substrate and product inhibition. As pointed out in Section 20.2.1 this may require the use of a denitrification reactor (Chapter 22) between two nitrification reactors. In addition, a large amount of bicarbonate will be required for cell synthesis and for neutralization of the hydrogen ions formed (Eq. 20.1). If it is not provided, the reactor will

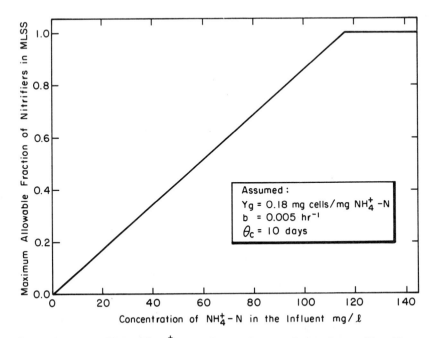

Figure 20.10. Effect of NH_4^+-N concentration on the maximum allowable fraction of nitrifiers in MLSS when the effluent suspended solids concentration is 20 mg/liter.

fail [43]. It is for these reasons that lab- or pilot-scale studies should always be performed on wastewaters containing concentrations of NH_4^+-N in excess of 300-400 mg/liter.

Although a single CSTR with cell recycle may be used satisfactorily for separate-stage nitrification, other reactor configurations may also be employed, such as several CSTR's in series with recycle around the entire chain or a long narrow aeration basin like that of TAAS. Because of the low K_s values associated with nitrification, reactors in series will achieve essentially complete nitrification and may be more stable than single CSTR's. The design can be approached in the same manner as for the single reactor, thus giving an additional safety factor. High purity oxygen systems may also be used, provided pH control is practiced [44], but a reactor system like CSAS should not be used.

Now let us consider an example of separate-stage nitrification of a wastewater that has been treated by CMAS.

EXAMPLE 20.3.2-1

Determine whether a CSTR with cell recycle can be used to nitrify an activated sludge effluent containing an average of 30 mg/liter NH_4^+-N to a concentration no greater than 0.75 mg/liter. The peak NH_4^+-N concentration is 45 mg/liter and the wastewater flow rate is 100 m^3/hr (10^5 liters/hr = 634,000 gal/day). It is estimated that the cold weather temperature will be 10°C. The growth of *Nitrosomonas* is assumed to be rate limiting with a μ_m value of 0.02 hr^{-1} and a K_s value of 2.0 mg/liter at 15°C. The combined Y_g for the two genera is assumed to be 0.10 mg cells/mg NH_4^+-N oxidized and b is 0.002 hr^{-1}.

Solution:

The first task is to correct the parameter values for the most severe conditions likely to be occurring in the reactor. Equation 20.4 can be used to estimate μ_{m10} by using a K value of 0.10:

$$\mu_{m10} = 0.02 \exp[0.10(10 - 15)]$$

$$\mu_{m10} = 0.012 \ hr^{-1}$$

Likewise K_s can be estimated with Eq. 20.5:

$$K_s = 2.0 \exp[0.118(10 - 15)]$$

$$K_s = 1.11 \ mg/liter$$

The value of μ_m must also be corrected for the effect of pH, using Eq. 20.3. For this purpose it will be assumed that the steady-state pH will be maintained at 7.0. Thus

$$\mu_m = 0.012[1 - 0.833(7.2 - 7.0)]$$

$$\mu_m = 0.010 \ hr^{-1}$$

The dissolved oxygen concentration will be kept at 3.0 mg/liter and thus will have little effect on the specific growth rate.

The MCRT required to achieve an effluent NH_4^+-N concentration of 0.75 mg/liter under severe conditions may be calculated with Eq. 16.23:

$$\theta_c = \frac{1.11 + 0.75}{0.75(0.010 - 0.002) - 1.11(0.002)}$$

$$\theta_c = 492 \text{ hrs} = 20.5 \text{ days}$$

The minimum MCRT may be calculated with Eq. 13.28:

$$\theta_{cmin} = \frac{1.11 + 30.0}{30.0(0.010 - 0.002) - 1.11(0.002)}$$

$$\theta_{cmin} = 131 \text{ hrs} = 5.4 \text{ days}$$

Thus the safety factor is 3.75. The minimum safety factor should be equal to the ratio of the peak mass flow rate of ammonia to the average mass flow rate, which is 1.5. Thus a design MCRT of 492 hrs should give stable operation while providing the required effluent quality.

Because the concentration of ammonia nitrogen in the influent is low, we suspect that it will be impossible to maintain the required MCRT with only nitrifying bacteria in the reactor. The first step in checking this is to calculate the mass of nitrifying bacteria formed hourly using Eq. 13.51:

$$P_x = \frac{(10^5)(0.10)(30.0 - 0.75)}{1 + (0.002)(492)}$$

$$P_x = 1.47 \times 10^5 \text{ mg/hr}$$

The maximum allowable concentration of nitrifiers in the final effluent is P_x divided by the average flow. This is 1.47 mg/liter and it is highly unlikely that a final clarifier will produce an effluent that good. Consequently a CSTR with cell recycle can't be used unless additional solids are provided to increase the mass of MLSS under aeration. If we assume that the final effluent will contain 20 mg/liter of suspended solids, then the maximum allowable fraction of nitrifiers in the MLSS will be

$$\psi_{max} = \frac{1.47}{20}$$

$$\psi_{max} = 0.0735$$

The rest of the MLSS must be made up of other suspended solids. A logical choice would be to divert a portion of the sludge wasted from the preceding activated sludge unit to the nitrification reactor. Since the MCRT of the nitrification reactor is 20.5 days it will serve as an aerobic digester for that sludge. If that were done the design would proceed as outlined in Section 17.3 with provision

made for the utilization of oxygen by the nitrifiers. The oxidation of the NH_4^+-N
in the influent will require a peak oxygen transfer rate of:

$$RO_{TS} = (45 - 0.75)(10^5)(4.33)$$
$$RO_{TS} = 1.92 \times 10^7 \text{ mg } O_2/hr = 19.2 \text{ kg } O_2/hr = 42.4 \text{ lb } O_2/hr$$

In addition, the NH_4^+-N released by the heterotrophs as they undergo digestion will
also be oxidized. Each kg/hr of cells destroyed will release 0.105 kg/hr of N
(see Section 16.3.4), thereby requiring that 0.45 kg/hr of O_2 be provided for its
oxidation by the nitrifiers.

Finally, to answer the original question, a CSTR with cell recycle can be used
to nitrify the effluent, but in the process it becomes an aerobic digester!

COMBINED CARBON OXIDATION-NITRIFICATION. Carbon oxidation and nitrification
will occur concurrently in any activated sludge reactor provided that the MCRT and
the environmental conditions are conducive to the growth of nitrifiers. Generally
the T_bOD:TKN ratio of untreated wastewaters is high so that the bulk of the microorganisms in the MLSS will be heterotrophic. Consequently, the system will be
designed primarily as an activated sludge system, with special provision for the
nitrifiers. Since the design of activated sludge systems has already been covered
in Section 16.3 emphasis will be given in this section to the special problems associated with nitrification.

The values of μ_m and K_s for the two nitrifying genera may be determined from
the effects of the MCRT on the concentrations of NH_4^+-N and NO_2^--N in the lab- or
pilot-scale reactors. Evaluation of Y_g and b for the nitrifiers requires that the
fraction of the MLSS contributed by them be estimated. This can be done by use of
a specific activity test which compares under constant environmental conditions the
specific rate of nitrification in the activated sludge MLSS to the specific rate
in an enrichment culture of nitrifying bacteria [8]. The ratio of the specific rates
is equal to the fraction of nitrifying bacteria in the MLSS. Estimation of that
fraction at each MCRT allows the concentration of nitrifying bacteria to be determined, thereby providing the data needed to estimate Y_g and b with the standard
techniques of Section 13.4.

After the kinetic parameters for the heterotrophs and the autotrophs are known,
the design may proceed as outlined in Section 16.3 and the preceding part of this
section. In order to ensure nitrification at all times, the MCRT must be selected
to give an adequate safety factor under extreme environmental conditions. Because
the level of dissolved oxygen in the reactor influences the kinetics of nitrification, considerable thought should be given to the selection of the design DO value.
For example, it may be more economic to design for a high DO level than to provide
for the additional cell residence time that would be required at a lower DO. Once

the MCRT has been selected, the Mτ product for the activated sludge (Eq. 13.67) may be added to the Xτ product for the nitrifiers (Eq. 13.38) to get the total Mτ product. When applying Eq. 13.38 it should be recognized that not all of the nitrogen in the influent wastewater will be available to the nitrifiers. Part of it will be used by the heterotrophs for the synthesis of cell material and the amount can be estimated by using Eq. 16.31 in conjunction with the predicted COD removal. Failure to do this will result in an over-estimation of the amounts of oxygen required and alkalinity destroyed by the nitrifiers [45]. Once the total MLSS concentration is known, the recycle ratio, α, may be selected with Eq. 16.25 and the wastage rate, F_w, may be calculated from Eq. 16.26 or 16.27 without regard to the MLSS composition. The amount of excess sludge to be disposed of is the sum of the heterotrophic sludge as calculated with Eq. 13.68 and the autotrophic sludge as calculated with Eq. 13.51, each with the appropriate parameters. Finally, the oxygen requirement may be calculated by adding the peak heterotrophic oxygen requirement as calculated with Eq. 15.3 to the autotrophic demand as calculated by multiplying the *peak available* TKN mass flow rate by a factor of 4.33 to 4.56 mg O_2 per mg N oxidized.

The constraints and factors discussed in Chapter 16 and in the preceding part of this section apply in this case also. Special attention should be given to the expected loss of solids in the final effluent to ensure that nitrification is indeed possible. The technique outlined earlier can be used to compute the maximum allowable fraction of nitrifiers in the MLSS, ψ_{max}. As long as it is greater than the actual fraction the system will function properly. If it is not, action should be taken to get better clarification in the final settler because the better its performance the larger ψ_{max} will be. If after doing that, the fraction is still not satisfactory, another MCRT must be tried. If that too fails, an alternative process must be used because there is no way to alter the ratio of autotrophs to heterotrophs in the MLSS in this type of system. A final factor which must be considered is the possibility that denitrification in the final settler will cause the sludge to rise. This can be minimized by using positive displacement sludge removal from the bottom of the settler to minimize the time that the sludge is under low DO conditions. This was discussed in Section 16.4.1.

Provided that the wastewater is free of inhibitory materials, nitrification may be accomplished concurrently with carbon oxidation in any of the alternative activated sludge systems discussed in Chapter 16. If TAAS, or any of the other modifications in which a concentration gradient exists, is to be used, care must be exercised to ensure that adequate DO is available at the influent end of the reactor or nitrification will be limited by the oxygen concentration. Under extreme conditions nitrification could be restricted to only a portion of the reactor. In that situation the MCRT would have to be increased by a factor equal to the reactor volume divided by the volume of that portion of the reactor in which nitrification could occur. CSAS may also achieve nitrification if the MCRT is long enough. Unlike the other types of activated sludge, however, complete nitrification cannot be achieved at reasonable contact tank space times. The equations for CSAS given in Section 16.3.1 can be used to estimate the extent of nitrification expected.

EXAMPLE 20.3.2-2

Consider the completely mixed activated sludge system described in Example 16.3.5-1. In addition to the constituents listed in that example, the wastewater contains an average TKN concentration of 47.6 mg/liter. Evaluation of the growth characteristics of nitrifying bacteria in that wastewater has indicated that μ_m for *Nitrosomonas* is approximately 0.02 hr^{-1} and K_s is 1 mg/liter NH_4^+-N under summer conditions. Furthermore, the lumped Y_g for the two genera together is 0.12 mg cells formed per mg NH_4^+-N oxidized to NO_3^--N, while the lumped decay rate (b) is 0.002 hr^{-1}. Is nitrification likely to occur during the summer if the MCRT is maintained at the design value of 7 days (168 hrs)? If so, how much will the presence of the nitrifiers increase the MLSS concentration if τ is held constant at 10 hours and how much additional oxygen must be provided under average conditions?

Solution:

To determine if nitrification will occur we must calculate θ_{cmin} using Eq. 13.28. Before we can do that, however, we must determine the concentration of NH_4^+-N available to the nitrifiers. In Example 16.3.5-1 we saw that the effluent T_bOD was 4.2 mg/liter. This tells us that the majority of the organic matter in the influent has undergone biodegradation, thereby releasing the organic nitrogen as NH_4^+-N. Consequently we may assume that the waste contains 47.6 mg/liter NH_4^+-N. Not all of this will be available for nitrification, however, because part of it will have been used by the heterotrophic bacteria to synthesize new cell material. The amount required can be estimated with Eq. 16.30:

$$NR = \frac{0.105 \, Y_g}{1 + b\theta_c} \tag{16.30}$$

where Y_g and b are for the heterotrophic bacteria, and are 0.5 and 0.0035 hr^{-1} respectively. Thus,

$$NR = \frac{(0.105)(0.5)}{1 + (0.0035)(168)}$$

$$NR = 0.033 \text{ mg N/mg } T_bOD \text{ removed}$$

Referring to Example 16.3.5-1 we see that 520 mg/liter T_bOD are removed, thus 17.2 mg/liter NH_4^+-N are required for cell synthesis. This means that 30.4 mg/liter NH_4^+-N are available for nitrification. Using this we can now calculate θ_{cmin} from Eq. 13.28:

$$\theta_{cmin} = \frac{1.0 + 30.4}{30.4(0.02 - 0.002) - 1.0(0.002)}$$

$$\theta_{cmin} = 57.6 \text{ hours} = 2.4 \text{ days}$$

Since the design MCRT is 168 hours, the safety factor is 2.92 and nitrification will occur. Thus the concentration of NH_4^+-N in the effluent can be estimated with Eq. 13.27:

$$S = \frac{1.0[1/168 + 0.002]}{0.02 - [1/168 + 0.002]}$$

$$S = 0.7 \text{ mg/liter } NH_4^+\text{-N}$$

The contribution of the nitrifiers to the MLSS can be calculated with Eq. 13.38:

$$X\tau = \frac{0.12(30.4 - 0.7)}{1/168 + 0.002}$$

$$X\tau = 448 \text{ mg/(liter·hr)}$$

Since τ is equal to 10 hours, the concentration of nitrifiers is 44.8 mg/liter, thereby incresing the design MLSS concentration from 3557 mg/liter to 3601 mg/liter.

The steady-state oxygen requirement for nitrification can be estimated by multiplying the mass of NH_4^+-N oxidized by 4.33.

$$RO_{SS} = (30.4 - 0.7)(10^5)(4.33)$$

$$RO_{SS} = 1.29 \times 10^7 \text{ mg } O_2/\text{hr} = 12.9 \text{ kg } O_2/\text{hr} = 28.5 \text{ lb } O_2/\text{hr}$$

The steady-state oxygen requirement for the removal of organic matter was 31.5 kg/hr, thus nitrification requires 41% more! Therefore, even though the presence of the nitrifiers only increases the MLSS concentration by 45 mg/liter it increases the average oxygen requirement from 31.5 kg O_2/hr to 44.4 kg O_2/hr.

20.3.3 Nitrification in Packed Towers

SEPARATE-STAGE NITRIFICATION. Most designers of packed tower nitrification reactors rely upon pilot plant data obtained on full height towers available from the media manufacturers. The period of pilot plant operation should encompass both winter and summer conditions since temperature influences system performance. As discussed in Section 20.2.3 a major design variable affecting performance is the surface area of media available for growth of the bacteria, and consequently one goal of the designer is to determine how the mass application rate of NH_4^+-N per unit of surface area will affect the degree of nitrification achieved. This is usually done by varying the application rate to a tower with a fixed area, and then extrapolating the results to other circumstances. One method of doing this would be to use the empirical model for trickling filters presented in Section 18.2.2. Since the use of that technique was discussed fully in Chapter 18 no further discussion will be presented here. If the influent NH_4^+-N concentration is relatively constant, an alternative approach would be to prepare design graphs directly from the pilot plant performance data. Parker *et al.* [6] have replotted the data of

Duddles and Richardson [35,36] to demonstrate how design graphs may be prepared for
packed towers. Their graph is reproduced as Fig. 20.11. Data from two other installations (not shown) were treated in a similar manner to demonstrate the general utility of the technique and to show that unique lines are obtained for each installation, thereby reflecting the nature of the wastewater undergoing treatment [6].

When a graph like Fig. 20.11 is available, the design procedure is as follows.
Using the effluent NH_4^+-N concentration desired, establish the area requirements from
the graph. Note that it is not possible to design for an effluent concentration
greater than 3.0 mg/liter using Fig. 20.11 because performance becomes independent of
area above that value. Using that requirement and knowledge of the average amount
of NH_4^+-N which must be oxidized daily, the surface area of media required can be
calculated. If the reactor will be subjected to diurnal loads the amount of surface
area determined from the graph must be multiplied by the ratio of the peak ammonia
load to the average load in order to ensure adequate performance during the peak
condition. The volume of media needed can be computed using its known specific
surface area. Then the depth and cross-sectional area of the tower can be selected
to meet the criteria of the media manufacturer. The use of recirculation may be considered if it is needed to maintain the minimum hydraulic application rate during

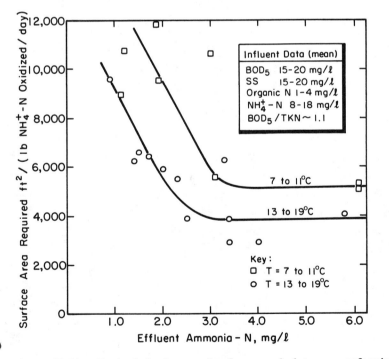

Figure 20.11.   Typical design graphs for a packed tower performing
                separate stage nitrification. *(Adapted from
                Parker et al. [6]).* The data is from Duddles and
                Richardson [35,36]. (1 ft^2/(lb/day) = 0.2 m^2/(kg/day)).

periods of low flow. In general, an amount of recirculation equal to the average
dry weather flow has been found to reduce fluctuations in performance [6]. The
necessity for a settler following the tower will depend upon the nature of the
waste being treated so that each situation must be judged on an individual basis.
As an example, if the influent contained 30 mg/liter of NH_4^+-N and the yield of
organisms were 0.18 mg cells/mg NH_4^+-N oxidized, only 5.4 mg/liter of cells would be
formed. Thus unless the influent contains either appreciable solids or sufficient
organic matter to cause the synthesis of heterotrophs within the tower, there would
be no need for a settler because settlers generally cannot reduce the solids concentration below that level anyway.

A factor likely to affect the performance of the tower is pH. Since fluid flow
is essentially plug flow, it is difficult to adjust the pH within the tower. Thus
consideration must be given to the amount of alkalinity that will be destroyed during
nitrification so that additional alkalinity can be made available in the infleunt
if needed.

EXAMPLE 20.3.3-1

Determine the surface area of media required in a packed tower to nitrify the wastewater used to obtain Fig. 20.11 to 1 mg/liter in summer (18°C) or to 2 mg/liter in
winter (8°C). The flow rate is 100 m^3/hr (10^5 liter/hr = 634,000 gal/day), and the
NH_4^+-N concentration is constant at 18 mg/liter. Also determine how many mg/liter of
suspended solids will be added to the liquid in the summer if the yield of nitrifiers
is 0.15 mg cells/mg NH_4^+-N oxidized.

Solution:

First calculate the mass of NH_4^+-N to be oxidized daily in summer and in winter.

Mass NH_4^+-N oxidized/day (summer) = $(18 - 1)(10^5)(24)$

$$= 40.8 \times 10^6 \text{ mg/day} = 40.8 \text{ kg/day} = 90 \text{ lb/day}$$

Mass NH_4^+-N oxidized/day (winter) = $(18 - 2)(10^5)(24)$

$$= 38.4 \times 10^6 \text{ mg/day} = 38.4 \text{ kg/day} = 85 \text{ lb/day}$$

Next enter Fig. 20.11 with the desired effluent concentrations in summer and winter
and read off the surface area requirement:

Ft^2 of area/lb NH_4^+-N oxidized/day (summer) = 9270

Ft^2 of area/lb NH_4^+-N oxidized/day (winter) = 9450

Multiply the area requirements by the mass of ammonia nitrogen to be oxidized to find
the actual area of media needed:

Area needed for summer = $(9270)(90) = 834,300 \text{ ft}^2$ $(77,500 \text{ m}^2)$

Area needed for winter = $(9450)(85) = 803,250 \text{ ft}^2$ $(74,600 \text{ m}^2)$

Thus summer operation is more severe and requires more media.

The increase in suspended solids during the summer may be calculated by multiplying the change in NH_4^+-N concentration by the yield:

Effluent suspended solids concentration = (18 - 1)(0.15)

$$= 2.55 \text{ mg/liter}$$

Thus a settler will not be needed unless other solids are present.

COMBINED CARBON OXIDATION - NITRIFICATION. It was seen in Fig. 20.6 that the degree of nitrification attained in a packed tower in which both carbon oxidation and nitrification are occurring is a function of the organic loading. Thus graphs like Fig. 20.6 can be generated from pilot plant data and then be used for design purposes. As seen in Chapter 18, organic loading is also used as the design parameter for trickling filters and thus the design procedure for a packed tower for combined carbon oxidation-nitrification is the same as for a trickling filter. Consequently, that material will not be repeated here, and the reader is instead referred to Section 18.3.

20.3.4 Nitrification in Rotating-Disc Reactors

SEPARATE-STAGE NITRIFICATION. Like a packed tower, the major design variable influencing the degree of nitrification obtained in a rotating-disc reactor is the surface area of media available for microbial growth. In this case, however, both mass loading and hydraulic loading have been found to be important. Consequently, design curves for a given influent concentration have been plotted as a function of the flow rate per unit area of disc. Just as with all of the other nitrification reactors, the safest approach is to develop the design curves from pilot plant data. Recently, however, sufficient data have become available to allow the development of generalized design curves for nitrification of typical domestic secondary effluents containing less than 15 mg/liter soluble BOD_5 and up to 30 mg/liter NH_4^+-N as shown in Fig. 20.12 [46]. It should be noted that the curves in that figure are applicable only to media manufactured by Autotrol Corporation. Because of differences in media configuration, mass transfer characteristics, etc., different curves must be used for other media. When curves like those in Fig. 20.12 are available and applicable, the design procedure is very simple. The curve corresponding to the average influent concentration is chosen and the design hydraulic loading corresponding to the desired effluent NH_4^+-N concentration is read from it. The design average daily flow rate in gallons per day is then divided by the hydraulic loading to determine the surface area of media required. As long as the ratio of the peak daily flow to the average daily flow does not exceed 2.5, no correction need be made for the normal diurnal flow pattern. This is because the effects of normal diurnal flow were incorporated into the curves. If the ratio exceeds 2.5, the design flow

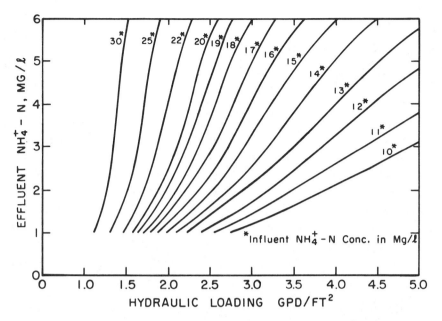

Figure 20.12. Design curves for nitrification in a rotating-disc reactor. *(Courtesy of Autotrol Corporation)*. (1 gpd/ft^2 = 0.091 m^3/(day·m^2)).

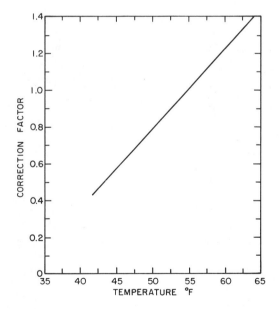

Figure 20.13. Temperature correction factors for nitrification in a rotating-disc reactor. *(Courtesy of Autotrol Corporation)*.

rate should be made equal to the peak divided by 2.5 [46]. If the design temperature is other than 55°F, then the temperature correction factor in Fig. 20.13 must be multiplied times the hydraulic loading obtained from Fig. 20.12 before the area is determined. The area obtained from this design procedure is the total wetted surface area for a four-stage system. The manufacturer's guidelines should be followed for determining the distribution of the area among the stages.

One problem frequently encountered in separate-stage nitrification is the maintenance of adequate alkalinity. Because the liquid in each stage of a rotating-disc reactor is well mixed, pH and alkalinity control can be practiced in each stage as needed. This can be particularly important when wastewaters containing a high concentration of NH_4^+-N are being treated. (Special design procedures must be used for wastewaters containing more than 30 mg/liter NH_4^+-N; see the manufacturer's literature [46]). Finally, just as with a packed tower, a final settler may not be needed if the concentration of NH_4^+-N being oxidized is low. The concentration of suspended solids formed may be estimated by the procedure used with the packed tower.

EXAMPLE 20.3.4-1

Determine the surface area of media required in a four-stage rotating-disc reactor to accomplish the same thing as the packed tower in Example 20.3.3-1.
Solution:

First consider summer operation when the temperature is 18°C (64°F). The influent NH_4^+-N concentration is 18 mg/liter and the desired effluent concentration is 1 mg/liter. Thus, from Fig. 20.12 the uncorrected hydraulic loading is 1.72 gal/(day·ft^2). Since the operating temperature is not 55°F, the temperature correction factor of 1.4 from Fig. 20.13 must be applied to the hydraulic loading, giving a design value of 2.4 gal/(day·ft^2). Since the flow rate is 634,000 gal/day, the required surface area for summer operation is:

Area = 634,000/2.4 = 264,000 ft^2

The same approach is used for winter operation when the temperature is 8°C (46°F). The influent NH_4^+-N concentration is the same, but the allowable effluent is now 2 mg/liter. Thus, from Fig. 20.12 the uncorrected hydraulic loading is 2.1 gal/(day·ft^2). Since the operating temperature is 46°F the temperature correction factor from Fig. 20.13 is 0.6. Thus the design hydraulic loading is 1.3, giving an area of:

Area = 634,000/1.3 = 488,000 ft^2

The winter area is larger and thus the reactor must be designed for winter conditions.

COMBINED CARBON OXIDATION-NITRIFICATION. Experience gained with full-scale rotating-disc reactors has shown that the nitrifying bacteria cannot compete effectively for space in the biofilm until the concentration of soluble organic matter is below 15 mg/liter as BOD_5 ($\sim$20 mg/liter T_bOD). Thus, because reactors in series are normally used, nitrification will occur only in the later stages after carbon oxidation has been accomplished. As a result, the design of a rotating-disc reactor system for both carbon oxidation and nitrification is a two step process. First, the procedures of Section 19.3 are used to size the reactors needed to reduce the concentration of soluble organic substrate to 20 mg/liter T_bOD and then the procedures outlined in the preceding part of this section are used to size the reactors needed to accomplish the desired degree of nitrification on that effluent. It should be noted that because reactors in series are used for both carbon oxidation and nitrification, the total number of reactors in the process train will most likely be large.

20.4 OPERATION

Operation of the activated sludge process, trickling filters, and rotating biological contactors was discussed in detail in Chapters 16, 18 and 19, respectively. The principles set forth therein are equally applicable here, but need not be repeated. There are a few specific problems which should be mentioned, however.

In Section 16.4 emphasis was given to the importance of correcting for the loss of microbial cells in the final effluent when determining the rate of sludge wastage from activated sludge reactors. Such correction is particularly important when nitrification is occurring because of the small amount of cell growth associated with nitrification and the necessity for maintaining the proper MCRT. In those corrections the assumption was made that the composition of suspended solids in the final effluent was the same as the composition in the reactor. Concern has been expressed by engineers that nitrifying bacteria may not settle as well as heterotrophic ones, thereby making their fraction in the effluent suspended solids greater than their fraction in the MLSS. Consequently, it would be prudent to apply a specific activity analysis [8] to both the effluent and mixed liquor suspended solids. If the fraction of nitrifiers in the effluent solids is found to be greater than the fraction in the MLSS, the amount of sludge wastage can be adjusted appropriately (using the techniques of Section 16.4) to maintain the MCRT of the nitrifying bacteria at the desired value. Two other operational factors of great importance to nitrifying CSTR's are the pH and dissolved oxygen concentration. As pointed out in Section 20.2.1 the rate of nitrification is quite sensitive to both. Thus steps should be taken to maintain the pH above 7.2 and the DO above 2.0 mg/liter throughout the reactor. All of the discussion in this chapter has categorized nitrification as either separate-stage or combined with carbon oxidation. Actually it may be advantageous to shift from one operating mode to another, depending upon the conditions existing at any given time.

Consequently Wilson [47] devised the MID (Mono, Intermediate, or Dual) step nitrification process which provides great flexibility for operation. Such flexible designs
have much to recommend them.

   Flexible design has also been recommended for nitrification in rotating-disc
reactors [48]. Generally such reactors are operated with the stages in series. When
this was done with a wastewater containing a high concentration of NH_4^+-N it was observed that most of the nitrogen was oxidized in the first stage during warm weather
so that the later stages had relatively little growth on the discs. As the wastewater temperature was reduced, however, the rate in the first stage decreased until
it was no longer large enough to oxidize all of the NH_4^+-N and some of it carried
over to the other stages. At that point, however, the growth rates of the nitrifiers
were so low that establishment of the required microbial film in the later stages
was difficult and the effluent quality deteriorated. To combat this it was recommended that the rotating-disc reactor be operated with the stages in parallel during
warm weather so that all discs would have growth established on them. Then during
cold weather the operation could be shifted back to series operation, thereby taking
advantage of the films on all discs.

20.5 EXAMPLES IN PRACTICE

Although Europeans have long been operating wastewater treatment plants to achieve
nitrification, experience with such plants within the United States has been fairly
recent. Parker et al. [6] summarized the characteristics of 22 nitrification facilities, but of those, only 9 were actual operating plants; four others were in the
design phase, and the remaining 9 were pilot plants. Table 20.4 summarizes the major
characteristics of the 13 plants that were either operating or in design, plus two
others discussed in reference [6]. Seven of the plants used combined carbon oxidation-nitrification while the rest used separate-stage nitrification. In all cases
where the MCRT was reported, the values reflect the slow growth characteristics of
the nitrifying bacteria. In almost all cases, however, the plants were able to
achieve nitrification within the range of MCRT's normally associated with activated
sludge. This implies that more activated sludge plants could achieve nitrification
if adequate oxygen transfer capacity were available and the pH were within the proper
range. Space does not permit a discussion of each of these plants. Parker et al. [6]
however, devote considerable space to several of them, and the reader is urged to
consult them for more details about the design features.

   One thing that has characterized the design of most nitrification facilities is
the extensive use of pilot plants for the development of design parameters. This has
been necessitated both by the uncertainty associated with the rates of nitrification
in any given wastewater and by the large number of options that are available. An
example of a detailed pilot plant study is that done for the city of Indianapolis,
Ind. [49]. Both combined carbon oxidation-nitrification and separate-stage

TABLE 20.4
Characteristics of Nitrification Facilities[a]

No.	Location	Flow MGD	$\dfrac{BOD_5}{TKN}$	MCRT Days	Classification Combined	Classification Separate	Type of Pretreatment	Reference No.
			Suspended Growth Reactors					
1.	Central Contra Costa San Dist., Ca.	30	2.4	20	*		Lime-Primary	50
2.	Livermore, Ca.	3.3	2.8	4.2-9.6		*	Roughing Filter	51
3.	Valley Community Services Dist., Ca.	3.8	10.8	8.1-15.2	*		Primary	50
4.	Whittier Narrows, LACSD, Ca.	12	6.6	9.4-40.2	*		Primary	52
5.	San Pablo San. Dist., Ca.	12.5	---	6.0-15.4		*	Roughing Filter	53
6.	Flint, Mich.	34	5.5	>6.0	*		Primary	54
7.	Jackson, Mich.	17	9	10.3-18.6	*		Primary	55
8.	El Lago, Texas	0.3	---	10		*	Roughing Filter	56
9.	Hyperion, Los Angeles, Ca.	46	7.3	---	*		Primary	57
10.	Manassas, Va.	0.2	1.2	---		*	Activated Sludge	58
11.	Blue Plains, D.C.	309	1.3-3.0	---		*	Activated Sludge	44
12.	Tampa, Fla.	60	3.0	---		*	Activated Sludge	59
13.	New Market, Ontario	2.4	2.6	---		*	Lime-Primary	60
			Packed Towers					
14.	Stockton, Ca.	58	5.3	---	*		Primary	61
15.	Allentown, Pa.	40	1.9	---		*	Trickling Filter	62

[a] Adapted from Parker et al. [6].

nitrification were studied using various combinations of rotating-disc reactors, plastic media packed towers, air activated sludge, and pure oxygen activated sludge. The process alternatives which could consistently achieve the desired degree of nitrification contained fixed-film roughing systems designed for about 50% soluble T_bOD removal followed by activated sludge operated at an MCRT appropriate for nitrification. Analyses were performed on the cost effectiveness of all combinations available using rotating-disc reactors and packed towers for roughing and air or oxygen activated sludge for nitrification. The best design was determined to be a packed tower followed by pure oxygen activated sludge which was capable of operating at a winter MCRT twice the summer value. No intermediate clarification was used between the towers and the activated sludge because the dense solids produced in the towers helped to improve the zone settling properties of the activated sludge. This study, as well as the ones summarized by Parker *et al.* [6], is an excellent example of the extensive engineering analyses required for the design of a nitrification facility. Although such studies are expensive, they represent the application of sound environmental engineering principles which will help ensure that the resultant facilities operate in the manner intended.

20.6 KEY POINTS

1. Nitrification involves the sequential conversion of NH_4^+-N to NO_2^--N and NO_3^--N by bacteria of the genera *Nitrosomonas* and *Nitrobacter*, respectively.

2. Nitrification may be performed in CSTR's with cell recycle, packed towers, and rotating-disc reactors.

3. Nitrification systems are designed to meet one objective, to convert NH_4^+-N to NO_3^--N.

4. The kinetics of nitrification can be expressed by the same equations used to describe heterotrophic growth, but because of the autotrophic nature of nitrification, the parameter values (μ_m, K_s, Y_g, b) are considerably different.

5. The growth kinetics of nitrifying bacteria are strongly influenced by the concentration of dissolved oxygen, the pH, and the temperature. Attempts have been made to quantify those effects so that they can be considered during design.

6. Although some organic compounds are inhibitory to nitrifying bacteria, in general the presence of organic compounds will have no deleterious effect upon their growth.

7. Because of the low K_s values associated with the growth of nitrifying bacteria, nitrification in CSTR's at steady state is an almost all-or-none phenomenon. Dynamic conditions, high concentrations of NH_4^+-N in the influent, or the presence of inhibitory compounds can alter that response, however.

8. The models presented in Chapter 13 for CSTR's with cell recycle satisfactorily predict the degree of nitrification obtained in reactor systems similar to those used for completely mixed activated sludge, conventional or tapered aeration activated sludge, and contact stabilization activated sludge.

9. Because of competition for space within the biofilm, nitrification will only occur in a packed tower when the organic loading is low. Consequently, the major parameter currently in use for predicting the extent of nitrification in a combined carbon oxidation-nitrification tower is the organic loading.

10. The degree of nitrification achievable in a packed tower performing separate-stage nitrification can be correlated with the hydraulic loading.

11. As long as the concentration of soluble organic matter exceeds 20 mg/liter T_bOD, nitrifying bacteria cannot compete effectively for space on the discs of a rotating-disc reactor and nitrification will be minimal.

12. For a given wastewater, once the concentration of soluble organic matter has been reduced to less than 20 mg/liter T_bOD, the degree of nitrification achieveable in a rotating-disc reactor can be correlated with the hydraulic loading.

13. An important decision which must be made by a design engineer is whether to use a separate-stage or a combined carbon oxidation and nitrification system. The information in Table 20.3 will aid in that decision.

14. The approach employed during design of a CSTR with cell recycle for nitrification is very similar to the one used for activated sludge, as discussed in Chapter 16.

15. Because of the low cell yields of nitrifying bacteria and the clarification inefficiencies normally associated with final settlers, it may be difficult to maintain the proper MCRT in a CSTR with cell recycle performing separate-stage nitrification. The problem can often be alleviated, however, by adding heterotrophic bacteria to the reactor.

16. If the T_bOD:TKN ratio in an untreated wastewater is high, the bulk of the microorganisms in a CSTR with cell recycle in which carbon oxidation and nitrification are occurring will be heterotrophic. Consequently, such a system will be designed primarily as an activated sludge system, with special provisions for the nitrifiers.

17. For separate-stage nitrification in a packed tower, pilot plant studies are normally performed using full height towers available from the media manufacturers. The resultant data can then be developed into a design graph like the one shown in Fig. 20.11.

18. The design procedure for a packed tower in which both carbon oxidation and nitrification are occurring is the same as the procedure for a trickling filter, as described in Chapter 18.

19. An estimate of the media area required to achieve separate-stage nitrification of typical secondary effluents in a rotating-disc reactor can be obtained from Fig. 20.12. Pilot plant studies are required to obtain more precise design values.

20. Because carbon oxidation and nitrification occur sequentially in a rotating-disc reactor the design of a combined system is a two step process. The initial stages are designed to remove the organic matter and the latter stages are designed for nitrification.

21. Because of the many factors which can influence the kinetics of the nitrifying bacteria, it is important that all nitrification systems be built to provide maximum operational flexibility.

20.7 STUDY QUESTIONS

The study questions are arranged in the same order as the text.

Section	Question
20.1	1
20.2	2-8
20.3	9-16
20.4	17

1. A wastewater contains 25 mg/liter of NH_4^+-N. How much oxygen will be required for its conversion to NO_3^--N? How much bicarbonate alkalinity will be destroyed?

2. State representative values of K_s for NH_4^+-N and dissolved oxygen for nitrifying bacteria and discuss the implications of their relative magnitudes.

3. Explain why pH control and the provision of adequate alkalinity are important to the proper operation of a nitrification reactor.

4. Explain how the presence of heterotrophic bacteria might have some effect upon the rates of nitrification.

5. Explain why the presence of a high concentration of NH_4^+-N in the influent to a CSTR in which nitrifying bacteria are growing can prevent the achievement of complete nitrification.

6. Explain the difference between separate-stage and combined carbon oxidation-nitrification systems using CSTR's with cell recycle, and the circumstances under which their performances are likely to be different.

7. Explain why the extent of nitrification achieved in a packed tower will depend upon whether biodegradable organic matter is in the influent whereas it will not in a CSTR with cell recycle.

8. Explain why the presence of organic matter in the influent to a rotating-disc reactor would make the degree of nitrification less than it would be if no organic matter were present.

9. Discuss the relative merits of separate-stage and combined carbon oxidation-nitrification systems.

10. Explain why it may be difficult to maintain the desired MCRT in a CSTR with cell recycle being used for separate-stage nitrification.

11. The effluent from an activated sludge system flows at a rate of 1000 m^3/hr and contains 15 mg/liter of bacteria, 5 mg/liter of inert solids, and 30 mg/liter of NH_4^+-N. Lab studies indicate that *Nitrosomonas* is rate limiting with $\mu_m = 0.011$ hr^{-1} and $K_s = 0.6$ mg/liter NH_4^+-N at the worst expected wastewater temperature. The combined Y_g for the two genera of nitrifying bacteria is 0.15 mg cells/mg NH_4^+-N oxidized and b is 0.003 hr^{-1}. The value of b for the heterotrophic bacteria in the influent is 0.005 hr^{-1} and it may be assumed that all of them are biodegradable and subject to decay. It has been suggested that a CSTR with cell recycle be used to reduce the NH_4^+-N concentration to 0.50 mg/liter by nitrification. What is the maximum concentration of suspended solids that could be lost in the effluent from the nitrification settler? State all assumptions made in your solution.

12. Assuming that a settler could be built to achieve the necessary effluent suspended solids concentration in Study Question 11, how many kg/hr of oxygen must be supplied to the reactor? State all assumptions.

13. Consider the situation described in Study Question 16 of Chapter 16. In addition to the constituents listed there, the wastewater contains 50 mg/liter TKN. Additional laboratory studies indicated that nitrification could occur concurrently with carbon removal in this wastewater and that μ_m for *Nitrosomonas* was 0.013 hr^{-1} while K_s was 1.0 mg/liter NH_4^+-N. Furthermore, Y_g for the two nitrifying genera combined was 0.20 mg cells/mg NH_4^+-N oxidized and b was 0.002 hr^{-1}. If a CMAS system is designed with a mean cell residence time of eight days, will nitrification occur? If it will, determine the following: (a) the concentration of NH_4^+-N in the effluent; (b) the concentration of NO_3^--N in the effluent; (c) the fraction of nitrifying bacteria in the MLSS; and (d) the additional quantity of oxygen which must be supplied. If it will not, determine the MCRT required to reduce the NH_4^+-N concentration to 1 mg/liter and answer (b), (c), and (d) above.

14. List and discuss the special provisions which must be made for the nitrifiers during design of a combined carbon oxidation-nitrification system using a CSTR with cell recycle.

15. A wastewater which has been treated by activated sludge must be nitrified before discharge to a stream. Consideration is being given to the use of a packed tower and pilot plant studies have yielded the data plotted in Fig. 20.11. The wastewater flows at a constant rate of 325 m^3/hr but the concentration of NH_4^+-N fluctuates about a mean value of 12 mg/liter, attaining peaks of 18 mg/liter frequently. If the two curves in the figure represent winter and summer operation, estimate the surface area of media required to attain an effluent NH_4^+-N concentration of 2.5 mg/liter in winter and 1.5 mg/liter in summer.

16. Rework Study Question 15 for a rotating-disc reactor containing four stages.

17. Explain why data relating to the fraction of nitrifying bacteria in the MLSS and in the final effluent suspended solids is important to proper operation of a CSTR with recycle within which nitrification is occurring.

REFERENCES AND FURTHER READING

1. H. A. Painter, "Microbial transformation of inorganic nitrogen," *Progress in Water Technology*, <u>8</u>, 3-29, 1977.

2. W. Verstraete and M. Alexander, "Heterotrophic nitrification in samples of natural ecosystems," *Environmental Science and Technology*, <u>7</u>, 39-42, 1973.

3. D. D. Focht and A. C. Chang, "Nitrification and denitrification processes related to waste water treatment," *Advances in Applied Microbiology*, <u>19</u>, 153-186, 1975.

4. D. P. Kelly, "Autotrophy: Concepts of lithotrophic bacteria and their organic metabolism," *Annual Review of Microbiology*, <u>25</u>, 177-210, 1971.

5. W. Gujer and D. Jenkins, "The contact stabilization process - Oxygen and nitrogen mass balances," Report No. 74-2, Sanitary Engineering Research Laboratory, University of California at Berkeley, Feb. 1974.

6. D. S. Parker, *et al*. *Process Design Manual for Nitrogen Control*, U.S. Environmental Protection Agency, Technology Transfer, Oct. 1975.

7. R. A. Poduska and J. F. Andrews, "Dynamics of nitrification in the activated sludge process," *Proceedings of the 29th Industrial Waste Conference,* Purdue University Engineering Extension Series No. 145, 1005-1025, 1974.

8. E. G. Srinath, *et al.* "Nitrifying organism concentration and activity," *Journal of the Environmental Engineering Division, ASCE,* 102, 449-463, 1976.

9. P. M. Sutton, *et al.* "Nitrogen control: A basis for design with activated sludge systems," *Progress in Water Technology,* 8, 467-481, 1977.

10. C. S. Huang and N. E. Hopson, "Nitrification rate in biological processes," *Journal of the Environmental Engineering Division, ASCE,* 100, 409-422, 1974.

11. M. J. Stankewich Jr., "Biological nitrification with the high purity oxygenation process," *Proceedings of the 27th Industrial Waste Conference,* Purdue University Engineering Extension Series No. 141, 1-23, 1972.

12. C. A. Nagel and J. G. Haworth, "Operational factors affecting nitrification in the activated sludge process," Paper presented at the 42nd Annual Conference of the Water Pollution Control Federation, Dallas, Texas, Oct. 1969.

13. A. L. Downing and G. Knowles, "Population dynamics in biological treatment plants," in *Advances in Water Pollution Research,* Proceedings of the Third International Conference held in Munich, Germany, Sept. 1966, Vol. 2, edited by S. H. Jenkins and L. Mendia, Water Pollution Control Federation, Wash., D.C., 117-136, 1967.

14. G. M. Wong-Chong and R. C. Loehr, "The kinetics of microbial nitrification," *Water Research,* 9, 1099-1106, 1975.

15. K. L. Murphy, *et al.* "Nitrogen control: Design considerations for supported growth systems," *Journal of the Water Pollution Control Federation,* 49, 549-557, 1977.

16. G. Knowles, *et al.* "Determination of kinetic constants for nitrifying bacteria in mixed culture, with the aid of an electronic computer," *Journal of General Microbiology,* 38, 263-276, 1965.

17. E. Boulanger and L. Massol, "Etudes sur les microbes nitrificateurs," *Annals Institute Pasteur, Paris,* 18, 181-196, 1904.

18. M. I. H. Aleem, "The physiology and chemoautotrophic metabolism of *Nitrobacter agilis,*" Ph.D. Thesis, Cornell University, Ithaca, NY, 1959.

19. B. Boon and H. Laudelot, "Kinetics of nitrite oxidation by *Nitrobacter winogradsky,*" *Biochemistry Journal,* 85, 440-447, 1962.

20. A. C. Anthonisen, *et al.* "Inhibition of nitrification by ammonia and nitrous acid," *Journal of the Water Pollution Control Federation,* 48, 835-852, 1976.

21. S. C. Rittenberg, "The roles of exogenous organic matter in the physiology of chemolithotrophic bacteria," *Advances in Microbial Physiology,* 3, 159-196, 1969.

22. H. Heukelekian, "The effect of carbonaceous materials on nitrification," *Proceedings of the 4th International Congress on Microbiology,* 460, 1947.

23. M. R. Hockenbury, *et al.* "Factors affecting nitrification," *Journal of the Environmental Engineering Division, ASCE,* 103, 9-19, 1977.

24. T. G. Tomlinson, *et al.* "Inhibition of nitrification in the activated sludge process of sewage disposal," *Journal of Applied Bacteriology,* 29, 266-291, 1966.

25. G. T. Austin, "The industrially significant organic chemicals, Parts 1-9," *Chemical Engineering,* 81, No. 2, 127; No. 4, 125; No. 6, 87; No. 8, 86; No. 9, 143; No. 11, 101; No. 13, 149; No. 15, 107; No. 17, 115, 1974.

26. M. R. Hockenbury and C. P. L. Grady, Jr., "Inhibition of nitrification - Effects of selected organic compounds," *Journal of Water Pollution Control Federation,* 49, 768-777, 1977.

27. T. B. S. Prakasam, *et al.* "Nitrogen removal from a concentrated waste by nitrification and denitrification," *Proceedings of the 29th Industrial Waste Conference,* Purdue University Engineering Extension Series No. 145, 497-509, 1974.

28. T. R. Bridle, *et al.* "Start-up of a full scale nitrification-denitrification treatment plant for industrial waste," *Proceedings of the 31st Industrial Waste Conference, 1976, Purdue University,* Ann Arbor Science Publishers, Ann Arbor, Mich., 807-815, 1977.

29. W. Gujer and D. Jenkins, "A nitrification model for the contact stabilization activated sludge process," *Water Research,* 9, 561-566, 1975.

30. A. R. Joel and C. P. L. Grady Jr., "Inhibition of nitrification - Effects of aniline following biodegradation," *Journal of the Water Pollution Control Federation,* 49, 778-788, 1977.

31. C. S. Huang, "Kinetics and process factors of nitrification on a biological film reactor," Ph.D. Thesis, University of New York at Buffalo, 1973.

32. E. L. Stover, *et al.* "Inhibiting nitrification in wastewater treatment plants," *Water and Sewage Works,* 123, No. 8, 56-59, 1976.

33. T. Stones, "Investigation on biological filtration at Salford," *Journal of the Institute of Sewage Purification,* 5, 406-417, 1961.

34. K. Williamson and P. L. McCarty, "A model of substrate utilization by bacterial films," *Journal of the Water Pollution Control Federation,* 48, 9-24, 1976.

35. G. A. Duddles, *et al.* "Plastic medium trickling filters for biological nitrogen control," *Journal of the Water Pollution Control Federation,* 46, 937-946, 1974.

36. G. A. Duddles and S. E. Richardson, "Application of plastic media trickling filters for biological nitrification," *USEPA, Environmental Protection Technology Series,* Report Number EPA-R2-73-199, June 1973.

37. F. F. Sampayo, "How to get low ammonia effluents," *Water and Sewage Works,* 121, No. 8, 92-94, 1974.

38. E. L. Stover and D. F. Kincannon, "One-step nitrification and carbon removal," *Water and Sewage Works,* 122, No. 6, 66-69, 1975.

39. R. L. Antonie, "Nitrification of activated sludge effluent: Bio-Surf process," *Water and Sewage Works,* 121, No. 11, 44-47; No. 12, 54-55, 1974.

40. H. D. Stensel, *et al.* "An automated biological nitrification toxicity test," *Journal of the Water Pollution Control Federation,* 48, 2343-2350, 1976.

41. W. C. Hutton and S. A. La Rocca, "Biological treatment of concentrated ammonia wastewaters," *Journal of the Water Pollution Control Federation, 47*, 989-997, 1975.

42. M. W. Tenney and W. F. Echelberger, "Removal of organic and eutrophying pollutants by chemical-biological treatment," *USEPA, Environmental Protection Technology Series,* Report Number EPA-R2-72-076, April 1972.

43. J. E. Barker and R. J. Thompson, "Biological removal of carbon and nitrogen compounds from coke plant wastes," *USEPA, Environmental Protection Technology Series,* Report Number EPA-R2-73-167, April 1973.

44. J. A. Heidman, "Experimental evaluation of oxygen and air activated sludge nitrification systems - With and without pH control," *USEPA Environmental Protection Technology Series,* Report Number 600/2-76-180, 1976.

45. J. H. Sherrard, "Destruction of alkalinity in aerobic biological wastewater treatment," *Journal of the Water Pollution Control Federation, 48*, 1834-1839, 1976.

46. "Design manual - Wastewater treatment systems," Autotrol Corporation, Bio-Systems Division, 1701 West Civic Drive, Milwaukee, Wisconsin, 53209, 1979.

47. T. E. Wilson, "Nitrogen control by means of the MID step biological nitrification process," *Progress in Water Technology, 8*, 495-507, 1977.

48. C. Lue-Hing, *et al.* "Biological nitrification of sludge supernatant by rotating-disks," *Journal of the Water Pollution Control Federation, 48*, 25-46, 1976.

49. R. E. Riemer, *et al.* "Pilot plant studies and process selection for nitrification, city of Indianapolis, Indiana," *Proceedings of the 31st Industrial Waste Conference, 1976, Purdue University,* Ann Arbor Science Publishers, Ann Arbor, Mich., 280-290, 1977.

50. G. A. Horstkotte *et al.* "Full-scale testing of a water reclamation system," *Journal of the Water Pollution Control Federation, 46*, 181-197, 1974.

51. W. E. Loftin, "Annual report, Livermore Water Reclamation Plant, 1970," City of Livermore, California, March, 1971.

52. County Sanitation Districts of Los Angeles County, "Monthly operating reports, Whittier Narrows Water Reclamation Plant," April 1973 to March 1974.

53. San Pablo Sanitary District, California, "Wastewater treatment plant operating reports," June 1973 to July 1974.

54. W. J. Beckman, *et al.* "Combined carbon oxidation-nitrification," *Journal of the Water Pollution Control Federation, 44*, 1916-1931, 1972.

55. R. A. Greene, "Complete nitrification by single stage activated sludge," Paper presented at the 46th Annual Conference of the Water Pollution Control Federation, Cleveland, Ohio, Oct. 1973.

56. Harris County Water Control and Improvement District No. 50, "Monthly progress report for October 1974, on Project 11010 GNM," prepared for the U.S. EPA.

57. City of Los Angeles, California, "Hyperion Treatment Plant West Battery operating reports," January through March, 1969.

58. M. C. Mulbarger, "Nitrification and denitrification in activated sludge systems,"
     Journal of the Water Pollution Control Federation, 43, 2059-2070, 1971.

59. D. Newton and T. E. Wilson, "Oxygen nitrification process at Tampa," In
     Applications of Commercial Oxygen to Water and Wastewater Systems, Edited by
     R. E. Speece and J. F. Malina, Jr., The Center for Research in Water Resources,
     Austin, Texas, 268-282, 1973.

60. S. A. Black, "Lime treatment for phosphorus removal of the New Market/East
     Gwillimbury WPCF," Paper No. W3032, Ontario Ministry of the Environment,
     Research Branch, May, 1972.

61. R. J. Steinquist, *et al.* "Carbon oxidation-nitrification in synthetic media
     trickling filters," *Journal of the Water Pollution Control Federation*, 46,
     2327-2339, 1974.

62. *Process Design Manual for Upgrading Existing Wastewater Treatment Plants*, U.S.
     EPA, Office of Technology Transfer, Washington, D.C. 1974.

63. S. Balakrishnan and W. W. Eckenfelder, "Nitrogen relationships in biological
     treatment processes - I - Nitrification in the activated sludge process,"
     Water Research, 3, 73-81, 1969.

64. K. Wuhrmann, "Nitrogen removal in sewage treatment processes," *XVth Verh.
     Internat. Verein. Limnol. Proc.*, 580, 1964.

65. W. K. Johnson and G. J. Schroepfer, "Nitrogen removal by nitrification and
     denitrification," *Journal of the Water Pollution Control Federation*, 36,
     1015-1036, 1964.

66. T. B. S. Prakasam and R. C. Loehr, "Microbial nitrification and denitrification
     in concentrated wastes," *Water Research*, 6, 859-869, 1972.

Further Reading

Benefield, L. D. and Randall, C. W., *Biological Process Design for Wastewater
     Treatment*, Prentice Hall, Inc., Englewood Cliffs, N.J., 1980. See Chapters
     4 and 7.

McCarty, P. L. and Haug, R. T., "Nitrogen removal from wastewaters by biological
     nitrification and denitrification," *Microbial Aspects of Pollution*, The
     Society for Applied Bacteriology Symposium Series, No. 1, edited by G. Sykes
     and F. A. Skinner, Academic Press, London, 215-232, 1971.

Painter, H. A., "A review of literature on inorganic nitrogen metabolism in micro-
     organisms," *Water Research*, 4, 393-450, 1970. A very complete review related
     to wastewater treatment.

Sharma, B. and Ahlert, R. C., "Nitrification and nitrogen removal," *Water Research*,
     11, 897-925, 1977. An overview of nitrification in wastewater treatment which
     emphasizes the literature which has appeared since the review by Painter,
     and integrates it with the older information.

Suzuki, I., "Mechanisms of inorganic oxidation and energy coupling," *Annual Review
     of Microbiology*, 28, 85-101, 1974.

CHAPTER 21

ANAEROBIC DIGESTION AND ANAEROBIC CONTACT

All of the biochemical operations discussed so far have been aerobic, i.e., the microorganisms in them use molecular oxygen as the terminal hydrogen acceptor. As discussed in Section 8.2.5, however, not all organisms must utilize oxygen, and in fact many can't. Metabolism in the absence of oxygen is called anaerobic. Anaerobic organisms have played a vital role in wastewater treatment through anaerobic digestion, which has been the primary means of degrading organic solids in aqueous systems. In addition, as shown in Table 1.2, anaerobic organisms have been used to remove soluble

organic matter, in an operation named anaerobic contact. Regardless of the objective, however, all anaerobic biochemical operations share a unique biological
feature - the formation of methane. This gas is rapidly discharged from the system
because of its low solubility in water resulting in a rather high degree of waste
stabilization. This unique feature governs the design and operation of anaerobic
biochemical operations to such a degree that the nature of the waste treatment
objective is less important than it is in aerobic operations. Consequently we will
consider anaerobic digestion and anaerobic contact together in this chapter.

21.1 DESCRIPTION OF ANAEROBIC OPERATIONS

Unlike aerobic operations, which contain diverse microbial communities and complex
food chains, anaerobic operations contain communities which are essentially totally
bacterial, as described in Section 7.4.1. In spite of this apparent simplicity,
interactions among the bacterial species have a severe effect upon system performance. Early attempts to design and control anaerobic operations were hampered by
our ignorance of these interactions, which caused the operations to acquire a reputation for being unreliable and difficult to control. As a better understanding of
the ecosystem was obtained it was recognized that the poor reputation was unwarranted,
leading to a resurgence of interest in anaerobic operations.

21.1.1 Microbiological and Biochemical Characteristics

General Nature of Anaerobic Operations. The multistep nature of anaerobic
operations is depicted in Fig. 21.1. Before insoluble organics can be consumed, they
must first be solubilized. In addition, large soluble organic molecules must be
reduced in size to facilitate transport across the cell membrane. The reactions
responsible for solubilization and size reduction are usually hydrolytic and are
catalyzed by enzymes which have been released to the medium by the bacteria.

The small molecules resulting from hydrolysis are used as carbon and energy
sources by bacteria which carry out fermentations. The oxidized end products of
those fermentations are primarily short-chain volatile acids such as acetic, propionic,
butyric, valeric, and caproic. Their production is referred to as acidogenesis and
the responsible organisms are called acid-producing bacteria. The reduced end products of the fermentation depend upon the nature of the culture and the environmental
conditions in the reactor. Some of the acid-producing bacteria possess a specialized
enzyme system which allows them to oxidize reduced coenzymes without passing the
electrons to an organic acceptor, thereby releasing hydrogen gas, H_2, to the medium.
As a result, these bacteria produce few reduced organic end products. In addition,
some of them utilize volatile acids larger than acetic, as well as reduced organic
compounds released by other bacteria, to produce acetic acid, CO_2, and H_2. The collective activity of these hydrogen-producing bacteria is called hydrogenogenesis.
Actually, the distinction between acid-producing and hydrogen-producing bacteria is

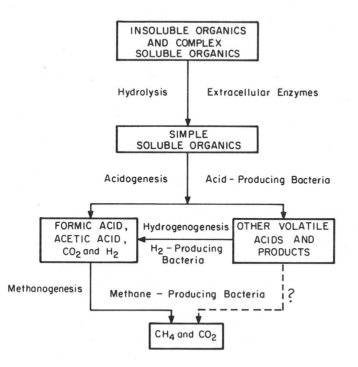

Figure 21.1. Multistep nature of anaerobic operations.

not clear. Since hydrogen-producing bacteria usually produce acids, but acid-producing bacteria do not all produce hydrogen, it is probably best to think of the hydrogen-producing bacteria as a subset of the acid-producing group. The combined groups of acid- and hydrogen-producing bacteria are generally referred to as non-methanogenic bacteria and their integrated metabolism results primarily in formic acid, acetic acid, CO_2, and H_2. If no hydrogen is formed, the nonmethanogenic phase results in insignificant reductions in COD because all electrons released in the oxidation of organic compounds are passed to organic acceptors which remain in the medium. Consequently, the energy level of the entire system is lowered only by losses due to microbial inefficiency. When hydrogen is formed, however, it represents a gaseous product which escapes from the medium thereby causing a reduction in the energy content, and thus the COD, of the liquid.

The products of the nonmethanogenic phase (i.e., formic acid, acetic acid, CO_2 and H_2) are utilized by methanogenic bacteria to produce methane gas. Although it is also possible that methane-producing bacteria exist which have the ability to use other volatile acids and organic end products to form methane, none have been isolated. Consequently that route to methane is shown as a dotted line in Fig. 21.1. With the exception of losses due to microbial inefficiency, almost all of the energy removed from the liquid is recovered in the methane. One mole of methane requires two moles of oxygen to oxidize it to CO_2 and water, consequently each 16 grams of methane produced and lost to the atmosphere corresponds to the removal of 64 grams of COD from the liquid [1]. At standard temperature and pressure, this corresponds to 5.62 ft^3 of methane for each pound of COD stabilized (0.34 m^3/kg) [2].

   NONMETHANOGENIC BACTERIA. The nonmethanogenic bacteria comprise a rather diverse group of facultative and obligately anaerobic bacteria. Although facultative bacteria were originally thought to be dominant, recent evidence indicates that the opposite is true [3,4,5,6], at least in sewage sludge digesters where the numbers of obligate anaerobes have been found to be over 100 times greater. This does not mean that facultative bacteria are unimportant, because their relative numbers can increase when the influent contains large numbers of them [7], or when the reactor is subjected to shock loads of easily fermentable substrates [8]. Nevertheless, it does appear that most important hydrolytic and fermentative reactions are carried out by strict anaerobes.

   Because of the diverse nature of anaerobic bacteria, a large number of end products can result from their metabolism. For example, Fig. 21.2 shows the many products which may arise from the fermentation of pyruvic acid, a key intermediate in

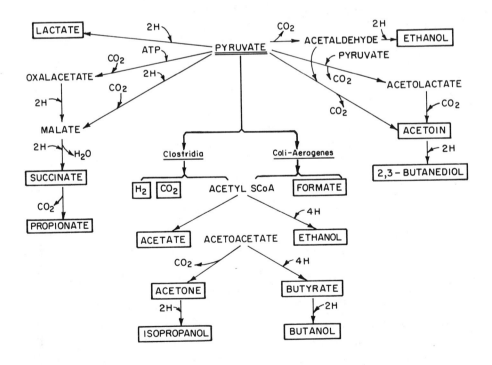

Figure 21.2. Fermentative product formation from pyruvate. *(Reprinted with permission from J.F. Andrews and E.A. Pearson, Int. J. Air & Wat. Poll., 9, 439-461, 1965, Pergamon Press Ltd.).*

the degradation of many compounds. Volatile acids may also arise from the fermentation of some amino acids and fats without passing through pyruvic acid as an intermediate. Although the major end products of the nonmethanogenic bacteria are the
short chain volatile acids (with acetic, propionic, and butyric being the most important) the relative concentrations of the various products are influenced by both
the environmental conditions (pH, temperature, electrode potential, etc) and the
specific growth rate imposed upon the culture [9]. This is due both to shifts in
the predominant species within the population, and to internal changes within individual organisms.

One of the most important internal changes can occur within the hydrogen-producing bacteria. If they are not producing hydrogen they must use organic compounds
as acceptors for the electrons removed during biological oxidation. As suggested
by Fig. 21.2, this results in the formation of reduced products such as butanol,
ethanol, lactic acid, and succinic acid, in addition to the oxidized product, acetate. If, on the other hand, a hydrogenase enzyme system is active, the electrons
may be transferred to hydrogen ions, forming molecular hydrogen (H_2) [10]

$$2 \; e^- + 2 \; H^+ \rightarrow H_2 \qquad\qquad\qquad\qquad\qquad\qquad (21.1)$$

Consequently, few reduced organic compounds need be formed and acetate becomes the
major end product [11,12,13]. Furthermore, hydrogen-forming bacteria can presumably
use the reduced organics produced by other organisms, thereby converting them to
acetate as well. Thus, when organisms with active hydrogenase systems are present,
the production of acetate is maximized while the production of reduced end products
is minimized.

The exact quantitative contribution of hydrogenogenesis to anaerobic operations
has not yet been determined because it has only recently been recognized that the
hydrogenase enzyme system in some organisms is under very strict metabolic control
by molecular hydrogen [11,12,13]. For example, *Selenomonas ruminantium* will produce
only trace amounts of hydrogen in batch culture if the gas is allowed to accumulate,
but will produce much more if it is continually removed [12]. One study [14], performed before this was known, found that 25% of the nonmethanogenic bacteria isolated
from a sewage sludge digester were capable of forming hydrogen in batch cultures in
which it was allowed to accumulate. However, because hydrogen is continually removed from operating digesters by the methanogenic culture, thereby maintaining its
partial pressure at a low level [15], it is likely that a higher percentage of the
nonmethanogenic bacteria are capable of producing hydrogen in such reactors. This
could help explain why acetate is the major organic end product of the nonmethanogenic
bacteria in mixed cultures.

Few kinetic studies have been performed with pure cultures of nonmethanogenic
bacteria, but in general they are thought to have μ_m values around 1 hr^{-1}. The kinetics of mixed cultures of nonmethanogenic bacteria will be discussed in
Section 21.2.1.

METHANOGENIC BACTERIA. The exact nature of the bacteria responsible for the
production of methane in anaerobic operations is not yet certain although it appears
that two distinct groups are involved. One group obtains its energy from the oxidation of molecular hydrogen whereas the other group oxidizes acetate.

For many years the only methanogenic bacteria which had been isolated in pure
culture were those which oxidize hydrogen [16]. They are all strictly obligate
anaerobes which obtain their energy from that oxidation and their carbon from carbon
dioxide. Because of this autotrophic mode of life, their cell yield is low. During
their metabolism they also use carbon dioxide as the terminal electron (hydrogen)
acceptor, forming methane gas in the process:

$$4 H_2 + CO_2 \rightarrow CH_4 + 2 H_2O \qquad\qquad\qquad (21.2)$$

Several of the species can utilize formic acid as the sole substrate [16] but this
may be because it breaks down easily to yield hydrogen and carbon dioxide. When the
partial pressure of hydrogen is high and the pH is near neutrality, the hydrogen-
utilizing methanogenic bacteria can grow with μ_m values around 0.04 hr^{-1} [15]. They
are very sensitive to pH, however, and appear to be inhibited by values outside the
range 6.7 - 7.4 [1].

Most information about the group which oxidizes acetate has been obtained with
enrichment, rather than pure, cultures. One such culture was unable to utilize
hydrogen and carbon dioxide or formate as energy substrates although its response to
pH was similar to that observed with hydrogen-utilizing bacteria [17]. Furthermore,
its μ_m was an order of magnitude smaller than hydrogen-utilizing methanogenic bacteria
and it was more tolerant of dissolved oxygen, although it was an obligate anaerobe.
The one organism capable of utilizing acetate which had been isolated in pure culture was a *Methanosarcina* [16], but its rates were so low that its use of acetate
was generally considered to be of little importance. Mah *et al*. [18], however, have
succeeded in isolating another strain of that genus which is capable of growing on
acetate with μ_m values up to 0.03 hr^{-1}, which are comparable to μ_m values reported
for hydrogen utilization. Although this genus is also capable of oxidizing hydrogen,
labeling studies indicate that hydrogen oxidation is not involved in acetate utilization [18]. In fact, addition of hydrogen and bicarbonate to cultures growing on
acetate resulted in a reduction in the amount of acetate metabolized to methane.
The significant differences between this organism and the enrichment culture mentioned above suggest that several types of bacteria may be responsible for acetate
utilization.

Although it has long been known that acetate is a major precursor of methane
in anaerobic operations, the relative importance of the two possible groups of
methanogenic bacteria in its utilization is unknown. The fact that most hydrogen-
utilizing methanogenic bacteria cannot use acetate for an energy substrate does not
mean that they cannot use it to form methane because they have been observed to use

it as a terminal electron acceptor [19]. The proposed reaction is:

$$CH_3COOH + 4\ H_2 \rightarrow 2\ CH_4 + 2\ H_2O \qquad\qquad (21.3)$$

Thus acetate could serve as a precursor for methane as the methanogenic bacteria remove the hydrogen formed by the nonmethanogenic bacteria, thereby preventing the
accumulation of hydrogen with its resultant inhibition of the hydrogen-forming bacteria. The role of the hydrogen-utilizing methanogenic bacteria in an enrichment
culture degrading propionate has been shown to be as a sink for the hydrogen formed,
thereby allowing propionate degradation to continue [20]. Thus they probably play a
similar role in anaerobic operations thereby removing acetate in the process. Ace-
tate-utilizing bacteria may play the major role in methane formation by mixed cultures degrading complex substrates, however, since radioactive labeling experiments,
including those with an enrichment culture growing on benzoate [21], have indicated
that even though a portion of the methane was formed by oxidation of hydrogen, the
majority was formed by cleavage of acetate:

$$CH_3COOH \rightarrow CH_4 + CO_2 \qquad\qquad (21.4)$$

A similar cleavage mechanism was responsible for methane formation from acetate by
Methanosarcina [18], although it was not present in the benzoate enrichment culture [21]. This suggests that other organisms, as yet unidentified, play an important role in the conversion of acetate to methane.

It is apparent that much work remains to be done before the production of methane
in anaerobic operations is fully understood. Regardless of the mechanism of methane
formation, however, acetate is known to be the major precursor of methane in anaerobic
operations and most kinetic models have been written with it as the major intermediate.

21.1.2 Physical Characteristics

Anaerobic reactors are usually covered, circular concrete tanks, 20 to 100 ft
(6.1 - 30.5 m) in diameter, with vertical sidewalls, ranging in depth from 25 to 45
ft (7.6 - 13.7 m). The bottoms are generally sloped at a rate of 1 vertical to 4
horizontal so that dense solids which settle may be removed easily. The covers may
be either fixed or floating. A typical reactor with a floating cover is shown in
Fig. 21.3. Mixing may be accomplished by any of several techniques, such as sludge
recirculation, gas recirculation, mechanical draft-tube mixers, or turbine and pro-
peller mixers. The reactors are normally heated to 85-95°F (30-35°C) either by recirculating the digester contents through an external heat exchanger or by using an
internal heat exchanger through which the reactor contents move at high velocity to
prevent fouling of the heat exchange surface. Static heat exchange surfaces should
be avoided because thick cakes of solids will build up on them, thereby reducing efficiency. Influent is pumped to the reactor continuously or by time-clock on a
30 minute to 2 hour cycle and effluent is removed by displacement. Provision should

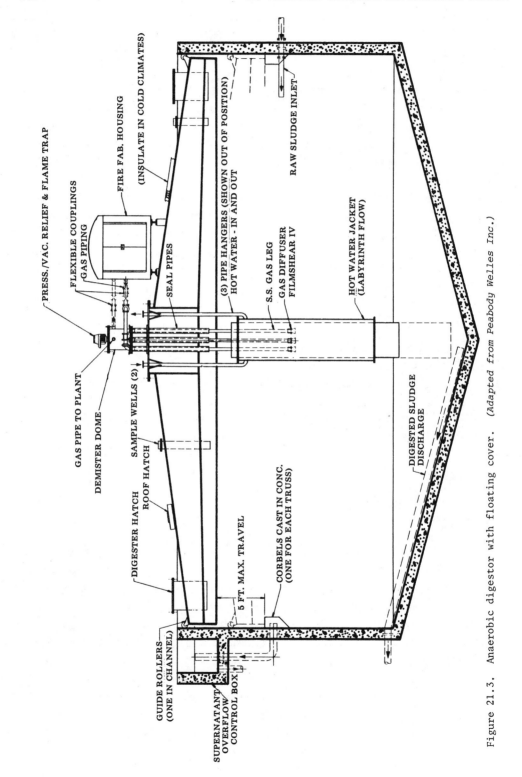

Figure 21.3.   Anaerobic digestor with floating cover.   (Adapted from Peabody Welles Inc.)

be made for withdrawing effluent at any of several levels within the reactor to help
ensure that the system is homogenously mixed.

In a simple CSTR without solids recycle, the effluent will flow either to a
holding tank or to a settler, depending upon the method of ultimate sludge disposal.
If a settler is employed, it is usually built like the reactor, with a space time of
several days but without heating or mixing. This allows gravity settling to occur,
so that supernatant may be continually withdrawn. The settled solids are withdrawn
periodically, depending upon the storage capacity for solids within the separator.

If solids recycle is to be employed, the design of the settler is more complicated. Generally high recycle ratios, such as 4:1, are used so that a positive displacement sludge return system is required. When gravity settling is employed, some
sort of degasifier is needed to remove gas bubbles entrapped in the solids so that
they will settle properly.

21.1.3 Applications

The traditional role of anaerobic operations has been the stabilization of organic
solids emanating from various points in the liquid processing train of a wastewater
treatment system. Anaerobic digestion is well suited to the stabilization of primary sludges, (i.e., those resulting from sedimentation of raw sewage) because the
large amount of nonmicrobial organic matter present can be reduced substantially with
a minimal production of biomass. It will also release the bound water, thereby
making the treated sludge easier to thicken. Secondary sludges can also be treated
anaerobically, but their relatively low solids concentration and high proportion of
microbial solids often makes them more amenable to aerobic digestion, as was discussed in Section 17.1.2.

Anaerobic operations have recently become accepted as an effective means of
treating high strength wastewaters. For example, it was pointed out in Section 16.1.2
that anaerobic contact is likely to be less expensive than activated sludge alone
for the treatment of wastewaters with a soluble T_bOD in excess of 4000 mg/liter.
Generally, anaerobic contact is used as a pretreatment technique prior to an aerobic
operation because the anaerobic effluent will contain too much residual organic matter
for direct discharge. As such, it has found application to wastewaters from the meat
packing, brewing, pharmaceutical, chemical, and food processing industries. As with
any biochemical operation, laboratory treatability studies should be performed to
ensure that the wastewater is amenable to treatment. This is particularly critical
with anaerobic operations because some organic compounds are resistant to anaerobic
degradation.

21.1.4 Merits and Defects

The main merits associated with anaerobic operations are directly attributable to the
special characteristics of anaerobic bacteria [2,22]. For example, because anaerobic
metabolism yields little energy for growth and because much of the energy in the

original substrate is released as product gases, only a small amount of the original
waste is converted to new cells. Furthermore, because of the low cellular yields,
relatively few inorganic nutrients are required making it possible to treat wastes
that have levels of nutrients which are marginal for aerobic operations. Of course,
an important benefit of anaerobic treatment is that the major product gas, methane,
is a valuable resource. The fact that no oxygen is required is also a merit. It
will be recalled from Section 16.3.3 that an upper limit was imposed on the MLSS
concentration (and consequently a lower limit was imposed on the reactor volume)
in an activated sludge reactor to prevent the power input per unit volume for
oxygen transfer from becoming so large as to disrupt the floc particles. Since
no oxygen is required in an anaerobic reactor decisions regarding the reactor volume
are not influenced by the limitations of oxygen transfer and proper mixing is the
only consideration when determining the required power input. Other merits of
anaerobic operations are that they are good at degrading natural organic compounds
and produce a residual sludge which is inoffensive and useful as a soil conditioner
and low grade fertilizer. Finally, anaerobic digestion of primary sewage sludge
effectively reduces the concentration of pathogenic microorganisms.

Anaerobic operations also suffer from a number of defects [2,22], the main one
being that the relationship between the major microbial populations is inherently
unstable thereby requiring that the system be kept under close supervision and control. Furthermore, the maximum specific growth rates of the organisms are relatively
low, making their response to perturbations sluggish and requiring the systems to be
designed with relatively long MCRT's, even when elevated temperatures are employed.
Some synthetic organic chemicals are resistant to anaerobic biodegradation, thereby
limiting the applicability of these operations, and when they can be applied, the
effluent generally contains enough soluble biodegradable organic matter to prevent
its direct discharge. Finally, because of the requirement for heat exchange, the
capital costs of anaerobic reactors can be high. Nevertheless, when high strength
wastes are being treated, the merits generally outweigh the defects so that anaerobic
operations find wide use.

21.2 PERFORMANCE OF ANAEROBIC OPERATIONS

Anaerobic digestion and anaerobic contact employ CSTR's, both with and without solids
recycle. Consequently we would expect the theory of Chapters 12 and 13 to be applicable in the general sense. However, it must be recognized that the models in those
chapters were developed for a single, homogeneous microbial community which responded
en masse to changes imposed upon the system, whereas the microbial community existing
in an anaerobic operation certainly does not meet that criterion. Thus the theory
cannot be expected to predict specific responses. Nevertheless, it does help us to
understand why anaerobic systems respond as they do.

21.2.1 Relationship between Observed Performance and Theory

The most significant finding to arise from the theory of Chapters 12 and 13 was that
the performance of a microbial community growing in a CSTR, both with and without
solids recycle, is determined primarily by the mean residence time of the cells within the reactor. This was shown to be true for the aerobic systems considered in
Chapters 15, 16, 17, and 20 and has also been shown to be true for anaerobic
systems [23,24,25,26,27]. Furthermore, it was shown in Chapter 13 and confirmed in
Chapter 16 that the performance of a CSTR with recycle is independent of the space
time and determined only by the MCRT. This too, has been found to be true for
anaerobic systems [21,27,28,29]. Consequently it is sufficient to talk only in
terms of the MCRT when discussing anaerobic process performance. Finally, as would
be expected from theory, higher degrees of waste stabilization and greater production
of methane are obtained as the MCRT is increased, and thus the MCRT becomes the primary parameter during design, although the exact relationship between it and performance must be determined experimentally for each particular waste. Let us therefore
review the effects of MCRT.

SOLID SUBSTRATE - PRIMARY SEWAGE SLUDGE. Anaerobic digestion of primary sludge
is a major use for anaerobic operations. McCarty [25] and O'Rourke [in 30] studied
the effect of MCRT on the decomposition of primary sewage sludge in a CSTR and the
results obtained by O'Rourke at 35°C are presented in Fig. 21.4. In a prior study,
McCarty [25] had shown that a portion of the organic material contained in municipal
sludge is relatively stable and not subject to biological treatment. Consequently,
the concentrations of the residual materials were subtracted off in order to present the data in terms of biodegradable material alone. Figure 21.4 shows that the
destruction of biodegradable volatile solids was relatively complete at an MCRT of
15 days for this particular waste, and thus operation at MCRT's in excess of that
provided little additional treatment. The three major components of the volatile
solids were protein, cellulose, and lipid. The hydrolysis and degradation of protein and cellulose were relatively rapid whereas lipid degradation was much slower,
so that destruction of the volatile solids appeared to be limited by it. Similar
results have been found by others [26]. The MCRT required to achieve relatively
complete destruction of volatile solids depends upon many factors, including the
composition of the solids and the particle sizes available for attack, consequently
other researchers have reported qualitatively similar, but quantitatively different
results [27,28,29].

Returning to Fig. 21.4, it can be seen that for MCRT's in excess of 15 days,
the amount of methane produced per unit of biodegradable COD added to the reactor
was relatively constant. Other authors have presented similar findings [26,27,28],
which suggest that waste stabilization occurs by the formation of methane and its
removal from the liquid. As the MCRT is reduced below 10 days, methane production

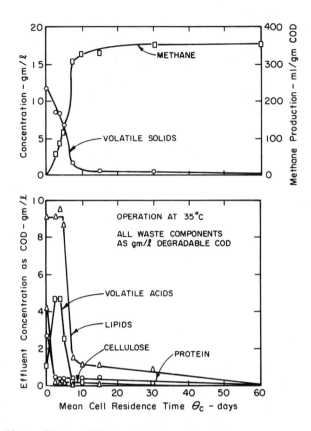

Figure 21.4.   Effect of MCRT on degradable components in domestic
               sewage sludge. *(From A.W. Lawrence, in Anaerobic
               Biological Treatment Processes, ACS Advances in Chem.
               Series #105, 163-189, 1971. Reproduced by permission
               of the American Chemical Society)*.

declines in a manner consistent with the washout of methane-forming bacteria. Concurrently with this decline there is a rise in soluble volatile acids implying that
they are formed from the degradation of the original substrate by the nonmethanogenic population and used in the production of methane by the methanogenic population.
Similar results have been presented by others [27]. Thus, the degree of destruction
of volatile solids is related to the MCRT until the limiting value is reached, at
which time washout of the methanogenic population occurs. If the MCRT is reduced
below that value, waste stabilization will cease because no more energy will be lost
in the gaseous form and the system will revert to a simple fermentation.

Examination of Fig. 21.4 shows that lipid degradation is incomplete even when
the MCRT is sufficiently long to prevent washout of the methanogenic population and
to give a low concentration of volatile acids. This implies that for solid substrates
the hydrolysis and degradation of the lipid component control the degree of

stabilization [29].   If the wastewater contains a soluble substrate, however, hydrolysis probably will not be the rate limiting step.

   SOLUBLE SUBSTRATE - COMPLEX.   Andrews and Pearson [9] studied the anaerobic
decomposition of a complex soluble synthetic waste containing tryptone, beef extract,
yeast extract, and glucose.   Figure 21.5 shows the effect of the MCRT on the performance of the system.   Most of the influent carbon was converted to volatile acids at
MCRT's too small to allow growth of the methanogenic population.   This suggests that
if a soluble waste is being treated, the rate limiting step will be the conversion
of volatile acids to methane.   Consequently, the concentration of organic matter in
the effluent will be determined by the rates at which the volatile acids are utilized by the methanogenic population, rather than by the rate at which the original
substrate is converted to volatile acids.   Because of this kinetic limitation, considerable research has been conducted on the conversion of volatile acids to methane
by enrichment cultures derived from active digesters.

   SOLUBLE SUBSTRATE - VOLATILE ACIDS.   As discussed in Section 21.1.1, acetic,
propionic, and butyric acids are the most important volatile acid precursors of
methane.   Consequently, the kinetics of methane formation from them by enrichment
cultures has received considerable attention.   Before considering the performance of

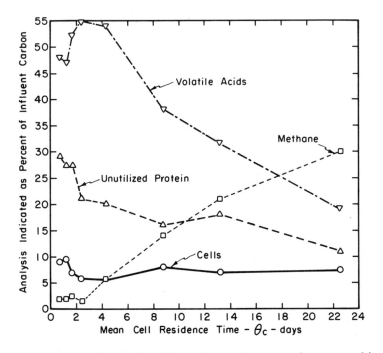

Figure 21.5.   Effect of MCRT on the performance of an anaerobic
               CSTR.   *(Adapted from Andrews and Pearson [9]).*

such cultures, however, it should be emphasized that the mechanism of methane formation from volatile acids is still unknown, and could well involve a mutualistic
interaction between hydrogen-producing and hydrogen-utilizing bacteria. None of the
kinetic studies reported to date have considered this possible interaction or the
effect that interphase hydrogen transfer may have on it. Instead they have all approached the problem as if the culture as a whole were using the volatile acids and
producing methane. Although this is probably an oversimplification, the data still
have utility because they tell us something about the rates that can be expected
from the total population. They should not, however, be interpreted as evidence
that all methanogenic bacteria *per se* are using the volatile acids directly for
energy metabolism.

   The impetus behind studies on the kinetics of volatile acid utilization is the
thought that if volatile acid utilization is rate limiting, then an anaerobic operation can be considered to be a single step process from the kinetic point of view,
thereby allowing direct application of the models of Chapters 12 and 13. Since the
major soluble, biodegradable constituents in the effluent are the volatile acids, if
the kinetic parameters for their utilization are known, then the T_bOD in the reactor
effluent at a given MCRT can be approximated by calculating the concentration of
each of the volatile acids individually using the techniques of Chapters 12 and 13,
and summing them [30]. In order to estimate the kinetic parameters for the utilization of acetate, propionate, and butyrate, Lawrence and McCarty [31] used CSTR's
operated at various MCRT's. The results obtained with acetic acid at 35°C are shown
in Fig. 21.6. The theoretical curve for acetic acid concentration tracks the data
well, indicating that the models of Chapters 12 and 13 can be used as suggested.

   A summary of the kinetic parameters calculated for acetic, propionic, and
butyric acids is presented in Table 21.1. There were slight differences observed
in the true growth yields (Y_g) and specific decay rates (b) for the three substrates

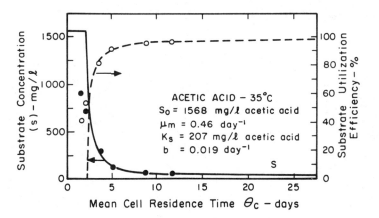

Figure 21.6. Effect of MCRT on acetate utilization.
(Adapted from Lawrence and McCarty [31]).

TABLE 21.1
Average Values of Kinetic Parameters from Growth of
Anaerobic Enrichment Cultures on Various Volatile Acids[a]

	35°C		30°C		25°C	
Volatile acid	μ_m hr^{-1}	K_s mg/liter COD	μ_m hr^{-1}	K_s mg/liter COD	μ_m hr^{-1}	K_s mg/liter COD
acetic	0.015	165	0.011	356	0.010	930
propionic	0.013	60	--	--	0.016	1145
butyric	0.016	13	--	--	--	--

[a] Adapted from Lawrence and McCarty [31].

and three temperatures studied, but Lawrence [30] felt that they were not significant for engineering purposes and recommended using a Y_g value of 0.044 mg cells produced per mg of COD removed from the medium (converted to methane) and a decay rate of $0.0008 \ hr^{-1}$ for all cases. The kinetic constants give minimum MCRT's of between 2 and 4 days, which are consistent with the total cessation of methane production in Figs. 21.4 and 21.5. Furthermore, use of the kinetic parameters suggests that at 35°C, MCRT's in excess of 10 days do not produce significant improvements in effluent quality. If a soluble wastewater containing long chain fatty acids is being treated, however, the degradation of those long chain acids will become rate limiting, requiring that longer MCRT's be employed [30]. Although the information gained from application of the models is helpful to our understanding of anaerobic systems, the best procedure is to determine experimentally the effects of MCRT upon performance.

In summary, observations of the performance of anaerobic reactors have revealed that the methanogenic population has a higher minimum MCRT than the nonmethanogenic population so that as the MCRT is reduced, process failure will occur because of washout of the methanogenic bacteria. Furthermore, if the wastewater being treated contains predominantly soluble material low in lipids, the degree of treatment achieved will be governed by the kinetics of the methanogenic population. On the other hand, if the waste contains primarily solid substrates and/or complex lipids, the degree of treatment will be governed by the conversion of those materials to short chain volatile acids.

21.2.2 Factors Affecting Performance

TEMPERATURE. Temperature has an effect upon anaerobic operations similar to all other biochemical operations, i.e., as the temperature is increased within a relatively narrow band the specific growth rates of the microorganisms will increase. This effect is particularly important in anaerobic systems because of the interacting populations. For example, different species of bacteria will respond to changes in temperature in qualitatively similar but quantitatively dissimilar ways.

Consequently a reactor which has been developed at one temperature is likely to have
a different balance of species than a reactor developed at another temperature. Unfortunately, little is known about the actual ecology of anaerobic populations at
various temperatures although it is known that a long time is required for a reactor
developed at one temperature to stabilize when even small changes are made in its
temperature. Furthermore, changes of only a few degrees can cause an imbalance between the two major populations which can lead to process failure. Consequently, the
maintenance of a uniform temperature is more important than the maintenance of the
temperature which gives the maximum possible rates [32].

     Little information has been reported about the effects of temperature upon the
nonmethanogenic bacteria. O'Rourke [in 30] observed that when he attempted to operate a sewage sludge digester at 15°C there was no significant degradation of the
lipid fraction, even when the MCRT was 60 days. The performance at 25 and 20°C,
however, was similar to that shown in Fig. 21.4, suggesting that the lipid-degrading
bacteria were quite sensitive to low temperature and were perhaps lost from the
population.

     More quantitative data are available concerning the effects of temperature upon
the utilization of volatile acids by the methanogenic population. For example, the
data in Table 21.1 show that μ_m for acetic acid decreases and K_s increases as the
temperature is decreased. Furthermore, analysis of the data for K_s revealed that it
fit the Arrhenius equation although the data for μ_m did not. The maximum specific
growth rate for propionic acid increased as the temperature decreased but microscopic
examination indicated that this may have been caused by changes in predominant species
within the population [31]. The importance of quantitative data is that it allows
an estimation to be made of the minimum acceptable MCRT. Lawrence [30] has estimated
that θ_{cmin} for the degradation of sewage sludge solids increased from a value of 4
days at 35°C to one of 10 days at 20°C, thereby suggesting that considerable reductions in reactor size can be achieved by operation at an elevated temperature.
Consequently, most mesophilic digesters are designed to operate at a temperature between 30 and 35°C. If the methane produced within the reactor is to be sufficient
to supply the heat required to maintain such a temperature, the amount of biodegradable COD available in the waste must exceed approximately 5000 mg/liter [2]. If it
doesn't, then the design engineer must decide whether to operate at a lower temperature with a longer MCRT or to supply heat from an external source.

     It is also possible to operate anaerobic reactors in the thermophilic range,
although different microbial species will be involved. However, experience in the
field with thermophilic digestion has not been highly satisfactory and there is still
considerable question as to whether the benefits received outweigh the disadvantages,
including the additional energy required to maintain high temperatures [33]. Consequently, most anaerobic operations are designed in the mesophilic temperature range
and the MCRT is adjusted to give the desired performance.

pH. The pH of the culture medium will exert an effect upon both of the main
microbial populations. Hydrogen-utilizing methanogenic bacteria are quite sensitive
to pH and generally show a relatively narrow range over which growth will occur. For
example, one species of methanogenic bacteria found in digesters showed a range of
growth from about pH 6.5 to 7.7 [34]. Generally, an optimum pH around neutrality
will exist, as demonstrated in Fig. 21.7. These data are for a mixed population of
methanogenic bacteria which had been grown in a laboratory reactor at neutral pH
and then removed for determination of the pH effect [35]. The substrate used during
the tests was formate, which implies that the effects observed are due primarily to
the hydrogen-utilizing methanogenic bacteria. There it can be seen that methanogenesis was almost totally inhibited at a pH of 6.2. As far as methanogenic bacteria
are concerned, the consensus is that the pH of anaerobic operations should be maintained near 7.0 and that severe problems can result if the pH is allowed to drop
below 6.5. The primary effect of pH upon the nonmethanogenic population is upon
the types of products formed [36,37]. This changes the substrates available to the
hydrogenogenic and methanogenic bacteria, which will, in turn, influence the rates
at which they can operate. It is not yet clear at what pH the best products are
formed by the nonmethanogenic bacteria, but as long as the two populations are grown
together, a pH near 7.0 is optimum for the system as a whole.

TOXICITY. Consider Fig. 21.8 which shows the effect of the concentration of any
material on the specific growth rate of bacteria when all other materials are present
in excess [2,38]. The left hand portion of the curve will be recognized as a classical Monod-type curve. That is, if absolutely none of a needed material is available
to a cell, it can't grow. As the concentration of that material is increased, the
specific growth rate will increase until μ_m is reached. The concentration range
over which μ increases as the concentration of the material is increased is called
the stimulatory region. The magnitude of that range will depend upon the particular

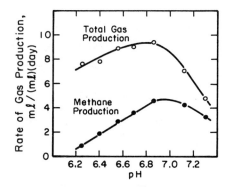

Figure 21.7.   Effect of pH on rates of methane and total
               gas production from formic acid. *(From
               L. van den Berg, et al., Biotech. Bioeng.
               16, 1459-1469, 1974. Reproduced by per-
               mission of John Wiley and Sons).*

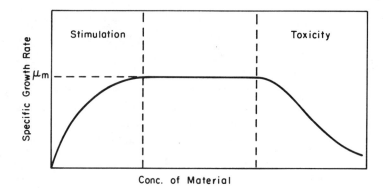

Figure 21.8. General nature of stimulation and toxicity.

material under consideration and can be anything from a few attograms per liter to
several grams per liter. As the concentration is increased further there will be a
region in which no effect is observed, but eventually a threshold value will be
reached at which the specific growth rate starts to decline. At that point, toxicity
is occurring and any concentration in excess of that is said to be toxic. Again,
that concentration can be anything from a few attograms per liter to several grams
per liter, depending upon the compound causing the effect. At concentrations above
the threshold value, the severity of the toxicity will increase as the concentration
increases. Let us now consider a few specific materials.

   Volatile Acids. Observations of the effects of volatile acids upon the microorganisms in anaerobic reactors are complicated by the fact that the acids can also
affect the pH of the medium. When the pH is held constant near neutrality, neither
acetic nor butyric acids have any significant toxic effects upon hydrogen-utilizing
methanogenic bacteria at concentrations up to 10,000 mg/liter [39]. Similar results
have been obtained for acetic acid in cultures from sewage sludge digestors [40].
Propionic acid, on the other hand, exhibits partial toxicity to methanogenic bacteria
at a concentration of 1000 mg/liter at neutral pH [39]. Furthermore, propionic acid
appears to retard acid-forming bacteria in sewage sludge digestion [38], which is consistent with studies performed in pure culture [41]. Thus it appears that at neutral
pH only propionic acid is likely to exhibit toxic effects in anaerobic operations,
and then only when the concentration is relatively high. There is still some question
concerning the response at other pH's. Andrews [42] presents preliminary evidence
which suggests that the nonionized form of propionic acid is the one which has the
deleterious effect, thereby making the inhibition stronger as the pH is decreased.
There is no evidence for this with acetic and butyric acids, so that conclusions concerning the generality of this pH-volatile acid interaction must await further study.
As far as operation of anaerobic reactors is concerned, little inhibition by volatile
acids will occur at neutral pH.

Ammonia. Ammonia is another material which arises as a natural result of sewage sludge digestion, yet which can be toxic to the bacteria. Most wastewater sludges contain substantial quantities of protein. As that protein is degraded, the nitrogen is released as ammonia but the form that the ammonia takes (i.e., either ammonium ion, NH_4^+, or dissolved free ammonia, NH_3) depends upon the pH of the system. Free ammonia is more toxic [2,38]. If its concentration exceeds 150 mg/liter severe toxicity will result, whereas the concentration of ammonium ion must be greater than 3000 mg/liter to have the same effect. As long as the pH is 7.2 or below, most ammonia will be in the form of ammonium ion so that total ammonia concentrations approaching 3000 mg/liter can be tolerated with little effect. Since one result of ammonia toxicity is a build up in volatile acids, it appears to be more toxic to the methanogenic bacteria.

Light Metal Cations. All of the preceding discussion indicates that pH control is very important to proper operation of an anaerobic reactor. Because volatile acids are produced during the normal course of anaerobic metabolism there is a tendency for the pH to drop during periods of imbalance between the two major populations. Consequently, pH control usually involves addition of a base to maintain a neutral pH. Care must be exercised in doing this, however, because the light metal cations associated with most bases can also exhibit toxic effects, presumably upon the entire microbial community, although the most detailed experiments have been performed with acetate enrichment cultures [2,38]. Sodium, potassium, calcium and magnesium are of particular concern because of their wide-spread usage and because their toxicity exhibits a complex interaction. For example, if the concentration of one cation is less than the concentration required to give maximum stimulation (see Fig. 21.8) then the toxicity exhibited by another cation will be more severe than it would be if the first cation were present at its maximally stimulatory concentration. In addition, if two cations are present at their toxic concentrations the effect will be larger than with either of the cations singly. In spite of these complications some generalities about the effects of various cation concentrations can be made, and these are shown in Table 21.2 [2]. The concentrations which are listed as stimulatory are those which will allow maximal reaction rates as shown in Fig. 21.8. Those concentrations will ensure optimum metabolic activity of the bacteria under normal conditions and will also help reduce toxic effects should the concentrations of one of the cations become excessive. Thus every effort should be made to maintain the stimulatory concentration in a reactor at all times. The concentrations listed as moderately inhibitory can be tolerated after a period of acclimation as long as they are applied steadily, however a sudden increase to those concentrations can be expected to retard the reactor significantly for several days. Concentrations listed as strongly inhibitory are those on the extreme right of the curve in Fig. 21.8 and will retard the growth of the bacteria so severely that extremely long MCRT's will be required to prevent process failure. If the toxic effects of a light metal cation

TABLE 21.2
Stimulatory and Inhibitory Concentrations of Light Metal Cations[a]

Cation	Concentrations in mg/liter		
	Stimulatory	Moderately Inhibitory	Strongly Inhibitory
Sodium	100-200	3500-5500	8000
Potassium	200-400	2500-4500	12000
Calcium	100-200	2500-4500	8000
Magnesium	75-150	1000-1500	3000

[a] Adapted from McCarty [2].

cannot be overcome by the addition of stimulatory concentrations of the others, then it will be necessary to dilute the waste.

Sulfide. Sulfides are produced in anaerobic reactors by reduction of sulfates present in the influent and by degradation of proteins. If the concentration of soluble sulfides exceeds 200 mg/liter then the metabolic activity of the methanogenic population will be strongly inhibited, leading to process failure [43]. Concentrations up to 100 mg/liter can be tolerated with little or no acclimation and concentrations between 100 and 200 mg/liter will have little effect upon reactor performance after acclimation has occurred. Only soluble sulfides exhibit toxicity because only they are available to the cells. Because heavy metals form highly insoluble precipitates with sulfide, the addition of a metal such as iron provides a simple means of reducing the soluble sulfide concentration, thereby reducing any toxicity effects. Sulfides can also be removed as gaseous hydrogen sulfide, therefore the concentration of soluble sulfides depends upon both the pH of the liquid and the composition of the gas space. Consequently, consideration must be given to all of these factors when determinations of the soluble sulfide concentration are made to check for possible toxic effects. Lawrence *et al.* [43] discuss this in detail.

Heavy Metals. Heavy metals are toxic to both major anaerobic populations at very low concentrations, as shown in Table 21.3 [44]. In spite of this extreme toxicity they need not cause a problem in anaerobic reactors because only soluble metals have

TABLE 21.3
Concentrations of Soluble Heavy Metals Exhibiting
50% Inhibition of Anaerobic Digesters[a]

Cation	Approximate Concentration in mg/liter
Fe^{++}	1-10
Zn^{++}	10^{-4}
Cd^{++}	10^{-7}
Cu^{+}	10^{-12}
Cu^{++}	10^{-16}

[a] Data extracted from Mosey and Hughes [44].

an effect and their soluble concentrations can be reduced to nontoxic values by precipitation with sulfides [44,45]. For example, the solubility product of heavy metal
sulfides ranges from 3.7×10^{-19} for FeS to 8.5×10^{-45} for CuS [38]. The approximate sulfide dose required to precipitate the heavy metals is 0.5 mg of sulfide per
mg of heavy metal. If the naturally occurring sulfides are not sufficient to prevent
heavy metal toxicity, then they should be supplemented by the addition of ferrous
sulfate [45]. In this way the excess sulfide formed will be held out of solution by
the iron. If additional heavy metals enter the reactor they will draw the sulfide
from the iron since iron sulfide is the most soluble heavy metal sulfide. As long
as the pH is above 6.4, the iron released will be precipitated as iron carbonate,
thereby preventing soluble iron toxicity.

RAPIDITY OF CHANGES. One thing that is evident from kinetic studies is that
the nonmethanogenic bacteria have much higher maximum specific growth rates than do
the methanogenic bacteria. Consequently they can respond more rapidly to environmental stress. Waste stabilization results from the proper functioning of the methanogenic culture and thus every precaution should be taken to ensure that their activity
is not impaired. This means that any changes in the reactor environment should be
made at a rate which can be tolerated by the methanogenic bacteria, i.e, they should
be made slowly or in small steps. As we will see in Section 21.4.4, if this is not
done the result is likely to be process failure.

ANAEROBIC BIOFLOCCULATION. Very little work has been done on bioflocculation
under anaerobic conditions, in spite of the fact that proper operation of the anaerobic
contact process is dependent upon it. In general, however, it appears that the same
factors which govern biological flocculation in aerobic systems also apply to anaerobic systems [27], although it is unlikely that protozoa play a large role. Most reports on the settling of anaerobic sludges indicate a considerable loss of solids in
the final effluent and the assumption is generally made that the proportions of the
various types of bacteria in those solids are the same as their proportions in the
return sludge, etc. Recent research, however, indicates that this assumption may not
be true, thereby suggesting that the overall MCRT in a system with solids recycle may
not be a reliable indicator of the specific growth rate of the methanogenic bacteria [46]. Until this question is resolved it would appear prudent to design anaerobic CSTR's with recycle using conservative MCRT's so that there will be minimal danger
of washout of the methanogenic bacteria.

LOADING. The loading on an anaerobic reactor is defined as the mass of organic
matter added per unit volume per unit time. For digesters receiving sewage sludge
the units are usually pounds of volatile solids per cubic foot per day [$kg/(m^3 \cdot day)$]
whereas for a reactor receiving soluble organic matter the units are pounds of $T_b OD$
per cubic foot per day. The loading is related to the space time of the reactor and
the concentration of the feed:

$$L = FS_o'/V = S_o'/\tau \tag{21.5}$$

For a reactor without cell recycle, the loading is related to the MCRT only because
the MCRT and τ are the same. For a reactor with recycle, the MCRT is independent of
τ, and thus the loading is not related to the MCRT. Equation 21.5 is depicted graphically in Fig. 21.9, which shows the inverse relationship between the loading and the
space time. The loading was used as an empirical design parameter before development
of the MCRT and recognition of its relationship to the specific growth rate. However,
since the loading is not related to the MCRT there is no direct correlation between
loading and process performance. Nevertheless the literature is replete with references to the fact that the loading on sewage sludge digesters should not exceed
0.4 lb VS/(ft^3·day) [6.4 kg VS/(m^3·day)]. Examination of Fig. 21.9 shows that in
order to achieve this loading at the normal space times of digestion, i.e., 10-15 days,
a highly concentrated sludge must be used. As that sludge undergoes degradation,
ammonia will be released and its concentration will be proportional to the concentration of the sludge entering the reactor. Thus high loadings will necessarily result in high ammonia concentrations, which could lead to toxicity problems. Consequently, experience dictated a realistic upper limit. If the limit on loading is
indeed caused by the concentration of ammonia and other metabolic products in the
liquid, then it would be reasonable that the upper limit on loading would decrease
as τ was increased, just as the loading itself does. Consequently, Clark and Orr [47]
suggested the probable loading limit curve in Fig. 21.10. Because they were considering only CSTR's without recycle, they did not propose limits for space times

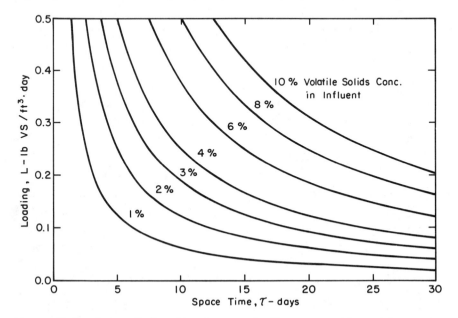

Figure 21.9.   Interrelationships between loading, space time, and
influent solids concentration for an anaerobic CSTR.

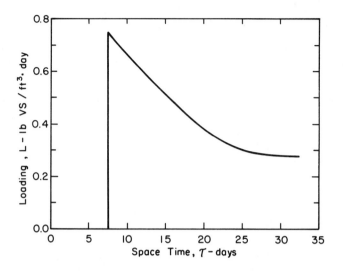

Figure 21.10. Probable loading limitation curve for
               anaerobic digestion of sewage sludge
               solids in CSTR's without sludge recycle.
               (Adapted from Clark and Orr [47]).

below 7.5 days. If reactors with recycle are used, Fig. 21.9 shows that very high
loadings can be obtained with relatively dilute sludges. This suggests that the
probable loading limit may be even higher than indicated in Fig. 21.10 when recycle
reactors with short space times are used. The limits in the figure are for digestion
of sewage sludges. There is no established loading limit for soluble wastes which
don't contain high concentrations of ammonia and other toxic materials, however.

21.2.3 Inadequacies of Models

From consideration of the performance of anaerobic reactors and the many factors affecting them, it is apparent that the simple models of Chapters 12 and 13 are woe-
fully inadequate for quantitative descriptions of anaerobic systems. First and fore-
most, those models consider only a single microbial population which is assumed to
respond *en masse* to changes in the concentration of a single growth limiting substrate. Figure 21.1 indicated, however, that anaerobic reactors contain at least
two and perhaps three microbial populations which are responsible for four distinct
reactions: hydrolysis, acidogenesis, hydrogenogenesis, and methanogenesis. Whereas
only one of these reactions is likely to be rate limiting, it is difficult to say
a priori which one it will be since the rates of all of the steps will be influenced
strongly by environmental conditions. Consequently the rate limiting reaction is
likely to change as environmental conditions are changed, thereby making the use of
a simple model with an assumed rate limiting step risky.

Another weakness of the simple models is that they do not quantitatively express the effects of pH upon the rates of microbial reactions. This is a significant weakness because not only are the metabolic rates of the bacteria strongly influenced by the pH, but the bacteria themselves affect the pH. The nonmethanogenic bacteria, for example, produce volatile acids which tend to depress the pH, whereas the methanogenic population removes them. In addition, both groups produce carbon dioxide which will be in equilibrium with the carbonic acid buffering system, thereby influencing the pH as well. The consideration of all of these interactions is quite important if a realistic picture of the system kinetics is to be obtained.

Finally, a number of materials can inhibit the anaerobic bacteria. The most important among these is propionic acid, because it is produced by the acidogenic bacteria and then used in the production of methane. Thus, in order to have a realistic model, its effects must be taken into consideration as well.

21.2.4 Modification of Models

A number of investigators have begun the task of developing more adequate models. For example, several groups [48,49,50] have developed models which use simultaneous equations to depict the production of volatile acids by the acidogenic bacteria and their utilization by the methanogenic population. One investigator [48] characterized the waste by chemical composition (protein, carbohydrate, and lipid) and wrote simultaneous equations for degradation of each component to the three major volatile acids (acetic, propionic, and butyric). An equation was also written for the utilization of each of the acids by the methanogenic population. Although no consideration was given to hydrogenogensis and the effects that it might have on methanogenesis, or to pH and inhibition effects, the resulting simulation compared favorably with data collected from a laboratory digester as shown in Fig. 21.11.

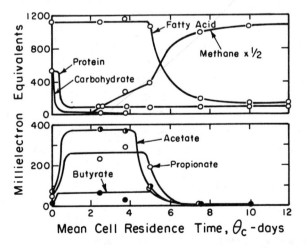

Figure 21.11.   Comparison of simulated and actual performance
of a laboratory digester. *(From P.L. McCarty,
in Anaerobic Biological Treatment Processes,
ASC Advances in Chem. Series #105, 91-107, 1971.
Reproduced by permission of the American Chemical
Society).*

Others [42,51] have concentrated on the methanogenic population and the effect
that pH plays on its kinetics. In so doing they assumed that volatile acids are
used by the methanogenic population but they made no distinctions among the various
acids. Furthermore, they assumed that the relationship between the specific growth
rate of the bacteria and the concentration of volatile acids is characterized by an
inhibition-type function, such as the curve in Fig. 21.8. Although data are available to indicate the concentration at which the toxicity region begins for propionic
acid, there are not yet data to indicate the concentration at which it starts for
acetic and butyric acids, although it is known to be at least 10 times higher [39].
Thus to assume that all volatile acids inhibit the system to the same degree appears
to be overly conservative. The most important part of their model, however, is the
assumption that only the nonionized portions of the volatile acids are available to
the cells as substrate, thereby making the specific growth rate associated with a
given total volatile acid concentration a function of pH. They also modeled the
carbonic acid-CO_2 equilibrium system so that the effects of carbon dioxide production
would be reflected and were thereby able to simulate effects that are known to occur
as system loading and pH are changed.

Although considerable progress has been made recently in the modeling of
anaerobic systems, the extreme complexity of the microbial interactions in them and
our lack of fundamental knowledge concerning those interactions suggests that it will
be some time before models are developed which have the simplicity and practical
utility of the models for aerobic systems. The development of such models presents
a challenge to the environmental engineering researcher of the 1980's comparable to
the challenge that activated sludge offered the researcher of the 1950's. In the
meantime, the design of anaerobic systems will depend upon the use of laboratory
studies and the presentation of their results in graphical form. Those graphs can
then be coupled with relatively simple mass balance and process design equations in
order to interpolate to the desired design condition.

21.3 DESIGN

As with the other slurry reactors considered in this text, the primary decision which
must be made during the design of an anaerobic reactor is the MCRT to be used. This
decision must be based upon the treatment objectives and how their attainment is influenced by the MCRT. If a solid substrate is being treated, two objectives are fore-
most in importance: the stabilization of the waste and improvement of its dewatering
characteristics. If a soluble substrate is being treated the major emphasis will be
upon its removal and the settling characteristics of the resulting sludge. All of
these factors can be assessed during laboratory treatability studies as can other
factors which relate to the potential stability of the process. In addition, the
mixing and heating requirements must also be estimated in order to complete the
process design.

21.3.1 Process Design from Lab Data

Process design data can be obtained from well mixed lab-scale anaerobic CSTR's operated at a variety of space times (MCRT's) between 5 and 30 days. Ideally, feeding
should be continuous but this is often difficult when sludge is being digested so it
is common practice to add feed and withdraw effluent once or twice a day. The frequency of feeding will be determined by the space time of the reactor since it is
best to remove not more than 5% of the reactor contents at any one time. The reactors
must be kept free of oxygen and provision should be made for collection and measurement of the gas produced. Each reactor should be run for several MCRT's to ensure
stability and samples should be taken over a period of at least one MCRT for determination of the concentrations of suspended solids, volatile suspended solids,
volatile solids, soluble COD, total COD, alkalinity, volatile acids, and ammonia
nitrogen in both the effluent and the influent. In addition the pH of the reactor
should be monitored and the effluent solids should be evaluated for their settleability and dewaterability. Detailed procedures for the running of such reactors
are available elsewhere [52].

   SOLID SUBSTRATE - ANAEROBIC DIGESTION. The main objective in the anaerobic
digestion of sewage sludge is the stabilization of the volatile solids. Often when
the percent volatile solids destruction is plotted as a function of the log of the
MCRT the result is a straight line as shown in Fig. 21.12 [29]. Another variable
of interest is the methane production per unit of volatile solids destroyed. This
will also be an apparent function of the MCRT as shown in Fig. 21.13 [28]. Different
types of volatile solids (i.e., proteins, lipids, etc.) will be destroyed to different degrees at various MCRT's. Since each type has a unique amount of energy associated with it, each will result in different quantities of methane, thereby making the

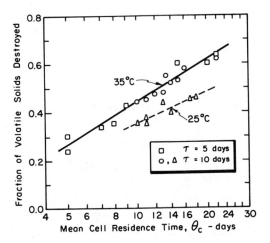

Figure 21.12. Effect of MCRT on the destruction
               of volatile solids during anaerobic
               digestion of primary and waste acti-
               vated sludge. *(Adapted from Pfeffer [29]).*

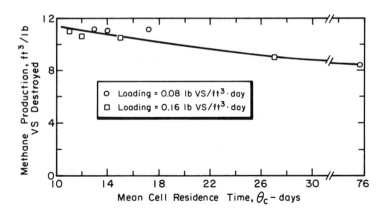

Figure 21.13.   Effect of MCRT on methane production.
                (*Adapted from Pfeffer et al. [28]*).

methane production per unit of volatile solids destroyed an apparent function of the
MCRT. Thus a graph such as Fig. 21.13 will be helpful in predicting the gas production rate. The data on the pH and concentration of soluble COD, volatile acids,
ammonia, and alkalinity will aid in estimating the stability of the digestion process
and in determining the best technique for disposal of the supernatant which will
normally be too high in COD and ammonia for discharge directly to the environment [53].
Finally, a plot should be made of the specific resistance (or some other appropriate
index of dewaterability) as a function of the MCRT.

Once the data have been collected the design decisions can be made. First it
will be necessary to choose the MCRT required to give the desired degree of volatile
solids destruction. In making this choice, a lower limit of around 10 days should be
adopted at a temperature of 35°C to ensure an adequate factor of safety against washout of the methanogenic population. As can be seen in Fig. 21.12, incremental changes
in percent destruction are relatively small for increases in MCRT above 15 days at
35°C. Consequently, the choice of MCRT is often dictated by the effects that it has
on the dewatering characteristics of the sludge. The effects of MCRT on dewaterability are influenced strongly by the characteristics of the raw sludge, and consequently no generalizations can be made. Nevertheless, it is an important consideration and should not be neglected, unless of course, the effluent sludge is to be
applied to land without the separation of supernatant and dewatering.

Most anaerobic digesters are designed as single pass reactors without solids
recycle. Thus, the space time is equal to the MCRT, thereby fixing the reactor volume
once the sludge flow rate is known (Eq. 4.31). After the volume has been selected,
the loading should be calculated using Eq. 21.5, and checked against the loading limit
associated with the space time as given in Fig. 21.10. If the loading is too high
and if the anticipated ammonia level in the reactor is high enough to indicate

860

possible toxicity problems it will be necessary to dilute the sludge prior to treatment. Given the limitations on current sludge thickening devices, this is an unlikely
possibility. It is more likely that the loading will be well below the limit curve.

If the loading is below 0.1 lb VS/(ft^3·day) [1.6 kg VS/(m^3·day)], effective utilization is not being made of the reactor volume and consideration should be given
either to thickening the influent sludge or to using solids recycle to make the MCRT
independent of τ. In making the decision regarding solids recycle it will be necessary to estimate the solids concentrations that would result with various space times.
This can be done by recognizing that the concentration of solids in a reactor with
recycle is equal to the concentration in one of equal MCRT without recycle, multiplied by the ratio of the MCRT to the space time. Thus

$$M = (\theta_c/\tau)[(1 - \xi_g)M_{go} + M_{fo}]$$ (21.6)

where ξ_g is the fractional destruction of volatile suspended solids at the desired
MCRT and M_{go} and M_{fo} are the concentrations of volatile and fixed suspended solids
in the influent, respectively. (It is generally assumed that fixed solids are con-
served in anaerobic digestion, thus no destruction of M_{fo} is assumed. The fractional
destruction of volatile suspended solids may be assumed equal to the fractional destruction of volatile solids with little error.) Once the solids concentration, M,
is known it can be used to size the settler. Generally, recycle ratios on the order
of 2 to 4 are employed. After the reactor size, solids concentration and recycle
ratio have been determined the sludge wastage rate required to maintain the desired
MCRT can be calculated using the techniques of Chapter 16.

After the required reactor volume has been determined, the amount of energy
required to heat it should be calculated to see if the methane generated during
digestion will be sufficient. The procedure for doing this is described elsewhere [52].
The heating value of the methane may be estimated from the fact that one cubic foot
of methane at standard temperature and pressure has a net heating value of 960 Btu.

SOLUBLE SUBSTRATE - ANAEROBIC CONTACT. Anaerobic contact is used to reduce the
concentration of a soluble substrate, generally before discharge to an aerobic
treatment process. Consequently, emphasis should be placed upon the soluble substrate
concentration although consideration should also be given to the concentration of suspended solids in the effluent so that the overall treatment efficiency can be predicted. Using the data collected on the lab-scale CSTR's, plots should be made of
the effluent soluble COD, C, the observed yield, Y, and the settleability of the
sludge as a function of the MCRT. In addition, a plot should be made of either the
volume of methane produced per unit mass of COD destroyed or β, the oxygen demand
coefficient of the cells formed, as a function of MCRT. The observed yield can be
calculated from

$$Y = (C_o - C)/X$$ (21.7)

where C_o and C are the soluble COD's in the influent and effluent, respectively and X is the cell concentration in the reactor. β may be calculated from

$$\beta = (T - C)/X \qquad\qquad\qquad (21.8)$$

where T is the total COD in the reactor.

The curve for the effluent soluble COD will probably look something like Fig. 21.6 and it can be used to select the MCRT that will give the desired degree of treatment (note, since the plot is of COD rather than T_bOD it will approach some positive value indicative of the nondegradable COD). Consideration should also be given to possible effects of MCRT on the settling characteristics of the sludge when choosing the design value. To guard against instability and washout of the methanogenic population, the design value should be at least three times as large as θ_{cmin}. If θ_{cmin} was not reached during the lab tests then a design value of 10 days should generally be sufficient if the temperature is 35°C.

As a first step in the design procedure a simple CSTR without recycle should be assumed and the volume calculated from the design space time (MCRT) and the known flow rate. Then the loading should be checked to see if it is greater than 0.1 lb biodegradable COD/(ft^3·day). Chances are, if the biodegradable COD (T_bOD) is less than 16,000 mg/liter it won't be. If it is not, then solids recycle can be used to decrease the volume, thereby increasing the loading. When solids recycle is used, the $X\tau$ product can be calculated from

$$X\tau = Y(C_o - C)\theta_c \qquad\qquad\qquad (21.9)$$

where Y is the yield observed at the design MCRT. Taking into account liquid-solids separation and a reasonable upper limit on loading, the appropriate values of X and τ may be determined. Again, high recycle ratios (2 to 4) are generally used and special precautions must be taken to ensure that the sludge will settle properly. Since little is known yet about the settling of anaerobic sludge, the design of the settler should be given considerable attention. After the recycle rate is known, the waste sludge flow rate may be estimated just as it was in Chapter 16. If the influent contains suspended solids they must also be corrected for in a manner analogous to that used in Chapter 16. Because the estimation of kinetic parameters is not required, it would be sufficient to estimate the inert suspended solids in the influent simply as the concentration of fixed suspended solids.

After the MCRT is known, the gas production rate may be estimated in either of two ways. If measurements were made of the quantity of methane produced per unit mass of COD removed by the biological process, then the mass removal rate of COD $[F(C_o - C)]$ need only be multiplied by the factor for the appropriate MCRT. The methane production rate can also be estimated from the amounts of soluble COD removed and cells formed [2]:

$$\text{Ft}^3/\text{day of methane at STP} = (5.62)(8.34)(F)[(C_o - C)(1 - \beta Y)] \qquad (21.10)$$

862   21/ ANAEROBIC DIGESTION AND ANAEROBIC CONTACT

where 5.62 is the cubic feet of methane produced per pound of COD destroyed, 8.34 is
a conversion factor, F is the flow in 10^6 gal/day, C_o and C are the influent and effluent soluble COD in mg/liter, and β and Y are the oxygen demand coefficient and the
observed yield at the appropriate MCRT. The heat value of this methane can then be
compared against the amount of heat required to maintain the desired temperature in
order to determine whether supplemental heating will be needed.

21.3.2 Factors Influencing the Design

MIXING. Adequate mixing is particularly important because it prevents the
development of microenvironments unfavorable to the methanogenic population (e.g.,
regions of low pH) which could reduce process efficiency [54]. Mixing also maintains
a uniform temperature in the reactor and can help to break apart sludge particles,
thereby exposing a greater surface area to biological attack. Finally, mixing helps
to prevent the formation of a scum layer on the reactor surface.

Because of the density and concentration of the solids in an anaerobic reactor,
care must be exercised in the selection of a mixer system. One important factor, the
mixing power input per unit volume, may be estimated using Eq. 17.23, but other
factors are also important. For example, because of convection, mixers which contain
heat exchangers will achieve more efficient mixing [55]. If they don't reach all the
way to the bottom of the reactor, however, thermal stratification is likely to occur,
thereby causing an unmixed region to form in the bottom. The importance of the configuration of the mixer system was demonstrated in one study in which it was found
that of three systems investigated, the one with the lowest power input per unit
volume actually had the best mixing characteristics [56].

Because of the special problems associated with the mixing of thick slurries,
a number of proprietary devices are available. They are described in detail elsewhere [57]. Some use gas recirculation whereas others use mechanical mixing. If
mechanical mixers are used in sewage sludge digesters care must be taken to ensure
that they are indeed designed for that particular purpose because rags and other
such items commonly found in sewage sludges can damage improperly designed mixers.
The engineer should set forth a performance specification for mixing and then check
the final facility to see that it has been met. One method of testing is to measure
the residence time distribution (Chapter 6) and special techniques have been developed
for doing this in anaerobic reactors [55,56].

TEMPERATURE. The effects of temperature upon the performance of anaerobic
reactors were discussed in Section 21.2.2 and Table 21.1 summarized kinetic data for
the methanogenic population. The decline in μ_m and increase in K_s associated with a
decrease in temperature cause θ_{cmin} to increase, thereby requiring longer MCRT's at
low temperatures to guard against washout of the methanogenic bacteria. The exact
effect of temperature upon system performance will depend upon the character of the
waste and can only be determined experimentally. For example, Fig. 21.12 shows that
for one sludge, an MCRT of 17 days was needed at 25°C to achieve the same degree of

volatile solids destruction as that achieved by an MCRT of 10.6 days at 35°C [29].
Thus it is very important that the effects of temperature be considered during design. The choice of the best temperature must balance the larger reactor size associated with a low temperature against the greater heating needs associated with a
high temperature.

ALKALINITY. As will be discussed in Section 21.4.4 the primary buffering system
in an anaerobic reactor is the carbonic acid-carbon dioxide system. Thus the alkalinity of the reactor will be indicative of its buffering capacity and the higher the
alkalinity the more stable the pH is likely to be. If the concentration of the waste
being treated is low, the amount of carbon dioxide produced will be small, which will
result in a low alkalinity, thereby giving a reaction system which may experience
stability problems. If a simple CSTR without recycle is being used, and the loading
is less than 0.1 lb VS/(ft^3·day) [1.6 kg VS/(m^3·day)] the alkalinity in the reactor
is likely to be less than desirable. This can be compensated for by concentrating
the waste prior to treatment or by adding additional alkalinity in the form of a
bicarbonate salt. Using recycle to decrease the reactor volume and increase the
loading will not solve the problem since it is caused by the concentration of the
waste and the associated carbon dioxide production.

NUTRIENTS. The chemical composition of anaerobic cells is quite similar to that
of aerobic cells, and consequently the amounts of nitrogen and phosphorus required
per unit mass of cells formed are the same. Much of the energy in the original substrate is lost from the liquid as methane, however, so that the mass of cells formed
per unit mass of COD removed anaerobically is much lower than it is aerobically.
Consequently, the amount of nitrogen and phosphorus needed per unit mass of COD removed will also be much smaller. When sewage sludge is being digested the possibility of a nutrient deficiency is remote; however, it may be necessary to add nutrients
to an industrial waste. The nitrogen requirement in mg N per mg COD removed can be
estimated by combining Eq. 16.31 with Eq. 13.39, the result being:

$$NR = 0.105 \ Y \tag{21.11}$$

where Y is the observed yield. Thus, once the design value of the MCRT is known the
nitrogen requirement may be estimated. The phosphorus requirement can be approximated
as one-fifth of the nitrogen requirement. During treatment of industrial wastes it
may also be necessary to add trace inorganic nutrients as well as organic nutrients
such as vitamins. This need should be investigated during lab-scale treatability
studies because the provision of proper nutrients may allow a reduction in the MCRT.

21.3.3 Example Problems

The easiest way to demonstrate the process design procedure is through examples.

EXAMPLE 21.3.3-1

A primary sewage sludge containing 4.0% dry solids, of which 65% are volatile, is produced at a rate of 100 m^3/day and is to be digested sufficiently to destroy 50% of the volatile solids. Laboratory tests gave the results indicated in Figs. 21.12 and 21.13.

(a) What volume must a digester have if the temperature is maintained at 35°C?

Entering Fig. 21.12 with a desired fractional destruction of 0.50, it can be seen that the required MCRT is 13 days. If a simple CSTR is used, then the MCRT is the same as the space time, thus τ is 13 days. This gives a volume of

$$V = (13 \text{ days})(100 \text{ } m^3/\text{day}) = 1300 \text{ } m^3 = 45900 \text{ } ft^3$$

(b) What will the loading on the digester be?

Using Eq. 21.5 and the fact that a concentration of 4% = 40 kg/m^3

$$L = \frac{(100 \text{ } m^3/\text{day})(40 \text{ } kg/m^3)(0.65)}{1300 \text{ } m^3}$$

$$L = 2 \text{ kg VS}/m^3 \cdot \text{day} = 0.12 \text{ lb VS}/(ft^3 \cdot \text{day})$$

This is sufficient for a single CSTR although it does not make the best possible use of digester volume. Thus consideration should be given either to thickening the sludge prior to treatment or to using a system with recycle.

(c) How much power will be required to mix the digester?

First we must know the concentration of solids.

Use Eq. 21.6, with θ_c set equal to τ.

$$M = (1 - \xi_g)M_{go} + M_{fo}$$

$$M_{go} = (40,000)(0.65) = 26,000 \text{ mg/liter}$$

$$M_{fo} = 40,000 - 26,000 = 14,000 \text{ mg/liter}$$

$$M = (1 - 0.5)26,000 + 14,000 = 27,000 \text{ mg/liter}$$

Then we may use Eq. 17.23 to estimate the minimum power per unit volume.

$$P/V = (0.935)(0.72)^{0.3}(27,000)^{0.298}$$

$$P/V = 17.7 \text{ kW}/1000 \text{ } m^3$$

$$P = (17.7)(1.3) = 23 \text{ kW} = 30.9 \text{ HP}$$

(d) How many Btu equivalents of methane will be produced daily by the reactor?

Entering Fig. 21.13 with an MCRT of 13 days, methane production is found to be 11 ft^3/lb VS destroyed. The loading on the reactor is 0.12 lb VS/($ft^3 \cdot$day) and of that 50% of the volatile solids are destroyed. Thus the total daily destruction of VS is

$$\text{lb VS destroyed} = (0.12)(0.50)(45,900) = 2754 \text{ lb VS/day}$$

Thus the methane production is

Methane = (2754 lb VS/day)(11 ft^3/lb VS) = 30,300 ft^3/day

The heat content of methane is 960 Btu/ft^3. Thus;

Available heat = (30,300)(960) = 29 x 10^6 Btu/day.

Once the digester has been sized, the heating requirement can be determined and compared to the heat available to see whether supplemental fuel will be required.

EXAMPLE 21.3.3-2

Returning to the sludge in Example 21.3.3-1, consider the use of a reactor with solids recycle as a means of increasing the loading to 0.3 lb VS/(ft^3·day), while still maintaining 50% VS destruction.

(a) What reactor volume and space time will be needed to achieve the desired loading?

Again we may use Eq. 21.5; in this case, however, we will use it to calculate the space time.

$\tau = S_o'/L$

If τ is in days and L is in lbs/(ft^3·day) then S_o' must be expressed as lbs/ft^3. However:

$S_o' = (4\%)(0.65) = 2.6\%$ VS

which must be converted to the proper units. The density of water is 62.4 lb/ft^3, and assuming that the density of sludge is similar, S_o' is found to be

$S_o' = (0.026)(62.4) = 1.62$ lb/ft^3

Therefore

$\tau = \dfrac{1.62}{0.3} = 5.4$ days

The reactor volume is

V = (5.4 days)(100 m^3/day)

V = 540 m^3 = 19,100 ft^3

(b) What will be the solids concentration in the reactor if the MCRT is maintained at 13 days to achieve 50% destruction of the volatile solids?

Again, use Eq. 21.6

M = (13/5.4)[(1 - 0.50)(26,000) + 14,000] = 65,000 mg/liter

This is a 6.5% sludge and it should be possible to mix, pump, and settle it with no difficulty.

(c) What recycle ratio will be needed if the sludge can be concentrated to 8%
(80,000 mg/liter) in the liquid-solids separator?

Use Eq. 16.25 to estimate that.

$$\alpha = \frac{1 - (\tau/\theta_c)}{(M_r/M) - 1}$$

$$\alpha = \frac{1 - (5.4/13)}{(80,000/65,000) - 1} = 2.5$$

Thus the recycle flow must be 2.5 times the sludge inflow rate. Care must be
exercised in the design of the settler (thickener) to ensure that the desired underflow solids can be achieved.

(d) At what rate must sludge be wasted from the underflow to maintain the desired
MCRT if no solids are lost in the supernatant (overflow)? Use Eq. 16.33

$$\theta_{ch} = \frac{VM}{F_w M_w}$$

$$F_w = \frac{V(M/M_w)}{\theta_{ch}}$$

$$F_w = \frac{(540 \text{ m}^3)(65,000/80,000)}{13 \text{ days}} = 33.8 \text{ m}^3/\text{day}$$

In practice the actual quantity wasted will be smaller than this because of the loss
of solids in the final effluent. That loss can be compensated for by using the
definition of MCRT.

(e) How much power will be required to mix this reactor?

Again, we may use Eq. 17.23 to estimate the minimum power per unit volume.

$$P/V = (0.935)(0.72)^{0.3}(65,000)^{0.298}$$

$$P/V = 23 \text{ kW}/1000 \text{ m}^3$$

$$P = (23)(0.54) = 12.4 \text{ kW} = 16.7 \text{ HP}$$

The higher sludge concentration requires a higher power input per unit volume. The
total power input is smaller, however, because the reactor is smaller.

EXAMPLE 21.3.3-3

A soluble wastewater with a COD of 12,000 mg/liter and a flow rate of 10 m^3/hr is to
be treated by anaerobic contact. Lab studies indicate that a MCRT of 15 days (30°C)
will give stable operation while maintaining a low level of volatile acids in the
effluent and reducing the soluble COD to 3000 mg/liter. During the studies the observed yield was 0.1 mg cells/mg COD removed and the oxygen demand coefficient of the
cells was 1.25 mg COD/mg cells when the MCRT was 15 days. Studies on the settling

characteristics of the resulting sludge indicated that the solids could be concentrated to 15,000 mg/liter in a settler of reasonable size. Perform a preliminary
design for an anaerobic contact process.

(a) The first thing that should be done is to check the loading for a CSTR without
recycle.

In that case τ would be 15 days and the loading can be calculated from Eq. 21.5.

$$L = \frac{12,000 \text{ mg/liter}}{15 \text{ days}}$$

$L = 0.8$ mg COD/(liter·day) $= 0.8$ kg COD/$(m^3 \cdot day) = 0.05$ lb COD/$(ft^3 \cdot day)$

The fact that this value is below 0.1 lb COD/$(ft^3 \cdot day)$ tells us two things: (1) that
a reactor with recycle must be used and (2) that the alkalinity produced in the process is likely to be too low to provide adequate buffering power. Consequently
consideration should be given to the addition of alkalinity in the final design.

(b) Next, we can prepare a table of alternative designs from which to choose by
making use of Eqs. 21.5 and 21.9.

Equation 21.5 allows us to calculate the space times required to obtain various
loadings:

$$\tau = S'_o/L$$

For our problem, conversion of S'_o to lb/ft^3 gives:

$$\tau = 0.75 \text{ lb/ft}^3/L \text{ lb/}(ft^3 \cdot day)$$

Equation 21.9 allows us to calculate the cell concentration that will develop for
any desired space time:

$$X = Y(C_o - C)(\theta_c/\tau)$$

For our problem this reduces to:

$$X = \frac{(0.1)(12,000 - 3,000)(15)}{\tau} = \frac{13,500}{\tau} \text{ mg/liter}$$

Using the two equations the values in Table E21.1 were calculated.

TABLE E21.1
Calculated Space Times and Solids Concentrations
Associated with Various Possible Loadings

Loading lb COD/$(ft^3 \cdot day)$	τ days	X mg/liter
0.1	7.5	1800
0.2	3.75	3600
0.3	2.5	5400
0.4	1.875	7200

Considering the settling characteristics of the cells, it appears reasonable to use a space time of 1.875 days which will give a cell concentration of 7200 mg/liter, a loading of 0.4 lb COD/(ft^3·day) and a reactor volume of 450 m^3. Because τ is lower than θ_{cmin} for methanogenic bacteria at 30°C, special care will be required during start up of the process. The system must be heavily seeded with an active sludge and then started slowly to ensure a stable methanogenic population.

(c) If a recycle ratio of 2.0 is used, what will be the concentration of solids in the underflow and at what rate must underflow sludge be wasted if no solids are lost in the effluent?

Use Eq. 16.25 to estimate M_r:

$$M_r = [\frac{1 - (\tau/\theta_c)}{\alpha} + 1] M$$

$$M_r = [\frac{1 - (1.875/15)}{2} + 1] \, 7200 = 10{,}350 \text{ mg/liter}$$

The settler must be designed to achieve this.

The wastage rate for sludge can be calculated from Eq. 16.33:

$$F_w = \frac{VM}{\theta_{ch}M_w}$$

$$F_w = \frac{(450)(7200)}{(15)(10{,}350)}$$

$$= 21 \text{ m}^3/\text{day}$$

In practice this quantity will have to be reduced to correct for the solids lost in the final effluent.

(d) How much methane will be produced per day?

This can be calculated from Eq. 21.10. In order to use that equation the flow rate must be expressed as 10^6 gal/day.

$$F = 10 \text{ m}^3/\text{hr} = 0.0634 \times 10^6 \text{ gal/day}$$

Thus

$$Ft^3/\text{day of methane} = (5.62)(8.34)(0.0634)[(12{,}000 - 3000)(1.0 - (1.25)(0.1))]$$

$$Ft^3/\text{day of methane} = 23{,}400$$

This has a heat value of 22.5 x 10^6 Btu/day. This can be compared with the heating requirements to see if supplemental heating will be required.

21.4 ANAEROBIC PROCESS OPERATION

Unlike some of the aerobic biochemical operations, anaerobic digestion and anaerobic contact are not tolerant of neglect. In fact, close operation and supervision are essential. Although the main factors concerning operation will be discussed herein, space does not permit a thorough discussion of all of the practical "tricks of the trade". Fortunately the U.S. Environmental Protection Agency has published an excellent manual [53] dealing with the practical aspects of operation and the reader is referred to that for additional information.

21.4.1 Unstable Nature of Anaerobic Processes

The statement has been made several times in this chapter that anaerobic processes are basically unstable. This results from the necessity for the two main groups of bacteria to coexist in one reactor and from the fact that under the environmental conditions normally imposed, the nonmethanogenic population can respond more rapidly to environmental changes than can the methanogenic one. Consequently if anything happens to upset the balanced condition between the populations, the two interacting fermentations become uncoupled and system performance deteriorates.

To amplify this point consider the following scenario. Suppose an additional load is placed upon an anaerobic reactor, thereby increasing the amount of substrate available to the nonmethanogenic population. It will respond fairly rapidly by increasing its specific growth rate, thereby increasing the rate at which volatile acids, carbon dioxide, hydrogen, and other end products are produced. As the concentration of end products increases they will act to stimulate the methanogenic population, which must also increase its specific growth rate. The maximum specific growth rate of cells in the methanogenic population is lower than that of cells in the nonmethanogenic population, however, so that if the original increase in loading is large, the methanogenic population may not be able to remove the volatile acid intermediates as rapidly as they are produced. This by itself need not cause a problem. However, if the buffering capacity of the system is not sufficient to maintain the pH near 7.0, the pH will drop and the methanogenic bacteria will begin to be inhibited, thereby causing an even greater imbalance between the two populations. Once this has happened, the reactor is on a course toward failure unless corrective action is taken to alleviate the original imbalance and chemicals are added to maintain the pH at 7.0. If corrective action is not taken, the pH will decline even more, which further inhibits the methanogenic population, making the imbalance worse, leading to an even lower pH, etc. If things get bad enough, hydrogen will begin to accumulate in the gas space, which will alter the metabolism of the nonmethanogenic bacteria, causing them to produce a higher proportion of propionic acid, which can inhibit both groups of bacteria. Eventually, the pH will drop to a very low value, volatile acids will build up within the system, methane production will stop and the digestion process will cease. When that occurs the reactor is referred to as "sour" (because of its odor) or "stuck". If the concentration of propionic acid is over 1000 mg/liter, restarting a stuck digester can be difficult.

As we will see in Section 21.4.3 a number of events can cause an imbalance between the two microbial populations. Since avoidance of a stuck reactor depends upon early detection of an imbalance it is important to monitor reactor performance frequently. Therefore let us consider some factors which can act as indicators of performance.

21.4.2 Indicators of Operational Performance

Unfortunately, there is no single indicator which will reliably signal an imbalance between the two major populations, so that a number of indicators must be considered simultaneously [2,57,58]. Among the most effective are: concentration of volatile acids, bicarbonate alkalinity, pH, and rate of methane production.

VOLATILE ACIDS CONCENTRATION. Although numbers such as 200-400 mg/liter as acetic acid are often quoted as being normal in good digestion, the magnitude of the volatile acids concentration is less important than its time rate of change [2,48]. Since the concentration is influenced by both microbial populations, its rate of change is a direct and early measure of their relative rates of activity. Consequently, a sharp rise in the volatile acids concentration indicates that something has happened either to retard the methanogenic population or to stimulate the acidogenic bacteria. Conversely, a sharp drop in the concentration could mean that either the methanogenic population has been stimulated or the nonmethanogenic bacteria have been selectively inhibited. The latter event could cause the whole process to stop without showing any of the symptoms of a stuck reactor. Knowledge of the types of volatile acids present can also be helpful. For example, if the propionic acid concentration is high (>1000 mg/liter), additional problems may result because it can inhibit the system, thereby magnifying the effect of the original problem. High levels of the other volatile acids will cause few problems as long as the pH is kept near 7.0, although they do indicate that something else has affected the system. Generally they are the result of an imbalance between the two populations and not the cause of it [2].

BICARBONATE ALKALINITY. Knowledge of the bicarbonate alkalinity tells us how much buffering capacity remains in the system. This is important because if the buffering capacity is low, relatively small increases in the concentration of volatile acids can have a severe effect upon the pH, with its concomitant adverse effect upon the methanogenic population. Conversely, if the system has adequate alkalinity, it can tolerate significant fluctuations in the concentrations of volatile acids without large changes in pH. It is difficult to specify a satisfactory alkalinity because the amount present will depend upon both the character of the carriage water and the concentration of the waste being treated. The latter determines the quantity of carbon dioxide produced per unit volume of liquid [54]. A more important factor for characterization of operation is the ratio of the concentration of volatile acids (as mg/liter of acetic acid) to the bicarbonate alkalinity (as mg/liter of $CaCO_3$) [53].

As long as this ratio is less than 0.4 the system should be able to accomodate moderate variations in the volatile acids concentrations with little fluctuation in pH. A rise in the ratio above that, however, is indicative of an imbalance within the system as well as a lack of reserve buffering capacity. If the ratio rises above 0.8 the system is likely to experience a severe drop in pH from even small changes in volatile acids. Consequently attention should be given both to the magnitude of the ratio and to its rate of change. Additional factors relating to pH and alkalinity will be discussed in Section 21.4.4.

pH. The pH is not a sensitive indicator of reactor behavior because significant environmental changes will have already occurred before the pH drops [2]. Knowledge of the pH is important to good operation, however, because of the rather narrow band within which the methanogenic bacteria can function properly. If the pH is not maintained near neutrality, inhibition of the methanogenic population will occur and the system will enter the failure spiral. In addition, maintenance of the proper pH is absolutely essential if the system is to recover from a metabolic imbalance.

METHANE PRODUCTION RATE. The rate of methane production is a direct measure of the metabolic activity of the methanogenic bacteria and as such has great potential as a diagnostic tool of digester performance [58]. Again, little can be said about the actual magnitude of the methane production rate, but any rapid change in it will indicate that something has happened to the methanogenic bacteria. The feed rate to the reactor must be quite uniform to prevent normal variations in the methane production rate from masking changes caused by toxicity or other problems. Any long-term downward trend, however, will indicate problems. Like the other indicators, this has maximum utility when used with one or more of the other techniques.

OTHER INDICATORS. Other parameters which are sometimes used to indicate the operational stability of anaerobic reactors are gas composition and total gas production rate. Both are influenced by the partial pressure of carbon dioxide in the gas space, which is in turn affected by interphase gas transfer with the liquid. Changes in either of these variables represent a complex interaction among several factors, thereby making small changes unreliable as indicators of reactor performance. An imbalance between the two major populations is likely to be manifested by a decrease in the production rate of methane and an increase in the rate of carbon dioxide production. Therefore a change in the composition is likely to show up before a change in the total production rate.

21.4.3 Causes of Operational Imbalance

As soon as an operational imbalance is detected every effort should be made to determine its cause and alleviate it. To aid in doing this the EPA operating manual [53] has a very complete trouble shooting guide.

One of the primary causes of an imbalance is an increase in loading, i.e., an increase in either the concentration or flow rate of feed to the reactor. Evidence for such a change would be an increase in the concentration of volatile acids in the reactor, accompanied by a constant or increasing rate of methane production. The fact that the rate of methane production has not decreased indicates that the methanogenic population has not been inhibited, therefore suggesting that the imbalance between the two populations was caused by an increase in the rate of activity of the acidogenic bacteria. This could be corrected by adjusting the feed rate.

An imbalance could also be caused by a change in reactor temperature. This is normally the result of a mechanical failure and would be detected during routine measurements of temperature.

Another factor which could cause an imbalance is a decrease in the MCRT. This could happen if solids deposition occurred, because it would reduce the effective reactor volume. The problem could be alleviated by repairing or improving the mixing system and removing the deposited solids. If the reactor has solids recycle, particular attention must also be paid to the quantity of sludge wasted and compensation made for solids lost in the effluent. Furthermore, the effluent should be tested to ensure that the percentage of methane formers in the effluent suspended solids is the same as the percentage in the reactor [35]. If it is higher, then the MCRT should be adjusted accordingly.

One of the most perplexing causes of an imbalance is toxicity because of the many potentially causative factors. (See Section 21.2.2). A decrease in the rate of methane production that is not accompanied by an increase in the concentration of volatile acids or a change in the pH indicates that something is inhibiting either the entire system or the nonmethanogenic population. This could be a heavy metal or sulfide. Every effort should be made to find the source and stop the input, but the immediate problem can be relieved by precipitation of the metal or sulfide ions. If a decrease in the rate of methane production is accompanied by an increase in the concentration of volatile acids, then inhibition of the methanogenic population should be suspected. This might be caused by ammonia or any of a number of specific chemicals. About all that can be done is to reduce the loading to help prevent total failure while the cause of the problem is detected and corrected.

No matter what the cause of the treatment imbalance, however, control of the pH to neutrality is essential for successful recovery.

21.4.4 pH Control

NATURE OF THE BUFFERING SYSTEM. As was indicated in Fig. 21.1, one of the end products of anaerobic metabolism is carbon dioxide. In aqueous systems carbon dioxide is in equilibrium with carbonic acid, which dissociates to give hydrogen and bicarbonate ions. Anaerobic reactors also contain other weak acid-base systems, such as ammonia and orthophosphoric, hydrosulfuric, and volatile acids, but the carbonic acid system is the most important in determining reactor pH [59]. This stems from several factors. First, in most anaerobic reactors the concentrations of the orthophosphoric and hydrosulfuric acid systems are too low to provide significant buffering capacity. Second, in the normal pH range of anaerobic reactors, the buffering actions of the volatile acids and ammonia are negligible; furthermore they are almost completely dissociated and thus act as a strong acid and base respectively. As a consequence, the interaction of the carbonic acid system and the net strong base, B_{ns}, controls the pH. The term net strong base, B_{ns}, refers to the summation of all strong acids and bases including volatile fatty acids and ammonia [59].

A proton balance on the system in the pH range 6.0 to 7.5 shows that the bicarbonate alkalinity will be approximately equal to the concentration of net strong base, so that the pH of the system is defined by those two species. Furthermore, there is a relationship between the bicarbonate alkalinity and the carbonic acid concentration, which is in turn related to the dissolved carbon dioxide concentration, so that through the equilibrium expressions for the first dissociation of carbonic acid and the solubility of carbon dioxide in water it is possible to construct a diagram showing the interrelationships among the pH of the reactor, the partial pressure of carbon dioxide in the gas space, and the concentration of bicarbonate alkalinity, HCO_3^- [59]. Such a diagram is shown in Fig. 21.14. Note that because the net strong base concentration and the bicarbonate alkalinity are approximately equal, both are given on the abscissa.

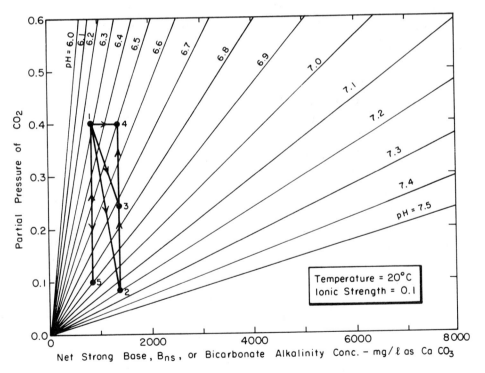

Figure 21.14. Theoretical relationships among CO_2 partial pressure, bicarbonate alkalinity, and pH in an anaerobic digester. *(Adapted from Capri and Marais [59]).*

When sewage sludge is being digested, the gas space contains about 30% carbon dioxide and 70% methane. Furthermore, because of the concentration of sludge in the feed and the quantity of ammonia released by degradation of the protein, it is common for the net strong base concentration to be 2000 mg/liter. This makes the bicarbonate alkalinity also around 2000 mg/liter. If the temperature of the system were 20°C (the temperature for which Fig. 21.14 was prepared [59]), the resulting pH would be 6.8. At a different temperature, the solubility of carbon dioxide and the equilibrium constants would be different, thereby giving a different pH for the same carbon dioxide partial pressure and alkalinity. Thus it can be seen that the pH of an anaerobic reactor is determined by the metabolic activity of the microorganisms (which determines the net quantities of carbon dioxide, volatile acids, and ammonia produced) as well as by the physical and chemical characteristics of the system. Because system pH has a strong effect upon the activity of the methanogenic population, however, it would be advantageous to control it to a desired value.

CONTROL OF pH. Several techniques can be used to control pH. Among them are addition of strong bases, bicarbonates, and carbonates to the liquid [59], or removal of carbon dioxide from the gas [58]. If a strong base (such as NaOH or NH_3) or a carbonate salt (such as Na_2CO_3) is added, ionic equilibrium occurs very rapidly and carbon dioxide is removed from the gas space to form the required bicarbonate alkalinity. For example, if the reactor conditions were originally at point 1 in Fig. 21.14 the addition of a strong base would reduce the carbon dioxide partial pressure and increase the alkalinity, thereby moving the equilibrium to point 2. However, as carbon dioxide was produced by the bacteria, the equilibrium would shift to point 4, thereby causing the pH to drop even though the alkalinity remained the same. If the same moles/liter of a carbonate salt had been added instead, the equilibrium would have shifted to point 3 before moving to point 4. On the other hand, had an equal amount of bicarbonate salt been added, the equilibrium would have shifted directly to point 4 because no carbon dioxide would have been dissolved to form the required alkalinity. As a final alternative, the gas could have been moved through an external scrubber to reduce the partial pressure of carbon dioxide to 0.1, which would have moved the equilibrium to point 5. As the bacteria produced carbon dioxide, however, its partial pressure in the gas would increase, thereby causing the equilibrium to drift back to point 1.

The preceding example has shown that chemicals for pH control fall into two major categories, those which add bicarbonate alkalinity directly, and those which trap carbon dioxide and convert it to bicarbonate. The question therefore arises as to which type of chemical is best. Two factors must be considered when answering that question. The first is the type of effect that the chemical has upon the system. A chemical which traps carbon dioxide makes the pH overshoot the desired value because of the time lag associated with carbon dioxide equilibrium. The first response is a rapid rise in pH, accompanied by a decline in carbon dioxide partial pressure.

If the chemical dose is sufficiently large this could raise the pH too much, thereby causing inhibition. Consequently, pH control by chemicals which trap carbon dioxide requires that their addition be in small steps to allow time for gas phase equilibrium to occur between each addition. The direct addition of bicarbonate, on the other hand, has no such effects on the system and thus can be done more precisely. The second factor which must be considered when evaluating the type of chemical is its solubility [2]. For example, sodium chemicals have very high solubilities under the normal conditions of anaerobic reactors, whereas calcium compounds are relatively insoluble, with $CaCO_3$ imposing the solubility limit in the pH range 6.9 to 7.5. As lime $(Ca(OH)_2)$ is added to a digester it traps carbon dioxide and converts it to bicarbonate. Calcium bicarbonate, however, has limited solubility. Consequently, when the bicarbonate alkalinity reaches 500 to 1000 mg/liter, the addition of more lime causes insoluble calcium carbonate to form. This removes carbon dioxide from the gas space, but it does not increase the alkalinity because insoluble $CaCO_3$ is being formed. The drop in carbon dioxide partial pressure will cause the pH to rise rapidly, but since the alkalinity has not been increased, this pH is unstable so that as soon as biological activity increases the pH drops rapidly. If more lime is added, the process repeats itself and an oscillation is set up. This not only causes a variation in pH, which is detrimental to the bacteria, it also increases the mass of sludge to be disposed of. Consequently, lime should only be added when the pH is below 6.5, and then only in sufficient quantity to raise the pH to about 6.7 to 6.8 [2]. This problem with calcium containing chemicals suggests that better pH control can be obtained with sodium containing chemicals. Furthermore, when coupled with the advantages associated with the direct addition of bicarbonate ion, it suggests that the use of sodium bicarbonate for pH control is the best procedure, despite the somewhat higher unit cost of chemical [33].

   CALCULATION OF QUANTITY OF CHEMICAL REQUIRED. The quantity of chemical required to change the pH of an anaerobic reactor may be calculated once the pH, alkalinity and total volatile acids concentrations in the reactor are known. The following procedure should be followed [60]:

1. Measure the pH of the reactor
2. Select the desired pH and determine the required pH change:

$$\Delta pH = \text{desired pH} - \text{measured pH} \tag{21.12}$$

3. Measure the alkalinity to pH 4.5 as mg/liter $CaCO_3$ $(A_{4.5})$
4. Measure the total volatile acids concentration (TVA) as mg/liter acetic acid.
5. Estimate the bicarbonate alkalinity $(HCO_3^-)_I$ in the reactor in mg/liter $CaCO_3$ by:

$$(HCO_3^-)_I = A_{4.5} - (0.833)(f_v)(TVA) \tag{21.13}$$

   where f_v is the fraction of the volatile acids salts titrated between the original pH of the reactor and pH 4.5:

$$f_v = \frac{K_I}{[H^+] + K_I} - \frac{K_I}{[H^+]_{4.5} + K_I} \qquad (21.14)$$

K_I = ionization constant representing the volatile acids in the system.

$[H^+]$ = hydrogen ion concentration at original reactor pH.

$[H^+]_{4.5}$ = hydrogen ion concentration at pH 4.5.

6. Determine required $(HCO_3^-)_R$ concentration in mg/liter $CaCO_3$ by

$$(HCO_3^-)_R = antilog \; [log(HCO_3^-)_I + \Delta pH] \qquad (21.15)$$

7. Determine required quantity of chemical in pounds by

$$lbs \; of \; chemical = (V/6) \, [(HCO_3^-)_R - (HCO_3^-)_I] E \qquad (21.16)$$

V = reactor volume in 10^6 gal.

E = equivalent weight of chemical added.

21.5 EXAMPLES IN PRACTICE

The major application of anaerobic reactors in practice is the stabilization of
sewage sludge. Originally anaerobic digestion was performed in unmixed reactors,
but because this created localized environments which were unfavorable to the methanogenic bacteria the loadings were kept very low, generally less than 0.1 lb VS/(ft^3·day)
[1.6 kg/(m^3·day)]. Now, however, digesters are usually designed as CSTR's without
recycle so that they may be operated at higher rates. For example, the single stage
digesters at Chicago [61] are operated at a space time of 14 days at 35°C, with
loadings up to 0.3 lb VS/(ft^3·day) [4.8 kg/(m^3·day)], and produce between 13 and 17
ft^3 of gas per lb of volatile solids destroyed (0.81 to 1.05 m^3/kg). Although such
loadings were not common in 1960 [62], they have found increasing use since the
establishment of MCRT as the major parameter in design. Thus as more reactors are
designed with MCRT's around 15 days, loadings around 0.3 lb VS/(ft^3·day)
[4.8 kg/(m^3·day)] will become the rule rather than the exception.

One result of our better understanding of the fundamental factors influencing
the performance of anaerobic reactors is an increased application of them to industrial wastewater treatment, where they are usually used to treat high strength soluble wastes and as such are applications of the anaerobic contact process. A summary
of the key factors of some such installations is given in Table 21.4. In addition
to those, examples can also be found of applications of anaerobic treatment to sugar
cane, textile, whey, pharmaceutical, milk, pineapple, synthetic organic chemical,
soap manufacturing, and animal wastes. Examination of Table 21.4 reveals some interesting facts. First, all of the wastes had a high organic content, which is consistent with the merits of anaerobic treatment. Second, all except one were operated

TABLE 21.4
Examples of Industrial Uses of Anaerobic Reactors

Example No.	Type of Wastewater	Waste Conc. mg/liter	Param.	Loading lb/(ft³·day)	τ days	MCRT days	Temp. °C	Removal %	Reference No.
1	Maize Starch	6300	BOD_5	0.11	3.3	--	23	88	69
2	Whiskey Distillery	25000	BOD_5	0.25	6.2	--	33	95	70
3a	Rum Distillery	65000	COD	0.24	16.7	--	36	78.5	63
3b	Rum Distillery	65000	COD	0.72	5.6	--	36	72	63
4	Molasses Distillery	30000	BOD_5	0.19	10	10	37	95.5	64
5	Wine Distillery	22000	COD	0.20	6.9	--	33	97.3	65
6	Wine	23400	VSS	0.73	2.0	--	33	85	71
7	Pomace Stillage	13740	COD	0.086	10	10	35	73	66
8	Brewery	3900	BOD_5	0.127	2.3	~33	--	96	67
9	Molasses	32800	VSS	0.546	3.8	--	33	69	71
10	Yeast	11900	VSS	0.372	2.0	--	33	65	71
11	Starch-Gluten	14000	VSS	0.10	3.8	--	35	80	72
12	Cotton Kiering	1600	BOD_5	0.074	1.3	--	30	67	73
13	Citrus	4600	BOD_5	0.214	1.3	--	33	87	74
14	Pear Waste	60000	COD	0.27	3	~108	35	91	46
15	Meat Packing	2000	BOD_5	0.11	1.3	--	33	95	73
16	Meat Packing	1380	BOD_5	0.156	0.5	~9	33	91	68

at a temperature between 30 and 36°C, attesting to the fact that it represents a
good compromise between maximum rates and heat requirements. Third, as expected,
there is no correlation between loading and performance. Unfortunately, the MCRT's
generally were not given, and in most cases there was not enough information available to estimate it. Fourth, relatively short space times have been used successfully
in practice, indicating that economic designs can be achieved if attention is paid to
process fundamentals. Perhaps the main point to be made by Table 21.4, however, is
that few generalizations can be made about the design of an anaerobic reactor, a
situation similar to that experienced with all of the other biochemical operations.

     In addition to the information in the table, a few other points should be made
which emphasize the need to investigate each case independently. For example, the
rum distillery waste listed as example 3 contained a very high level of sulfates.
Consequently, it was necessary to continually scrub hydrogen sulfide from the reactor
gas space in order to keep the dissolved sulfide level below the toxic limit. During
another study of a distillery waste [69], however, no provision was made for sulfide
removal, and the reactor failed when the loading was only 0.22 lb $BOD_5/(ft_3 \cdot day)$
[3.5 kg/(m^3·day)] in spite of the fact that the space time was 10 days and the MCRT
was even greater. This points out the necessity for considering all of the factors
which can affect the performance of an anaerobic system, because quite often some
environmental factor will govern the results that can be achieved. A crucial unknown
facing the designer of an anaerobic contact reactor is how the settling characteristics
of the sludge will be influenced by the operational characteristics of the reactor.
Little is known yet about anaerobic flocculation and sedimentation. Thus the designer
must rely heavily upon data obtained during treatability tests. In spite of that,
anaerobic contact is a valuable tool which can do an excellent job when properly
applied. Consequently, it will undoubtedly be used more often in the future.

21.6 KEY POINTS

1. The degradation of organic matter in anaerobic operations occurs in multistep
    fashion as depicted in Fig. 21.1.

2. The nonmethanogenic bacteria constitute a rather diverse group of facultative
    and obligately anaerobic bacteria which produce primarily formate, acetate,
    hydrogen and carbon dioxide as end products.

3. All methanogenic bacteria which have been isolated in pure culture are auto-
    trophic, obligate anaerobes which obtain their energy from the oxidation of
    molecular hydrogen and their carbon from carbon dioxide. Although only one
    species has been isolated in pure culture, evidence indicates that methanogenic
    bacteria which can grow and produce methane from acetate play an important
    role in anaerobic operations.

4. Anaerobic reactors are usually covered, circular concrete tanks which are mixed
    by any of several techniques, such as sludge recirculation, gas recirculation,
    mechanical draft-tube mixers, or turbine and propeller mixers.

5. When an anaerobic reactor is used for the degradation of organic solids the
    process is called anaerobic digestion, but when it is used to treat high
    strength soluble wastes it is called anaerobic contact.

6. The performance of an anaerobic reactor, either with or without solids recycle,
    is a function of the MCRT.

7. Even though the mechanism of methane formation from acetic, propionic, and
    butyric acids is unknown the kinetics of the utilization of those acids can be
    modeled by the techniques presented in Chapters 12 and 13.

8. Most anaerobic digesters are designed to operate at a uniform temperature
    between 30° and 35°C.

9. The maintenance of a constant pH near 7.0 is important to stable operation of
    an anaerobic reactor.

10. There is no direct correlation between the loading applied to an anaerobic
    reactor and its process performance. However, high loadings require high con-
    centrations in the influent, which may cause toxicity problems, thereby in-
    directly affecting performance.

11. The most important variable which must be considered during the design of an
    anaerobic operation is the MCRT. Consequently laboratory treatability studies
    must be performed to determine its effect upon the attainment of the treatment
    objectives.

12. The nutrient requirements for anaerobic operations are less than those for
    aerobic operations because the cell yields are less.

13. Nonmethanogenic bacteria can normally respond to environmental changes more
    rapidly than methanogenic bacteria, thereby producing conditions which are
    deleterious to the methanogenic bacteria. Consequently, anaerobic processes
    are naturally unstable and require close supervision to achieve reliable per-
    formance.

14. There is no single indicator which will reliably signal an imbalance between
    the two major populations in an anaerobic reactor. Consequently a number of
    indicators (such as concentration of volatile acids, bicarbonate alkalinity,
    pH, and rate of methane production) must be considered simultaneously.

15. Some causes of operational imbalances in anaerobic reactors are an increase in
    loading, a change in reactor temperature, a decrease in MCRT, and the presence
    of toxic materials.

16. Recovery of an anaerobic reactor from an imbalance requires elimination of the
    cause and maintenance of a neutral pH.

17. The pH in an anaerobic operation is a result of the interaction of the carbonic
    acid buffering system with the net strong base (i.e., the summation of all
    strong acids and bases including volatile fatty acids and ammonia) present.

18. Chemicals for pH control fall into two major categories, those which add bi-
    carbonate alkalinity directly and those which trap carbon dioxide from the gas
    space and convert it to bicarbonate. Greater care is required when the latter
    are used.

21.7 STUDY QUESTIONS

The study questions are arranged in the same order as the text

Section	Questions
21.1	1-4
21.2	5-10
21.3	11-13
21.4	14-16

1. Describe briefly the multistep nature of an anaerobic operation.

2. Discuss the role that hydrogen plays in determining the nature of the soluble end products formed by the nonmethanogenic bacteria.

3. Discuss the role that hydrogen production plays in the production of methane in mixed cultures.

4. List the merits and defects associated with anaerobic operations.

5. Discuss, with the aid of sketches, the effects of MCRT upon the degradation of sewage sludge in an anaerobic reactor.

6. Discuss the concept of the rate limiting step in an anaerobic operation and tell how it will depend upon the nature of the substrate.

7. Explain why investigators have been interested in the application of the kinetics of Chapters 12 and 13 to the utilization of volatile acids by the methanogenic population.

8. Discuss briefly the effects of the following factors upon anaerobic reactors: volatile acids, ammonia, sulfide, light metal cations, and heavy metals.

9. Explain why the mathematical models of Chapters 12 and 13 are inadequate for describing the performance of anaerobic reactors.

10. Explain why adequate mixing is important to the proper operation of anaerobic reactors.

11. Laboratory tests on anaerobic digestion of a primary sewage sludge gave the results shown in Figs. 21.12 and 21.13. The sludge is produced at a rate of 75 m^3/day at a concentration of 5.0% solids, of which 70% are volatile. What volume of reactor will be required to destroy 60% of the volatile solids and how much power will be required to mix it? How much gas will be produced? What fraction of the residual solids are volatile?

12. Assume that you wish to digest the sludge in the preceding problem in an anaerobic CSTR with sludge recycle at a loading of 0.32 lb VS/(ft^3·day). What volume of reactor will be required to destroy 60% of the volatile solids and how much power will be required to mix it?

13. A soluble wastewater with a COD of 15,000 mg/liter and a flow rate of 25 m^3/hr is to be treated by anaerobic contact. Lab studies indicate that a MCRT of 20 days will give stable operation while maintaining a low level of volatile acids in the effluent and reducing the soluble COD to 2000 mg/liter. During the studies the observed yield was 0.1 mg cells formed/mg COD removed when the MCRT was 20 days and β was 1.25 mg COD/mg cells. Studies on the settling characteristics of the resulting sludge indicated that the solids could be concentrated to 14,000 mg/liter in a settler of reasonable size. Assuming a recycle sludge concentration of 12,000 mg/liter and a loading of 0.35 lb COD/(ft^3·day) determine

the following: (a) the reactor volume, (b) the cell concentration, (c) the recycle ratio, (d) the maximum sludge wastage rate, and (e) the heat content of the methane produced daily.

14. Explain what is meant by the term stuck digester.

15. Explain what each of the following observations might signify about an anaerobic reaction: (a) a rapid rise in the concentration of volatile acids, (b) a volatile acids to alkalinity ratio of 1.0, (c) a rapid rise in the concentration of volatile acids accompanied by a decrease in the methane production rate, (d) a rise in the concentration of volatile acids accompanied by a slow increase in the methane production rate, (e) a decrease in the rate of methane production and in the concentration of volatile acids.

16. Discuss the relative merits of lime and soda ash for pH control in an anaerobic reactor.

REFERENCES AND FURTHER READING

1. P. L. McCarty, "The methane fermentation," *Principles and Applications of Aquatic Microbiology,* edited by H. Heukelekian and N. C. Dondero, J. Wiley and Sons, Inc., New York, 314-343, 1964.

2. P. L. McCarty, "Anaerobic waste treatment fundamentals," *Public Works,* 95; No. 9, 107-112; No. 10, 123-126; No. 11, 91-94; No. 12, 95-99, 1964.

3. D. F. Toerien *et al.* "The bacterial nature of the acid-forming phase of anaerobic digestion," *Water Research,* 1, 497-507, 1967.

4. P. G. Thiel *et al.* "Interrelations between biological and chemical characteristics in anaerobic digestion," *Water Research,* 2, 391-408, 1968.

5. E. J. Kirsch, "Studies on the enumeration and isolation of obligate anaerobic bacteria from digesting sewage sludge," *Developments in Industrial Microbiology,* 10, 170-176, 1969.

6. R. A. Mah, and C. Susman, "Microbiology of anaerobic sludge ferementation - 1 - Enumeration of the nonmethanogenic anaerobic bacteria," *Applied Microbiology,* 16, 358-361, 1968.

7. P. N. Hobson and B. G. Shaw, "The bacterial population of piggery waste anaerobic digesters," *Water Research,* 8, 507-516, 1974.

8. R. A. Mah, ESE Notes, Univ. of North Carolina, 6, 1, 1969.

9. J. F. Andrews and E. A. Pearson, "Kinetics and characteristics of volatile acid production in anaerobic fermentation processes," *International Journal of Air and Water Pollution,* 9, 439-461, 1965.

10. C. T. Gray and H. Gest, "Biological formation of molecular hydrogen," *Science,* 148, 186-192, 1965.

11. E. L. Iannotti, *et al.* "Glucose fermentation products of *Ruminococcus albus* grown in continuous culture with *Vibrio succinogenes* - Changes caused by interspecies transfer of H_2," *Journal of Bacteriology,* 114, 1231-1240, 1973.

12. C. C. Scheifinger, *et al.* "H_2 production by *Selenomonas ruminantium* in the absence and presence of methanogenic bacteria," *Applied Microbiology,* 29, 480-483, 1975.

13. C. A. Reddy, *et al.* "Characteristics of S organism isolated from *Methanobacillus omelianskii*," *Journal of Bacteriology,* 109, 539-542, 1972.

14. R. M. Sykes and E. J. Kirsch, "Fermentative patterns in the initial stages of sewage sludge digestion," *Developments in Industrial Microbiology,* 11, 357-366, 1970.

15. T. G. Shea, *et al.* "Kinetics of hydrogen assimilation in the methane fermentation," *Water Research,* 2, 833-848, 1968.

16. R. S. Wolfe, "Microbial formation of methane," *Advances in Microbial Physiology,* 6, 107-146, 1971.

17. L. van den Berg, *et al.* "Factors affecting rate of methane formation from acetic acid by enriched methanogenic cultures," *Canadian Journal of Microbiology,* 22, 1312-1319, 1976.

18. R. A. Mah, *et al.* "Studies on an acetate-fermenting strain of *Methanosarcina*," *Applied and Environmental Microbiology,* 35, 1174-1184, 1978.

19. J. G. Ziekus, *et al.* "Bacterial methanogenesis: Acetate as a methane precursor in pure culture," *Archives for Microbiology,* 104, 129-134, 1975.

20. P. J. Shuba, "Propionic acid fermentation in sludge," *Dissertation Abstracts B,* 34, 5582, 1974.

21. J. G. Ferry and R. S. Wolfe, "Anaerobic degradation of benzoate to methane by a microbial consortium," *Archives for Microbiology,* 107, 33-40, 1976.

22. A. F. Weland and P. N. Cheremisinoff, "Anaerobic digestion of sludge," *Water and Sewage Works,* 122, No. 10, 80-83; No. 11, 45-47, 1975.

23. H. Schmidt and C. N. Sawyer, "High-rate sludge digestion," *Journal of the Boston Society of Civil Engineers,* 42, 1-17, 1955.

24. W. N. Torpey, "Loading to failure of a pilot high-rate digester," *Sewage and Industrial Wastes,* 27, 121-133, 1955.

25. P. L. McCarty, "Kinetics of waste assimilation in anaerobic treatment," *Developments in Industrial Microbiology,* 7, 144-155, 1965.

26. C. N. Sawyer, "An evaluation of high-rate digestion," in *Biological Treatment of Sewage and Industrial Wastes, Vol II, Anaerobic Digestion and Solids-Liquid Separation,* edited by J. McCabe and W. W. Eckenfelder, Jr., Reinhold Publishing Corp., New York, 48-60, 1958.

27. R. R. Dague, *et al.* "Solids retention in anaerobic waste treatment systems," *Journal of the Water Pollution Control Federation,* 42, R29-R46, 1970.

28. J. T. Pfeffer, *et al.* "Population dynamics in anaerobic digestion," *Journal of the Water Pollution Control Federation,* 39, 1305-1322, 1967.

29. J. T. Pfeffer, "Increased loadings on digesters with recycle of digested solids," *Journal of the Water Pollution Control Federation,* 40, 1920-1933, 1968.

30. A. W. Lawrence, "Application of process kinetics to design of anaerobic processes," in *Anaerobic Biological Treatment Processes,* American Chemical Society Advances in Chemistry Series, *105, 163-189, 1971.*

31. A. W. Lawrence and P. L. McCarty, "Kinetics of methane fermentation in anaerobic treatment," *Journal of the Water Pollution Control Federation,* 41, R1-R17, 1969.

32. R. R. Dague, "Application of digestion theory to digester control," *Journal of the Water Pollution Control Federation,* 40, 2021-2032, 1968.

33. E. J. Kirsch and R. M. Sykes, "Anaerobic digestion in biological waste treatment," *Progress in Industrial Microbiology,* 9, 155-237, 1971.

34. P. H. Smith and R. E. Hungate, "Isolation and characterization of *Methanobacterium ruminantium* n.sp.," *Journal of Bacteriology,* 75, 713-718, 1958.

35. L. van den Berg, *et al.* "Assessment of methanogenic activity in anaerobic digestion: Apparatus and method," *Biotechnology and Bioengineering,* 16, 1459-1469, 1974.

36. P. N. Hobson and R. Summers, "The continuous culture of anaerobic bacteria," *Journal of General Microbiology,* 47, 53, 1967.

37. E. P. Willimon, Jr. and J. F. Andrews, "Multi-stage biological processes for waste treatment," *Proceedings of the 22nd Industrial Waste Conference,* Purdue University Engineering Extension Series, No. 129, 645-660, 1967.

38. I. J. Kugelman and K. K. Chin, "Toxicity, synergism, and antagonism in anaerobic waste treatment processes," in *Anaerobic Biological Treatment Processes,* American Chemical Society Advances in Chemistry Series, 105, 55-90, 1971.

39. P. N. Hobson and B. G. Shaw, "Inhibition of methane production by *Methanobacterium formicicum*," *Water Research,* 10, 849-852, 1976.

40. P. L. McCarty and R. E. McKinney, "Volatile acid toxicity in anaerobic digestion," *Journal of the Water Pollution Control Federation,* 33, 223-232, 1961.

41. N. D. Ierusalimsky, "Bottle-necks in metabolism as growth-rate controlling factors," in *Microbial Physiology and Continuous Culture,* edited by E. O. Powell, *et al.,* Her Majesty's Stationery Office, London, 23-33, 1967.

42. J. F. Andrews, "Dynamic model of the anaerobic digestion process," *Journal of the Sanitary Engineering Division, ASCE,* 95, No. SA1, 95-116, 1969.

43. A. W. Lawrence, *et al.* "The effect of sulfides on anaerobic treatment," *Proceedings of the 19th Industrial Waste Conference,* Purdue University Engineering Extension Series, No. 117, 343-357, 1964.

44. F. E. Mosey and D. A. Hughes, "The toxicity of heavy metal ions to anaerobic digestion," *Water Pollution Control,* 74, 18-39, 1975.

45. A. W. Lawrence and P. L. McCarty, "The role of sulfide in preventing heavy metal toxicity in anaerobic treatment," *Journal of the Water Pollution Control Federation,* 37, 392-409, 1965.

46. L. van den Berg and C. P. Lentz, "Anaerobic digestion of pear waste; Factors affecting performance," *Proceedings of the 27th Industrial Waste Conference,* Purdue University Engineering Extension Series, No. 141, 313-323, 1972.

47. R. H. Clark and V. D. Orr, "Digestion : Concentration - loading time limits," *Journal of the Sanitary Engineering Division, ASCE,* 98, 809-811, 1972.

48. P. L. McCarty, "Energetics and kinetics of anaerobic treatment," in *Anaerobic Biological Treatment Processes,* American Chemical Society Advances in Chemistry Series, 105, 91-107, 1971.

49. S. Ghosh and F. G. Pohland, "Kinetics of substrate assimilation and product formation in anaerobic digestion," *Journal of the Water Pollution Control Federation,* 46, 748-759, 1974.

50. S. Ghosh, *et al.* "Anaerobic acidogenesis of wastewater sludge," *Journal of the Water Pollution Control Federation,* 47, 30-45, 1975.

51. J. F. Andrews and S. P. Graef, "Dynamic modeling and simulation of the anaerobic digestion process," in *Anaerobic Biological Treatment Processes,* American Chemical Society Advances in Chemistry Series, 105, 126-162, 1971.

52. Metcalf and Eddy, Inc., *Wastewater Engineering, Treatment, Disposal, Reuse.* McGraw-Hill Book Co., New York, N.Y., 424, 1979.

53. C. Zickefoose and R. B. J. Hayes, "Operations manual - Anaerobic sludge digestion," Publication #EPA 430/9-76-001, Office of Water Program Operations, U.S. Environmental Protection Agency, Washington, D.C., Feb. 1976.

54. R. C. Loehr, "Design of anaerobic digestion systems," *Journal of the Sanitary Engineering Division, ASCE,* 92, No. SA1, 19-29, 1966.

55. F. H. Verhoff, *et al.* "Mixing in anaerobic digestion," *Biotechnology and Bioengineering,* 16, 757-770, 1974.

56. J. Zoltek Jr. and A. L. Gram, "High-rate digester mixing study using radio-isotope tracer," *Journal of the Water Pollution Control Federation,* 47, 79-84, 1975.

57. J. F. Malina Jr. and E. M. Miholits, "New developments in the anaerobic digestion of sludges" in *Advances in Water Quality Improvement,* edited by E. F. Gloyna and W. W. Eckenfelder Jr., University of Texas Press, Austin, 355-379, 1968.

58. S. P. Graef and J. F. Andrews, "Stability and control of anaerobic digestion," *Journal of the Water Pollution Control Federation,* 46, 666-683, 1974.

59. M. G. Capri and G. V. R. Marais, "pH Adjustment in anaerobic digestion," *Water Research,* 9, 307-313, 1975.

60. W. Murray, "pH Control in anaerobic sewage sludge digestion," MSCE Thesis, Purdue University, West Lafayette, IN, 1970.

61. S. P. Graef, "Anaerobic digester operation at the Metropolitan Sanitary Districts of Greater Chicago," paper presented at the National Conference on Municipal Sludge Management, Pittsburgh, Penn., June 1974.

62. A. A. Estrada, "Design and cost consideration in high-rate sludge digestion," *Journal of the Sanitary Engineering Division, ASCE,* 86, No. SA3, 111-127, 1960.

63. W. C. Hiatt, *et al.* "Anaerobic digestion of rum distillery waste," *Proceedings of the 28th Industrial Waste Conference,* Purdue University Engineering Extension Series, No. 142, 966-976, 1973.

64. B. P. Sen and T. R. Bhaskaran, "Anaerobic digestion of liquid molasses distillery wastes," *Journal of the Water Pollution Control Federation,* 34, 1015-1025, 1962.

65. G. G. Cillie, *et al.* "Anaerobic digestion - IV - The application of the process in waste purification," *Water Research,* 3, 623-643, 1969.

66. T. H. Chadwick and E. D. Schroeder, "Characterization and treatability of pomace stillage," *Journal of the Water Pollution Control Federation,* 45, 1978-1984, 1973.

67. D. Newton, *et al.* "Pilot plant studies for the evaluation of methods of treating brewery wastes," *Proceedings of the 16th Industrial Waste Conference,* Purdue University Engineering Extension Series, No. 109, 332-350, 1962.

68. A. J. Steffen and M. Bedker, "Operation of full-scale anaerobic contact treatment plant for meat packing waste," *Proceedings of the 16th Industrial Waste Conference,* Purdue University Engineering Extension Series, No. 109, 423-437, 1962.

69. J. Hemens, *et al.* "Full-scale anaerobic digestion of effluents from the production of maize-starch," *Water and Waste Treatment,* 9, No. 3, 16-18, 1962.

70. H. A. Painter, *et al.* "Treatment of malt whiskey distillery wastes by anaerobic digestion," *The Brewer's Guardian,* August 1960.

71. G. J. Stander and R. Synders, "Effluents from fermentation industries, Part V," *Proceedings of the Institute of Sewage Purification,* 447-458, 1950.

72. J. T. Ling, "Pilot investigation of starch-gluten waste treatment," *Proceedings of the 16th Industrial Waste Conference,* Purdue University Engineering Extension Series No. 109, 217-231, 1962.

73. A. E. J. Pettet, *et al.* "The treatment of strong organic wastes by anaerobic digestion," *Journal of the Institution of Public Health Engineers,* 58, 170-191, 1959.

74. R. R. McNary, *et al.* "Experimental treatment of citrus waste water," *Proceedings of the 8th Industrial Waste Conference,* Purdue University Engineering Extension Series No. 83, 256-274, 1954.

Further Reading

Benefield, L. D. and Randall, C. W., *Biological Process Design for Wastewater Treatment,* Prentice-Hall, Inc., Englewood Cliffs, N.J., 1980. See Chapter 8.

Bryant, M. P., "Microbial methane production - Theoretical aspects," *Journal of Animal Science,* 48, 193-201, 1979. This review summarizes the literature relating to inter-species hydrogen transfer.

Buhr, H. O. and Andrews, J. F., "The thermophilic anaerobic digestion process," *Water Research,* 11, 129-143, 1977. This review summarizes many of the current mathematical models.

Hobson, P. N. and Shaw, B. G., "The role of strict anaerobes in the digestion of organic material," in *Microbial Aspects of Pollution,* edited by G. Sykes and F. A. Skinner, Academic Press, New York, NY, 103-121, 1971.

Hobson, P. N., *et al.* "Anaerobic digestion of organic matter," *CRC Critical Reviews of Environmental Control,* 4, 131-191, 1974. A detailed and comprehensive review of anaerobic digestion with emphasis upon microbiology and biochemistry.

Kotze, J. P., *et al.* "Anaerobic digestion - II - The characterization and control of anaerobic digestion," *Water Research,* 3, 459-494, 1969.

Pohland, F. G. (Editor), *Anaerobic Biological Treatment Processes,* Advances in
    Chemistry Series, #105, American Chemical Society, Washington, D.C., 1971.
    Contains papers presented at a symposium held at the 159th meeting of the
    ACS, in Houston, Texas, Feb. 1970.

Pohland, F. G. and Ghosh, S., "Developments in anaerobic treatment processes,"
    Biotechnology Bioengineering Symposium No. 2, 85-106, 1971.

Pretorius, W. A., "Anaerobic digestion - III - Kinetics of anaerobic fermentation,"
    Water Research, 3, 545-558, 1969.

Toerien, D. F. and Hattingh, W. H. J., "Anaerobic digestion - I - The microbiology
    of anaerobic digestion," *Water Research,* 3, 385-416, 1969.

CHAPTER 22

DENITRIFICATION

As discussed in Chapter 20, nitrification is used to reduce the amount of ammonia nitrogen discharged to a receiving body of water, thereby reducing the oxygen demand associated with its oxidation. This does not reduce the mass of nitrogen discharged, however; it merely changes its state. Consequently, nitrification alone will do nothing to alleviate the problem of eutrophication because that requires a reduction in the availability of nutrients in the aquatic environment. One way of minimizing the availability of nitrogen is to discharge it to the atmosphere as nitrogen gas through the application of biological denitrification. Actually, like nitrification, denitrification is really a type of biochemical reaction or conversion, rather than a named biochemical operation, and as such it can be carried out in any of a number of reactor types. In this chapter we will consider the requirements for performing denitrification in CSTR's with cell recycle and in packed towers.

22.1 DESCRIPTION OF DENITRIFICATION SYSTEMS

22.1.1 Microbiological and Biochemical Characteristics

As defined in Section 7.1.3 denitrification is the reduction of nitrate nitrogen (NO_3^--N) as it serves as the terminal hydrogen acceptor for microbial respiration in the absence of molecular oxygen. As such, it is an alternative to the reduction of oxygen, and thus is called anaerobic respiration (see Section 9.1.3). The bacteria responsible for denitrification are facultative and utilize the same basic biochemical pathways during both aerobic and anaerobic respiration. The only major differences are in the enzymes catalyzing the terminal electron transfer and their sites in the electron transport chain, as shown in Fig. 9.2. Denitrification can be accomplished by a large number of microbial genera commonly found in wastewater treatment systems, including *Achromobacter, Aerobacter, Alcaligenes, Bacillus, Flavobacterium, Micrococcus, Proteus,* and *Pseudomonas* [1,2], thereby making the establishment of a denitrifying culture relatively easy.

There are two types of enzyme systems involved with the reduction of NO_3^--N: assimilatory and dissimilatory. Assimilatory nitrate reduction converts NO_3^--N to ammonia nitrogen for use by the cells in biosynthesis, and functions when NO_3^--N is the only form of nitrogen available. Dissimilatory nitrate reduction results in the formation of nitrogen gas from NO_3^--N, and is the one responsible for the denitrification of wastewater. The term denitrification will be restricted herein to dissimilatory nitrate reduction and assimilatory nitrate reduction will not be discussed further.

Because dissimilatory nitrate reduction serves as an alternative means of microbial respiration, there has been considerable interest in the influence of oxygen upon the responsible enzyme system [3,4,5]. Although there are exceptions, as a general rule the presence of oxygen in the medium (and/or its active utilization as the terminal hydrogen acceptor) represses the synthesis of the nitrate reducing enzyme system. When oxygen is absent, or is present in amounts which are insufficient to meet the needs of the culture, derepression occurs and the enzymes are synthesized. The effect of oxygen upon the activity of the enzymes depends upon the bacterial species involved. In some, the activities are diminished in the presence of oxygen, whereas in others they are not. Thus denitrification probably can occur at diminished rates in the presence of oxygen, provided that anoxic conditions had previously existed during which enzyme synthesis could occur. One factor complicating the determination of the effects of oxygen upon denitrification in wastewater treatment systems is the flocculent nature of the cultures involved. Because of the large size of microbial floc particles there is likely to be a region in the interior which is devoid of oxygen. Thus denitrification could occur in the interior of the floc

even when oxygen was present in the medium [6]. Nevertheless, considering all of the
factors known about the synthesis and activity of the enzymes responsible for denitrification, it is generally agreed that the level of dissolved oxygen should approach zero in order to achieve consistently good performance.

The steps in the reduction of nitrate were given in Eq. 9.2:

$$NO_3^- \rightarrow NO_2^- \rightarrow NO \rightarrow N_2O \rightarrow N_2 \tag{9.2}$$

Any of the last three substances can be released as a gaseous end product, but for
minimal environmental degradation the release of N_2 would be preferred. Relatively
few studies have been performed on the factors influencing the type of end products
formed, but it is known that the type of organism is important. The pH of the medium
is also important, with values below 7.3 causing an increase in the production of
N_2O [7,8]. Generally, however, N_2 appears to be the major product formed by the
mixed cultures used in wastewater treatment [9,10]. Consequently, it will be assumed herein that N_2 is the only product formed in significant amounts.

The perspective from which denitrification must be viewed is just the opposite
of that from which most wastewater treatment systems are viewed. In most systems the
wastewater contains organic matter which serves as an electron donor and the designer
must provide for the addition of the proper quantity of oxygen (the electron acceptor)
to allow complete conversion of that organic matter to cell material and carbon
dioxide. The objective during the design of a denitrification system, however, is the
removal of an electron acceptor (nitrate) and to do this, a sufficient amount of
electron donor (organic matter) must be made available. This requires consideration
of both the stoichiometric and the kinetic requirements of the system. In our consideration of the supply of oxygen to an activated sludge system the stoichiometric
requirement was determined through the use of Eq. 13.47. The kinetic requirement was
met by specifying an oxygen transfer rate which would provide the stoichiometric requirement while maintaining a dissolved oxygen concentration of 2.0 mg/liter, a value
large enough to ensure that oxygen would not be rate limiting. With denitrification,
the same approach can be used to determine the stoichiometric amount of organic matter
required. The concentration which must be provided in the reactor to meet the kinetic
requirement, however, will depend upon the specific design. Thus these factors will
be considered in detail in Section 22.2.1. Let us now turn our attention to the
factors influencing the selection of an electron donor.

Throughout this text the concentration of organic matter has been measured in
terms of its oxygen demand. Because NO_3^--N serves as the electron acceptor its concentration can also be expressed on an oxygen equivalence basis, thereby allowing
continuation of the use of oxygen demand as the measure of the amount of electron
donor (organic matter) required. Looking at the half reactions in Table 9.4 it can
be seen that 1/5 mole of NO_3^- is equivalent to 1/4 mole of O_2. When converted to a
mass basis each mg of NO_3^--N can accept the same number of electrons as 2.86 mg of O_2.

In other words, each mg of NO_3^--N is equivalent to 2.86 mg O_2 in its ability to bring about the oxidation of organic matter.

If all of the organic matter added to a denitrification reactor were converted to carbon dioxide and water it would be easy to calculate the amount (expressed as T_bOD) required to reduce all of the NO_3^--N to N_2. It would be equal to 2.86 times the NO_3^--N concentration. However, some organic matter is always converted into cell material and thus is not oxidized completely so that the amount required is always greater than that value. Furthermore, the higher the yield (mg cells formed/mg T_bOD removed), the greater the fraction of organic matter converted to cells and the greater the amount required. This suggests that the electron donor should be one with a low yield if the quantity to be added is to be minimized. This has the added benefit of reducing the quantity of sludge produced as well. From Chapter 9 it will be recalled that single-carbon compounds generally have low yields because the energy required to synthesize cell constituents from them is large. Thus a single-carbon compound would be an ideal candidate for the electron donor. In addition, if the electron donor is highly reduced, the T_bOD per unit mass will be high, thereby minimizing the quantity which must be purchased for addition. The most highly reduced single-carbon compound is methane, but the practicality of its use is questionable. Another highly reduced single-carbon compound is methanol (CH_3OH; $T_bOD = 1.5$ mg/mg), which is widely available, of consistent quality, and relatively inexpensive. Furthermore, its yield is low, thereby minimizing the amount required [11]. For all of these reasons, methanol is the compound usually chosen when an external electron donor must be added [12] and it is the one to which primary consideration will be given herein. Other electron donors can be used, however, particularly when they are already present in the wastewater.

22.1.2 Physical Characteristics

Denitrification reactors fall into two broad categories: slurry reactors and fixed-film reactors. The slurry reactors are quite similar in appearance to those used for activated sludge (Chapter 16) except that the mixing systems are designed to minimize oxygen transfer while maintaining the sludge in suspension. The most effective way of excluding oxygen is to use a closed reactor, but it is not uncommon to see open ones as well. There is a great deal more variety among packed tower reactors. This makes it difficult to establish a general approach to design [1], although one method looks promising [13]. One aspect of the variation is in the type of media employed, which covers the range from corrugated plastic sheets to fine sand grains. Another is in the void spaces, which may be filled with either liquid or nitrogen gas. A third is in the fluid regime, which can range from downward in thin films to upward at a velocity sufficient to fluidize sand particles [14]. A detailed discussion of all of the reactor configurations which have been proposed is beyond the scope of this text, and consequently, the interested reader is referred to Chapter 5 of the Nitrogen Control Manual by Parker *et al.* [1].

Another method of classifying denitrification systems is with respect to the
source of the electron donor: (1) it may be added as needed; (2) it may come from
inside the organisms themselves, such as when microbial decay occurs; and (3) it may
be the organic matter in the wastewater itself. The first source is the most common
in the United States because of the treatment which the wastewater undergoes prior
to denitrification. The nitrogen in most untreated wastewaters is in the form of
either ammonia or organic nitrogen and is only converted to the nitrate form as the
wastewater undergoes treatment. As a natural result of the treatment conditions required to do this (see Chapter 20) almost all of the organic matter is removed from
solution. Thus it will be necessary to add organic matter to serve as the electron
donor. Methanol is usually chosen. It has been argued, however, that it is uneconomic to add an external electron donor when the organisms used in the activated
sludge process contain organic compounds which will be utilized for maintenance energy
if they are left without exogenous energy supplies. Consequently, a number of workers,
particularly in Europe, have advocated denitrification techniques which use the activated sludge developed during carbon removal to remove the nitrate by endogenous decay (see references in [1] and [2]). Other workers have argued that it should be
possible to make use of the organic matter present in the untreated wastewater. The
problem lies in how to do this while still achieving a high degree of nitrogen removal across the entire system. Several ingenious flow schemes have been proposed
but a discussion of them is beyond the scope of the text. Thus, the interested reader
is referred to the discussions in [1] and [2].

The purpose of this chapter is to provide an understanding of the fundamental
concepts governing denitrification systems. This can be done most easily by considering only a single reactor receiving a wastewater which contains predominantly
NO_3^--N and little organic matter, in other words, the effluent from a nitrification
reactor. Consequently, the remainder of this chapter will be limited to that situation. It is important, however, that the reader not interpret this limitation as
a criticism of other denitrification systems. To the contrary, the reader is urged
to consult references [1] and [2] for further reading about them.

22.1.3 Applications

Denitrification systems are designed to meet one objective, the conversion of NO_3^--N
to nitrogen gas. Relatively few denitrification systems have been built for normal
domestic wastewaters, because the levels of NO_3^--N resulting from their treatment are
not yet considered high enough to cause significant problems in most situations.
Thus denitrification is generally applied to them only when they are to be reused or
discharged to impoundments within which eutrophication is likely to be a problem.
Denitrification is finding wider application in the treatment of industrial wastewaters. For example, as discussed in Section 20.1.3, it is being used in conjunction
with nitrification to reduce the concentration of nitrogen in wastewaters which are
high in ammonia. It is also being used to remove nitrogen from industrial wastes

which contain high concentrations of nitrate, such as those from munitions manufacturing. Nonbiological nitrogen removal systems are also available [1], but to date denitrification systems are the most popular.

22.1.4 Merits and Defects

Like nitrification and activated sludge, the major merit associated with biological denitrification is its capability for producing an effluent of high quality at reasonable cost. Its major defect is our limited knowledge about it. Among the problems still to be solved are how to: (1) select the proper media size for packed towers, (2) minimize the increase in head loss through such systems without disrupting their performance; and (3) ensure reliable settling in CSTR's with cell recycle. There is a considerable research effort underway on denitrification so that most of the major problems will soon be overcome.

22.2 PERFORMANCE OF DENITRIFICATION SYSTEMS

The models in Chapters 12, 13, and 14 were based upon the concept of a single rate limiting substance, e.g., soluble organic matter. This was possible because all other materials, including oxygen could be provided in excess without causing a deterioration in effluent quality. The situation is more complicated for denitrification, however, because the amount of electron donor added to the wastewater must balance the amount of nitrate (electron acceptor) to be removed. If organic matter is added in excess, it will pass to the effluent where it will exert an oxygen demand, thereby reducing the quality of the effluent. If insufficient organic matter is added, some NO_3^--N will remain and the treatment objective will not be attained. If exactly the right amount is added, the concentrations of both the organic matter and the nitrate are likely to be rate limiting, a situation which was not covered by the performance models presented earlier. Thus before the theoretical performance of denitrification reactors can be discussed, it will be necessary to consider the kinetics of two rate limiting materials.

22.2.1 Kinetics of Denitrification

Conceptually, limitation by both NO_3^--N and the organic electron donor in a denitrification system is similar to limitation by both oxygen and the organic matter in an aerobic system. Such a dual limitation has been of concern to biochemical engineers desiring to maximize the production rate of biomass in single-cell protein systems and thus has been the subject of several papers [15,16,17,18]. One technique which has been used successfully to model this dual limitation expresses the specific growth rate as a double Monod-type function of the concentrations of the two limiting substances, S_1 and S_2:

$$\mu = \frac{\mu_m S_1 S_2}{(K_{s1} + S_1)(K_{s2} + S_2)} \tag{22.1}$$

A similar equation has been proposed for denitrification [1], and the success of Eq. 22.1 in modeling the performance of systems limited by both oxygen and organic carbon suggests that it is appropriate. Therefore let us consider briefly the implications of Eq. 22.1.

As discussed by Bader et $al.$ [19], Eq. 22-1 can be considered as defining the contours of constant μ on a graph of S_1 versus S_2 as shown in Fig. 22.1. This graph emphasizes that any of the pairs of values of S_1 and S_2 defining a particular contour could occur in a CSTR operated to give that particular specific growth rate. There is no single value of S_1 or S_2 associated with that value of μ as there would be if only one of them were rate limiting. How then does one determine the actual values of S_1 and S_2 existing for any particular situation? To do this an additional relationship between the two substrates is needed and it is provided by stoichiometry. The cell concentration in the CSTR is given by

$$X = \frac{Y_{g1}(S_{1o} - S_1)}{1 + b\tau} = \frac{Y_{g2}(S_{2o} - S_2)}{1 + b\tau} \qquad (22.2)$$

where S_{1o} and S_{2o} are the concentrations of S_1 and S_2 in the influent and Y_{g1} and Y_{g2} are the true growth yields for each substrate. Simplifying Eq. 22.2 reveals that it is just the equation for a straight line which has slope Y_{g1}/Y_{g2} and which must pass through the point (S_{2o}, S_{1o}).

$$S_2 = (Y_{g1}/Y_{g2})S_1 - (Y_{g1}/Y_{g2})S_{1o} + S_{2o} \qquad (22.3)$$

This "stoichiometric line" is shown in Fig. 22.1. The intersection of the stoichiometric line with the curve corresponding to μ for the reactor defines the actual

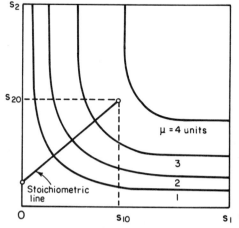

Figure 22.1.   Contour plot of specific growth rate as a function of two substrate concentrations. Both substrates are rate-limiting. (From F.G. Bader et al., Biotech. Bioeng., 17, 279-283, 1975 [19]. Reproduced by permission of John Wiley and Sons, Inc.).

values of S_1 and S_2 existing under the influent conditions characterized by S_{1o} and S_{2o}. In the situation depicted in Fig. 22.1, substrate 2 is being fed in excess of the stoichiometric amount because a finite concentration of it would exist in the reactor when the concentration of substrate 1 was zero (the stoichiometric line intersects the S_2 axis). Nevertheless, both substrates are rate limiting because a slight change in either S_{1o} or S_{2o} will cause a change in both S_1 and S_2. Now consider the situation depicted in Fig. 22.2 in which the value of S_{2o} has been increased considerably, thereby shifting the stoichiometric line upwards. In this case it can be seen that S_1 is rate limiting because it remains relatively constant when slight changes are made in either S_{1o} or S_{2o}. Before it can be assumed that only one substrate is rate limiting it must be shown that the stoichiometric line intersects the μ contours in the regions where they are essentially linear. This requires knowledge of three things: (1) the ratio of the yields, Y_{g1}/Y_{g2}, which determines the slope of the stoichiometric line; (2) the influent concentrations S_{1o} and S_{2o}, which determine its position; and (3) the relative values of K_{s1} and K_{s2}, which determine the shapes of the μ contours.

The importance of dual substrate limitation to denitrification arises from the desire to achieve low concentrations of both NO_3^--N and the electron donor in the effluent because such a situation is likely to cause the stoichiometric line to intersect the μ contour in the curved region. If slight changes were made in either S_{1o} or S_{2o} during experimentation to determine the kinetic parameters, the result would be changes in both S_1 and S_2 even though μ was held constant. This could be confusing

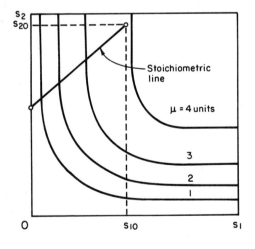

Figure 22.2. Contour plot of specific growth rate as a function of two substrate concentrations. Substrate 2 is fed in excess so that only substrate 1 is rate-limiting. *(From F.G. Bader et al. Biotech. Bioeng., 17, 279-283, 1975 [19]. Reproduced by permission of John Wiley and Sons, Inc.).*

if the possibility of dual substrate limitation were not recognized. If the investigator were aware of this possibility, however, correct interpretation of the data
would be more likely.

Although no studies have been performed to confirm the applicability of Eq. 22.1
to denitrification systems, the similarity between anaerobic and aerobic respirations
suggests that it can be used. Thus it will be adopted herein for illustrative purposes. To emphasize the nature of the two potential rate-limiting substances,
Eq. 22.1 will be rewritten as

$$\mu = \frac{\mu_m SN}{(K_s + S)(K_n + N)} \tag{22.4}$$

where S refers to the concentration of electron donor expressed as T_bOD, N the concentration of electron acceptor expressed as NO_3^--N, and K_s and K_n are the saturation
constants for each. The value of μ_m will depend upon the type of electron donor employed. Calculations made from data presented in one study [20] indicate a value of
0.33 hr^{-1} on methanol. Similarly, K_s will also depend upon the type of carbon
source. One study [in 1] done with methanol found it to be 0.15 mg/liter as COD
whereas another found it to lie between 4 and 13 mg/liter, depending upon the temperature [21]. The saturation constant for NO_3^--N, K_n, has also been found to be low
with values around 0.1 - 0.2 mg/liter NO_3^--N being reported [1,2,20]. Thus the curved
portions of the contours in figures like 22.1 are likely to be small. The true
growth yield with respect to the electron donor, Y_{gs}, is smaller when NO_3^--N serves
as the electron acceptor than it is when oxygen serves [3]. Thus the value for a
given system must be determined under the appropriate conditions. McCarty et al.
[11] have calculated a theoretical value for Y_{gs} on methanol of 0.16 mg cells
formed/mg T_bOD removed and others have reported experimental values which agree
[21,22,23]. Because NO_3^--N serves as the electron acceptor during biosynthesis, the
true growth yield with respect to NO_3^--N (Y_{gn}) must be in proportion to Y_{gs}. If
enough NH_4^+-N is present to make the amount of NO_3^--N incorporated into cell material
small, the relationship is:

$$Y_{gn} = \frac{2.86\, Y_{gs}}{1 - \beta Y_{gs}} \tag{22.5}$$

where Y_{gs} is expressed as mg cells/mg T_bOD removed and Y_{gn} is expressed as mg cells/mg
NO_3^--N removed. If the growth medium contains no NH_4^+-N nitrogen, a portion of the
NO_3^--N must be reduced to NH_4^+-N for incorporation into the cell material. This will
make the relationship between Y_{gn} and Y_{gs} different from that expressed by Eq. 22.5.
The exact relationship will depend upon what fraction of the NO_3^--N must be incorporated into the cells. Thus under those circumstances where incorporation of NO_3^--N
will occur, the relationship between Y_{gn} and Y_{gs} should be determined experimentally.
Equation 22.5 suggests that if Y_{gs} is 0.16 and β is 1.25, Y_{gn} will be 0.57 mg
cells/mg NO_3^--N removed. Experimental values of Y_{gn} [20] are consistent with values

calculated from Eq. 22.5 using Y_{gs} values reported in the literature [1,21]. Microbial decay occurs during denitrification just as it does in aerobic systems and the values of b which have been reported are around 0.002 hr^{-1} [1,21].

The values of the kinetic parameters presented in the previous paragraph are average for typical conditions, i.e., neutral pH, temperature around 20°C, etc. In general pH and temperature affect the kinetics of denitrification just as they do any other biochemical operation. Parker *et al.* [1] have summarized the effects of pH observed by a number of workers and these are shown in Fig. 22.3. It is apparent from the figure that denitrification rates (and thus μ_m) are depressed below pH 6.0 and above pH 8.0. In general, the highest rates occur within the range of pH 7.0 to 7.5 which is consistent with the previously cited effect of pH upon the gaseous end products from denitrification. Figure 22.4 summarizes the effects of temperature on μ_m for denitrifying systems as compiled by Parker *et al.* [1]. These data are consistent with other reports which suggest that temperature exerts a larger effect below 15°C than above it [12]. It can be seen from the figure, however, that the effects are very system specific and thus pilot studies will be required to determine the exact effects for a given system. The presence of dissolved oxygen can affect the rate of denitrification, although little quantitative data exist. Additional electron donor (organic material) must be supplied to remove any oxygen which enters the system. Thus good engineering practice dictates that every reasonable effort be made to exclude oxygen. Finally, it should be mentioned that methanol acts as an inhibitory substrate at high concentration (>3000 mg/liter) [24]. Such concentrations will seldom be reached in most systems.

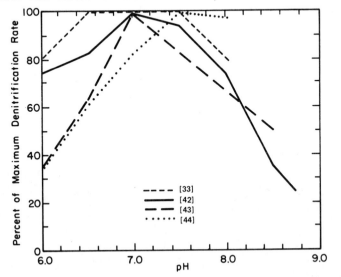

Figure 22.3.   Effects of pH upon the rate of denitrification
as observed by several workers and summarized
by Parker *et al.* [1]. The reference numbers
refer to the sources of the data.

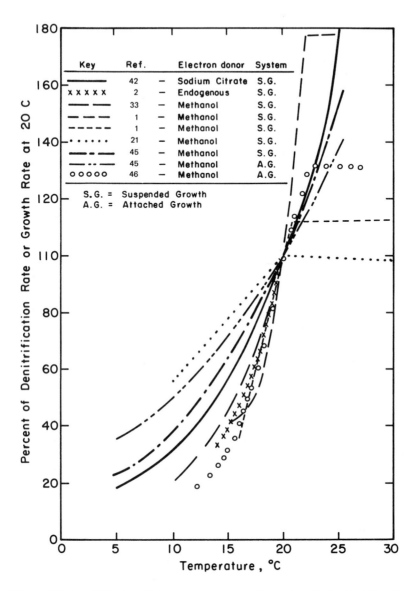

Key	Ref.		Electron donor	System
———	42	—	Sodium Citrate	S.G.
x x x x x	2	—	Endogenous	S.G.
—— ——	33	—	Methanol	S.G.
— — —	1	—	Methanol	S.G.
- - - -	1	—	Methanol	S.G.
· · · · ·	21	—	Methanol	S.G.
—·—·	45	—	Methanol	S.G.
—··—··	45	—	Methanol	A.G.
o o o o o	46	—	Methanol	A.G.

S.G. = Suspended Growth
A.G. = Attached Growth

Figure 22.4. Effects of temperature upon the rate of denitrification
as observed by several workers and summarized by
Parker et al. [1]. The reference numbers refer to the
sources of the data.

22.2.2 Denitrification in CSTR's

THEORETICAL MODEL. The kinetic concepts of the previous section can be used to
develop a model for denitrification in a CSTR. For simplicity this will be done for
a reactor without recycle, although from the principles of Chapters 12 and 13 it is
apparent that it is also applicable to one with recycle. Consider the situation
shown in Fig. 22.5 in which a reactor with volume V receives influent at rate F

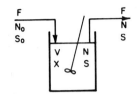

Figure 22.5. Denitrification CSTR.

containing NO_3^--N at concentration N_o and a soluble electron donor at concentration
S_o (as T_bOD). For simplicity it will be assumed that the influent is devoid of dissolved oxygen and contains sufficient ammonia nitrogen to supply the biosynthetic
requirements of the culture. Thus the only role of the NO_3^--N is as the electron
acceptor. Determine the concentration of cells, X, soluble T_bOD, S, and NO_3^--N in
the reactor as a function of τ. If cell death is neglected (nothing is known yet
about cell death in denitrification reactors), a mass balance on cells leads to:

$$\mu = 1/\tau + b \tag{22.6}$$

A mass balance on the electron donor yields:

$$X = \frac{Y_{gs}(S_o - S)}{1 + b\tau} \tag{12.16}$$

The mass balance on NO_3^--N is:

$$FN_o - FN - (\mu XV/Y_{gn}) - (bXV/Y_b) = 0 \tag{22.7}$$

The term $-\mu XV/Y_{gn}$ represents the reduction of NO_3^--N which occurs as it acts as
the electron acceptor during cell synthesis. Because the specific growth rate is a
function of the concentrations of both S and N this term will change whenever either
concentration changes. The term $-bXV/Y_b$ represents the reduction of NO_3^--N by acceptance of the electrons released during microbial decay. During denitrification it
is likely that the specific decay rate, b, will be affected in some manner by the
NO_3^--N concentration. The nature of this dependence is unclear, however, because
some decay could go on in the absence of N by using fermentative electron acceptors.
Because of this, and because the inclusion of a functional relationship between b
and N would complicate the model significantly, it will be assumed herein that b is
constant and independent of N. The parameter Y_b gives the mg cells destroyed/mg
NO_3^--N reduced and is equal to $2.86/\beta$ if no fermentative electron acceptors are used
during decay. In practice, Y_b should be determined experimentally. Rearrangement
of Eq. 22.7 and substitution of Eqs. 22.6 and 12.16 for μ and X, respectively, lead
to:

$$N = N_o - Y_{gs}(S_o - S)[\frac{1}{Y_{gn}} + \frac{1}{Y_b}(\frac{b\tau}{1 + b\tau})] \tag{22.8}$$

The concentration of soluble organic matter in the reactor comes from application of Eq. 22.4:

$$\mu = \frac{\mu_m SN}{(K_s + S)(K_n + N)} \tag{22.4}$$

Substitution into Eq. 22.4 of Eq. 22.6 for μ and Eq. 22.8 for N, and rearrangement yields a quadratic equation:

$$S = [-B \pm (B^2 - 4AC)^{\frac{1}{2}}]/2A \tag{22.9}$$

where:

$$A = GHY_{gs} \tag{22.10}$$

$$B = K_n I + N_o G + Y_{gs}H(K_s I - S_o G) \tag{22.11}$$

$$C = K_s I(N_o - Y_{gs}S_o H + K_n) \tag{22.12}$$

and

$$G = 1 + b\tau - \mu_m\tau \tag{22.13}$$

$$H = \frac{1}{Y_{gn}} + \frac{1}{Y_b}(\frac{b\tau}{1 + b\tau}) \tag{22.14}$$

$$I = 1 + b\tau \tag{22.15}$$

Examination of this set of equations shows that S and N depend upon S_o and N_o as well as upon μ, which is consistent with Figs. 22.1 and 22.2.

A factor which is of importance to the design and operation of denitrification systems is the amount of $T_b OD$ utilized (ΔS) per unit mass of NO_3^--N reduced (ΔN). Since ΔS is just ($S_o - S$) and ΔN is just ($N_o - N$), the equation for $\Delta S/\Delta N$ can be obtained by rearranging Eq. 22.8:

$$\frac{\Delta S}{\Delta N} = \frac{1}{Y_{gs}[\frac{1}{Y_{gn}} + \frac{1}{Y_b}(\frac{b\tau}{1 + b\tau})]} \tag{22.16}$$

Equation 22.16 shows that $\Delta S/\Delta N$ is a function of τ and the various parameters describing the culture. If dissolved oxygen and NO_2^--N are present in the influent, the measured value of $\Delta S/\Delta N$ will be larger than that predicted by Eq. 22.16. The same is true if NO_3^--N must be used as the nitrogen source for cell synthesis in addition to its use as the electron acceptor. Nevertheless Eq. 22.16 will serve to illustrate the general trends that will occur. All of these equations (except 12.16) can be generalized to the case of a CSTR with cell recycle by substitution of θ_c for τ. In that case, Eq. 12.16 would be replaced by Eq. 13.38.

PREDICTIONS FROM MODEL. In order to establish in general terms how a CSTR would perform for denitrification, consider the use of methanol as the electron donor. Typical parameter values are given in Table 22.1. The value of Y_{gn} was calculated from Eq. 22.5 and Y_b was calculated by dividing 2.86 by a β value of 1.25. The rest are from the literature. The effects of MCRT upon the concentrations of methanol T_bOD and NO_3^--N in the reactor are shown in Fig. 22.6 for three different S_o/N_o ratios.

TABLE 22.1
Parameter Values for Denitrification Using Methanol

Parameter	Value	Parameter	Value
μ_m	0.333 hr^{-1}	Y_{gs}	0.18 mg cells/mg T_bOD removed
K_s	4.0 mg/liter T_bOD	b	0.002 hr^{-1}
K_n	0.21 mg/liter NO_3^--N	Y_{gn}	0.664 mg cells formed/mg NO_3^--N reduced
		Y_b	2.288 mg cells destroyed/ mg NO_3^--N reduced

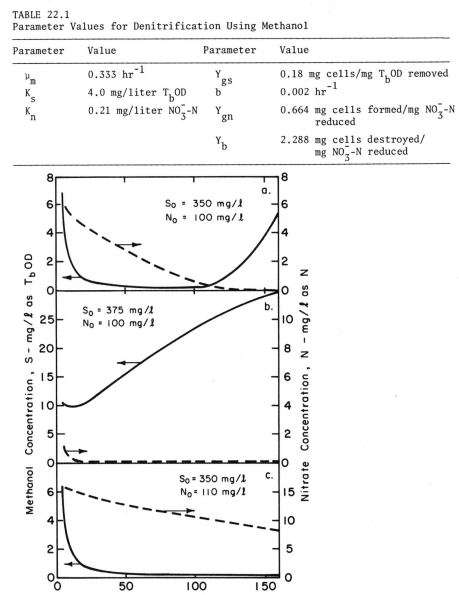

Figure 22.6. Effect of MCRT on the concentration of NO_3^--N and methanol T_bOD in a CSTR as predicted by the theoretical model. The influent conditions are given on the graphs. The parameter values are given in Table 22.1.

Figure 22.6a represents the case where the S_o/N_o ratio is nearly stoichiometric. At low MCRT's little cell decay occurs so that the main requirement for NO_3^--N reduction is as the electron acceptor during cell synthesis. This leaves a relatively large residual concentration in the effluent. As the MCRT is increased, however, cell decay becomes a more significant source of electrons and thus more acceptor is required. Consequently the concentration of NO_3^--N decreases, even though the amount of methanol in the feed has not been changed. Up to an MCRT of approximately 100 hours, the concentration of methanol in the reactor imposes the kinetic limitation on the system and thus remains relatively constant. After that, however, a region is reached in which dual limitation is occurring and the concentrations of both consti-tuents respond to a change in the MCRT. Finally, for MCRT's in excess of 130-140 hours, the concentration of NO_3^--N imposes the kinetic limitation. The concentration of methanol T_bOD in the reactor begins to increase because cell decay is contributing a larger fraction of the electrons needed to reduce the NO_3^--N and thus less exogenous electron donor is required. In contrast, Fig. 22.6b shows that when excess methanol is provided, NO_3^--N will be the limiting substance at all MCRT's. This allows com-plete removal of the NO_3^--N, but releases considerable organic matter to the effluent. Finally, Fig. 22.6c illustrates that insufficient methanol causes incomplete removal of NO_3^--N at all MCRT's.

Most denitrification reactors are run at a constant MCRT and thus it would be interesting to know what effect changes in the S_o/N_o ratio would have under that condition. Figure 22.7 is the prediction of the model. It shows that ratios less

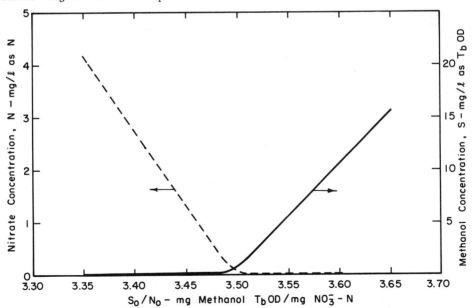

Figure 22.7.   Effect of the S_o/N_o ratio on the concentrations of NO_3^--N and methanol T_bOD in a CSTR with a MCRT of 120 hours receiving flow with a NO_3^--N concentration of 100 mg/liter, as predicted by the theoretical model. The parameter values used are given in Table 22.1.

than stoichiometric will leave residual NO_3^--N in the reactor whereas ratios greater than stoichiometric will leave excess methanol T_bOD. More importantly, however, this figure shows that the ratio range over which relatively complete removal of both material occurs can be quite small. This suggests that it would be exceedingly difficult to control the addition of the electron donor accurately enough to achieve complete removal of both the donor and the acceptor. It should be noted that the exact shapes of the curves and the positions of the crossover points will depend upon the MCRT and upon the amount of oxygen and/or ammonia nitrogen in the influent.

A final point of interest concerning theoretical performance is the effect of MCRT on the $\Delta S/\Delta N$ ratio. Figure 22.8 is a plot of Eq. 22.16 using the parameter values in Table 22.1. There it can be seen that the $\Delta S/\Delta N$ ratio decreases as the MCRT increases. This is a direct result of the increased importance of cell decay as an electron donor at longer MCRT's. Again, if oxygen or ammonia nitrogen were present in the influent the actual values would be different.

OBSERVED PERFORMANCE. Engberg and Schroeder [20] studied the effects of MCRT upon the concentrations of methanol and NO_3^--N in a CSTR with cell recycle and their results follow the general trends predicted by the model as shown in Fig. 22.6. They also presented data on the effect of MCRT upon the $\Delta S/\Delta N$ ratio and this is shown in Fig. 22.9. Included in the figure are data from a previous paper [22]. Even though the values of $\Delta S/\Delta N$ are quantitatively different from those in Fig. 22.8, the qualitative similarity is apparent. Similar curves have been reported by others [25], thereby verifying the generality of the trends. Several researchers have shown an interest in the effects of the S_o/N_o ratio upon the residual carbon and nitrogen remaining after denitrification. Figure 22.10, taken from the work of Dawson and Murphy [26], is typical of the results obtained. Comparison of it with Fig. 22.7 reveals that the trends predicted by the model are correct. Furthermore, the scatter associated with the data in Fig. 22.10 further emphasizes the futility

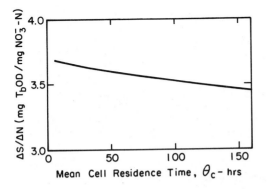

Figure 22.8.   Effect of MCRT on the $\Delta S/\Delta N$
ratio as calculated with
Eq. 22.16 using the parameters
in Table 22.1.

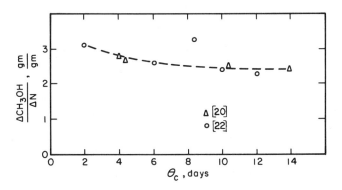

Figure 22.9. Effect of MCRT on the $\Delta S/\Delta N$ ratio as reported by D.J. Engberg and E.D. Schroeder, *Water Research*, 9, 1051-1054, 1975 [20]. *(Reproduced by permission of Pergamon Press, Ltd.)* The reference numbers refer to the sources of the data.

of trying to maintain close control over the concentrations of both S and N remaining in the effluent. This will have an important effect upon the design approach which must be used for a CSTR with recycle, as discussed in Section 22.3.2.

22.2.3 Denitrification in Packed Towers

THEORETICAL MODEL. Theoretical models for packed towers were presented in Chapter 14 where it was seen that the overall performance could be controlled either by the kinetics of substrate removal or by the rates of transport of materials into and out of the biofilms. Harremoës [13] has considered the effects of each of the potentially rate controlling factors and has concluded that the nature of the kinetic constants associated with denitrification allows simplified models to be developed.

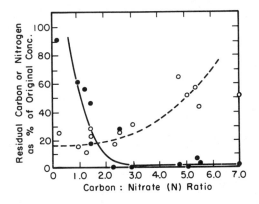

Figure 22.10. Effect of S_0/N_0 (expressed as C/N ratio) upon the removal of carbon (o) and nitrogen (●) in a denitrification reactor. *(From R.N. Dawson and K.L. Murphy, Advances in Water Pollution Research, 1973, p. 681 [26]. Reproduced by permission of Pergamon Press Ltd.)*

While space does not permit a detailed presentation of the development of his model
a brief discussion of the concepts involved would be instructive. Consider first
the effects of diffusion into the biofilm by assuming that the film contains a large
number of pores within which reaction occurs. Substrate must be transported into
the pores by diffusion and the rate limiting step can be either reaction or transport. If the reaction within the pores is first order with respect to the reactant
concentration at the reaction site, it can be transformed into an overall reaction
that is of apparent first order with respect to the concentration of reactant in the
bulk of the fluid at the external surface of the film. If the reaction within the
pores is zero order with respect to the concentration of reactant at the reaction
site, the situation is more complex and depends upon the relative magnitudes of the
diffusivity and the reaction rate constant. If the diffusivity is large enough in
relation to the reaction rate to allow penetration of substrate to the end of the
pore, then the entire pore is effective and the overall reaction behaves as zero
order with respect to the concentration of reactant in the bulk of the liquid. If
the diffusivity is too small to allow this, however, then the overall reaction behaves as if it were one-half order with respect to the bulk reactant concentration.

Now consider the specific case of denitrification for which K_n and K_s (for
methanol) are quite low. Because a packed tower approximates plug flow, the concentrations of NO_3^--N and methanol will be much larger than K_n and K_s throughout most of
the tower. Thus, the rate of NO_3^--N reduction will approximate a zero-order reaction.
In addition, it is likely that the attached biomass will build up on the media to
such an extent that the inner part will become starved, thereby making the film behave like partially effective pores. Thus the NO_3^--N reduction rate throughout the
major portion of the tower should be one-half order with respect to the NO_3^--N concentration in the bulk of the liquid. If the tower is of sufficient length to remove
essentially all of the NO_3^--N, its concentration will approach the value of K_n, thereby forming a transition region leading eventually into a zone within which the
reaction rate is of apparent first order.

Application of these concepts to a packed tower of height L and cross-sectional
area A_c receiving flow at rate F, yields [13]:

$$N = N_o \left(1 - \frac{0.5\, k_{\frac{1}{2}}\, a_v \tau}{N_o^{0.5}} \right)^2$$ (22.17)

where

$k_{\frac{1}{2}}$ = a ½-order surface reaction rate constant

a_v = area of biofilm/unit volume of tower

τ = empty bed space time (LA_c/F)

More complete equations have been developed for the situation in which the reaction
order changes within the tower, but in the interest of brevity they will not be

presented here. Instead, the reader is referred to the work of Harremoës [13] for
them.

   COMPARISON OF MODEL TO OBSERVED PERFORMANCE. An equation defining the $NO_3^- $-N
profile down through a packed tower can be obtained by taking the square root of
each side of Eq. 22.17 [13]:

$$N^{0.5} = N_o^{0.5} - 0.5\ k_{\frac{1}{2}}a_V \tau \tag{22.18}$$

Thus a plot of the square root of the concentration of NO_3^--N remaining at any depth,
L, as a function of the empty bed space time to that depth should be a straight line.
Figure 22.11 is a plot of experimental data from a pilot-scale downflow reactor containing gravel media. Five different hydraulic application rates (F/A_c) were used
but all of the data fall along a single straight line indicating that Eq. 22.18 does
a reasonable job of simulating nitrate reductions. The only significant deviation
was at very low NO_3^--N concentrations where the assumption of half-order kinetics was
no longer valid. On the basis of information presented earlier in this chapter, we

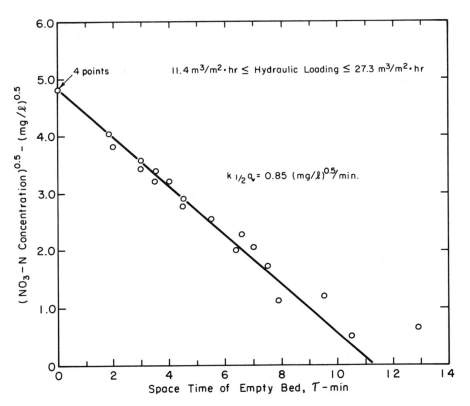

Figure 22.11. Performance of a down-flow denitrification packed tower when
         plotted according to Eq. 22.18. Five hydraulic loadings were
         used. *(Adapted from Harremoës [13])*.

would expect the kinetics of denitrification to shift to first order at NO_3^--N concentrations below 1 mg/liter. Plots made with data from a variety of reactor types showed similar trends, with very good fits until the NO_3^--N concentration was well below 1 mg/liter. Table 22.2 is a summary of the results obtained from those plots [13]. Values of the half-order surface reaction rate constant, $k_{\frac{1}{2}}$, were estimated from the slopes of the lines and the known specific surface areas of the media. Examination of those values indicates that in spite of the broad range of reactor sizes, media sizes, media types, and temperatures employed, most of them lie between 1×10^{-3} and 10×10^{-3} $mg^{0.5}/(liter^{0.5} \cdot min)$. This suggests that most denitrification towers function by the same principle: zero-order reaction in pores that are only partly effective, leading to a half-order reaction.

22.3 DESIGN OF A DENITRIFICATION SYSTEM

22.3.1 Denitrification in a CSTR with Cell Recycle

The intent of this section is to outline a design procedure which focuses on the main points which should be considered. In order not to obscure those points only the simplest case will be investigated; a CSTR with cell recycle receiving a wastewater which is devoid of dissolved oxygen and which contains sufficient ammonia nitrogen to meet the biosynthetic needs of the microorganisms. More complicated situations may be encountered in practice, and they may be handled by application of these principles coupled with the use of experimentally determined parameter values. Furthermore, as more is learned about the kinetics of denitrification, other reactor types may be found to be better suited for the job. The equations describing them are more complex, however, and it was feared that their use here would obscure the concepts which should be presented.

   PROCESS DESIGN EQUATIONS. The first decision that must be made is the concentration of NO_3^--N (N) and $T_b OD$ (S) desired in the final effluent. Once they have been chosen, the MCRT can be chosen by combining Eq. 22.4 with Eq. 22.6 (in which θ_c has been substituted for τ):

$$\theta_c = \frac{(K_s + S)(K_n + N)}{\mu_m SN - b(K_s + S)(K_n + N)} \tag{22.19}$$

Usually the value of N will be fixed by the effluent quality criteria but the designer may choose the value of S. If S is low, the effluent from the reactor will be of high quality and no additional treatment will be required to remove the excess organic matter. However, a large MCRT will be required to achieve a low value of N. In addition, Fig. 22.1 demonstrates that attempts to attain low values of both S and N place the reactor in the curved region of the constant μ (i.e., constant MCRT) contours so that small changes in S will have large effects upon N. An effluent with a high value of S will require additional treatment before it can be discharged. However, the high S value may allow a lower MCRT to be used and may make N relatively

TABLE 22.2
List of Data on Half-Order Surface Removal Rates for Denitrification Filters[a]

Type of Media	Media Size (mm)	Porosity	Filter Diam. (cm)	a_v $\dfrac{dm^2}{dm^3}$	$k_{\frac{1}{2}}$ $\dfrac{mg^{\frac{1}{2}}}{dm^{\frac{1}{2}} \cdot min}$ $\times 10^{-3}$	Temperature (°C)	Reference
Gravel, crushed	4.0	0.45	18.6	112	2.5-8.0	6-19	35
Gravel, round	3.5	0.39	18.6	130	1.8-7.4	6-19	35
Ceramic saddle	13	0.73	12.5	62	1.3-2.2	20	36
Sand	3-4	0.40	180	82.5	3.7	26	31
Flex rings	15x15	0.92	300	34.6	2.6	27	31
Stones	25	0.40	45	13.9	1.6-4.1	12-22	37
Flex rings	25x25	0.96	45	21.4	0.3-1.7	10-26	38
Gravel	3.4	0.28	10	80	10.4	27	39
Stones	5.9	0.34	10	58	9.0	27	39
Stones	14.5	0.37	10	28	7.1	27	39
Glass beads	3	----	3	67	>4	5-30	40
Plexiglass cylinders	16-28	0.80	10	26	4-9	25	41

[a] Adapted from Harremoës [13].

insensitive to changes in S (see Fig. 22.1). The magnitude of S at which N becomes insensitive to S depends upon K_s. For methanol, K_s is relatively small, so that values of S of less than 10 mg/liter as COD should be sufficient. Contour plots like those in Fig. 22.1 can be prepared from Eq. 22.19 by assuming a constant MCRT and calculating N for various values of S. They can then be used to help make the decision regarding S.

After the MCRT has been chosen, the $\Delta S/\Delta N$ ratio can be calculated using Eq. 22.16 (with substitution of θ_c for τ). Since both N_o and N are known (i.e., ΔN is known), ΔS is fixed. Furthermore, since S is also known, S_o is fixed:

$$S_o = S + \Delta S \tag{22.20}$$

Organic matter must be added to the influent at concentration S_o (as T_bOD) as long as N_o is the same. If N_o changes, S_o must be changed in order to maintain the same $\Delta S/\Delta N$ ratio. This may be difficult to accomplish in practice, and thus fluctuations in S may occur. That is why it is important to choose S so that small changes in it have little effect upon N. Once S_o has been established, Eq. 13.38 may be used to calculate the $X\tau$ product:

$$X\tau = \frac{Y_{gs}(S_o - S)}{1/\theta_c + b} \tag{13.38}$$

Either X or τ may then be chosen just as is done during CMAS design. Finally a re-cycle ratio should be chosen that results in an economic settler design and the rate of sludge wastage required to maintain the desired MCRT should be determined. The mass wastage rate of sludge, P_x, should also be calculated. All of these compu-tations should be made in exactly the same manner that they were for activated sludge.

Once a system has been designed, the effects of changes in N_o, S_o, or θ_c must be estimated by using Eqs. 22.8 and 22.9. Of course θ_c should be substituted for τ in those expressions.

DETERMINATION OF PARAMETER VALUES. As with the other biochemical operations, either lab- or pilot-scale experiments are required to establish the values of the kinetic parameters. Ideally CSTR's with cell recycle should be used because in ad-dition to the parameter values they will provide information about the settling characteristics of the sludge. It has been demonstrated, however, that when methanol is the electron donor the parameters determined in simple CSTR's are the same as those obtained from CSTR's with recycle [20]. The first two parameters which must be determined are Y_{gs} and b. The techniques outlined in either Section 12.4 or 13.4 may be used, depending upon the reactor type. The values of μ_m and K_s can be determined from data obtained while operating the reactor under carbon limited conditions, i.e., N_o must be made so large that $N \gg K_n$, so that $N/(K_n + N)$ ap-proaches 1.0. Under that condition, Eq. 22.4 reduces to the Monod equation (Eq. 10.42) allowing μ_m and K_s to be estimated from data on S as a function of θ_c

as outlined in Chapter 12 or 13. Likewise, the value of K_n can be determined from data of N as a function of θ_c obtained while operating the reactors under nitrate limited conditions, i.e., with S_o so large that $S \gg K_s$ thereby causing $S/(K_s + S)$ to approach 1.0. Note that the value of μ_m should be independent of the type of material which is growth rate limiting. If it is not, then the data are either incorrect or the equations are not applicable to the situation in question. The values of Y_{gn} and Y_b can be estimated from experimental data of $\Delta S/\Delta N$ as a function of θ_c. Rearranging Eq. 22.16 yields:

$$\frac{\Delta N}{\Delta S} = \frac{Y_{gs}}{Y_{gn}} + \frac{Y_{gs}}{Y_b} \left(\frac{b\theta_c}{1 + b\theta_c}\right) \tag{22.21}$$

Since Y_{gs} and b are known, a plot of $\Delta N/\Delta S$ versus $b\theta_c/(1 + b\theta_c)$ will yield a straight line with slope Y_{gs}/Y_b and intercept Y_{gs}/Y_{gn}. If the wastewater contains dissolved oxygen or does not contain sufficient ammonia nitrogen to meet the biosynthetic requirements of the cells then the plot of Eq. 22.21 may deviate somewhat from linearity and the value of Y_{gn} obtained may not be exactly equal to the value which could be estimated with Eq. 22.5. Nevertheless, it should still be possible to obtain a reasonable estimate of the parameter from Eq. 22.21 and its value should be given preference over the one obtained from Eq. 22.5.

   REQUIRED CONSTRAINTS. The constraints which must be applied to the design of a denitrification reactor result from the need for cell recycle and thus are quite similar to those for activated sludge. Although no studies of the effects of MCRT on sludge settling have been reported, it seems likely that they will be similar to those observed with activated sludge (see Section 16.2.1). Thus, it appears reasonable to use 3 days (72 hours) as the minimum practical MCRT. Consideration must also be given to θ_{cmin} when choosing the MCRT in order to achieve a stable response to variations in loading and to minimize the possibility of sludge washout. Once the average values of S_o and N_o are known, θ_{cmin} can be estimated from Eq. 22.19 by substituting S_o for S and N_o for N. The design MCRT should then be checked to ensure that the ratio of θ_c to θ_{cmin} exceeds the ratio of the peak to average NO_3^--N loading entering the reactor [1]. The choice of the MLSS concentration (once X_T is known) in an activated sludge reactor was influenced in part by the requirement for oxygen transfer (see Section 16.3.3). Since that requirement does not exist in a denitrification reactor it is likely that the economic upper limit on the MLSS concentration will be higher than it was for activated sludge, although it is not yet well established. The main factors influencing the choice of that concentration will probably be sludge settling and recycle so that consideration must be given to the effects of the choice upon the design of the final settler. It does not seem likely, however, that sludge concentrations below 1500 mg/liter would be economic.

   FACTORS INFLUENCING DESIGN. Because anoxic conditions are required for the synthesis and maximal activity of the enzymes responsible for denitrification, it is important that oxygen be excluded from denitrification reactors. Furthermore, because the amount of electron donor (methanol, etc.) required is proportional to the total amount of electron acceptor (nitrate, nitrite, and oxygen) present, economics also dictates that oxygen be excluded. For example, approximately one unit mass of methanol will be required to remove each unit mass of oxygen which enters the reactor.

As discussed in Section 22.2.1, temperature and pH affect the kinetics of denitrification. Figure 22.4 showed that low temperatures can cause a considerable reduction in rate. Consequently, the design should be performed in a manner that will ensure adequate performance at the lowest anticipated sustained temperature. Figure 22.3 showed that the pH should be near 7.0. Low pH's will generally not be a problem because approximately 3.5 mg of alkalinity (as $CaCO_3$) are produced for each mg of NO_3^--N reduced to N_2 gas, thereby tending to make the pH rise. In one situation in which a wastewater containing a high concentration of NO_3^--N was being treated in a lab-scale reactor, the pH rose above 9.0 [27]. Consequently consideration should be given to pH control, particularly if the concentration of NO_3^--N being removed is high.

A factor which was an important consideration in the design of a slurry nitrification reactor was the loss of suspended solids in the final effluent and its effect upon the maintenance of the proper MCRT. This can also be an important factor in denitrification design. The mass of cells formed during denitrification can be estimated by Eq. 13.51

$$P_x = \frac{FY_{gs}(S_o - S)}{1 + b\theta_c} \qquad (13.51)$$

and the maximum concentration which can escape in the final effluent without disrupting the process can be found by dividing P_x by the effluent flow rate, (which may be approximated by F)

$$(X_e)_{max} \cong \frac{Y_{gs}(S_o - S)}{1 + b\theta_c} \qquad (22.22)$$

Noting that $(S_o - S)$ is ΔS, Eq. 22.22 may be put in terms of the NO_3^--N removal (ΔN) by substituting Eq. 22.21 for $(S_o - S)$. This yields:

$$(X_e)_{max} = \frac{\Delta N}{\dfrac{1 + b\theta_c}{Y_{gn}} + \dfrac{b\theta_c}{Y_b}} \qquad (22.23)$$

Furthermore, when methanol is the electron donor the values of the kinetic parameters given in Table 22.1 are representative, allowing Eq. 22.23 to be reduced to:

$$(X_e)_{max} = \frac{\Delta N}{1.51 + 0.004\theta_c} \qquad (22.24)$$

Thus, for an MCRT of 5 days (120 hrs), only 0.5 mg/liter of cells may be lost for each mg/liter of NO_3^--N reduced. A nitrified domestic wastewater typically contains

25 mg/liter NO_3^--N, consequently only 12.5 mg/liter of cells can be lost in the effluent if methanol is the carbon source. Special care would be required during the
design of the final settler to meet this limitation.

One problem which can occur during denitrification is flotation of the sludge
in the final settler due to the entrapment of nitrogen gas bubbles [1,2]. Thus the
maintenance of a low effluent suspended solids concentration requires that those
bubbles be removed before the mixed liquor enters the final settler. Two alternatives are available, vacuum degasification, such as is used with anaerobic contact,
and aeration in a reactor with short space time. Aeration is generally used because
it is also effective in removing any residual electron donor which may be present,
thereby reducing the concentration of soluble organic matter in the effluent [1,2].
Because short periods of aerobiosis have no effect upon denitrifying enzymes which
have been formed under anoxic conditions, the passage of the sludge through the
aerated reactor has no effect upon the rate of nitrate reduction in the denitrifying
reactor [28]. However, there is a tendency for sludges which are exposed to alternating anoxic and aerobic conditions to bulk [29]. Consequently special attention
must be given to thickening limitations during design of the settlers. The space
time required to prevent the sludge from floating in the settler is approximately
10 minutes [30]. Longer times may be required to remove the residual organic matter,
however, with the exact amount depending upon its concentration as well as that of
the cells. The aerated basin will also increase the amount of cell decay that occurs [1]. The effects of the aerated basin may be estimated by modeling the system
as two reactors in series with cell recycle around the entire chain, using the
techniques outlined in Chapter 13. In so doing the values of the kinetic parameters
in each of the reactors should reflect the particular reaction environment. Simulation studies in which the MCRT and the individual reactor space times are varied
will allow an appropriate system design to be selected.

EXAMPLE 20.3.1-1

Consider denitrification of the effluent from the completely mixed activated sludge
process discussed in Examples 16.3.5-1 and 20.3.2-2 which had a flow rate of
634,000 gal/day (10^5 liters/hr = 100 m^3/hr) and a NO_3^--N concentration of 29.7 mg/liter.
Size a completely mixed denitrification reactor to reduce the NO_3^--N concentration to
0.5 mg/liter or less. Use methanol as the electron donor. Assume that the temperature is 20°C and that the parameters in Table 22.1 apply.
Solution:

The first task is to choose the concentration of methanol T_bOD in the effluent
and the associated MCRT. Using Eq. 22.19, prepare contour lines of constant MCRT
for a plot of S versus N. The contours for MCRT's of 3 and 4 days are shown in
Fig. E22.1 where it can be seen that very low concentrations of NO_3^--N can be achieved

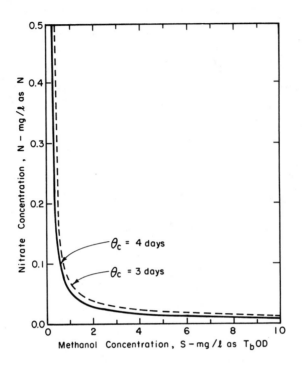

Figure E22.1.   Contour plots for θ_c = 3 days and
θ_c = 4 days for Example 22.3.1-1.

at either MCRT so long as the methanol T_bOD concentration is in excess of 3 mg/liter.
After considering the rises in N that might be caused by small variations in S, a
design S concentration of 7 mg/liter as T_bOD and an MCRT of 4 days are chosen because they should give stable performance.

The next task is to calculate the $\Delta S/\Delta N$ ratio using Eq. 22.16.

$$\frac{\Delta S}{\Delta N} = \frac{1}{0.18[\frac{1}{0.664} + \frac{1}{2.288}(\frac{(0.002)(96)}{1 + (0.002)(96)})]}$$

$\frac{\Delta S}{\Delta N}$ = 3.52 mg methanol T_bOD/mg NO_3^--N reduced.

Since essentially all of the NO_3^--N is reduced, ΔN is 29.7 mg/liter, making ΔS equal
to 104.5 mg/liter, which implies that

S_o = 7 + 104.5 = 111.7 mg/liter methanol T_bOD

Thus 111.7 mg/liter of methanol T_bOD must be added to the influent, as long as N_o
is 29.7 mg/liter NO_3^--N.

The $X\tau$ ratio may be calculated with Eq. 13.38:

$$X\tau = \frac{(0.18)(104.5)}{(1/96) + 0.002}$$

$X\tau = 1515$ (mg·hr)/liter

Either X or τ may be chosen, thus fixing the other. A space time of 1 hour seems
reasonable, as it will give a cell concentration of 1515 mg/liter. The recycle
ratio, maximum rate of sludge wastage, and the mass of sludge to be disposed of can
all be calculated in the way that they were for activated sludge.

22.3.2 Denitrification in a Packed Tower

A number of types of packing materials and flow regimes are being tried in packed
towers but experience is limited so that none has yet proven to be superior to the
others. Furthermore, no design procedure has yet been accepted as being applicable
to all situations. Consequently, the only reliable design technique is to conduct
pilot plant studies with a particular packing and then use the data to scale up.

PROCESS DESIGN. The task of scale up is simplified if performance can be correlated with the major process variables. Table 22.2 indicated that the half-order
reaction theory of Harremoës [13] was applicable to a large number of reactor types,
thereby suggesting that the space time based upon the empty reactor volume is an
important process variable. Consequently Eq. 22.18 can be used to correlate performance and space time by plotting the square roots of the NO_3^--N concentrations at
various depths as a function of the empty reactor space times to those depths. The
half-order reaction rate constant, $k_{\frac{1}{2}}$, can then be obtained from the slope of the
line and the specific surface area of the media, a_v. This technique will probably
give good correlations down to NO_3^--N concentrations of 1 mg/liter, but below that it
may be necessary to use a first-order model similar to the one used in Chapter 18.
Once the correlation equations are available, process design becomes a matter of
choosing the reactor space time required to reduce the NO_3^--N concentration to the
desired value. This will normally require sequential application of the half-order
and first-order equations. Equation 22.18 can be used to calculate the τ required
to reduce the NO_3^--N concentration from N_0 to 1 mg/liter whereas Eq. 18.4 can be used
to find the additional τ required to continue the reduction to the desired effluent
concentration.

FACTORS INFLUENCING THE DESIGN. The concentration of electron donor must not
be rate limiting if the concentration of NO_3^--N is to control the reaction rate. For
a slurry reactor the $\Delta S/\Delta N$ ratio could be estimated thereby allowing estimation of
the amount of donor which must be added. The equation cannot be applied to a packed
tower, however, because little is known about its cell residence time. Thus $\Delta S/\Delta N$
must be determined experimentally. Figure 22.12 relates the performance of towers
receiving methanol as the donor to that ratio [2]. There it can be seen that the

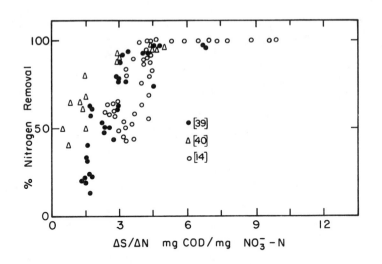

Figure 22.12.   Relationship between observed ΔS/ΔN ratio and
                performance of denitrification packed towers.
                (Adapted from Christensen and Harremoës [2]).
                The reference numbers refer to the sources of
                the data.

values needed for complete denitrification are higher than those required for slurry
systems, and are actually higher than theoretical. Even though no explanation has
yet been provided concerning this, it may be taken as an empirical fact that ΔS/ΔN
values in excess of 4.5 mg COD/mg NO_3^--N are required to achieve more than 90% removal [2]. Consequently, the methanol requirements for packed towers are likely to
be higher than for slurry reactors. Temperature and pH affect packed towers in much
the same manner that they do CSTR's, and their effects must be anticipated and provided for through changes in the half-order or first-order reaction rate constants.

     Perhaps the most important factor influencing denitrification in packed towers
is the build up of biomass on the media and its subsequent removal. When the reactor
contains corrugated plastic media whose spaces are filled with nitrogen gas, sloughing of the biomass is continuous, just as it is in a trickling filter. Consequently
the tower must be followed by some type of physical operation to remove the microorganisms prior to discharge. For domestic wastewaters (in which the biomass production during denitrification is approximately 12 mg/liter) multimedia filtration
may be used. For high NO_3^--N wastewaters, however, sedimentation may be needed. When
the media in a tower is submerged, the liquid velocity is normally not high enough to
remove the excess biomass; consequently the head loss tends to increase with time.
To prevent clogging of these reactors they must be periodically backwashed, much like
a sand filter, to remove the excess solids. If the media is fine, like a sand filter,
the tower will tend to retain all of the biosolids until it is backwashed. Consequently the normal discharge will require no additional treatment. If the media is

coarse, on the other hand, the reactor will continually lose biosolids to the effluent so that filtration will be required to prepare it for discharge. Even those
reactors require backwashing, however, because the normal rate of discharge of the
biomass is less than the rate of build up and clogging will eventually occur. One
type of tower, employing fluidized sand grains to which the denitrifying bacteria ad-
here, is operated at a steady state by continually removing and cleaning a portion of
the media. Consequently, its effluent can be discharged without filtration. Interested readers should see Parker *et al.* [1] for a detailed discussion of the effects
of biomass build up on each of the reactor types.

22.3.3 Selection of the Reactor Type

A natural question which arises during the design of a denitrification system concerns the type of reactor which should be employed. Unfortunately, there is no easy
answer to that question since each type is capable of reliably achieving a high degree
of denitrification. Each also has a potential problem associated with it. For a
CSTR with cell recycle the problem lies in the ability of the final settler to remove
a sufficient amount of the organisms from the effluent to allow the desired MCRT to
be maintained. This problem can generally be solved through careful design of the
settler. For a packed tower the major problem is the excess electron donor which
must be added to achieve complete denitrification. With a CSTR the excess donor can
easily be removed by passing the mixed liquor through an aerobic stage prior to sedimentation and recycle. This technique cannot be applied to a packed tower, however.
To date, systems for the removal of excess electron donor from the effluent of packed
towers have not been developed [1] and thus much closer control of donor addition is
required. This can be difficult.

22.4 OPERATION OF DENITRIFICATION SYSTEMS

Addition of the correct amount of electron donor (methanol, etc.) is essential to the
proper operation of a denitrification reactor. If the quantity is insufficient, not
all NO_3^--N will be reduced, whereas, if it is excessive, the effluent T_bOD will be un-
acceptable. In practice, addition of exactly the right amount is difficult to achieve.
Consequently, the usual procedure is to add a slight excess and then use an aerated
reactor to remove it prior to discharge. The most promising system for the addition
of methanol appears to be feedforward control based upon both the wastewater flow rate
and the NO_3^--N concentration [1]. If additional fine tuning is required it can be
achieved by a feedback loop based upon the ORP or COD in the reactor.

   Operation of a CSTR with cell recycle is exactly the same as operation of the
activated sludge process, which was described in Section 16.4. As with nitrification,
the most crucial concern is with the loss of suspended solids in the effluent. If it
is greater than the quantity of sludge produced, maintenance of the desired MCRT will
be impossible unless additional solids are added to the reactor, for example, waste
activated sludge. This would make the reactor behave like a CSTR receiving microbial

cells and the performance equations would have to be altered accordingly using the techniques presented in Chapters 12 and 13.

Operation of a packed tower may be like operation of either a trickling filter (Section 18.4) or a sand filter. When the tower is filled with plastic media whose voids contain nitrogen gas, operation is like a trickling filter. When the media is submerged, on the other hand, operation is much like a sand filter in that a primary concern is with the control of head loss. Backwashing, in which the liquid flows upward through the reactor at a rate fast enough to carry away entrapped parti- cles of biomass, is the most common method of reducing head loss. The frequency with which it must be done depends upon the type of media and the method of operation. If the media is of high porosity, like rings or saddles, there is a continual loss of biomass from the reactor so that the build up of head loss is slow. Consequently backwashing is performed only infrequently. In one such installation, backwashing was performed every four weeks, although that time interval was chosen to prevent the concentration of suspended solids in the effluent from rising to too high a level rather than to prevent excessive head loss [31]. Reactors containing small diameter media of low porosity are operated like sand filters because they entrap the biomass and prevent its escape to the effluent. Consequently the increase in head loss is rapid, requiring frequent backwashing. A common interval is once a day. Entrapment of nitrogen gas bubbles in fine media can also contribute to head loss and thus one system routinely backwashes for one or two minutes every four to twelve hours to re- move the gas [32]. A more detailed description of the operational techniques employed with packed towers is available elsewhere [1].

22.5 EXAMPLES IN PRACTICE

Relatively few full-scale denitrification reactors have been built; consequently, most experience with separate stage denitrification using methanol has come from pilot- scale facilities. Studies conducted by the EPA on a 0.2 MGD (760 m^3/day) suspended growth denitrification system demonstrated that the performance of the process is quite stable and capable of producing an effluent of high quality [1,33]. For example, over a four month period the process reduced the NO_3^--N concentration in nitrified domestic sewage to an average of 0.8 mg/liter. The Central Contra Costa Sanitary District (CCCSD) in California performed studies on a 0.5 MGD (1990 m^3/day) pilot plant over a 23 month period in which the NO_3^--N concentration was consistently reduced from 27 to 0.5 mg/liter [34]. Furthermore, the 90th percentile performance level did not vary widely from the median level, thereby providing statistical evi- dence of process stability.

As a result of the CCCSD studies a full-scale separate stage denitrification plant was designed using CSTR's in series with cell recycle around the entire chain [29]. Each reactor system consists of nine open tanks in series, with pro- vision for aeration in the last four. All nine are equipped with low-shear turbine

mixers which are turned just rapidly enough to keep the MLSS in suspension. The design calls for an MLSS concentration of 3000 to 4000 mg/liter with a space time of
102 minutes for the entire chain. The fraction of the space time devoted to reaeration
can be adjusted as needed by using any number of the last four tanks for that purpose.
The methanol feed rate will be automatically controlled in proportion to the wastewater
flow rate and the NO_3^--N concentration, although it is anticipated that the minimum
methanol COD:NO_3^--N ratio will be 4:1.

A nitrogen gas filled denitrification column was also tested at the CCCSD pilot-
plant facility and the results obtained led to the design of a full-scale plant for
Canberra, Australia [1,34]. The medium employed had a specific surface area of
42 ft^2/ft^3 (12.8 m^2/m^3) and the design removal rate was 9.9 x 10^{-4} lb NO_3^--N/(day·ft^2)
of media surface area (4.83 x 10^{-3} kg/(m^2·day)). This required towers 19 ft (5.8 m)
deep with a hydraulic application rate of 2 gal/(min·ft^2) of tower cross-sectional
area (0.8 m^3/(min·m^2)).

Both fine and coarse media packed tower pilot plants were run at El Lago, Texas
on nitrified municipal sewage [1,31]. The coarse media consisted of Koch Flexirings
with 92% void space and a specific surface area of 105 ft^2/ft^3 (346 m^2/m^3). The
hydraulic application rate was 2.5 gal/(min·ft^2) of tower cross-sectional area
(1 m^3/(min·m^2)) and the empty bed space time was 1 hour. The tower was able to reduce the NO_3^--N concentration from 13.6 to 0.9 mg/liter while only requiring backwashing every four weeks. As would be expected with coarse media, the effluent suspended solids concentration was high, 19 mg/liter, thereby requiring additional treatment prior to discharge. The fine media was uniform sand with a diameter between 3
and 4 mm, a specific surface area of 250 ft^2/ft^3 (825 m^2/m^3), and a void space of 40%.
The hydraulic application rate was 7.4 gal/(ft^2·min) (2.9 m^3/(m^2·min)) and the empty
bed space time was 0.25 hr. In spite of the higher loading, this tower was able to
reduce the NO_3^--N concentration from 15.2 to 2.6 mg/liter while requiring backwashing
only once a day. The backwashing was done with nitrified effluent (which contained
suspended solids) and consequently the suspended solids in the packed tower effluent
averaged 17 mg/liter. Tertiary filtration was performed prior to discharge.

Although other examples of denitrification systems can be found in the literature they will not be presented here because the preceding are sufficient to provide the reader with an appreciation of what denitrification systems can do. As effluent guidelines are made more stringent, more and more use will be made of this
biochemical operation, thereby increasing our knowledge and allowing us to do an
even better job of design.

22.6 KEY POINTS

1. Denitrification, which is the reduction of nitrate nitrogen as it serves as the
   terminal hydrogen acceptor for microbial respiration, is performed by facultative bacteria as an alternative to the use of oxygen when the concentration
   of dissolved oxygen approaches zero.

2. Because methanol is a highly reduced compound with a low microbial yield it is commonly employed as an electron donor for denitrification.

3. Both slurry and fixed-film reactors may be used for denitrification; the electron donor may be the organic matter in the untreated wastewater, organic reserves inside the microorganisms, or an added compound such as methanol.

4. Denitrification systems are used when it is necessary to reduce the concentration of nitrogen in a wastewater prior to discharge.

5. Either the electron donor (organic matter) or the electron acceptor (NO_3^--N) can be rate limiting during denitrification. Thus a dual substrate equation should be used to depict the relationship between their concentrations and the specific growth rate.

6. The true growth yield based on the electron acceptor (Y_{gn}) must be in stoichiometric balance with the true growth yield based on the electron donor (Y_{gs}).

7. Like other biological reactions, temperature and pH affect the rate of denitrification. Temperature exerts a stronger effect below 15°C than above it. In general, for any given temperature, the highest rates occur within the range of pH between 7.0 and 7.5.

8. For a fixed S_0/N_0 ratio, the concentrations of NO_3^--N and T_bOD in the effluent from a CSTR depend upon the MCRT employed. If the S_0/N_0 ratio is near stoichiometric, there will be a relatively narrow range of MCRT's over which both constituents will be reduced to low concentrations.

9. For a fixed MCRT, the concentrations of NO_3^--N and T_bOD in the effluent from a CSTR will be strongly dependent upon the S_0/N_0 ratio. The range of S_0/N_0 values over which both constituents will be reduced to low concentrations will be small.

10. The $\Delta S/\Delta N$ ratio for a CSTR will decrease as the MCRT is increased because cell decay plays an increasingly important role in supplying electrons for the reduction of the NO_3^--N.

11. It appears that most packed towers used for denitrification function by the same principle: zero-order reaction in pores that are only partly effective, leading to a performance equation which is based upon a half-order reaction. Thus a plot of the square root of the NO_3^--N concentration remaining at any depth as a function of the empty bed space time to that depth is often a straight line.

12. The design of a CSTR with cell recycle for denitrification is similar to the design of CMAS. The MCRT is chosen to give the desired values of S and N and the $\Delta S/\Delta N$ ratio is thereby fixed, which allows determination of S_0. Once θ_c, S_0 and S are known, the $X\tau$ product is fixed and the design may proceed as it did for CMAS.

13. Several factors and constraints must be considered during the design of a CSTR with cell recycle for denitrification. Among them are the loss of suspended solids in the final effluent and the removal of residual organic matter. Both are influenced by a short period of aeration between the denitrification reactor and the settler.

14. The only reliable way to design a packed tower for denitrification is to scale up from pilot-plant data.

15. An important limitation on the application of packed towers for denitrification is the build up of biomass within the media, which in some cases leads to excessive head losses.

16. Operation of a CSTR with cell recycle is exactly the same as operation of the activated sludge process. An additional step is required, however: maintenance of the correct amount of electron donor in the feed.

17. The major operational procedures associated with packed towers are the addition of the proper amount of electron donor and the removal of excess biomass.

18. Experience with pilot-scale denitrification reactors has shown that their performance is reliable and stable.

22.7 STUDY QUESTIONS

The study questions are arranged in the same order as the text.

Section	Questions
22.1	1-3
22.2	4-9
22.3	10-12
22.4	13

1. Explain what is meant by the statement that "The perspective from which denitrification must be viewed is just the opposite of that from which most wastewater treatment systems are viewed."

2. Explain why the properties of methanol make it an advantageous electron donor for denitrification.

3. Explain why it is necessary to supply an electron donor for denitrification of the effluent from a nitrification reactor.

4. Draw contour plots for μ when μ_m is 0.5 hr^{-1} and K_{s1} and K_{s2} are 20 and 40 mg/liter, respectively. If Y_{g1} is 0.4 and Y_{g2} is 0.5 use the stoichiometric line to determine which substrate is rate limiting for each of the following feed situations:

(a) S_{1o} = 120 mg/liter; S_{2o} = 140 mg/liter

(b) S_{1o} = 120 mg/liter; S_{2o} = 100 mg/liter

(c) S_{1o} = 80 mg/liter; S_{2o} = 150 mg/liter

5. If the value of Y_{gs} on methanol is 0.14 mg cells/mg T_bOD and the value of β is 1.25, what is the value of Y_{gn} if no NO_3^--N is incorporated into cell material?

6. Using the parameter values in Table 22.1 calculate the concentrations of cells, NO_3^--N, and T_bOD in the effluent from a CSTR with a space time of 60 hrs receiving 200 mg/liter NO_3^--N and 715 mg/liter T_bOD. What would happen to the concentrations if the value of S_o were reduced to 700 mg/liter? What would happen if it were increased to 730 mg/liter?

7. Repeat Study Question 6 using a space time of 180 hours.

8. Calculate the $\Delta S/\Delta N$ ratio for each case in Study Questions 6 and 7.

9. Using data from Table 22.2 and Eq. 22.18 prepare graphs comparing the performance of two denitrification packed towers.

10. Explain why the loss of suspended solids in the effluent may be an important limitation in the design of a CSTR with cell recycle for denitrification.

11. A wastewater flowing at a rate of 634,000 gal/day (10^5 liters/hr = 100 m^3/hr)
    and containing 50 mg/liter of NO_3^--N is to be denitrified in a CSTR with cell
    recycle using methanol as the electron donor. Assuming that the parameters in
    Table 22.1 apply, choose an MCRT to reduce the NO_3^--N concentration to
    0.5 mg/liter or less and justify your choice. Also determine S, S_0, X and τ.
    Justify the values. Finally, determine the maximum allowable effluent sus-
    pended solids concentration. You may assume that the influent is devoid of
    oxygen and nitrite nitrogen and that it contains sufficient NH_4^+-N to meet the
    biosynthetic needs of the culture. How much would that be?

12. Enumerate and explain the major points which must be considered during design
    of a packed tower for denitrification.

13. Explain why it is necessary to backwash denitrification packed towers employing
    submerged media.

REFERENCES AND FURTHER READING

1. D. S. Parker, et al. "Biological denitrification," Chapter 5 in Process Design
    Manual for Nitrogen Control, U.S. Environmental Protection Agency, Technology
    Transfer, October, 1975.

2. M. J. Christensen and P. Harremoës, "Biological denitrification of sewage: A
    literature review," Progress in Water Technology, 8, 509-555, 1977.

3. A. H. Stouthamer, "Biochemistry and genetics of nitrate reductase in bacteria,"
    Advances in Microbial Physiology, 14, 315-375, 1976.

4. C. C. Delwiche and B. A. Bryan, "Denitrification," Annual Review of Micro-
    biology, 30, 241-262, 1976.

5. W. J. Payne, "Reduction of nitrogenous oxides by microorganisms," Bacteriologi-
    cal Reviews, 37, 409-452, 1973.

6. J. M. Krul, "The relationship between dissimilatory nitrate reduction and oxygen
    uptake by cells of an Alcaligenes strain in flocs and in suspension and by
    activated sludge flocs," Water Research, 10, 337-341, 1976.

7. C. C. Delwiche, "Denitrification" in Inorganic Nitrogen Metabolism ed. by
    W. D. McElroy and B. Glass, Johns Hopkins Press, Baltimore, 233-256, 1956.

8. J. Wijler and C. C. Delwiche, "Investigations of the denitrifying process in
    soils," Plant and Soil, 5, 155-169, 1954.

9. P. J. du Toit and T. R. Davies, "Denitrification studies with laboratory-scale
    continuous flow units," Water Research, 7, 489-500, 1973.

10. W. K. Johnson and G. J. Schroepfer, "Nitrogen removal by nitrification and
    denitrification," Journal Water Pollution Control Federation, 36, 1015-1036,
    1964.

11. P. L. McCarty, et al. "Biological denitrification of wastewaters by addition
    of organic materials," Proceedings of the 24th Industrial Waste Conference,
    Purdue University Engineering Extension Series, No. 135, 1271-1285, 1969.

12. D. D. Focht and A. C. Chang, "Nitrification and denitrification processes re-
    lated to waste water treatment," Advances in Applied Microbiology, 19, 153-186,
    1975.

13. P. Harremoës, "The significance of pore diffusion to filter denitrification," *Journal Water Pollution Control Federation,* 47, 377-388, 1976.

14. J. S. Jeris and R. W. Owens, "Pilot-scale, high-rate biological denitrification," *Journal Water Pollution Control Federation,* 47, 2043-2057, 1975.

15. D. N. Ryder and C. G. Sinclair, "Model for the growth of aerobic microorganisms under oxygen limiting conditions," *Biotechnology and Bioengineering,* 14, 787-798, 1972.

16. C. G. Sinclair and D. N. Ryder, "Models for the continuous culture of microorganisms under both oxygen and carbon limiting conditions," *Biotechnology and Bioengineering,* 17, 375-398, 1975.

17. M. Shoda, *et al.* "Simulation of growth of methane-utilizing bacteria in batch culture," *Journal of Applied Chemistry and Biotechnology,* 25, 305-318, 1975.

18. S. Nagai, *et al.* "Growth kinetics based on energetics of methane utilizing bacteria," in *Microbial Growth on C_1-Compounds,* Proceedings of the International Symposium on Microbial Growth on C_1-Compounds, Sept. 5, 1974, Tokyo, Japan, The Society of Fermentation Technology, Tokyo, Japan, 221-230, 1975.

19. F. G. Bader, *et al.* "Comments on microbial growth rate," *Biotechnology and Bioengineering,* 17, 279-283, 1975.

20. D. J. Engberg and E. D. Schroeder, "Kinetics and stoichiometry of bacterial denitrification as a function of cell residence time," *Water Research,* 9, 1051-1054, 1975.

21. H. D. Stensel, *et al.* "Biological kinetics of suspended-growth denitrification," *Journal Water Pollution Control Federation,* 45, 249-261, 1973.

22. S. F. Moore and E. D. Schroeder, "An investigation of the effect of residence time on anaerobic bacterial denitrification," *Water Research,* 4, 685-694, 1970.

23. R. P. Michael and W. J. Jewell, "Optimization of denitrification process," *Journal Environmental Engineering Division, ASCE,* 101, 643-657, 1975.

24. B. J. Chen, *et al.* "A model for bacterial growth on methanol," *Biotechnology and Bioengineering,* 18, 1629-1633, 1976.

25. D. J. R. Dodd and D. H. Bone, "Nitrate reduction by denitrifying bacteria in single and two stage continuous flow reactors," *Water Research,* 9, 323-328, 1975.

26. R. N. Dawson and K. L. Murphy, "Reply to discussion of 'Factors affecting biological denitrification of wastewater' by R. N. Dawson and K. L. Murphy," in *Advances in Water Pollution Research* edited by S. H. Jenkins, Pergamon Press, Oxford, England, 682-683, 1973.

27. W. J. Jewell and R. J. Cummings, "Denitrification of concentrated nitrate wastewaters," *Journal Water Pollution Control Federation,* 47, 2281-2291, 1975.

28. P. M. Sutton, *et al.* "Nitrogen control: A basis for design with activated sludge systems," *Progress in Water Technology,* 8, 467-481, 1977.

29. D. S. Parker, *et al.* "Development and implementation of biological denitrification for two large plants," *Progress in Water Technology,* 8, 673-686, 1977.

30. K. Mudrack, "Experiments on the use of microbial denitrification for biological purification of industrial waste water," *Veröffentlichungen des Instituts für Siedlungswasserwirtschaft der Technische Universität Hannover* (in German), Heft 36., Hannover, 1970.

31. "Description of the El Lago, Texas advanced wastewater treatment plant," Harris County Water Control and Improvement District Number 50, Seabrook, Texas, March, 1974.

32. E. S. Savage and J. J. Chen, "Operating experiences with columnar denitrification," *Water Research,* 9, 751-757, 1975.

33. M. C. Mulbarger, "The three sludge systems for nitrogen and phosphorus removal," Paper presented at the 44th Annual Conference of the Water Pollution Control Federation, San Francisco, California, October 1971.

34. G. A. Horstkotte, *et al.* "Full-scale testing of a water reclamation system," *Journal Water Pollution Control Federation,* 46, 181-197, 1974.

35. P. Harremoës and M. Riemer, "Report on pilot experiments on down-flow filter denitrification," Report from Dept. of Sanitary Engineering, Technical University of Denmark, Lyngby, 1975.

36. P. M. Sutton, "Continuous biological denitrification of wastewater," Thesis presented to McMaster University, Hamilton, Ontario, Canada, in 1973 in partial fulfillment of the requirements for the degree of Master of Engineering.

37. P. P. St. Amant and P. L. McCarty, "Treatment of nitrate rich waters," *Journal of the American Water Works Association,* 61, 659-662, 1969.

38. J. R. Jones, "Denitrification by anaerobic filters and ponds - Phase II. Bio-engineering aspect of agricultural drainage. San Joaquin Project, California," U.S. EPA Report #13030 UBH 6/71-14, June 1971.

39. J. M. Smith, *et al.* "Nitrogen removal from municipal waste water by columnar denitrification, *Environmental Science and Technology,* 6, 260-267, 1972.

40. S. G. Dholakia, *et al.* "Methanol requirements and temperature effects in waste-water denitrification," Water Pollution Control Research Series, ORD-17010 DHT 09/70, U.S. Federal Water Quality Administration, 1970.

41. D. A. Requa, "Kinetics of packed bed denitrification," Thesis presented to the University of California at Davis, in 1970 in partial fulfillment of the requirements for the degree of Master of Science.

42. R. N. Dawson and K. L. Murphy, "Factors affecting biological denitrification in wastewater" in *Advances in Water Pollution Research* edited by S. H. Jenkins, Pergamon Press, Oxford, England, 671-680, 1973.

43. E. D. Renner and G. E. Becker, "Production of nitric oxide and nitrous oxide during denitrification by *Cornybacterium nephridii,*" *Journal of Bacteriology,* 101, 821-826, 1970.

44. G. W. Clayfield, "Respiration and denitrification studies on laboratory and works activated sludges," *Water Pollution Control,* 73, 51-76, 1974.

45. K. L. Murphy and P. M. Sutton, "Pilot scale studies on biological denitrification," *Progress in Water Technology,* 7, 315, 1975.

46. Ecotrol, Inc., "Biological denitrification using fluidized bed technology," August, 1974.

Further Reading

Bailey, D. A. and Thomas, E. V., "The removal of inorganic nitrogen from sewage ef-
    fluents by biological denitrification," *Water Pollution Control*, 74, 497-515,
    1975.

Barnard, J. L., "Nutrient removal in biological systems," *Water Pollution Control*,
    74, 143-154, 1975.

Benefield, L. D. and Randall, C. W., *Biological Process Design for Wastewater Treat-
    ment*, Prentice-Hall, Inc., Englewood Cliffs, N.J., 1980. See Chapters 4 and 7.

Christensen, M. H. and Harremoës, P., "Biological denitrification in water treatment -
    A literature study," Report No. 2-72, prepared by the Department of Sanitary
    Engineering, Technical University of Denmark, Building 115, 2800 Lyngby, Denmark,
    1972. A very complete review of the literature through 1971.

McCarty, P. L. and Haug, R. T., "Nitrogen removal from wastewaters by biological
    nitrification and denitrification," in *Microbial Aspects of Pollution*, edited
    by G. Sykes and F. A. Skinner, Academic Press, New York, 215-232, 1971.

Painter, H. A. "Microbial transformations of inorganic nitrogen metabolism in
    microorganisms," *Water Research*, 4, 393-450, 1970. An excellent review of the
    microbiology of nitrogen transformations.

Painter, H. A., "Microbial transformation of inorganic nitrogen," *Progress in
    Water Technology*, 8, 3-29, 1977.

Parker, D. S., *et al. Process Design Manual for Nitrogen Control*, U.S. Environmental
    Protection Agency, Technology Transfer, Oct. 1975, pp 3-29 - 3-43 and Chapter 5.
    An excellent source of material on the engineering aspects of denitrification.

AUTHOR INDEX

A

Adams, C.E., (68) 329, 348*; (8) 441,
    506*; (53) 642, 684*; (2) 690,
    692, 693, 696, 699, 701, 702,
    709, 712*; 714*
Adelberg, E.A., 208*; 212*
Ahlberg, N.R., (33) 708, 709, 714*
Ahlert, R.C., 831*
Aiba, S., (1) 27, 36*; 349*; 430*
Alberda, G., (5) 172, 192*
Albertson, J.G., (19) 627, 657, 683*
Alberty, R.A., (5) 27, 36*; (3) 50, 59*
Aleem, M.I.H., (18) 793, 828*
Alexander, M., (2) 788, 827*
Ali, M.E.A.R., (34) 732, 752*
Amberg, H.R., (25) 603, 612, 613, 617*;
    (34) 611, 612, 618*
American Public Health Association,
    361*
Andrews, J.F., (13) 288, 302*; (19) 325,
    345*; 430*; 507*; (78) 669, 686*;
    (89) 675, 687*; 687*; (44) 744,
    752*; (7) 790, 793, 795, 796, 797,
    828*; (9) 836, 837, 845, 881*;
    (37) 849, 883*; (42) 850, 857,
    883*; (51) 857, 884*; (58) 870,
    871, 874, 884*; 885*
Anthonisen, A.C., (20) 793, 796, 828*
Antonie, R.L., (26) 573, 583*; (1) 756,
    760, 780, 785*; (3) 759, 760,
    785*; (4) 760, 779, 785*; (5) 760,
    785*; (6) 760, 785*; (11) 762,
    763, 765, 768, 774, 785*; (17)
    764, 786*; (25) 780, 781, 782,
    786*; (39) 804, 805, 829*
Ardern, E., (1) 620, 681*
Aris, R., 125*; (10) 541, 582*
Arrhenius, S., (11) 54, 59*; (50) 336,
    347*
Asano, T., (22) 491, 507*
Atkinson, B., (6) 530, 582*; (7) 530,
    582*; 687*; (34) 732, 752*; (35)
    732, 752*
Austin, G.T., (25) 794, 795, 829*

Autotrol Corp., (22) 772, 773, 774, 775,
    780, 786*; (46) 818, 819, 820, 830*

B

Bader, F.G., (19) 893, 894, 921*
Bailey, D.A., 923*
Bailey, J.E., 349*
Balakrishnan, S., (19) 724, 726, 727, 733,
    734, 737, 751*; (43) 739, 752*;
    (63) 796, 831*
Balasha, E., (7) 592, 616*
Ball, J.E., (47) 639, 684*
Bamford, C.H., 125*; 161*
Banks, N., (48) 639, 684*
Barker, J.E., (43) 810, 830*
Barnard, J.L., (49) 329, 335, 347*; 923*
Barnhart, E.L., (3) 589, 605, 616*; (22)
    602, 617*; (1) 689, 692, 693, 695,
    712*; (48) 746, 753*
Barson, G., 587, 616*
Bartsch, E.H., (30) 605, 617*
Bauchop, T., (1) 278, 285, 302*
Becker, G.E., (43) 896, 922*
Beckman, W.J., (54) 823, 830*
Bedker, M., (68) 877, 885*
Benedek, P., (66) 328, 348*; (10) 692,
    701, 712*
Benedict, A.H., (53) 337, 347*; (58) 338,
    347*
Benefield, L.D., 618*; (56) 644, 685*;
    (73) 662, 686*; 687*; (22) 694,
    713*; 714*; 753*; 786*; 831*;
    885*; 923*
Benjes, H. Jr., (24) 602, 617*
Benson, S.W., 59*; 161*
Bentley, T.L., (23) 727, 751*
Beran, K. 430*
Berger, H.F., (28) 604, 613, 617*
Bess, F.D., (6) 591, 612, 613, 616*
Bettenhausen, C., (7) 285, 302*
Bewtra, J.K., (8) 692, 712*
Beychock, M.R., (21) 602, 610, 612, 617*

925

A

α (*see* Recycle ratio α)
A (*see* Area)
Acetic acid 248-249, 834-839, 846-848, 850
Acidogenesis 834-837
Actinomycetes 667
Activated sludge 9-11, 619-688 (*see also* Completely mixed activated sludge, Contact stabilization activated sludge, Extended aeration activated sludge, Pure oxygen activated sludge, Step aeration activated sludge, Tapered aeration activated sludge)
  aeration 659-660
  appearance 667-669
  applications 628, 675-676
  cell concentration 651
  control 667
    computer aided 675
    hydraulic 670-672, 675
    long term 669-674
    short term 674-675
  defects 629-630
  and denitrification (*see* Denitrification: CSTR)
  description 619-630
  design 649-666
    checklist 676-677
    constraints 656-658
    equations 650-655
    factors affecting 658-662
    standards 676
  example problems 663-666
  kinetic parameters, assessing 655-656
  mathematical models 631, 644-649
    verification 655-656
  mean cell residence time (*see also* Mean cell residence time: activated sludge)
    constraints 631-634
    effects of 634-642
    equation 650
  merits 629-630

[Activated sludge]
  microbiology 221-223, 621
  microorganisms, excess 652
  mixing 658-660
  and nitrification (*see* Nitrification, combined carbon oxidation-:CSTR)
  nutrient addition 662
  operation 667-675
  oxygen requirement 653
  performance 631-649
    observed 642-644
    theoretical 634-642
  physical characteristics 619-628
  reactor configuration, choice of 659-660
  recycle ratio 651-652
    constraints 657
  space time 651
  substrate concentration 651
  suspended solids concentration 651
    constraints 656-658
    excess 653
  temperature 660-662
  volume 651
  wastage flow rate 652
Activation energy E 54-56
Activation, enzyme 264
Active sites 306
Activity test, specific 812
Adenosine triphosphate (*see* ATP)
Aerated lagoons (*see* Completely mixed aerated lagoons)
Aeration
  activated sludge 620, 656-660
    completely mixed 625, 653
    contact stabilization 655
    conventional 621-622, 629
    pure oxygen 627-628
    tapered aeration 622
  aerobic digestion 690, 700-701
  CMAL 589, 598, 602-603
  denitrification 911
  heat loss from 661-662
  ice on systems 610
  RBC 756
  trickling filters 739, 745

Y

Z